PHYSIOLOGY OF THE HUMAN BODY

PHYSIOLOGY OF THE HUMAN BODY

SECOND EDITION

J. ROBERT McCLINTIC, Ph.D.
CALIFORNIA STATE UNIVERSITY, FRESNO

WILEY INTERNATIONAL EDITION

JOHN WILEY & SONS

New York • Chichester • Brisbane • Toronto

Library of Congress Cataloging in Publication Data:

McClintic, J. Robert, 1928–
Physiology of the human body.

Includes bibliographies and indexes.
1. Human physiology. I. Title. [DNLM:
1. Physiology. QT4 M127p]
QP34.5.M32 1977 612 77-27066
ISBN 0-471-02664-6

Printed in the United States of America

10 9 8 7 6 5 4 3

PREFACE TO THE SECOND EDITION

The second edition of *Physiology of the Human Body* continues the major aim of the first edition: to present an up-to-date, relevant discussion of how the human body functions in health and disease.

It has been organized to appeal to all levels of readers, from the general student to the health-profession oriented reader who needs a discussion of how health and disease are interrelated.

The unifying theme of the second edition is homeostasis, the mechanisms that operate to ensure its maintenance, and what the body must do to restore homeostasis in case of disease or disorder.

A new feature is the student objectives at the beginning of each chapter. These objectives are correlated with the chapter summaries and with the end of chapter questions. Thus, guidance and a means of checking knowledge are provided. Fifty-seven new pieces of art have been added in the second edition, and major changes were made in over half the chapters. Explanations of function have been greatly expanded to lead the reader into the mechanisms of how the body operates. Chapters dealing with the nervous system have been completely rewritten and placed early in the book. The chapters on cellular function, muscular tissue, body fluids, acid-base balance, immunity, the heart, blood vessels, respiration, digestion, excretion, endocrines, and the last chapter on the human cycle are other areas in which there are major additions and reorganizations.

Where appropriate, fetal and neonatal physiology are discussed to show how function changes with age, and to emphasize that infants are not "small adults" with regard to body function.

Clinical orientation is provided by carefully chosen examples of disorders that illustrate not only the disease itself but how it developed.

Readings are extremely current, with those from 1974 onward; also included are significant earlier works.

In-text summaries, tables, and "flow-sheet" diagrams ensure the "workability" of the text. *Physiology of the Human Body* is meant to be not only informative, but enjoyable to read.

Responsibility for errors in this text are mine.

JRM

ACKNOWLEDGMENTS

I am indebted to my wife Peggy, MSN, whose patience, constructive criticism, typing, and proofreading contributed greatly to the production of this text.

To the staff of John Wiley and Sons, Publishers, and especially to Robert L. Rogers, Biology Editor, I extend my appreciation for encouragement, assistance, and excellent production during the preparation of this text.

I also wish to thank the following individuals for the many hours they spent reviewing the manuscript.

John P. Harley. Ph.D., Department of Biological Sciences, Eastern Kentucky University, Richmond, Kentucky.

August W. Jaussi, Ph.D., Department of Zoology, Brigham Young University, Provo, Utah.

Richard G. Pflanzer, Ph.D., Department of Physiology, Indiana University School of Medicine, Indianapolis, Indiana.

David W. Saxon, Ph.D., Department of Biological Sciences, Morehead State University, Morehead, Kentucky.

CONTENTS

19

BODY DEFENSES AGAINST DISEASE; MECHANISMS OF PROTECTION, ANTIGEN-ANTIBODY REACTIONS, IMMUNITY, BLOOD GROUPS 291

20

THE HEART 314

21

THE ROLE OF THE BLOOD VESSELS IN CIRCULATION 344

1

INTRODUCTION

OBJECTIVES

After studying this chapter, the reader should be able to:

- ☐ Define physiology.
- ☐ Define and explain the meaning of homeostasis.
- ☐ Define what is meant by cybernetics.
- ☐ Distinguish between a mechanistic and vitalistic approach to the study of physiology.
- ☐ List some other disciplines and their contributions to the understanding of physiology.
- ☐ Relate the role of animal experimentation to the understanding of human physiology.
- ☐ Define and explain the several levels of body organization.
- ☐ Define what the internal environment is and explain its importance as an integrator of body function.
- ☐ Comment on the importance and basic types of control mechanisms of the body.
- ☐ Relate the concept of pathophysiology to homeostasis.

WHAT IS PHYSIOLOGY?

PHYSIOLOGY is one of the life sciences and deals with HOW LIVING THINGS FUNCTION. A more modern definition describes physiology as the study of HOMEOSTASIS and of the CONTROL MECHANISMS (cybernetics) operating to attain and maintain homeostasis.

HOMEOSTASIS (*homio,* like; *stasis,* fixed condition) is a term coined by the American physiologist Walter B. Cannon (1871–1945). He used it to refer to the stability maintained within the body of a living organism, even though many forces may attempt to upset that stability. Today, knowing that constant activity is required on the part of the organism to maintain stability, the terms *homeokinesis* (*kinesis,* work) and *homeodynamics* (*dynamis,* power) have been suggested by some as appropriate substitutes for the term homeostasis. Homeostasis and how it is maintained forms an appropriate integrating theme for this book, inasmuch as health reflects the proper operation of these principles, while pathology indicates some breakdown in the mechanisms.

Researchers in physiology believe that all explanations of physiological phenomena may ultimately be made in terms of chemical, physical, and mathematical principles; they thus hold to what is termed a MECHANISTIC VIEW of life. Opposed to mechanism is the VITALISTIC VIEW, which holds that unexplained or often supernatural "vital forces" are involved in the creation and operation of living organisms. In many areas such as what constitutes intelligence, what constitutes morality, and many aspects of brain function, phenomena cannot be visualized mechanistically, and vitalistic views may still be retained until replaced by mechanistic experimental data.

In addition to utilizing chemistry, physics, and mathematics to explain physiological phenomena, physiology draws upon ANATOMY to set the structural basis for understanding function; upon BIOCHEMISTRY to provide knowledge of the chemical reactions occurring within and outside of cells; upon BIOPHYSICS to aid in the explanation of electrical and mechanical events occurring in the body; and upon GENETICS and EMBRYOLOGY to explain hereditary potential, growth, differentiation, and development. Physiology thus brings together, into an integrated whole, the information provided by many other disciplines.

HOW IS PHYSIOLOGY STUDIED?

Physiology, by definition, is concerned with living organisms, for one cannot investigate an ongoing phenomenon in a subject whose activities have been terminated by death. It is fortunate that most mammalian forms function similarly, so that much information gathered from animal experiments proves to be applicable to humans. The field of medicine provides information on humans, through the study of those who are diseased or who have suffered derangements of body function, and on those who give their "informed consent" for research procedures to be carried out on their bodies.

Within the area of human physiology, it is important to point out that body function changes with age, and that what may be considered "normal" for one person may not be normal for another. Where appropriate, this book indicates and discusses such differences.

THE ORGANIZATION OF THE HUMAN BODY

CELLS

Living organisms are typically composed of one or more units that carry on the minimum activities required for their own survival, as individuals and as a species. Such basic units of structure and function are usually referred to as CELLS. Among the minimum activities a cell must carry out to ensure its survival are those listed below.

INTAKE (*ingestion*). This usually occurs in the acquisition of materials for use as energy sources or as raw materials for the synthesis of new substances. Glucose and amino acids are examples of substances "imported" as raw materials.

OUTPUT (*egestion*). This refers to the *elimination* from cells of metabolic wastes, and to the *release* of synthesized materials for use elsewhere in the body. Carbon dioxide, enzymes, and hormones are examples of "exported" materials.

METABOLISM. This term describes the sum total of the chemical reactions occurring in the unit. *Anabolic* (energy storing) reactions are usually associated with the synthesis of new materials by a cell, while *catabolic* (energy releasing) reactions are associated with destruction of materials.

REPRODUCTION of new units. Replacement of cells and healing occurs (in multicellular forms) by the process of *mitosis*. Continuance of many organisms as species is assured by the production of sex cells in the process of *meiosis* and their fusion to form new individuals.

GROWTH. Growth implies an increase in size of cells, by the process of forming new material, or in the size of the whole organism by increase in cell size or numbers.

MOVEMENT. Materials may move about within individual cells or the entire organism may move through its environment.

EXCITABILITY. This property refers to the ability of cells to respond to changes in their environment, and is a fundamental attribute of all living material.

Since multicellular organisms are composed of individual cells, the activity of the whole organism is the result of activity in its individual cells, and the list just presented thus applies to the organism as well. In an organism such as the human being, other properties are assumed by particular cells, and an INTERDEPENDENCE between cells develops as a result of specialization and division of labor. For example, muscle becomes specialized to contract or shorten, and movement of materials through the body (food through the gut, blood through the vessels) depends on muscular activity; erythrocytes or red blood cells specialize in transporting oxygen to all body cells. It is thus not surprising that when certain body functions fail ("heart attack"), the entire organism suffers.

TISSUES

The next level of body organization beyond cells is the TISSUE. A tissue is a group of cells that are similar in structure and function, together with all associated intercellular material.

EPITHELIAL TISSUES form effective coverings

and linings for the body and its parts, because the cells of an epithelium fit very closely together with minimal amounts of intercellular substance between the cells. They do not contain blood vessels of their own, but are nourished by vessels in the connective tissue that underlies the epithelium. The cells in an epithelium may possess special modifications (e.g., the keratin in the cells of the skin; cytoplasmic extensions called *microvilli*) that increase their ability to act as protective and absorptive tissues.

CONNECTIVE TISSUES are just the opposite of epithelial tissues in that their cells are widely scattered in large amounts of intercellular substance. The intercellular substance may contain fibers that give the tissue strength (tendons) or elasticity (walls of large arteries); in bone, the intercellular material has been *calcified,* making it effective as a supporting and protective tissue. It has been said, with ample justification, that without the connective tissues of the body, the human organism would be a shapeless mass of several trillion cells.

MUSCULAR TISSUES possess contractility, or the ability to shorten. They can thus create pressure or tension, and assure propulsion of materials through the body, and movement of the body through space.

NERVOUS TISSUES form and transmit *nerve impulses* through the body. These impulses cause response of effectors (muscle and glands) that are concerned with maintenance and control of body function.

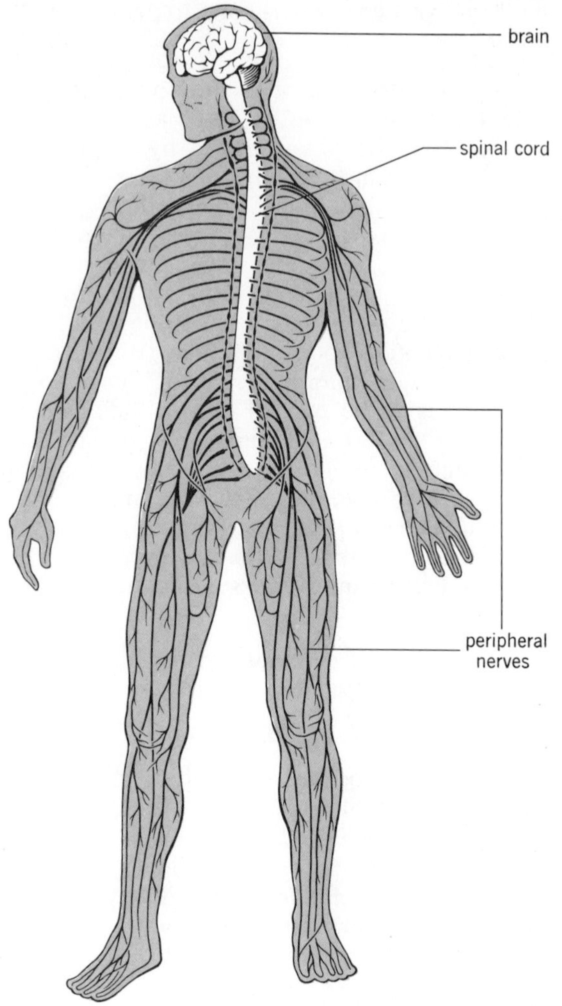

FIGURE 1.1
The nervous system.

ORGANS

ORGANS are composed of two or more tissues put together in a particular way to carry out a particular function. Thus the kidney, composed of epithelium, connective tissue, and provided with muscular blood vessels, regulates the composition of the body fluids; the liver, with its hepatic epithelium, and connective tissue framework, manufactures new materials, stores others, and detoxifies harmful substances in the body. Space precludes a complete listing of all organs and their functions; the reader is encouraged to add more organs to the examples just presented.

SYSTEMS

SYSTEMS are composed of several organs organized to carry out a larger process in the body. The heart and blood vessels comprise a system for circulation of the blood; the mouth, esophagus, stomach, intestines, liver, and other organs form a digestive system designed to reduce large food molecules to forms usable by cells and to absorb those products. Body systems form a primary focus in the study of physiology, and their integration of activity forms one of the bases for understanding homeostasis.

A functional grouping of the body systems may further emphasize the interrelationships of those systems.

Controlling systems. These systems provide a means for controlling and integrating body activity, and its early development.

The NERVOUS SYSTEM (Fig. 1.1) provides a means of monitoring and detecting changes within and outside the body, a means of interpreting and integrating the information received, and a means of initiating responses to maintain or restore homeostasis.

The ENDOCRINE SYSTEM (Fig. 1.2) is, to a large degree, controlled by the nervous system, and provides chemicals known as hormones. These hormones are placed into the bloodstream and are distributed to all cells, including those of the nervous system, where they influence such processes as growth, metabolism, reproduction, and differentiation.

Vegetative systems. Under this heading are those systems responsible for the intake, processing, absorption and distribution of nutrients essential for body function. They also provide for elimination or excretion of metabolic wastes so that the constancy of the internal environment may be maintained.

FIGURE 1.2
Locations and names of the major endocrine organs

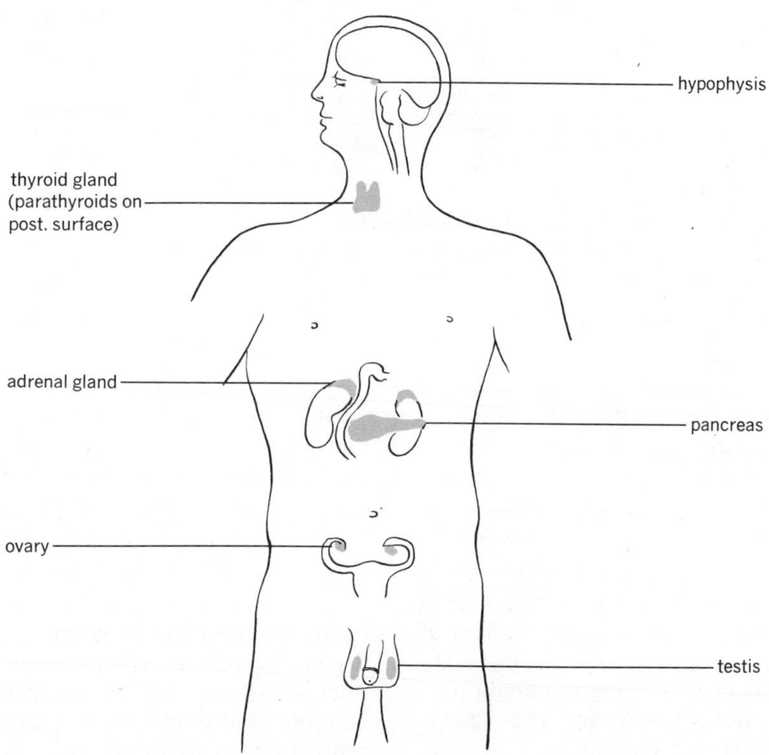

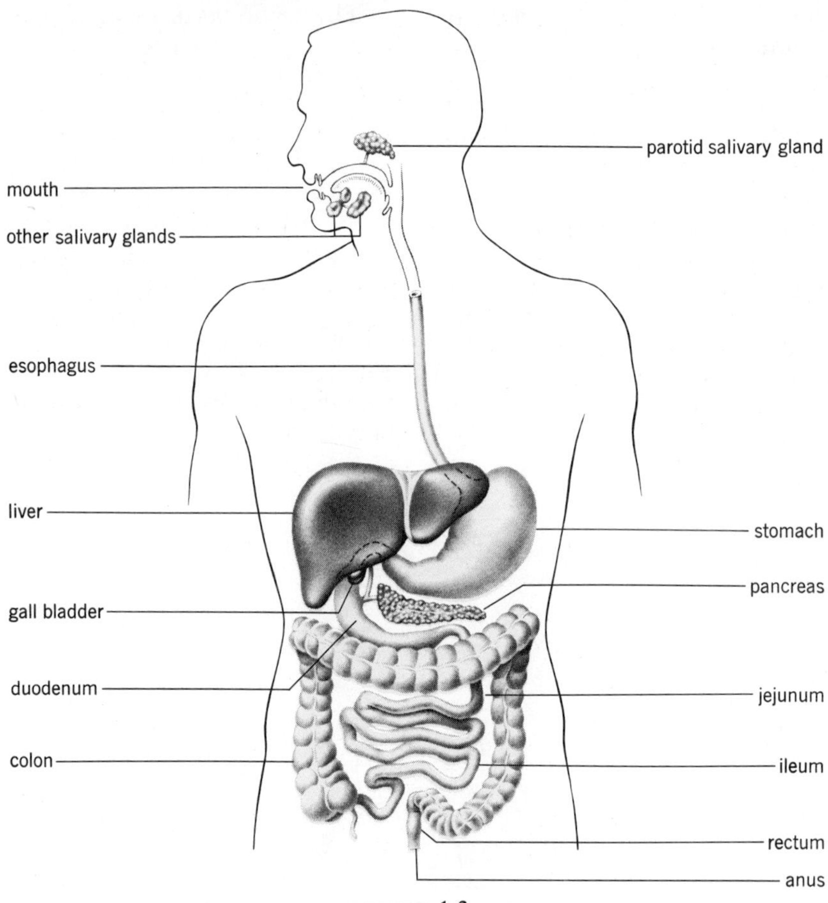

FIGURE 1.3
The digestive system.

The DIGESTIVE SYSTEM (Fig. 1.3) serves as the normal route of intake of food and drink, and breaks down large food molecules to absorbable end products. Absorption of those products and the elimination, via the feces, of unusable materials are also part of its tasks.

The RESPIRATORY SYSTEM (Fig. 1.4), along with the muscles of respiration, provides a means for intake of oxygen and elimination of carbon dioxide. Acid–base balance of the body relies to a large degree on lung elimination of carbon dioxide, again illustrating the dependence of systems on one another.

The URINARY SYSTEM (Fig. 1.5) excretes the solid wastes of metabolism, and water, and regulates the composition of the blood and hence all body fluids.

The CIRCULATORY SYSTEM (Fig. 1.6) provides a means of interconnecting all body systems by transporting materials to all parts of the body.

Protective systems. Protective systems include the integumentary system (skin and associated structures) and the reticuloendothelial system. The lymphatic system and blood also provide protection for the body.

The SKIN (Fig. 1.7) provides a mechanical barrier to the entry of microorganisms and toxic substances from outside the body, and prevents loss of vital body constituents from inside the body. Its sweat glands and blood vessels aid in control of body temperature.

The RETICULOENDOTHELIAL SYSTEM (*RES* or *mononuclear phagocyte system*) is composed of the free and fixed macrophages (phagocytes) of the body, and the plasma cells that produce antibodies against foreign chemicals. These cells form a functional system only, and are found in many different organs and systems. The basis of the body's immunity to various agents resides in this system.

The LYMPHATIC SYSTEM (Fig. 1.8) acts as a source of many RE cells, produces certain blood cells, returns tissue fluid to the blood vessels, and cleanses tissue fluid.

The BLOOD contains many phagocytic cells that form a last line of defense against microorganisms and foreign chemicals.

FIGURE 1.4
The respiratory system.

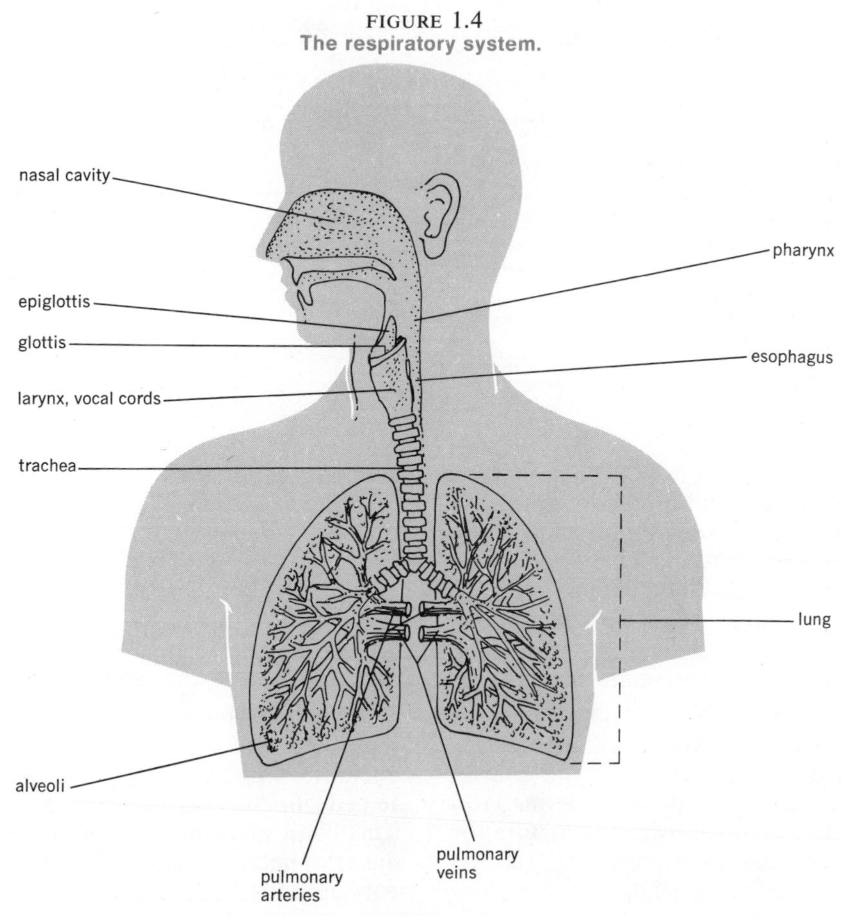

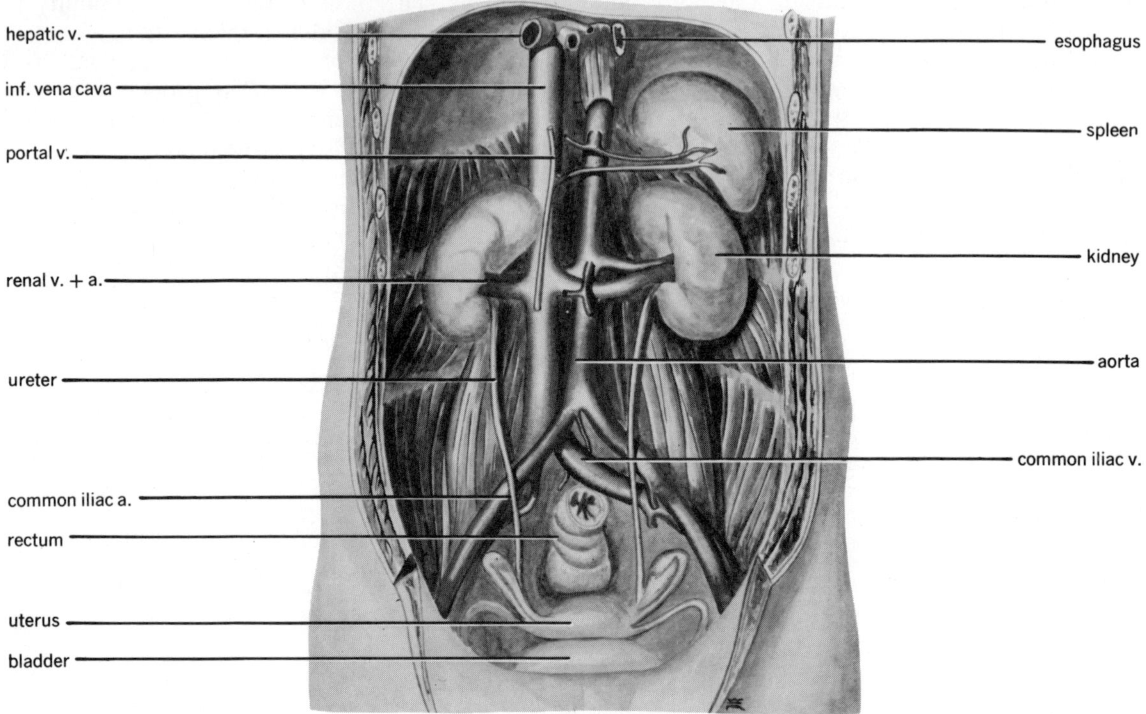

FIGURE 1.5
The urinary system and some associated structures.

Musculoskeletal system (Fig. 1.9). The SKELETON provides support and protection for many body organs, forms the levers for muscles to pull upon to cause body movement, acts as a storehouse for minerals, and contains the bone marrow that produces a wide variety of blood cells.

The MUSCULAR SYSTEM refers to the muscles attaching to and moving the skeleton.

Reproductive systems (Fig. 1.10). These systems provide for perpetuation of the species through production of eggs and sperm. Hormones essential to body growth and maturation are produced by certain organs of these systems. Protection and nourishment of offspring are provided by certain organs of the female reproductive system.

THE INTERNAL ENVIRONMENT

Life is said to have originated in the sea. That sea provided nutrients for the organisms living within it, and acted as a means of removing metabolic wastes from the organism producing them. As multicellular forms developed, they enclosed within their bodies fluids that continued the functions of the aquatic medium. This fluid enabled a water–land transition to occur. The body fluids, especially those lying outside of the cells, constitute the INTERNAL

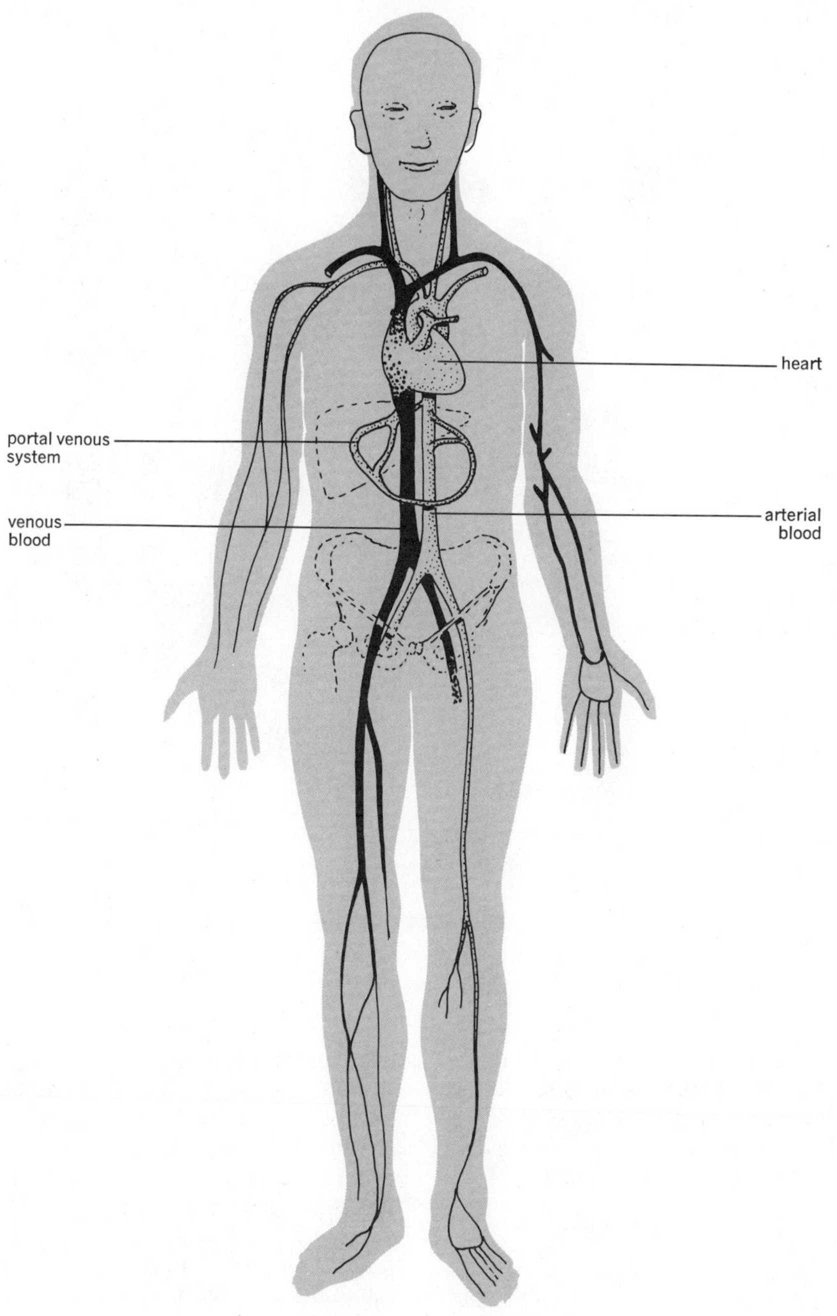

FIGURE 1.6
The circulatory system.

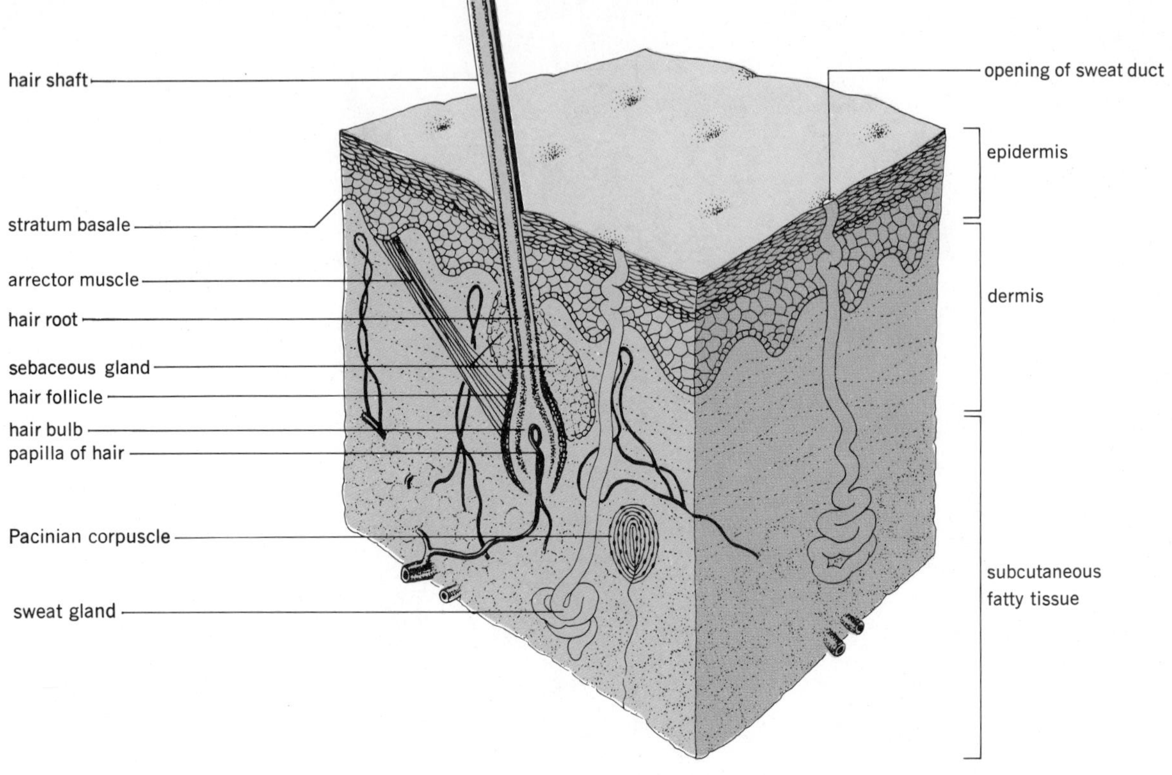

FIGURE 1.7
The structure of hairy skin.

ENVIRONMENT of the organism. Since these fluids are in direct contact with body cells, their composition and characteristics must be maintained within narrow limits, a prime example of the necessity for homeostasis.

CONTROL MECHANISMS

If CONTROL is to be achieved on the internal environment and on body organs and systems, changes must be recognized and appropriate responses initiated. Recognition is provided by nervous system structures that monitor pressure, temperature, pH, organ output, and a host of other factors inside and outside the body. Analysis, usually within the central nervous system (brain and spinal cord), is followed by signals (nerve impulses or hormone secretion) that stimulate or inhibit the activity of an organ concerned with homeostasis of that particular factor. These relationships are diagrammed in Figure 1.11. The stimulation or inhibition of activity through the intermediary of a monitoring device constitutes FEEDBACK. The operation of a thermostat (monitor) in controlling a furnace illustrates the operation of *negative feedback*. A fall of temperature

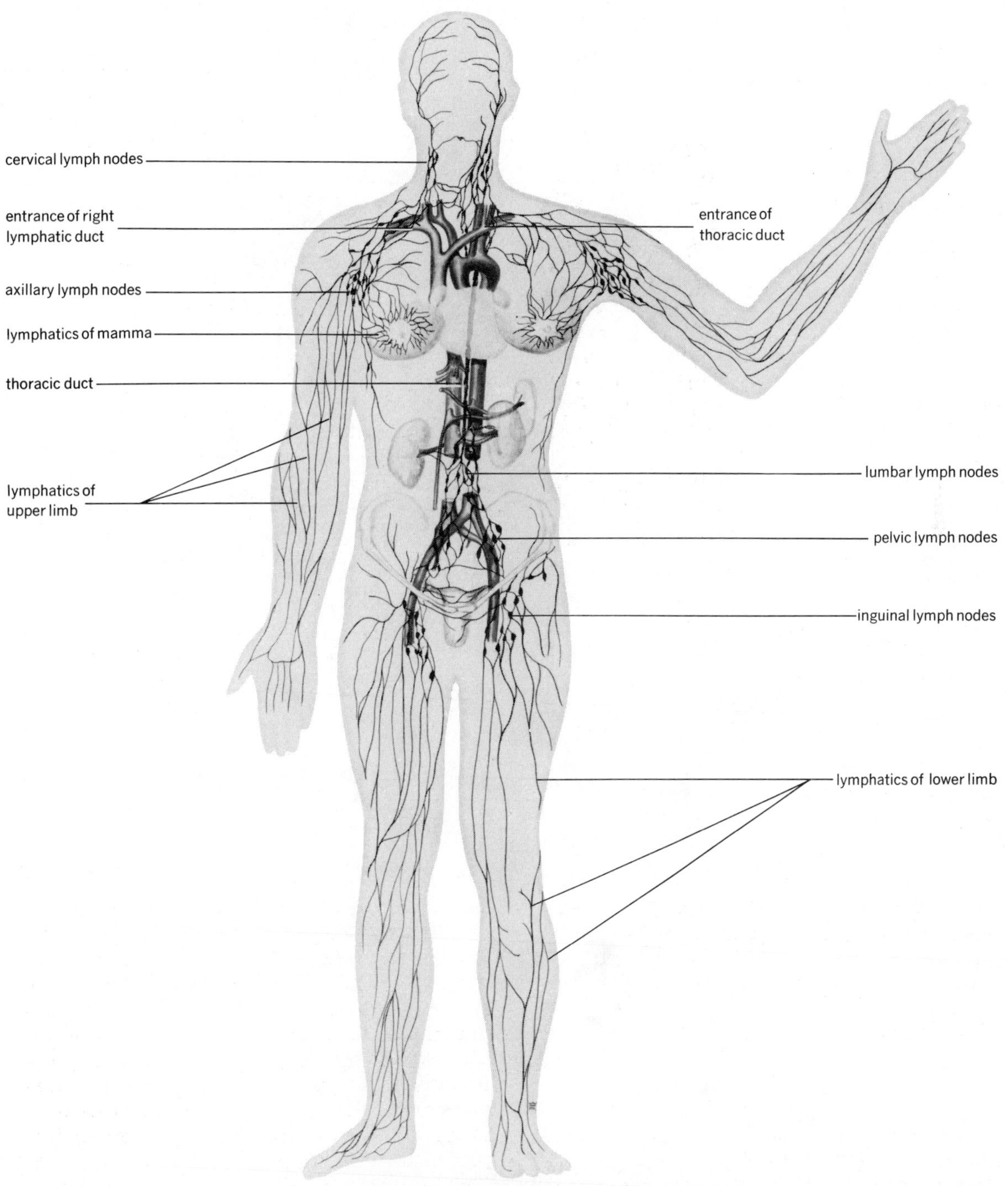

FIGURE 1.8
Some organs of the lymphatic system.

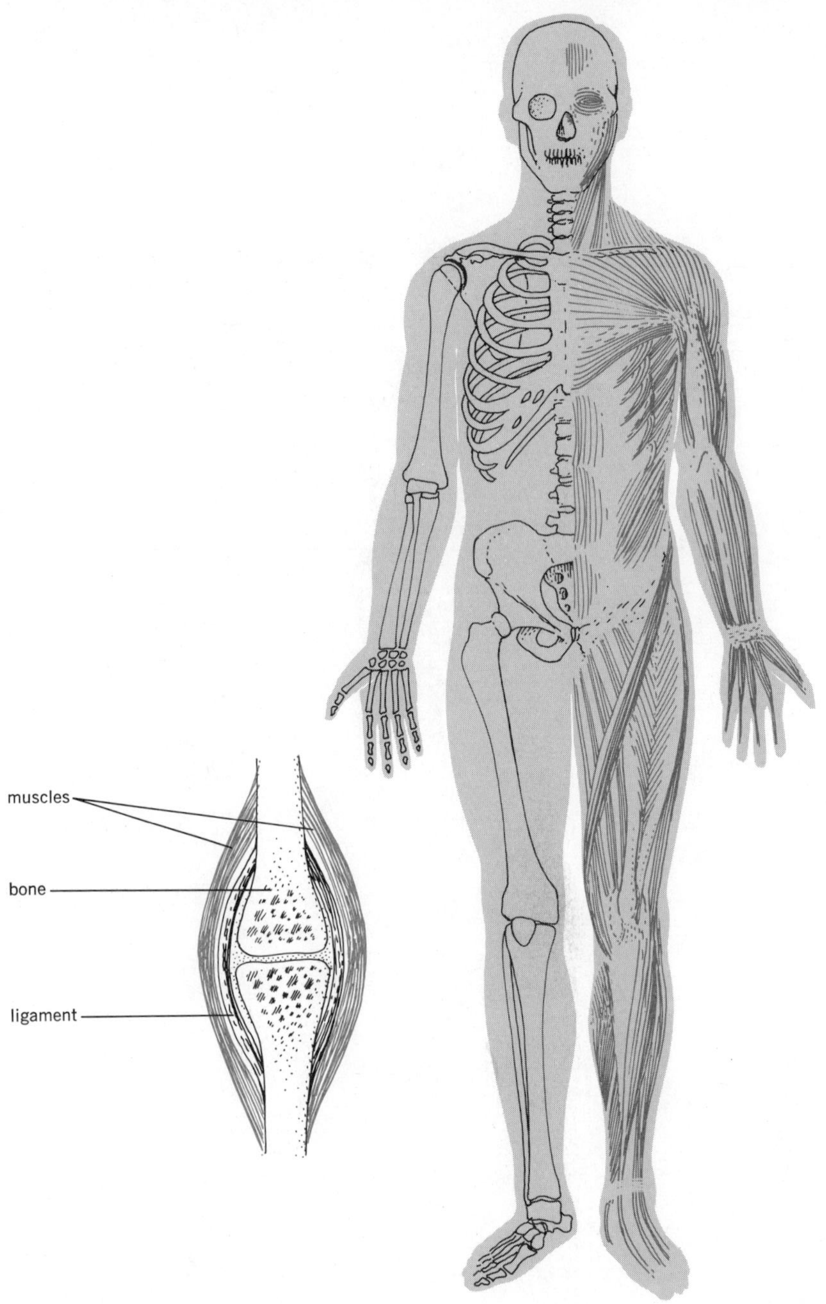

FIGURE 1.9
The musculoskeletal system.

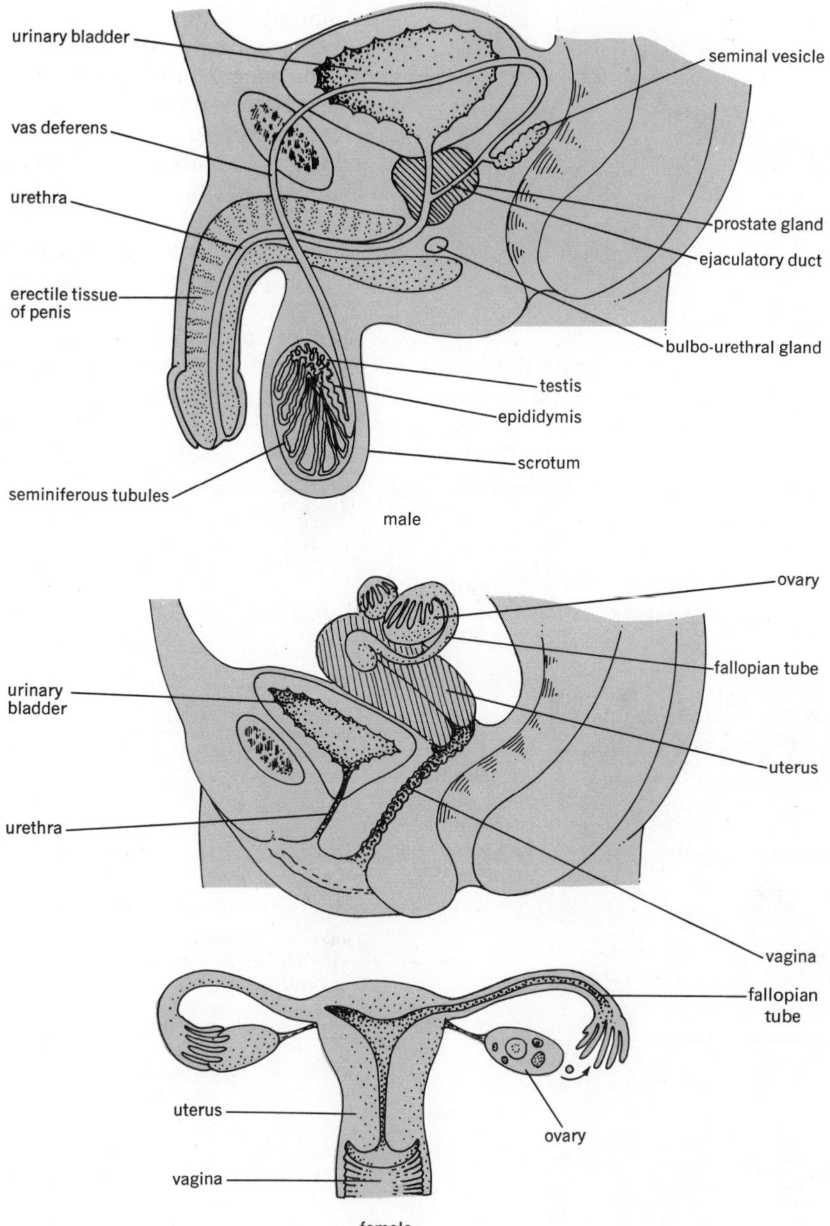

FIGURE 1.10
The male and female reproductive systems.

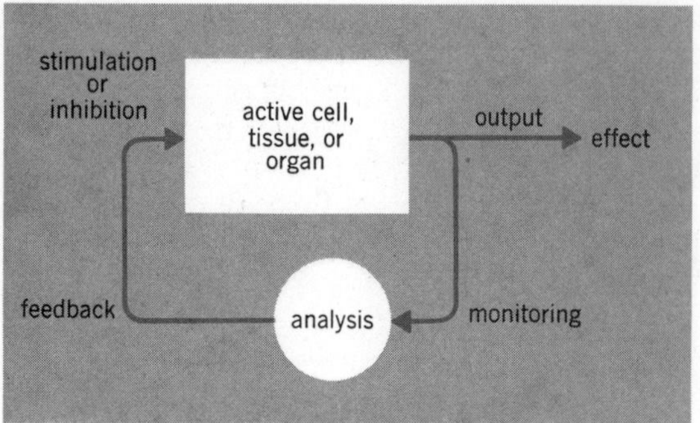

FIGURE 1.11
The basis of cybernetic control.

triggers the furnace to go on, it heats the air, and when the temperature rises to a particular point, the furnace shuts off (it is inhibited). The air temperature thus fluctuates within limits set by the thermostat and "temperature homeostasis" has been achieved. *Positive feedback* occurs when output or activity is increased (stimulated) by the monitoring device, as in the events occurring during depolarization of a cell membrane. In this situation, depolarization results in increased permeability to sodium ions that causes further depolarization that results in greater permeability, and so forth. Positive feedback does not result in fluctuation within limits and is not generally employed in homeostatic control mechanisms. Figure 1.12 illustrates the principles of negative and positive feedback.

Control mechanisms are basically CHEMICAL, with the chemical messenger or transmitter being produced by nervous, endocrine, or other body systems. The nucleic acids—chemicals themselves—provide the ultimate genetic control for all body functions.

FIGURE 1.12
A diagram illustrating the principles of positive and negative feedback.

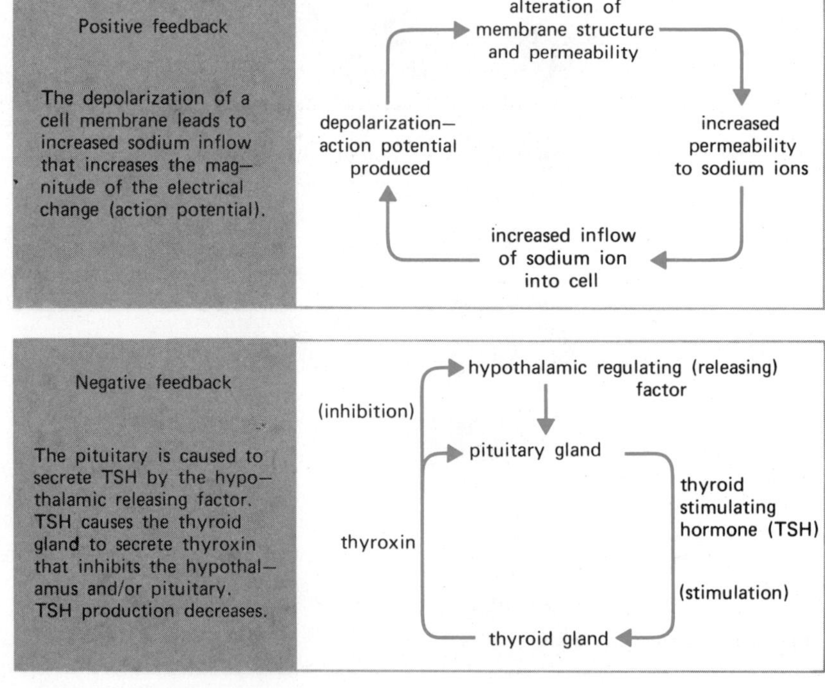

THE BASIS OF PATHOPHYSIOLOGY

If we accept that health represents normal or adequate homeostatic and cybernetic activity, then disorder and disease may be understood as a malfunction of that activity. This is the approach taken in this text, that is, understand the basic mechanism, understand what went wrong, and understand what must be done to restore normal function. In this way, we should be able to see a "whole package" of how the body maintains itself and its operations. The reader may also be able to judge the truth or fiction in certain health claims reaching us through the mass media, knowing the basic mechanisms the body utilizes to survive.

EPILOGUE

This chapter is designed to introduce the reader to the basic structure of the body and to point out that the functioning of the body is and must be controlled. Control mechanisms are basically similar regardless of the particular function being controlled. The concept that control may break down and that "the chain is only as strong as its weakest link" should provide us with a stimulus for body care that begins with conception, and ends only with the death of that particular individual.

SUMMARY

1. Physiology or its researchers:
   a. Is the science of homeostasis and the control mechanisms that operate to attain and maintain homeostasis.
   b. Believe that function may ultimately be explained in terms of chemical, physical, and mathematic laws.
   c. Studies how living things function.
   d. Utilizes input from many other scientific fields to explain body function.

2. Physiology is studied
   a. By animal experimentation.
   b. On sick persons.
   c. On those who give their informed consent for experimental procedures.
   d. In those of all ages, because function is different at different ages.

3. Several levels of organization are found in the human body, and functions at one level are often repeated at higher levels.
   a. Cells are the basic structural and functional units of the body. They show intake, output, metabolism, reproduction, growth, movement, and excitability.
   b. In multicellular organisms, some cells become specialized for specific functions, and interdependency develops.
   c. Tissues, four groups (epithelial, connective, muscular, and nervous), form body organs.
   d. Organs have specific functions in the body.
   e. Systems are composed of organs and carry out larger processes in the body.

4. Body systems include those for control, acquisition, processing, and distribution of products, regulation of body composition, protection, support and movement, and perpetuation of the species.

5. The internal environment
   a. Consists of the body fluids.
   b. Is controlled as to its composition and characteristics; cellular activity is involved in this control.

6. Control
   a. Requires the presence of detecting or monitoring devices, interpretation and integration of information, and initiation of appropriate response to maintain homeostasis.
   b. Most often employs negative feedback.
   c. Is basically chemical in nature, with the chemicals produced by various body organs or systems.

7. Pathophysiology may be viewed as a derangement of a particular homeostatic control mechanism.

QUESTIONS

1. Name four scientific fields that contribute to the study of physiology, and indicate the importance of each. How, or what sort of contribution could a worker in each of these fields make to the discipline of physiology?
2. Explain, and give examples of, homeostasis.
3. Speculate on the problems involved in experimentation on humans, and how it is possible to gain information about humans from other mammals.
4. List the levels of organization of the human body and the types of activities that occur at each level.
5. What is meant by "internal environment"; how did it evolve, and what is its importance to the human body?
6. Why does the body need control mechanisms?
7. What are the components of an effective control system? What does each contribute to the control process?
8. Give examples of positive and negative feedback systems other than those that appear within this chapter. Analyze each that you mention as to its components and result. (Note that the question does not specify *body* mechanisms. Seek as many "everyday examples" as you can.)

READINGS

Barber, Bernard. "The Ethics of Experimentation with Human Subjects." *Sci. Amer. 234:*25 (Feb) 1976.

Gordon, Richard. "Image Construction from X-ray." *Sci. Amer. 233:*56 (Oct) 1975.

Gray, Bradford H. *Human Subjects in Medical Experimentation.* Wiley. New York, 1975.

2

CELL STRUCTURE AND FUNCTION

OBJECTIVES

After studying this chapter, the reader should be able to:

☐ Discuss some of the factors involved in determining the size and shape of cells.

☐ Explain the structure of cell membrane systems, and discuss the role of these membranes in controlling what the cell takes in and releases.

☐ Describe the factors involved in determining to what substances a membrane is permeable.

☐ List the formed bodies of the cytoplasm, describe their structure, and give a function for each.

☐ Discuss, in general terms, the structure of the nucleus and its role in control of cellular activity.

☐ Describe: the differences between passive and active methods of movement of substances across membranes; the methods within each category, with emphasis on how each process occurs; the characteristics each exhibits.

☐ Compare and contrast mitosis and meiosis as to how each process occurs, and the effects of each process on the cell chromosome number and genetic potency.

☐ Give some of the general characteristics of cancers, list some possible causes for cancer, and discuss how cancers are classified.

DETERMINANTS OF CELLULAR MORPHOLOGY

DETERMINANTS OF SIZE

Individual cells of living organisms vary in size from about 200 nm* to the size of an ostrich egg. The minimal size a cell can be is set by the fact that the cell must be large enough to contain the MACROMOLECULES (nucleic acids and proteins) required to CONTROL and SUSTAIN CELLULAR ACTIVITY. Most cells lie within the size range of 0.5 to 20 μm in diameter, but there are very large cells (e.g., muscle and nerve cells) that carry on vital activities.

* m = meter
mm = millimeter = 1/1000m or 10^{-3}m
μm = micrometer or micron (μ) = 1/1,000,000m or 10^{-6}m
nm = nanometer or millimicron (mμ) = 1/1,000,000,000m or 10^{-9}m
Å = angstrom = 1/10nm or 10^{-10}m
pm = picometer = 1/100Å or 10^{-12}m

How large a cell may become, and still function adequately, depends primarily on two factors.

The first factor is the relationship between the nucleus of the cell and the amount of cytoplasm it can influence or control. Since the nucleus utilizes chemicals that pass from nucleus to cytoplasm to exert this control, a problem of getting the controlling chemicals to the entire cytoplasmic mass may arise. This problem is solved in some cells by having multiple nuclei, each imparting identical instructions to a part of the cytoplasm, as in skeletal muscle. Another method for distributing controlling chemicals is by movement (streaming) of the cytoplasm, as in the axoplasmic flow of materials in the processes of nerve cells.

The second factor is the surface available through which nutrients and wastes must pass to service the volume of living material. What is being referred to here is the ratio of plasma (cell) membrane surface area to cytoplasmic volume. As size of a cell increases, the surface area of the plasma membrane increases as the square of the change, while volume increases as the cube of the change. In short, the volume of material to be serviced rises more rapidly than the area available for contact with the internal environment. Cells may develop foldings, become elongated or flattened to help solve this problem.

DETERMINANTS OF SHAPE

The shape of body cells is very variable (Fig. 2.1), and is related to the cell surface-volume ratio mentioned above, and to the function the cell serves. Muscle cells are capable of shortening to at least two-thirds of their original length, and must therefore have an initial elongated form in order to shorten effectively. Secretory and absorptive cells are generally cuboidal or columnar in shape to provide room for cytoplasmic constituents, and commonly have numerous small extensions of the cell surface called *microvilli*. Such microvilli vastly increase the membrane surface across which materials may be passed. Secreting cells are often cuplike to contain their secretions, and nerve cells may reach several feet in length in order to carry nerve impulses from, for example, the spinal cord to a muscle in a toe.

Regardless of specialization of form or function, almost all cells of the body have certain features in common that allow cellular activity to proceed normally.

FIGURE 2.1
Diverse forms of mammalian cells. (Not to the same scale.)

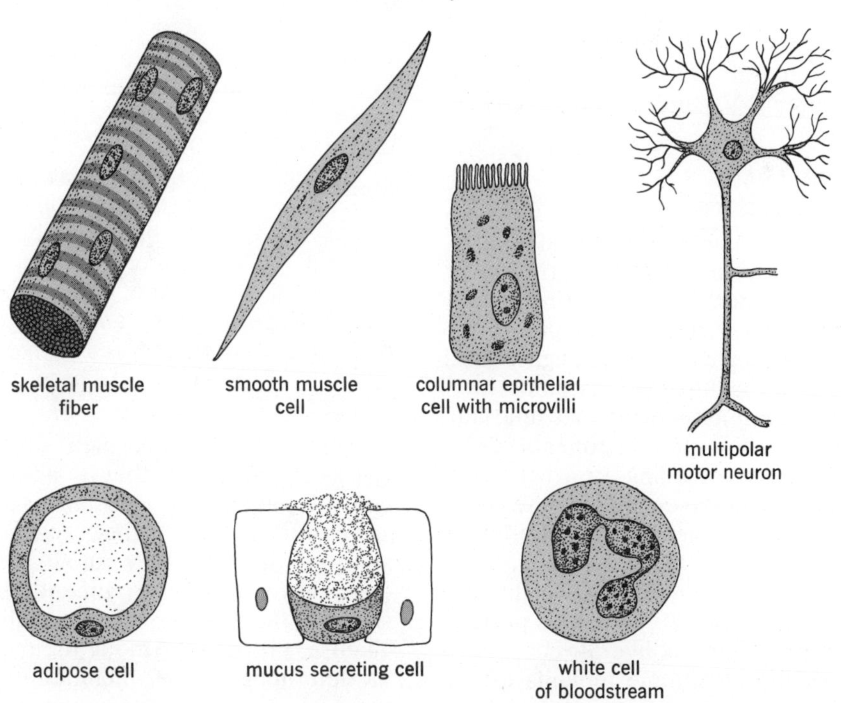

TABLE 2.1
COMPOSITION OF REPRESENTATIVE MEMBRANES

MEMBRANE	PROTEIN-LIPID RATIO NUMBER : 1	LIPIDS PREDOMINATING
Erythrocyte (plasma membrane)	1.5	Phospholipid (55%) Cholesterol (25%)
Mitochondrial (inner membrane)	3.0	Phospholipids (95%) Cholesterol (5%)
Myelin (nerve fibers)	0.25	Phospholipids (32%) Cholesterol (25%) Sphingolipids (31%)

CELL STRUCTURE AND FUNCTION

In general, almost any body cell may be said to have three basic parts: MEMBRANE SYSTEMS that surround the cell and compartmentalize the cell interior so that chemical reactions for particular schemes of metabolism do not interfere with one another; the CYTOPLASM or material inside the cell membrane (excluding the nucleus); and the NUCLEUS itself, which serves as the directing agent of cellular activity.

MEMBRANE SYSTEMS

Theories of structure. The outer boundary of a cell is formed by the PLASMA MEMBRANE. Membranes surround or form the periphery of many cellular organelles, including mitochondria, the endoplasmic reticulum, lysosomes, and Golgi apparatus. Although not all membranes may be said to have the same structure, there is general agreement that lipid and protein form the two major components of any membrane system. According to what membrane one selects for investigation, the lipids involved in the membrane may be different as indicated in Table 2.1. On the average, lipids account for about 40 percent by weight of a membrane system, about 50 percent is protein, with about 10 percent carbohydrate.

Structure of the membrane remains a continuing source of investigation and debate. Again, it may be stated that structure and organization of the membrane components depends to a large degree on what membrane is selected for study. The lipid component appears to be arranged as a bimolecular (two layers of molecules) layer with protein covering these layers, or in some manner "floating" in or lying around the lipid component. Figure 2.2 indicates some of the current theories of arrangement of the lipid and protein components in various cellular membranes. The configuration of the components of the membrane may create channels or "pores" 4 to 8 Å in diameter in plasma membranes, and several hundred Å in diameter in nuclear membranes. These areas represent regions through which substances may pass with greater ease than through the rest of the membrane; they are not areas of free passage, for they appear to be lined with molecules that exert a restrictive effect on the passage of substances. These pores are also indicated in Figure 2.2.

Functions. Because of their structure and the types of molecules they contain, membranes control passage of substances across themselves. They are SELECTIVE and SEMIPERMEABLE, in that they allow some materials (e.g., water) to cross unhindered, while restricting the passage of other substances (e.g., glucose, proteins). Among the factors involved in determination of membrane permeability, or involved in changes in permeability are the following.

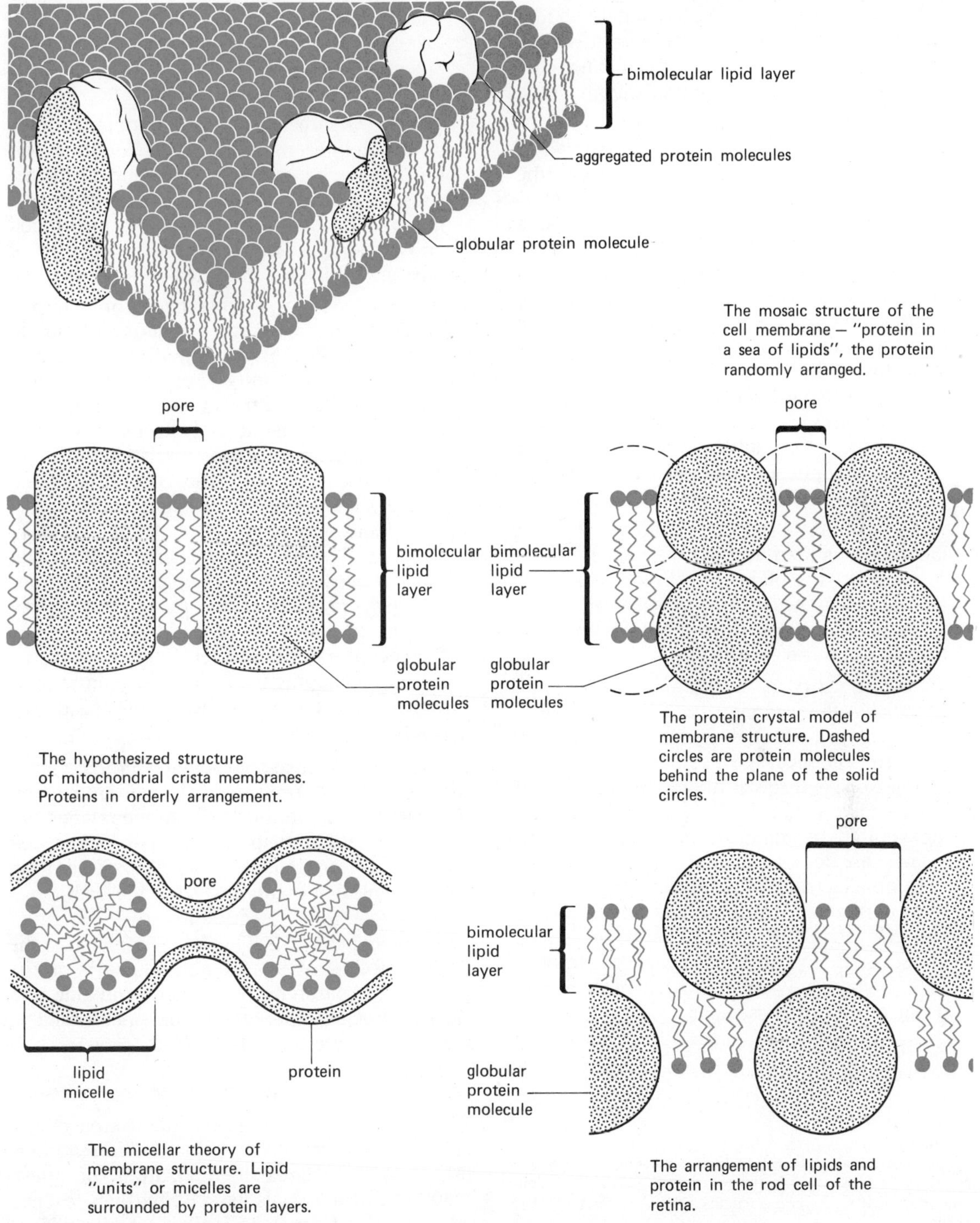

FIGURE 2.2

Various arrangements of the lipid and protein components of the cell membrane. Small round circles indicate lipid polar (charged) portions; wavy lines indicate the nonpolar lipid tails.

MEMBRANE THICKNESS. A thicker or wider membrane represents a greater distance through which a substance must pass, and thus the time required to pass through would be increased.

SIZE. This factor may operate in two ways. First, substances trying to pass through the membrane have a diameter or length determined by their molecular structure, and a weight determined by the number of atoms in their molecules. Second, the orientation of lipids and proteins in the membrane creates spaces (pores?) of different sizes in the membrane. Substances having diameters less than the size of a glucose molecule (8.6 Å) will pass relatively easily through most membranes. Urea, glycerol, and ethyl alcohol are other substances that are near the upper limit of size for easy passage through pores (7 to 8 Å diameter).

LIPID SOLUBILITY. Because almost half of a membrane is lipid, it might be expected that substances dissolving in lipids, such as steroid hormones, would pass more easily or rapidly through a membrane, and this proves to be the case.

ELECTRICAL CHARGE. Membrane components, especially the proteins, may acquire electrical charges, through ionization of certain of their chemical groups. If a substance trying to pass through the membrane is also electrically charged, its passage may be accelerated or inhibited. In general, membranes are less permeable to electrolytes* than nonelectrolytes of similar size, and are less permeable to ions carrying multiple charges than to those carrying single charges. Entry of singly charged (monovalent) ions may be ranked in descending order of rate of passage as

$$K^+ > NH_4^+ > Na^+ > Li^+ \text{ and } SCN^- \text{ (thiocyanate)} > I^- > Br^- > NO_3^- > Cl^-$$

* The term electrolyte refers to the presence of charged particles or substances in the solution. Solutions that contain electrically charged particles (ions) conduct electric current.

Charge may also determine how many layers of water molecules (hydration layers) are associated with the ion and this may influence its overall diameter or size.

ACTIVE TRANSPORT SYSTEMS. As detailed later in this chapter, membranes may contain systems that can combine with and transport substances through membranes relatively independently of the factors mentioned above. Such systems control both rate of passage and type of material being carried.

BINDING SITES. Many substances, for example, alanine (an amino acid), glucose (a simple sugar), vitamin B_{12}, antibodies (chemicals involved in immunity) and insulin (a hormone), in order to pass through or influence a membrane, must actually attach to specific sites on the membrane. Such sites are referred to as receptor sites or binding sites. These sites appear to be genetically determined and may be actual molecules or parts of molecules of the membrane itself. It has been suggested, for example, that the metabolism of cholesterol (a lipid) requires binding to a membrane, and that absence of the receptor for cholesterol may be a factor in the development of high blood lipid concentrations and such processes as atherosclerosis.

Membranes also participate in enzymatically controlled processes. For example, mitochondrial membranes contain enzymes involved in electron transfer. Also, liver cell membranes contain enzymes important in glucose entry into and exit from the cell.

The functions described above are summarized in Table 2.2. This table emphasizes the degree of control any membrane exerts on the passage of materials through it, and one can thus appreciate the role of the membrane in maintaining internal cellular homeostasis.

THE CYTOPLASM

The CYTOPLASM acts as the main area of the cell in which catabolic and anabolic reactions take place. Compartmentalization of these metabolic reactions is assured by the presence of permanent, metabolically active, self-re-

TABLE 2.2
SUMMARY OF FUNCTIONS OF THE CELL MEMBRANES

FUNCTION	COMMENTS
Limiting the cell	Separates cytoplasm from internal environment
Controlling passage of materials, such as	
Water	Freely permeable
Charged substances	Like charges repel, opposites attract. Relationship is between charge on membrane and charge of particle attempting to cross the membrane
Large molecules	Larger molecules pass more slowly unless a transport system is present
Lipid soluble substances	Pass more rapidly than materials having low lipid solubilities
Digestion	Protein, carbohydrate, and lipid digesting enzymes are present in most membranes
Transport	"Carriers" for active transport are found in the membrane
Binding	Sites are described for alanine, glucose, vitamin B_{12}
Energy transformations	Adenosine triphosphate (ATP) is split in the membrane

producing formed structures known as ORGANELLES (Fig. 2.3). Also found in the cytoplasm are structures or masses of material that serve to store products, or which are crystals, droplets, or other forms of wastes of, and raw materials for, cellular activity. Such structures are termed INCLUSIONS.

The cytoplasm is divided into a thin outer portion that lies just beneath the plasma membrane. This area is usually free of large numbers of organelles and inclusions, and is called the EXOPLASM. The remaining area of the cytoplasm and the greatest portion, is called the ENDOPLASM.

Organelles. Most body cells contain the following organelles that ensure their metabolic reactions, and their survival and perpetuation.

Endoplasmic reticulum (ER). The ER lies in and ramifies through all parts of the endoplasm, and extends from plasma membrane to nucleus. It consists of FLUID-FILLED TUBULAR CHANNELS that connect with one another and are bounded by membranes having a structure and composition similar to that of the plasma membrane. The fluid is derived from the internal environment in which the cells lie. Substances derived from outside the cell, from the nucleus, or from the cytoplasm itself, may "circulate" through these channels and thus be distributed to all parts of the cell. Control of activity of all parts of the cytoplasm may thus be achieved. Two varieties of ER are distinguished: GRANULAR ER has ribosomes (see below) attached to the outer surface of its walls, and the ribosomes are responsible for protein synthesis within the cell; AGRANULAR ER lacks ribosomes and is involved in the synthesis of lipids and certain large carbohydrate molecules.

Ribosomes. Ribosomes are small granules of ribonucleoprotein (a combination of ribonucleic acid or RNA, and a protein) that are attached to the ER or may be found free in the cytoplasm. "Chains" of ribosomes may form what are called polyribosomes. Cellular PROTEINS ARE SYNTHESIZED on ribosomes.

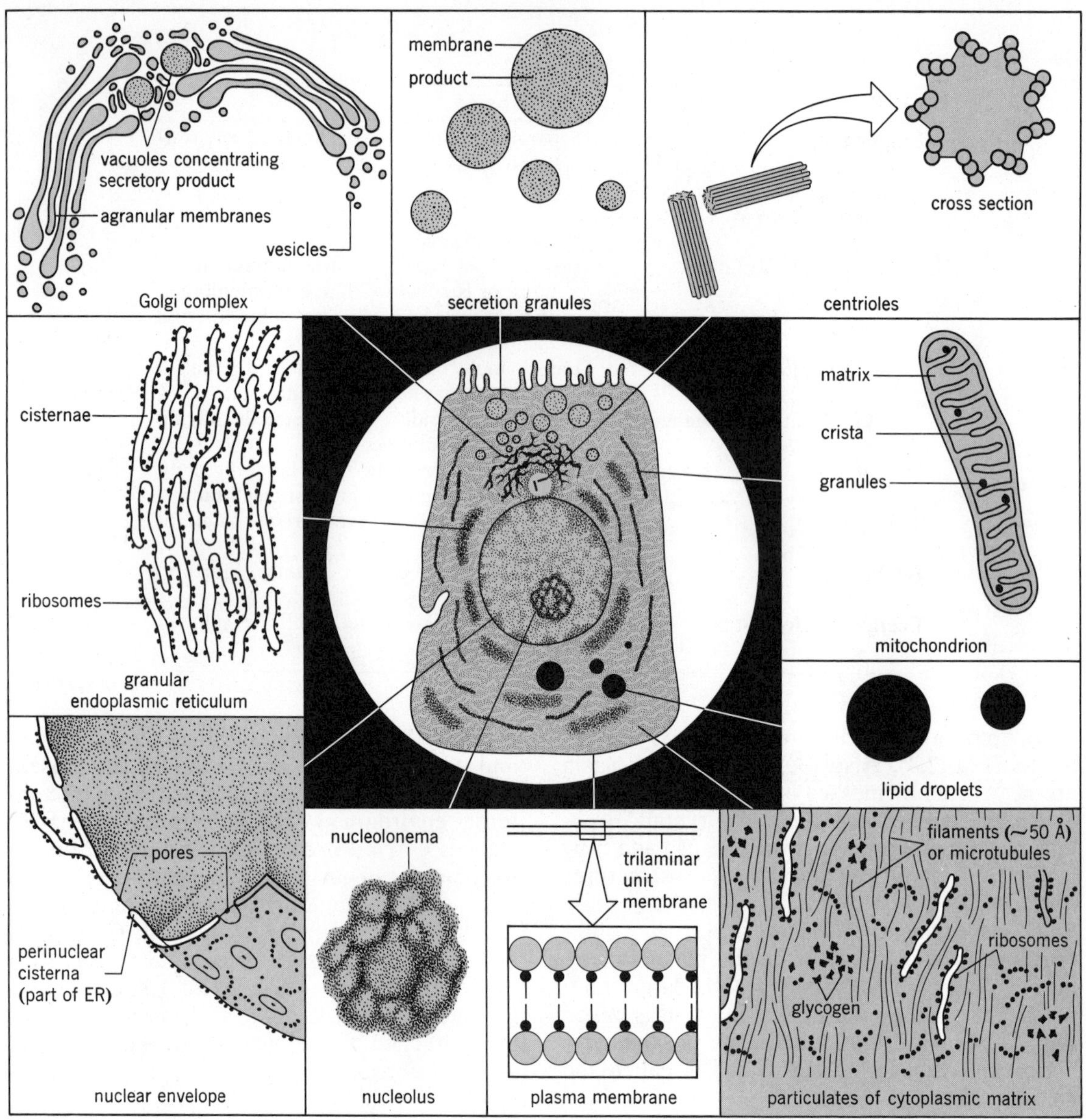

FIGURE 2.3
In the center of this figure is a diagram of the cell as it might appear viewed by light microscopy. The peripheral diagrams present the finer structure of the cell as it would appear in electron micrographs.

The Golgi body, complex, or apparatus. This organelle is usually situated near the nucleus in one part of the cell. It consists of flattened hollow channels surrounded by membranes *(Golgi membranes)* that may contain enlarged areas *(Golgi vesicles)*. The organelle serves as the site of SYNTHESIS of glycoproteins (combinations of carbohydrates and protein), glycolipids (combinations of lipid and protein), and mucus. It is also involved in PLACING MEMBRANES around enzymes, hormones, or other products of cellular activity so that they cannot influence the activity of the cell producing them. Such "packaged" products may then be released from the cell for use elsewhere in the body. A final function of the Golgi body is to act as a region where substances may accumulate and react with one another *(condensation membrane)*.

Lysosomes. Lysosomes are membrane-bounded "sacs" of powerful HYDROLYTIC ENZYMES capable of digesting large molecules that have been taken into the cell. Some ten enzymes acting on about twenty different materials have been identified in the lysosomes. Some of the more important ones are presented in Table 2.3. Lysosomes have been implicated in causing cell death and digestion ("suicide packets") when their enzymes are released into the cell itself. Such activity occurs when cells are damaged or when they die from "natural causes," and clears the way for new cells to be formed by mitosis. Lysosomal release of enzymes may also contribute to the inflammation of joints that occurs in rheumatoid arthritis. Genetically determined lack of one of the lysosomal enzymes results in the accumulation of its corresponding substrate (material acted upon), and forms the basis for several "storage diseases" occurring in humans. Several of these disorders are presented in Table 2.4.

TABLE 2.3
SOME LYSOSOMAL ENZYMES

ENZYME	SUBSTRATE (MATERIAL ACTED UPON)
Collagenase	Collagen, a connective tissue protein
Cathepsins	Cartilage, protein polysaccharides
Elastase	Elastin, a connective tissue protein
Hyaluronidase	Connective tissue, protein polysaccharides
Protease	Protein polysaccharides
Acid peptidase	Fibrin
Glucosidase	Glucocerebroside
Sphingomyelinase	Sphingomyelin
Glucosaminidase	Ganglioside

TABLE 2.4
SOME HEREDITARY DISORDERS ASSOCIATED WITH LACK OF LYSOSOMAL ENZYMES

DISEASE	ACCUMULATED MATERIAL	MISSING OR DEFECTIVE ENZYME
Gaucher's disease	Glucocerebroside (carbohydrate-lipid compound)	β-Glucosidase
Niemann-Pick disease	Sphingomyelin (a lipid)	Sphingo-myelinase
Glycogen storage disease II	Glycogen	α-Glucosidase
Tay-Sachs disease	Ganglioside	Gluco-saminidase

Mitochondria. Mitochondria have a double walled structure, with the inner wall folded into shelves or CRISTAE. Within the outer membrane and on the cristae are systems of enzymes that are responsible for PRODUCTION OF ADENOSINE TRIPHOSPHATE *(ATP)*. ATP is an energy-rich compound that is used to operate cellular activity, and has been termed "the coin of energy exchange." Transport systems in membranes, muscle contraction, and many other activities require ATP. We eat foods to provide the nutrient energy to synthesize ATP.

The central body. Two CENTRIOLES and a surrounding area of differentiated cytoplasm comprise the central body or centrosome. The centrioles give rise to the SPINDLE at the time of cell division, and also form CILIA and

FLAGELLAE that move materials across cell surfaces or propel cells (e.g., sperm) through a fluid environment.

Microtubules. Tiny (about 250 Å diameter) hollow tubules permeate the entire cytoplasm. These microtubules appear to act as a CYTOSKELETON that imparts stiffness to the cell and may direct the flow of cytoplasm. The spindle formed during cell division is composed of microtubules.

Peroxisomes (microbodies). Peroxisomes are spherical cytoplasmic bodies about 0.5 μm in diameter that are surrounded by a single membrane. They are found in kidney and liver cells and in the nonciliated cells of bronchioles in the respiratory system. They contain ENZYMES (urate oxidase, amino acid oxidase) capable of reducing oxygen to hydrogen peroxide (H_2O_2) and an enzyme (catalase) that breaks down H_2O_2 to water. Hydrogen peroxide is produced in certain white blood cells, aids the destruction of phagocytosed bacteria, and is a by-product of metabolism in other cells. Peroxisomes may protect the cell from excessive H_2O_2 production, and may be involved in the conversion of fats to carbohydrates.

Other generalized organelles include protein strands, commonly called MICROFILAMENTS. In muscle, these fibers confer contractility on the cells. In other instances, such fibers may support the cell.

Inclusions. Vacuoles – membrane surrounded fluid droplets – serve as areas of storage for water soluble substances. Fat droplets, crystals or molecules of glycogen, and pigment granules are other examples of materials that are not metabolically active and serve as raw materials or sources of energy for cellular activity, or as products of activity.

The cellular organelles and their functions are summarized in Table 2.5.

THE NUCLEUS (FIG. 2.4)

The method by which the nucleus governs cellular activity is considered in greater detail in Chapter 4. Here, we may state that DIRECTION OF CELLULAR ACTIVITY is provided by the nucleus. Two MEMBRANES lie around the nucleus. The outer membrane is derived from the ER, and the inner one from the nucleus itself. At intervals, the two membranes appear to be fused to create "nuclear pores," actually minute areas of greater permeability. These pores may be 100 Å or more in diameter to allow passage of large nucleic acid molecules from the nucleus. A NUCLEOLUS, composed chiefly of ribonucleic acid (RNA) floats in the NUCLEAR FLUID. Deoxyribonucleic acid (DNA) forms the CHROMATIN MATERIAL of the nucleus. The chromatin material forms the visible chromosomes when the cell divides.

The DNA of the chromatin and chromosomes contain sequences of nucleotides that form the GENES. The genes determine the individuality of the organism. The basic method by which control of activity is achieved is by the DNA of the nucleus directing ultimately the synthesis of a protein, which, as an enzyme or hormone, influences a particular chemical reaction within the cell. Figure 2.5 shows some postulated interrelationships between DNA, the RNA it directs to be produced, and the proteins that result.

PASSAGE OF MATERIALS ACROSS MEMBRANES OR THROUGH FLUIDS

PASSIVE PROCESSES

Passive processes cause materials to move through membranes or fluids according to pressure or concentration differences, and do not require the cell to expend energy or do work to cause the movement to occur.

Bulk flow. If a pressure or concentration difference causes the rapid movement of considerable volumes of fluid *(solvent)* through a membrane in a given direction, and the contained particles (*solutes*, dissolved or suspended in the solvent) are swept through the membrane with the solvent; bulk flow has occurred. Such flow occurs through the most permeable membranes of the body, such as the blood capillaries in the body tissues and organs. This type of flow must be distinguished from the

TABLE 2.5

A SUMMARY OF CELL STRUCTURE AND FUNCTION

CELL PART	STRUCTURE	FUNCTION	COMMENTS
Plasma membrane	Three-layered sandwich of lipid and protein, or units (micelles)	Protects, limits, controls entry and exit of materials	Selectively permeable
Cytoplasm	Contains many organelles and inclusions	Factory area; synthesizes, metabolizes, packages	Bulk of cell
ER	Hollow channels in whole cytoplasm; filled with fluid	Transport	"Circulatory system" of cell
Ribosome	Granules of nucleic acid and protein	Protein synthesis	Free or attached to ER
Mitochondria	Double layered with cristae	Production of ATP	"Powerhouse" of cell
Golgi apparatus	Flattened sacs with vacuoles	Secretion; condensation membrane	Packaging of materials
Central body	Centrioles and centrosome	Cell division	Lacking or nonfunctional in mature nerve cells
Lysosome	Membrane surrounded sacs of enzymes	Reduces large molecules to smaller units	"Digestive system" of cell
Fibrils	Protein strands	Support and movement	Extreme development in muscle cells
Microtubules	Tiny hollow tubes	Support and conduction	Universal occurrence
Peroxisomes (microbodies)	Spherical; contain oxidative enzymes	Form hydrogen peroxide from oxygen; change peroxide to water	Protect cell from hydrogen peroxide excess
Inclusions	Fats, sugar, wastes	Fuels and wastes	"Nonliving"
Vacuoles	Membranous sacs of fluid	Storage and excretion	Are not active metabolically
Nucleus	Membrane, nucleolus, chromatin (DNA), and fluid	Directs cell activity	"Brains" of cell

passage of a randomly moving particle that may strike the membrane and pass through without affecting the movement of other particles in the solution.

Permeability constants. The rate at which a material crosses a membrane, or the rate at which it accumulates on one side of a membrane reflects how permeable the membrane is to that substance. Recall that permeability of a membrane depends on its thickness, chemical composition and other factors mentioned previously. The symbol k_p, or permeability constant, is used to express permeability of the membrane to a particular substance, and it takes into account the influence of such factors as thickness, temperature, and type of particle moving. It may be calculated using a mathematical formula*, and the larger the value of k_p, the more permeable is the membrane to that substance. Such calculations for a variety

* $dS/dt = \frac{-DA\ (C_i - C_o)}{d} = -\left(\frac{D}{d}\right) A\ (C_i - C_o)$ where dS = number of moles of a substance that will pass through a given area (A), in a given time (dt) under a concentration gradient or difference represented by the difference in the concentration of the material inside (C_i) and outside (C_o) the cell or membrane. d represents the thickness of the membrane, and is assumed to be 75 to 100 Å, although this may not be true. The ratio of D to d is k_p, and it has the dimensions of a velocity (cm/sec). D is the diffusion constant.

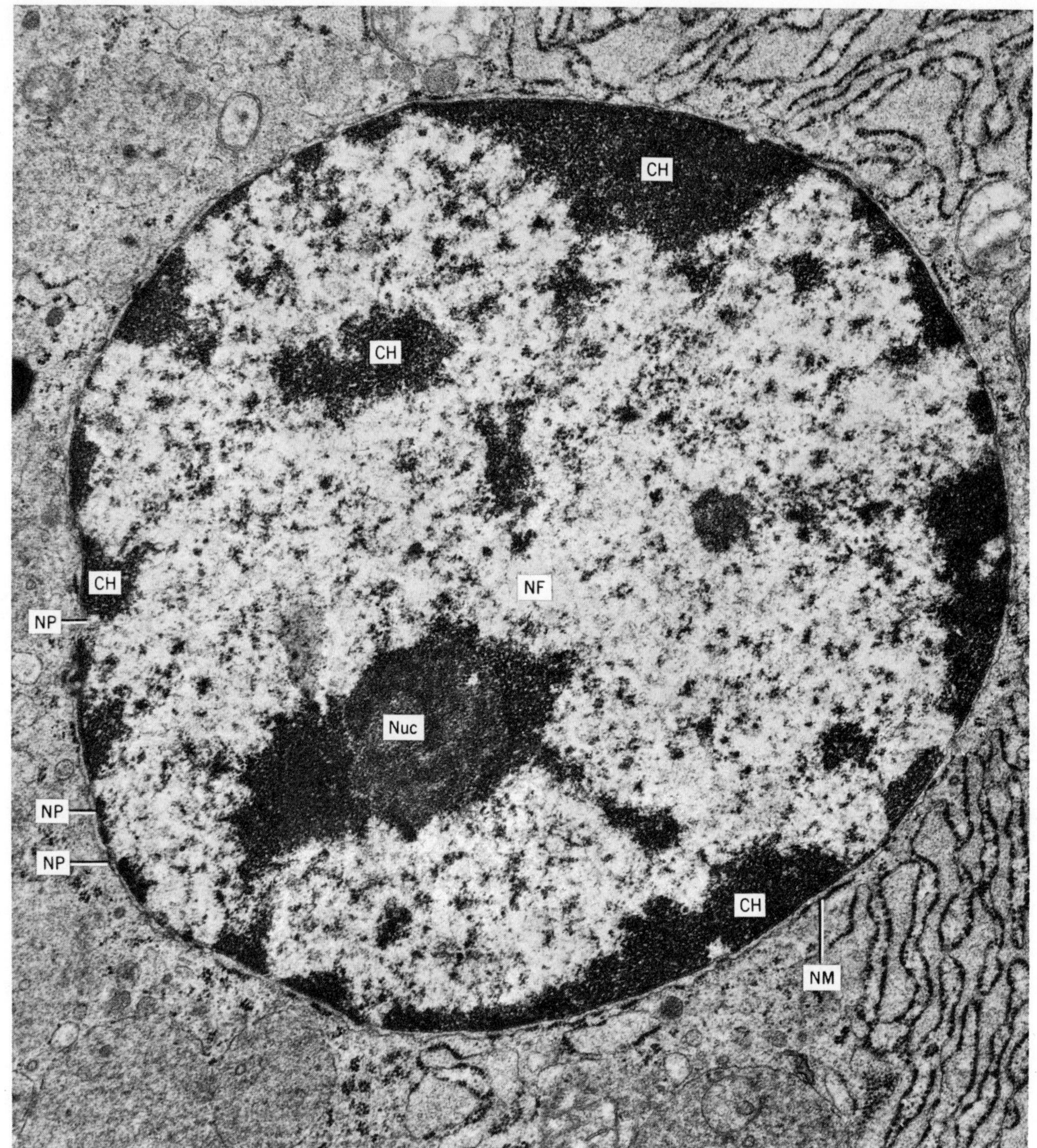

FIGURE 2.4
Nucleus from laboratory mouse trachea (mucous cell) NM = nuclear membrane; Nuc = nucleolus; NP = nuclear pores; CH = chromatin; NF = space containing nuclear fluid. ×15,000. (Courtesy of Norton B. Gilula, The Rockefeller University.)

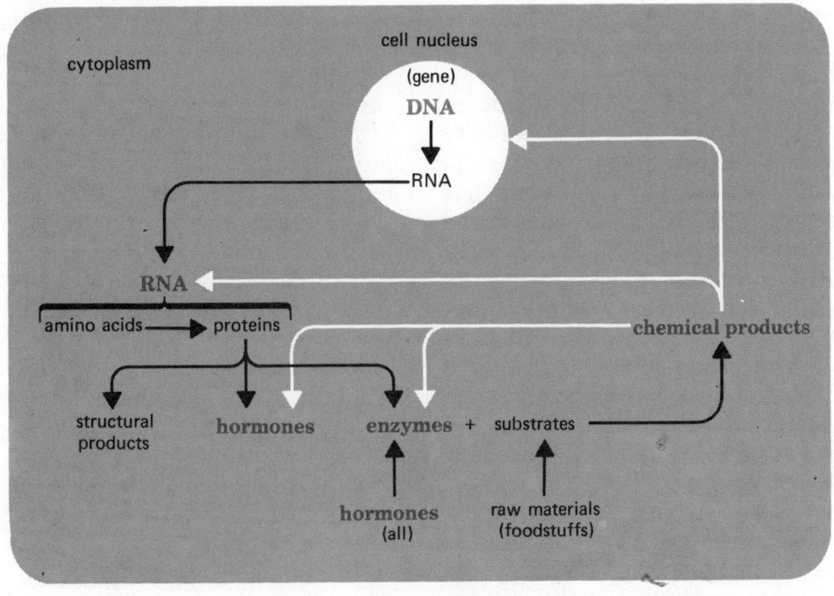

FIGURE 2.5
Control of cellular activity by genes and hormones, with controlling homeostatic mechanisms indicated by white lines. The control mechanisms may either stimulate or inhibit the activity of the structure controlled.

FIGURE 2.6
Diffusion. Molecules of dye move from area of higher concentration to area of lower concentration.

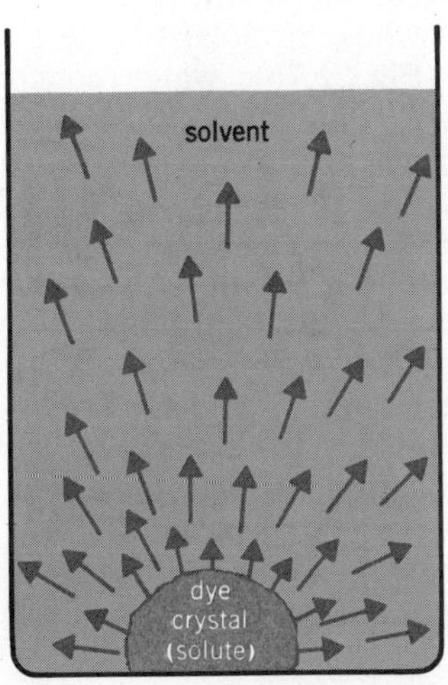

of substances indicate that each has its own k_p; in other words, the membrane "treats" each substance differently. Water appears to be a substance that enjoys a large k_p.

Diffusion (Fig. 2.6). All molecules are in constant motion, with their speed of movement depending on temperature. The warmer an object or solution is, the faster the movement of the constituent particles (molecules or ions). Movement of the particles causes them to collide with one another and with other solute or solvent molecules that are present. If there is a greater concentration of a particle (solute) in a given area of a solution, or on one side of a membrane, more collisions will occur per unit of time among particles and with the membrane. The solute will therefore distribute itself rapidly in the solvent initially, and then the distribution will slow as the solute concentration tends to equalize in the solvent. If a membrane is involved, the greater number of collisions on one side of the membrane will, if the membrane is permeable to the solute, result in a net passage of more solute particles, *from* the side of greatest concentration *to* the side of lesser concentration. As concentrations approach equality on the two sides of the membrane, passage will be equal in both directions, and equilibrium has been attained. Distribution of materials in the manner described is called diffusion. It may also be noted that large particles move more slowly than smaller ones at a given temperature, and are shifted less by collisions. In other words, they diffuse more slowly. Diffusion is thus a process that occurs more rapidly when small molecules are involved, as in the movement of oxygen and carbon dioxide across lung surfaces. Diffusion emphasizes two of the basic characteristics of any passive process: first, movement is from high to low (in this case concentrations of diffusing material); second, rate of movement slows as equality is approached. It may also be noted that passive processes occur from an area of higher kinetic energy to one of lower kinetic energy; flow of heat, solutes, and solvents all move because of this difference.

If a substance is capable of diffusing through a membrane, but its passage requires a carrier (see below), it is said to have undergone FACILITATED DIFFUSION. In this process, the carrier can react with a substance at either side of the membrane; thus, net transport in a given direction occurs only so long as there is a concentration difference on the two sides. Glucose may enter the cell partly by this process.

Osmosis. OSMOSIS is a type of diffusion, but the term is used to refer to a net movement of ONLY WATER *(solvent)* through a membrane that restricts the passage of solutes while permitting the movement of water molecules. Movement of water occurs from the area of higher water concentration to one of lower water concentration. What establishes a difference in water concentration on two sides of a membrane is the relative number of solute molecules, to which the membrane is not permeable, or permeable only to a slight degree. Thus, the greater the solute concentration, the less the water concentration as solute molecules displace water molecules in a given volume of solution. The total solute concentration of a solution is known as the OSMOLARITY of the solution, and the difference in osmolarity of the solutions on either side of a membrane gives an idea of the magnitude of the force that will cause water movement across the membrane. Osmolarity, and thus the rate of movement of water molecules, depends only on numbers of particles in the solution, not on their type. A one molar (1M) solution of glucose has an osmolarity of 1 osmol; a 1M solution of sodium chloride has an osmolarity of about 2 osmols, because the sodium chloride ionizes in solution to give both sodium ions and chloride ions (actual osmolarity is 1.8 osmols, since only about 90 percent of the ions act as if they are 2 osmotic particles).

Cells behave like osmotic systems because their membranes permit movement of water molecules, but restrict the passage of solute molecules. To illustrate the operation of the process, consider placing cells in solutions containing varying amounts of solute (Fig. 2.7). If a cell is placed in a solution that has an osmolarity equal to that of the cell (Fig. 2.7a), water *and* solute concentrations are equal inside and outside the cell, and water movement will be at equal rates into and out of the cell. The cell will neither shrink nor swell and is said to have been placed in an ISOTONIC (*iso,* equal) or ISOSMOTIC SOLUTION. If the cell is placed in a solution whose osmolarity is less than that of the cell (Fig. 2.7b), water concentration is greater than within the cell, and a *net flow* of water will occur into the cell; it will swell. In this situation, the cell is said to have been placed in a HYPOTONIC (*hypo,* less) or HYPOSMOTIC SOLUTION. If the cell is placed in a solution whose osmolarity is greater than that of the cell (Fig. 2.7c), the water concentration is now greater inside the cell, and a net flow of water will occur out of the cell; it will shrink. In this situation, the cell is said to have been placed in a HYPERTONIC (*hyper,* more) or HYPEROSMOTIC SOLUTION. Note that it is the solution *around* the cell that is designated iso-, hypo-, or hyper-tonic (-osmotic).

Three things may be reemphasized about osmosis:

Water flow occurs from the area of greater water concentration (the area of lesser solute concentration) to the area of lesser water concentration (the area of greater solute concentration).

There is always water movement in *both* directions across the membrane, but there will be a net flow that is greater in one direction than another if there is a difference in osmolarity on the two sides of a membrane.

As with all passive processes, the flow is from higher to lower areas of water and energy concentration, *as determined by the solutes.*

Obviously, one problem the body faces is to maintain water and solute concentrations in the extracellular fluids within narrow ranges, so that large shifts of water into and out of cells do not occur. The kidneys perform a large part of this regulatory task.

Dialysis. DIALYSIS occurs when there are two or more solutes on one side of a membrane, and the membrane is permeable to only certain ones. If a concentration difference exists for a solute, and the membrane is permeable to it, that solute will diffuse through the membrane, while solutes to which the membrane is not permeable will be retained. Thus, a separation of solutes will occur. This principle is utilized in the artificial kidney, where the patient's blood (about 500 ml at a time) is passed through a series of dialysing tubes. The tubes are surrounded by a solution that has the same concentration of diffusible solutes as the blood, except for zero concentration of solutes to be removed from the bloodstream. There is no net flow of necessary solutes, but molecules such as urea move from the blood in the tube to the solution (solution is hypotonic or hypo-osmotic for that particular solute). Renewing the solution as these "wastes" diffuse out keeps the concentration gradient high and can cause the removal of nearly all the toxic substance.

Filtration. FILTRATION occurs when there is a

FIGURE 2.7
Osmosis. Changes in cell size in (*a*) isotonic, (*b*) hypotonic, and (c) hypertonic solutions. Arrows indicate flow of water with length denoting greatest direction of flow.

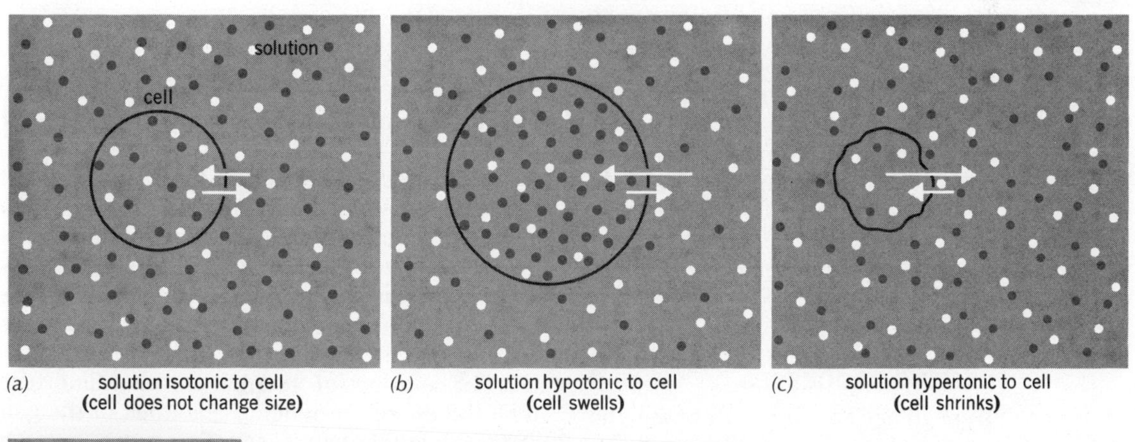

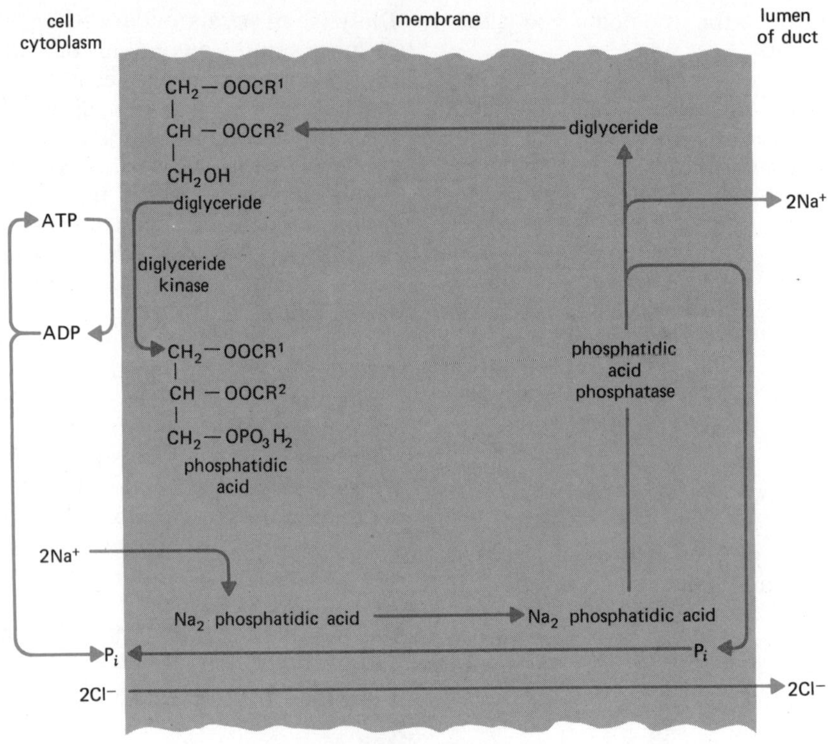

FIGURE 2.8
A theory of active transport of Na^+ involving the use of a carrier (diglyceride), an energy source (ATP), and enzymes (kinase and phosphatase). (After Hokin and Hokin, *J. Gen. Physiol, 44,* 1960.)

PRESSURE OR ENERGY DIFFERENCE on the two sides of a membrane. The membrane acts much like a sieve, allowing molecules to be forced through the membrane by the pressure according mainly to size. Rate of passage is determined by the magnitude of the pressure difference. The movement of solutes and solvent through capillary walls due to blood pressure is an example of filtration as it occurs in the body.

ACTIVE PROCESSES

If, to move a substance across a membrane, the cell is required to participate by supplying energy or specific molecules for the process, the movement is called active.

Active transport. ACTIVE TRANSPORT, sometimes called *mediated transport,* is one distinguished from a passive process in several ways.

Molecules of a SPECIFIC type are handled by the transport system. Glucose, amino acids, and inorganic ions are transported, all by different systems. The evidence for different systems revolves around the fact that each moves at a rate independent of the others, suggesting separate transport systems.

The addition of INHIBITORS that interfere with the energy producing reactions of the cell generally stops the transport process.

Transport tends to occur at more or less CONSTANT RATES, providing temperature does not change, suggesting some rate-limiting chemical reaction is involved.

Most transport systems have a MAXIMUM RATE at which they can carry materials. The system can be *saturated* by presenting it with more material than it can handle. Thus, a curve depicting entry of a substance into a cell by active transport as a function of concentration outside the cell will show a plateau when the transport maximum is reached.

COMPETITION exists between materials for the same transport mechanism. For example, a red dye, chlorphenol red, and an X-ray contrast medium called Diodrast, are transported by the same kidney mechanism. If the dye is present alone, it will be transported at a given rate. If the Diodrast is added to the solution, the rate of dye transport will decrease. This is evidence that two molecules are competing for the same transport system. Competitive inhibition (on dye transport) has occurred.

Active transport can move substances *against* concentration gradients if energy supply is not diminished. For this reason, active transport mechanisms are often referred to as "pumps."

Explanations of how active transport occurs have centered mainly around hypotheses that suggest the presence in the membrane of a "carrier molecule." This molecule is activated, probably via ATP and attaches to a substance that has formed a bond with the cell membrane on one of the membrane's binding sites. The reaction between the carrier and the molecule to be transported may require enzymes. The carrier then delivers the substance to the other side of the membrane.

The nature of the carrier has not been determined. Some years ago, Hokin and Hokin (Fig. 2.8) related active transport to the metabolism of certain lipids. A protein may be involved (Fig. 2.9), or a "mobile complex" (Fig. 2.10) may be involved. If a specific molecule *is* involved as a carrier, it may diffuse from one side of the membrane to the other, rotate, or "unwind" to move the material across the membrane. Lastly, a "gate" hypothesis has been advanced suggesting that the binding of a substance to the membrane causes induction of a pore or "opens a gate" through the membrane.

In summary, we may indicate that the characteristics of active transport have been determined, but the actual mechanism remains undetermined. A primary need that bears reemphasizing is the requirement for an energy source for active transport, provided by ATP.

Endocytosis. Cells take in not only individual ions or molecules by active processes, but may acquire macromolecules (nucleic acids, viruses), particulates (bacteria), and liquid droplets. The general name for such methods of acquisition is ENDOCYTOSIS. It is subdivided into pinocytosis and phagocytosis.

FIGURE 2.9
The protein-coiling hypothesis of active transport. (Diagrammatic.)

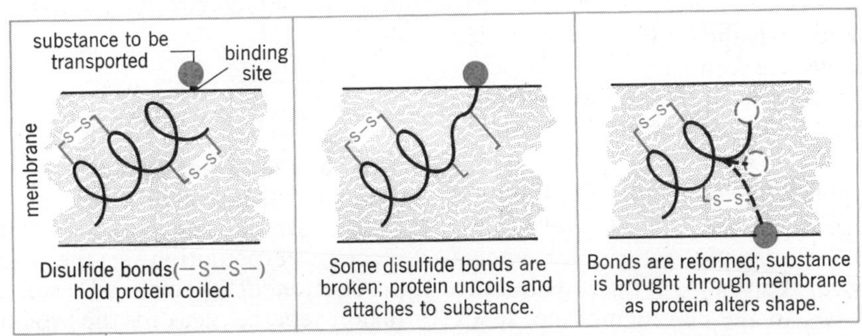

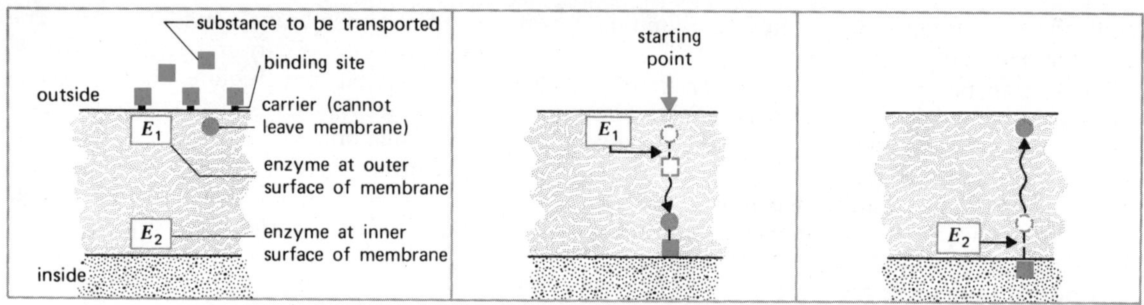

Substance to be transported is joined to carrier by E_1. Many such complexes are formed, and diffuse to the inner boundary of the membrane.

E_2 splits substance from carrier, and it enters cell. Carrier diffuses back to outer boundary of the membrane to repeat the process.

FIGURE 2.10
The mobile-complex theory of active transport. It is similar to the mechanism depicted in Figure 2.8, but not tied to lipid metabolism.

PINOCYTOSIS, or "drinking" by cells, results when the cell membrane "sinks" or invaginates and a membrane-surrounded intracellular vesicle containing the fluid and any dissolved or suspended material is formed within the cell. PHAGOCYTOSIS occurs in similar fashion, but takes in particulate matter, even bacteria.

In phagocytosis, several steps are usually involved:

There is attraction of the phagocytic cells to the site of injury, invasion, or inflammation, probably by the release of chemicals by the injury or inflammation.

The particles are coated by gamma globulin (opsonization) to increase attractiveness to the phagocyte.

The phagocyte ingests the particle as described above. The vesicle formed by either process may fuse with a lysosome, and the molecules or microorganism(s) may be digested and the end products released into the cell cytoplasm as nutrients. Some of the relationships described are depicted in Figure 2.11.

Exocytosis. A "reverse pinocytosis" or emeiocytosis can occur if a vesicle formed within the cytoplasm moves to the cell membrane, fuses with it, and discharges its contents to the exterior of the cell. Such activity is commonly referred to as EXOCYTOSIS. This process may be utilized by the cell to eliminate wastes or materials that might prove toxic to the cell, or to remove excess substances (salts), deactivated products of cell activity, and products (e.g., hormones).

Considered together, the mechanisms described in this section enable a cell to exert a great degree of control over what it accepts from its environment. That such control mechanisms exist at the cellular level is to further emphasize the importance of control in maintaining body homeostasis. Control *must* exist at the cellular level, for organism survival depends on cell survival, and survival involves control.

CELL REPRODUCTION

The body does not spend its life with the same number or same cells with which it was born. Cells wear out and die, and often must be replaced. Perpetuation of the species requires production of specialized "sex cells," whose fusion sets the basis for the production of a new

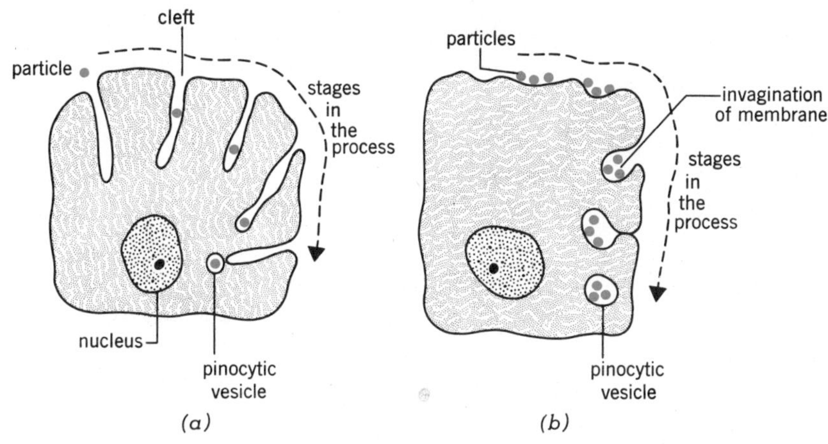

Pinocytosis. (*a*) The particle enters a cleft and becomes enclosed in a vesicle. (*b*) The particle is adsorbed on the surface of the membrane and is enclosed in a vesicle. (*c*) Electron micrograph of formation of vesicles in skeletal muscle capillary (× 22,000); pv = pinocytic vesicle.

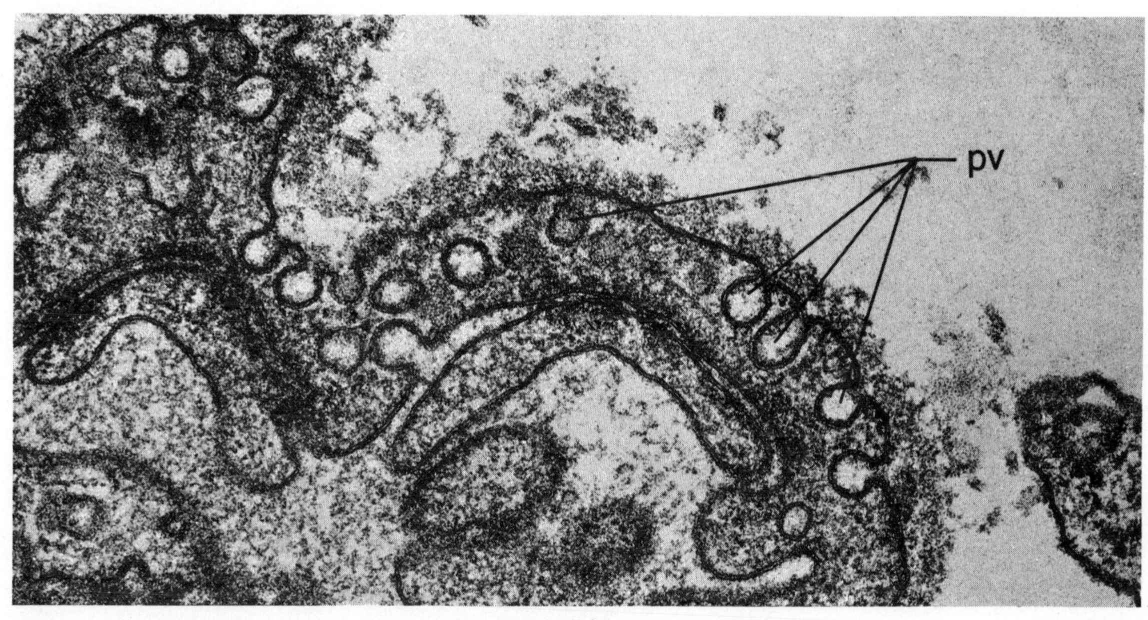

(*c*)

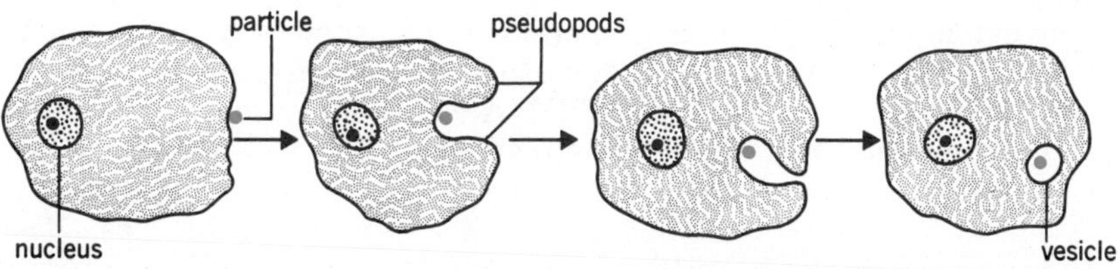

Phagocytosis. Pseudopods are formed that engulf the particle.

FIGURE 2.11
Pinocytosis (upper) and phagocytosis (lower).

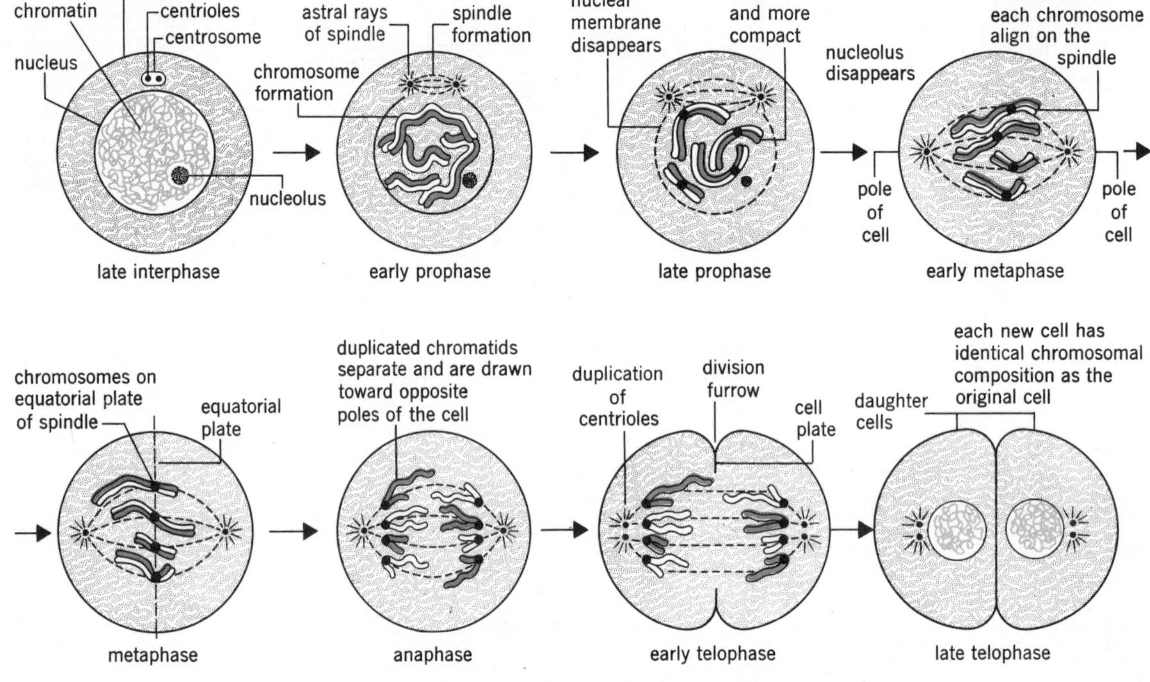

FIGURE 2.12
The major events occurring in mitosis as depicted in a cell containing four chromosomes.

organism. Two processes occur in the body to ensure cellular replacement and perpetuation of the species.

MITOSIS

MITOSIS (Fig. 2.12) is the type of cell reproduction the body utilizes as a method of growth and to replace dead or lost cells. In this type of division, the nucleus undergoes a complicated series of changes that ensure the daughter cells will receive the same complement of DNA (chromosomes) as possessed by the parent cell. The cytoplasm and its organelles undergo an approximately equal division between the daughter cells by simply cleaving in half. Mitosis is described as occurring in STAGES, each designated by events that are characteristic for that stage.

In INTERPHASE, the cell appears typical for its type, and is not undergoing division. It is engaged in accumulating chemicals for its own use, synthesis of proteins, metabolism of fuels, replication of DNA, and RNA production. The chromatin material, and therefore each chromosome, is duplicated. (Strands of nucleic acid attach to a portion of a chromosome called a centromere. One centromere is equivalent to one chromosome regardless of how many strands of DNA are attached to it.)

In PROPHASE, the centrioles migrate to opposite poles of the cell, and a spindle of microtubules forms between the centrioles. The nuclear membrane disappears and the chromatin resolves itself into visible chromosomes. At this stage, each chromosome consists of two threads (chromatids) that are copies of one another.

In METAPHASE, the chromosomes align on the equatorial plate of the cell, and attach to the spindle.

In ANAPHASE, centromeres divide and the duplicated chromosomes separate, possibly by repulsion or being pulled by the spindle fibers. The centromeres "lead" the separating chromosomes and they present a typical "V" or "J" shape.

In TELOPHASE, the nucleus reorganizes, chromosomes return to chromatin material, centrioles divide, and the cytoplasm undergoes division. The cell then returns to its interphase appearance.

Studies of the timing of mitosis in the human indicates that it has a cyclical duration 12 to 24 hours in length, from interphase to interphase. Actual division takes only about one hour.

MEIOSIS

MEIOSIS (Fig. 2.13) occurs in the gonads (ovaries and testes) and produces gametes (sex cells: ova and sperm). Each daughter cell formed contains one-half the chromosome number present in the parent cell. If this process did not occur, the next generation would have twice the normal species number of chromosomes, as egg and sperm combined during fertilization. Such a situation is not compatible with human survival. In meiosis, *two* cell divisions occur, one without previous duplication of chromosomes. Also, exchange of portions of chromosomes may occur during the process. This exchange produces genetic variability in the members of a given pair of chromosomes. Again, stages are described in the process.

In MEIOTIC INTERPHASE, duplication of chromatin occurs.

In FIRST MEIOTIC PROPHASE, chromosomes appear, the nuclear membrane disappears, centrioles migrate, and a spindle is formed as in mitosis. Homologous chromosomes (a pair of chromosomes, one from each parent that have the same genes in the same order) pair, and crossing over of arms occurs. These points of CROSSING OVER (*chiasmata*) may also be sites of breakage, whereby a part of one homologous chromosome is exchanged for a part of another. This creates a variability in the "new chromosome" that results.

In FIRST MEIOTIC METAPHASE, chromosomes align on the equatorial plate of the cell, but the *centromeres do not divide.*

In FIRST MEIOTIC ANAPHASE, reductional division occurs, with one member of each homologous pair being drawn to the opposite poles of the cell. Thus, the chromosome number is divided in half (from 46 to 23 for the human).

FIRST MEIOTIC TELOPHASE sees a cleavage of the cell cytoplasm and the formation of two daughter cells, each with a haploid chromosome number.

A SECOND MEIOTIC DIVISION then occurs that is basically a *mitotic division;* that is, the centromeres divide and the duplicated chromosomes are separated without further change. Two daughter cells are formed from each previous daughter cell, giving a total of four haploid cells from each diploid parent cell.

Formation of sperm in man has a time cycle of about 74 days. In woman, ova production proceeds to first meiotic prophase during intrauterine life, and then remains "suspended" until puberty sees commencement of ova maturation and ovulation. The time period for egg maturation may be 9 to 50 years.

NEOPLASMS (CANCERS)

Cell division by mitosis is usually balanced to the needs of the body for cell replacement and repair. The mechanisms controlling such replacement are virtually unknown, but are undoubtedly related to chemical factors and blood supply. Uncontrolled growth constitutes a NEOPLASM (literally, new growth), commonly called a cancer.

Two major types of neoplasms are recognized. BENIGN NEOPLASMS usually grow slowly, are limited from surrounding normal tissue by connective tissue capsules, and are not considered grave threats to life unless they develop in areas where they are difficult to detect. For example, in the brain, mechanical pressures develop as the size of the neoplasm increases.

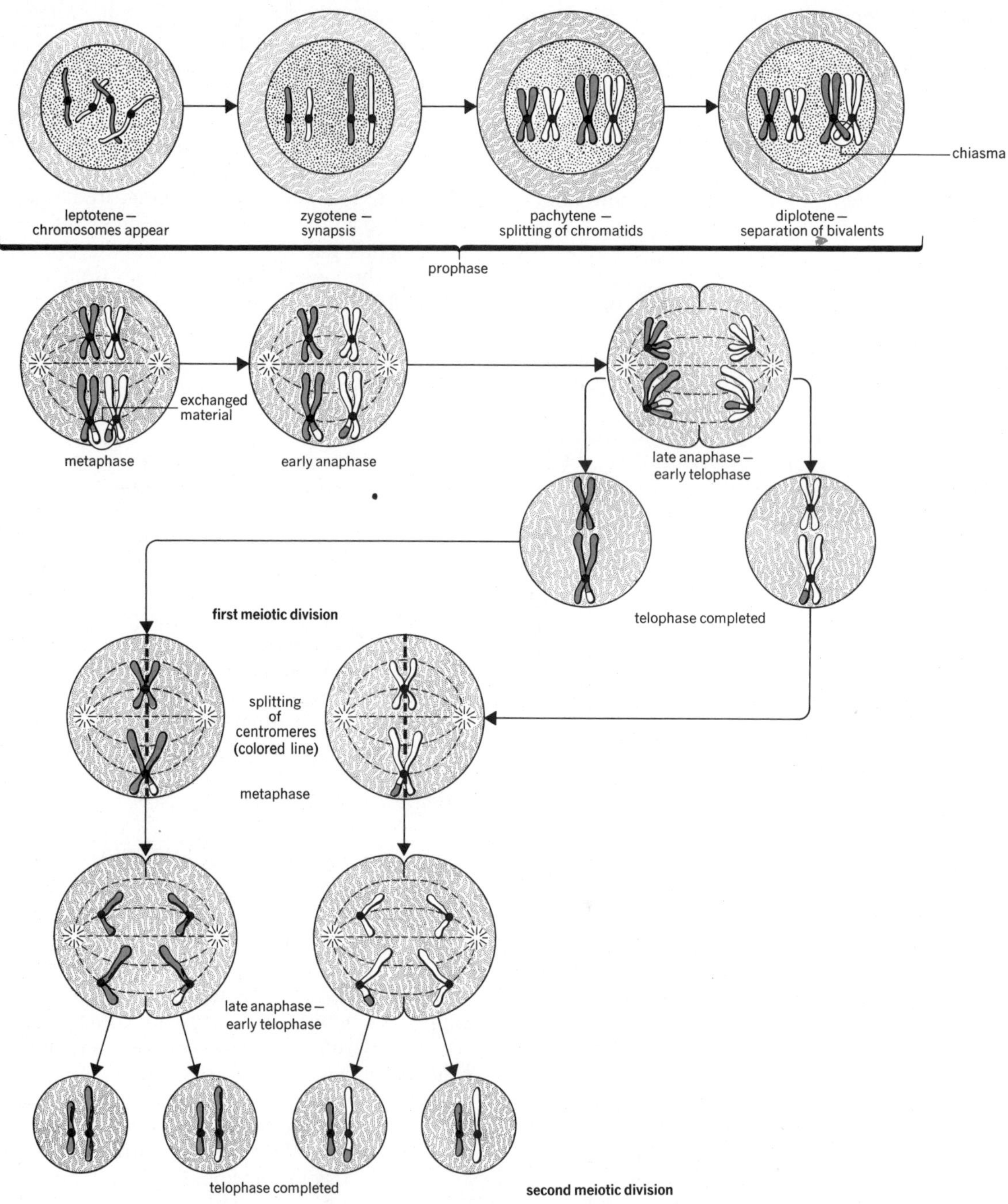

FIGURE 2.13
The major events of meiosis. Only two of the 23 human chromosome pairs are shown. Chromosomes from one parent are solid color, those from the other are white.

MALIGNANT NEOPLASMS grow rapidly, are not usually limited by capsules, and shed cells (*metastasis*) into blood and lymph vessels. New sites of "infection" may be created as these cells are trapped in different body areas.

Causes of neoplasms may be genetic (Wilm's tumor of the kidney), or environmental. The latter involves cancer-causing agents (*carcinogens*) that increase the odds that the cells may grow uncontrolled. Viruses, radiation, and chemicals in food and drink have been shown to be carcinogenic. Prognosis depends on early discovery and adequate treatment.

Terminology used to describe neoplasms is often confusing. The word TUMOR (*swelling*) is nondescriptive of either cause or characteristics of the neoplasm. Benign tumors are usually classified by adding the suffix *-oma* to the cell or tissue of origin; for example, *osteoma* (bone neoplasm), *lipoma* (adipose neoplasm) or *fibroma* (connective tissue neoplasm). Malignant neoplasms are designated CARCINOMAS if they arise from epithelial cells, or SARCOMAS if they arise from connective tissues. They are further described as to GRADE (I to IV according to the degree the neoplastic cells differ in organization and structure from normal cells); by STAGE (extent of spread from original site of development—Stage I implies limited to original site, IV indicates widespread metastasis); by TYPE (A,B,C) according to depth of penetration through the wall of an organ (A implies inner layers only involved, B implies the whole wall is involved, and C implies lymph nodes in the area are invaded).

A neoplasm sustains itself at the expense of the host, and emaciation, loss of weight, anemia, and disfigurement may result.

The aim of treatment of a neoplasm is to remove or inhibit the growth while promoting normal tissue responses and patient comfort and well-being.

SUMMARY

1. Cell size depends on nuclear-cytoplasmic relationships and ability to transmit instructions from nucleus to cytoplasm. Ability of cell surface to supply cytoplasmic volume with nutrients and remove its wastes also determines cell size.
2. Cell shape is determined mainly by function and requirements of the cell for nutrients.
3. Cells consist of membrane systems, cytoplasm, and nucleus.
   a. Membrane systems differ in structure according to particular membrane, but all have a lipid-protein basis. Membranes control entry and exit of materials into and out of cells and its parts, aid in enzymatic reactions and bind substances.
   b. The cytoplasm contains organelles that carry out anabolic and catabolic reactions. Table 2.5 summarizes these organelles and their functions. The cytoplasm also contains inclusions that represent storage facilities, energy sources, or wastes.
   c. The nucleus directs cell activity by way of DNA, RNA, and synthesized proteins.
4. Cells acquire materials by passive and active processes.
   a. Bulk flow moves solvent and solutes as a mass.
   b. Rate of movement of a substance is reflected by its permeability constant. The higher the constant, the more permeable the membrane is to it.
   c. Diffusion refers to movement of solutes, depends on molecular motion, and may occur with or without the presence of a membrane.
   d. Facilitated diffusion utilizes carriers and is used to speed the passage of a substance in a direction it would normally pass by diffusion alone.
   e. Osmosis is diffusion of solvent molecules through a semipermeable membrane due to a concentration difference for water. Osmolality refers to solute concentration in a solution, and cells retain shape in a solution (isotonic) that has osmolality equal to that of the cell, swell in a solution (hypotonic) that has a lower osmolality and shrink in a solution (hypertonic) that has a greater osmolality than the cell.
   f. Dialysis separates solutes because of differential membrane permeability to the solutes.
   g. Filtration is movement by hydrostatic pressure with the membrane acting like a sieve to pass molecules mainly on the basis of size.
   h. Active transport requires the cell's participation, transports specific substances, may be

inhibited by metabolic "poisons," has a maximum rate of transport, and may be competitively inhibited. The carrier hypothesis has been advanced as a device to explain active transport.

i. Endocytosis refers to the taking in of liquids (pinocytosis) and solids in that liquid, and particles (phagocytosis) by cells.

j. Exocytosis is a "reverse pinocytosis" by which wastes and other materials may be removed from cells.

5. If a cell is capable of dividing, reproduction of cells involves division of the nucleus and cytoplasm. Two processes occur.
   a. Mitosis results in cells whose nuclei are genetically identical to the original cell, and whose cytoplasm has undergone nearly equal division between the two daughter cells. Healing of wounds and replacement of cells occurs by mitosis.
   b. Meiosis produces four daughter cells for each parent cell, reduces the chromosome number in half for each of those cells, and introduces variation in the genetic potential of the daughter cells (by crossing over).

6. Neoplasms represent uncontrolled cell division of no value to the host. They may be benign or malignant, and are due to genetic and environmental factors. Neoplasms are classified according to structural changes, degree of spread, and layers of an organ involved.

QUESTIONS

1. If you wished to make an "ideal cell" that would carry out the body's basic processes, what would you include in it? Justify each item you include in your cell. How big would it have to be, and what shape would you propose for it?
2. What properties would an ideal cell membrane possess? What functions would you expect it to serve in order to maintain the cell homeostasis against stressors?
3. Contrast diffusion and filtration as to driving force, selectivity, and what determines when or if the process ceases to move materials.
4. Define the terms isotonic (iso-osmotic), hypotonic (hypo-osmotic), and hypertonic (hyperosmotic) solutions, and discuss what would happen and why to a human cell placed in each type of solution.
5. Outline the carrier hypothesis of active transport, and discuss several characteristics of active transport.
6. In what ways are phagocytosis and pinocytosis alike? Different?
7. Compare and contrast mitosis and meiosis as to the effects of each process on nuclear chromatin (chromosomes), and the value of each to the body.
8. Speculate as to the value of meiosis to the evolutionary process and Darwin's principle of survival of the fittest.
9. In what ways do neoplasms differ from normal cells or tissues? How may a neoplasm cause death of the host?

READINGS

Baltimore, David. "Viruses, Polymerases, and Cancer." *Science 192*:632. 14 May 1976.

Berlin, R. D., J. M. Oliver, T. E. Ukena, and H. H. Yin. "The Cell Surface." *New Eng. J. Med. 292*:515. March 6, 1975.

Brown, Michael S., and Joseph L. Goldstern. "Receptor-Mediated Control of Metabolism." *Science 191*:150. 16 Jan 1976.

Capaldi, Roderick A. "A Dynamic Model of Cell Membranes." *Sci. Amer. 230*:26. (Mar) 1974.

Claude, Albert. "The Coming of Age of the Cell." *Science 189*:433. 8 Aug 1975.

Cone, Clarence D., Jr., and Charlotte M. Cone. "Induction of Mitosis in Mature Neurons in Central Nervous System by Sustained Depolarization." *Science 192*:155. 9 April 1976.

Culliton, Barbara J. "Breast Cancer: Second Thoughts About Routine Mammography." *Science 193*:555. 13 Aug 1976.

DeDuve, Christian. "Exploring Cells with a Centrifuge." *Science 189*:186. 18 July 1975.

Glynn, I. M., and S. J. D. Karlish. "The Sodium Pump." *Annual Rev. Physiol. 37*:13. Palo Alto, Cal. 1975.

Hoover, Robert, et al. "Cancer by County: New Resource for Etiologic Clues." *Science 189:*1005. 19 Sept 1975.

Kolata, Gina Bari. "Ribosomes (I): Genetic Studies with Viruses." *Science 190:*136. 10 Oct 1975.

Kolata, Gina Bari. "Water Structure and Ion Binding: A Role in Cell Physiology?" *Science 192:* 1220. 18 June 1976.

Luria, Salvador E. "Colicins and the Energetics of Cell Membranes." *Sci. Amer. 233:*30. (Dec) 1975.

Marx, Jean L. "Cell Biology: Cell Surfaces and the Regulation of Mitosis." *Science 192:*455. 30 April 1976.

Marx, Jean L. "Estrogen Drugs: Do They Increase the Risk of Cancer?" *Science 191:*838. 27 Feb 1976.

Marx, Jean L. "Immunology: Role of Immune Response Genes." *Science 191:*277. 23 Jan 1976.

Maugh, Thomas H. "Ribosomes (II): A Complicated Structure Begins to Emerge." *Science 190:*258. 17 Oct 1975.

Mazia, Daniel. "The Cell Cycle." *Sci. Amer. 230:*54. (Jan) 1974.

Nicholson, Garth L., and George Poste. "Cell-Surface Organization and Modification with Cancer." *New Eng. J. Med. 295:*197. July 22, 1976. *295:*253. July 29, 1976.

Simpson, Ian, Birgit Rose, and Werner R. Loewenstein. "Size Limit of Molecules Permeating the Junctional Membrane Channels." *Science 195:*294. 21 Jan 1977.

Weber, George. Medical Progress: "Enzymology of Cancer Cells." *New Eng. J. Med. 296:*486. March 3, 1977. *296:*541. March 10, 1977.

3

CHEMICAL ORGANIZATION OF THE CELL

OBJECTIVES

After studying this chapter, the reader should be able to:

☐ Explain the differences between elements, radicals, ions, molecules, and compounds.

☐ Differentiate between a solution and a colloidal suspension.

☐ List the most common elements and molecular constituents of the body.

☐ Show the molecular configuration or give examples of the formulae, and give functions for each of the following constituents of living material: water, electrolytes, proteins, enzymes, lipids, carbohydrates, nucleic acids, and trace substances.

☐ Describe briefly the subdivisions or categories of materials included in the proteins, carbohydrates and lipids and give examples of and functions for each. Specifically: amino acids, polypeptides, monosaccharides, disaccharides, polysaccharides; simple, compound, and derived lipids.

CHEMICAL ORGANIZATION OF THE CELL

Living material is composed of free elements, ions, radicals, molecules, and compounds. ELEMENTS, numbering over 100, are substances in atomic form and are the simplest chemical units that retain the properties ascribed to them. Elements, alone or in combinations known as RADICALS, are usually capable of gaining or losing electrons, and thus may become IONS or electrically charged units. A MOLECULE is formed of two or more atoms or ions to create a substance having properties different from the substances composing it. The term COMPOUND is used to refer to large complex molecules or the combination of molecules into a unit having properties different from the molecules composing it.

These materials may be found in the cell as a SOLUTION, in which the materials exist in the dissolved or unaggregated state. They may also exist as a COLLOIDAL SUSPENSION, in which the particles are large and form aggregates that do not diffuse easily through membranes. Such materials may also form part of the cellular structure (cell membranes and organelles).

The living material is thus a heterogenous mixture of chemical substances in various states exhibiting different properties.

THE COMPOSITION OF THE CELL

Four elements compose over 95 percent of the body. These elements and their percent of body weight are: hydrogen (H) 10 percent, oxygen (O) 65 percent, carbon (C) 18 percent, and nitrogen (N) 3 percent.

Additional elements common to the body include calcium (Ca), sodium (Na), potassium (K), phosphorus (P), sulfur (S), and magnesium (Mg).

Hydrogen, oxygen, carbon, nitrogen, phosphorus, and sulfur are usually found linked into larger molecules and compounds to form the basis of body structure and function. These six elements form water, carbohydrates, lipids, proteins (some of which are enzymes), nucleic acids, hormones, and other substances essential for life. Analysis of the adult body for major molecular constituents indicates the following.

MATERIAL	PERCENT OF BODY WEIGHT MALE	FEMALE
Water	62	59
Protein	18	15
Lipid	14	20
Carbohydrates	1	1
Other (nucleic acids, radicals)	5	5

MOLECULES AND SUBSTANCES FOUND IN THE CELL

WATER composes, on the average, 55 to 60 percent of the cell substance. It is a good SOLVENT, is NOT TOXIC to the cell when isotonic, and is an excellent MEDIUM FOR HEAT TRANSFER. In water solution many substances IONIZE or assume electrical charges, making them more reactive chemically. Most water in the body is available for movement by osmosis between fluid compartments. About 4 percent of the total body water appears to be associated with cell membranes, electrolytes, and other materials as hydration layers. Such layers consist of water molecules attached by bonding forces to the cellular constituents.

ELECTROLYTES are substances that can dissociate into electrically charged particles. They may be ELEMENTS and RADICALS (two or more atoms behaving as a single unit) and are known as IONS. Positively charged materials are attracted to a negatively charged electrical pole *(cathode)* and are thus called CATIONS; negatively charged materials are attracted to a positively charged electrical pole *(anode)* and are thus called ANIONS. A solution of electrolytes will conduct electrical current. The chief body electrolytes are presented in Table 3.1.

These electrolytes are necessary for establishment of OSMOTIC GRADIENTS, influence enzyme activity, function as components in BUFFER SYSTEMS, and aid in the establishment of EXCITABILITY (membrane potentials) in all cells.

PROTEINS are complex molecules composed primarily of C (50 to 55 percent), H (6 to 8 percent), O (20 to 23 percent), N (15 to 18 percent), and S (0 to 4 percent). The basic building units of proteins are some 20 AMINO ACIDS (Fig. 3.1) linked together by peptide bonds (Fig. 3.2). Intermediate in complexity between amino acids and proteins are the PEPTIDES. A peptide containing less than 10 amino acids is known as an *oligopeptide;* one with more than 10 amino acids is known as a *polypeptide*. Con-

TABLE 3.1
MAIN ELECTROLYTES OF THE BODY

SUBSTANCE	CHARGE	ELEMENT	OR RADICAL	ANION	OR CATION
Sodium (Na^+)	+1	X			X
Potassium (K^+)	+1	X			X
Calcium (Ca^{++})	+2	X			X
Magnesium (Mg^{++})	+2	X			X
Chloride (Cl^-)	−1	X		X	
Phosphate ($HPO_4^=$)	−2		X	X	
Sulfate ($SO_4^=$)	−2		X	X	
Bicarbonate (HCO_3^-)	−1		X	X	

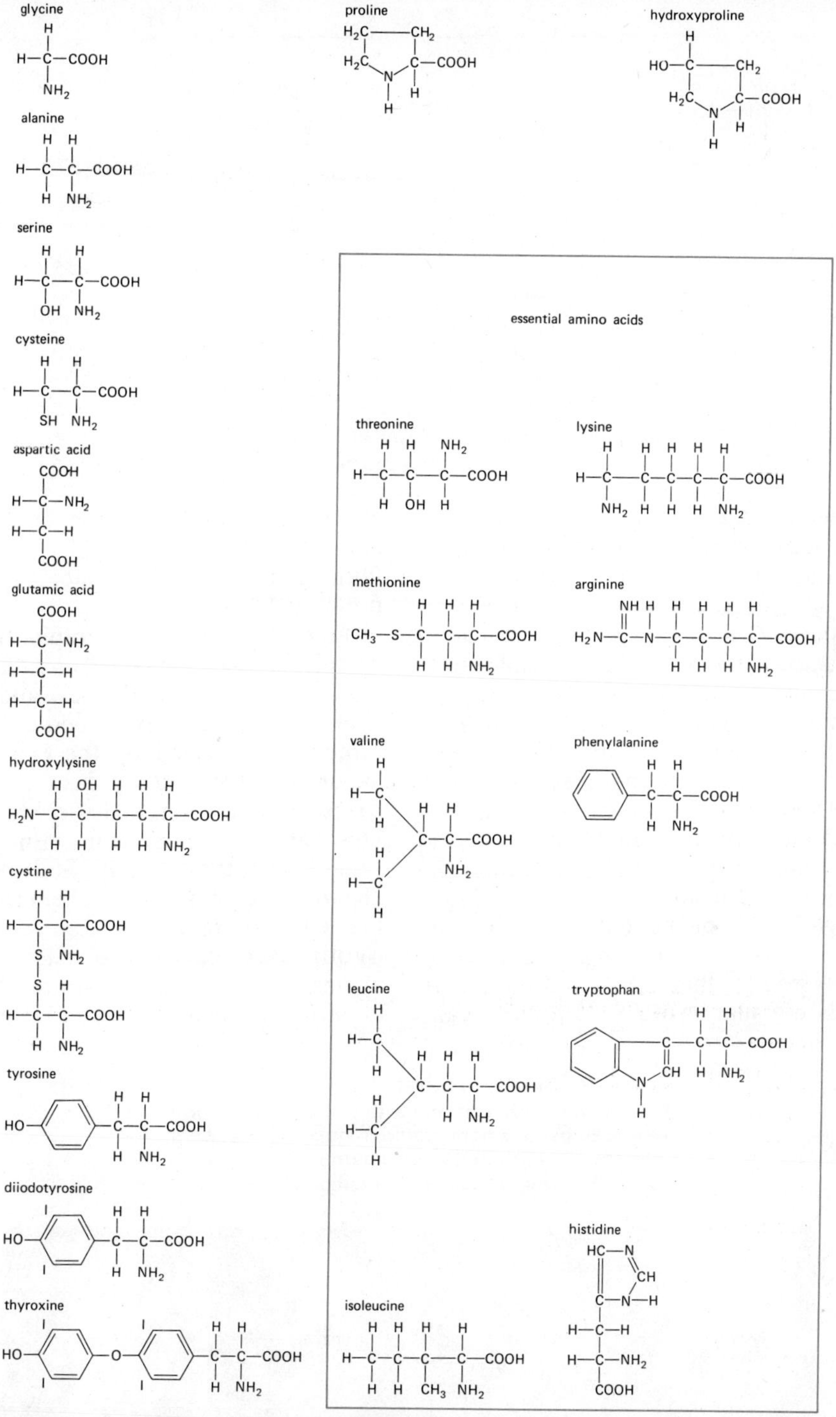

glycine

proline

hydroxyproline

alanine

serine

cysteine

aspartic acid

glutamic acid

hydroxylysine

cystine

tyrosine

diiodotyrosine

thyroxine

essential amino acids

threonine

lysine

methionine

arginine

valine

phenylalanine

leucine

tryptophan

histidine

isoleucine

FIGURE 3.1

Amino acids. *Essential* amino acids *cannot* be synthesized by body cells and must be ingested as such in the diet. The remaining amino acids (*nonessential*) are produced by body cells. Arginine and histidine are synthesized, but not in amounts sufficient to meet body needs.

$$\text{R—CH(NH}_2\text{)—C(=O)—}\boxed{\text{OH + H}}\text{—N(H)—CH(R)—COOH} \longrightarrow \text{R—CH(NH}_2\text{)—}\boxed{\text{C(=O)—N(H)}}\text{—CH(R)—COOH} + H_2O$$

amine group (NH_2)

carboxyl group (COOH)

peptide bond

FIGURE 3.2
The linking of two amino acids by a peptide bond.

jugated proteins are a combination of a protein and a nonprotein substance (e.g., nucleoproteins, lipoproteins).

Two broad categories of proteins are described. FIBROUS PROTEINS consist of elongated filamentous chains that are stable and relatively insoluble. Fibrous proteins are found as contractile elements in muscle (actin filaments), as the collagen molecules in connective tissues, and as the keratin molecules in the epidermis of the skin. GLOBULAR PROTEINS are more soluble and exhibit folding, which makes them more compact than the fibrous proteins. Hemoglobin (the pigment within red blood cells), myoglobin (a pigmented substance in muscle cells), cytochrome (a protein serving to transport electrons within cells), and the albumins are all globular proteins. Proteins can change from fibrous to globular states, as in the conversion of the soluble globular protein fibrinogen, to the insoluble fibrous protein fibrin, as blood clots.

ENZYMES permit the rapid progression of chemical reactions and act as catalysts. Without enzymes, these reactions would proceed too slowly to be useful to the body. An enzyme is a protein synthesized by the cell and is inactivated or destroyed by heat, strong acids or bases, products of cellular activity (metabolites), and organic solvents. Enzymes are SPECIFIC in that they usually catalyze only a single chemical reaction. The chemical with which the enzyme reacts is termed its *substrate*. GROUP SPECIFICITY also exists for some enzymes in that several substrates that are similar in structure (e.g., hexoses) may be reactive

FIGURE 3.3
The reaction of an enzyme with a substrate to form an enzyme-substrate combination. The combination then gives the final products and returns the enzyme to its original state. S = Substrate.

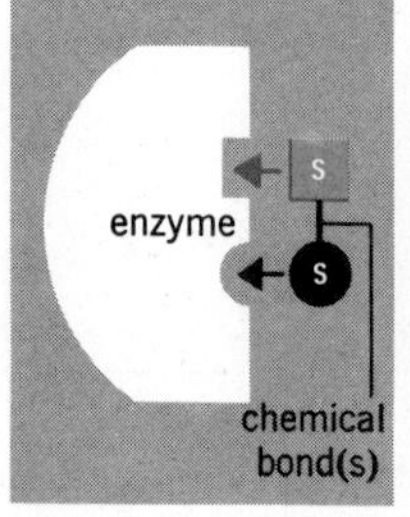

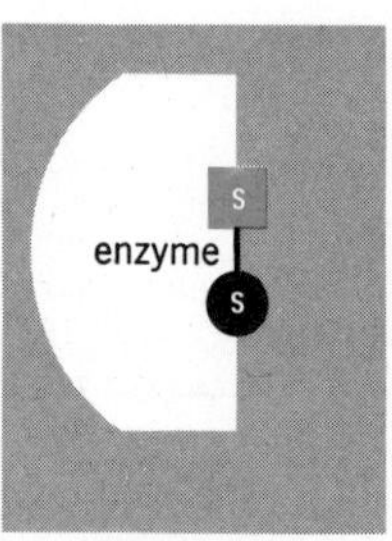

combination

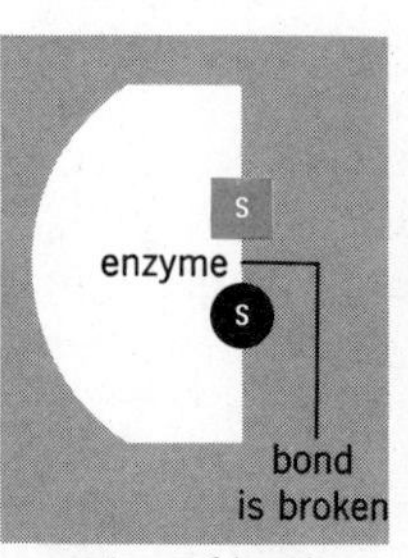

rupture of bonds

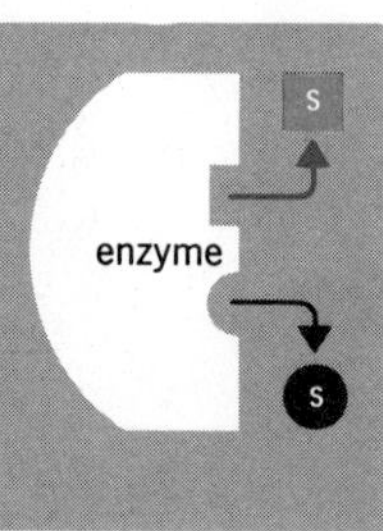

release of products

TABLE 3.2
A PARTIAL CLASSIFICATION OF ENZYMES

NAME	SUBSTRATE (MATERIAL WITH WHICH ENZYME REACTS)	ACTION	EXAMPLES
Hydrolases	None specific. Fats, phosphate containing compounds, nucleic acids, many others	By adding water (hydrolysis), splits larger molecules into smaller units	Lipases, phosphatases, nucleases
Carbohydrases	Carbohydrates; starches, disaccharides	Fragment carbohydrates by hydrolysis to smaller units	Salivary amylase, pancreatic amylase, maltase, lactase, sucrase
Proteases	Proteins, smaller units containing fewer amino acids	Fragment proteins by hydrolysis to smaller units	Pepsin, trypsin, erepsin
Phosphorylases	Many molecules	Split or activate molecule by addition of phosphate group	Muscle phosphorylase, glucokinase, fructokinase
Dehydrogenases	Any compound capable of losing 2 H^+	Oxidizes compound by removal of 2 H^+	Succinic dehydrogenase, fumaric dehydrogenase
Oxidases	Any compound that may add oxygen	Oxidizes compound by addition of oxygen	Peroxidase
Aminases	Amine-containing compounds and keto acids	Addition or removal of amine groups	Transaminase, deaminase
Decarboxylases	Acids containing carboxyl group	Remove CO_2 from carboxyl group	Ketoglutaric decarboxylase
Isomerases	Chiefly carbohydrates	Moves a radical to different part of molecule	Isomerase

with one enzyme. Specificity may depend on a "lock and key" type of arrangement between the enzyme and its substrate. The relationship between an enzyme and its substrate is depicted in Figure 3.3.

Enzymes may be INHIBITED by other compounds that are not the enzyme's primary substrate. The inhibition is usually the result of combination of the compound with the enzyme, reducing its activity.

Enzymes are NAMED in various ways, including the substrate with which they react, and the type of reaction catalyzed. The suffix -ASE is commonly employed to designate the enzyme. Some important categories of enzymes are presented in Table 3.2.

LIPIDS are insoluble in water, but are soluble in fat solvents such as ether, alcohol, and chloroform. Lipids are composed primarily of C, H, and O and contain relatively few atoms of oxygen. Three categories of lipids are recognized.

SIMPLE LIPIDS (Fig. 3.4) include the fats, waxes, and oils. The TRIGLYCERIDES, compounds of three fatty acids with glycerol, are widespread in the body. They are the most common storage form of fat in adipose cells, and are the primary lipids metabolized for energy by the body. There are basically two types of fatty acids, the *saturated* and *unsaturated.* A saturated fatty acid contains all the hydrogens it can hold, that is, it is saturated with hydrogens. An unsaturated fatty acid does not contain all the hydrogens that could be incorporated into the molecule, and shows one or more double bonds in the molecule (—C═C—) between carbon atoms. Saturated fatty acids, the predominant type in

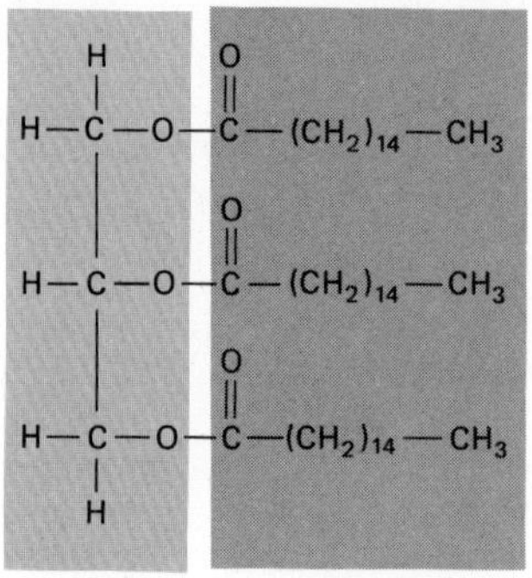

glycerol palmitic acid

FIGURE 3.4
A triglyceride, tripalmitin, composed of three palmitic acid molecules attached to glycerol.

animal fats, are reported to have a role in stimulating atherosclerosis. Unsaturated fatty acids predominate in plants. Therefore, dietary reduction of animal fats, and utilization of vegetable oils may slow the development of atherosclerosis and cardiovascular disease.

COMPOUND LIPIDS (Fig. 3.5) consist of a lipid portion combined with a nonlipid portion. PHOSPHOLIPIDS, found in cell membranes, GLYCOLIPIDS found in photosynthetic membranes, and SPHINGOLIPIDS found in nervous tissue, are examples of compound lipids. Lipoproteins (see Chap. 24) are important factors in the generation of atherosclerosis.

DERIVED LIPIDS (Fig. 3.6) include the sterols and intermediate compounds in lipid metabolism. Sterols are derived from cholesterol, and include several steroid hormones (testosterone, estradiol, progesterone, adrenal cortical hormones). Cholesterol has been implicated as a lipid deposited in the walls of blood vessels leading to *atherosclerosis* and *heart attack*. Synthesis of cholesterol in the body is believed to be inversely related to dietary intake, and the consumption of low cholesterol diets is thought to retard deposition of the compound in blood vessel walls, at least until the low dietary intake is compensated for by an increase in synthesis by the body. Also, a genetic lack of membrane receptor sites for cholesterol and fatty acids that prevents intake and metabolism of these substances may contribute to high blood levels and sclerosis.

Lipids form important SOURCES OF ENERGY, releasing on combustion over twice the calories per gram as do carbohydrates. In their storage depots in skin, they INSULATE and thus are important in control of body temperature. They give form to the body, and cushion organs. They form, as phospholipids, an important CONSTITUENT OF CELL MEMBRANES.

FIGURE 3.5
The basic chemical structures of some representative compound lipids. (*a*) A phospholipid (lecithin, found in membranes), (*b*) A glycolipid (monogalactosyl diacyl glycerol, found in photosynthetic tissue), (*c*) A sphingolipid (sphingomyelin, found in nervous tissue). R = additional chemical groups (not specified).

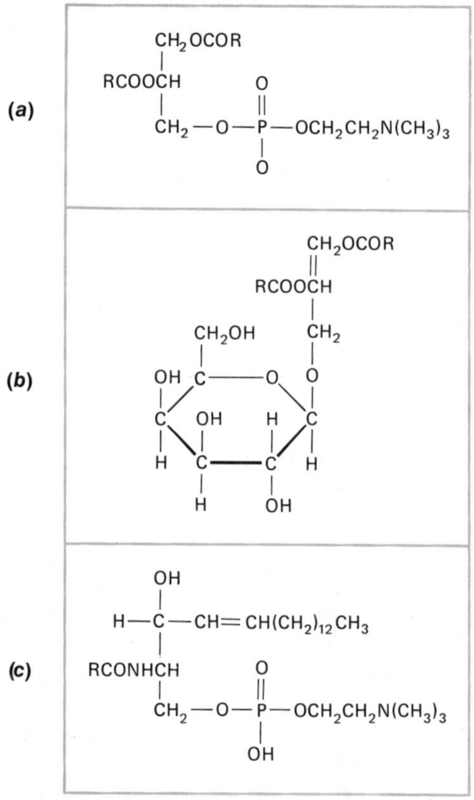

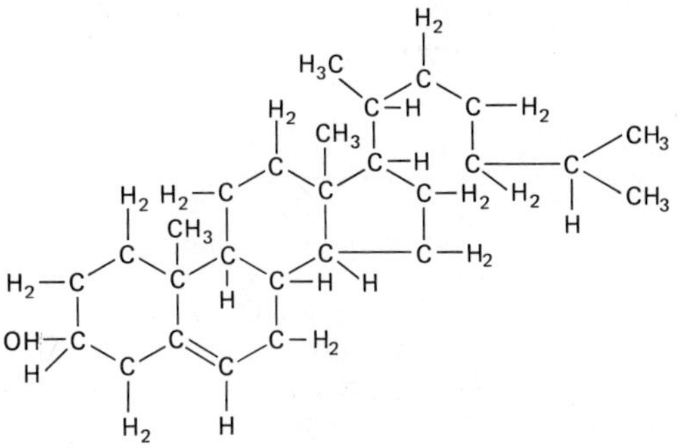

FIGURE 3.6
Cholesterol, a derived lipid.

CARBOHYDRATES include the starches and sugars. They are composed of carbon, hydrogen, and oxygen, with hydrogen and oxygen usually in a 2:1 ratio. Three categories of carbohydrates are recognized.

MONOSACCHARIDES *Simple sugars* (Fig. 3.7). Pentoses (ribose) with five carbons, and hexoses (glucose, fructose, galactose) with six carbons are the most common examples.

DISACCHARIDES *Double sugars* (Fig. 3.8). These are two simple sugar units linked together. They usually represent intermediate stages in the synthesis or breakdown of other carbohydrates.

POLYSACCHARIDES (Fig. 3.9). These are hundreds or thousands of simple sugar units linked into storage forms such as glycogen (animal starch) and plant starch. Polysaccharides may also be joined with other substances to form structural materials such as the protein polysaccharides.

Simple sugars are the cell's PREFERRED SOURCE OF ENERGY for ATP synthesis, and other chemical reactions requiring energy.

NUCLEIC ACIDS (Fig. 3 10) are composed of purine and pyrimidine BASES in combination with a SUGAR and PHOSPHORIC ACID. According to the bases present and the sugar present, two types of nucleic acids are distinguished. DEOXYRIBONUCLEIC ACID (DNA) contains the bases *adenine, guanine, cytosine,* and *thymine,* plus the sugar *deoxyribose;* RIBONUCLEIC ACID (RNA) contains *adenine, guanine, cytosine,* and *uracil,* plus the sugar *ribose*. Both compounds also contain phosphoric acid. ATP, which functions as an intermediate of energy exchange in the cell, NAD (nicotinamide adenine dinucleotide), and FAD (flavine adenine dinucleotide), which function as hydrogen acceptors, are examples of nucleotides. A NUCLEOSIDE is a unit containing only the base and sugar components, while a NUCLEOTIDE consists of the base, sugar, and acid. The

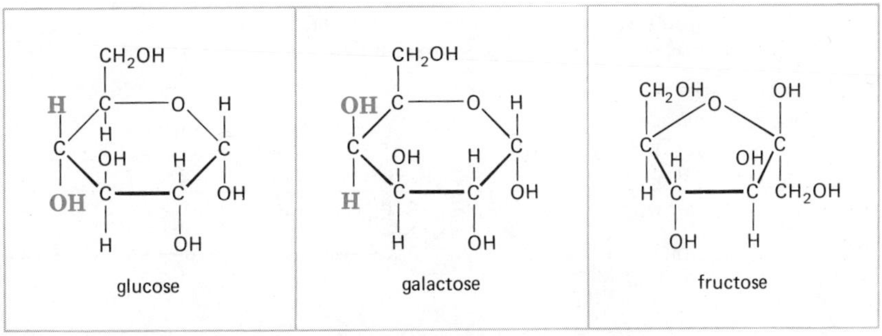

glucose

galactose

fructose

FIGURE 3.7
Monosaccharides (simple sugars). Glucose and galactose differ only in the *position* of the boldfaced chemical groups. Fructose has only five atoms in the ring.

TABLE 3.3
A SUMMARY OF THE CHEMICALS OF LIVING MATERIAL

SUBSTANCE	LOCATION IN CELL	FUNCTION
Water	Throughout	Dissolve, suspend, and ionize other materials; regulate temperature
Electrolytes	Throughout	Establish osmotic gradients, pH, buffer capacity, and membrane potentials
Proteins	Membranes, cytoskeleton, ribosomes, enzymes	Give form, strength, contractility, catalysts, buffering
Lipids	Membranes, Golgi apparatus, inclusions	Reserve energy source, form, shape, protection, insulation
Carbohydrates	Inclusions	Preferred fuel for activity
Nucleic Acids		
DNA	Nucleus, in chromosomes and genes	Direct cell activity
RNA	Nucleolus, cytoplasm	Carry instructions, transport amino acids
Trace Materials		
Vitamins	Cytoplasm, nucleus	Work with enzymes
Hormones	Cytoplasm, nucleus	Work with enzymes, activate or deactivate genes
Metals	Cytoplasm, nucleus	Specific functions in various synthetic schemes (e.g. synthesis of insulin, maturation of red cells)

FIGURE 3.8
Some important disaccharides. Maltose is malt sugar. Lactose is milk sugar. Sucrose is cane or beet sugar.

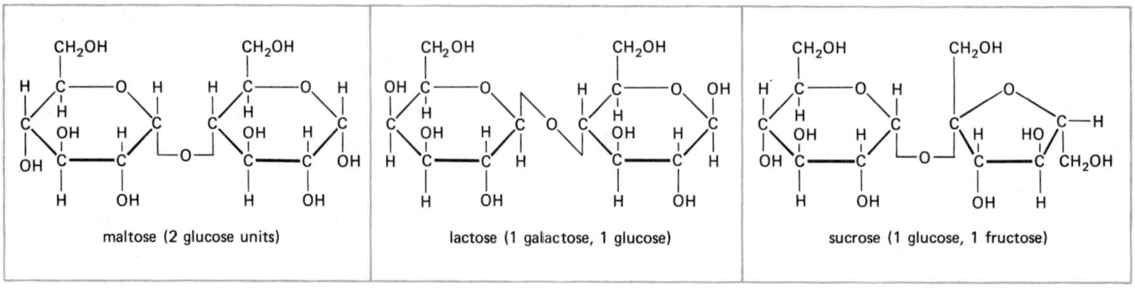

maltose (2 glucose units)

lactose (1 galactose, 1 glucose)

sucrose (1 glucose, 1 fructose)

nucleic acids control cellular activity through protein synthesis.

TRACE MATERIALS are essential for cell function but are required in only very small amounts. This category includes vitamins, certain metals, and hormones. The actions of such materials are considered in greater detail in appropriate chapters later in the text.

A summary of the chemicals of the cell is presented in Table 3.3.

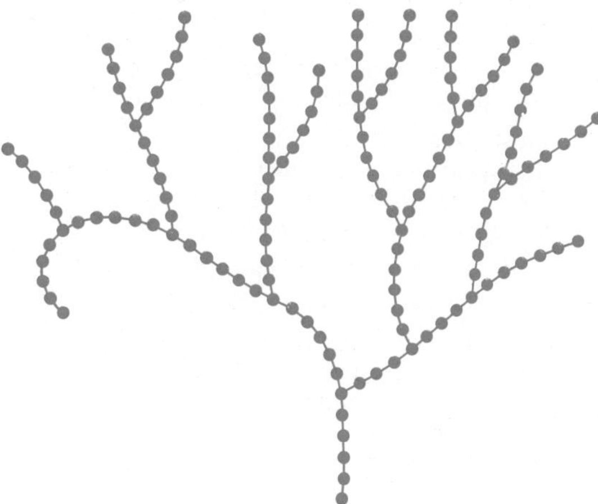

FIGURE 3.9
The glycogen molecule. Each circle represents a glucose molecule.

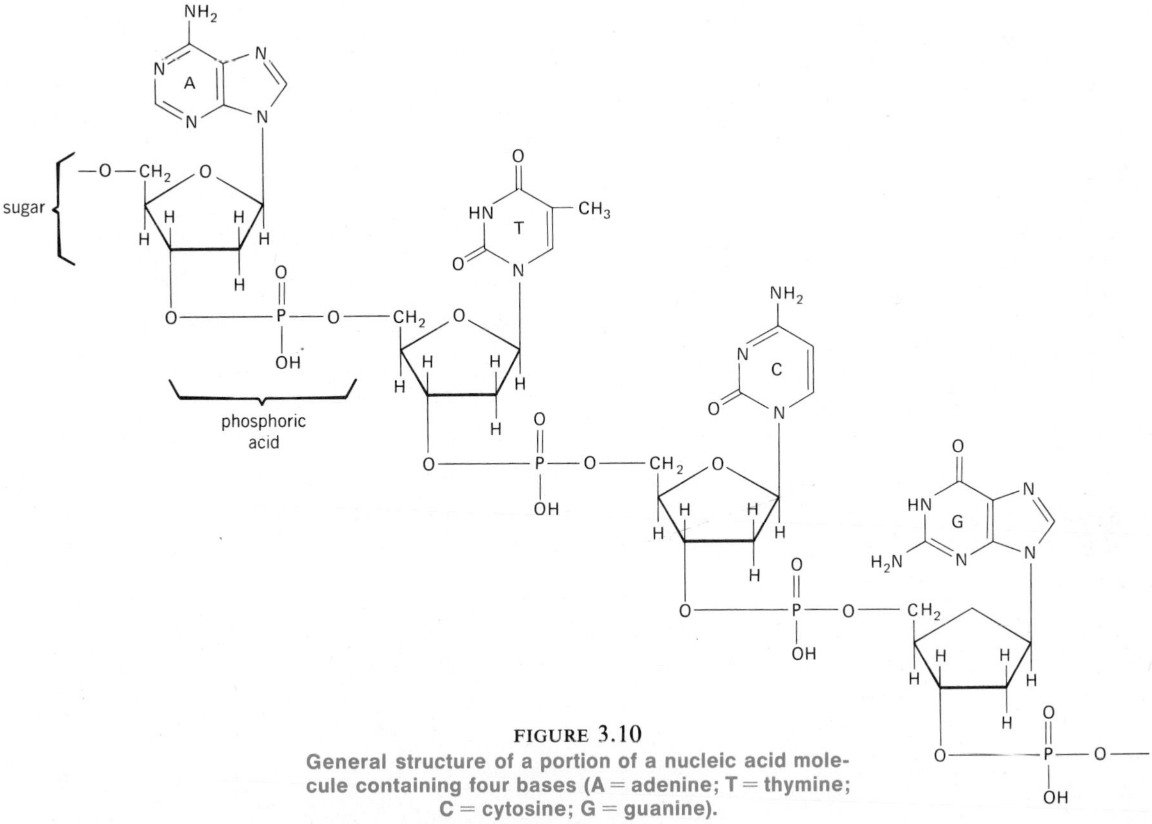

FIGURE 3.10
General structure of a portion of a nucleic acid molecule containing four bases (A = adenine; T = thymine; C = cytosine; G = guanine).

SUMMARY

1. Cells are composed of elements, ions, radicals, molecules, and compounds.
2. Substances may exist in the cell as solutions (dissolved), or in the form of colloidal suspensions. The latter consist of larger molecules that do not settle out of suspension.
3. Water serves as a solvent and as a medium in which chemical reactions occur.
4. Electrolytes are electrically charged elements or radicals and are known as ions. They aid in establishing and maintaining the osmotic and pH homeostasis of the body fluids. They are also necessary for the establishment of the excitable state in cells.
5. Proteins are structural materials, form enzymes, and are composed of amino acids linked by peptide bonds.
6. Enzymes are catalysts for chemical reactions and are specific as to the materials they work on. They require rather specific conditions of pH and temperature to exhibit their optimum activity.
7. Lipids are fuel sources, form some hormones, and give form and insulation to the body.
8. Carbohydrates are the sugars and starches and are the preferred sources of energy for cellular activity.
9. Nucleic acids provide direction of cellular activity, particularly protein synthesis.
10. Trace substances include vitamins and metals. Such substances are required in only small amounts but are vital to normal cellular function.

QUESTIONS

1. What are some characteristics or properties of a colloidal suspension that distinguish it from a true solution?
2. Why do you suppose water was "chosen" as the fluid medium of the body?
3. What are the chief electrolytes of the body?
4. What characteristics differentiate globular and fibrous proteins? Give examples of each.
5. If enzymes were not present in the body, what sorts of disruptions might occur in body function? Name several classes of enzymes, indicating substrate and manner of action.
6. List the categories of lipids, give an example of each, and explain the functions served by each you list.
7. How may lipid metabolism be related to circulatory disorders?
8. What are carbohydrates used for in the body? Give an example from each major group, and a specific function each example serves.
9. What is a nucleoside? A nucleotide? A nucleic acid molecule?
10. What distinguishes "trace materials" from other cellular chemicals? Give examples.

READINGS

Sharon, Nathan. "Glycoproteins." *Sci. Amer. 230:*78. (May) 1974.

Smith, J. Cecil, Jack A. Zeller, Ellen D. Brown, and S. C. Ong. "Elevated Plasma Zinc: A Heritable Anomaly." *Science 193:*496. 6 Aug 1976.

4

CELLULAR CONTROL MECHANISMS AND ENERGY SOURCES FOR CELLULAR ACTIVITY

OBJECTIVES

After studying this chapter, the reader should be able to:

☐ Explain the structure of a nucleic acid, and relate it to the synthesis of a protein.

☐ Describe the differences between DNA and RNA.

☐ Distinguish between the three types of RNA as to function and form.

☐ Describe the processes of replication, transcription, and translation as to product and mechanism.

☐ Describe the genetic code and its significance.

☐ Explain how mutations cause alterations in body structure and function.

☐ Describe the various mechanisms by which protein synthesis and gene activity are controlled by nucleic acids and hormones.

☐ Interpret cellular homeostasis of protein synthesis and activity in terms of nucleic acid and hormone controls.

☐ Describe the major metabolic schemes as related to ATP production, energy release, and value to the body in general.

☐ Describe the importance of the "metabolic mill" to the body.

Much of the cellular activity described in the preceding sections depends on the production of chemical compounds that exert control over synthetic and degrading processes in terms of their rates and efficiencies. Control, as stated in Chapter 1, is basically chemical in nature. Two important mechanisms appear to be available for governing such reactions: GENETIC CONTROL is exerted by way of nucleic acid molecules (deoxyribonucleic acid or DNA, and ribonucleic acid or RNA) and their eventual involvement in the production of body proteins; HORMONAL CONTROL is exerted by specific protein, glycoprotein, steroid and small molecular weight molecules produced by the endocrine glands of the body. Both mechanisms complement one another to achieve control over cellular and therefore body processes. The intent of this chapter is to indicate the manner in which the control is established and not to imply that only two mechanisms exist.

CONTROL BY NUCLEIC ACIDS

STRUCTURE OF NUCLEIC ACIDS

The nucleus of the cell contains nearly all of the cell's DNA, and about 10 percent of the cell's RNA Both types of nucleic acid are composed of subunits of structure known as NUCLEOTIDES, and these in turn are composed of NITROGENOUS BASES (*purines* and *pyrimidines*), a PENTOSE SUGAR (*ribose* or *deoxyribose*), and PHOSPHATE (Fig. 4.1).

Phosphate-to-sugar bonds link nucleotides into the nucleic acid molecules that form the double helix of DNA and the single-stranded RNA.

In DNA, four common bases are found: ADENINE (A), GUANINE (G), CYTOSINE (C), and THYMINE (T). In 1953, James D. Watson and Francis H. C. Crick proposed that DNA consisted of two sugar-phosphate chains, twisted into a helix with the two chains bound together

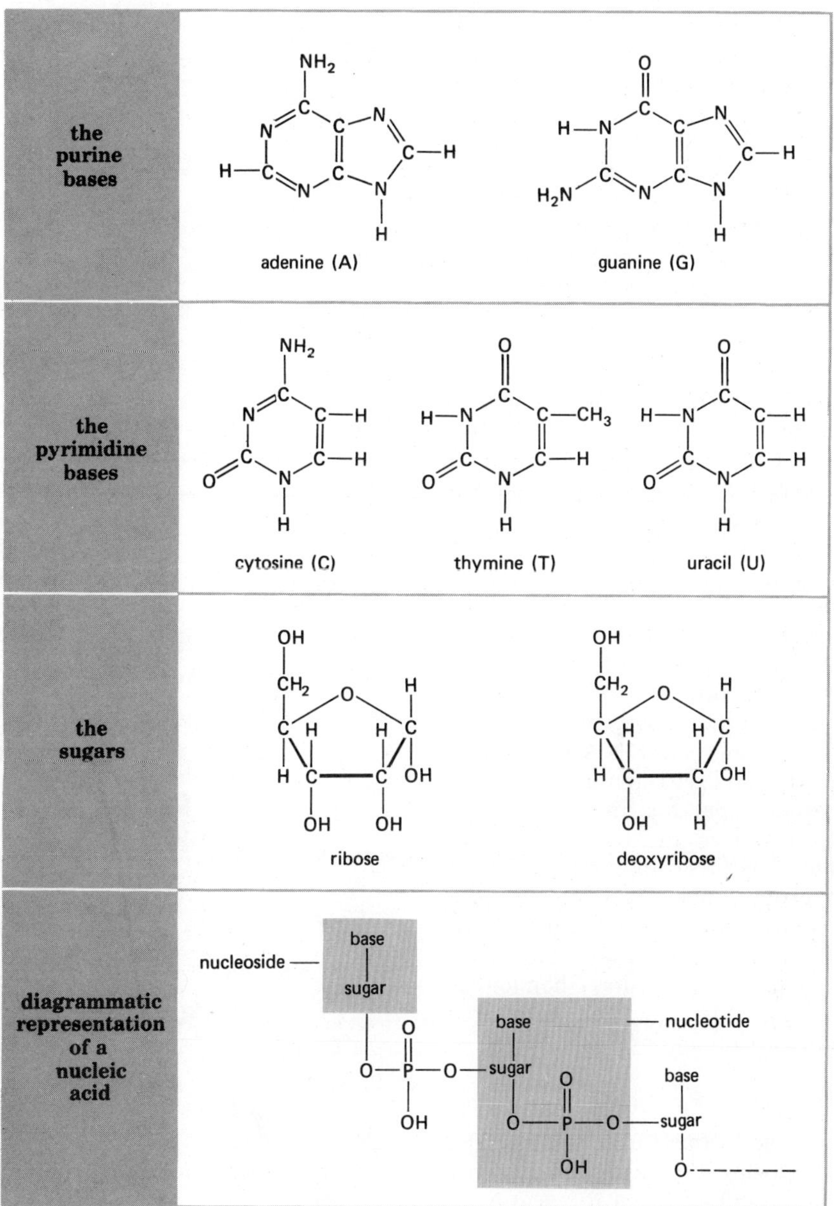

FIGURE 4.1
The constituents of the nucleic acids.

by purine–pyrimide pairings held by hydrogen bonds (Fig. 4.2). Specifically A–T, and C–G linkages were proposed to occur. The sugar in DNA is deoxyribose ($C_5H_{10}O_4$) and the presence of this compound accounts for the name of the nucleic acid.

In RNA, the sugar is RIBOSE ($C_5H_{10}O_5$), and the pyrimidine URACIL (U) is substituted for thymine. Thus, linkages may be made as A–U, and C–G. RNA consists of a single rather than a double strand of sugar-phosphate linked nucleotides which may be twisted upon itself (Fig. 4.3) to give the appearance of a double helix. Where the twisting of the molecule causes the one section of the strand to lie alongside another section of the strand, these portions are bound together by A–U, C–G bondings. RNA exists in several forms designated as messenger RNA (m-RNA), transfer-RNA (t-RNA), and ribosomal-RNA (r-RNA).

Production of new DNA molecules from preexisting DNA, as happens, for example, during mitosis, occurs by the process known as REPLICATION (Fig. 4.4). The DNA helix is believed to unwind and separate at one end. New nucleotides are attracted to the separated strands according to the pairing properties described above. The enzyme DNA-polymerase then joins the nucleotides to create a new DNA molecule that then separates from the original one. A newly advanced theory of aging suggests that failure of DNA–polymerase to incorporate the proper nucleotides into DNA in aging human cells results in a breakdown of cellular information transfer via the DNA-RNA-protein sequence of events. Chemical reactions within the cells may then be altered, causing cellular, tissue, and organ damage.

THE GENETIC CODE

Watson and Crick further proposed that the four bases composing DNA carried information that eventually determined the sequences of amino acids in proteins. It was suggested that a sequence of three of these bases (a "triplet") carried the information about the amino acid. In turn, sequences of triplets form genes that direct the production of a protein. The protein may then become part of the body

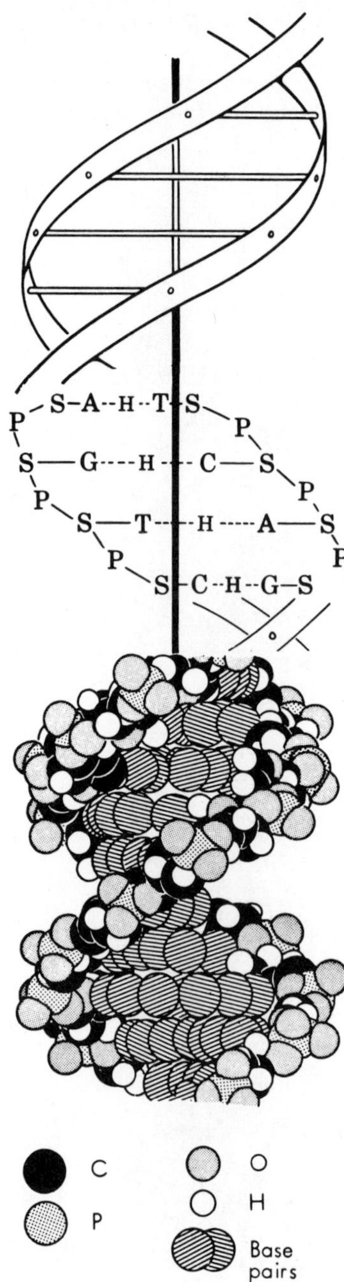

FIGURE 4.2
The double-strand helix of DNA represented three ways. P = phosphoric acid; S = sugar. Other letters indicate specific bases.

structure or an enzyme or hormone influencing a chemical reaction.

A three-base sequence corresponding to a particular amino acid is termed a CODON. With four bases available to form codons, it is possible to form 64 "three-letter words," as shown below. It was calculated that three-letter words were needed because there are about 20 amino acids to be coded for by the four bases. One base as a code for one amino acid yields only four possibilities ($4^1 = 4$); two bases yields 16 possibilities ($4^2 = 16$); three bases yields 64 possibilities ($4^3 = 64$), enough to handle all the amino acids.

AAA	AAG	AAC	AAT
AGA	AGG	AGC	AGT
ACA	ACG	ACC	ACT
ATA	ATG	ATC	ATT
GAA	GAG	GAC	GAT
GGA	GGG	GGC	GGT
GCA	GCG	GCC	GCT
GTA	GTG	GTC	GTT
CAA	CAG	CAC	CAT
CGA	CGG	CGC	CGT
CCA	CCG	CCC	CCT
CTA	CTG	CTC	CTT
TAA	TAG	TAC	TAT
TGA	TGG	TGC	TGT
TCA	TCG	TCC	TCT
TTA	TTG	TTC	TTT

FIGURE 4.3
The single strand "cloverleaf" helix of RNA.

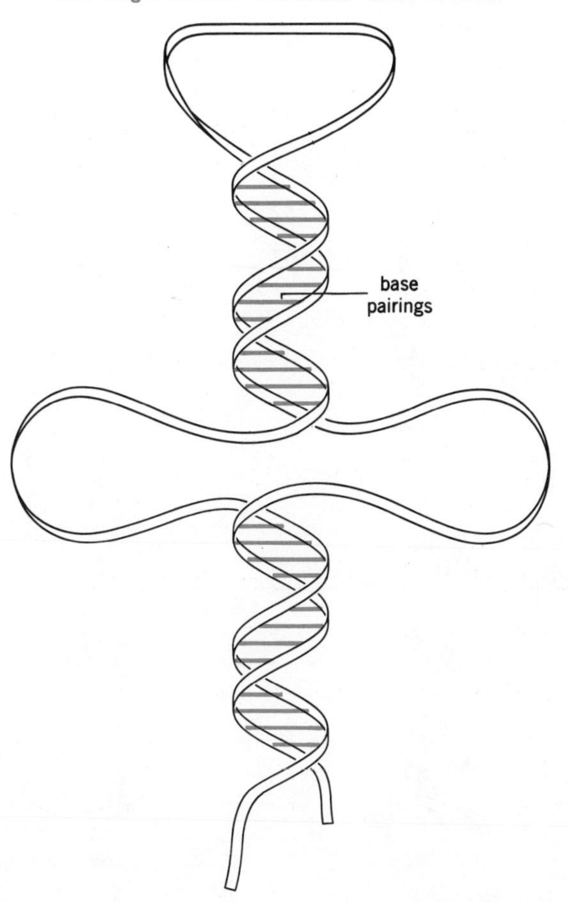

Because there are about 20 amino acids but 64 codons, some codons may specify the same amino acids, or may act as "punctuation marks" to signal the beginning or end of genes. Codons that do not specify particular acids are termed "nonsense codons." Research has shown that the DNA codons specify particular amino acids as indicated below.

Isoleucine	Valine	Leucine	Phenylalanine
ATT	GTT	CTT	TTT
ATC	GTC	CTC	TTC
ATA	GTA	CTA	
		CTG	
Methionine	**Valine/ Methionine**	**Proline**	**Leucine**
ATG	GTG	CCT	TTA
		CCC	TTG
		CCA	
		CCG	
Threonine	**Alanine**	**Histidine**	**Serine**
ACT	GCT	CAT	TCT
ACC	GCC	CAC	TCC
AGA	GCA		TCA
ACG	GCG		TCG
Asparagine	**Aspartic acid**	**Glutamine**	**Tyrosine**
AAT	GAT	CAA	TAT
AAC	GAC	CAG	TAC

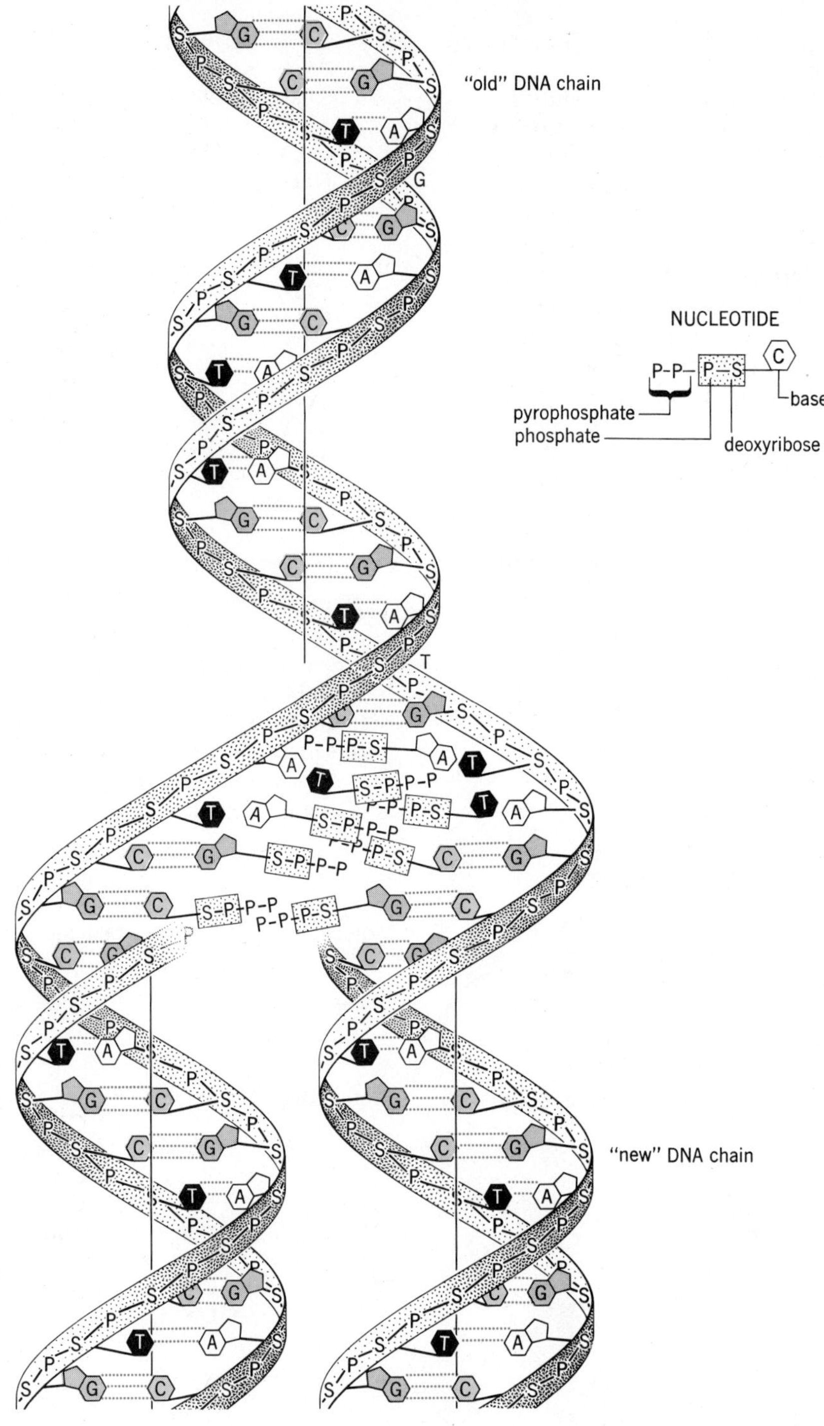

FIGURE 4.4
The replication of DNA.

Lysine	Glutamic acid	Arginine	Cysteine
AAA	GAA	CGT	TGT
AAG	GAG	CGC	TGC
		CGA	
		CGG	
Serine	Glycine		"punctuation marks"
AGT	GGT		TAA
AGC	GGC		TAG
	GGA		TGA
	GGG		TGG
Arginine			
AGA			
AGG			

Recalling that in RNA, uracil (U) replaces thymine (T), there is a similar set of complementary RNA codons, also specifying particular amino acids, that could be set up.

PROTEIN SYNTHESIS

The synthesis of a protein requires that m-RNA and t-RNA be present. DNA participates in m-RNA production by the process known as TRANSCRIPTION. In this process, the two joined bases at a given level in the double DNA helix are believed to determine which nucleotide will be attracted to a particular point along the DNA chain. As different nucleotides are "lined up" on the DNA molecules, RNA-polymerase joins them together into a nucleic acid molecule called m-RNA. The m-RNA then leaves the nucleus through one of the large "nuclear pores," moves to the cytoplasm, and becomes associated with the ribosomes.

T-RNA is believed to be synthesized on a chromosome by DNA, and is then collected on the nucleolus. From here it moves to the cytoplasm. In the cytoplasm, the ANTICODON, a series of three bases in the top of the middle loop of the molecule (Fig. 4.5) specifies the particular amino acid to which the t-RNA will attach. Inosine, I, is an additional base found in t-RNA. It pairs as does guanine. The free end of the molecule, ACCA, serves as the point of attachment for specific amino acids. Each t-RNA attaches to a specific amino acid. Because the anticodon is attracted to a particular codon on m-RNA (associated with the ribosomes) as determined by base pairing, the t-RNA brings a particular amino acid to a specified position in the protein as designated by m-RNA codon sequence. Sequences of amino acids in the protein are thus determined.

The process whereby specific amino acids are placed in a particular sequence as determined by m-RNA, is called TRANSLATION, a sort of decoding of m-RNA's instructions. These relationships are shown in Figure 4.6. In order for an amino acid to be picked up by t-RNA, it must be activated by ATP, a process requiring enzymes. Once at the site of the ribosome and m-RNA, the ribosome moves down the m-RNA strand joining the acids together into the protein chain (Fig. 4.7).

The complete series of events, from DNA to protein synthesis is presented in a "flow sheet" in Figure 4.8.

MUTATIONS

Alterations in base sequences in the DNA molecules (codons) or RNA molecules (anticodons or elsewhere) must result in the production of faulty or abnormal proteins. Such alterations constitute MUTATIONS. J. C. Kendrew illustrates the ways in which mutations arise by the following passages:

> A mutation, interpreted in the light of the genetic code may be compared to a misprint in a line of newspaper type.
>
> "Say it with glowers." In this example, there has been a SUBSTITUTION of an incorrect for a correct letter.
>
> "The prime minister spent the weekend in the country, shooting p_easants." In this case a letter is missing; a DELETION has occurred.
>
> "The treasury controls the public monkeys." Here, an additional letter is present in what is called an INSERTION.
>
> "We put our trust in the Untied Nations." Two letters have been reversed in what is termed an INVERSION.
>
> "Each gene consists of xgmdonrsd ad zbqpt ytrs." Suddenly, a sequence becomes gibberish, forming NONSENSE words.

Most mutations usually affect only one member of a pair of ALLELES (one of two or more

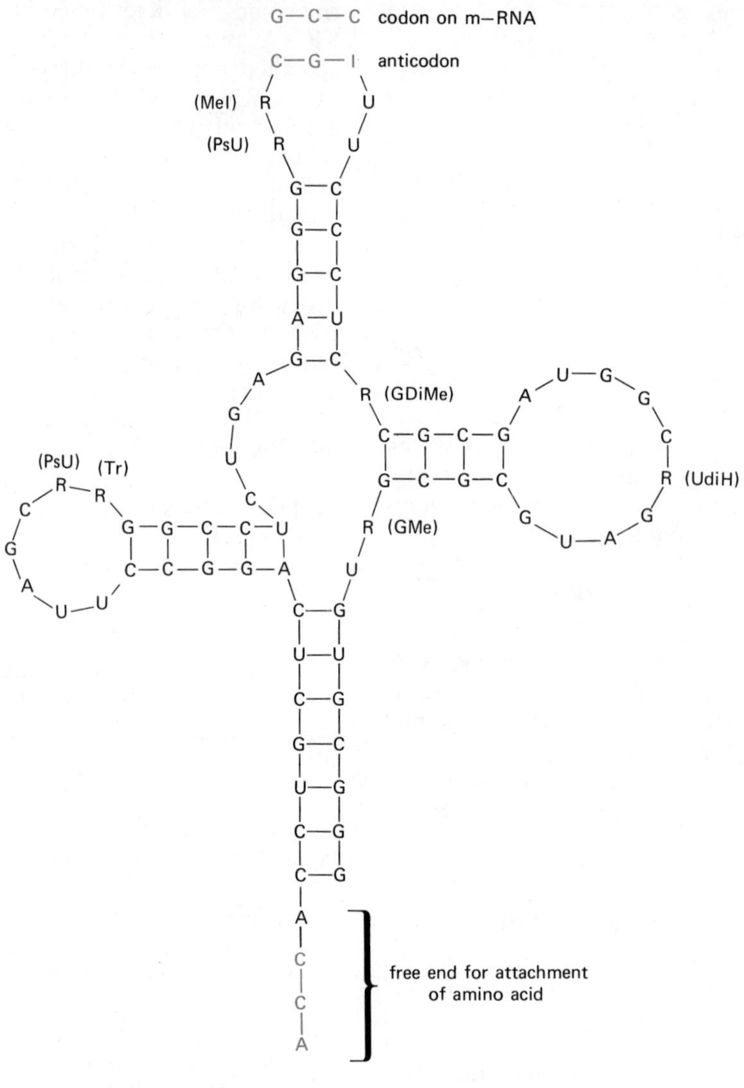

FIGURE 4.5
Diagram for alanine t-RNA. Each letter represents a nucleotide containing a nitrogenous base. R indicates the presence of one of the rare or unusual bases whose formulas are not given in this book. The rare nucleotides are thus indicated: MeI = methylinosine; PsU = pseudouridylic acid; Tr = ribothmidylic acid; Me + base letter = methylated derivative of common base (di Me = two methyl groups); DiH + base letter = di hydro (2–OH) derivative of common base.

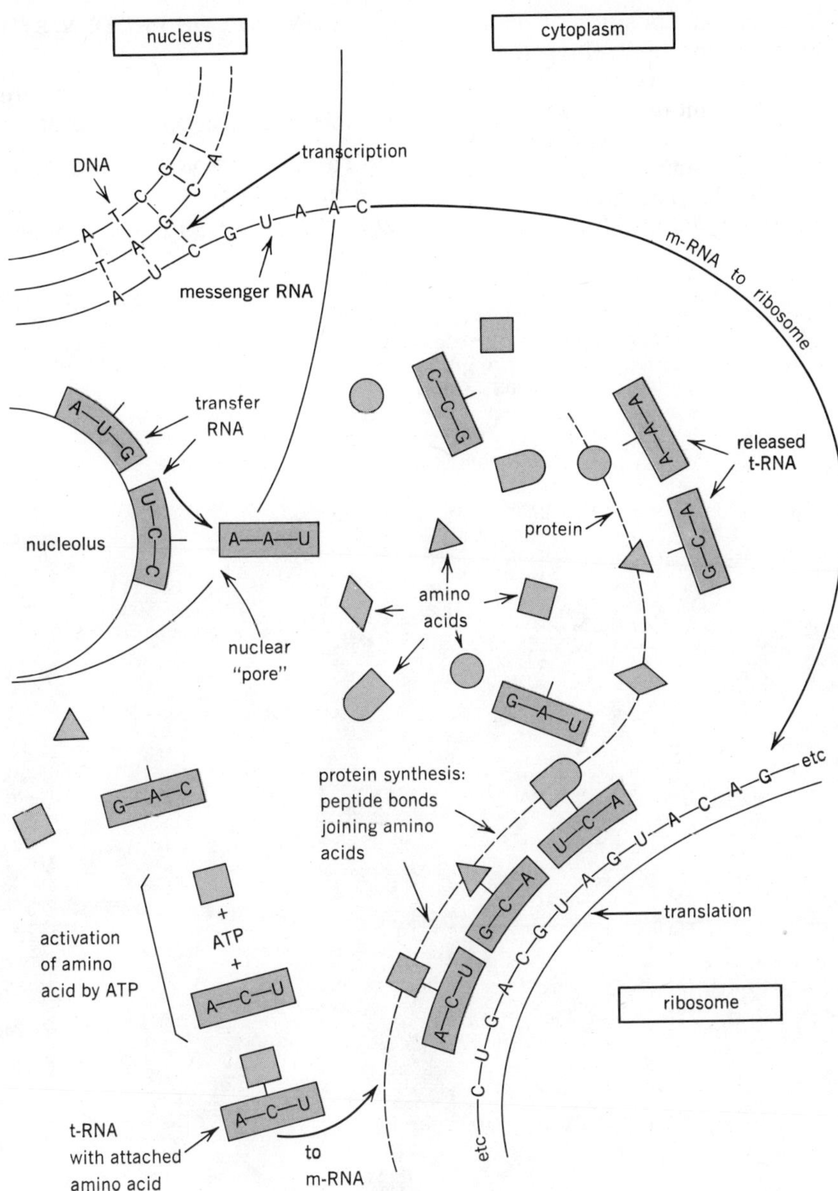

FIGURE 4.6
Transcription, translation, and protein synthesis in a cell.

genes occurring at the same position on a given pair of chromosomes, and controlling the expression of a given characteristic). The normal allele exerts its effect more strongly (*dominance*), and the mutation is not expressed. The defect is still carried and may be transmitted to new cells as they divide. The abnormal characteristic may be expressed at a future time.

CONTROL OF PROTEIN SYNTHESIS

The scheme by which proteins are synthesized is liable to alteration at at least four points:

AT TRANSCRIPTION, the formation of m-RNA from DNA.

When m-RNA and the ribosome ASSOCIATE.

FIGURE 4.7
The synthesis of a protein as the ribosome moves relative to an m-RNA strand. A ribosome consists of a smaller portion (30S) and a larger portion (50S). The smaller portion "locks" the ribosome to the m-RNA; the larger portion attracts t-RNA molecules with attached amino acids.

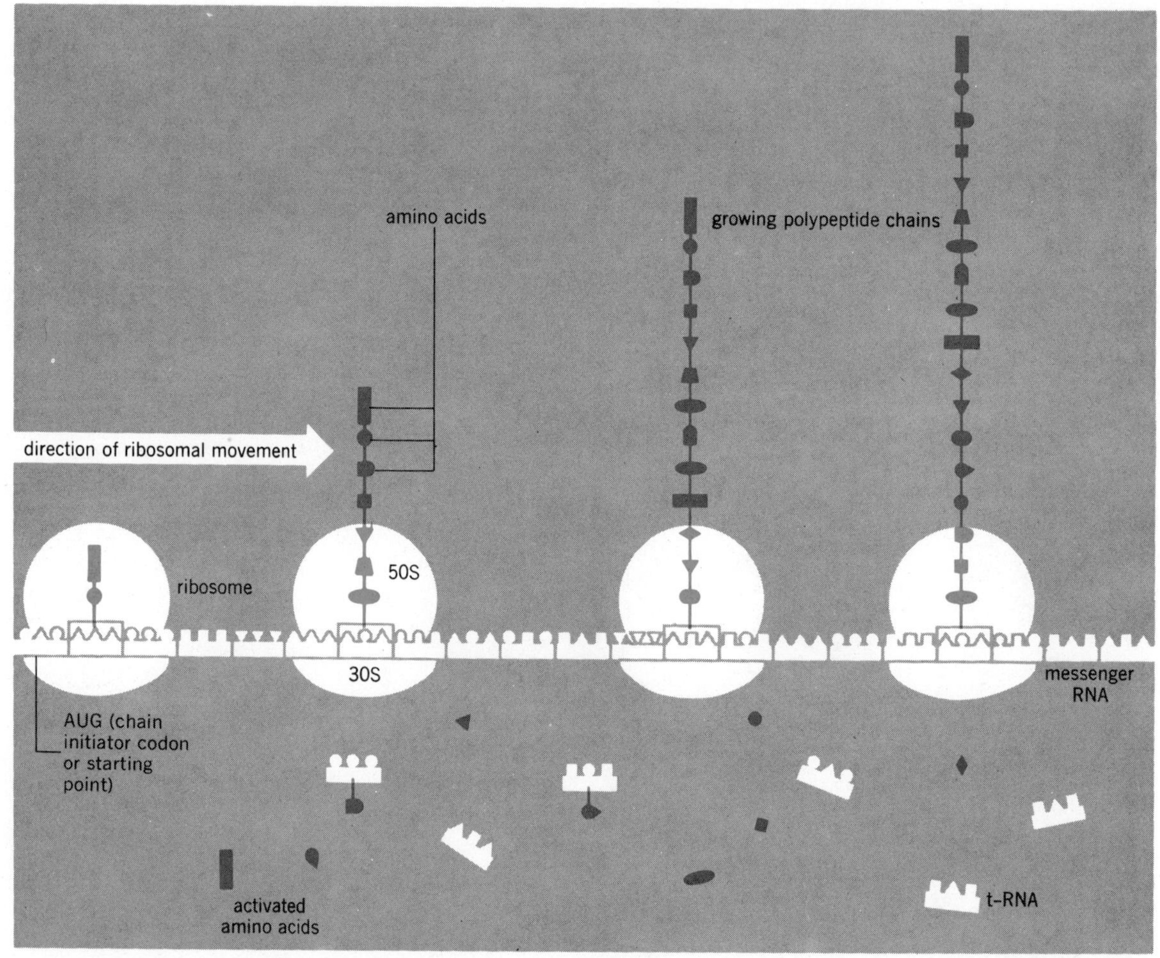

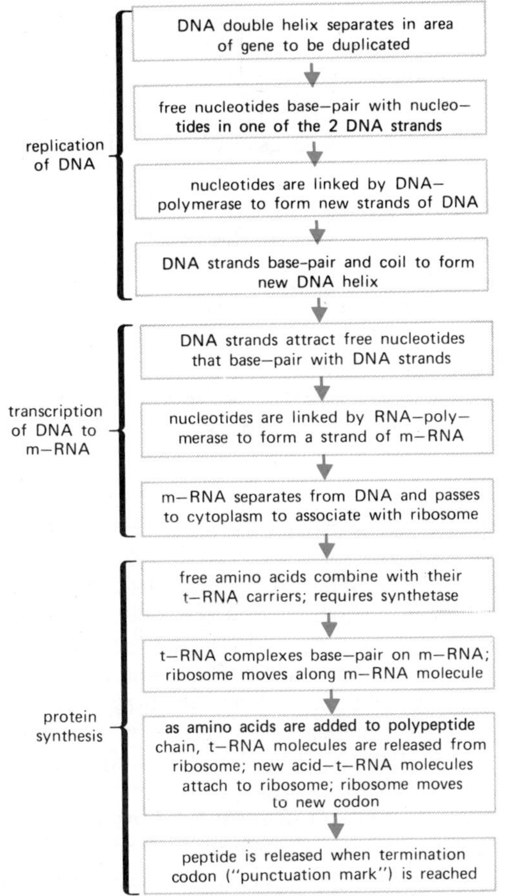

FIGURE 4.8
The events leading from DNA to the synthesis of a protein.

If M-RNA IS DESTROYED in the cytoplasm, and must be replaced.

At the time of T-RNA SYNTHESIS.

Although the exact point at which control *is* exerted is not known, the following mechanisms are postulated to be methods by which control occurs.

INDUCTION is a process by which the presence of a chemical (e.g., lactose) causes (induces) the production of enzymes to metabolize the lactose. Capacity to form the enzymes was already present in the cell, and was merely "turned on" by the presence of the chemical. This mechanism conserves cellular resources until the need for the enzyme has arisen.

REPRESSION involves the ability of a chemical (e.g., histidine) to inhibit the production of the enzymes required to synthesize the chemical. Thus, as the product accumulates, its formation is slowed until its levels are reduced by metabolism and synthesis begins to rise. This mechanism prevents the accumulation of a particular substance, and could conceivably prevent toxic levels of a material from being reached.

Theories of gene function embodying such mechanisms as described above are included in the OPERON CONCEPT. This concept further suggests that there are STRUCTURAL GENES and OPERATOR GENES. Structural genes direct the synthesis of specific proteins via the m-RNA mechanism; operator genes turn the structural genes "off or on." The operon is the *combination* of structural and operator gene.

REGULATOR GENES in turn control the operator genes through the synthesis of proteins called REPRESSORS.

This whole sequence of events has emphasized the role of nucleic acids in producing chemicals that control cellular activity.

HORMONAL CONTROL

Hormones, produced by the endocrine glands, are chemical messengers that are proteins, glycoproteins, steroids, or small molecules. They influence chemicals or processes already available in the body. An additional means of altering the rates of cellular reactions or processes is thus made available.

Hormones appear to exert their control at one or more of the following points:

They may INFLUENCE MEMBRANES by changing their permeability to specific substances or by affecting the binding of materials to membrane sites.

They may INFLUENCE ENZYME ACTIVITY directly.

They may INFLUENCE GENE ACTIVITY.

They may CAUSE THE PRODUCTION AND/OR RELEASE OF SMALL MOLECULES that may then exert effects on cellular function.

HORMONAL EFFECTS ON MEMBRANES

Insulin, a product of the pancreatic islet tissue, is postulated to increase passage of glucose into certain cells, by increasing plasma membrane permeability to glucose. The permeability change results from the reaction of insulin with its membrane receptor sites. Certain cells (e.g., brain, red blood cells, testis, and kidney tubules) are not affected by insulin, and presumably have no receptor sites for the hormone. Thus, glucose enters these cells unaffected by the presence of the hormone.

Parathyroid hormone alters (increases) mitochondrial membrane permeability to magnesium and potassium, and stimulates mitochondrial activity (Krebs cycle, oxidative phosphorylation).

HORMONAL EFFECTS ON ENZYMES

Hormones have been shown to alter the activities of many enzymes *in vitro* (in the "test tube"), but their effects *in vivo* (in the living organism) are more difficult to prove. As examples of known *in vivo* effects, we may cite: the ability of *cortisone* (an adrenal cortical hormone) to stimulate gluconeogenesis (a process by which glucose may be made from amino acids and lipids) by increasing the synthesis of enzymes required in the conversion; the ability of *aldosterone* (another adrenal cortical hormone) to induce the formation of enzymes necessary for sodium and potassium transport through cell membranes and to stimulate their activity; *estrogens* (female sex hormones) influence certain enzymes in the Krebs cycle.

HORMONAL EFFECTS ON GENES

Many hormones (cortisone, aldosterone, testosterone) appear to act to cause alterations in rates of DNA and RNA synthesis, or to act as inducers or repressors of gene activity. There appear to be four points at which an effect could be exerted.

ON THE DNA MOLECULES and their replication and/or transcription.

BY COMBINING WITH PRODUCTS OF THE REGULATOR GENES, and thus *acting* as inducers or repressors.

BY ALTERING ACTIVITY OF AN ENZYME required for synthesis of an inducer or repressor.

BY ALTERING MEMBRANE TRANSPORT of a substance essential for the synthesis of an inducer, suppressor, or for production of the nucleic acid itself.

HORMONAL EFFECTS ON SMALL MOLECULES AND IONS

Certain hormones increase the production of a molecule known as CYCLIC ADENOSINE MONOPHOSPHATE (c-AMP) from ATP by increasing or turning on the activity of the enzyme adenylcyclase. C-AMP in turn activates enzymes that alter the rates of chemical reactions. For example, *epinephrine* (an adrenal medullary hormone) increases c-AMP production, which stimulates phosphorylase, an enzyme increasing glucose release from glycogen.

IONS, particularly calcium and sodium, may be released from areas of binding or storage. *Oxytocin,* a posterior pituitary hormone, stimulates a release of calcium ions that causes uterine contraction. This mechanism is largely responsible for the ability of the uterus to contract after childbirth and prevent hemorrhage. *Aldosterone* causes release of sodium ions that aid in active transport of amino acids into cell nuclei.

All of these control methods are interrelated within the cell to govern its activities. Some of these interrelationships are diagrammed in Figure 4.9.

This figure indicates the role of the regulator gene and its control of the operator gene by synthesis of an inducer or repressor substance. The operator gene influences the structural gene that is involved in DNA and RNA synthesis, and ultimately a protein. This protein may be an enzyme or hormone. If the protein is a hormone, Figure 4.9 shows how it can influence a membrane, small molecule formation, or genes themselves.

ENERGY SOURCES FOR CELLULAR ACTIVITY

ADENOSINE TRIPHOSPHATE

To support the chemical reactions necessary for continuance of cellular activity, sources of energy are necessary. Energy is typically provided by the ultimate oxidation or combustion of basic foodstuffs such as glucose and fatty acids. The sequential steps in the degradation of such molecules to form specific end products is carried out by specific metabolic pathways, schemes, or cycles. Most degradative metabolic pathways produce energy that is utilized to synthesize ADENOSINE TRIPHOSPHATE (ATP). This compound serves as a means of transferring energy from the foodstuff to the physiological mechanism such as muscular contraction, membrane transport systems, or secretion of products by a gland. Among the more im-

FIGURE 4.9
A representation of the action of genetic and hormonal controls on cellular activity (+ = stimulation; − = inhibition).

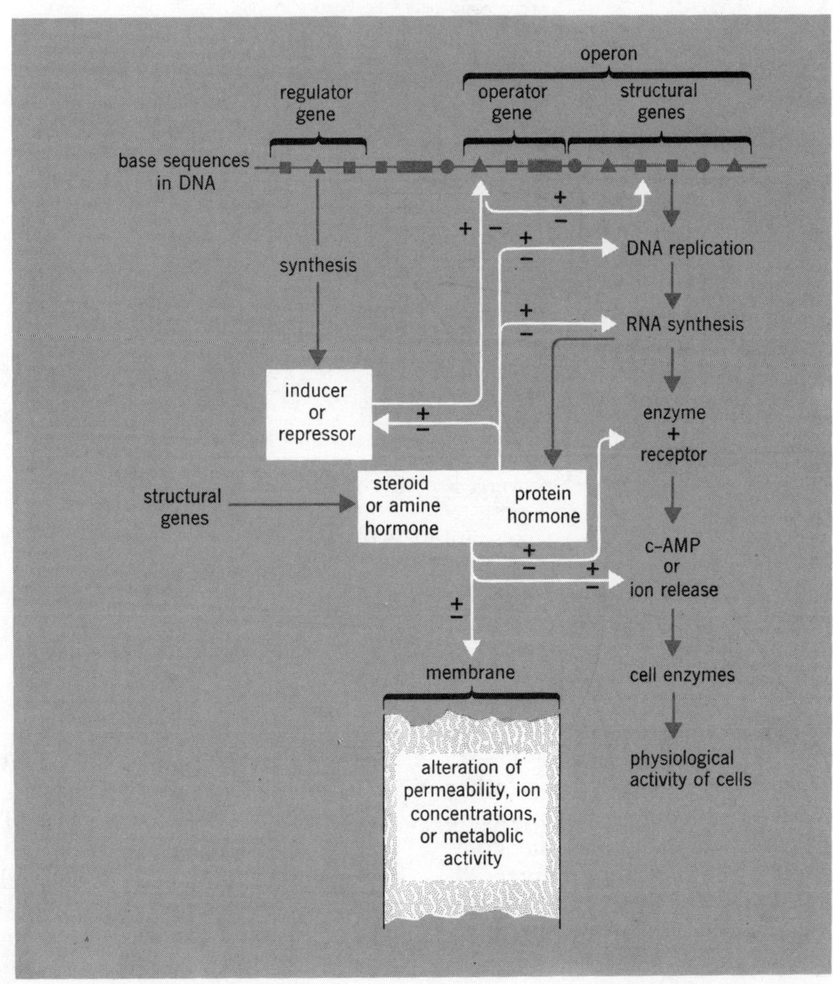

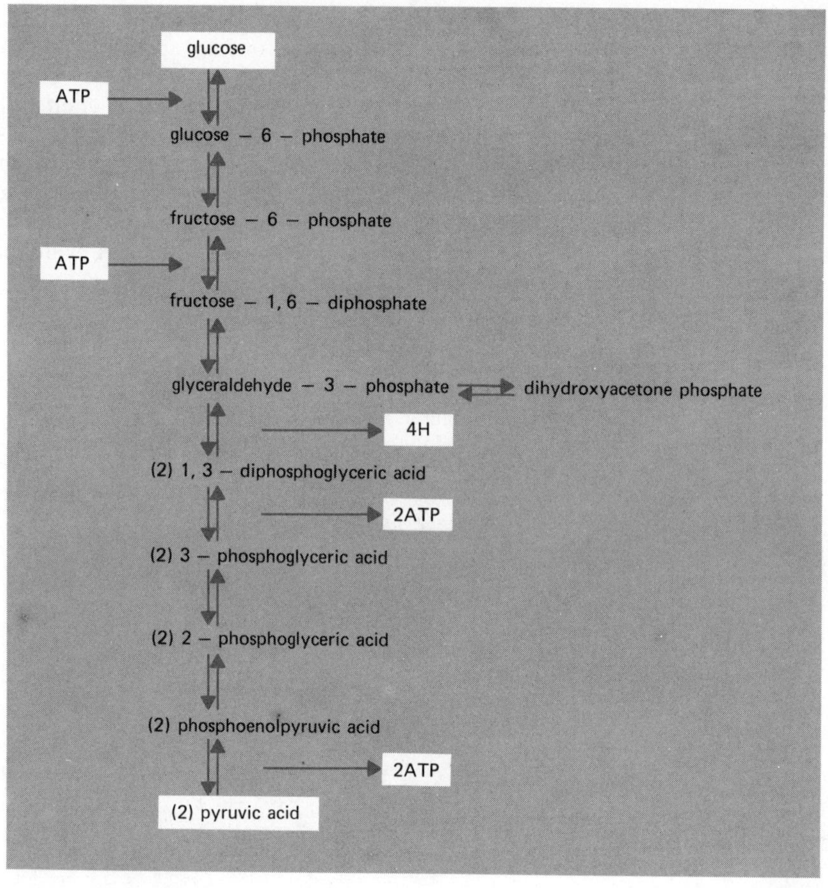

net reaction:

glucose → 2 pyruvic acid + 2ATP + 4H + 56 kcal/mole

FIGURE 4.10
Glycolysis.

portant cycles that provide energy for the synthesis of ATP are glycolysis, Krebs cycle, beta oxidation, and oxidative phosphorylation. Other cycles are available for reactions of other compounds. The basic reactions in each scheme are presented in this chapter, and the complete schemes may be found in Chapter 24. Many of the schemes are carried out in the cytoplasmic organelles, particularly the mitochondria.

GLYCOLYSIS

The *anaerobic* pathway of glycolysis (Fig. 4.10) commences with glucose and results in the production of two molecules of pyruvic acid and two new ATP molecules. In the course of the reactions, four hydrogens are liberated and attach to large molecules known as *hydrogen acceptors* (in this case, NAD). In glycolysis, a total of 56,000 calories/mole of glucose are evolved, of which 27.5 percent of this energy is captured in the two new ATP molecules produced by the pathway. Further degradation of the pyruvic acid requires the removal of CO_2 and 2H from each pyruvic acid molecule resulting from glycolysis. Acetic acid is produced, which is directed to the Krebs cycle.

THE KREBS CYCLE

This *aerobic* cycle (Fig. 4.11) takes the two-carbon acetic acid molecules produced from decarboxylation (removal of CO_2) of the pyruvic acid, and produces two CO_2 molecules, one ATP, and eight H for each acetic acid channeled through the cycle. The cycle is aerobic and releases 629,000 calories of heat energy per mole of pyruvic acid. Hydrogens released in the conversion of pyruvic acid to acetic acid and in the Krebs cycle are also bound to hydrogen acceptors.

FIGURE 4.11
The conversion of pyruvic acid to acetyl CoA and combustion of acetyl CoA in the Krebs cycle.

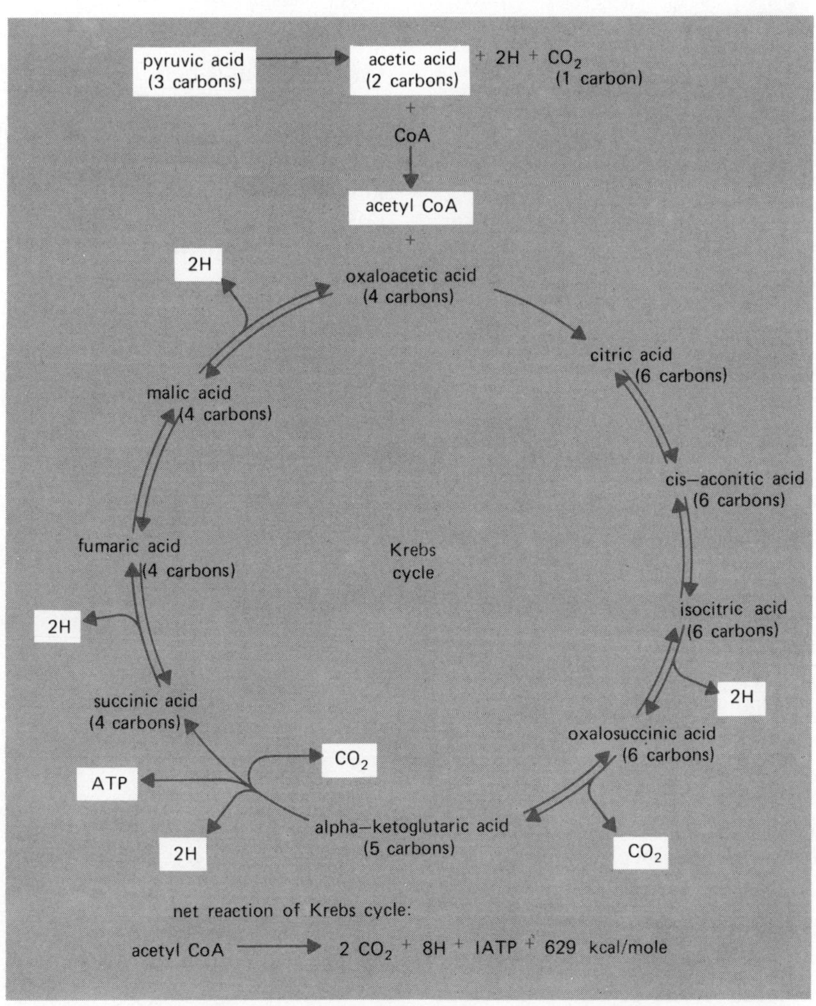

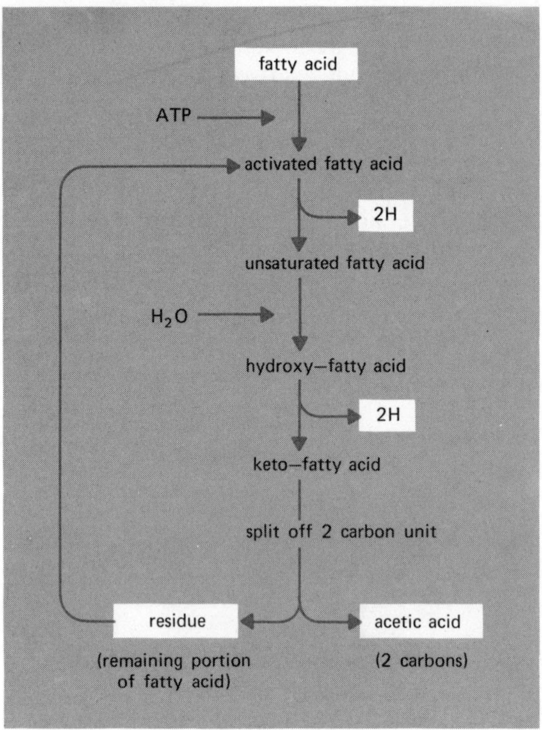

net reaction per split:

fatty acid ⟶ acetic acid + 4H + residue

to other cycles
(e.g. Krebs)

FIGURE 4.12
The oxidation of fatty acid.

BETA OXIDATION

Beta oxidation (Fig. 4.12) reduces long chain fatty acids to two-carbon units of acetic acid. Each cleavage of a two-carbon unit from the fatty acid also releases four H ions that, again, go to hydrogen acceptors. The acetic acid molecules then enter the Krebs cycle.

ELECTRON TRANSPORT AND OXIDATIVE PHOSPHORYLATION (FIG. 4.13)

Hydrogens gathered from glycolysis, Krebs cycle, beta oxidation, and other sources are picked up by large molecules known as hydrogen acceptors. The acceptor has the hydrogens and their accompanying electrons catalytically removed. The electrons are transported over a series of coupled oxidation-reduction reactions utilizing compounds known as cytochromes. The electrons are ultimately transferred to molecular oxygen. Hydrogen ions combine with that oxygen to form water. Energy released during the electron transfer is used to phosphorylate (add phosphate) to ADP, creating ATP. Because the energy is stored by synthesis of ATP, the overall coupled process is termed OXIDATIVE PHOSPHORYLATION.

OTHER SCHEMES OF METABOLISM

Glycolysis and beta oxidation are reversible if energy is supplied to the reactions. Thus, glucose or fatty acids may be synthesized from acetic acid. Additionally, by polymerization of glucose units, glycogen, a storage form of glucose, may be synthesized. Production of glycogen from glucose is termed GLYCOGENESIS

FIGURE 4.13
Biological oxidation of hydrogen; oxidative phosphorylation.

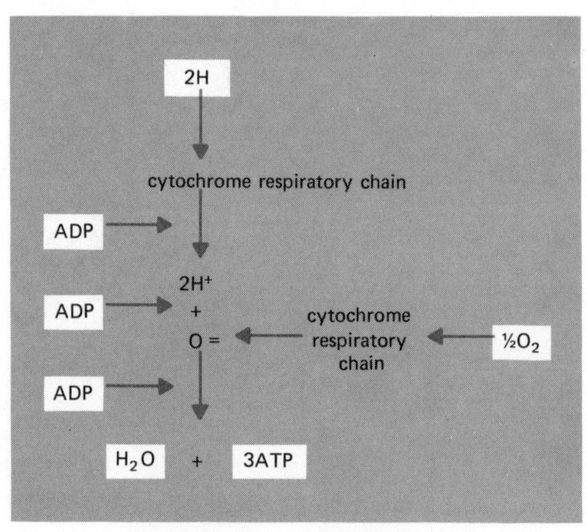

net reaction:

$2H + \frac{1}{2}O_2 + 3ADP \longrightarrow H_2O + 3ATP$

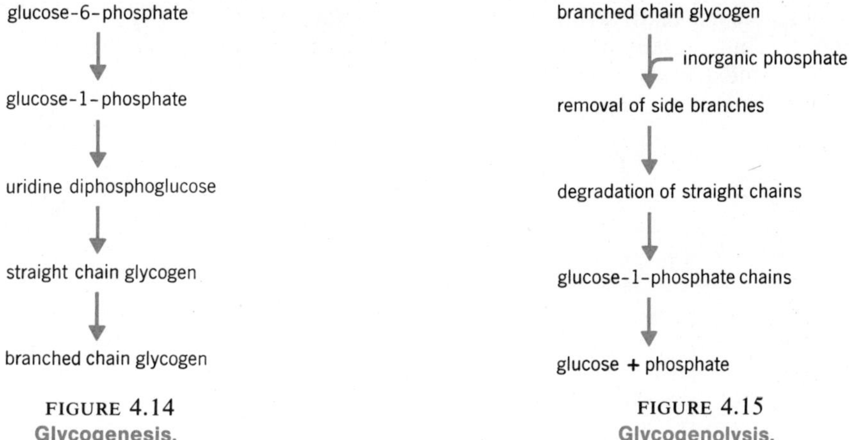

FIGURE 4.14
Glycogenesis.

FIGURE 4.15
Glycogenolysis.

(Fig. 4.14). Recovery of glucose from glycogen constitutes GLYCOGENOLYSIS (Fig. 4.15).

Amino acids may become energy sources for cellular activity if they are DEAMINATED (Fig. 4.16). This reaction produces ammonia and what is termed a keto acid. Certain keto acids are found in the schemes of glycolysis and beta oxidation and may thus suffer the same fate as compounds resulting from metabolism of glucose. In short, a given cycle treats the same compound similarly regardless of its source. Synthesis of glucose from the keto acids resulting from deamination of amino acids or fatty acid metabolism is termed GLUCONEOGENESIS.

Conversely, TRANSAMINATION of a keto acid (Fig. 4.17) produces amino acids by transferral of an amine group from an amine donor.

Lastly, the three basic foodstuffs, represented by glucose, fatty acids, and the non-

FIGURE 4.16
Deamination, using alanine as the example.

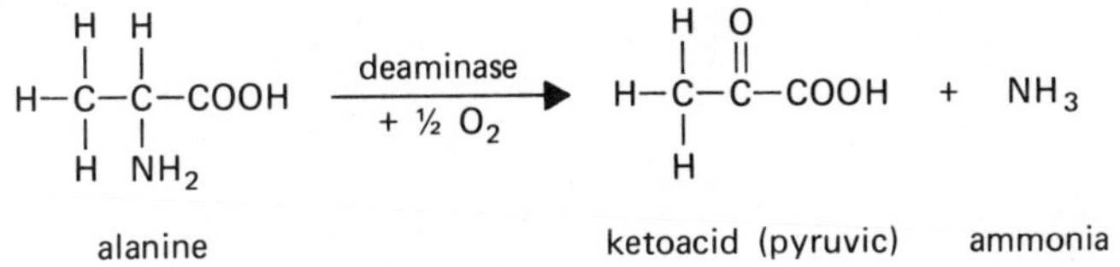

$$\mathrm{H{-}\underset{H}{\overset{H}{C}}{-}\underset{NH_2}{\overset{H}{C}}{-}COOH} \xrightarrow[+\ \frac{1}{2}\,O_2]{\text{deaminase}} \mathrm{H{-}\underset{H}{\overset{H}{C}}{-}\overset{O}{\overset{\|}{C}}{-}COOH} + NH_3$$

alanine — ketoacid (pyruvic) — ammonia

FIGURE 4.17
Transamination, using pyruvic acid as the example.

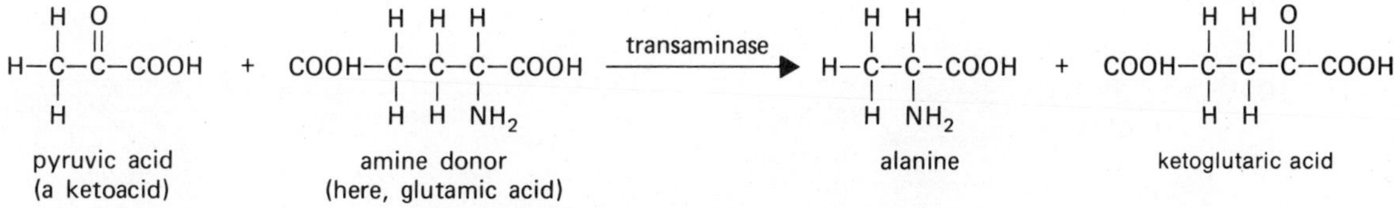

$$\mathrm{H{-}\underset{H}{\overset{H}{C}}{-}\overset{O}{\overset{\|}{C}}{-}COOH} + \mathrm{COOH{-}\underset{H}{\overset{H}{C}}{-}\underset{H}{\overset{H}{C}}{-}\underset{NH_2}{\overset{H}{C}}{-}COOH} \xrightarrow{\text{transaminase}} \mathrm{H{-}\underset{H}{\overset{H}{C}}{-}\underset{NH_2}{\overset{H}{C}}{-}COOH} + \mathrm{COOH{-}\underset{H}{\overset{H}{C}}{-}\underset{H}{\overset{H}{C}}{-}\overset{O}{\overset{\|}{C}}{-}COOH}$$

pyruvic acid (a ketoacid) — amine donor (here, glutamic acid) — alanine — ketoglutaric acid

essential amino acids are, to a large degree, *interconvertible*. Thus, relatively independent of dietary intake, most essential materials for synthesis of body substances are assured. The "metabolic mill" (Fig. 4.18) illustrates the interrelationships in the basic metabolism of the three basic foodstuffs.

FIGURE 4.18
The interrelationships in the metabolism of protein, fatty acids, and glucose; the "metabolic mill."

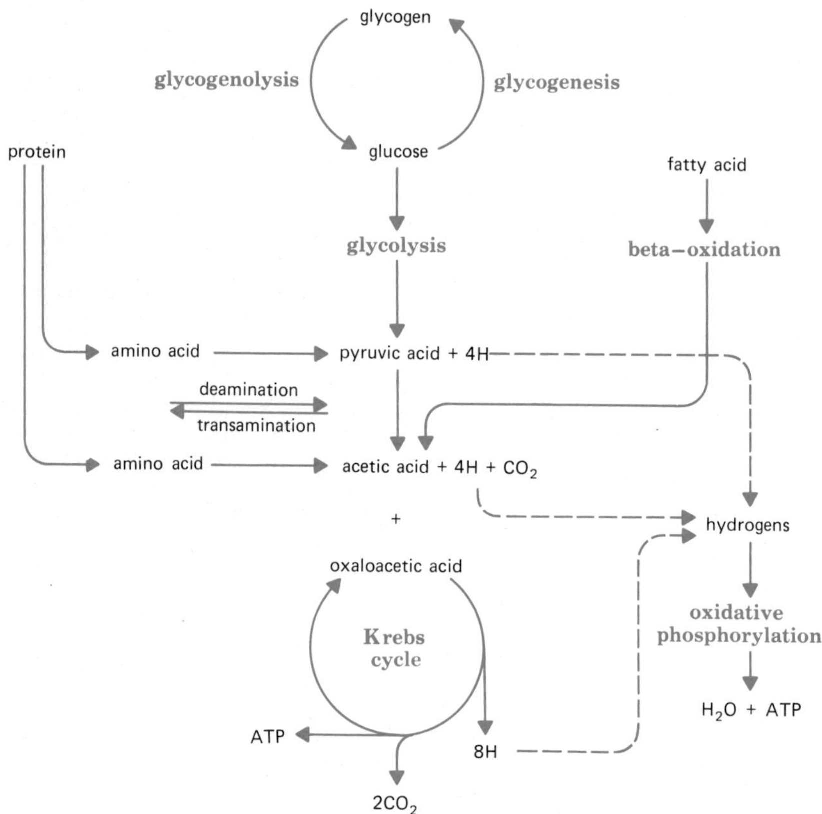

SUMMARY

1. Control of cellular activity is basically chemical in nature, and involves the use of DNA and RNA to synthesize specific proteins and hormones of diverse chemical nature.
2. Nucleic acids are composed of purine and pyrimidine bases, sugars, and phosphate. The bases characteristic of DNA are adenine, guanine, cytosine, and thymine. Uracil replaces thymine in RNA. The DNA sugar is deoxyribose, the RNA sugar, ribose.
3. Production of new DNA from preexisting DNA is called replication, and involves separation of DNA strands. The DNA carries a genetic code that ensures the proper sequencing of amino acids in a protein.
4. RNA is produced by transcription from DNA. Messenger-RNA and transfer-RNA are produced and, respectively, carry instructions for amino acid insertion into a protein, and the amino acids themselves to the insertion site on ribosomes.
5. Translation utilizes m-RNA instructions and t-RNA amino acid transport to actually synthesize a specific protein, which, as an enzyme or hormone, influences a body reaction.
6. Mutations involve alterations of several types (substitutions, deletions, insertions, inversions) that result in faulty or abnormal proteins being produced by body cells.
7. Control of protein synthesis may occur at transcription, at the ribosome level, with metabolism of m-RNA in the cytoplasm, and at the time of RNA synthesis. Induction and repression are additional methods of control.
8. The operon concept suggests that there are structural genes (to direct the production of proteins) and operator genes that signal structural genes to turn on or off. Regulator genes produce repressors that shut off operator genes.
9. Hormones exert control by possibly acting on membranes, enzymes, genes themselves (structural, operator, or regulator), and on the production/release of small molecules or ions.
10. Energy for cellular activity is provided by the degradation of carbohydrates, fats, and proteins, and the energy released is used to synthesize ATP.
11. Several schemes are utilized in the synthesis of ATP.
    a. Glycolysis is an anaerobic cycle that degrades glucose to pyruvic acid and produces two new ATPs.
    b. The Krebs cycle degrades acetic acid, from whatever source, to CO_2 and produces one new ATP per acetate molecule.
    c. Beta oxidation degrades fatty acids.
    d. Electron transfer and oxidative phosphorylation produces water and the bulk of new ATP for the cell.
    e. Glycogenesis is a process that forms glycogen from glucose.
    f. Glycogenolysis is a process that releases glucose from glycogen.
    g. Deamination produces keto acids that may enter other cycles.
    h. Gluconeogenesis is the production of glucose from noncarbohydrate precursors.
    i. Transamination produces amino acids from keto acids.
    j. The three basic foodstuffs (carbohydrates, fats, proteins) are interconvertible by way of the "metabolic mill."

QUESTIONS

1. In general terms, describe the structure of a nucleic acid molecule.
2. Of what does the "genetic code" consist? What does this code determine?
3. What steps are involved in the synthesis of a protein? Place them in proper order of their occurrence.
4. What alterations in the genetic code may result from mutations?
5. Explain the operon concept.
6. By what means may chemical substances, particularly hormones, control cellular activity?
7. What metabolic cycles would be involved in the use of glycogen as a source of energy for ATP synthesis? What would each cycle contribute to the synthesis?
8. Discuss the concept of the "metabolic mill" and its significance to body function.

READINGS

Brown, Donald D. "The Isolation of Genes." *Sci. Amer. 229*:20. (Aug) 1973.

Cohen, Stanley N. "Gene Manipulation." *New Eng. J. Med. 294*:883. April 15, 1976.

Cohen, Stanley N. "The Manipulation of Genes." *Sci. Amer. 233*:24. (July) 1975.

Erbe, Richard W. "Current Concepts: Principles of Medical Genetics." *New Eng. J. Med. 294*:381. Feb 12, 1976. *294*:480. Feb 26, 1976.

Koshland, D. E. Jr. "Protein Shape and Biological Control." *Sci. Amer. 229*:52. (Oct) 1973.

Maniatis, Tom, and Mark Ptashne. "A DNA Operator-repressor System." *Sci. Amer. 234*:64. (Jan) 1976.

Maugh, Thomas H. "The Artificial Gene: It's Synthesized and It Works in Cells." *Science 194*:44. 1 Oct 1976.

Miller, O. L., Jr. "The Visualization of Genes in Action." *Sci. Amer. 228*:34. (Mar) 1973.

O'Malley, Bert W., and W. T. Schrader. "The Receptor of Steroid Hormones." *Sci. Amer. 234*:32. (Feb) 1976.

Ruddle, Frank H., and Raju H. Kucherlapati. "Hybrid Cells and Human Genes." *Sci. Amer. 231*:36. (July) 1974.

Stein, Gary S., Janet S. Stein, and Lewis J. Kleinsmith. "Chromosomal Proteins and Gene Regulation." *Sci. Amer. 232*:46. (Feb) 1975.

Sussman, Joel L., and Sung-Hou Kim. "Three-Dimensional Structure of a Transfer RNA in Two Crystal Forms." *Science 192*:853. 28 May 1976.

5

MUSCULAR TISSUE

OBJECTIVES

SKELETAL MUSCLE

PHYSIOLOGICAL ANATOMY

EXCITATION OF SKELETAL MUSCLE

THE NEUROMUSCULAR JUNCTION

CLINICAL CONSIDERATIONS

EXCITATION-CONTRACTION COUPLING; THE SLIDING-FILAMENT MODEL

SOURCES OF ENERGY FOR CONTRACTION

OTHER CONSIDERATIONS OF SKELETAL MUSCLE CONTRACTION

ELECTRICAL EVENTS

HEAT PRODUCTION

PHYSIOLOGICAL PROPERTIES

CHARACTERISTICS OF THE SEVERAL TYPES OF CONTRACTIONS EXHIBITED BY SKELETAL MUSCLE

THE TWITCH

TETANUS

TREPPE

TONE

SMOOTH MUSCLE

PHYSIOLOGICAL ANATOMY

TYPES OF SMOOTH MUSCLE

SPONTANEOUS ACTIVITY

CHEMICAL FACTORS

NERVOUS FACTORS

CLINICAL CONSIDERATIONS

SUMMARY

QUESTIONS

READINGS

OBJECTIVES

After studying this chapter, the reader should be able to:

- ☐ Describe the microscopic and ultramicroscopic anatomy of skeletal muscle essential to understanding its activity.
- ☐ List and relate to muscle activity the four major muscle proteins.
- ☐ Describe the anatomy and functioning of the neuromuscular junction.
- ☐ Relate certain muscle disorders to the operation of the neuromuscular junction or to the muscle itself.
- ☐ Describe, in stepwise fashion, the events that lead to muscular contraction and relaxation.
- ☐ Describe the energy sources utilized to initiate and sustain contraction, and to replenish muscle ATP.
- ☐ Relate electrical events and heat production to the events occurring during muscular activity.
- ☐ Describe the physiological properties unique to skeletal muscle and relate them to the tasks skeletal muscle performs.
- ☐ Describe a twitch and what it demonstrates.
- ☐ Account for the genesis of tetanus.
- ☐ Explain treppe and tone and their value to the body.
- ☐ Describe the physiological anatomy of smooth muscle, and the types of smooth muscle.
- ☐ Describe how smooth muscle is stimulated and the effects of chemical and nervous factors on its operation.

It is the function of muscular tissue to CONTRACT or shorten, and so perform work. Contraction then causes MOVEMENT of body parts or PROPULSION of materials through the body. There are three types of muscle in the body, each possessing particular properties that fit the muscle for its particular jobs.

SKELETAL MUSCLE attaches to and moves the skeleton, allowing the organism to move through space and adjust itself to its external environment. It also forms the type of muscle causing breathing. Skeletal muscle is normally caused to contract by impulses delivered to it by way of nerves. Skeletal muscle accounts for about 40 percent of the body weight.

SMOOTH OR VISCERAL MUSCLE is located around hollow organs of the body, such as the organs of the digestive system, reproductive system, urinary system, and blood vessels. Contents of such hollow organs may be propelled through the organ by smooth muscle contraction (peristalsis) or the size (diameter) of the organ may be altered. The latter phenomenon is of particular importance in adjusting blood pressure and blood flow. Contraction of most smooth muscle is basically due to inherent factors, but contraction may be controlled or altered by nerves and chemicals.

CARDIAC MUSCLE is located only in the heart, and its contraction propels blood through the body. Its contraction, like that of smooth muscle, is due to inherent factors, but may be altered by nervous and chemical stimuli.

Cardiac and smooth muscle together constitute 10 percent of the body weight.

The activity and properties of skeletal and smooth muscle form the topics in this chapter. Cardiac muscle is considered in Chapter 20, as a part of the study of the heart.

SKELETAL MUSCLE

PHYSIOLOGICAL ANATOMY

Skeletal muscle (Fig. 5.1) is composed of multinucleated, long cylindrical FIBERS or cells. The diameter of an individual fiber varies from

FIGURE 5.1
The structure of skeletal muscle, (*F*), (*G*), (*H*), and (*I*) are cross sections of the filaments at the points indicated.

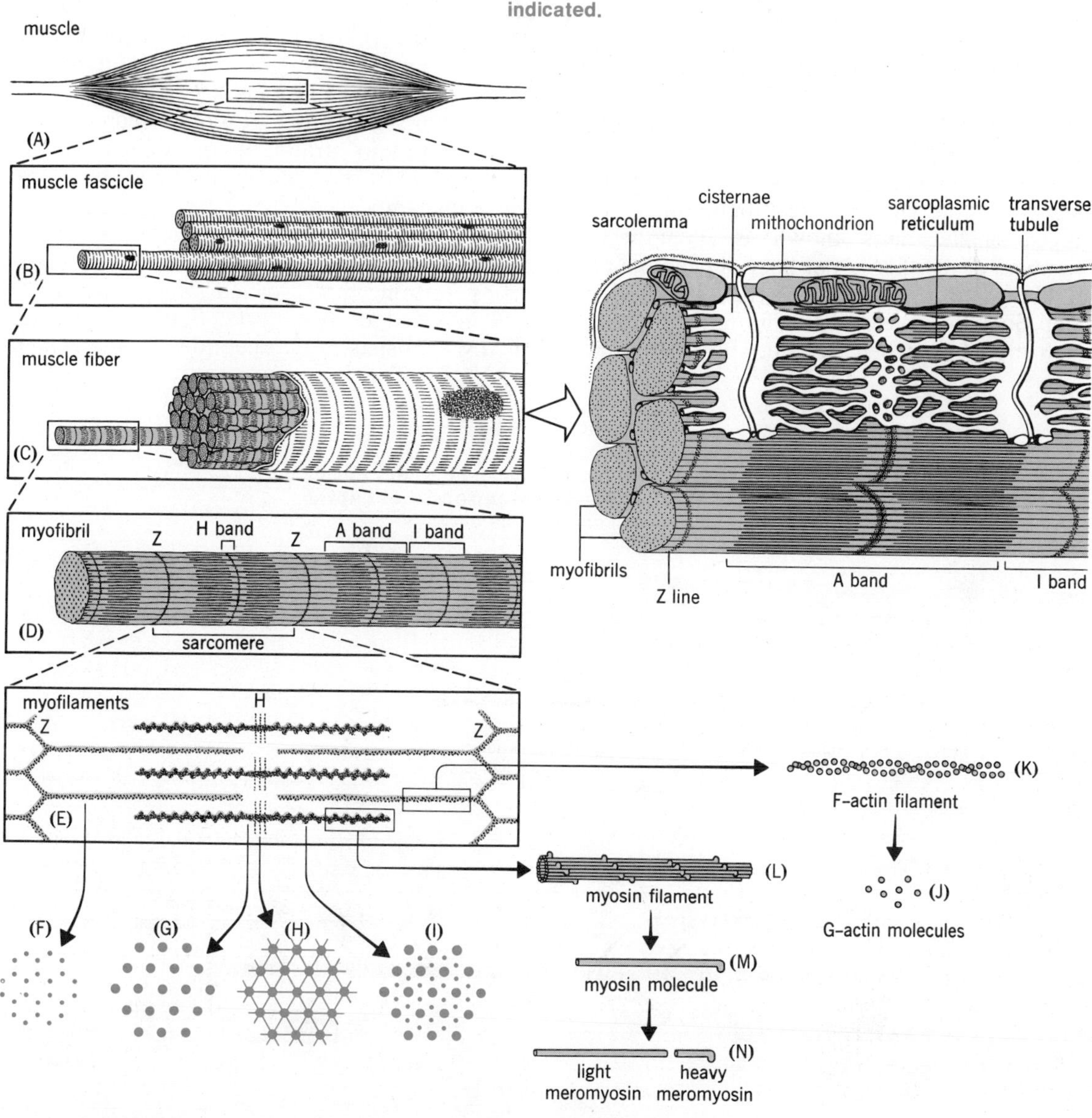

10 to 100 μm, while the length varies from 2 to 3 mm to 7.5 cm or more. In general, small fibers are found in small muscles, such as the muscles turning the eyeball, and larger fibers are found in larger muscles, such as those of the appendages. Each fiber is surrounded by a membrane that serves as the cell membrane. It is called the SARCOLEMMA, and consists of an inner plasma membrane similar in structure to plasma membranes elsewhere in the body, and an outer layer of polysaccharide material that contains fine protein fibrillae (collagen). Where a muscle connects to a tendon, the fibrillae fuse with the tendon fibers, the latter connecting with the periosteum of a bone to form the basis for bone movement.

Inside the fiber, lying in the muscle cytoplasm or SARCOPLASM, are longitudinally arranged smaller units known as MYOFIBRILS. Numbers of myofibrils per fiber depend mainly on fiber size, and may be a few hundred in small fibers to several thousand in large fibers. Each myofibril, when stained, shows alternating light and dark regions in it. The LIGHT BANDS are called the I (*isotropic*) bands or lines, the DARK BANDS the A (*anisotropic*) bands or lines. Careful observation of the I band shows it to be divided by a thin Z LINE, and the A band may be shown to be divided by an H LINE. In the entire fiber, all bands on the myofibrils line up above one another and impart obvious STRIATIONS to the entire fiber. (An alternative name for skeletal muscle is *striated muscle*, given because of this obvious cross banding.) The distance between two Z LINES is termed a SARCOMERE, a unit that becomes narrower when the muscle contracts.

Myofibrils, in turn, are composed of FILAMENTS, protein strands involved in muscular contraction. Four major proteins compose these filaments.

MYOSIN (Fig. 5.2) is a large (MW about 500,000) protein possessing enzymelike properties, forming the so-called *thick filaments*. It is composed of two parts: *light meromyosin* (LMM) consists of two peptide strands wound

FIGURE 5.2

(*a*) The myosin molecule. "Hinges" are areas of greater flexibility within the molecule where bending may occur. (*b*) The formation of cross bridges between myosin and thin filaments.

(*a*)

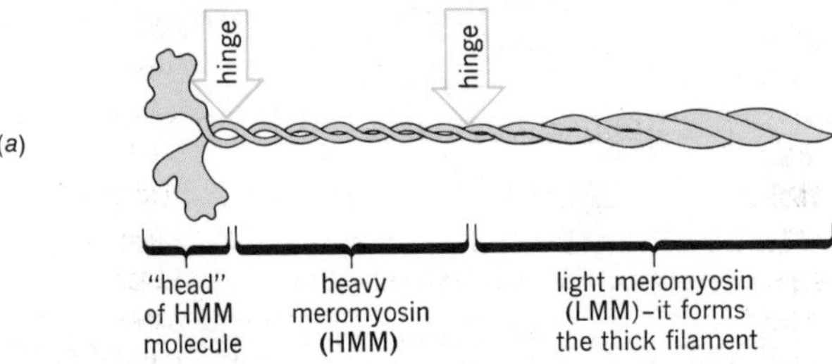

(*b*)

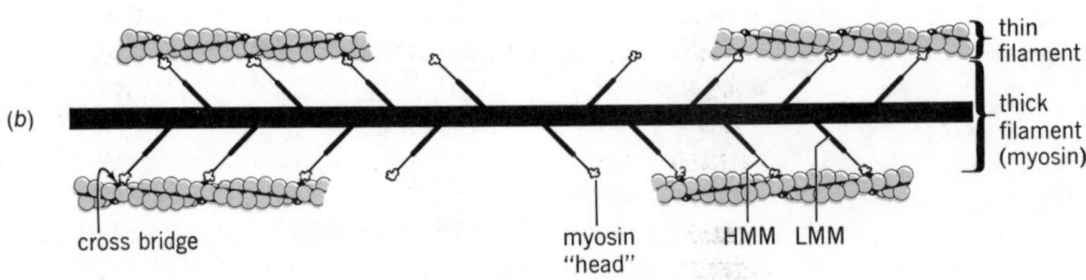

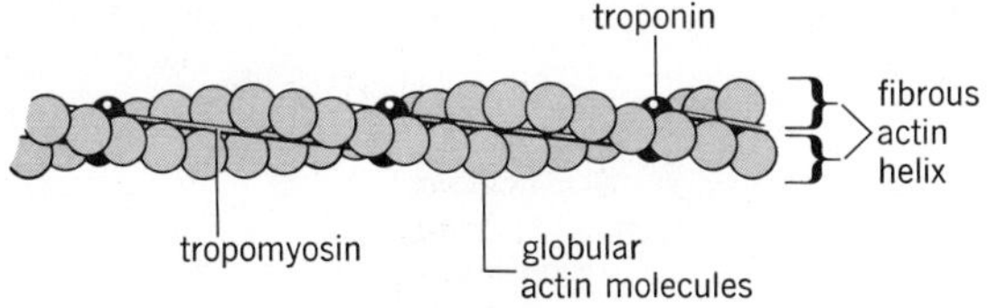

FIGURE 5.3
The structure of the thin filaments.

around one another in a helix; *heavy meromyosin* (HMM) is also a double stranded helix with a "head" or enlarged end on it. It has the enzymelike properties ascribed to the myosin molecule in general. Myosin is restricted to the A line.

ACTIN (Fig. 5.3), MW 60,000, forms the *thin filaments* and consists of individual G-actin (G, globular) units polymerized into F-actin (F, fibrous) strands. F-actin strands are also wound into a double helix that either attach to or pass through the Z lines. Thin filaments are separated from one another in the H line and are continuous between H lines (*see* Fig. 5.1), through the I band. Actin has "reactive sites" on it that may represent where ATP or ADP is attached, or where myosin-actin bonds can form.

TROPOMYOSIN is a filamentous protein having a molecular weight of about 70,000. It is believed to lie twisted around the F-actin helix (*see* Fig. 5.3).

TROPONIN is a globular protein with a molecular weight of about 50,000. It binds to tropomyosin, but not actin, at only certain points along the tropomyosin molecule (*see* Fig. 5.3). Troponin has an especially strong affinity for binding Ca^{++}.

Other proteins (e.g., M-protein and α-actinin) in the muscle probably maintain the organization of the thick and thin filaments.

Other substances in the fiber include Ca^{++}, Na^{+}, Cl^{-}, K^{+}, ATP, phosphocreatine, and glucose. *Myoglobin* is a colored substance found in varying quantities in muscle. It can bind oxygen. "Red muscle" contains much myoglobin, "white muscle" does not.

The SARCOPLASM also contains numerous mitochondria, usually alternating in rows with myofibrils, and the sarcoplasmic reticulum and T-tubules.

The SARCOPLASMIC RETICULUM (*SR*) corresponds to the endoplasmic reticulum of other cells, and consists of longitudinally arranged tubules between the myofibrils. The ends of each longitudinal tubule terminate in expanded CISTERNAE associated with the T-tubules (*see* Fig. 5.1). Two adjacent cisternae and the intervening T-tubule form a TRIAD.

T-TUBULES are transversely oriented hollow tubes reaching entirely across a fiber. They form ringlike ANNULI around each myofibril, and allow tissue fluid around the fibers to flow into the fiber. In mammalian skeletal muscle, the T-tubules are located at the junction of A and I bands; in frog muscle, they are found at Z lines. Thus, mammalian muscle has two annuli per sarcomere.

The reticulum and T-tubule systems play a major role in muscle excitation and in excitation-contraction coupling.

EXCITATION OF SKELETAL MUSCLE

The neuromuscular junction. As stated earlier, the excitation of skeletal muscle is normally handled by nerve stimulation. The connection between a nerve and the muscle fibers is handled by the NEUROMUSCULAR JUNCTION (Fig. 5.4). The junction consists of the *terminal branches of a motor nerve* and a specialized portion of the muscle sarcolemma known as a *motor end plate*.

Each individual nerve fiber in a motor nerve branches at its end and each branch forms an enlarged bulb. In the bulb are many mitochondria that are believed to supply the energy necessary to continuously synthesize a chemical called ACETYLCHOLINE (*Ach*). The Ach is thought to be stored in the enlarged nerve ends in the form of membrane-surrounded VESICLES. The end bulb lies in an invagination of the sarcolemma called a SYNAPTIC GUTTER, and is separated from the sarcolemma by a SYNAPTIC CLEFT about 200 to 300 Å wide that is filled with a gelatinous ground substance (an extracellular or tissue fluid material). The bottom of the gutter has many FOLDS extending into the fiber and which greatly increase the surface

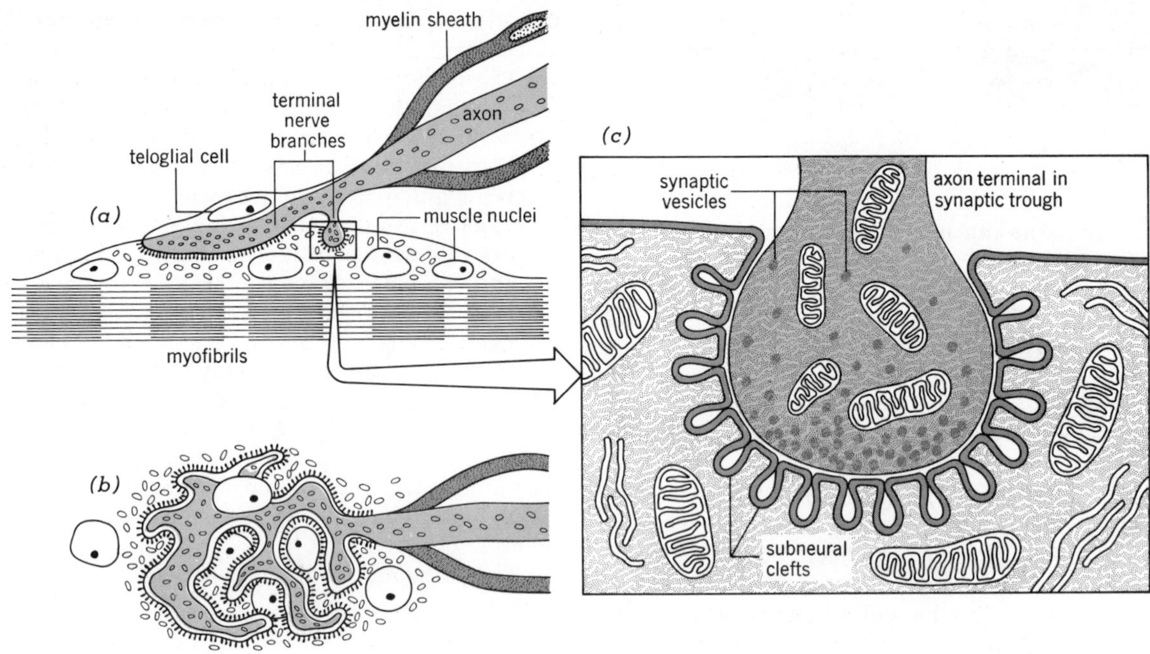

FIGURE 5.4
The neuromuscular junction as seen by light and electron microscopy. (*a*) A section of the junction. (*b*) Surface view of the junction as seen by light microscopy. (c) Details of the junction as seen by the electron microscope.

area of the fiber. The enzyme CHOLINESTERASE, which destroys Ach after the Ach has exerted its effect, is found in the folds.

A single nerve fiber forms branches that may innervate as many as 1000 muscle fibers. A single nerve fiber and its associated muscle fibers constitute a MOTOR UNIT. A given muscle contains hundreds or thousands of motor units, and the muscle fibers of given motor units do not lie adjacent to one another, but are intermingled with the muscle fibers of other motor units.

When a nerve impulse is conducted down a nerve fiber, it arrives at the terminal end of a branch and is there believed to increase the fiber's permeability to Ca^{++}, located in the extracellular fluid. Calcium enters the end bulb and causes the release of Ach by releasing vesicles or by rupturing the vesicles in which the Ach is being stored. It has been shown that the number of vesicles that are ruptured or released is a function of duration of action of the nerve impulse on the membrane of the bulb. After release from the vesicles, Ach diffuses from the terminal bulb across the synaptic cleft to act on the muscle membrane of the gutter. Within 2 to 3 milliseconds (msec) after contacting the gutter membrane, Ach serves as the substrate for cholinesterase, is cleaved into acetic acid and choline, and is no longer an excitant to the muscle membrane. Because Ach is effective and its components are not suggests that the muscle membrane has a series of receptor sites specifically for Ach. Binding of Ach to the muscle membrane is believed to cause the muscle membrane to become very permeable to Na^{+}, which is actively maintained in high concentration outside the muscle fiber. Sodium ions then move rapidly into the fiber by diffusion and cause a change in the electrical properties of the fiber's membrane systems (sarcolemma, sarcoplasmic reticulum, T-tubules), which in turn leads to contraction of the muscle fiber.

It may thus be said that transducing (changing) a nerve impulse into excitation of a muscle fiber is carried out chemically.

Clinical considerations. MYASTHENIA GRAVIS is a condition characterized by abnormal fatiguability of the skeletal muscles. Muscles are weak, contracting with little force or speed. No neural lesions may be demonstrated. Changes in the thymus gland are usually seen with development of the condition.

Myasthenia has, in the past, been suggested to be due to low vesicle concentrations in the end bulbs at neuromuscular junctions, indicating a possible reduction in Ach synthesis, or to overactive or excessive cholinesterase activity that destroys Ach before enough molecules can bind to the muscle membrane to trigger depolarization. More recently, an immunologic cause for the disease has been advanced. According to this hypothesis, antibodies are developed to one's own muscle fibers, perhaps secondary to some inflammatory disorder of the muscles or after invasion by lymphocytes. The antibodies then react with the muscle fiber, causing it to become damaged and incapable of response. The thymus may be the source of the cells producing antibodies. Also, gutter fold damage may occur, which destroys Ach receptor sites.

PARALYSIS refers to inability to voluntarily contract a skeletal muscle, and may usually be traced to nerve damage. The muscle responds to direct stimulation.

FATIGUE may, in the light of a chemical method of neuromuscular transmission, be interpreted as utilization of Ach at a greater rate than it is synthesized. Thus, transmission across the junction is slowed or stops until vesicles are replaced. A "safety factor" is thus built into the junction, one that prevents the muscle from reaching a nonfunctioning condition as a result of overactivity—that is, the junction "quits" before the muscle does.

MUSCULAR DYSTROPHY is regarded as an inherited disorder involving enzymes and/or proteins in the muscle. Muscle fibers appear degenerated with obvious variations in size. Fibers may be replaced with fibrous tissue or fat and the overall size of the muscle may increase because of this accumulation. Muscular fiber loss may eventually confine the patient to a wheelchair. Death, if it occurs, is most often due to pulmonary infection as respiration becomes shallow and weak, or the heart muscle may become affected. At present, there is no specific treatment for the disorder.

EXCITATION–CONTRACTION COUPLING; THE SLIDING-FILAMENT MODEL

This term refers to the series of events by which excitation of the muscle fiber's membrane systems leads to muscle contraction. Steps in the process are presented below.

In a muscle at rest, there is an electrical potential difference maintained between the inside of the fiber and the extracellular fluids around it, and the membranes are said to be POLARIZED. The potential difference is due largely to active transport systems that keep sodium outside the fiber and potassium inside. The potential has a magnitude of −70 millivolts (the negative sign indicates that the inside of the fiber is electrically negative to the outside).

Excitation of the muscle fiber membrane occurs first at the neuromuscular junction when Ach acts on the membrane. Increased permeability of that membrane to Na^+ and K^+ is followed by a loss of the electrical potential and the membrane is said to be DEPOLARIZED.

The depolarized state is transmitted to the interior of the fiber over the membranes of the T-tubule system. Some evidence suggests that sarcoplasmic reticulum membranes are also polarized and may also be depolarized by the spread of the excitation wave, but the manner by which the disturbance is passed from T-tubule to SR (which lies 120 Å from the T-tubule) is not clear.

In any event, it is believed that SR membranes undergo depolarization and that this results in the release to the sarcoplasm of calcium ions that have been "stored" in the SR.

The tropomyosin-troponin complex, described earlier, is believed to prevent the binding of actin to myosin, perhaps by the tropomyosin-troponin complex inhibiting or physically covering the active sites on the actin molecules. Calcium ions, released from the SR, are believed to combine with troponin, which then undergoes a change in shape and "pulls" the entire tropomyosin strand so that the actin's active sites are now "uncovered" or activated.

Then, it is believed that the enlarged heads of the myosin molecules become attracted to the active sites on the actin molecules, and a bond between the myosin and actin results. The bond also causes a change in the shape of the head, causing it to tilt and pull on the actin filaments, making them draw closer together; this causes the Z lines to draw together.

Once the head has tilted, another reactive site on the myosin head is exposed and ATP now binds to that region. This binding causes detachment of the head from its binding site on the actin molecule.

Now, the ATP itself is cleaved to ADP by the ATPase (enzyme) activity of the heavy meromyosin; the energy released "cocks" the head back to its original position, and it binds to another actin active site further down the actin molecule.

These events are repeated, and the actin filaments are drawn toward one another. The events have been compared to a rowboat being moved by oars that represent the oscillating myosin heads, or to a person pulling an outstretched rope toward the body using each hand in alternate fashion. The essential fact is that the actin filaments are drawn toward one another, pulling the Z lines together and shortening the muscle fiber. The actin filaments overlap as the shortening occurs.

Figure 5.5 shows some of the events described above.

As long as the calcium ions remain in the sarcoplasmic fluid, muscle contraction will continue. However, if depolarization does not continue as a result of nerve impulses, a continually active calcium pump, activated by ATP, returns the Ca^{++} to the SR. Removal of Ca^{++}

FIGURE 5.5
Excitation-contraction coupling in skeletal muscle. (Redrawn from Guyton.)

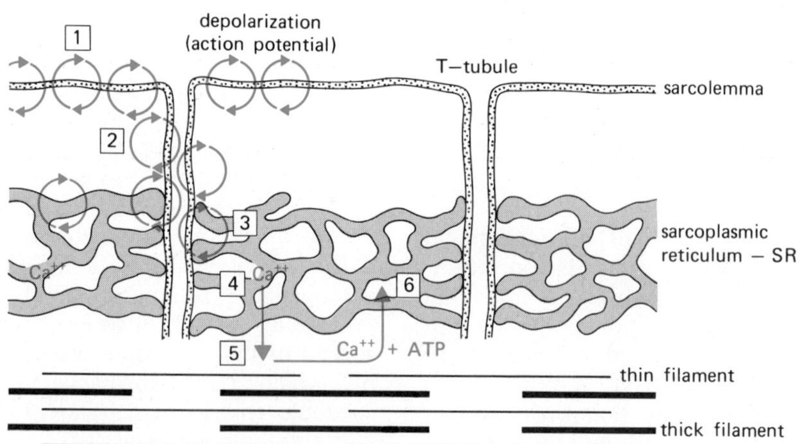

allows the troponin-tropomyosin to "recover" the actin active sites and myosin can no longer bind with actin. The actin filaments return to their original state, perhaps by an "elastic effect" exerted by the helical nature of the coils and by the pull of the connective tissues within the muscle. The muscle relaxes.

The entire sequence of events is summarized in Figure 5.6.

SOURCES OF ENERGY FOR CONTRACTION

The role of ATP in the contraction process has been described above. It should be evident that without ATP, cocking of the myosin head cannot occur and therefore maintenance of supplies of ATP becomes a prime factor in maintaining muscle activity.

Sources of energy for ATP synthesis in a muscle are the same as those described in Chapter 4. They are summarized below.

Anaerobic degradation of glucose (glycolysis):

$$\text{glucose} + 2\ \text{ATP} \rightarrow 2\ \text{pyruvic acid} + 4\ \text{ATP}$$
$$\downarrow +4\text{H}$$
$$2\ \text{lactic acid}$$

Aerobic degradation of glucose (Krebs cycle and oxidative phosphorylation):

$$\text{glucose} + 2\ \text{ATP} \rightarrow 6\ CO_2 + 6\ H_2O + 38\ \text{ATP}$$

Aerobic degradation of fatty acids (β-oxidation, Krebs cycle, oxidative phosphorylation):

$$\text{fatty acid} \rightarrow CO_2 + H_2O + \text{ATP}$$
(amounts dependent on length of fatty acid chain)

This source is not normally utilized as a primary energy source unless carbohydrate supplies are depleted.

A source of energy for synthesis of ATP that is unique to muscle is the breakdown of phosphocreatine (creatine phosphate) utilizing the enzyme creatine phosphokinase (CPK):*

$$\text{phosphocreatine} \xrightarrow{\text{CPK}} \text{phosphate} + \text{creatine} + \text{energy}$$

$$\text{energy} + \text{ADP} + \text{phosphate} \rightarrow \text{ATP}$$

or

$$\text{Phosphocreatine} + \text{ADP} \rightarrow \text{ATP} + \text{creatine}$$

ATP may be synthesized much more rapidly using phosphocreatine as an energy source than it can through glycolytic pathways.

* Elevated blood CPK levels may be used to evaluate the extent of muscle damage during heart attacks, and in muscular dystrophy.

FIGURE 5.6
A summary of the events in muscular activity.

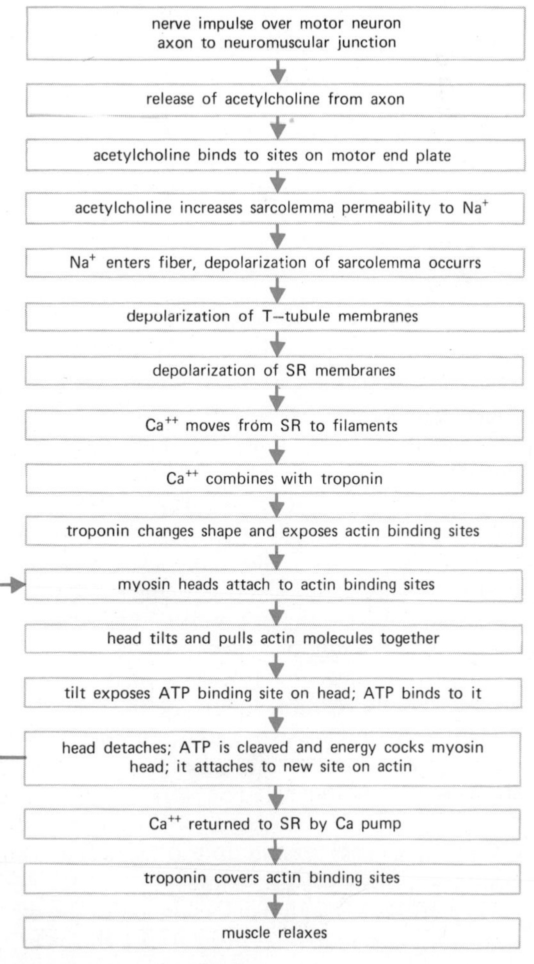

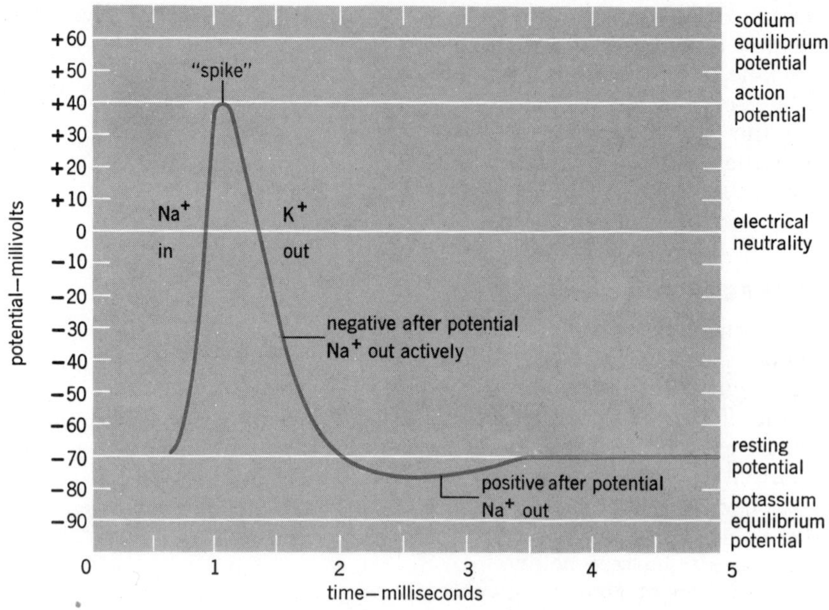

FIGURE 5.7
Changes in membrane potential during development of an action potential.

OTHER CONSIDERATIONS OF SKELETAL MUSCLE CONTRACTION

Electrical events. An ACTION POTENTIAL (Fig. 5.7) always accompanies depolarization of the muscle membranes, and is recorded as an ELECTROMYOGRAM (*EMG*) from skeletal muscle. In a *single* muscle fiber, the potential lasts about 5 msec, with the spike potential about 2 msec long. The electrical and mechanical events, however, do not occur synchronously (Fig. 5.8). Spread of the depolarization wave through the fiber, the release and diffusion of Ca^{++}, and the establishment of the bonds between actin and myosin require about 2 msec, and until these events have been completed, no shortening will occur. The term LATENT PERIOD is used to designate the time lag that occurs between stimulation of the fiber and commencement of shortening.

The EMG is often used as an aid to the diagnosis of muscular disorders. Both the rate of discharge of action potentials and their magnitude may be altered in such disorders as *myasthenia, fibrillation* (rapid "spontaneous" discharge of motor units not visible through the skin), and *fasciculation* (a "spontaneous" discharge of motor units that *is* visible through the skin). The two latter conditions represent neuromuscular junction disorder, and disease associated with the neurons supplying skeletal muscle respectively.

Heat production. Heat production accompanies muscular activity and is associated with the chemical reactions of that activity. The term ACTIVATION HEAT refers to the total heat released during one complete contraction–relaxation cycle of activity. It has several subdivisions:

INITIAL HEAT appears when the muscle is stimulated, and represents the heat released by the processes that lead to contraction; myosin activation, and ATP splitting required for cocking the myosin heads. Oxygen is not required for the production of initial heat.

RELAXATION HEAT is produced when stimulation ceases and relaxation occurs. Its magnitude correlates with the load on the muscle, and does not appear if the load is removed from the muscle before it relaxes. It may be associated with the use of ATP to fuel the calcium pump that returns Ca^{++} to the SR.

RECOVERY HEAT is produced for some time after relaxation has occurred, and represents aerobic metabolism designed to replenish ATP stores and rid the muscle of metabolites accumulated during the activity. It does not appear in the absence of oxygen.

Utilizing heat production and knowing about how much energy is available for contraction, the efficiency of muscular activity can be calculated to be about 25 percent. That is, about one-quarter of the energy available is directly used for contraction, the remaining three-quarters appears as heat.

FIGURE 5.8
Electrical and mechanical events correlated for skeletal muscle. Note asynchrony of events. *a* = latent period; *b* = contraction; *c* = relaxation.

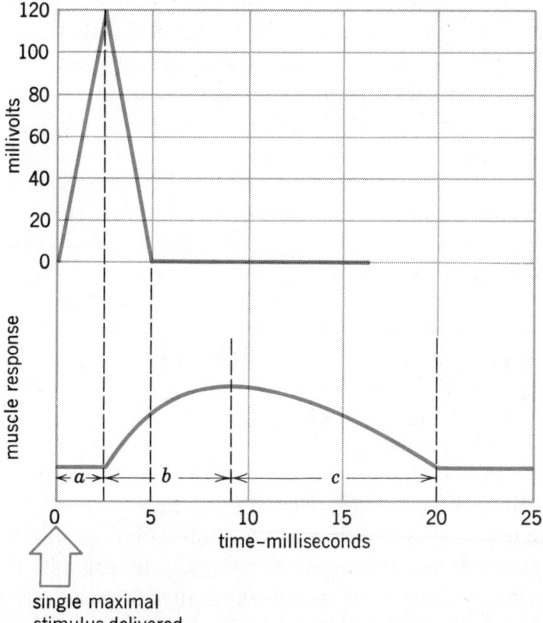

The formula used for the calculation is:

$$\text{efficiency} = \frac{W}{A + 0.16\, P_o x + 1.18W}$$

where

W = work (load × distance)
A = initial heat, constant at about 0.1 gm cm
P_o = maximum tension developed
x = distance shortened

Physiological properties.

Stimulus properties. It is possible to deliver measurable electrical stimuli to a muscle that are too weak to cause a response on the part of the muscle. Such a stimulus is called a SUBLIMINAL (*subthreshold*) stimulus. Increasing the strength of the stimulus will result in a point being reached when a barely perceptible response is obtained. The strength of the stimulus at this point is said to be LIMINAL (*threshold*). It is just strong enough to cause depolarization of those fibers having the lowest thresholds (a point at which a physiological effect is produced) for depolarization. From this point, further increase of stimulus strength produces an increasing response, until the point is reached when *all* fibers in the muscle are active. At this time, the stimulus is said to be MAXIMAL, as is the response of the muscle.

Grading of strength of contraction. From the preceding discussion, it is obvious that stronger stimuli will result, to a point, in stronger contractions. Remembering that skeletal muscles are served by nerves, that nerves are composed of many smaller nerve fibers, and that each nerve fiber supplies a number of muscle fibers (motor units), we can say that nerve stimulation of different intensities will activate different numbers of motor units, thereby grading or adjusting contraction strength to the job at hand. Experience and training play a large part in knowing how much force is required for a particular task. We have all had the experience of nearly "throwing something through the roof" when we put out more force than was necessary.

All-or-none law. Even though the number of active motor units can be altered, it may be demonstrated that if a stimulus is just strong enough to activate a motor unit, that unit will give a maximal response. If not strong enough to activate, no response will occur. These phenomena are stated in the ALL-OR-NONE LAW, which further stipulates: response will not occur or will be maximal *according to the conditions at the time.* The latter takes into account such factors as pH, temperature, load on the muscle, and so on, all of which may change the muscle's response to stimulation.

Length-tension relationships. If a muscle fiber is stretched, it exhibits elasticity much in the same manner as a rubber band. This elasticity may be due to the helical arrangement (like springs) of the muscle's major proteins, and to the connective tissue components holding fibers together in the muscle. If the muscle is fixed at the length it normally has in the body (resting length), and it is stimulated maximally, it will develop its greatest contractile tension (Fig. 5.9). It appears to contract most efficiently when placed under tension by the skeletal structures to which it attaches, and a length greater or lesser than resting results in less development of tension when stimulated. Microscopically, when the sarcomere has a length of 2.0 to 2.2 μm, all bridges are pulling on the actin filaments, and maximum tension is developed (Fig. 5.10). Shorter than this, and some myosin heads have no actin sites to attach to. It has been suggested that degree of overlapping of actin filaments on the myosin cross bridges within the sarcomere (a function of length) is the factor determining contractile strength. Maximum contraction occurs when there is maximum overlap between the actin filaments and the cross bridges of the myosin filaments, and supports the idea that the number of cross bridges pulling the actin filaments determines strength of contraction.

FIGURE 5.9
A length-tension curve. Length is measured in arbitrary units, with 100 representing the resting length of the muscle in the body. Note that maximum tension is developed when muscle is at its normal body length.

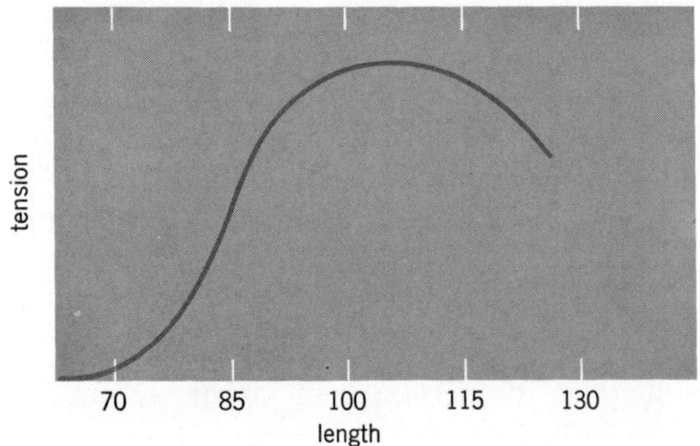

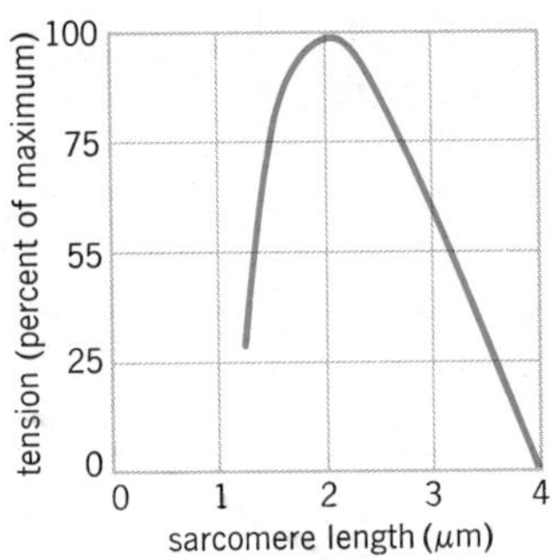

FIGURE 5.10
A length-tension curve for one sarcomere. Maximum tension is developed when sarcomeres are at a length of about 2.1 μm.

Force-velocity relationships. Speed of muscular contraction varies inversely with the weight the muscle must move, and becomes zero when the load is too heavy to move (Fig. 5.11). Contractions in which tension develops *and* shortening occurs are termed ISOTONIC CONTRACTIONS, while those in which tension develops but no shortening occurs are termed ISOMETRIC CONTRACTIONS. Shortening is most rapid, in an isotonic contraction, when the muscle is loaded by a weight equivalent to that imposed by body structures, that is, the skeleton.

The term isometric is probably familiar to those who have undertaken training programs with a view to increasing muscular strength. Isometric exercises, in which one muscle or group of muscles is pitted against another with no movement being allowed, have been shown

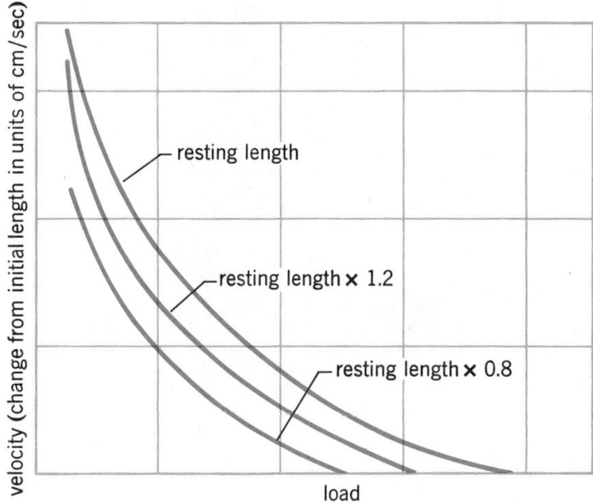

FIGURE 5.11
Several force-velocity curves. Shortening is most rapid when muscle is loaded by a weight equivalent to that normally imposed by body structures.

to cause a more rapid increase in muscular strength than isotonic exercises (e.g., lifting bar bells). Although a complete physiological explanation for this phenomenon is lacking, several things occur more rapidly under an isometric program: muscle fibers increase in size more rapidly; fibers that are smaller or not normally active are brought into play more rapidly; faster increase in capillary networks occurs; and neuromuscular junctions increase in size and chemical content, bringing stronger stimuli to the muscle.

Independent irritability. It is possible to cause a muscle to contract by direct placement of stimulating electrodes on the skin. This method of stimulation bypasses the nerve fibers normally responsible for stimulation. Independent irritability refers to the capacity to respond to direct stimulation. If a muscle is denervated it undergoes atrophy (decrease in size), or if it is immobilized, as in a cast, it may also atrophy. Direct stimulation of a denervated muscle may prevent atrophy until regrowing nerve fibers reestablish connections. If connections cannot be reestablished, the fibers will eventually degenerate and be replaced by scar and fatty tissue. Atrophy of a muscle when a limb is placed in a cast is an *atrophy from disuse*. Muscle health is, in part, maintained by utilization—in short—use it or lose it.

Refractory periods. Immediately after stimulation, a muscle cannot be stimulated to contract again until a definite time period has passed. This is followed by a period of time where a stronger than normal stimulus is required to initiate a mechanical response. The first time period is termed the ABSOLUTE REFRACTORY PERIOD, the second, the RELATIVE REFRACTORY PERIOD. The time after which a skeletal muscle *will* respond to a second stimulus is about 1/500 second, a time that permits addition of the effects of the second contraction to that of the first. Such addition of contractile effects is termed WAVE SUMMATION, and allows the development of TETANIC and SUBTETANIC CONTRACTIONS (see below).

This section has demonstrated and discussed some of the unique properties skeletal muscle has. It may be emphasized that such properties fit this type of muscle especially well for the jobs that it must do, such as moving the body through space and maintaining its posture.

CHARACTERISTICS OF THE SEVERAL TYPES OF CONTRACTIONS EXHIBITED BY SKELETAL MUSCLE

The twitch. A single maximal stimulus applied to a muscle elicits a single isotonic response called a TWITCH (Fig. 5.12) (not the "twitching" we sometimes feel in a muscle if it is tired or has some chemical imbalance). Such a contraction demonstrates the time relationships of the reaction. For a frog muscle, for example, latent period, period of contraction, and relaxation take about 0.1, 0.4, and 0.5 seconds. In human skeletal muscle, differing fiber size results in "fast" and "slow" types of twitches. In general, smaller fibers are "faster," larger fibers are "slower." Figure 5.13 shows twitches obtained isometrically (so that tension or force of contraction may be measured). Fast muscles ("white" muscles) react most rapidly, and are found in those body areas where speed is important (moving eyes, upper appendage); slow muscles ("red" muscles) react more slowly, and are found in body areas where prolonged ac-

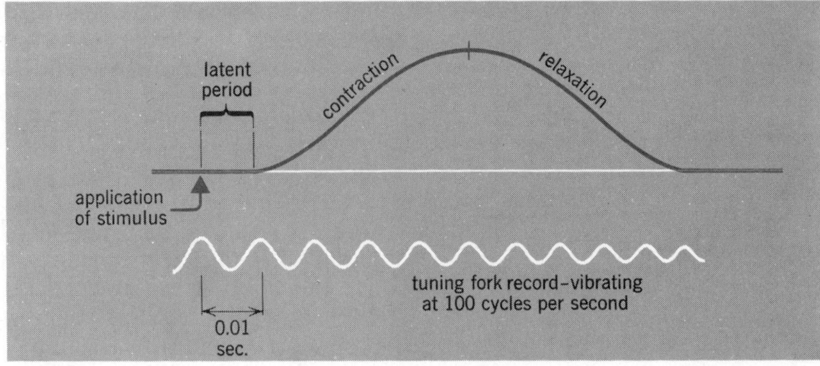

FIGURE 5.12
A single muscle twitch.

tivity is required, as in the postural muscles of the back and hip area. Again we see modifications of muscle according to function.

Tetanus (Fig. 5.14). As indicated above, repeated maximal stimulation of skeletal muscle with *increasing frequency* of stimulation results in summation of twitches. The *lowest* frequency that causes fusing of contractions so that they appear as a single smooth response is called the CRITICAL FREQUENCY, and *when* it occurs, TETANIZATION has taken place. This phenomenon may result, in part, from the continuous presence of Ca^{++} in the area of the filaments—it is being released at a rate equal to or faster than the pump can return it to the SR. The increased strength of contraction that oc-

FIGURE 5.13
Several isometric contraction curves showing the duration of the twitch in several different muscles of the body.

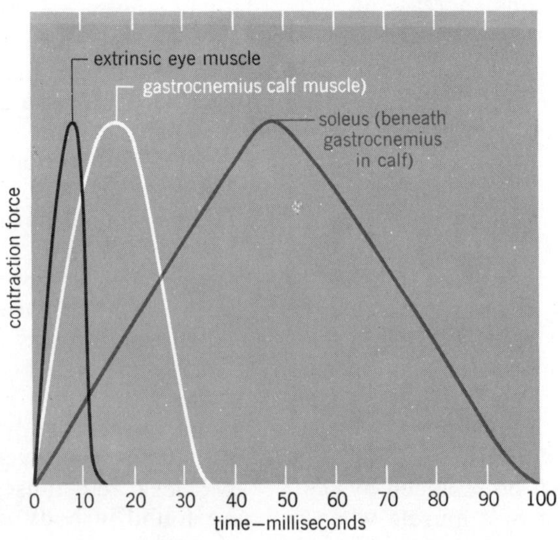

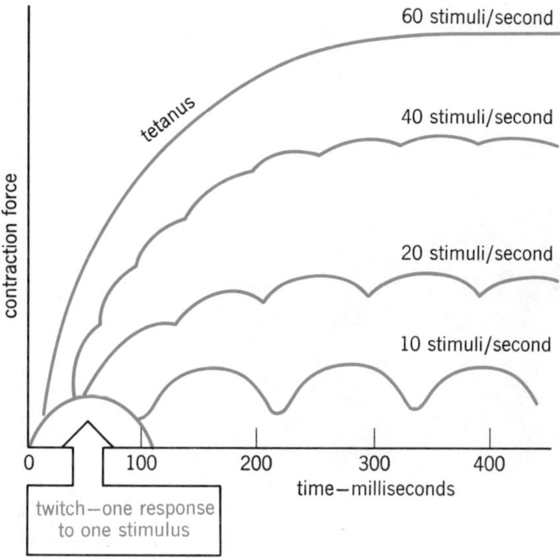

FIGURE 5.14
Wave summation and the genesis of tetanus in a "slow" muscle, as a function of stimulus frequency.

curs in the body with development of tetanus is the result of nervous mechanisms that increase *both* the rate of impulse stimulation and the number of motor units firing synchronously.

Treppe (Fig. 5.15). Even with maximal stimulation of a muscle, the first few stimuli delivered to it may result in an increase of contraction strength called TREPPE. This phenomenon is partly attributed to release of heat from ATP splitting and other chemical reactions that causes "thinning out" of the sarcoplasm, much in the same way that oil in an auto engine becomes less viscous as the engine "warms up." Internal resistance of the muscle is decreased, and more energy can be directed to shortening and less to overcoming internal resistance. Additionally, it has been shown that there is an increased influx of Ca^{++} into the muscle cell during treppe, as compared to efflux. Eventually, influx and efflux become equal and the strength of contraction "levels off."

Tone. TONE in a muscle refers to a continuous submaximal contraction that maintains a continual though not necessarily constant degree of tension in the muscle. Tone appears to depend on an intact nerve supply to the muscle and may be largely reflex in nature. The "relay theory" suggests that alternating motor units are caused to respond by some type of central nervous system mechanism, with another motor unit being caused to contract as the first is about to relax. Thus, a certain number of, *but not the same,* motor units will be stimulated to contract. This maintains a certain level of tension without fatigue and keeps the muscle in a state of readiness for activity.

SMOOTH MUSCLE

PHYSIOLOGICAL ANATOMY

SMOOTH MUSCLE (Fig. 5.16) consists of spindle-shaped fibers 2 to 5 μm in diameter and 50 to 100 μm in length. Each fiber possesses a single nucleus located near its center in the widest portion of the cell. A sarcolemma surrounds the fiber. Myofibrils are few and nonstriated. Smooth muscle is usually said to be involuntary, although its activity may be modified and in some cases controlled by nervous stimulation. The fibers are capable of maintaining a more-or-less constant contraction (tone) regardless of length. Concentrations of actin and myosin are about seven times lower in smooth muscle than in skeletal, as are concentrations of ATP and phosphocreatine. The muscle fibers appear to lack T-tubules and have a poorly developed sarcoplasmic reticulum.

FIGURE 5.15
Treppe.

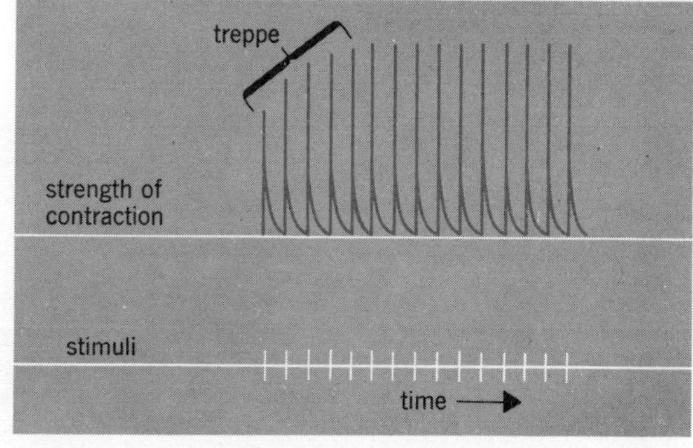

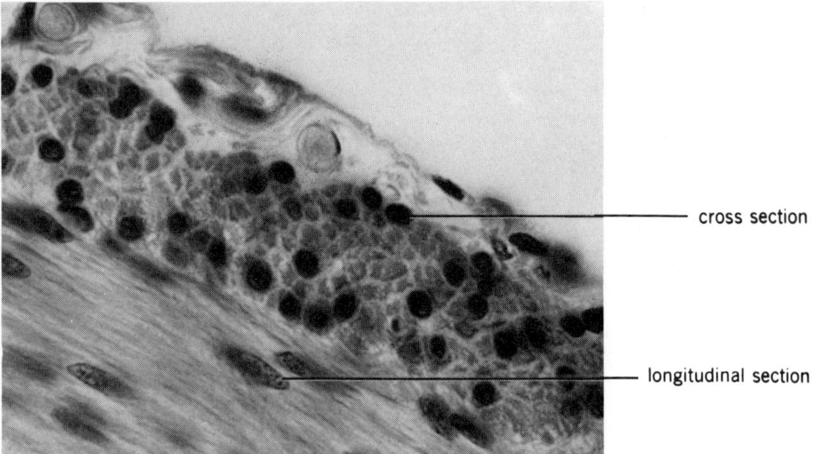

View in light microscope X440.

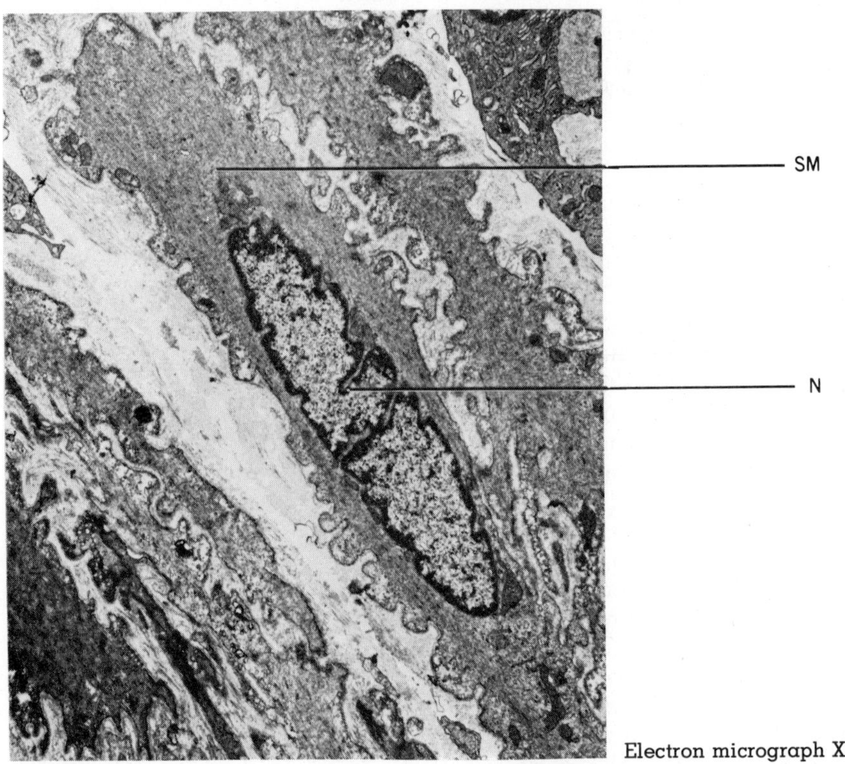

Electron micrograph X8,000.

FIGURE 5.16
The morphology of smooth muscle. SM = smooth muscle cell; N = nucleus of smooth muscle cell. (Electron micrograph, courtesy Norton B. Gilula, The Rockefeller University.)

TYPES OF SMOOTH MUSCLE

Two categories of smooth muscle may be distinguished.

UNITARY SMOOTH MUSCLE exhibits spontaneous activity and conducts impulses from one muscle cell to another as though the entire muscle mass was a single cell. This type of muscle is found in the alimentary tract, uterus, ureters, and small blood vessels.

MULTIUNIT SMOOTH MUSCLE does not exhibit spontaneous contraction, usually requires stimulation by nerves, and can grade its strength of contraction. This type of muscle is found in the ciliary muscles and iris of the eye and in larger blood vessels.

Some smooth muscle does not fall clearly into either group. For example, in the bladder, stretch and/or nerve stimulation can cause contraction. Unitary smooth muscle has been studied most extensively and the comments that follow are based primarily on investigation of this type.

SPONTANEOUS ACTIVITY

Spontaneous activity depends on local depolarization similar to that occurring in cardiac nodal tissue. The membrane potential is about two-thirds that of skeletal muscle and nerve (about 60 mv) and decays so that spontaneous depolarization occurs. This depolarization does not arise in a constant area in the muscle but "migrates" from one place to another. The waves or areas of depolarization are then conducted from cell to cell by way of low resistance "tight junctions" between fused adjacent cell membranes. Contraction of smooth muscle is believed to occur by a process similar to that described for skeletal muscle. Actual contraction time is long, presumably because of the time required for the few myofibrils to develop tension. Strength of smooth muscle contraction depends on the frequency of spontaneous discharge from individual cells and the number of cells contracting together. Smooth muscle contraction, and the degree of tension developed, may be influenced by several factors.

STRETCHING usually serves as an adequate stimulus for depolarization and is followed by contraction.

Chemical factors. ACETYLCHOLINE stimulates most unitary smooth muscle to increased activity. The mechanism of its action is not clear but may be to increase membrane permeability to Ca^{++} or Na^{+}, similar to its effect on skeletal muscle. EPINEPHRINE effects depend on type of muscle and species. Epinephrine usually inhibits contraction of gut musculature and stimulates activity of uterine and vascular muscle. This chemical probably exerts its action by changing membrane permeability to Ca^{++} or Na^{+}. NOREPINEPHRINE is similar to epinephrine in action and effect but requires about 100 times the concentration of an epinephrine "dose" to achieve the same degree of effect. ESTROGENIC HORMONES and OXYTOCIN also modify activity of smooth muscle, particularly of the uterus. Estrogens stimulate activity, as does oxytocin. PROGESTERONE tends to quiet or inhibit activity. OTHER CHEMICALS affecting smooth muscle are reserpine, ephedrine, and botulinus toxin. Reserpine, a chemical employed as an antihypertensive agent, causes relaxation of arteriolar smooth muscle. Ephedrine, a substance employed in nasal sprays, constricts vessels in the repiratory system. Botulinus toxin blocks acetylcholine release and leads to muscular paralysis.

Nervous factors. Autonomic nerves supply most smooth muscle and, by secretion of chemicals at their endings, exert effects similar to those produced by acetylcholine and norepinephrine. Most smooth muscle receives parasympathetic, acetylcholine-secreting nerves and sympathetic, norepinephrine-secreting nerves. No specialized end plate exists to transmit the effects of nerve stimulation to the muscle. Instead, the axons themselves may be seen to contain vesicles that presumably contain the chemicals that are released to influence muscle activity.

We may again note that a double nerve supply enables activity to be regulated to maintain homeostasis of such activities as blood pressure and peristalsis.

CLINICAL CONSIDERATIONS

Nerve plexuses* and fibers are found in the walls of such organs as the small and large intestines. These structures aid in coordinating activity that may originate within the muscle and/or serve to transmit the effects of extrinsic control to the muscle.

CONGENITAL MEGACOLON (*Hirschsprung's disease*) is enlargement of the colon because of failure of the nervous structures to develop in a section of the organ. The section without nervous structures undergoes a persistent contraction and presents a functional obstruction to passage of materials. The colon proximal to the obstruction enlarges as material accumulates. If the section without nervous structures is short and near the anus, laxatives may provide a means of aiding passage of feces. Most cases, however, require surgical treatment in which the abnormal section is removed and the ends of the colon anastomosed (joined together).

This disorder emphasizes the importance of the intrinsic nervous elements serving smooth muscle in the gut.

SUMMARY

1. Muscular tissue is contractile, and can shorten and perform work.
2. There are three types of muscular tissue:
   a. Skeletal muscle forms 40 percent of the body weight, attaches to the skeleton and is served by nerves that cause it to contract.
   b. Smooth (visceral) muscle lies around hollow body organs. It may or may not depend on nerves to contract and controls propulsion of materials through organs.
   c. Cardiac muscle is found only in the heart, and pumps the blood through the body. Its contraction is controlled by inherent factors.
3. Skeletal muscle cells are called fibers. Each fiber is surrounded by a sarcolemma, and contains sarcoplasm, myofibrils (composed of filaments), mitochondria, T-tubules, and sarcoplasmic reticulum (SR). The fibers show characteristic striations.
4. The myofibrils are composed of four major muscle proteins:
   a. *Myosin.* Composed of light and heavy portions, myosin acts to form bonds to actin during contraction.
   b. *Actin.* A double helix of protein molecules, actin is the molecule drawn together when the muscle shortens.
   c. *Tropomyosin.* This protein lies imbedded in the grooves around the actin helix.
   d. *Troponin.* This protein binds Ca^{++} and is located at repeating distances along the tropomyosin helix.
5. Muscle also contains ATP, Ca^{++} (in the reticulum), phosphocreatin, glucose, and other substances.
6. The neuromuscular junction serves to transmit nerve impulses to the muscle as follows:
   a. Nerve impulses cause increase of permeability of nerve fibers to Ca^{++}.
   b. Ca^{++} causes release of acetylcholine (Ach) stored in nerve endings.
   c. Ach causes depolarization of muscle membrane systems.
   d. An enzyme, cholinesterase, rapidly destroys Ach.
7. Muscle contraction and relaxation involves
   a. Depolarization of sarcolemma, T-tubule membranes, and possibly SR membranes.
   b. Ca^{++} moves from SR to filaments.
   c. Ca^{++} binds to troponin, which alters shape and uncovers active sites on actin.
   d. Myosin forms bonds to actin.
   e. ATP attaches to myosin, causing head to oscillate.
   f. Head attaches repeatedly to different actin sites, drawing actin filaments together.
   g. Ca^{++} is pumped back to SR.
   h. Tropomyosin recovers actin sites.
   i. No more binding of myosin to actin.
   j. Muscle relaxes.
8. ATP forms the immediate source of energy for muscle contraction and return of Ca^{++} to the SR. Metabolism of creatine phosphate, glucose, and fatty acids furnish energy for ATP synthesis.

* Meissner's plexus (submucosal) and Auerbach's plexus (between smooth muscle layers).

9. Electrical events precede a muscle contraction. An action potential forms, recordable as an EMG.
10. Heat production occurs; activation heat refers to all heat produced during one cycle and is subdivided into initial, relaxation, and recovery heats.
11. Some physiological properties of skeletal muscle are
    a. It won't contract until its threshold for depolarization is reached.
    b. Its strength of contraction can be graded according to requirement for tension.
    c. Its motor units follow the all-or-none law.
    d. It contracts more strongly when slightly stretched.
    e. It contracts more rapidly when slightly stretched.
    f. It may be stimulated directly; it shows independent irritability.
    g. It has refractory periods that are short, enabling it to be tetanized.
12. Contraction forms that a skeletal muscle shows include
    a. Isotonic. Shortening occurs; causes movement.
    b. Isometric. No shortening occurs; maintains posture.
    c. Twitch. One cycle as a result of one stimulus. Shows phases and timing of activity.
    d. Tetanus. Sustained maximal contraction from fusion or summation of twitches.
    e. Treppe. Increased strength of contraction as muscle "warms up."
    f. Tone. Sustained partial state of contraction maintained without fatigue by alternate contraction of different motor units.
13. Smooth muscle is slow contracting, contains one nucleus per spindle-shaped unstriated cell, and occurs in unitary and multiunit varieties.
    a. Unitary exhibits spontaneous activity and is found primarily in digestive, reproductive, and urinary systems.
    b. Multiunit depends on nerves to contract and is found in the iris of the eye and larger muscular blood vessels.
14. Spontaneous activity originates in a "pacemaker" within the muscle and spreads via tight junctions between cells.
15. Smooth muscle is very sensitive to chemicals
    a. Acetylcholine usually causes unitary muscle to contract.
    b. Epinephrine may cause contraction *or* relaxation depending on location and species.
    c. Other chemicals and their effects are presented.
16. Most smooth muscle has a dual nerve supply: one set causes it to contract; the other causes its relaxation. Activity can therefore be closely regulated.

QUESTIONS

1. Compare skeletal and smooth muscle as to structure and the properties that fit each for the jobs it does.
2. How does a nerve impulse "get to" a muscle to cause it to react?
3. What would happen to muscle contraction, and why, if Ca^{++} was not available?
4. What are the roles of the following in muscle contraction/relaxation?
    a. ATP
    b. Myosin
    c. Troponin
5. List the steps involved in muscle contraction.
6. What roles do glucose and phosphocreatine play in muscle metabolism?
7. What effect would you expect a low environmental temperature to have on muscle activity? Explain.
8. Describe one disorder of skeletal muscle and how it interferes with normal function.
9. How do skeletal and smooth muscle react to changes in their chemical environments? Which do you speculate, and why, is more controlled by chemicals?
10. Curare is a chemical that blocks the action of acetylcholine at the neuromuscular junction. Explain, in terms of the junction, why curare is chosen by certain South American Indian tribes as an arrow poison.

READINGS

Cohen, Carolyn. "The Protein Switch of Muscle Contraction." *Sci. Amer. 233:*36. (Nov) 1975.

Felig, Philip, and John Wahren. "Fuel Homeostasis in Exercise." *New Eng. J. Med. 293:*1078. Nov 20, 1975.

Fuchs, Franklin. "Striated Muscle." *Ann. Rev. Physiol.* Vol 36. Palo Alto, Cal., 1974.

Huddart, Henry, and Stephen Hunt. *Visceral Muscle, Its Structure and Function.* Halstead (Wiley). New York, 1975.

Lester, Henry A. "The Response to Acetylcholine." *Sci. Amer. 236:*106. (Feb) 1977.

Prosser, C. Ladd. "Smooth Muscle." *Ann. Rev. Physiol.* Vol 36. Palo Alto, Cal., 1974.

THE BASIC ORGANIZATION AND PROPERTIES OF THE NERVOUS SYSTEM

OBJECTIVES

After studying this chapter, the reader should be able to:

- ☐ Define what is meant by excitation and conduction.
- ☐ Explain some of the mechanisms by which cells are rendered excitable and polarized.
- ☐ Explain how a membrane becomes depolarized, what ion(s) move(s), and how an impulse is created.
- ☐ Explain what repolarization is, and how it occurs.
- ☐ Diagram and explain an action potential.
- ☐ Explain how an impulse is conducted.
- ☐ Diagram a "typical neuron."
- ☐ List and explain neuronal properties in terms of de- and repolarization.
- ☐ Give functions and properties of glial cells, as well as listing important types.
- ☐ Describe the structure of a synapse.
- ☐ Describe how a synapse transmits impulses.
- ☐ List and explain synaptic properties in terms of a chemical mode of transmission of impulses.
- ☐ List the parts of a reflex arc.
- ☐ Explain the characteristics of reflex activity, and the value of such activity to the maintenance of homeostasis.

MEMBRANE POTENTIALS

It may be demonstrated that electrical potentials exist across cell membranes in practically all body cells. The magnitude of the potential is not the same across all membranes, however, and certain cells, such as muscle and nerve cells, have transmembrane potentials that are among the highest in the body. If a cell can establish an electrical potential across its membrane, it becomes EXCITABLE, or capable of responding to changes (stimuli) in its internal or external environments with formation of an action potential. Some cells, again especially muscle and nerve cells, are capable of transmitting the effects of stimuli over the length and depth of the cell, and this forms the basis of CONDUCTION of impulses.

The ability to render a cell excitable, and to form and transmit disturbances, revolves primarily around the electrolytes of the fluids outside (*extracellular fluid,* ECF) and inside (*intracellular fluid,* ICF) the cells. Of the various ions available in these fluids, as shown in Table 6.1, three are present in greatest concentration and are likely candidates for involvement in production of a membrane potential.

It may also be recalled that cell membranes contain active transport systems capable of causing unequal distributions of electrolytes on two sides of a membrane, and that the membrane itself is not equally permeable to all ions. Such factors as size, electrical charge, and whether an ion has a single or multiple charge all enter into its ability to cross a membrane. Because of such factors as mentioned above,

TABLE 6.1
AVERAGE CONCENTRATIONS OF MAJOR ELECTROLYTES IN BODY FLUIDS

ELECTROLYTES	CONCENTRATION IN EXTRACELLULAR FLUID meq/L[a]	CONCENTRATION IN INTRACELLULAR FLUID meq/L
Sodium	145	12
Potassium	4	155
Chloride	120	4

[a] One equivalent weight of a substance is its atomic weight divided by its valence. It represents the number of grams of an element that will combine with 8 grams of oxygen or 1.008 grams of hydrogen. A milliequivalent is 1/1000 of an equivalent weight.

typical plasma membranes are, when at rest or not reacting to stimuli, much more permeable to potassium than to sodium ions. Also, there are many large nondiffusible ions, chiefly proteins and organic phosphate and sulfate ions inside the cell. These are also involved in creation of a membrane potential.

With these facts in mind, let us next examine how an electrical difference or potential is established across a plasma membrane.

MECHANISMS OF ESTABLISHMENT OF MEMBRANE POTENTIALS

If a difference of electrical potential is to be created across a membrane, and the items available for establishing it are charged particles, it seems reasonable to assume that a difference of potential can only be created if there is an *unequal distribution* of charges on the two sides of the membrane. There would appear to be two basic methods by which unequal ionic separation can be achieved: differential permeability of membranes to different ions, and active transport.

Differential permeability of membranes to ions can cause a slight separation or unequal distribution of ions because membranes are not *equally* permeable to all ions. Sodium, although it appears earlier in the periodic table of elements than potassium, passes with greater difficulty through most membranes. If a concentration gradient exists for a diffusible ion, it will move toward the side of the membrane with the lowest concentration of that ion. This mechanism can therefore cause an unequal distribution of ions on two sides of a membrane.

In living cells, the presence on the inside of the cell of large nondiffusible ions also influences ionic distribution by what is called the DONNAN-GIBBS EQUILIBRIUM. When the process reaches equilibrium, three criteria must be met:

The *products of the diffusible ions must be equal* on the two sides of the membrane.

The side containing the nondiffusible ions must have a *greater concentration of diffusible cations* than the other side.

There will be *electrical neutrality* between ions on a given side of the membrane.

To illustrate the operation of the Donnan-Gibbs principle, consider the following diagram:

SEMIPERMEABLE MEMBRANE		SEMIPERMEABLE MEMBRANE	
$5\ Na^+$	$10\ Na^+$	$5\ Na^+$	$6\ Na^+$
$5\ A^-$	$10\ Cl^-$	$5\ A^-$	$6\ Cl^-$
		$4\ Na^+$	
		$4\ Cl^-$	
side 1	side 2	side 1	side 2
Initial condition		At equilibrium	

A^- = large, nondiffusible anion

At the start, there is a diffusion gradient for Na^+ from side 2 to side 1, and it will move toward side 1. As positively charged sodium ions diffuse, they attract negatively charged chloride ions and the latter will move across the membrane in equal numbers with sodium ions. Movement of these charged ions causes the production of an electrical potential on side 1 that tends to oppose the further movement of Na^+ (more Na^+ on side 1 creates a *repulsion* to the passage of positive charges), and the diffusion of Na^+ stops. The membrane potential that just stops the net diffusion of an ion is called its EQUILIBRIUM POTENTIAL; the potential may be calculated by the use of the Nernst equation (*see* Appendix). Thus, at equilibrium, the ionic distribution would appear as shown on page 95. Have the conditions stated earlier been met?

On side 1, the products of *diffusible* ions are:

$$9\ Na^+ \times 4\ Cl^- = 36$$

On side 2, the products are:

$$6\ Na^+ \times 6\ Cl^- = 36$$

The nondiffusible ion is on side 1; side 1 contains a greater concentration of diffusible cation ($Na^+ = 9$) than side 2 ($Na^+ = 6$).

On side 1, positive charges (9 Na^+) are equal to negative charges (5 A^- + 4 Cl^-); on side 2, positive charges (6 Na^+) equal negative charges (6 Cl^-).

The conditions are thus met, *but,* there is an *unequal* distribution of charges on the two sides.

We thus have seen that purely passive forces may operate to create a potential. If only such passive forces were operating, the measurable membrane potential would be about 30 millivolts (mv). Actual potentials on highly excitable cells such as nerve cells are 70 to 90 mv. Something else must be operating to achieve an even greater separation of ions.

ACTIVE TRANSPORT SYSTEMS play the role of achieving large ionic separations. A coupled Na-K pump, such as the one illustrated in Figure 6.1, can achieve the degree of separation of ions shown in Table 6.1, and the calculation of potential by the Nernst equation using these values accords nicely with the actual measured potential. We may therefore conclude that active mechanisms contribute the majority of the force causing ionic separation and the bulk of the membrane potential.

EXCITABILITY

The resting potential. By such passive and active mechanisms as just described, ions may be distributed unequally, and a membrane potential results. The name given to the potential

FIGURE 6.1

Sodium combines with a specific carrier (y) and is transported to the interstitial fluid. Carrier y is enzymatically transformed (E_1) into carrier x, which brings potassium in. Again, at the inside of the membrane, enzymatic transformation (E_2) changes x to y and the process is repeated.

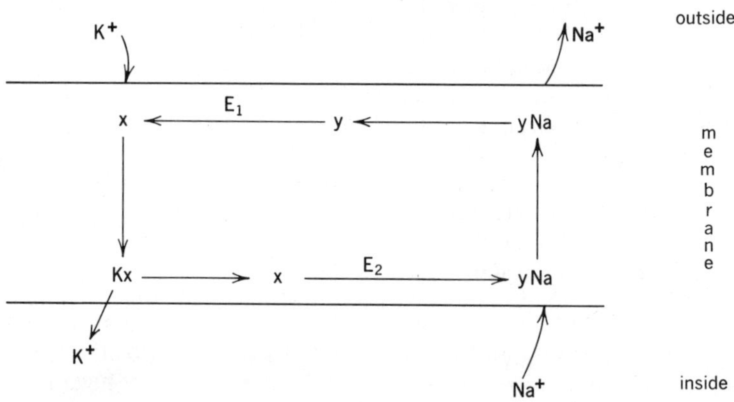

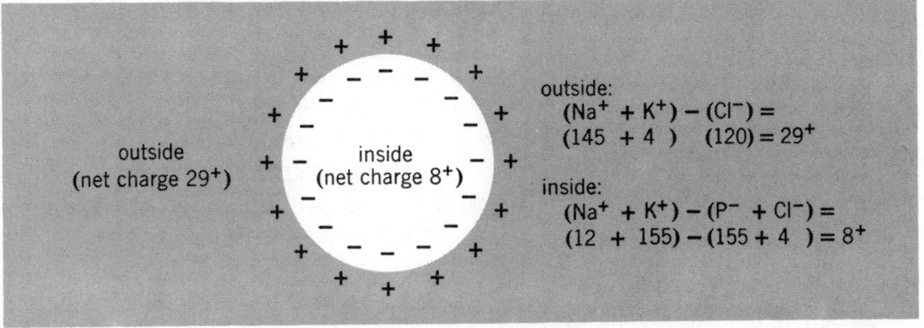

FIGURE 6.2
The separation of ions and charges resulting during attainment of the excitable state in a cell.

resulting from such activity is the RESTING POTENTIAL. In a nerve fiber, the ion believed most responsible for that potential is K^+. The reasons are as follows:

Mainly by active transport, sodium is being kept outside the cell, while potassium is being placed inside the cell.

The resting cell membrane is more permeable to K^+ than to Na^+ (50 to 100 times more), so that K^+ diffuses *out* of the fiber and tends to accumulate on the outside of the cell.

More positive charges, because of K^+ diffusion, are found on the outside of the fiber relative to the inside, and a "potassium potential" is established; the membrane is POLARIZED.

These relationships are presented in Figure 6.2. By convention, if the outside of a cell is electrically positive to the inside, a negative sign is placed in front of the measured membrane potential. Applying the Nernst equation (see Appendix) to calculate the membrane potential using the values of Table 6.1 shows that if the potential were due to Na alone it would have a value of −90 mv. Thus, the actual resting potential, being closer to that of K, is believed to be due to K more than to Na.

FORMATION OF TRANSMISSIBLE IMPULSES

If the permeability of the membrane to sodium should suddenly increase, sodium ion could move into the cell, by following its diffusion gradient. When a cell is stimulated, such an increase in permeability may actually be recorded; the membrane becomes about 500 times more permeable to sodium than it was before, followed by a 30 to 40 fold increase in permeability to potassium. The actual mechanism by which the increased sodium permeability is achieved is unknown. Suggestions have been advanced involving interruption of the Na-K pump, or the opening of "pores" for sodium movement in the membrane. It does not matter whether the stimulus leading to increased sodium permeability is thermal, mechanical, electrical, or chemical in nature; all it must do is lead to increased permeability.

If the fiber is polarized,

then stimulation, by causing a greater increase in permeability to sodium than potassium at the point of stimulation, allows inflow of sodium of a magnitude sufficient to reverse the original ionic relationships. The fiber becomes DEPOLARIZED, with the membrane now a "sodium membrane," as sodium flows into the fiber due to its increased permeability.

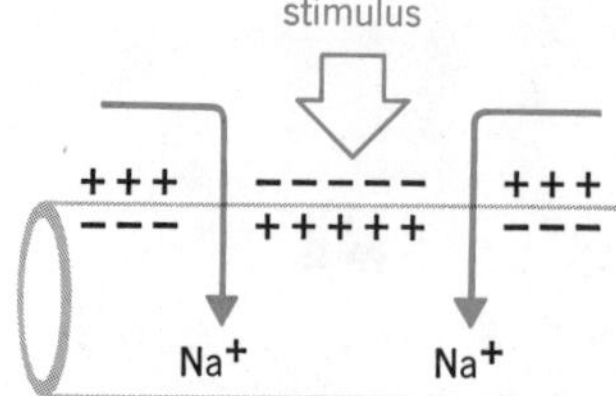

A depolarized area now lies adjacent to a still polarized area. Utilizing the ion-laden ECF and ICF, a current flow begins between the polarized and depolarized regions, as though two batteries were connected to one another.

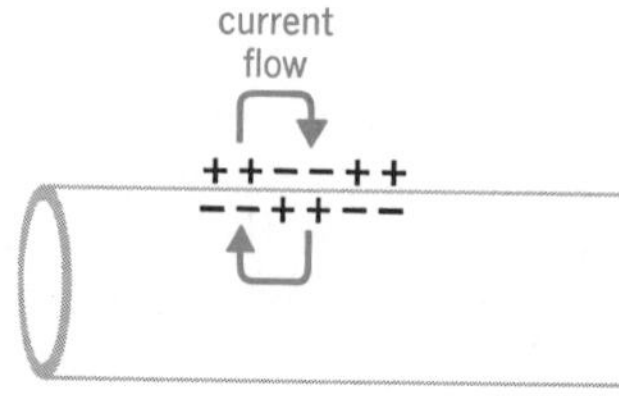

Indeed, the term "battery effect" is often used to describe this situation. In any event, the current flow may be said to represent a measurable or recordable phenomenon equivalent to a NERVE IMPULSE.

If no continuation of the stimulus occurs, permeability to sodium returns to original resting values, and there is a great increase of potassium permeability. Presumably, the Na-K pump moves sodium back again to the outside of the fiber, and recaptures some of the lost potassium. The fiber has become REPOLARIZED as the original condition is restored.

The action potential. As membrane permeability changes and ionic movement occurs, the associated alterations of electrical potential may be recorded as an ACTION POTENTIAL (Fig. 6.3). The action potential has several parts.

The SPIKE POTENTIAL is the initial large change in membrane potential. In large nerves, it lasts about 0.4 msec and reflects the vast inflow of sodium as membrane permeability changes to that ion. The degree by which the spike reverses beyond electrical neutrality is the REVERSAL POTENTIAL.

A NEGATIVE AFTER POTENTIAL reflects the return of membrane permeability to resting values for both ions, the active movement of Na^+ out of, K^+ back into the cell, with a buildup of K^+ immediately outside of the membrane.

A POSITIVE AFTER POTENTIAL occurs as the resting potential becomes a bit more positive than it was originally; it is as if the active Na^+ pump "overshot" its mark, or if the K^+ returned more slowly to the cell than Na^+ was pumped out, allowing excessive positive charges to accumulate outside the fiber. In any event, this after potential may last from 50 msec to several seconds as the pump equilibrates.

CONDUCTILITY

If the current flow and electrical disturbance described above can be transmitted or caused to move along the fiber, CONDUCTILITY results, and messages may then be sent long distances through the body over nerve cells. Conductility is believed to occur as follows.

The "battery effect" creates an electrical field which, while its strength decreases exponentially from its center, is nevertheless strong enough at some point from its center to cause depolarization of the next section of the fiber. Depolarization of *that* section allows another flow of current that distance further along the fiber. In effect, the area of depolarization has been advanced a distance along the fiber. These events are repeated and the impulse "travels" down the fiber. Behind the advancing impulse, repolarization is occurring, and the fiber is ready to respond to another stimulus. The events involving depolarization–repolarization occur in a very short time, about 1 msec in large fibers. The transmitted impulse does not decrease in strength as it passes down the nerve cell; conduction is *decrementless.*

THE CELLS OF THE NERVOUS SYSTEM

DEVELOPMENT

The nervous system develops from that germ layer known as ECTODERM, which forms the outer layer of cells of the embryo. From ecto-

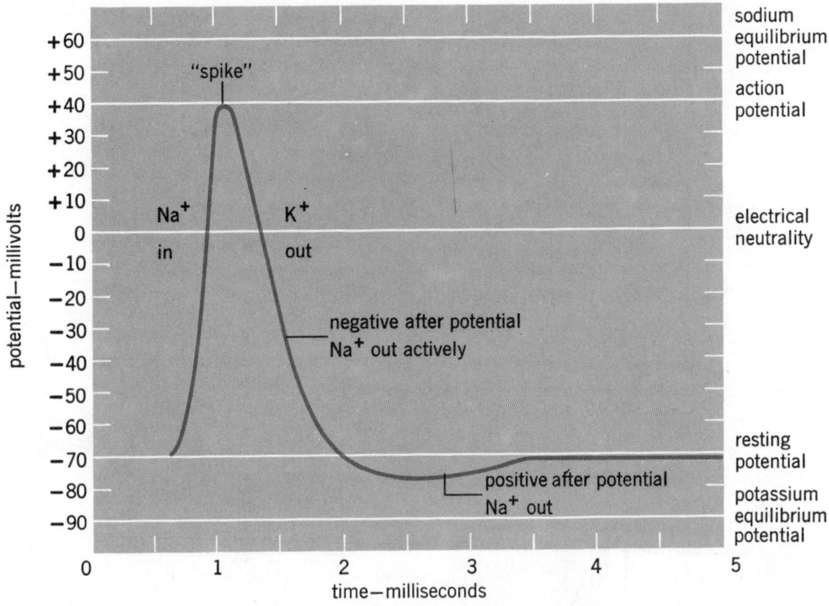

FIGURE 6.3
Changes in membrane potential during development of an action potential.

derm, two types of cells involved in formation of the nervous system differentiate: NEUROBLASTS will give rise to the NEURONS, which are highly excitable and conductile cells of the system; SPONGIOBLASTS will give rise to a variety of GLIAL CELLS, classically regarded as connective and supportive cells of the system.

NEURONS

Neurons in different parts of the nervous system have different sizes and shapes. Anatomically, there are enough common features between all neurons to permit a description of structure of a multipolar motor neuron and a sensory neuron of the peripheral nervous system to serve as a basis for all neuron anatomy.

Physiological anatomy (Fig. 6.4). A CELL BODY is surrounded by a typical plasma membrane. The membrane surrounds the NEUROPLASM or neuron cytoplasm. The cytoplasm contains the usual cell organelles (mitochondria, ER, etc.), but appears, after about five years of age to lack or develop a nonfunctioning cell center. Thus, neurons are not replaced mitotically if lost. In the neuroplasm are NISSL BODIES (masses of granular ER) that are involved in protein synthesis, and hollow microtubules known as NEUROFIBRILS. The latter may act as supportive structures or routes for distribution of substances through the neurons. A large NUCLEUS, surrounded by a thin nuclear membrane, contains the KARYOPLASM or nuclear protoplasm. In the karyoplasm, composed largely of NUCLEAR FLUID, floats a large NUCLEOLUS and granular CHROMATIN MATERIAL.

The cell body gives rise to one or more extensions or PROCESSES. AXONS are usually long (up to several feet in length), generally single, sparsely branched, uniform-diameter processes that conduct impulses away from (*efferent*) the cell body. DENDRITES are shorter, usually multiple, irregular, and highly branched processes that conduct impulses toward (*afferent*) the cell body. A multipolar neuron shows both types of

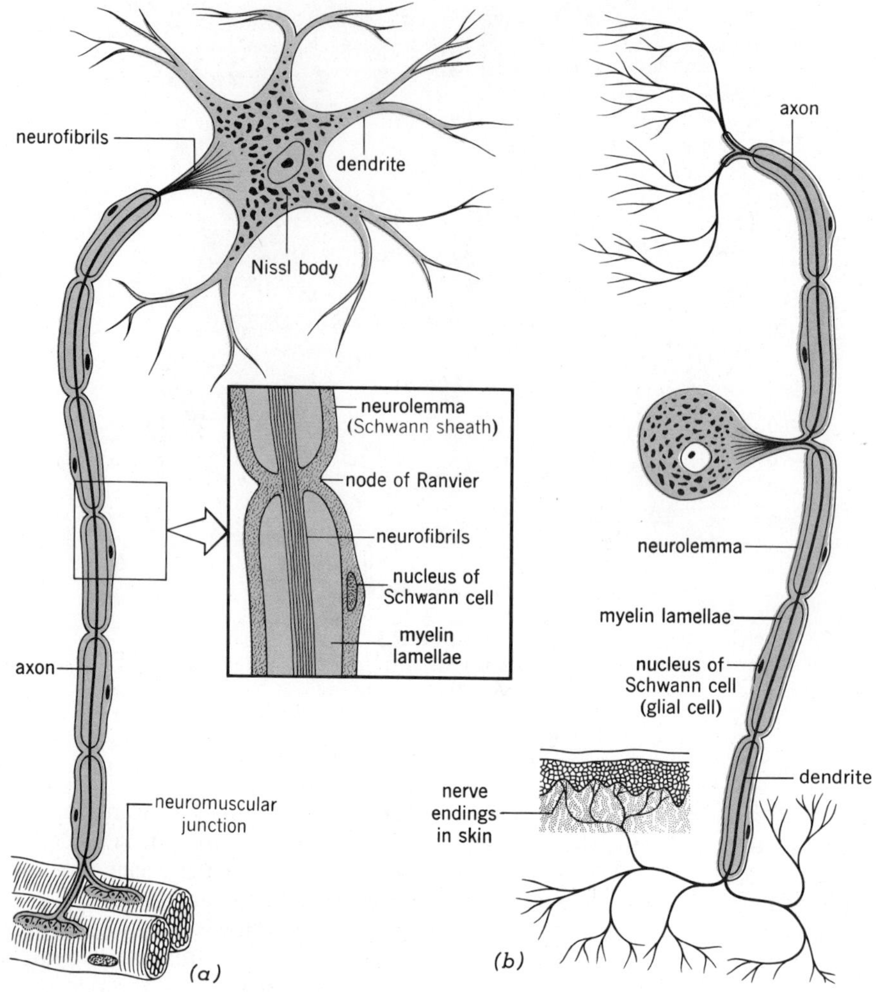

FIGURE 6.4
The two major types of neurons in the body (spinal nerve as the example). (*a*) Motor neuron. (*b*) Sensory neuron.

processes; a sensory neuron such as that shown in Figure 6.4, has processes that are structurally axons, but one will *act* as a dendrite and the other as an axon. Both processes are often termed *nerve fibers;* axons or dendrites are bound together by connective tissue to form the large visible *nerves* that run throughout the body.

Both types of processes may have SHEATHS or coverings on them, if they are structurally built like axons. A MYELIN SHEATH (Fig. 6.5) is a segmented fatty covering acting to insulate the fiber against current loss, and which is responsible for *saltatory conduction* (see page 102) in nerve fibers. Saltatory conduction results in faster conduction rates in myelinated fibers. A NEURILEMMA (*Schwann sheath*) is a thin membranous covering forming part of the myelin

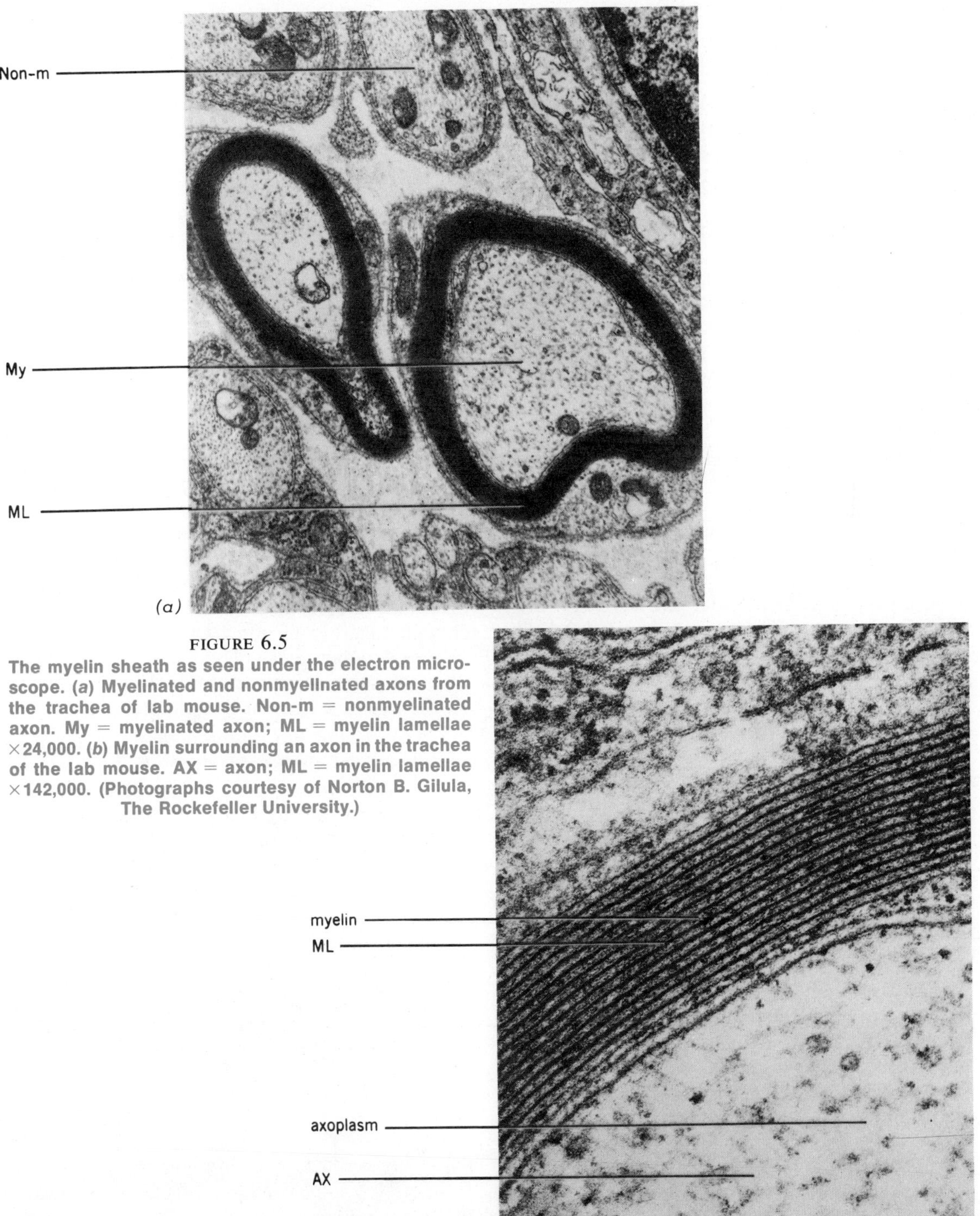

FIGURE 6.5
The myelin sheath as seen under the electron microscope. (*a*) Myelinated and nonmyelinated axons from the trachea of lab mouse. Non-m = nonmyelinated axon. My = myelinated axon; ML = myelin lamellae ×24,000. (*b*) Myelin surrounding an axon in the trachea of the lab mouse. AX = axon; ML = myelin lamellae ×142,000. (Photographs courtesy of Norton B. Gilula, The Rockefeller University.)

TABLE 6.2
CHARACTERISTICS AND TYPES OF NERVE FIBERS ACCORDING TO DIAMETER

TYPE	DIAMETER (μm)	MYELINATED	CONDUCTION VELOCITY (m/SEC)	OCCURRENCE
A fibers	1–20	Yes	5–100	Motor and sensory nerves
B fibers	1–3	Some	3–14	Autonomic system
C fibers	<1	No	<3	Skin, viscera

sheath (the lamellae) and acting as a guiding tube for regeneration of damaged fibers. SCHWANN CELLS, whose membranes form the neurilemma and myelin lamellae, are the myelin producing cells of this sheath.

Speed of conduction of an impulse along a nerve fiber depends on overall fiber diameter, and on the presence or absence of a myelin sheath. Table 6.2 presents data related to size and sheaths.

A myelin sheath permits SALTATORY CONDUCTION to occur. The processes involved are the same as those described earlier; that is, depolarization is followed by ion movement and a current flow. However, if a myelin sheath is present, the only regions where current flow can occur between fiber and ECF is at the NODES of the sheath, where the myelin is extremely thin. Because nodes may lie 1 to 3 mm apart in the sheath, the depolarization "jumps" from node to node over longer sections of the fiber than if a myelin sheath is absent (Figure 6.6).

Basic physiological properties

Threshold. Neuronal membranes have a voltage at which depolarization will occur. A stimulus, to be effective, must possess a certain strength. The term THRESHOLD is used to describe both quantities. The fiber threshold may vary according to the chemical and physical environment of the neuron, and the neuron may respond to stimuli of a given strength at one time but not at another.

Summation. A stimulus not strong enough to cause depolarization is termed a SUBTHRESHOLD (*subliminal*) STIMULUS. Although not strong enough to cause depolarization by itself, it does lower the membrane potential slightly. If several subthreshold stimuli arrive simultaneously at one neuron over several other neurons that connect to it (*spatial summation*) the additive effect of these stimuli may be strong enough to depolarize the neuron. Also, if several subthreshold stimuli are rapidly given to *one* neuron that connects to another neuron, they may arrive at that second neuron close enough together to add (*temporal summation*) to give depolarization. The basis for both these types of summation is described in the section on the synapse later in this chapter.

Strength-duration relationships. Even if a stimulus is of threshold strength, it must act for a certain *time* to be effective. The relationship between strength and duration is an inverse one. A strength-duration curve (Fig. 6.7) depicts these relationships. The term RHEOBASE

FIGURE 6.6
Saltatory conduction in a myelinated nerve fiber.

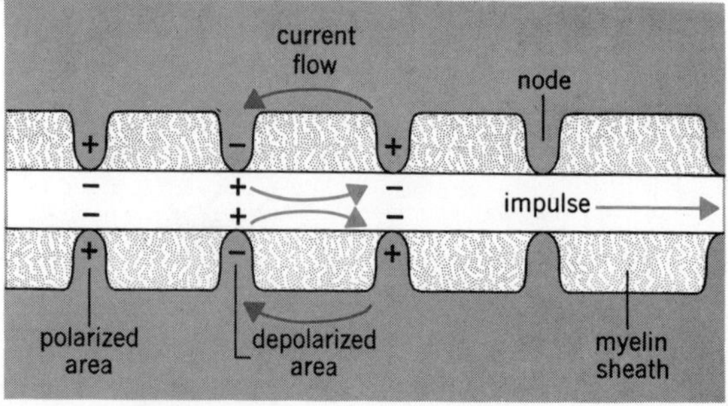

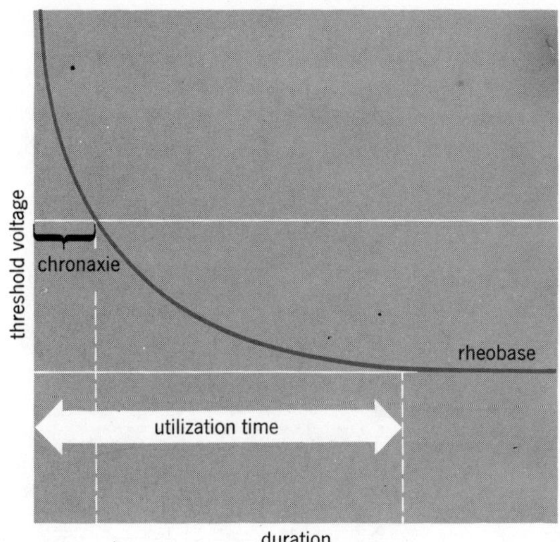

FIGURE 6.7
A strength-duration curve.

(*threshold*) defines the minimum current strength required to cause depolarization, and the term CHRONAXIE is the duration required, for a current that is twice rheobase strength, to cause depolarization. The latter quantity (chronaxie) provides a convenient way of measuring rapidity of neuron response. It is preferable to rheobase in determinations of excitability, because rheobases are not constant.

Accommodation. The application of a continual or *chronic* rheobase current to a nerve fiber causes a rise of threshold in the process known as ACCOMMODATION. It thus becomes more difficult to cause depolarization; put another way, it will take a stronger current to depolarize a fiber that has undergone accommodation.

All-or-none response. For *conducted* impulses, the law of a maximal response or no response (all-or-none) is followed. The events leading to depolarization usually go "all the way" or not at all.

Refractory period. During the time the fiber is depolarized, it is not capable of responding (it is *refractory*) to a second stimulus, no matter how strong that stimulus is. This period of time is called the ABSOLUTE REFRACTORY PERIOD. As repolarization progresses, stronger-than-normal stimuli may become effective. *This* time period is designated as the RELATIVE REFRACTORY PERIOD. Absolute refractory periods in large axons last about 1 msec, and, theoretically, such an axon can conduct 1000 impulses per second.

These physiological properties can be largely explained if one understands how nerve impulses are formed and transmitted. The properties also indicate that there is considerable ability of the neuron to exercise control as to whether it responds to stimuli.

GLIAL CELLS

Glial cells far outnumber (about five fold) the number of neurons in the nervous system. They have processes, but do not have axons and dendrites, and do not conduct impulses as do neurons. They possess the ability to continue to divide throughout life. They possess the organelles of all actively metabolizing cells (mitochondria, ER, ribosomes, etc.).

Physiological anatomy. Figure 6.8 illustrates several of the basic types of glial cells. Seven types of glial cells are recognized:

Astrocytes – protoplasmic and fibrous types
Oligodendroglia
Microglia
Satellite cells
Ependyma
Schwann cells

Astrocytes, oligodendrocytes, microglia, and ependyma are found in the central nervous system (brain and spinal cord). Schwann cells are the cells of the neurilemma in the peripheral nervous system (all nervous structures other than brain and cord), and satellite cells form capsulelike coverings around groups of nerve cells (ganglia) outside the brain and cord.

Properties and functions. Glial cells, although they possess membrane potentials, *do not participate in transmission of nerve impulses* through the nervous system. An important functional relationship exists between glial

FIGURE 6.8
Some types of glia in the central nervous system. (*a*) Ependyma and neuroglia in the region of the central canal of a child's spinal cord: *A*, Ependymal cells; *B* and *D*, fibrous astrocytes; *C*, protoplasmic astrocytes. Golgi method. (*b*) Interstitial cells of the central nervous system: *A*, Protoplasmic astrocyte; *B*, fibrous astrocyte; *C*, microglia; *D*, oligodendroglia. (Courtesy W. B. Saunders. From J. W. Ranson and S. L. Clark, *The Anatomy of the Nervous System*, tenth edition, W. B. Saunders, 1964.)

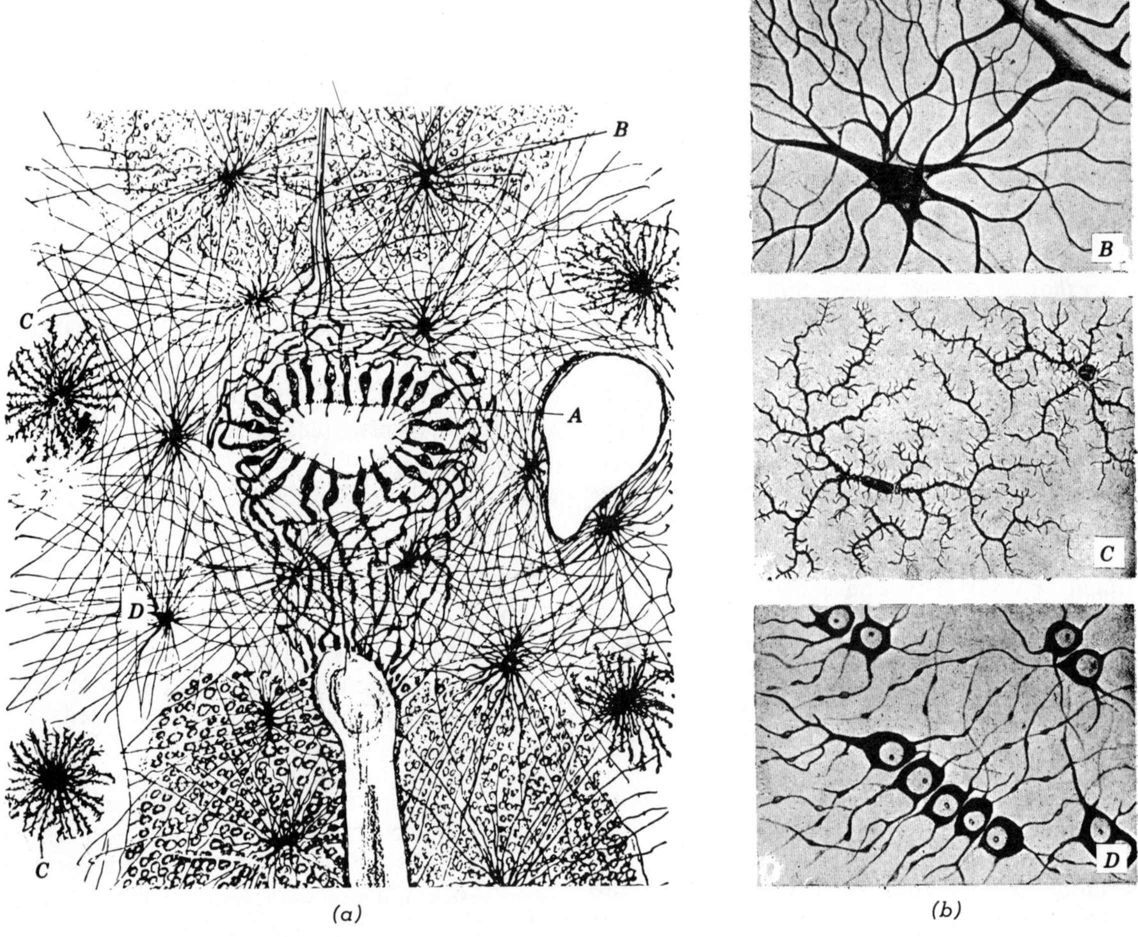

cells and neurons. Neurons do not survive long in tissue culture if deprived of their associated glial elements, suggesting nutritionally oriented roles for glia. Glia, especially astrocytes, commonly lie between blood vessels and neurons, and may transfer substances to the neurons. Oligodendrocytes synthesize the myelin for fibers in the central nervous system. Ependyma are ciliated cells lining the cavities inside the brain and cord, and are possibly responsible for movement of certain molecules within those cavities. Microglia are ameboid, phagocytic, mesodermally derived cells that "cleanse" the central nervous system of dead cells and products of disease.

Glia also maintain high levels of RNA and proteins. Evidence suggests a direct relationship between levels of these chemicals and neuron activity. In short, activity (experience?) may be "recorded" in the nervous system (memory?) by synthesis of RNA and protein. This relationship is explored in greater depth in Chapter 9. In any event, glia, while they do "hold things together" in the nervous system, have additional functions related to neuron health and activity.

THE SYNAPSE

A synapse is defined as an area of functional but not anatomical continuity between the axonal terminations of one neuron (*presynaptic neuron*) and the dendrites, cell body, or axon of another neuron (*postsynaptic neuron*). It is an area where much *control* over impulses is exerted in terms of blockage, passage, or alteration in nature of the impulse. The synapse also governs direction of passage of impulses across itself.

STRUCTURE

The terminal ends of axons may be enlarged to form end feet, end knobs, or other specialized structures (Fig. 6.9). Within these structures are SYNAPTIC VESICLES containing chemicals that may be different in different synapses. A nerve impulse arriving at the end structures increases membrane permeability to Ca^{++}, which causes release of whole vesicles or of their contained chemical. The chemical diffuses across a synaptic cleft to cause depolarization of the postsynaptic neuron. A method for destroying the chemical is provided so that it does not exert a continual effect. If all this sounds familiar, recall that the neuromuscular junction, described in Chapter 5, operates in essentially the same manner as does a synapse.

A variety of chemicals acting as synaptic transmitters and methods of destroying them are presented in Table 6.3. A chemical method of enabling an impulse to cross a synapse endows that synapse with properties different from those of the neurons composing it.

PHYSIOLOGICAL PROPERTIES

One-way conduction. Because only the axon terminals of the presynaptic neuron contain the synaptic vesicles, an impulse must travel a presynaptic neuron–synapse–postsynaptic neuron course. An impulse can be conducted the "wrong way" up a postsynaptic neuron, but is blocked at the synapse. Prevention of "short circuits" in neuronal pathways is the obvious advantage of 1-way conduction.

Facilitation. Synaptic transmitters may lower synaptic threshold (*hypopolarization*) by producing an *excitatory postsynaptic potential* (EPSP) that may allow another impulse traveling the same pathway to more easily induce a change in the postsynaptic neuron (*facilitation*). Repeated use of such a pathway may render it one of least resistance, something of value in the learning or conditioning process.

Inhibition. If the transmitter chemical makes it more difficult to induce a change in the postsynaptic neuron, it is said to produce an *inhibitory postsynaptic potential* (IPSP) by raising the threshold of depolarization (*hyperpolarization*). Thus, trivial or nonessential information may be "filtered out" of the system.

Synaptic delay. The events leading to synaptic transmission (release and diffusion of the chemical transmitter, depolarization) all require time in which to occur. It thus takes *more time* for an impulse to traverse a pathway of a given length, if there is a synapse in the path, than if there was no synapse. The extra time represents SYNAPTIC DELAY, and amounts to 0.5 to 1.0 msec per synapse.

Summation. Both TEMPORAL and SPATIAL SUMMATION occur at synapses. A subthreshold stimulus releases a small amount of chemical from the presynaptic axons. If another such stimulus releases more chemical into the synapse before the first chemical is destroyed, the additive effect may produce enough chemical to depolarize the postsynaptic neuron. This is temporal summation. Spatial summation occurs if several neurons terminating on a single

FIGURE 6.9
Various forms of synaptic junctions.

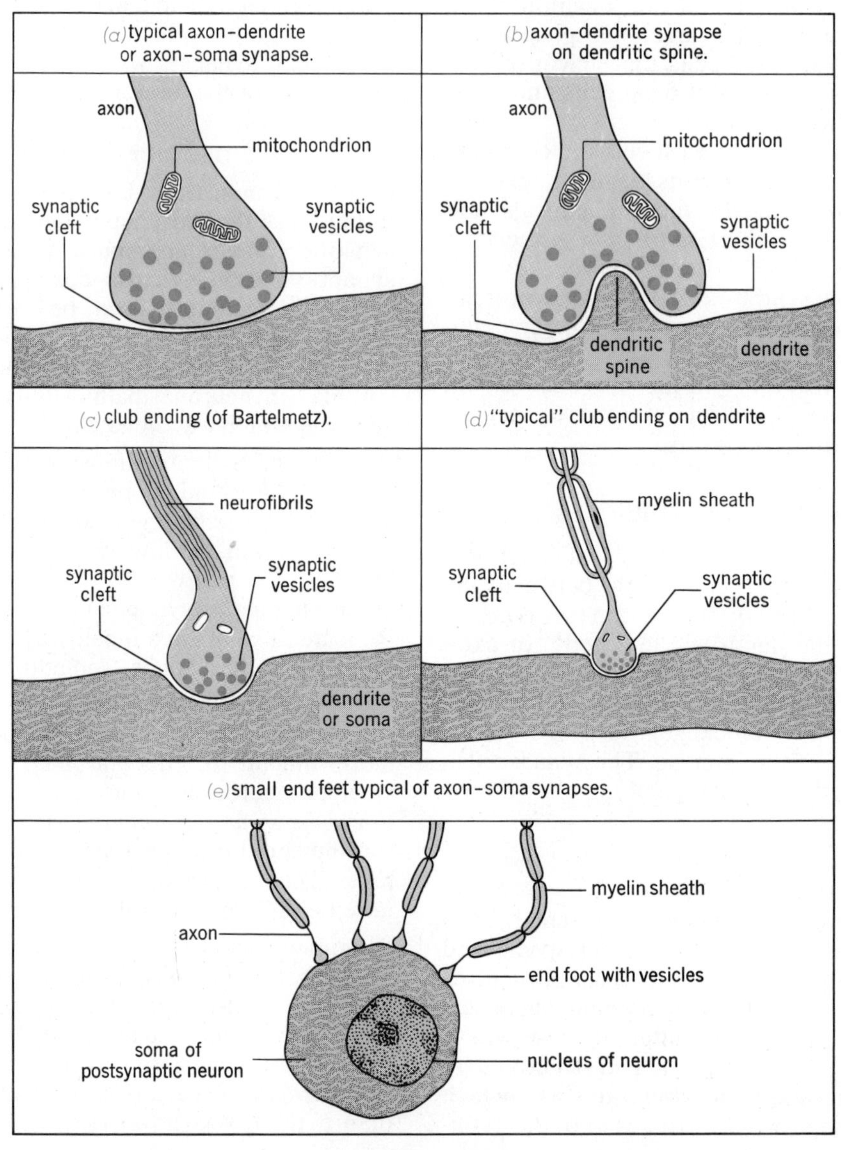

TABLE 6.3
SYNAPTIC TRANSMITTERS

SUBSTANCE OR AGENT	ENZYME OR PROCESS DESTROYING THE AGENT	WHERE FOUND, HIGHEST CONCENTRATION	COMMENTS
Acetylcholine	Cholinesterase	Neuromuscular junctions; peripheral synapses between neurons; brain stem, thalamus, cortex, retina	In CNS, not distributed evenly. Perhaps serves specific functions in each area
Norepinephrine	Monamine oxidase (MAO) or catechol-o-methyl transferase (COMT)	Postganglionic sympathetic neurons; midbrain, pons, medulla	Low concentration in cerebrum
Serotonin	MAO	Hypothalamus, basal ganglia, spinal cord	May be involved in expression of behavior
Histamine	Imidazole-N-methyl transferase, histaminase	Hypothalamus	In mast cells (tissue basophils)
Amino acids	Oxidation or conversion	Cortex, visual area	GABA is inhibitory to most synapses (GABA = gamma amino butyric acid)

neuron each simultaneously release small amounts of chemical, which, added together, amount to enough to depolarize the postsynaptic neuron.

Again it may be noted that the synapse EXERTS CONTROL over what happens to the impulses reaching it, leading to much control over body activity in general.

Enlargement of the axonal end feet in the presynaptic neuron, as that synapse is used repeatedly, has been observed. Such enlarged feet contain more chemical than smaller ones and presumably release more as impulses reach the synapse. This may contribute to the establishment of the habitual or preferred pathways essential to learning.

DRUGS have great effects on synapses. Hypnotics, analgesics (pain killers), and anesthetics such as bromides, aspirin, morphine, and opium, create IPSP's and decrease synaptic activity. Curare blocks neuromuscular transmission by preventing acetylcholine from reacting with postsynaptic structures. Physostigmine and strychnine inactivate cholinesterase and permit continued action of the transmitter agent.

CHANGES IN PH alter synaptic activity. Increase of pH increases ease of transmission, while decrease of pH depresses it.

HYPOXIA (lowered O_2 levels) produces cessation of synaptic activity.

The synapse thus seems to be quite responsive to many factors in its environment, a fact that may become extremely important during disease states that may alter the synaptic environment. Such disorders are discussed in later chapters.

THE REFLEX ARC: PROPERTIES OF REFLEXES

The reflex arc (Fig. 6.10) is the simplest functional unit of the nervous system capable of detecting change and causing a response to that change. Many body activities are controlled reflexly. This term implies an *automatic* adjustment to maintain homeostasis without conscious effort. A reflex arc always has five basic parts to it:

A *receptor* to detect change.

An *afferent neuron* to conduct the impulse, re-

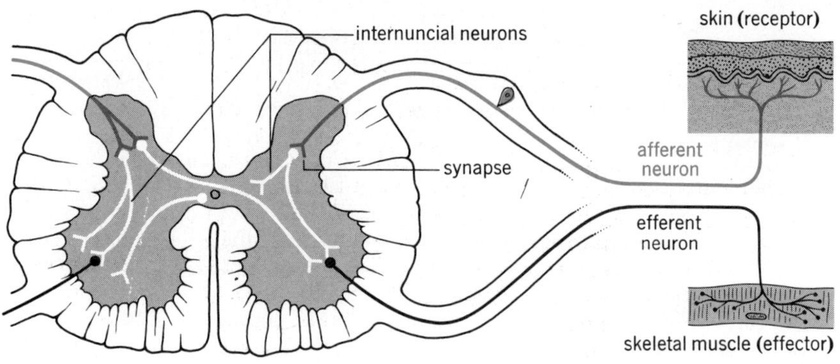

FIGURE 6.10
The components of a reflex arc.

sulting from stimulation of the receptor, to the central nervous system.

A *center* or *synapse* where a junction is made between neurons. An internuncial neuron (a neuron interposed between two other neurons) is commonly found at this point, which is often within the central nervous system.

An *efferent neuron* to conduct impulses for appropriate response to an organ.

An *effector* or *organ* that does something to maintain homeostasis.

Functionally, reflexes may be classed as exteroceptive, if the receptor is at or near a body surface; interoceptive, if the receptor is within a visceral organ; proprioceptive, if the receptor is in a muscle or tendon.

Activity that results from passage of impulses over a reflex arc is called the REFLEX ACT or the REFLEX. Reflex acts, regardless of what the stimulus and response are, have several characteristics in common.

Activity of the effector depends on stimulation of a receptor. In short, the arc serving the act must be complete: if any part of the arc is not working, the effector will not respond.

Reflex activity is *involuntary,* but can often be stopped by an act of will; if we expect a certain stimulus we need not respond to it. Reflex activity is *stereotyped* (repetitious); stimulation of a given receptor always causes the same effector response. Thus, if the patellar tendon is tapped, the knee *always* extends. It is thus *predictable.*

Reflex activity is *specific* in terms of the stimulus and the response. A particular stimulus causes a particular response; the response is usually *purposeful.*

The value of these reflex properties to the body is obvious. A particular stimulus causes a predictable and appropriate response, which will usually result in or aid in maintenance of homeostasis. If responses were different to the same stimulus, the body would have a difficult time maintaining its posture, balance, and overall functions.

Examples of reflexes, especially as they are involved in homeostatic control are presented in subsequent chapters.

SUMMARY

1. Cells, by achieving an unequal separation of ions, may create an excitable state and an electrical potential across their membranes. The ions responsible for these potentials are sodium and potassium.
   a. Diffusion, of differential nature for different ions, plus a semipermeable membrane, creates a part of the membrane potential.
   b. Active transport, of sodium and potassium primarily, creates a larger part of the membrane potential.
2. Polarization results when ions are separated.
   a. A resting potential of −70 to −90 mv is created by the processes indicated in 1a, 1b. It is due primarily to "leakage" of K^+ from inside to outside of the cell.
3. Impulses are formed when membrane permeability to Na^+ is greatly increased.
   a. A stimulus causes increased (500 ×) permeability to sodium ion, sodium enters the fiber, the membrane potential is reversed and the fiber is depolarized.
   b. Current flows between the polarized and depolarized regions and this is the impulse.
   c. Repolarization occurs on return of Na^+ permeability to the original state and pumping out of Na^+. Potassium conductance increases as Na^+ conductance decreases, and this and active K^+ transport return K^+ to the fiber, and slow its outward movement.
4. An action potential is produced as ions move through a depolarized membrane.
   a. The spike potential correlates with sodium entry.
   b. The reversal potential correlates with the reversal of membrane potential as large quantities of Na^+ enter the fiber.
   c. After–potentials correlate with active movement of Na^+ and K^+, and return of membrane permeability to normal.
5. Conductility occurs as the current flow depolarizes the next section of the fiber and moves, by repeating the process, down the fiber.
6. Neurons, the highly excitable and conductile cells of the nervous system, arise from neuroblasts of the ectodermal germ layer. Glia arise from spongioblasts of ectoderm.
7. The structure of a neuron is described. The involvement of a fatty myelin sheath in speeding impulse transmission (saltatory conduction) is discussed.
8. Physiological properties of neurons are presented.
   a. Threshold. The minimum strength of stimulus or the minimum voltage required to depolarize a neuron.
   b. Summation. Subthreshold stimuli may "pool their influence" to cause depolarization.
   c. A stimulus must last a certain time (duration) to cause depolarization. Weaker stimuli must last longer and vice versa.
   d. Continued stimulation causes a rise of neuron threshold called accommodation.
   e. Response of a neuron is either maximal or none (all-or-none response).
   f. Refractory periods exist during which the neuron cannot be stimulated or is more difficult to stimulate.
9. Several types of glial cells are described.
   a. Glia play no part in transmission of impulses.
   b. They may be involved in neuron nutrition, form supporting networks for neurons, synthesize myelin, and are phagocytic. They may be involved in formation of memory traces.
10. A synapse is a functional junction between neurons. Its properties are different from the neurons composing it.
    a. Synaptic transmission is by means of chemical substances, which are then destroyed to prevent continued action.
    b. One-way conduction is assured by having the transmitter substance on only one side of the synapse.
    c. Facilitation means an easier passage for an impulse resulting from production of an EPSP.
    d. Inhibition means more difficult passage for an impulse due to production of an IPSP.
    e. Synaptic delay means a slower passage of an impulse across a synapse than down a nerve fiber due to time required for chemical release and diffusion.
    f. Temporal and spatial summation occur as small amounts of chemical are added together.
    g. Drugs may affect synapses to increase or decrease impulse transmission. pH and oxygen levels also influence synaptic activity.
    h. Synapses control their activity to a large degree.

11. The reflex arc is the simplest unit of the nervous system that can automatically maintain homeostasis.
    a. Its parts include: a receptor to respond to change; an afferent neuron to conduct impulses to the central nervous system; a synapse, with the possibility of an internuncial neuron, in the central nervous system; an efferent neuron to conduct impulses away from the central nervous system; an effector to respond.
    b. Reflex acts are involuntary, stereotyped, predictable, and usually purposeful. They automatically control homeostasis.

QUESTIONS

1. How do excitation and conduction differ in the mechanisms by which they occur?
2. How are ions separated so as to create the excitable state?
3. Explain why a resting potential is a "potassium potential."
4. Explain why an action potential is a "sodium potential."
5. How does saltatory conduction differ from conduction in a nonmyelinated fiber? How are they similar?
6. Compare and contrast neuronal and synaptic properties. Account for them in terms of the methods of neuronal and synaptic transmission.
7. Of what value are reflex arcs to the control of homeostasis?
8. What are the parts of a reflex arc, and what does each contribute to the reflex?
9. From your own knowledge or experience, give some examples of reflex activity that contribute to your homeostasis.

READINGS

Axelrod, Julius. "Neurotransmitters." *Sci. Amer. 230:*58. (June) 1974.

Lim, Ramon, David E. Turriff, Shauang S. Troy, Blake W. Moore, and Laurence F. Eng. "Glia Maturation Factor: Effect on Chemical Differentiation of Glioblasts in Culture." *Science 195:*195. 14 Jan 1977.

McEven, Bruce S. "Interactions Between Hormones and Nerve Tissue." *Sci. Amer. 235:*24. (Aug) 1976.

Sonyen, George C. "Electrophysiology of Neuroglia." *Ann. Rev. Physiol.* Vol 37. Palo Alto, Cal. 1975.

7

THE BRAIN

OBJECTIVES
THE DEVELOPMENT OF THE NERVOUS SYSTEM
THE TELENCEPHALON (CEREBRUM)
THE CEREBRAL CORTEX
STRUCTURE
FUNCTIONS
CLINICAL CONSIDERATIONS
THE MEDULLARY BODY AND BASAL GANGLIA
STRUCTURE
FUNCTIONS
CLINICAL CONSIDERATIONS
THE DIENCEPHALON
THE THALAMUS
STRUCTURE
FUNCTIONS
CLINICAL CONSIDERATIONS
THE HYPOTHALAMUS
STRUCTURE
FUNCTIONS
CLINICAL CONSIDERATIONS
THE MESENCAPHALON (MIDBRAIN)
STRUCTURE
FUNCTIONS
CLINICAL CONSIDERATIONS
THE METENCEPHALON AND MYELENCEPHALON; PONS AND MEDULLA (MEDULLA OBLONGATA)
STRUCTURE
FUNCTIONS
CLINICAL CONSIDERATIONS
THE RETICULAR FORMATION
THE IMMATURE BRAIN
BIOCHEMICAL CHANGES
PHYSIOLOGICAL CHANGES
SUMMARY
QUESTIONS
READINGS

OBJECTIVES

After studying this chapter, the reader should be able to:

☐ Give, in general terms, an outline of development of the nervous system.

☐ Name the major parts of the brain.

☐ Name the fissures and lobes of the cerebrum.

☐ Describe or give the names and/or numbers of the major functional areas of the cerebral cortex, and the functions centered therein.

☐ Describe the medullary body, name the basal ganglia, and discuss the functions of both.

☐ Describe the structure and list the functions of the thalamus and hypothalamus.

☐ Describe the structure and list the functions of the midbrain.

☐ Describe the structure and list the functions of the pons and medulla.

☐ Describe the location of the reticular formation and indicate its functions.

☐ For all brain areas, indicate the expected symptoms that would result from either stimulation of, or damage to, that area.

☐ List and briefly discuss some of the biochemical and physiological changes that occur as the brain matures.

It is difficult to know where to begin a study of the physiology of the nervous system. Some say that all experience depends on sensory impressions received by the body and this is what should form the starting point of study. Others believe that the role of the brain as the integrator and analytic component of the nervous system should form the initial focus for study. All parts of the nervous system cooperate to assure optimum levels of body activity, and to begin with the brain will perhaps underline the homeostatic and integrative functions served by this part of the nervous system.

THE DEVELOPMENT OF THE NERVOUS SYSTEM (FIG. 7.1)

At about 18 days of embryonic development, there appears on the embryonic disc an area of thickened ectoderm known as the NEURAL PLATE. A groove, the NEURAL GROOVE, appears in the plate, and the lips of the groove enlarge to form two NEURAL FOLDS. The folds grow, contact one another over the groove and form a hollow NEURAL TUBE. From the upper and lateral regions of the neural tube, cells of the NEURAL CREST separate from the tube and come to lie alongside the tube. The anterior end of the neural tube closes and forms a series of three enlargements that undergo subdivision and refinement as development proceeds (Table 7.1), and which will give rise to specific parts of the central nervous system (brain and cord). The reader should note what organs compose the three major parts of the brain; the cerebrum, brain stem, and cerebellum.

The walls of the neural tube develop a three layered structure. An *outer* MARGINAL ZONE or LAYER is the forerunner of the white matter

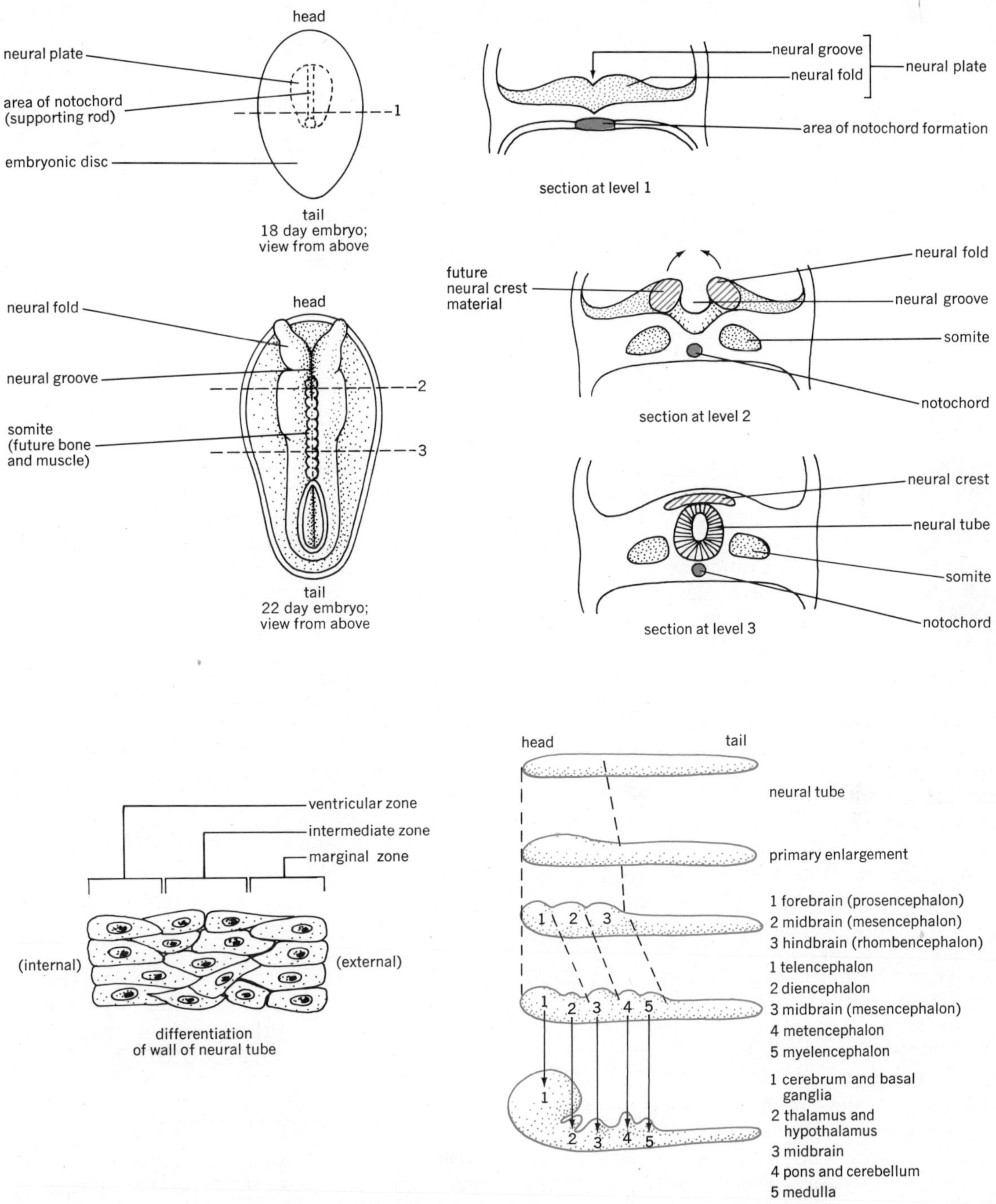

FIGURE 7.1
The early development of the nervous system.

TABLE 7.1
DERIVATIVES OF THE PRIMARY NEURAL TUBE ENLARGEMENTS

PRIMARY ENLARGEMENT	SUB-DIVISION(S)	STRUCTURE(S) ARISING FROM SUBDIVISION
Forebrain (Prosencephalon)	Telencephalon	Cerebrum Basal ganglia
	Diencephalon	Thalamus Hypothalamus
Midbrain (Mesencephalon)	None	Midbrain
Hindbrain (Rhombencephalon)	Metencephalon	Pons Cerebellum
	Myelencephalon	Medulla

of the central nervous system. In the spinal cord, the layer remains essentially free of neuron cell bodies and lies peripherally; in the telencephalon and cerebellum, the marginal layer is invaded by migrating neuroblasts that will form an *outer* layer of nerve cells as the cortices (sing., *cortex*) of these two organs. An INTERMEDIATE ZONE or LAYER (*mantle layer*) develops neuroblasts that will become the gray matter of the central nervous system. This zone remains internally to the white matter in the spinal cord, and serves as the source of the cells that will migrate to form the cerebral and cerebellar cortices. An inner layer of epithelial cells called the VENTRICULAR ZONE (*ependymal layer*) remains as the lining of the cavities of the brain and cord. Within the various regions that will form the brain, groups of nerve cells will remain to form the nuclei of cranial and some peripheral nerves, and the basal ganglia and nuclei of thalamus, hypothalamus, and other parts of the brainstem.

The fibers of the peripheral nervous system (Fig. 7.2) develop from several sources. Afferent (sensory) neurons of the cranial and spinal nerves develop from NEURAL CREST material whose neuroblasts send one fiber peripherally (dendrite) and another (axon) to the central nervous system. The peripheral GANGLIA of the system also develop from neural crest material. The efferent (motor) fibers of the peripheral nervous system originate from the neuroblasts within the neural tube, and send processes (axons) that terminate eventually on muscular and glandular structures.

Though brief, this section on development should acquaint the reader with some of the basic terminology employed later in this chapter.

THE TELENCEPHALON (CEREBRUM)

The cerebrum is a derivative of the primitive forebrain, and, as it grows, it becomes convoluted or folded. This folding creates more surface for housing nerve cells. Upfolds of the cerebrum are called GYRI (sing., *gyrus*), and shallow downfolds are called SULCI (sing., *sulcus*). At several places on the cerebrum, deep FISSURES divide the organ into two HEMISPHERES and several LOBES (Fig. 7.3).

The gray matter of the cerebrum forms an outer covering or CORTEX varying between 2.5 and 4.0 mm in thickness, having a volume of about 600 cm³, a surface area of about 2000 cm², and containing an estimated 12–15 billion neurons.

THE CEREBRAL CORTEX

Structure. Many years ago, Korbinian Brodmann attempted to correlate cellular differences in the cerebral cortex with functional localization. His numbered maps (Fig. 7.4) of areas of the cortex are widely quoted today when re-

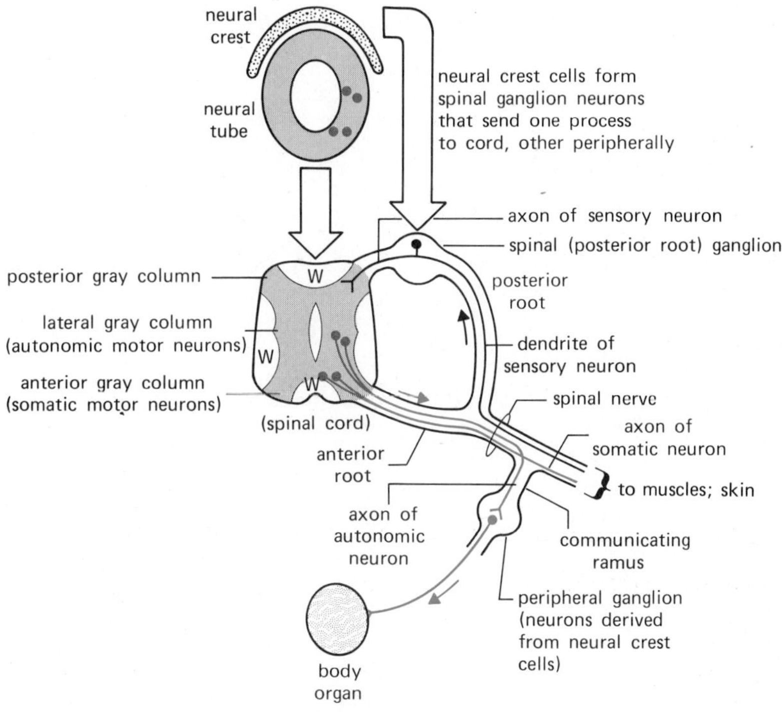

FIGURE 7.2
The development of the spinal portion of the peripheral nervous system (w = white matter).

ferring to specific regions of the cortex, even though there may not be indisputable structural–functional relationships. What *is* implied by such regions is that some areas, because of the connections they make, *are involved with one particular aspect of function.*

Within the cortex, the neurons appear to be arranged in vertical columns (Fig. 7.5) that allow lateral spread of impulses mainly through the most superficial layer; in short, chains of neurons are formed that analyze and store information, and exert effects on adjacent columns by connections through the surface layer of the cortex.

Functions. Following the designation of cortical areas as devised by Brodman, functional regions of the cerebral cortex may be described in each hemisphere. Further evidence of cortical function is provided by direct stimulation of the brain such as may be necessary to determine function before attempting operations on the brain (our brains are no more identical than our faces, even though both contain a basic set of components).

The frontal lobes. Lying in the precentral gyrus (*see* Fig. 7.3), is area 4, designated as the PRIMARY MOTOR AREA. Stimulation in area 4 brings about movement of individual muscles or functional groups of muscles on the body. The body is represented upside down and with those parts of the body where complex movements are required (e.g., face, hands) receiving larger areas of representation (Fig. 7.6). A SUPPLEMENTARY MOTOR AREA lies on the medial surface of a hemisphere (areas 24, 31). Both the primary and supplementary areas

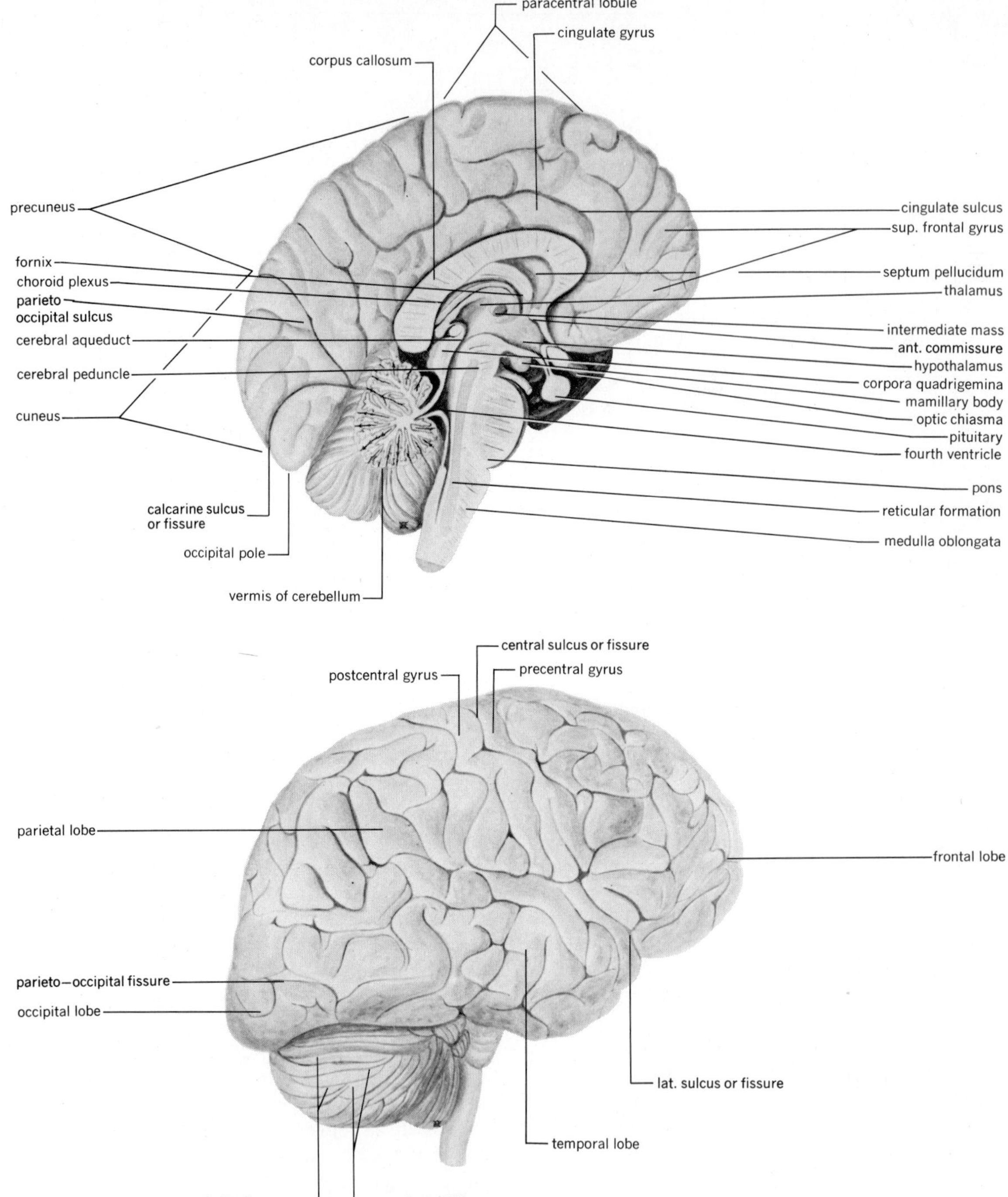

FIGURE 7.3
(*Above*) Medial view of the left half of the brain. (*Below*) Lateral view of the right side of the brain.

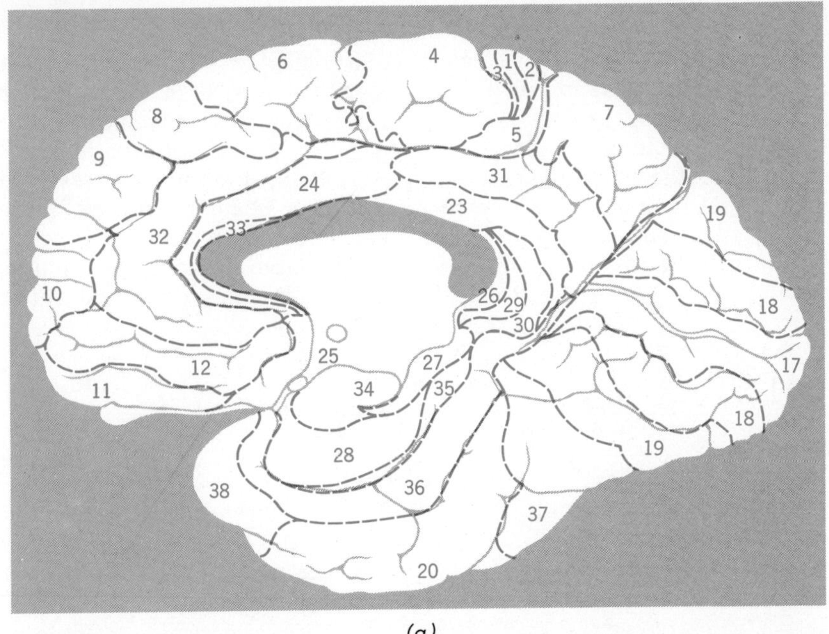

(a)

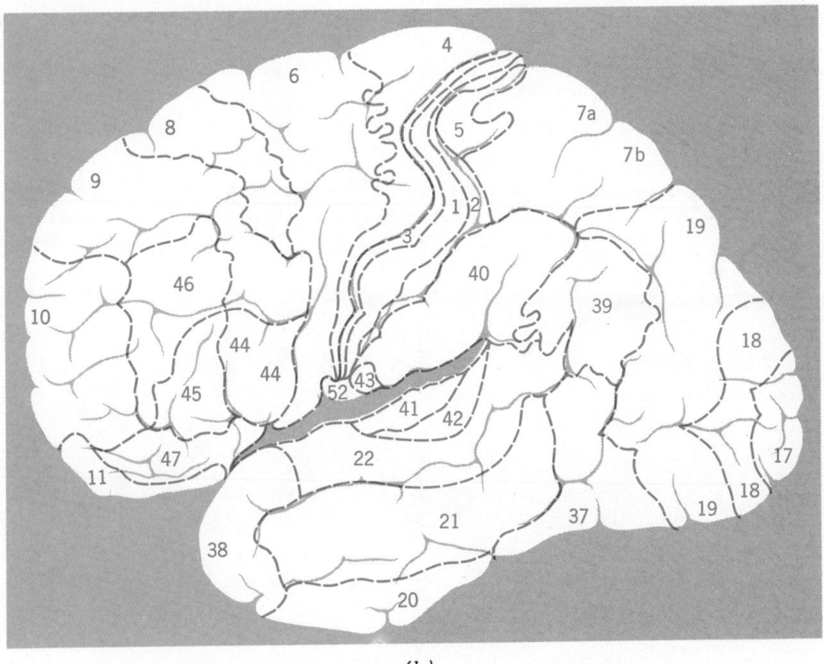

(b)

FIGURE 7.4
The cytoarchitectural areas according to Brodmann. Within certain numbered regions are specific functional areas (see text). (*a*) Medial aspect of cerebrum. (*b*) Lateral aspect of cerebrum.

control muscles primarily on the opposite side of the body. Efferents from these motor areas form about 40 to 45 percent of the fibers giving control over voluntary muscle movement.

Area 6, designated as the PREMOTOR AREA, provides input to area 4. Stimulation of area 6 causes contractions of muscle groups only if area 4 is intact, and also causes movement of the head, neck, and trunk. *Learned* motor activity may reside in area 6, because lesions (damage) placed in this region interfere with *performance* of movement even though there is no voluntary motor paralysis.

Area 8, if stimulated, results in eye movements of scanning nature, and is called the FRONTAL EYE FIELD.

The large expanse of frontal lobe anterior to the motor areas described above (specifically

FIGURE 7.5
The cellular arrangement of the cerebral cortex. A = axon; D = dendrite.

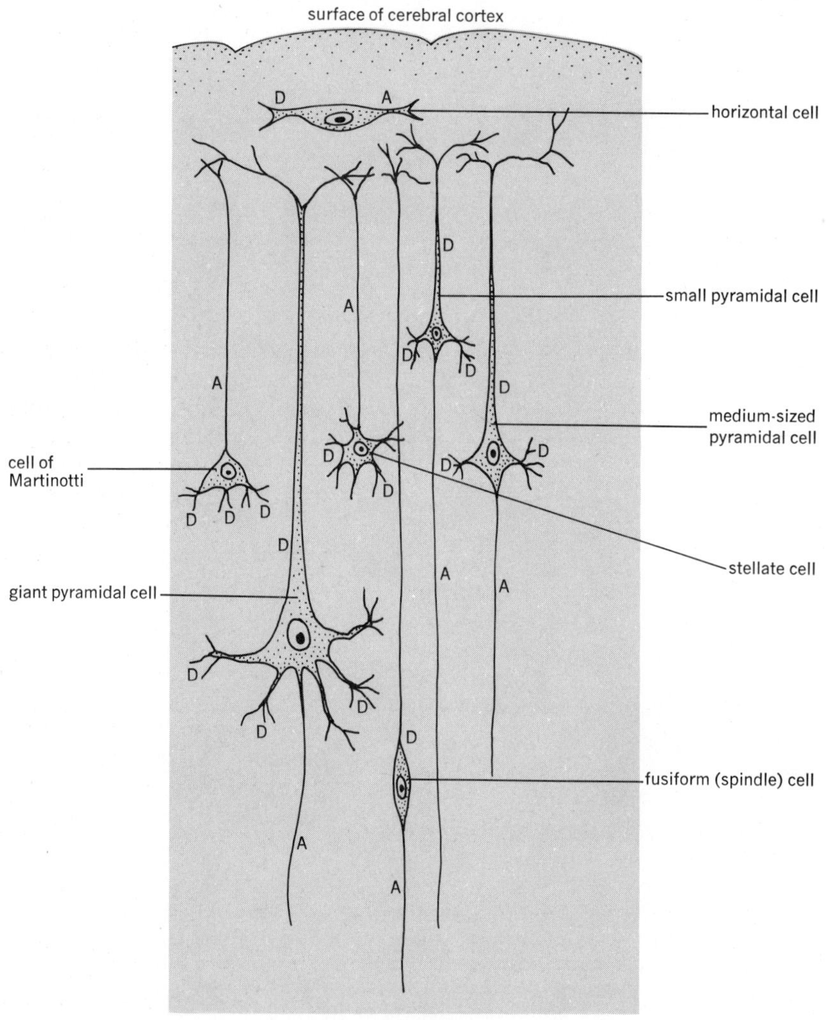

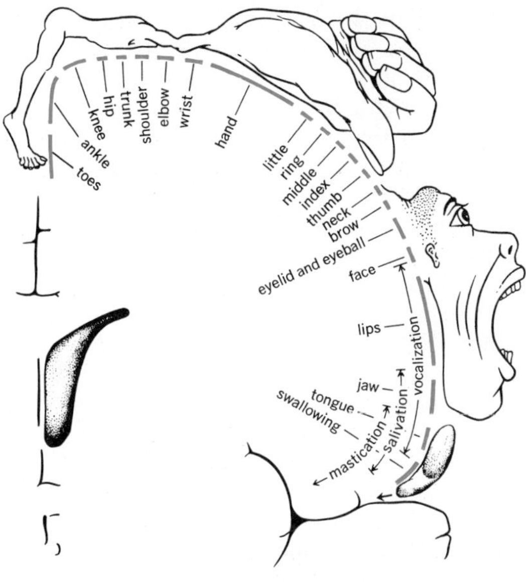

FIGURE 7.6
Distribution of motor function in area 4. A motor homunculus.

areas 9, 10, 11, 12), is designated as the PREFRONTAL CORTEX. Because stimulation in these areas produces no clearcut movement or sensation, and because they receive fibers from other brain areas, they are also sometimes called the *frontal association areas.* In animals, removal of the prefrontal cortex may cause *hyperactivity,* characterized by aimless incessant pacing, or *delayed response reactions may be eliminated.* In the latter phenomenon, a normal animal seeing food or other reward being placed in a cup or other hiding device will remember where the reward was after delays of up to 90 seconds. In a lesioned animal, 5 seconds delay reduces the ability of the animal to find the reward to mere chance. Also, *emotional display becomes excessive,* with rage predominating. This suggests that the prefrontal areas may inhibit such display. Lastly, some reactions typical of neurosis, such as emotional outbursts, anxiety, and rage, may be inhibited by severing the prefrontal regions from the rest of the brain. Such results were used to justify, in the past, the *prefrontal lobotomy* in the human.

In humans, there is great diversity and variability in symptoms displayed as a result of frontal lobe damage, by accident or intent (lobotomy). In humans, as in animals, changes are most consistently seen in personality, emotional reactions, ability to accept the responsibilities of life, and moral and social concepts. The famous "crowbar case" of Phineas P. Gage, described as "an efficient and capable foreman," who had a tamping iron blown through the prefrontal regions of his brain, illustrates some of the activities carried on by this part of the brain. Gage's attending physician, one J. M. Harlow, described Gage as follows:

> He is fitful, irreverant, indulging in the grossest profanity (which was not previously his custom), manifesting but little deference to his fellows, impatient of restraint or advice when it conflicts with his desires, at times . . . obstinate, capricious, and vacillating. . . . His mind was radically changed, so that his friends said he was no longer Gage.

The parietal lobes. In the postcentral gyrus (*see* Fig. 7.3) lie areas 3, 1, 2, designated as the GENERAL SENSORY or SOMESTHETIC AREA. Some motor responses may be obtained if this area is stimulated, but they are generalized and unskilled in type. The areas serve as the termination for the somesthetic senses, such as pain, heat, cold, touch and pressure, and sensations of tension, and limb position. Within the area, the body is represented, as it was in area 4, upside down and with more sensitive regions of the body given more room (Fig. 7.7). Additionally, skin sensations (touch, pressure, pain, thermal) are found anteriorly, and those called *deep sensibility* (remaining after the skin is anesthetized) such as tension and position, are placed posteriorly.

The PARIETAL ASSOCIATION AREAS (areas 5, 7a,b) provide interpretation of textures, shapes of objects when handled, and appreciation of degrees of changes as in the thermal sensations.

The occipital lobes. The VISUAL AREA (17) occupies the greater part of the occipital lobe and receives fibers from the retina. The VISUAL ASSOCIATION CORTEX occupies areas 18 and 19 and handles the relationships of past to present visual experiences as well as integrating eye

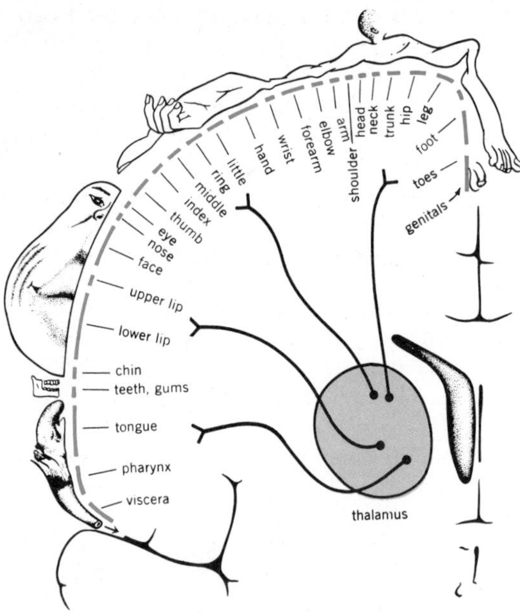

FIGURE 7.7
The distribution of sensory fibers from thalamus to cortex, and the orientation of the sensory homunculus.

movements to fix the eyes directly and involuntarily upon objects viewed.

The temporal lobes. Areas 41 and 42 constitute the AUDITORY AREA and AUDITORY ASSOCIATION AREA respectively. The auditory area receives fibers from the cochlea. The association area is an important component of the system responsible for language functions, and stores memories of both auditory and visual experience.

Aiding in language functions is BROCA'S SPEECH AREA, areas 44 and 45. The area includes portions of the motor area, and ensures appropriate and coordinated use of the organs of speech.

Clinical considerations. Lesions or damage to any motor area will interfere with the ability to perform movements. A primary motor area lesion leads to *paralysis*, or the inability to voluntarily move muscles. Sensory area lesions destroy the ability to perceive sensations. Prefrontal lesions were described in connection with that area. Visual and auditory area lesions will destroy ability to see in particular areas of the visual field, and the ability to perceive sounds of particular pitches.

Of particular interest are the effects of damage to the temporal and parietal association areas. Temporal lesions produce loss of memory of things learned, and occurs in one or more of three forms.

AGNOSIA is the loss of ability to recognize familiar objects. Failure may manifest itself through lack of ability to distinguish things touched (*astereognosis*), heard (*auditory agnosia*), or seen (*visual agnosia*), or in failure to distinguish right from left (*autotopognosis*).

APRAXIA refers to an inability to perform voluntary movements, particularly of speech, when no true paralysis is present.

APHASIA is an inability to use or comprehend ideas communicated by spoken or written symbols. It may be motor, resulting in inability to *express* oneself, or sensory, in that there is a *deficit in understanding*.

ALEXIA, a form of aphasia, refers specifically to inability to comprehend the written word (*word blindness*) or to speak words correctly. It is usually discovered first in school-age children, who exhibit reading defects.

Parietal lobe lesions produce agnosia if the lesion is on the "dominant side" of the cerebrum (usually left side in right-handed persons and vice versa). If on the "nondominant side," a lesion produces AMORPHOSYNTHESIS, characterized by defective perception of one side of the body.

THE MEDULLARY BODY AND BASAL GANGLIA

Structure. Each cerebral hemisphere includes a large mass of white matter consisting of three types of fibers.

COMMISSURAL FIBERS connect the two cerebral hemispheres through the *corpus callosum* (*see* Fig. 7.3). These fibers allow one side of the cerebrum to communicate with the other.

PROJECTION FIBERS are fibers that come *to* the cortex from noncerebral regions, or which pass *from* the cortex to go to other brain or body areas.

ASSOCIATION FIBERS stay within a given hemisphere, connecting one cortical area with another. Many of these fibers form bundles, to which specific names have been assigned.

Buried deep within the white matter of each cerebral hemisphere are several masses of gray matter collectively known as the BASAL GANGLIA (Fig. 7.8). The largest of these is called the CORPUS STRIATUM, which consists of the *caudate nucleus, putamen,* and *globus pallidus.*

Lying beneath the thalamus are a group of SUBTHALAMIC NUCLEI, including the *substantia nigra,* and *red nucleus.* The *amygdaloid nucleus* lies in the temporal lobe at the tip of the tail of the caudate nucleus and while it is considered a basal ganglion, it is involved in the limbic system discussed in Chapter 9.

Functions. The basal ganglia represent, in such animals as fish, amphibians, and reptiles, *the* highest motor center of their brain. In birds and mammals, cerebral development has resulted in much motor function being moved out of the ganglia into the cerebral cortex. The question results then as to what contributions to motor function remain in the human basal ganglia. Perhaps the most information concerning function comes from degenerative disorders that involve the ganglia; in such disorders, involuntary movements figure prominently.

Stimulation of the caudate nucleus produces inhibition of movement and muscle tone. Lesions of caudate, putamen, globus pallidus, or substantia nigra produce involuntary and uncontrollable tremor, suggesting that normal movement requires an inhibitory function of the ganglia.

Clinical considerations. Lesions of the various types of medullary fibers cause deficiencies characteristic of the fiber tracts interrupted.

Interruption of the commissural fibers results in the creation of essentially "two brains" within the same skull. Each hemisphere can still accumulate and analyze information, but information received by one half of the cerebrum cannot be transferred directly to the other half. Nevertheless, through "lower" centers in the brain, such as the brainstem, compensation for such losses can occur to a great degree.

Interruption of incoming projection fibers results primarily in loss of ability to perceive sensations, while interruption of outgoing motor projection fibers produces signs of an *upper motor neuron* lesion: paralysis of voluntary movement with muscles in a contracted state (*spastic paralysis*); also exaggeration of reflexes.

Interruption of association fibers results in inability to integrate activity on one side of the cerebrum. For example, motor responses to sensory input (sensorimotor integration) become defective.

CHOREIFORM MOVEMENTS are rapid, jerky, and purposeless movements most pronounced in the trunk and appendages. *Sydenham's chorea* (acquired) and *Huntington's chorea* (congenital) are specific examples of choreiform disorders. Lesions in the corpus striatum are most commonly associated with such disorders.

ATHETOID MOVEMENTS are slow sinous movements most common in the outer (distal) parts of the appendages. Globus pallidus lesions may cause such movements.

The substantia nigra is the most commonly involved nucleus in PARKINSON'S DISEASE (*paralysis agitans, shaking palsy*). This disease makes its appearance most often in persons between 50 and 65 years of age. There is fixation of facial expression, tremor of the limbs at rest, muscular rigidity, slow movements, and postural abnormalities. The use of levodopa (L-DOPA, levorotatory dihydroxyphenylalanine), an amino acid, may reduce the severity of the disease's symptoms by increasing the synthesis of certain synaptic transmitters in the central nervous system.

THE DIENCEPHALON (FIGS. 7.9, 7.10, 7.11)

The diencephalon forms the superior end of the brainstem, and is part of the original forebrain. It develops into larger superiorly placed paired THALAMI, and an inferiorly placed, single HYPOTHALAMUS.

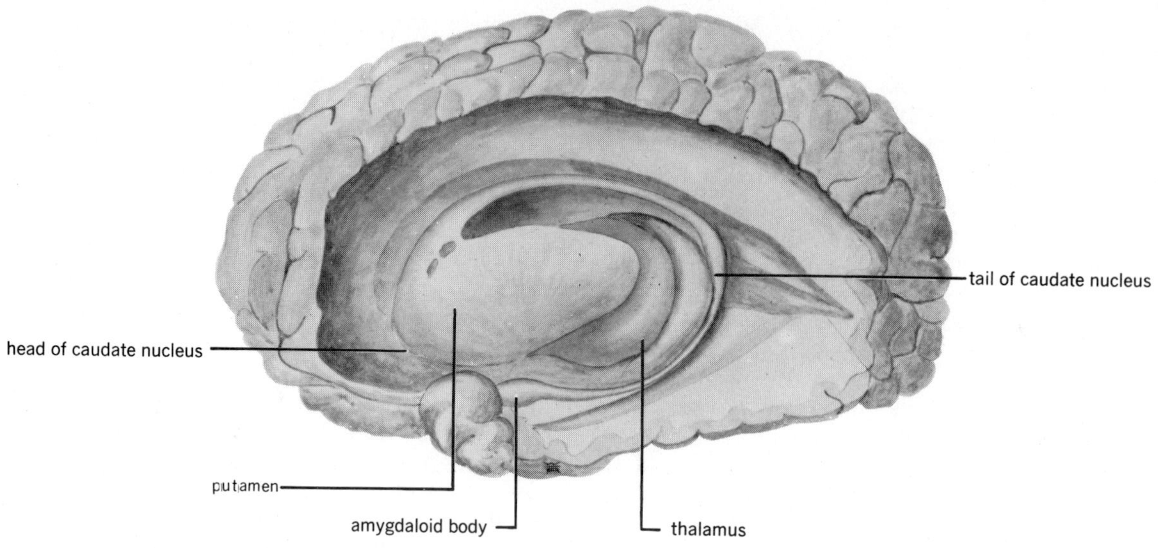

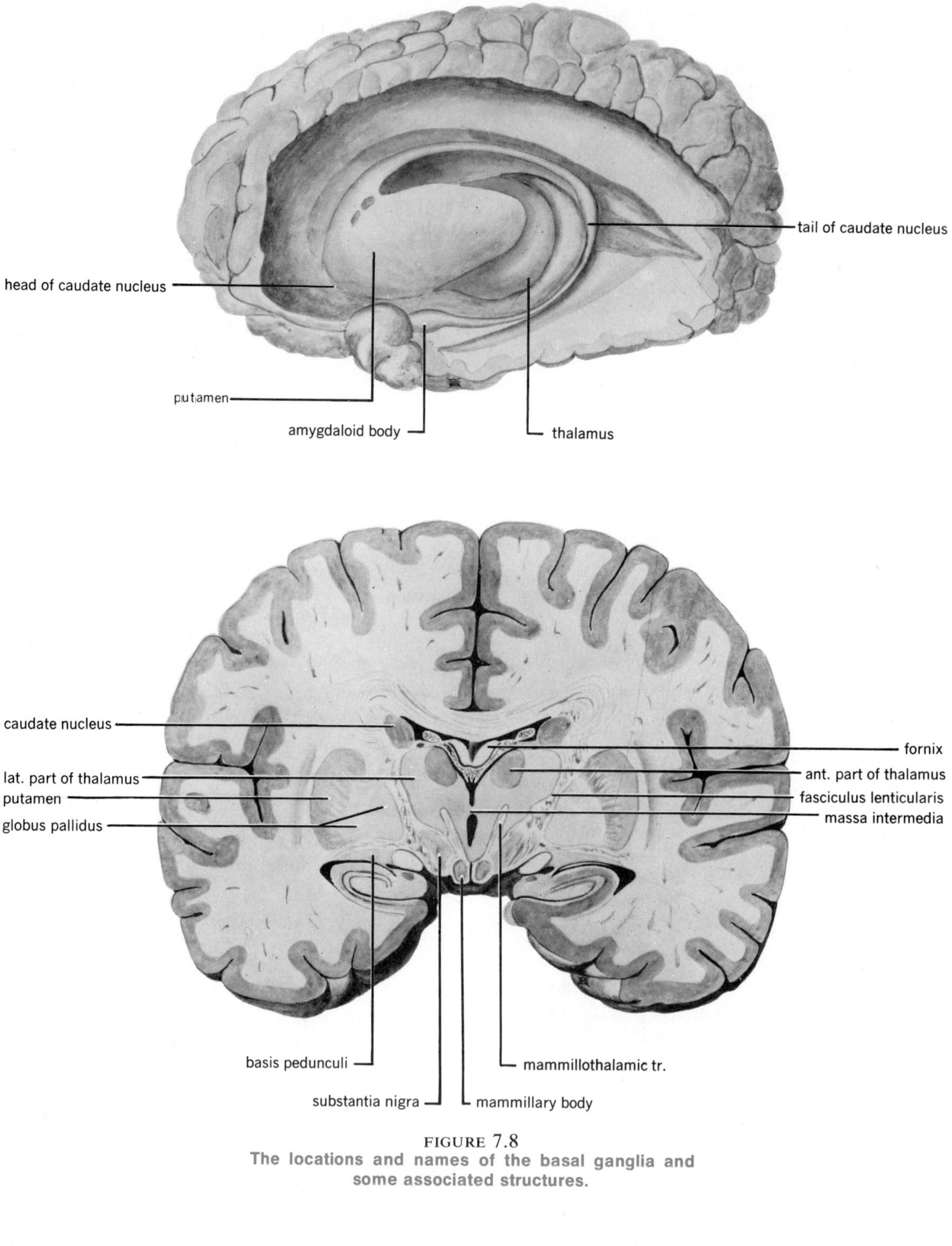

caudate nucleus
fornix
lat. part of thalamus
ant. part of thalamus
putamen
fasciculus lenticularis
massa intermedia
globus pallidus
basis pedunculi
mammillothalamic tr.
substantia nigra
mammillary body

FIGURE 7.8
The locations and names of the basal ganglia and some associated structures.

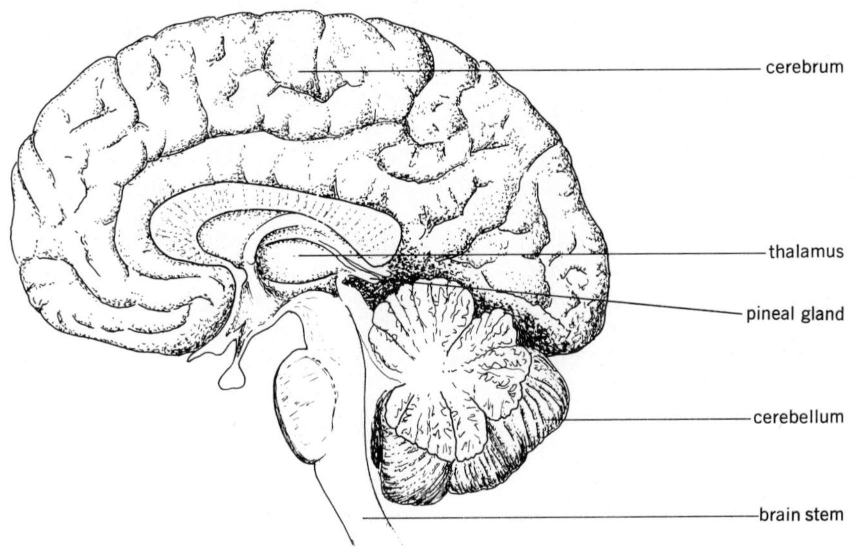

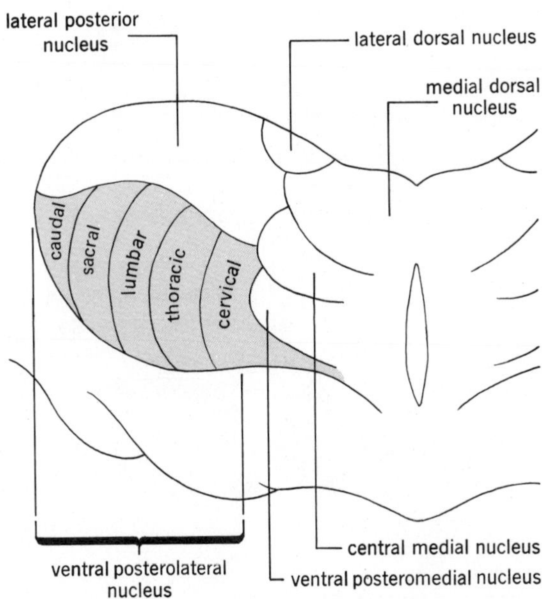

FIGURE 7.9
The location and body representation in the thalamus.

THE THALAMUS (*SEE* FIGS. 7.9, 7.10)

Structure. Each thalamus measures about 3 cm from front to back and 1.5 cm in height and width, and together they compose about four-fifths of the diencephalon. They are connected across the midline by the *intermediate mass*. Developmentally, three portions differentiate within the thalamus primordia to give rise to an upper *epithalamus*, a central *dorsal thalamus*, and a lower *ventral thalamus*. Several nuclei, or groups of cell bodies are found in each area (*see* Fig. 7.10). Fibers entering the thalamus are primarily sensory in nature, and outgoing fibers from the dorsal and ventral thalami pass primarily to the sensory and motor regions of the cerebral cortex.

Functions. The epithalamus consists of the PINEAL GLAND or BODY (*epiphysis*) and the paired habenular nuclei located anterior to the pineal gland. The epithalamus receives fibers from the olfactory regions of the cerebrum and from the limbic system. (The latter is concerned with emotions and their expression.) These fibers synapse in the habenular nuclei. The epithalamus sends fibers to the midbrain that then pass to the autonomic nuclei of the brainstem and to the reticular formation. Such connections allow olfactory stimuli to influence

FIGURE 7.10
The major thalamic nuclei. (*a*) View from the dorso-lateral direction. (*b*) Posterior view.

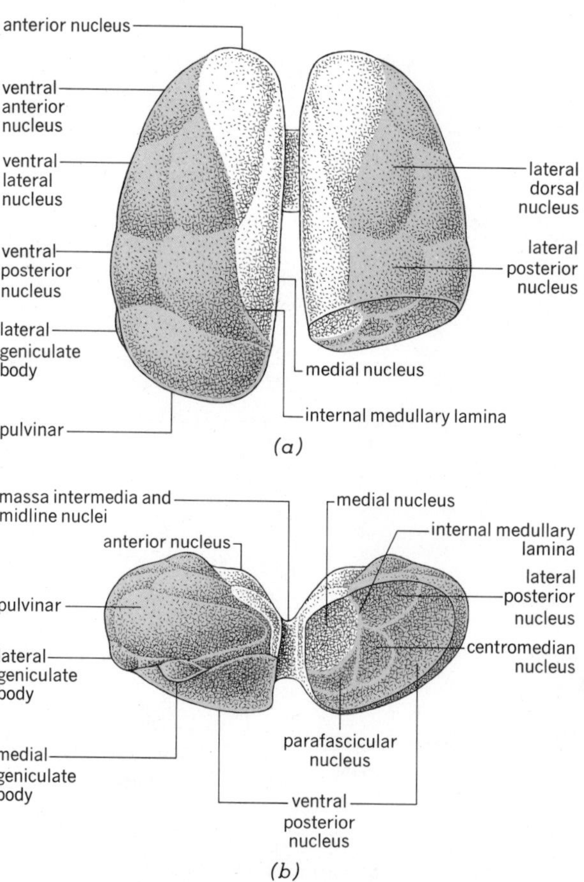

emotional behavior, as in mating or food seeking, and also to influence visceral activity. The pineal gland is a cone-shaped structure measuring about 7 by 5 mm (in length and diameter); in the human it appears to be concerned with onset of puberty and with the establishment of circadian rhythms.

The dorsal and ventral thalamus (commonly called simply the *thalamus*) contains the following groups of nuclei.

The reticular nucleus. This portion receives fibers from the reticular formation, and sends fibers to the cerebral cortex. It is part of the mechanisms that alert and maintain the waking state of the organism.

The midline nuclei. These nuclei receive sensory information from receptors in the body viscera and taste buds, and send fibers to other parts of the diencephalon. There are no connections directly to the cerebral cortex.

Specific thalamic nuclei. This term designates nuclei projecting (sending fibers) to sensory and motor areas of the cerebral cortex. Named nuclei include the following.

The MEDIAL GENICULATE BODY that receives fibers from the cochlea and inferior colliculi and that sends fibers to the auditory cortex for relay of auditory information.

The LATERAL GENICULATE BODY that receives fibers from the retina and superior colliculus and sends fibers to the visual cortex.

The VENTRAL POSTERIOR NUCLEUS (*ventral posterolateral nucleus*) that receives the *spinothalamic tracts* that carry sensations of pain, heat, and cold; the *medial lemniscus* conveying sensations of touch and pressure, and sensory fibers of cranial nerve V (trigeminal) conveying sensory information from the face. The body is represented in topographical fashion in this nucleus (*see* Fig. 7.9), with the head medially, and lower body laterally. Some interpretation of crude pain may occur here. Fibers are projected to the sensory cortex for discrimination and localization.

The VENTRAL LATERAL and VENTRAL ANTERIOR NUCLEI that receive fibers from several of the basal ganglia and the cerebellum. Fibers are projected to the motor and premotor areas of the frontal cortex. These nuclei are thus part of a system influencing voluntary motor activity.

Nonspecific thalamic nuclei. This term is used to describe nuclei that project to the association areas of the cerebral cortex. Specific nuclei include the PULVINAR, LATERAL POSTERIOR, LATERAL DORSAL, MEDIAL, and ANTERIOR nuclei. As a group, these nuclei are concerned with reception of sensory information and analysis of that information leading to expression of emotion, intellectual function, memory storage, and "moods" and "feelings."

In summary, the thalamus operates primarily in the reception and sorting out of sensory information, at both the basic *and* intellectual levels.

Clinical considerations. Damage to the sensory portions of the thalamus produces the THALAMIC SYNDROME. A lesion here is commonly vascular in origin (a vessel blocked or ruptured) and results in little or no awareness of sensation until a high and critical point is reached. Then the sensation may "burst" on the consciousness and become intolerable (especially the sense of pain). Selective destruction of thalamic nuclei (e.g., ventral posterior nucleus) may ameliorate the condition. In older individuals, vascular problems may result in damage to the thalamic motor areas and interrupt pathways connecting cortex and basal ganglia. Tremor and muscular rigidity may develop (Parkinson's disease). L-DOPA may alleviate such symptoms.

THE HYPOTHALAMUS (*SEE* FIGS. 7.9, 7.11)

Structure. The hypothalamus weighs about 4 grams, and contains many groupings of cell bodies or nuclei. It receives fibers from the thalamus, cerebral cortex, brainstem, and limbic system, and projects fibers or sends chemicals to the pituitary gland and autonomic nuclei in the brainstem.

Functions. The hypothalamus is concerned with the maintenance of HOMEOSTASIS of a variety of body functions.

Temperature regulation. Human body temperature is normally maintained within a degree or two of 98.6°F or 37°C. A proper balance between heat production plus conservation and heat loss is maintained by HEAT LOSS and HEAT GAIN CENTERS in the hypothalamus. Receiving fibers from the thalamus concerned with heat and cold, and by monitoring blood temperature, the heat loss center controls skin blood vessels (dilation for greater radiation of heat), sweating (increase for evaporation and cooling), and decreased muscle tone (less muscle contraction for less fuel consumption and lowered heat production). The heat gain center causes skin vessel constriction, shivering, and cessation of marked sweating, all of which conserve body heat or increase heat production.

Physical means of heat loss occurs by several methods.

RADIATION is the transfer of heat, as infrared radiation, from a warmer to a cooler medium. Rate of radiation is influenced by the surface area of the radiating structure, by the volume of active cells, and by the heat gradient (or temperature difference) between the radiator and the medium around it. For the human, the radiant surface is the skin and the gradient depends on the amount and temperature of blood circulating through the skin versus the environmental temperature. In cool environments, radiation is the major pathway of heat loss.

CONDUCTION is the transfer of heat from molecule to molecule, as from the radiator to another body in direct contact with the radiator. Transfer of heat to a piece of clothing from the body is an example of conduction.

CONVECTION is the transfer of heat to a moving medium (fluid or air) that carries the heat away and tends to maintain the gradient, as swimming in cold water.

VAPORIZATION or evaporation of water (e.g., perspiration) on the body surface requires heat. Cooling is achieved as vaporization proceeds.

As environmental temperature approaches and exceeds body temperature, radiation decreases and stops, and vaporization assumes increasing importance as the main route of heat loss. Vaporization decreases as the humidity of the warm environment increases.

FEVER is a term used to describe a body temperature that is above the usual limits of normal. It may result from damage to the brain, from chemicals that affect the hypothalamic thermostat, from failure of the normal cooling mechanisms of the body (heat stroke), or from dehydration. In many diseases caused by bacteria or viruses, the toxic products produced by the microorganism, or the products of cell destruction, "reset" the thermostat to a higher value. Chemicals that have this effect are called PYROGENS (*pyro,* fire + *gen,* to produce). Brain tumors may affect the thermostat. Dehydration results in a smaller volume of body water to get rid of metabolic heat, and body temperature rises. HEAT STROKE occurs when normal methods of ridding the body of heat fail, and body temperature rises. Elevated body temperature stimulates chemical reactions, more heat is produced and a "vicious circle" is established. Heat stroke most commonly occurs when heavy work is performed in a hot humid environment. To prevent this condition, one must cool off by whatever means are available.

HYPOTHERMIA (low body temperature) is often created artificially during surgical operations (e.g., heart surgery). The combination of an anesthetic and cooling of the body can reduce core temperature by 10 to 15°F. Such low core temperatures result in less demand for nutrients by the brain and promote the ability of the patient to tolerate the surgery. Core temperatures cannot be reduced below about 84°F without running the danger of damaging the hypothalamic centers.

Regulation of water balance. Hypothalamic neurons continually monitor blood osmotic pressure and adjust the tonicity of the body extracellular fluid compartment by ADH (*antidiuretic hormone*) that is passed to the posterior lobe of the pituitary gland for storage and release. The hormone permits greater water reabsorption from the fluids contained within the kidney tubules.

Control of pituitary function. At last count, nine chemicals influencing pituitary function have been shown to originate in the hypothalamus. Properly termed PITUITARY REGULATING FACTORS, these chemicals are produced as a result of blood-borne and nervous stimuli, are passed to the anterior lobe of the pituitary gland over a system of blood vessels, and there stimulate or inhibit production and/or release of anterior lobe hormones.

Control of food intake. Initiation of feeding is controlled by a FEEDING CENTER. When the animal is "full," a SATIETY CENTER causes cessation of feeding. The stimulus that triggers activity of either center has been debated at length. Three theories have been advanced.

The GLUCOSTAT THEORY suggests that feeding is initiated by a lowered blood sugar level to the hypothalamic feeding center. Feeding and absorption of glucose raises blood sugar and stimulates the satiety center. Feeding ceases. The seemingly paradoxical hyperphagia of diabetes mellitus is explained by suggesting that the cells of the centers, in spite of high blood-sugar levels, do not take in glucose because of insulin absence, and are thus "fooled" into believing the blood glucose level is low.

The THERMOSTAT THEORY suggests that the centers are sensitive to body temperature. Between feedings, body temperature is lowered, triggering feeding. The absorption and processing of this food is associated with accelerated metabolism and heat production that stimulates the satiety center.

Brobeck's ABDOMINAL STRETCH RECEPTOR THEORY suggests that as the stomach fills, receptors are stimulated, resulting in inhibitory impulses being conducted to the feeding center.

Regulation of gastric secretion. Hypothalamic stimulation causes increased gastric secretion. Emotions may thus cause increased gastric secretion when there is no food in the stomach and lead to ulcer development.

Emotional expression. The hypothalamus is part of the system necessary for expressions of rage and anger, and sexual behavior. Emotions may also be expressed as changes in heart rate, blood pressure, and pupillary changes.

CLINICAL CONSIDERATIONS

A wide variety of symptoms may develop with hypothalamic damage. Relating the functions just described to the areas shown in Figure 7.11 gives an idea of what regulation would be lost. Disorders associated with disturbances of water balance, temperature regulation, and emotions appear most commonly. *Diabetes insipidus* refers to the production of a large volume of dilute urine as a result of diminished

FIGURE 7.11
The nuclei of the hypothalamus.

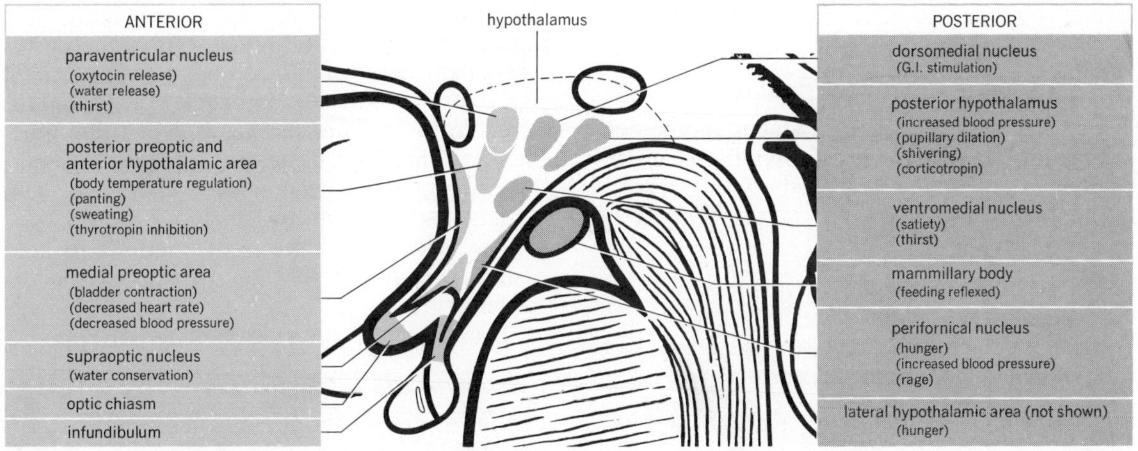

ADH secretion. Inability to regulate body temperature (in an adult) indicates hypothalamic damage. Because of immaturity of their nervous system, infants do not maintain as good a control of their body temperature as do adults, and this is a normal condition.

THE MESENCEPHALON (MIDBRAIN)

STRUCTURE

The midbrain is a wedge-shaped portion of the brain stem, about 1.5 cm in length, lying between the thalamus and pons (*see* Fig. 7.9). Its anterior portion consists chiefly of two large bundles of fibers called the CEREBRAL PEDUNCLES. These bundles carry most of the efferent motor fibers from cerebrum to the cerebellum and spinal cord. The posterior portion contains the paired SUPERIOR and INFERIOR COLLICULI and the nuclei of cranial nerves III and IV.

FUNCTION

The midbrain, by way of the peduncles, serves as a *motor relay station* for fibers from cerebrum to cerebellum and spinal cord. The colliculi integrate a variety of *visual* (superior colliculi) and *auditory reflexes* (inferior colliculi), including those concerned with avoiding objects that are seen, and turning the head to achieve the greatest benefit from sights and sounds. Reflexes that maintain the body in proper orientation to its environment (clued mainly by visual impressions) are in part controlled by the midbrain. These latter reflexes are termed *righting reflexes*.

CLINICAL CONSIDERATIONS

Damage to the midbrain produces motor deficits if the peduncles are involved, and disturbances in auditory, visual, and righting reflexes if the colliculi are involved.

THE METENCEPHALON AND MYELENCEPHALON: PONS AND MEDULLA (MEDULLA OBLONGATA)

The pons and medulla develop from the hindbrain.

STRUCTURE

The pons is about 2.5 cm long and part of it forms a conspicuous bulge on the anterior surface of the brainstem. A BASAL or ANTERIOR PORTION receives fibers from a cerebral hemisphere and sends fibers to the opposite cerebellar hemisphere over the CEREBELLAR PEDUNCLES. It is the fibers of the peduncles that form the bulge on the anterior aspect of the pons. The posterior portion or TEGMENTUM contains ascending (sensory) and descending (motor) fibers and the nuclei of cranial nerves V to VIII. A respiratory center called the PNEUMOTAXIC CENTER is also located in the posterior pons.

The medulla forms the inferior 3 cm or so of the brainstem. It is slightly wedge-shaped, wider at the superior end. It joins the spinal cord at the level of the foramen magnum of the skull. Internally, the medulla shows an extensive rearrangement of ascending and descending fibers, many of which cross over (*decussate*) from one side to the other. Groups of nerve cell bodies known as nuclei or centers are scattered throughout the medulla. Some of these groups form the VITAL and NONVITAL CENTERS of the medulla. Others serve cranial nerves IX to XII, whose fibers emerge from the medulla.

FUNCTIONS

The basal portion of the pons (peduncles) acts as a synaptic or relay station for motor fibers from cerebrum to cerebellum. The cranial nerve nuclei provide for sensory and motor fibers to the skin of the head and certain muscles, for hearing, taste, and eye movement. The pneumotaxic center is part of a system designed to regulate breathing.

The VITAL CENTERS of the medulla include centers necessary for the survival of the organism.

The CARDIAC CENTERS include *cardioacceleratory* and *cardioinhibitory centers* that receive fibers from blood vessels, heart, lungs, cerebrum and other areas. These centers provide for the reflex control of heart rate according to body activity levels and levels of carbon dioxide and oxygen in the bloodstream. These activities are considered in greater detail in Chapter 20.

Two of three paired RESPIRATORY CENTERS (discussed more fully in Chapter 22), including the *inspiratory* and *expiratory centers,* lie in the medulla. The inspiratory centers provide stimulation to the muscles (diaphragm, external intercostals) that ultimately cause the drawing of air into the lungs. The expiratory centers provide part of the influence necessary to interrupt inspiratory activity and permit air to be expired or driven out of the lungs.

The VASOMOTOR CENTERS include *vasoconstrictor* and *vasodilator centers* (described more fully in Chapter 21). These centers reflexly control the diameter of muscular blood vessels and hence blood pressure.

All of these vital centers are integrated so that breathing, heart rate, and blood pressure work to achieve the desired end result. For example, activity speeds heart rate, raises blood pressure, and accelerates breathing to speed the flow of oxygenated blood to the active areas.

NONVITAL CENTERS of the medulla include those for *swallowing, vomiting, sneezing,* and *coughing.*

CLINICAL CONSIDERATIONS

Damage to the pons may produce *motor deficits* through involvement of the peduncles; movements are less precise and posture is difficult to maintain. Involvement of the pneumotaxic center leads to *apneustic breathing,* characterized by long sustained inspiratory activity.

Damage to the medulla may occur with blows to the back of the head or upper neck; such blows may involve the vital centers, leading to respiratory paralysis, loss of blood pressure, and death. One area that poliovirus may attack is the medulla. *Bulbar polio,* characterized by impaired respiration, results in the victim being confined to an "iron lung" (a negative-pressure apparatus) that alternately compresses and decompresses the chest to cause air movement into and out of the lungs.

THE RETICULAR FORMATION

Located in the brainstem, extending from the medulla to the lower border of the thalamus, is a column of gray matter distinct from that described in preceding sections called the RETICULAR FORMATION. Within this column are the *reticular nuclei,* specifically designated by position and what part of the brainstem they are located within. The formation receives fibers from brain motor regions and from most of the sensory systems of the body. Its outgoing fibers pass to the thalamus and ultimately to the cerebral cortex in general. Stimulation of most parts of the formation causes immediate and marked activation of the cerebral cortex, a state of alertness and attention, and, if the subject is sleeping, immediate awakening. The upper portion of the formation *plus* its pathways to thalamus and cortex have been designated the RETICULAR ACTIVATING SYSTEM (RAS) reflecting the alerting function it has. Maintenance of the waking state is achieved by a positive feedback mechanism in which RAS stimulation of the cortex causes the cortex to raise the activity level of the RAS, stimulate visceral activity and muscle tone, and cause epinephrine (adrenaline) release from the adrenal medulla. Coma, or a state of unconsciousness from which even the strongest stimuli cannot awaken the individual, develops when there is brainstem reticular formation damage. Drugs that make us sleepy or tranquilized act, in part, on the RAS, depressing synaptic conduction through it and reducing the positive feedback mechanism mentioned above.

THE IMMATURE BRAIN

The fetal and newborn brains may be characterized as immature in structural, biochemical, and physiological ways. The basic structure is there, but there is a paucity of synaptic connections as compared to the adult brain. Myelination is incomplete, motor control is incomplete, and much programming, by sensory input, is required. It is obviously impossible to detail all of these changes in this book. The discussion to follow indicates some of the milestones in biochemical and physiological development. The reader should be aware that central nervous system development is incomplete at birth and that this is reflected in many body functions as well as that of the brain itself.

BIOCHEMICAL CHANGES

The central nervous system depends almost entirely on glucose as its energy source. The fetal brain is especially dependent on carbohydrate metabolism. It has the capability to utilize both anaerobic and aerobic pathways for glucose utilization, and is thus more resistant to hypoxia (anoxia) than is the adult brain, which is more dependent on the aerobic pathways. In rats, whose developing brains have been extensively studied during their entire development, glucose utilization increases from 36 to 56 percent of the body glucose consumption, while anaerobic degradation (glycolysis) of glucose decreases from 45 percent of total utilization to 33 percent. The immature brain may utilize glucose derivatives that the adult brain cannot because of the development of the blood-brain barrier (*see* Chapter 10). The amount of glycogen in the brain decreases as maturity proceeds, and may reflect the development of enzymes necessary for its metabolism as maturation proceeds.

Lipid concentrations generally increase in the brain as maturation proceeds. A good bit of this increase reflects the advancement of myelination of fiber tracts. The remainder appears to be correlated with changes in the metabolism of lipids, such as a shift toward the metabolism of longer chain fatty acids with maturation.

Changes in proteins, amino acids, and amines generally follow an increasing curve as development proceeds. Protein synthesis is more active in immature animals reflecting the more rapid early growth of the brain. Amino acids or their derivatives may increase with age (glutamic acid, glutamine, gamma-aminobutyric acid, aspartic acid), decrease with age (taurine, ethanolamine, cystathione) or show no specific pattern of change (alanine, glycine, threonine, serine). Gamma-aminobutyric acid (GABA) is regarded as an inhibitory synaptic transmitter, and its increase appears to be correlated with the increasing degree of inhibitory functions the CNS acquires as it develops. The amines mentioned include serotonin, dopamine, epinephrine, and norepinephrine, most of which are involved in synaptic transmission. All tend to increase with age, perhaps correlated with the increasing number of synapses that develop with maturation, and with the increasing excitability that neurons exhibit.

Nucleic acid concentrations in brain generally decrease with age, DNA at a more rapid rate than RNA. Ribosomal and messenger RNA species do not share this decrease with age, perhaps because much information is being recorded as memory traces with maturation and experience.

Water and extracellular electrolytes (Na^+, Cl^-) progressively decrease as maturation proceeds, in part due to increased amounts of organic substances (lipid, protein) as maturation occurs. Intracellular potassium increases with age, as neurons acquire their full excitable capability.

Enzymes involved in metabolism of carbohydrates, lipids, and synthesis of synaptic transmitters increase with maturity. This appears to be correlated with increasing bulk of tissue, number of synapses, and activity; this enzyme production is one point at which "metabolic errors" due to mutations in the DNA responsible for their appearance may appear, with the possibility arising for mental retardation and other effects.

PHYSIOLOGICAL CHANGES

Immature neurons do not develop as strong an action potential as do mature ones, and it lasts longer. Synaptic transmission is slower, and facilitation and EPSP's are not exhibited to the degree seen in mature neurons. Inhibitory activity and IPSP's are not seen until several weeks after birth. Electrical activity in the fetal and infant brain is more variable and probably is related to immaturity of the sense organs that provide much of the brain's input. Fatigue of synapses is more pronounced in immature brains and EEG (electroencephalogram) patterns do not assume adult forms until puberty.

TABLE 7.2
SOME FETAL AND NEONATAL REFLEXES

MENSTRUAL AGE (WEEKS)	PROTECTIVE AND/OR AVOIDING; LOCOMOTION	FEEDING	GRASPING AND PLANTAR	GENITAL	RESPIRATION-LIKE
Stimulation site[a]	*Perioral*[1] *or eyelid*[2]	*Perioral*[1] *or palmar*[2]	*Palmar*[1] *or plantar*[2]	*Genital*[1] *or thigh*[2]	*Chest*[1] *or abdominal*[2]
7.5	Contralateral neck flexion				
8–8.5	Contralateral neck and trunk flexion[1]	Ipsilateral neck and trunk flexion			
9.5	Head and trunk extension[1]	Mouth opening[1]			
10–10.5	Orbicularis oculi contraction[2] ("squint"); Rotation of face to contralateral side, usually with trunk flexion[2]	Ventral head flexion[1]	Incomplete finger closure[1]; Plantar flexion of all toes[2]	Bilateral flexion of thighs on pelvis	
11–11.5	Corrugator supercilli contraction[2] ("scowl")	Ipsilateral head rotation[1]			
12–12.5	Squint and scowl combined[2]	Lip closure and swallowing[1]	Principally dorsiflexion of big toe and toe fanning; flexion at foot, knee, hip[2]		
13–16.5	Squint combined with neck and trunk extension or with contralateral head rotation[2]	Mouth opening, closure, swallowing, ventral head flexion; Tongue movements[1]	Complete finger closure;[1] Maintained finger closure[1]		Isolated respiratory contractions;[1] Abdominal muscle contractions[2]
17–23.5		Protrusion of lips[1]	Effective, but weak true grasp[1]		Temporary diaphragmatic contraction[1] Temporary effective respiratory chest and phonation if delivered[1]

(continued on page 132)

TABLE 7.2
SOME FETAL AND NEONATAL REFLEXES (continued)

MENSTRUAL AGE (WEEKS)	PROTECTIVE AND/OR AVOIDING; LOCOMOTION	FEEDING	GRASPING AND PLANTAR	GENITAL	RESPIRATION-LIKE
24		Sucking[1]			
27			Maintained grasp to support most of body weight momentarily		Permanent respiration established on delivery
32				Cremasteric reflex[2]	
Premature and newborn	Primitive swimming movements	Babkin reflex[2]	Grasp weak, but stronger while sucking[1]		

[a] Varies for each reflex as indicated in column headings.
Source: Timiras—Develop. Physiol. & Aging, Table 9.4, pg. 154, Macmillan, 1972.

Reflex activity proceeds in a predictable pattern as maturation proceeds as shown in Table 7.2. Reflex development requires both maturation of the receptors involved and of the pathways the impulses traverse.

Integrative activity begins in the spinal cord and brainstem, and is followed by cortical control. In the human, stimulation of the fetal cortex up to the seventh fetal month causes no response; beyond this time, the cortex assumes increasing control over inhibitory and motor activity.

Finally, it may be pointed out that strictly environmental influences, such as hormones, hypoxia, and ionizing radiation are concerned with the direction and rate of genetically determined maturation processes. The effects of these factors are explored in more detail in Chapters 26 and 28.

SUMMARY

1. Development of the nervous system begins at about 18 days of embryonic life.
   a. A neural plate develops on the embryonic disc.
   b. A neural groove forms, whose lips, the neural folds, close over the midline to form the neural tube.
   c. Neural crest cells split off from the posterior aspect of the neural tube.
   d. The anterior end of the tube forms a series of enlargements that will give rise to the several parts of the brain (*see* Table 7.1).
   e. The walls of the neural tube contain cells that will give rise to neurons and glia. In the cerebrum, neurons lie externally; in the cord internally.
   f. The peripheral nerves develop from neural crest and outgrowths from the central nervous system.
2. The cerebrum (telencephalon) is subdivided by several fissures into hemispheres and lobes (frontal, parietal, occipital, temporal).
   a. The outer gray matter of the cerebrum is the cortex, and it is divided into numbered (Brodmann) areas according to cellular differences.
   b. The cortical functional areas are described as follows:
      1) Frontal lobes: Area 4, primary motor area for voluntary movement; Areas 24, 31, supplementary motor area for voluntary movement; Area 6, the premotor area for

contraction of muscle groups; Area 8, for eye movements; Areas 9 to 12, the pre-frontal area for intelligence, moral and social senses, and emotions.

2) Parietal lobes: Areas 3, 1, 2, the general sensory areas for general motor activity and reception of somesthetic senses; Areas 5, 7ab, the parietal association areas for sensory interpretation.

3) Occipital lobes: Area 17, the visual area where retinal pathways ultimately terminate; Areas 18, 19, the visual association areas for interpretation of visual experience.

4) Temporal lobes: Area 41, the auditory area where cochlear fibers end; Area 42, the auditory association area for interpretation of auditory experience, language function, and memory.

5) Broca's speech area occupies Areas 44, 45, and is essential for formation of words.

3. The white matter of the cerebrum (medullary body) contains fibers and the basal ganglia.
   a. Association fibers connect different parts of a hemisphere.
   b. Commissural fibers connect hemispheres.
   c. Projection fibers enter or leave the cerebrum.
   d. The basal ganglia are listed. They control motor functions, muscle tone, and rhythmical movements.
4. The diencephalon includes the thalamus and hypothalamus.
   a. The thalamus is primarily a sensory relay center to the cerebrum (specific nuclei and their functions are described).
   b. The hypothalamus is a center for homeostasis of temperature, water balance, pituitary function, food intake, and for gastric secretion and emotional expression.
5. The midbrain
   a. Contains cerebral peduncles transmitting fibers from cerebrum to cerebellum and spinal cord.
   b. Carries the colliculi integrating visual, auditory, and righting reflexes.
6. The pons serves as a means of transmitting motor fibers from cerebrum to cerebellum and contains a respiratory center ensuring normal expiration.
7. The medulla (oblongata) contains vital centers for respiration, heart rate and blood vessel diameter, nonvital centers for sneezing, coughing, vomiting and swallowing, and crossing fiber tracts.
8. The reticular formation is gray matter in the medulla, pons, and midbrain, and deals with arousal of the organism.
9. The immature brain is characterized
   a. There are increases in: ability to metabolize materials, lipid content, protein, amino acid and amine content, and enzymes; and decreases in nucleic acid content as the brain matures.
   b. Neurons do not realize their full potential of excitability, facilitation, and electrical activity for variable periods after formation. Reflex and integrative activity are minimal at birth and become more advanced after birth.

QUESTIONS

1. What are the "milestones" in the development of the nervous system?
2. How do peripheral nerves and ganglia form?
3. What are the derivatives of the three primary enlargements of the neural tube?
4. What functions are centered in each of the following brain areas?
   a. Frontal lobes
   b. Temporal lobes
   c. Midbrain
   d. Thalamus
   e. Medulla
   f. Hypothalamus
5. Where and what is the reticular formation?
6. How does an immature brain differ from an adult brain?
7. What sorts of symptoms could be expected to develop with damage in the
   a. Area 4?
   b. Area 17?
   c. Hypothalamus?
   d. Frontal association area?
   e. Temporal association area?
8. Compare the contributions of cerebral cortex and basal ganglia to movement.

READINGS

Brady, Roscoe O. "Inherited Metabolic Diseases of the Nervous System." *Science 193:*733. 27 Aug 1976.

Culliton, Barbara J. "Psychosurgery: National Commission Issues Surprisingly Favorable Report." *Science 194:*299. 15 Oct 1976.

Evarts, Edward V. "Brain Mechanisms in Movement." *Sci. Amer. 229:*96. (July) 1973.

Kolata, Gina Bari. "Brain Biochemistry: Effects of Diet." *Science 192:*41. 2 April 1976.

Stricker, Edward M., Wilson G. Bradshaw, and Robert H. McDonald, Jr. "The Renin-Angiotensin System and Thirst: A Reevaluation." *Science 194:*1169. 10 Dec 1976.

THE CEREBELLUM

OBJECTIVES

After studying this chapter, the reader should be able to:

☐ Describe the generalized gross and microscopic structure of the cerebellum.

☐ Describe what constitutes the "functional unit" of the cerebellum.

☐ Describe the sources of input of information to the cerebellum, and its efferent pathways.

☐ Describe the functions of the cerebellum and its involvement in motor activity.

☐ Predict some of the signs of damage to the cerebellum, given knowledge of its functions.

☐ Give a generalized integrated scheme by which cerebral cortex, subcortical structures, and the cerebellum produce and control movement.

THE CEREBELLUM

The cerebellum develops from the primitive hindbrain, from the same area giving rise to the pons. It is second in size only to the cerebrum, and comes to lie on the posterior aspect of the brain stem above the pons and medulla (Fig. 8.1).

STRUCTURE

Grossly, the cerebellum is composed of a centrally placed VERMIS, and two CEREBELLAR HEMISPHERES. It is divided into two major LOBES, *anterior* and *posterior,* and several smaller LOBULES, by fissures and sulci. Some of these features are depicted in Figure 8.2. The surfaces of the vermis and hemispheres are folded into numerous small *folia,* each of which contains an outer CORTEX of gray matter, and an inner MEDULLARY BODY (*medulla*) composed of white matter. The tree-like arrangement of cerebellar white matter is called the *arbor vitae*.

FIGURE 8.1
The location of the cerebellum.

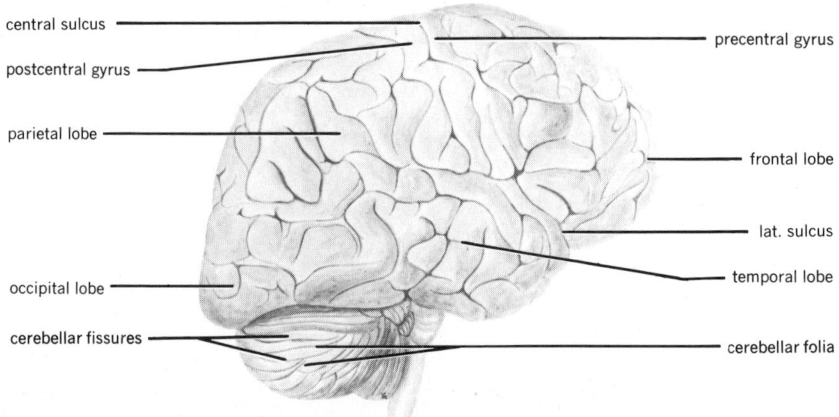

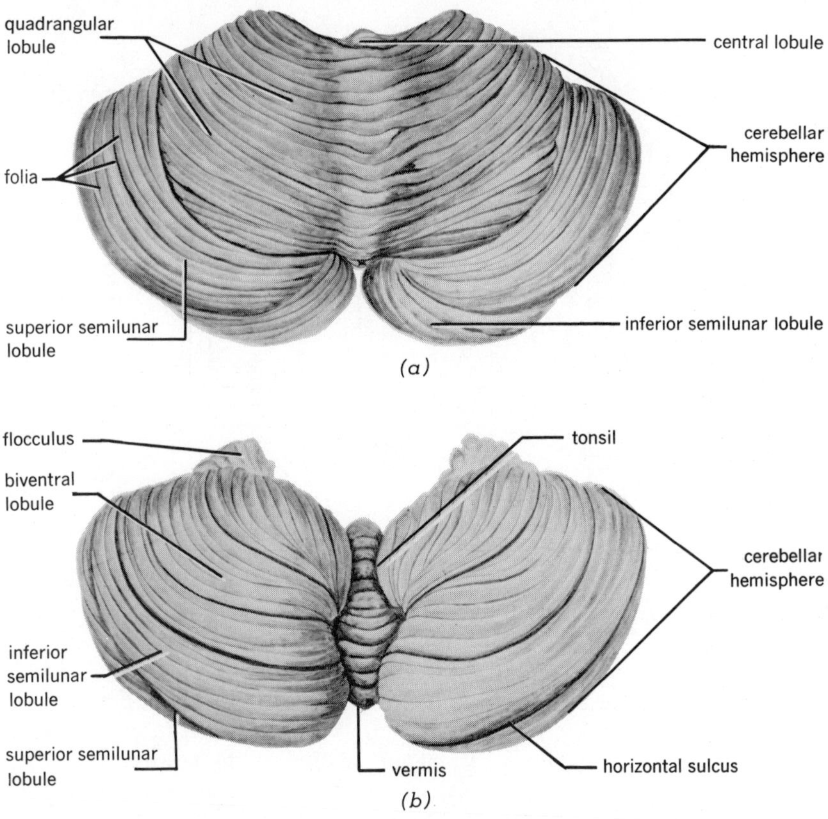

FIGURE 8.2
The cerebellum. (*a*) Dorsal view; (*b*) Ventral view.

Microscopically, a section of a folium shows the same basic structure anywhere in the organ. The characteristic cell of the cerebellum is the PURKINJE CELL, and these, together with the granule cells, basket cells, and stellate cells, form some 30 million functional units in the organ. The microscopic anatomy and functional unit are depicted in Figure 8.3. Deep within the white matter of the cerebellum are its CENTRAL NUCLEI (Fig. 8.4). The axons of the Purkinje cells terminate in these nuclei, and the nuclei in turn send impulses to the brainstem.

Input to the cerebellum occurs over three major pathways:

Fibers from the semicircular canals and maculae of the inner ear (equilibrial and motion detectors)

Fibers from sensory receptors in general, including the muscle spindles and tendon organs, and the eye

Fibers from the cerebral cortex, basal ganglia, thalamus, and reticular formation

Output from the cerebellum passes to the brainstem and from there to the skeletal muscles, basal ganglia, thalamus, and cerebral cortex. Connections such as those described provide a series of interlocking circuits that can compare intent with actual movement.

FUNCTIONS

In general terms, the cerebellum, while operating totally at the subconscious level, and initiating no movements of its own, continually

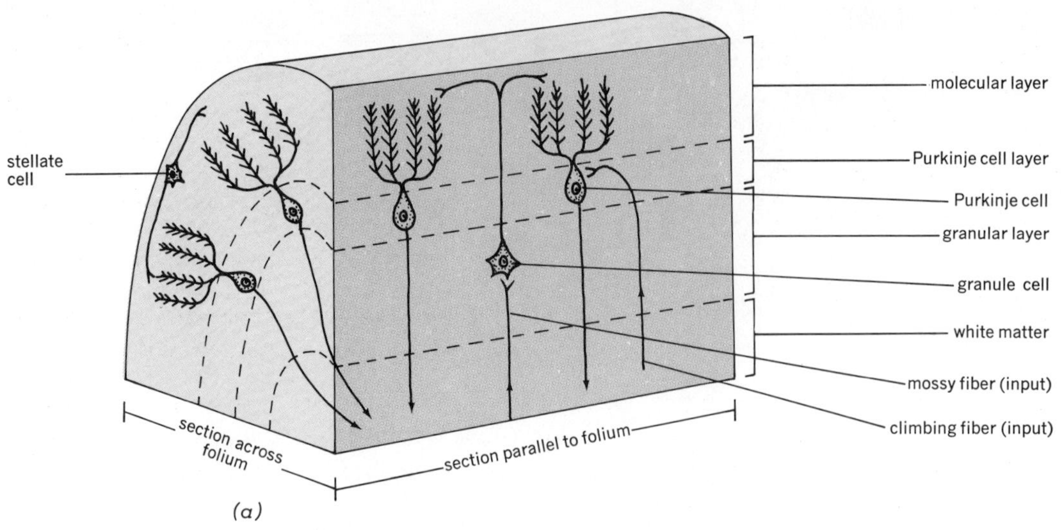

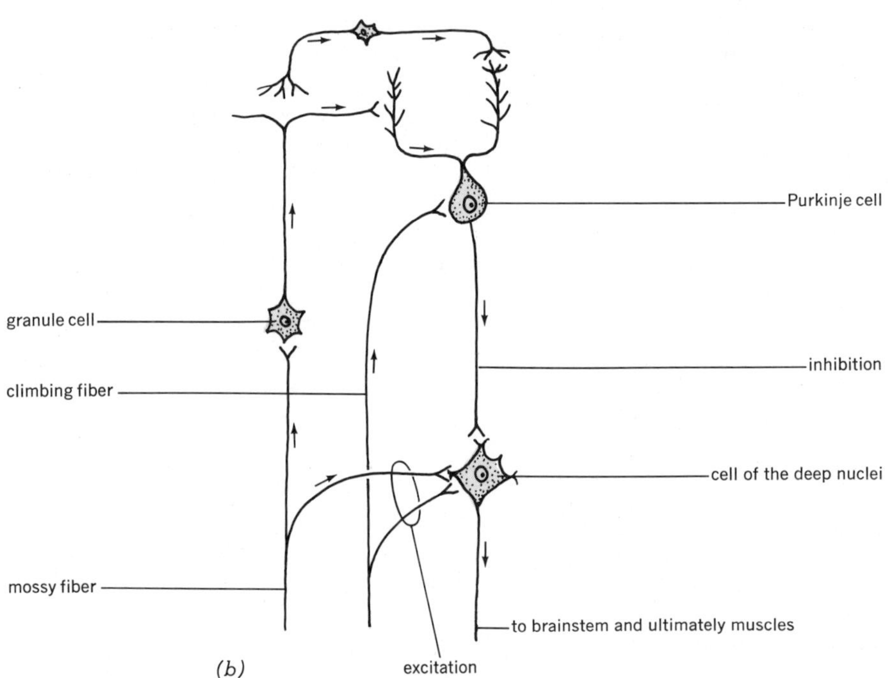

FIGURE 8.3
(*a*) The basic histology of the cerebellum. (*b*) The functional unit of the cerebellum. [(*a*) redrawn from Barr.]

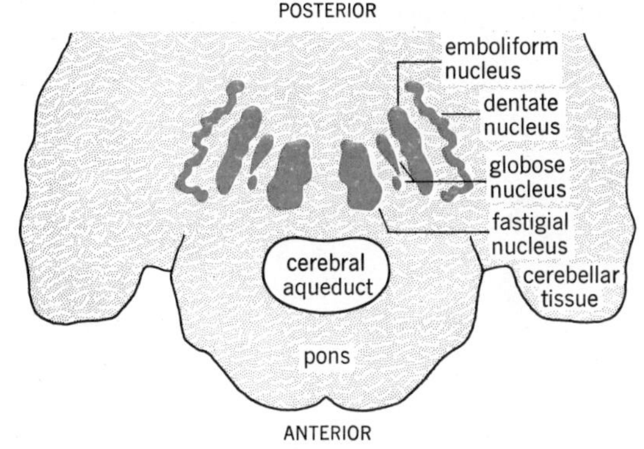

FIGURE 8.4
The deep nuclei of the cerebellum as seen in a cross section of the pons.

monitors and adjusts the motor activities originating elsewhere in the brain. Utilizing information provided by body sense organs, the cerebellum adjusts rate of movements, direction of movement, and force required, to what the motor system intended.

Functions in voluntary movements. Actual movement of the body due to skeletal muscular contraction probably originates in the basal ganglia, cerebral cortex, or as a response to sensory input. As impulses pass from the cortex to skeletal muscles, signals also go to the cerebellum. These signals constitute "intent" of the movement. As the muscles respond to the impulses from the cerebral cortex and a movement begins, information as to that movement's force, speed, and direction is transmitted from muscle, tendon, and skin receptors to the cerebellum. Intent and motion are compared or integrated in the cerebellum, and signals will be sent from there back to basal ganglia and cerebral cortex that correct the movement if required. Such a system provides several things.

Error control. By comparing intent and performance, the cerebellum ensures that a motion goes where it is supposed to, at the proper rate and with the amount of force appropriate to the resistance being overcome. Such action involves initial strong contraction by muscles causing the desired motion, and subsequent contraction of antagonistic muscles to slow or stop the action at the proper time. The second component of this activity is "involuntary" and is a major cerebellar function.

Damping functions. Most body movements are pendular in nature. A movement in one direction must be opposed by a force applied in the opposite direction if the movement is to be slowed or stopped. If too much opposing force is applied, motion in the opposite direction may occur that will have to be opposed by force applied in the original direction. Thus, oscillating movements may occur that are called *tremors.* Cerebellar signals cancel out this potential tremor and stop the motion at the proper point.

Prediction. Using information provided by direction and rate of movement of body parts, and also utilizing visual information, the cerebellum calculates or predicts *when* a motion should be slowed. Such activity obviously requires that the cerebellum provide a time base against which performance may be compared. Lastly, proper progression of the components of a given motion is necessary if the movement is to be directed and purposeful. Prediction of what muscle group must next act to give proper activity is also provided by the cerebellum.

An outline of the pathways for such actions is shown in Figure 8.5.

FIGURE 8.5
A diagrammatic representation of cerebellar error control of voluntary movements.

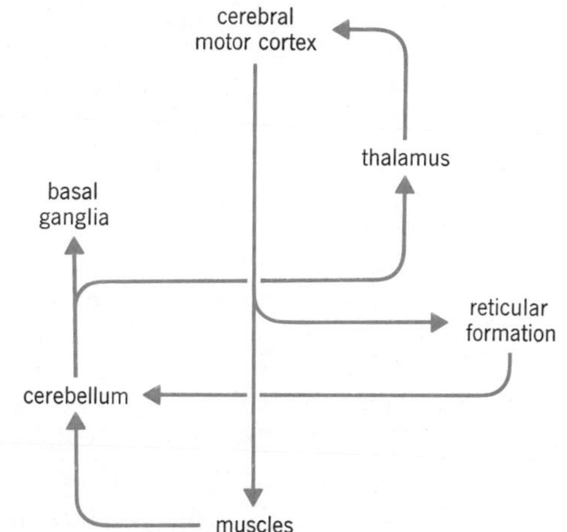

Functions in involuntary movements. These functions are concerned with maintenance of muscle tone, posture, equilibrium, and orientation in the face of forces tending to upset our relationship to the external world. Essentially, the same "services" are provided here by the cerebellum as were provided for voluntary movement; error control, damping, and prediction. This time, the result is involuntary reflex activity that maintains posture, causing us, for example, to "lean into a turn," or go "forwards on acceleration" and "backwards on deceleration." The pathways for such control are different from those for voluntary control and are depicted in Figure 8.6.

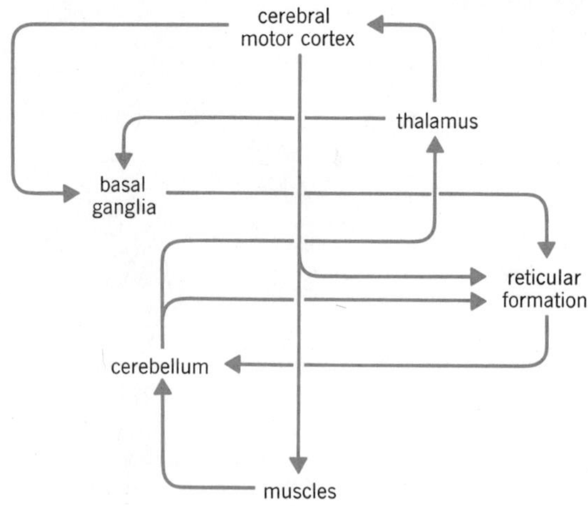

FIGURE 8.6
A diagrammatic representation of cerebellar error control of involuntary movements.

CLINICAL CONSIDERATIONS

Because the structure of the cerebellum is essentially the same in all parts, the severity of the effects of damage in the organ depend primarily on how much tissue is destroyed. The body is represented in the cerebellum as shown in Figure 8.7, and thus *where* in the body the effects of loss will be appreciated depends on where in the cerebellum the damage occurs. The ability of the cerebellum to overcome the effects of small losses of tissue is called COMPENSATION. The larger the amount of tissue destroyed, the more slowly a given movement must be performed for it to remain accurate, as though the "computer" was still functioning, but without as many resources or "memory banks" available.

As should be suspected from the discussion above, signs of cerebellar dysfunction should and do appear in the integration of motion, its accuracy, and in muscle tone and equilibrium. Some signs of cerebellar disorder are presented below.

Asthenia, fatigability, and slowness of movement. Asthenia refers to lack of or loss of muscular strength. In cerebellar disease, muscles on the same side where damage has occurred are weak, tire easily, and have delayed contraction and relaxation times. These effects are believed to be due to involvement of the cerebellar nuclei.

Hypotonia. The muscles feel flabby (like half-filled hot water bottles), offer little resistance to stretching, and reflexes may show oscillating qualities. Such abnormalities indicate that the cerebellum is not responding to information supplied to it from stretch receptors (spindles) in muscles.

Dysmetria. This term refers to "overshoot" or "past-pointing" occurring when a movement is not stopped at its proper point. It indicates failure of the predictive function of the cerebellum.

Ataxia. This term indicates muscular incoordination when muscles are in use. Several more definitive terms are included under the heading of ataxia: *intention tremor* is an oscillating movement occurring at the end of fine movements; *asynergia,* or lack of cooperation between muscle groups for accurate performance of motion; *decomposition of movement,* in which a motion is done in its parts and not smoothly; *dysdiadochokinesia,* in which rapid alternating movements (e.g., tapping a finger) cannot be performed rapidly and smoothly.

Nystagmus. In cerebellar disease, the eyeballs may undergo a tremor when the victim attempts to fix the eyes on an object to the side of the head.

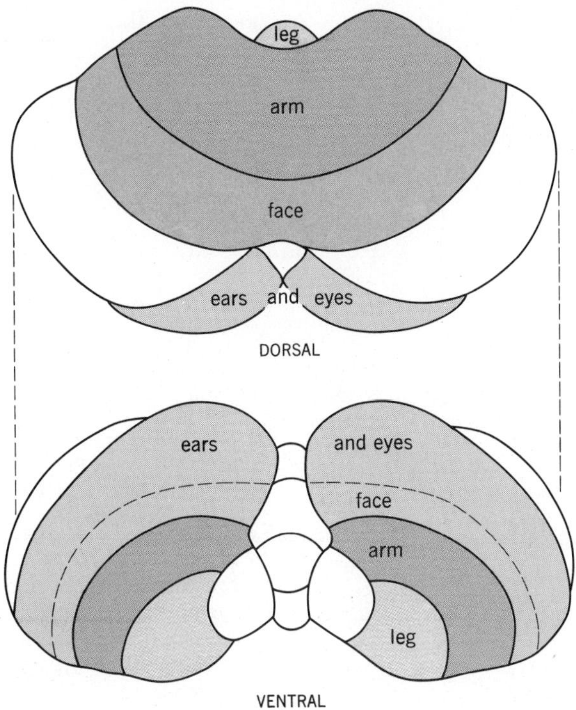

FIGURE 8.7
The representation of the body in the cerebellum. (Compare with Figure 8.2.)

INTERRELATIONSHIPS OF BRAIN STRUCTURES IN THE PRODUCTION AND CONTROL OF MOVEMENT

In this chapter and the preceding one, muscular movement has been discussed. It is perhaps appropriate at this time to look at movement and its control as a whole, at least insofar as brain structures are involved.

There is no evidence to suggest that the motor areas of the cerebral cortex *spontaneously* generate nerve impulses that lead to muscular contraction. They are triggered into activity by impulses reaching them from other body areas. From where do such impulses come? Partly from sensory receptors that supply visual, auditory, equilibrial, and proprioceptive information, partly from subcortical areas such as the brainstem reticular formation, the basal ganglia, and brainstem nuclei. Such mechanisms provide a postural background for fine movements and ensure the smoothness of actions. Every bit of input and output is sooner or later passed to the cerebellum where intent is compared to performance, and the movement is integrated and coordinated as to power required, direction and speed of the movement, and when to stop the motion.

Output of impulses to the muscles leaves the brain over two major pathways. The PYRAMIDAL SYSTEM consists of fibers originating in the various cortical motor areas (40 percent from Area 4, the rest from motor areas other than 4) that convey information leading to precise voluntary movements. The EXTRAPYRAMIDAL SYSTEM conveys fibers from subcortical areas such as reticular formation, basal ganglia, and the cerebellum. These impulses are concerned with maintenance of posture, muscle tone, and automatic reflex adjustments designed to maintain equilibrium. Both "systems" synapse within the spinal cord with LOWER MOTOR NEURONS that pass from cord to muscles. These neurons form then a FINAL COMMON PATH to convey instructions of all types to the muscles themselves.

SUMMARY

1. The cerebellum develops from the hindbrain, and lies on the posterior aspect of the pons and medulla.
2. It has a vermis, two hemispheres, lobes and lobules, and is folded into folia, each of which has a cortex and medulla. Purkinje cells are the characteristic cell of the cortex, and the white matter (medulla) contains central nuclei that send impulses to the brainstem.
3. Input to the cerebellum is from the equilibrial structures of the inner ear, sensory receptors in general, and from cerebral cortex, basal ganglia, thalamus, and reticular formation. Output is to brainstem, from there to skeletal muscles directly or indirectly.
4. The cerebellum operates entirely at the subconscious level to integrate and coordinate information into coordinated and purposeful movement.
   a. Error control is exerted by comparing intent with performance.
   b. Damping of oscillatory movements is exerted.
   c. Prediction of when to stop movements is provided.
   d. Maintenance of posture and equilibrium is provided against acceleratory, deceleratory, and change in direction of motions.
5. Signs of cerebellar damage are associated with slow movements, incoordination of movement, development of tremor, and inability to perform rapid rhythmical movements.
6. The integration of motor activity between cerebral cortex, subcortical structures, and the cerebellum is discussed.
   a. The pyramidal system conveys information from the brain concerned with fine voluntary movements to synapses in the spinal cord.
   b. The extrapyramidal system conveys information from the brain concerned with posture, muscle tone, and equilibrial reflexes, to synapses in the spinal cord.
   c. A lower motor neuron serves as the pathway from spinal cord to muscles to convey instructions for all types of muscular activity.

QUESTIONS

1. What is meant by the "functional unit" of the cerebellum, and what does it do?
2. From where does the cerebellum receive its information? What type of information is provided from each source?
3. What is meant by error control, damping, and prediction? Discuss the role of the cerebellum in each of these processes.
4. Discuss the effects of small and large lesions in the cerebellum. Is or would there be any localization of effects in particular body areas according to where the damage was? Explain.
5. What do the cerebral cortex, basal ganglia, and other subcortical structures, and the cerebellum contribute to muscle activity? Where does it all begin?

READINGS

Barr, Murray L. *The Human Nervous System*. Harper and Row. New York, 1975.

Enna, Salvatore J. "Huntington's Chorea: Changes in Neurotransmitter Receptors in the Brain." *New Eng. J. Med. 294:*1305. June 10, 1976.

Llinás, Rodolfo R. "The Cortex of the Cerebellum." *Sci. Amer. 233:*56. (Jan) 1975.

ELECTRICAL ACTIVITY IN THE BRAIN; WAKEFULNESS AND SLEEP; EMOTIONS, LEARNING, AND MEMORY

OBJECTIVES

After studying this chapter, the reader should be able to:

☐ Define the two basic types of electrical activity occurring in the cerebral cortex.

☐ List and explain the basic electrical wave patterns a normal brain exhibits, and with what type of activity each is associated.

☐ Tell how continuous and evoked activity differ from one another.

☐ Describe epilepsy as a brain disorder, and mention how the EEG changes in the disorder.

☐ Describe some of the mechanisms employed that maintain the waking state of the body.

☐ List and briefly explain some of the attributes of the waking state.

☐ Discuss some of the theories as to how sleep occurs and where "sleep centers" may be located in the brain.

☐ Describe what REM sleep is, and its value to the body.

☐ Explain how coma may develop, and how it differs from the normal nonwaking state.

☐ Define and list the attributes of emotion.

☐ Explain the involvement of the limbic system in emotional expression.

The brain exhibits recordable electrical activity that is the result of millions of action potentials in individual neurons. Such recordings give clues about the status of the brain, and, if abnormal, may lead to localization of an abnormal process in the brain. Electrical activity changes according to the level of consciousness of the individual and according to emotional state. Emotion, learning, and memory are interwoven in that emotion requires previous experience by which to measure need for and degree of expression of emotion.

ELECTRICAL ACTIVITY IN THE BRAIN

Two general types of electrical activity may be recorded from the cerebral cortex by the simple expedient of attaching recording electrodes to the scalp.

A continuous rhythmical activity designated as SPONTANEOUS ACTIVITY, because its origin is not known.

Activity that appears in the cerebrum as a result of sensory stimulation and is known as EVOKED ACTIVITY.

SPONTANEOUS ACTIVITY

The continuous variety of activity exhibited by the cortex is recorded as an ELECTROENCEPHALOGRAM (*EEG*). Several basic rhythms may be distinguished according to frequency, voltage, and area from which the rhythm may be recorded. These are summarized in Table 9.1 and are shown in Figure 9.1.

ALPHA WAVES are recorded with greatest *amplitude* in the parietal and occipital regions, and are thought to originate from a conscious and alert brain, but one in which no active processing of information is occurring. Alpha

TABLE 9.1

CHARACTERISTICS OF NORMAL EEG WAVES

WAVE	FREQUENCY /SEC (Hz)	VOLTAGE (μV)	WHERE AND WHEN RECORDED
Alpha	10–12	50	Parietal and occipital, at rest with eyes closed. Brain is alert but "unoccupied"
Beta	13–25	5–10	Frontal, when brain is stimulated by sensory input or mental activity
Delta	1–5	20–200	Sleep, brain damage
Theta	5–8	10	Temporal-occipital, emotional stress; noxious stimuli

FIGURE 9.1

Normal EEG patterns from different regions of the cortex. Alpha waves predominate in parietal and occipital areas; beta waves in precentral area. Alpha waves are blocked when eyes are opened.

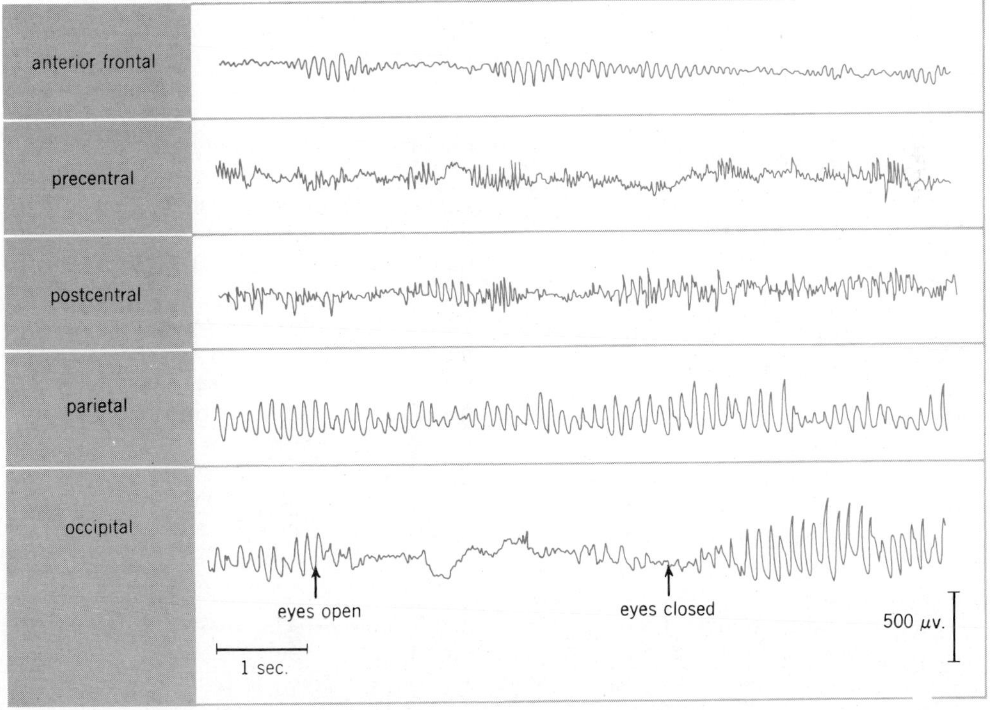

wave frequency is normally 10 to 12 Hz, has a voltage of about 50 μV, and is remarkably constant in both frequency and voltage. Because this type of wave represents relaxation in the brain and body, interest has arisen in the possibility that one can train oneself to voluntarily "bring up" the alpha rhythm to achieve a relaxed state at will. To the extent that one can train oneself to relax the muscles and clear the mind of anxiety, such training will increase the changes of the brain reflecting such a state.

BETA WAVES are recorded best from the precentral regions of the frontal lobes and are associated with sensory stimulation and/or conscious effort associated with vision and mental activity. They have a frequency of 13 to 25 Hz, and a voltage of 5 to 10 μV. Such activity is commonly referred to as a *desynchronized pattern;* that is, a pattern replacing the relaxed alpha rhythm ("desynchronization of the alpha rhythm").

DELTA WAVES and THETA WAVES are "slow rhythms"; delta waves have a frequency of 1 to 5 Hz, and a voltage of 20 to 200 μV, while theta waves have a frequency of 5 to 8 Hz and voltages of about 10 μV. If either wave occurs other than in newborn infants or during severe emotional stress or during sleep, they are to be regarded as signs of disease or injury to the brain.

THE EEG CHANGES ACCORDING TO AGE. Newborn infants show continuous irregular waves from all regions of the cortex. Development of an adult type of pattern occurs by adolescence.

EVOKED ACTIVITY

Potentials that may be recorded from the cortex in response to stimulation of receptors or their afferent pathways constitute EVOKED ACTIVITY. They differ from continuous activity in that they are superimposed on the continuous activity, and bear a definite time relationship to the stimuli that cause their appearance. Furthermore, they typically appear in restricted areas of the brain, those associated with the terminations of the particular sensory pathway involved. Thus, evoked visual potentials are best recorded in the occipital area, those associated with touch in the parietal area, and so forth. Evoked potentials may lead to the accumulation of information in the brain, particularly if the stimulation is repeated. They may also be used to study the distribution of sensory terminations if one is certain that only one particular type of sense organ is being stimulated.

ABNORMAL ELECTRICAL ACTIVITY

In some individuals, spontaneous abnormal electrical discharge may occur in the brain, with the development of convulsive disorders (seizures), alterations in the state of consciousness, and motor and/or sensory disturbances. Such activity may be the result of areas of lowered threshold in the brain or areas of damage (lesions).

Among the conditions that may be associated with the development of *isolated* instances of convulsive disorder are head injuries, high fever, infections (meningitis, rabies), hypoglycemia (low blood sugar levels), excessive alcohol intake, lead and mercury poisoning, cerebral hypoxia (low blood oxygen levels), withdrawal from hypnotics, and tranquilizers.

Epilepsy. *Recurrent patterns* of seizure are termed EPILEPSY. If an organic cause, that is, a lesion, is found to account for the seizures, the epilepsy is termed SYMPTOMATIC EPILEPSY. If no organic cause can be demonstrated, it is termed IDIOPATHIC EPILEPSY. In adults, about 75 percent of convulsive attacks are of the idiopathic variety, and it has been suggested that there may be an hereditary disposition for this variety of epilepsy. Epileptic attacks cause varied symptoms according to where they originate in the brain and the extent of the disturbance. In some cases, an attack that originates in one hemisphere of the cerebrum may involve the same area in the opposite hemisphere by passing impulses over the cerebral commissural fibers, contributing to the spread of the disturbance. In any event, five types of epilepsy are usually distinguished, based primarily on the symptoms that develop.

Grand mal. The most severe type of seizure, grand mal is subdivided into *focal* or *Jacksonian seizures,* and *typical grand mal seizures.*

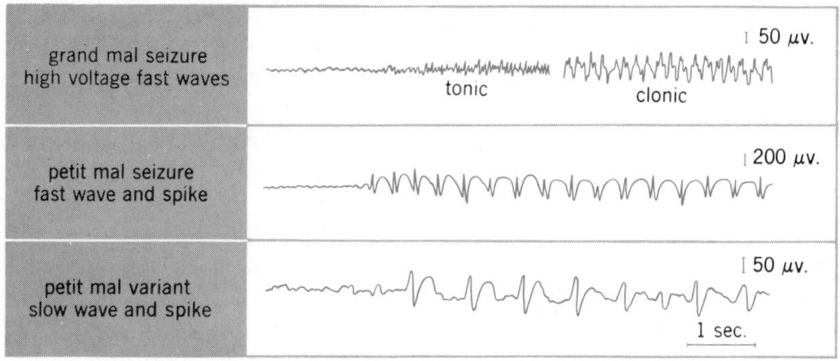

FIGURE 9.2
Some abnormal EEGs. Note large spikes and fast rhythms.

Focal seizures are precipitated by specific events or focal disturbances, such as a sudden noise, a flashing light, hyperventilation, or drug taking. These stimuli apparently cause a lowering of threshold in a particular brain area (a locus), and impulses are discharged that are normally inhibited. Typical grand mal seizures lack a locus, and are of more diffuse origin. Both types are associated with muscular twitching, loss of consciousness, falling, and outcry by the subject.

Petit mal. This type is characterized by lapses of attention, and clouding of consciousness. No muscular symptoms are present. Petit mal is most common in children, and rarely occurs after 20 years of age.

Psychomotor epilepsy (epileptic automatism). In this type, there is loss of contact with the environment for one to two minutes, and the subject performs automated or stereotyped movements, such as chewing, swallowing, and twitching of arms and legs. The attack is often preceded by an unpleasant olfactory hallucination. This type is often associated with temporal lobe damage.

Infantile spasms. In this type of epilepsy, there is arm and trunk flexion and leg extension. Attacks last only a few seconds, but may be repeated several times a day. This type usually occurs before three years of age.

Epileptic equivalents. This type is associated with periods of mental confusion lasting for several hours.

Epilepsy must be considered as a disease and, as such, may be treated with drugs to control the seizures, or, in the case of a demonstrated lesion, surgery may be effective.

Many of the epileptic states produce characteristic alterations in the EEG pattern. Figure 9.2 shows the large spikes and fast rhythms characteristic of several states.

WAKEFULNESS AND SLEEP

WAKEFULNESS

To a large degree, it would appear that the state of alertness or nonalertness of the organism depends on cerebral cortical activation. The reticular activating system was described in Chapter 7, and serves as one mechanism for maintaining the waking state. A thalamocortical pathway receiving input mainly from the skin also appears to be present, and has led to the suggestion that wakefulness depends on continual afferent impulses.

Regardless of what the activating pathways are, consciousness (the waking state) has several attributes.

It is a neural phenomenon. Organisms without

nervous systems do not exhibit the observable aspects of consciousness such as:

Ability to demonstrate and shift attention.

Ability to think abstractly, both in the use of tools and in social behavior.

Self-awareness, and awareness of other organisms.

Creation and use of ethical and esthetic values.

Perception and *memory,* involved in judgment of future action.

It is variable from time to time in the individual, and is different in different individuals.

All of these attributes indicate a state of brain function where thought and activity are proceeding. Most organisms exhibit a cyclical pattern in their conscious state, with alternating periods of wakefulness and unconsciousness, the latter often termed sleep.

SLEEP

Sleep is an interruption of consciousness and the mental processes associated with the waking state. It may provide one of the most essential needs of higher organisms. Length of sleep is an individual matter, varying from five or six hours for some adults to the "all-day" sleep some infants exhibit. An "internal clock" or CIRCADIAN RHYTHM (*see* also Chapter 27) appears to be involved in determining time and habits of sleeping. What sets this "clock" is still a matter of debate, but appears to involve light–dark cycles that may affect the pineal gland or hypothalamus-thalamus complex.

Theories of sleep. Onset of sleep may be a passive process in that withdrawal of the activity of the reticular activating system and a decrease of sensory input are the major events in genesis of sleep. Thus, loose-fitting clothing, an environment devoid of turmoil and noise, and activities designed to relax the muscles and brain all contribute to a sense of drowsiness.

On the other hand, there appears to be increasing evidence that sleep may be an active process, one associated with the production of a "hypnotoxin" or sleep chemical that "puts us to sleep." Serotonin, gamma-aminobutyric acid (GABA), and dihydroxy-phenylalanine (DOPA) have been suggested as possible hypnotoxins. More recently, a polypeptide with a molecular weight between 350 to 500 has been isolated from the cerebrospinal fluid of goats whose ventromedial thalamic nucleus had been stimulated, and which, when infused into wakeful animals, puts them into a state indistinguishable from true sleep.

If there is a "sleep center" in the brain that responds to decreased sensory input or a chemical, it would appear to lie in one of three primary areas, as determined by experiments involving sectioning, cooling, or stimulation.

In the pons, where stimulation produces alpha synchronization and drowsiness, while cooling produces arousal

In the thalamus, where stimulation produces sleep

In the area between hypothalamus and the frontal lobes, where sectioning results in insomnia, presumably due to interruption of a sleep inducing pathway.

Stages of sleep. As sleep approaches, several physiological parameters undergo alteration. Body temperature declines, and adrenal steroid production decreases (true for those who normally sleep during the dark and remain awake during the day; this may be reversed by 12 hours in those whose activity is opposite to this). As sleep progresses, there are four stages through which one passes.

STAGE 1 sleep is signalled by a slowing and decrease in amplitude of the EEG alpha rhythm. There is a sensation of drifting or floating, and, if awakened during this stage, humans will usually assert they were not asleep, but merely had closed their eyes.

STAGE 2 sleep is characterized by the appearance of "sleep spindles" in the EEG record. These are bursts of 14 to 15 Hz waves lasting several seconds. During this stage, the sleeper may be easily awakened, and will not describe dreaming as having occurred.

STAGE 3 sleep is associated with the appearance of delta waves in the EEG. Breathing is slow and even, and pulse rate is about 60 beats per minute; temperature and blood pressure continue to decline. It is a stage of intermediate depth sleep.

STAGE 4 sleep is deep or oblivious sleep. External stimuli, such as clicks, may cause evoked activity in the brain, so the brain is still receptive. The sleeper, however, is not aroused by such stimuli.

A sleeper appears to descend into Stage 4 sleep, and then "emerges" into a modified Stage 1 sleep. Such emergence occurs throughout the entire sleep cycle in intervals that are 80 to 120 minutes long. During the emergent phase, the eyes may be seen to undergo rapid movements (REM, rapid eye movement), and the subject may make sounds and show muscular movement. REM sleep is associated with dreaming.

Deprivation of sleep is associated with the development of bizarre behavior and temporary neurosis and/or psychosis. After four to five days of sleep deprivation, there is a decrease of energy mobilization by the body. ATP stores are decreased, there are elevated levels of adrenal cortical "stress hormones" in the bloodstream, and a chemical similar to LSD-25. Memory fails, normal mental agility is impaired, and attention span is severely limited. Hallucinations are common, perhaps brought on by the LSD-like chemical mentioned above. Hallucination and reality may merge into a continuum from which the only release is sleep. Release is accomplished by sleeping, and the subject may sleep 12 to 14 or more hours. During sleep, time spent in REM activity is increased. Thus, REM sleep may be essential for both release of psychological tension and for "exercising" the brain. Up to 10 days may be required for body functions to return to normal, and sleep may be concerned with the prevention of such disorders as noted above.

Coma. Coma is defined as a state of unconsciousness from which even the most powerful external stimuli cannot arouse the subject. It seems to require suppression of neuronal function such as might occur by compression (tumors, cerebrovascular accident or *CVA* with clot formation), deprivation of nutrients ("stroke," rupture of vessels), poisoning (hepatic coma, uremia), hyperglycemia (diabetic coma), and infections (meningitis).

Emergence from a coma depends on removal of the cause (if possible), and on the nature and extent of damage that has occurred to the involved neurons. If cell destruction has occurred, chances for recovery are less than if the cells have been merely compressed. The brain area most commonly involved in production of coma is the brainstem between the anterior end of the third ventricle and the medulla, an area that contains sleep centers, and the reticular activating system. Thus the cortex cannot be aroused. Cerebral injuries do not in and of themselves necessarily produce coma; there must be brainstem involvement as a result of the action of the causative agent.

All things considered, preservation of brain function becomes the overriding concern in treatment of coma or other disorders. The human brain is an organ that makes enjoyment of life possible, and any procedure that "lets it wait" while another problem is solved is not worth the effort.

EMOTIONS

An emotion may be defined as a mental state or strong feeling that is usually accompanied by physical changes in the body such as heart rate, vasomotor reactions, breathing, and muscle tone. According to the particular emotion, positive or negative drives may be provided that lead to the satisfaction of needs, or to the avoidance of noxious stimuli.

EMOTIONAL EXPRESSION

The limbic system; aspects of emotion. Emotional *expression* appears to be associated with the structures of the LIMBIC SYSTEM (Fig. 9.3). Included as parts of this system are the amygdaloid nuclei, the hypothalamus (e.g., mammillary bodies), fornix, and several subcortical nuclei (e.g., habenula). Stimulation of the amygdala produces aggressive responses, while their destruction causes violent animals to become passive, to exhibit no fear of objects that formerly produced fear, and to exhibit hypersexuality. Hypothalamic integrity is required for outward expression of emotion, particularly

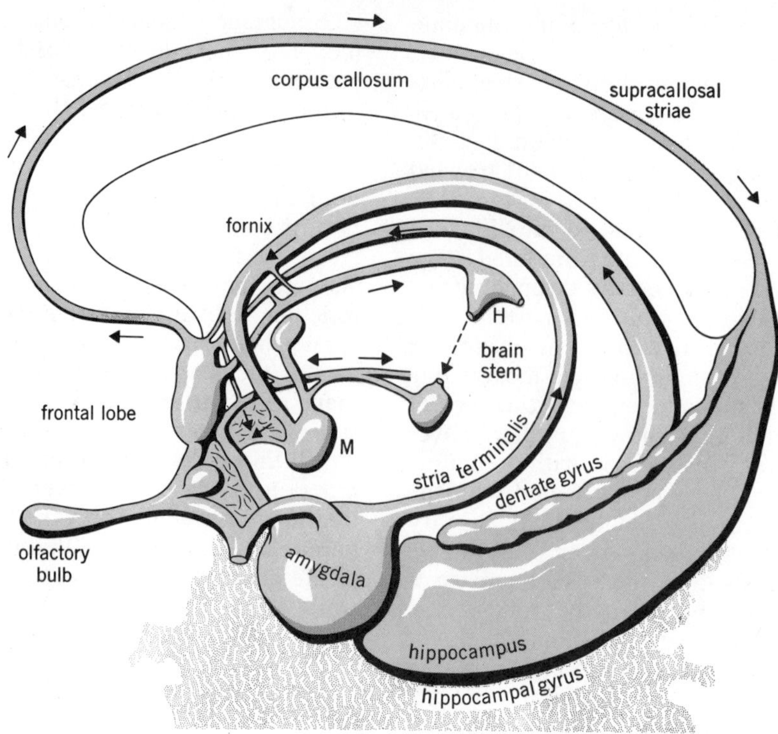

FIGURE 9.3
The structures of the limbic system. M = mammillary body; H = habenula.

the visceral components. *Directed* emotional expression, such as attack on a specific object, requires the cerebral cortex. Also included as part of the limbic system are areas that, when stimulated, produce feelings of pleasure and positive drives.

Emotion itself has four aspects.

COGNITION, or perception and evaluation of a particular situation before a reaction is made.

EXPRESSION, or external sign of emotion.

EXPERIENCE, or the subjective evaluation of an incident as compared to previous incidents.

EXCITEMENT, or degree of vividness of the emotion.

These aspects may be seen to require both cortical and infracortical regions. Experience would seem to require memory, and excitement results from reticular activating activity.

Of interest is the development of emotions. Although there is no definite agreement as to whether a newborn exhibits emotions or not, several general statements may be made.

A newborn appears to have a comfort–discomfort reaction. It may cry if something makes it hurt, cold, or lonely. It may turn up the corners of the mouth (a smile?) in response to internal stimuli (gas, a need for defecation). By age two to eight weeks, it may exhibit a true smile in response to comfort, satisfaction of needs, or the appearance of a parent in the field of view, and a rudimentary form of love develops.

If there is a hereditary factor in emotions, it appears to lie in the area of setting basic neural pathways that can be "filled" with experience as life progresses. The infant learns very rapidly that certain activities on its part will cause reactions by its caretakers, and so positive or negative reinforcement of behavior and emotional expression occurs. By two years, the child has probably acquired a full set of emotional expressions. It may have shown frustration at the time of weaning; it may have developed guilt about toilet training and curiosity as new items appear in its world; it may show anxiety as anticipation of danger. Fear does not seem to be inherent in the child, but is learned.

CLINICAL CONSIDERATIONS

Emotional disturbance is too broad a field to discuss in a textbook of this sort; nevertheless, some mention of what it is needs to be made. The reader is referred to the readings section for references to emotional disorder.

As disease-causing agents, there is considerable attention paid to psychological causes of malfunction. The term PSYCHOSOMATIC MEDICINE stresses a broad approach to illness from the psychological, organic, and social viewpoints. Basically, mental health depends on the individual's abilities to cope with living in general, and the many stresses associated with life. Inability to successfully cope results in a change in one's own self image or as viewed by another, and may lead to gastrointestinal, musculoskeletal, skin, cardiovascular, respiratory, and nervous system reactions.

Although all individuals have different personalities, behavior and method of emotional expression must remain within certain limits as set by peers, society, and other factors. The term NEUROSIS implies maladjustment to life that interferes with normal functioning, but in which contact with reality is still maintained. PSYCHOSIS results in loss of contact with reality, and usually is associated with behavior that results in society's viewing the victim in unfavorable terms.

MENTAL RETARDATION generally implies failure to advance intellectually or in abilities as measured against one's peers. Neurosis and psychosis cannot usually be related to organic causes, except when toxic reactions or poisons may be suspected. Retardation may be inherited as a metabolic disorder that allows accumulation of toxic chemicals, or as a result of birth injury. All in all, it requires a trained individual to evaluate causes of emotional disturbance, and self-treatment, beyond admitting the existence of disorder, should be discouraged. Similarly, treatment personnel should not assume that a tranquilizer is a panacea for emotional disturbance.

LEARNING AND MEMORY

Learning and memory are perhaps inseparable qualities of the nervous system, for one usually must have a base for comparison (provided by memory) against which to measure whether some new bit of information shall be permanently incorporated into the nervous system.

Learning may be produced through conditioning, as illustrated by Pavlov's experiments with dogs, or by reinforcing either negatively or positively a particular activity such as bar pressing to obtain food.

There appear to be several areas of the nervous system that are involved in learning. The spinal cord may be involved in the learning of simple conditioned responses such as those involving withdrawal reactions to painful stimuli; the cortex appears to speed the rate of learning; the reticular formation may be "trained" to arouse the cortex since less stimulation is required to produce the same evoked cortical potential as learning proceeds. In short, there appears to be no *specific area* that is involved to the exclusion of others in the learning process. Learning may proceed through the establishing of "preferred pathways." If a particular circuit is used again and again, enlargement of synaptic endings, increase in numbers of vesicles, and increase in number of endings occur. Thus, the next time the same stimulus occurs, it tends to follow a given path and learning has taken place.

Memory is another problem. Two questions come to mind immediately: *where* is the information stored, and *how* is it stored?

The question of where is not completely answerable today. Stimulation of points within the temporal lobes evokes recall of memories of long past experiences. The memories themselves are probably not stored here, the stimulation merely activates pathways that cause retrieval from brain stem or frontal and parietal storage areas. The hippocampal area appears to be necessary in the storage of information as memories for recent events since destruction of that area impairs recent memory.

The process of memory formation appears to require the involvement of three mechanisms: one for retention momentarily (remembering a phone number only long enough to dial it); memory for events occurring minutes to hours before (short-term memory); memory for events occurring in the past (long-term memory). Both the momentary and short-term memories appear to depend on some electrical phenomenon such as impulses circling around neuronal chains in the cerebral association areas, because they may be abolished by electroshock. The latter procedure imposes an external electrical field on the entire brain and causes all neurons to simultaneously assume a refractory state. This is thought to result in blockage of electrical impulses circling the neuronal chains. After some four hours beyond the experience, electroshock does not affect the memory. Between about 10 minutes and four hours, administration of an antibiotic, puromycin (which is also an inhibitor of protein synthesis), prevents the establishment of long-term memory. It prevents "consolidation of the memory trace" or the creation of a permanent record of the memory as a chemical of some sort. It is tempting then to suggest that long-term memory involves the production of a protein. Protein synthesis requires one or more forms of RNA. A theory has thus evolved suggesting that consolidation of memory traces results from production of specific RNA "templates" that cause production of specific proteins for each bit of information stored. Retrieval of the memory is associated with destruction of the protein but not of its template so that, once recalled, the memory is not lost forever, since more protein may be synthesized using the RNA template. Supporting this theory are investigations that show increases of RNA in the brain as learning and experience increases, and the fact that RNAase inhibits learning (memory?) in certain worms. Nevertheless, we are a long way from a "smart pill." Some drugs (*see* Appendix) however, do increase rate of learning. For example, stimulants such as caffeine, physostigmine, amphetamines, nicotine, and strychnine improve learning by speeding consolidation of the trace. Pemoline (Cylert), a mild stimulant, is also a drug increasing RNA synthesis, and it too speeds learning.

SUMMARY

1. Electrical activity in the brain is the result of "spontaneous" activity or evoked activity.
   a. Electrical activity may be recorded as an EEG (electroencephalogram).
   b. Continuous or spontaneous electrical activity is seen as the basic alpha, beta, delta, and theta waves.
   c. Evoked activity is the result of specific sensory input to the brain and is reflected as changes in the EEG.
   d. Abnormal electrical activity is associated with convulsive disorders such as epilepsy.
2. The waking state of the organism is maintained by input from sensory receptors acting through the reticular formation of the brain stem.
3. Sleep is an interruption of the waking state.
   a. Sleep occurs usually once in a 24-hour period and is associated with depression of body temperature, heart rate, blood pressure, respiration, and metabolism.
   b. Sleep occurs in four stages, in cycles of 80 to 120 minutes. Emergence to REM sleep occurs after Stage 4 sleep and is associated with dreaming. Dreaming appears to serve psychological and physiological needs.
   c. Sleep deprivation is associated with disorders of personality and behavior. Chemicals produced during deprivation may account for these symptoms.

4. Coma is a state of unconsciousness from which the subject cannot be aroused.
   a. Coma is usually associated with damage to brain cells in the brainstem, or deprivation of glucose and oxygen.
   b. Recovery from coma depends on extent and duration of damage.
5. Emotion has four aspects and appears to be learned.
   a. Cognition or perception of a situation.
   b. Expression or external sign.
   c. Experience or evaluation.
   d. Excitement or vividness of the emotion.
6. The limbic system involves the amygdaloid nuclei, hypothalamus, and several other nuclei and connecting tracts.
   a. The amygdaloid nuclei are concerned with aggression.
   b. The hypothalamus is concerned with outward expression of emotion.
   c. The hypothalamus is also the site of areas concerned with motivation.
7. Emotional disturbances reflect inability to cope successfully with life, and may be associated with the development of organic symptoms.
   a. Psychosomatic medicine refers to a total approach to illness from psychological, organic, and social viewpoints.
   b. A neurosis implies maladjustment but maintaining contact with reality.
   c. A psychosis implies loss of contact with reality.
   d. Mental retardation most commonly has an organic cause.
8. Certain drugs accelerate learning, perhaps by stimulating RNA and/or protein synthesis.
9. Learning is the ability to acquire a bit of knowledge or a skill. It may be retained, in part, as a memory.
   a. Short-term memory appears to be served by electrical phenomena.
   b. Long-term memory appears to require the production of a chemical, possibly a protein, mediated by RNA.

QUESTIONS

1. What are the dominant rhythms of the EEG? With what state of the organism is each correlated?
2. What, if any, is the common denominator of epileptic seizure? What justifies inclusion of the disorder as a disease?
3. How is the waking state of the organism maintained? What allows sleep to ensue?
4. How is sleep produced? What characterizes each stage?
5. How can sleep deprivation affect the organism?
6. What is coma? What part(s) of the brain may be involved in production of coma? Can coma be produced by a cause other than injury to the brain? Explain.
7. What parts of the brain are involved in emotional expression, and what does each contribute to the process?
8. How, and in what parts of the nervous system, can learning occur?
9. What "types" of memory are there? What are some theories to explain each type?

READINGS

Allison, Truett, and Domenic V. Cicchetti. "Sleep in Mammals: Ecological and Constitutional Correlates." *Science 194:*732. 12 Nov 1976.

Dement, William C. *Some Must Watch While Some Must Sleep*. Freeman. San Francisco, 1974.

Hart, Leslie A. *How the Brain Works: A New Understanding of Human Learning, Emotion, and Thinking*. Basic Books. New York, 1975.

Marx, Jean L. "Learning and Behavior: Effects of Pituitary Hormones." *Science 190:*367. 24 Oct 1975.

Meyer, David E., and Roger W. Schvaneveldt. "Meaning, Memory Structure, and Mental Processes." *Science 192:*27. 2 April 1976.

Pappenheimer, John R. "The Sleep Factor." *Sci. Amer. 235:*24. (Aug) 1976.

Wallace, Patricia. "Neurochemistry: Unraveling the Mechanism of Memory." *Science 190:*1076. 12 Dec 1975.

Witkin, Herman A., et al. "Criminality in XYY and XXY Men." *Science 193:*547. 13 Aug 1976.

BLOOD SUPPLY OF THE BRAIN AND SPINAL CORD; THE VENTRICLES AND CEREBROSPINAL FLUID

OBJECTIVES

VESSELS OF THE CENTRAL NERVOUS SYSTEM

SUPPLY TO THE BRAIN

SUPPLY TO THE SPINAL CORD

THE BLOOD–BRAIN BARRIER

VASCULAR DISORDERS OF THE CENTRAL NERVOUS SYSTEM

CLASSIFICATION OF VASCULAR DISORDER

SYMPTOMS OF VASCULAR DISORDER

THE VENTRICLES AND MENINGES OF THE BRAIN AND SPINAL CORD

CEREBROSPINAL FLUID

FORMATION, FLOW, AND COMPOSITION

FUNCTIONS

CLINICAL CONSIDERATIONS

SUMMARY

QUESTIONS

READINGS

OBJECTIVES

After studying this chapter, the reader should be able to:

☐ Describe the pattern of blood flow to and from the brain and spinal cord.

☐ Describe the importance of the circle of Willis to brain circulation.

☐ Describe the anatomical basis of the blood-brain barrier, what it is, and its physiological importance.

☐ Give some causes of vascular disorder of the CNS.

☐ Describe those symptoms that may occur with vascular disorders of the CNS.

☐ Describe the organization of the cavities of the brain and cord, and their relationships to the meninges.

☐ Describe the formation, flow, and composition of cerebrospinal fluid.

☐ Give functions served by cerebrospinal fluid.

☐ Account for the development of hydrocephalus.

Vascular lesions, including rupture, narrowing, and occulsion, account for more neurological disorders than any other category of pathological processes. Permeability in the cerebral circulation is different from that elsewhere in the body. Vascular flow is also essential for the production of the cerebrospinal fluid, a fluid classically regarded as cushioning the brain, but which has been recently assigned additional functions.

Though the physiology of the spinal cord has not yet been described, its blood supply is discussed in connection with that of the brain.

VESSELS OF THE CENTRAL NERVOUS SYSTEM

SUPPLY TO THE BRAIN

The paired INTERNAL CAROTID ARTERIES, that arise from the common carotid arteries of the neck, and the paired VERTEBRAL ARTERIES that arise from the subclavian arteries (beneath the clavicles) form the arterial supply to the brain (Fig. 10.1). The carotids give rise to the anterior and middle cerebral arteries and to the posterior communicating arteries. The vertebrals fuse to form the basilar artery that lies on the anterior aspect of the brainstem, and from which vessels supply the stem, cerebellum, the posterior cerebrum, and other areas. The two arterial supplies communicate with one another at the CIRCLE OF WILLIS, composed of carotids, anterior cerebral arteries, the anterior communicating artery, the posterior communicating arteries, and the proximal portions of the posterior cerebral arteries. Although there is normally little exchange of blood through the small communicating vessels, the circle provides for maintenance of blood flow when one of the major vessels is occluded. Capillary beds to the various parts of the brain are formed from these vessels, and from these the venous drainage arises. The DURAL SINUSES (Fig. 10.2) form the largest vessels draining the brain, and they ultimately empty into the internal jugular vein that carries blood eventually to the superior vena cava.

SUPPLY TO THE SPINAL CORD

A single ANTERIOR SPINAL ARTERY arises in a Y-shaped manner (*see* Fig. 10.1) from the vertebral arteries, and paired POSTERIOR SPINAL

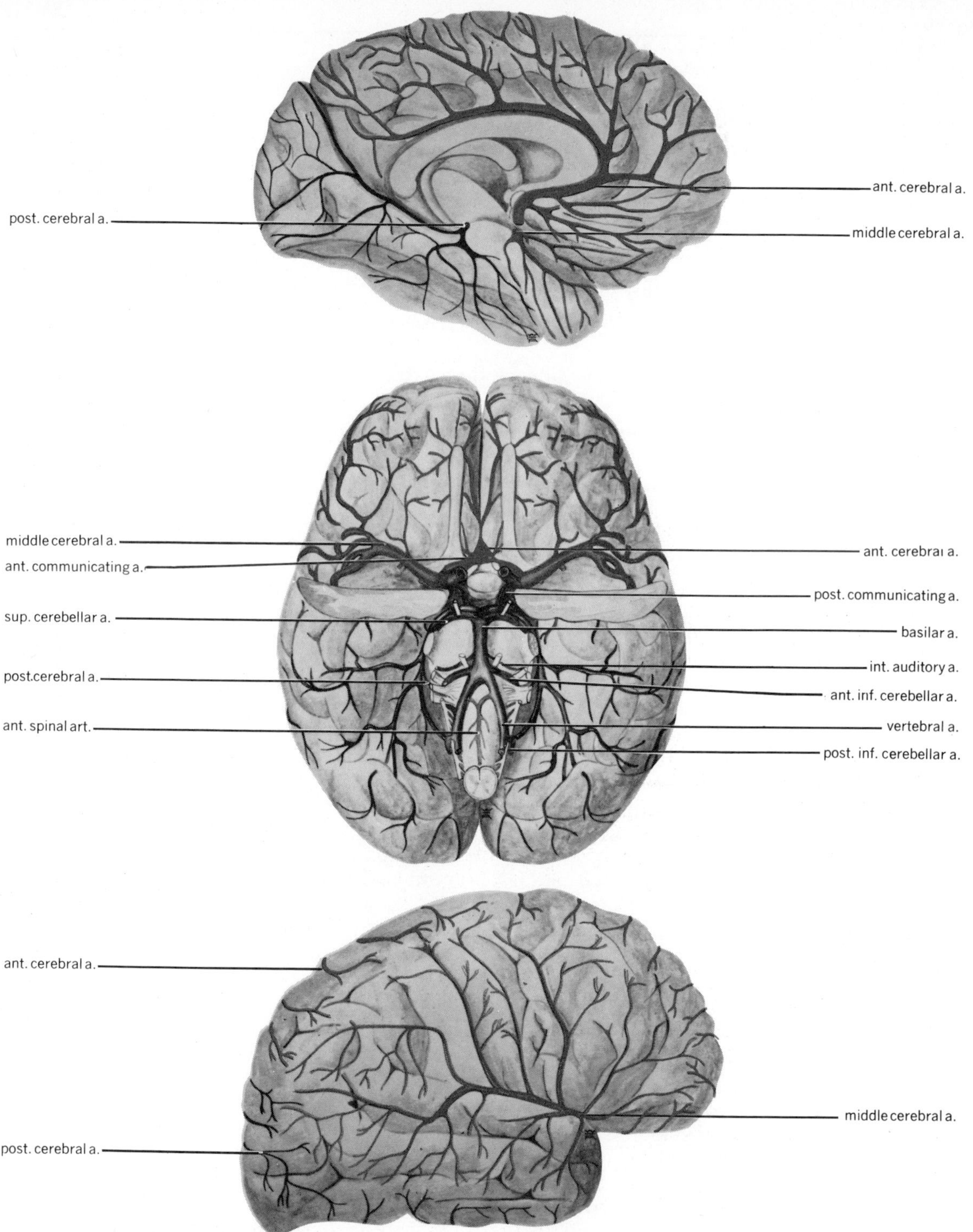

FIGURE 10.1
The blood vessels of the brain.

ARTERIES arise separately from the vertebrals or posterior cerebellar arteries. These three vessels run the length of the cord. Additional vessels are supplied from the thoracic and abdominal aorta and are called SPINAL ARTERIES. Major venous drainage of the cord accompanies the arterial supply and is supplemented by smaller vessels that empty into veins lying alongside the vertebrae.

THE BLOOD–BRAIN BARRIER

The passage of a variety of substances from the cerebral capillaries of mature brains into the cellular portion of the cerebrum is either prevented or is slowed as compared to passage through capillary beds elsewhere in the body. Examples of such substances include trypan blue (a dye), protein molecules, lipids, and most

FIGURE 10.2
The dural sinuses and their connected vessels.

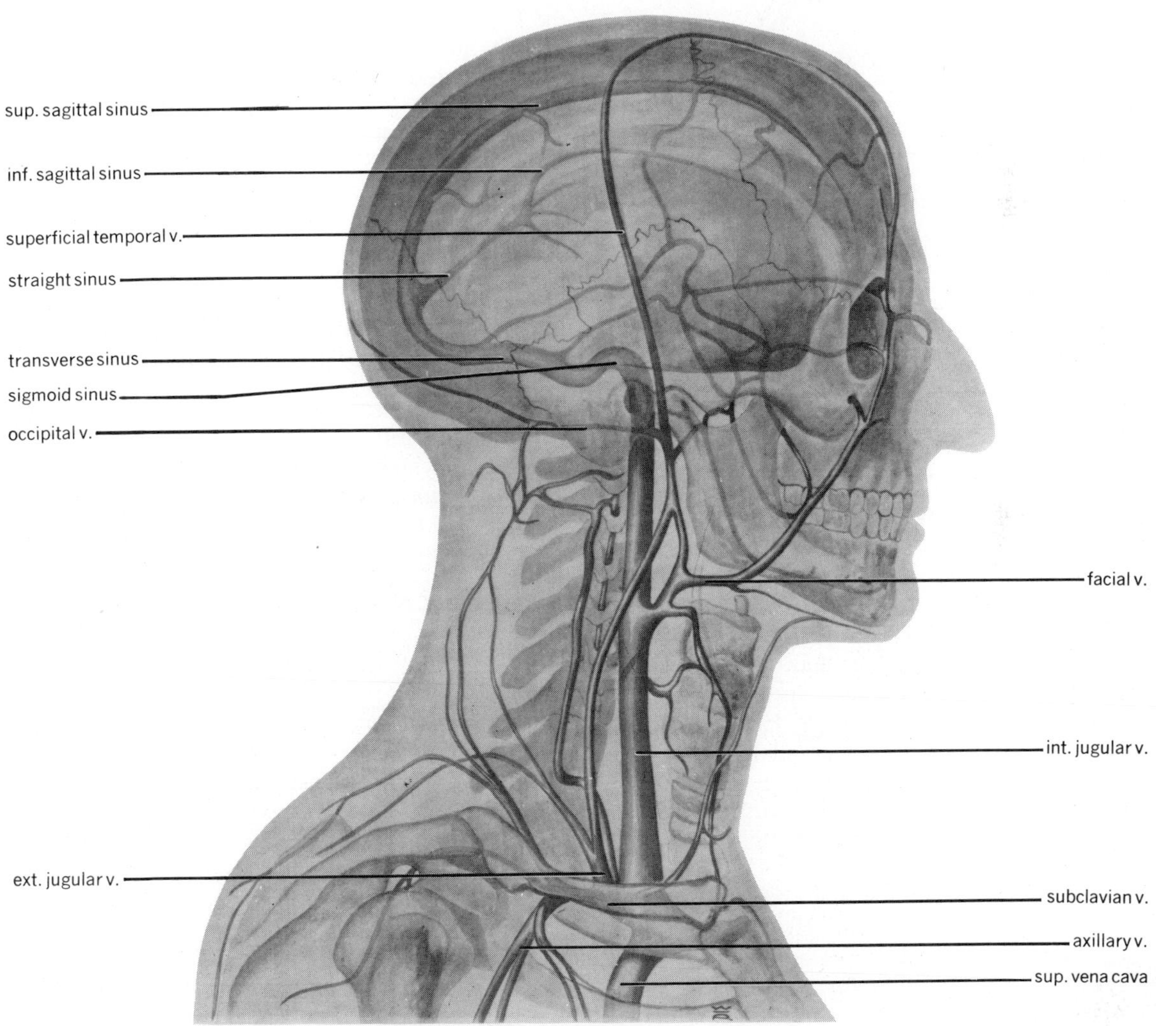

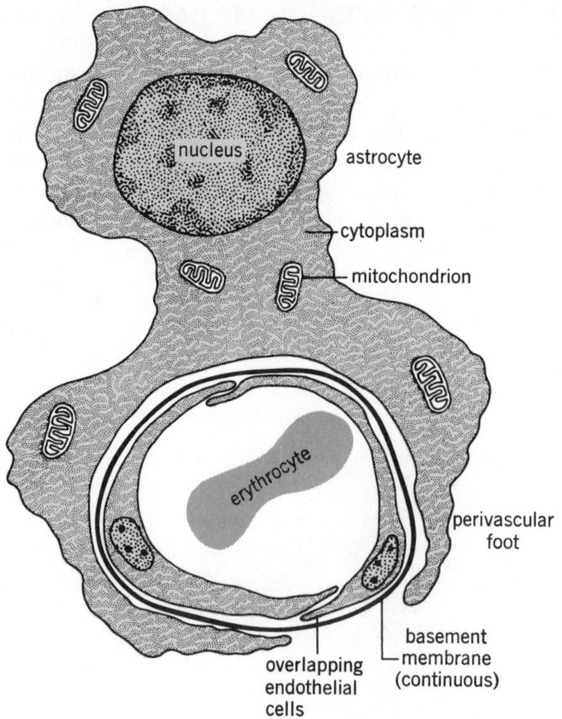

FIGURE 10.3
The anatomy of the blood brain barrier.

antibiotics, whose passage is blocked, and creatinine, urea, chloride ion, inulin (a polysaccharide) and sucrose, all of whose passage is restricted. Glucose, oxygen, and potassium and sodium pass very rapidly through cerebral capillary walls. These facts have led to the hypothesis that a BLOOD–BRAIN BARRIER exists that is responsible for the restriction of passage of certain substances. Electron microscope studies of cerebral capillaries show the anatomical features presented in Figure 10.3. Several differences are evident in the structure of the cerebral capillaries as compared to other capillaries.

Capillary endothelial cells either lack the "pores" found in other capillaries, or the pores are fewer and smaller.

Lining cells (*endothelial cells*) overlap and are sealed by tight junctions (fusion of cell membranes) that prevent passage of materials through spaces between the cells.

A continuous basement membrane surrounds each capillary and, outside of this, processes of glial elements (chiefly astrocytes) cover about 85 percent of the capillary surface. These features combine to produce a series of membranous barriers to the passage of solutes. Curiously, the barrier appears to be lacking in immature brains, and the immature brain accepts nutrients not available later. There is no particular age at which the barrier may be said to develop, because determination of the presence of the barrier depends on what substance is used to demonstrate it. Suffice it to say that it develops in fetal life, possibly as blood vessels invade the brain (at about 8 to 10 weeks of intrauterine life). The barrier, once developed, may protect the cerebral neurons from potentially harmful substances; at least it preserves the environment of the neurons in a state of constancy that is necessary for their normal function.

VASCULAR DISORDERS OF THE CENTRAL NERVOUS SYSTEM

No attempt is made to present all possible disorders of blood vessels involving the central nervous system. General categories by cause are described with common symptomatology indicated to give some acquaintance with symptoms of central vascular disorder.

CLASSIFICATION OF VASCULAR DISORDER

There are three major categories of vascular disorders, based on cause.

NARROWING or OCCLUSIVE DISORDERS are characterized by diminution of vessel diameter as in: the deposition of lipids in vessel walls (*atherosclerosis*); hardening of the artery walls (*arteriosclerosis*) with inability to enlarge to permit greater blood flow; infections (e.g., syphilis) that thicken vessel walls and reduce diameter. (Flow is proportional to the fourth power of the radius of the tube, so small alterations in size result in large alterations of blood flow).

EMBOLIC DISEASE, in which there is blockage of vessels by floating materials such as fats, blood clots, or air bubbles. Sudden decompression (the "bends") or entry of air into the circulation after surgery may create an air embolism. Clots that lodge in the cerebral vessel after left heart surgery may block vessels (clots originating in veins are typically trapped in the lungs, not the brain). Atherosclerotic fatty plaques may release from vessel walls and become floating hazards.

WEAKENING of vessel walls may lead to dilation (aneurysm) and possible vessel rupture. Weakening may be secondary to congenital defects such as failure to develop muscular or fibrous coats in the vessel wall, to infectious processes (syphilis), toxic processes (alcohol, arsenic poisoning), or hypertension (high blood pressure).

Although the results of such processes are most obvious when they involve the brain, damage to the cord arteries can also occur.

SYMPTOMS OF VASCULAR DISORDER

Symptoms are produced by gradual or sudden reduction of blood flow to nervous structures. In the brain, symptoms of *any* vascular disorder are essentially the same.

HEADACHE occurs, which is vague and transient.

DIZZINESS and FAINTNESS occur, which is exaggerated by postural changes.

NAUSEA and VOMITING are common.

UNEQUAL REFLEXES, such as pupillary response to light, or myotatic reflexes, are common.

MUSCULAR WEAKNESS, usually unilateral (one-sided) is common in lesions involving the cerebral motor areas or cerebellum.

MENTAL CONFUSION, evidenced by aphasia, speaking deficits, and inability to focus attention.

Involvement of the spinal cord may show up as exaggerated and/or unequal muscular reflexes, and disturbances of motion and sensation.

It must be emphasized that the foregoing list of symptoms is not complete and will not lead to specific diagnosis of a specific condition. It is intended only to indicate those symptoms that should lead to the suspicion that a CVA (cerebrovascular accident) has occurred. "STROKE" is the term used in the total picture of a CVA, regardless of type.

THE VENTRICLES AND MENINGES OF THE BRAIN AND SPINAL CORD

The VENTRICLES or cavities within the brain, and the CENTRAL CANAL of the spinal cord, are structures arising from the cavities developed as the neural tube formed. Also, as the CNS develops, it becomes surrounded by three membranes collectively known as the MENINGES. An outer, tough *dura mater* ("hard mother") serves as the major protective membrane. A middle *arachnoid* (like a spider web) is a delicate membrane enclosing a *subarachnoid space* that is filled with *cerebrospinal fluid*. An inner *pia mater* ("tender mother") is a thin vascular membrane carrying blood vessels into brain and cord.

Each cerebral hemisphere contains one LATERAL VENTRICLE that communicates through an INTERVENTRICULAR FORAMEN with the THIRD VENTRICLE that lies in the diencephalon. The CEREBRAL AQUEDUCT (of Sylvius) is a small channel running through the midbrain and pons to connect the third ventricle with the FOURTH VENTRICLE of the medulla (Fig. 10.4). Three openings are found in the roof of the fourth ventricle: two lateral *foramina of Luschka,* and a single midline *foramen of Magendie*. These foramina communicate with the subarachnoid space of the meninges around cord and brain. Vascular derivatives of the pia mater, known as CHOROID PLEXUS, are located in the lateral, third, and fourth ventricles, and these are the major sites of formation of the cerebrospinal fluid. The central canal of the spinal cord does not participate in the circulation of the fluid.

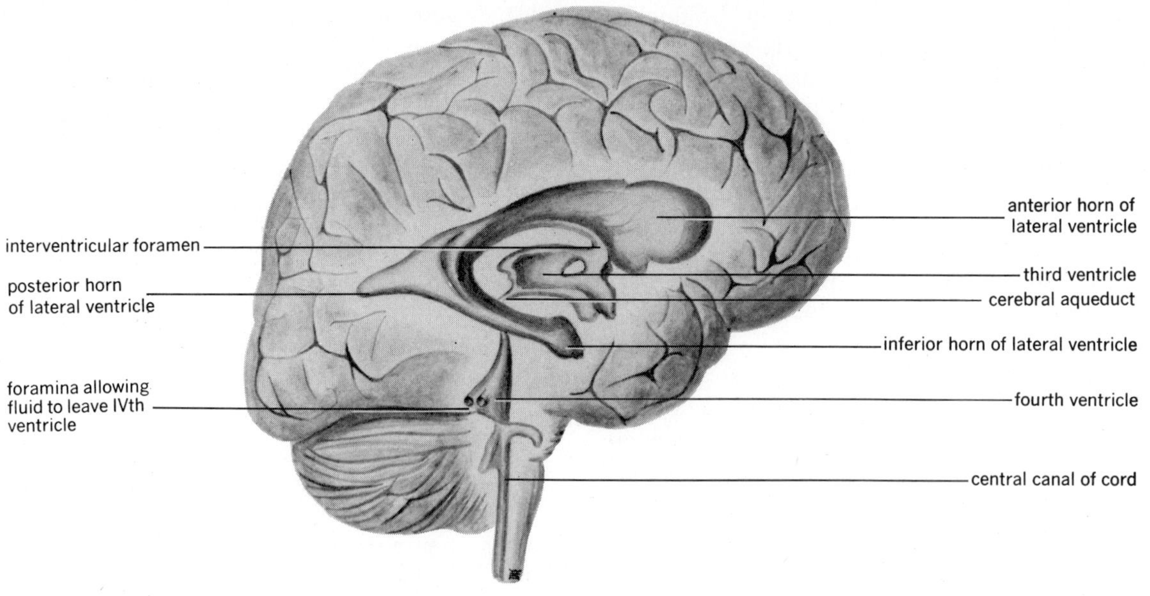

FIGURE 10.4
The ventricles of the brain.

CEREBROSPINAL FLUID

FORMATION, FLOW, AND COMPOSITION

Formation of the cerebrospinal fluid (CSF) is believed to occur by active secretory processes from the choroid plexuses and to a small degree from ependymal lining cells. A comparison of plasma and CSF (Table 10.1) shows differences suggestive of secretory processes. If a passive process was responsible, concentrations of smaller molecules should be nearly equal in plasma and CSF. About 25 ml of CSF is contained within the ventricles, while about 120 ml is found in brain and cord subarachnoid spaces. Rate of CSF formation is continuous and is about 3/4 ml per gram of choroid plexus per minute (this is about one-quarter of the blood flow through the plexuses). Flow occurs from lateral to third, to fourth ventricles, and then to subarachnoid spaces via the foramina in the fourth ventricle roof. In the brain, fluid is absorbed into blood vessels by way of the SUBARACHNOID GRANULATIONS (Fig. 10.5), and from the cord via spinal venous vessels. In the various cavities, CSF has a pressure of about 150 mm H_2O, or about twice that of the body veins.

FUNCTIONS

The fluid acts as a shock absorber for the brain and cord. In the brain, CSF volume tends to vary inversely with blood volume and thus the volume of the cranial contents is kept nearly constant (this avoids compression of neurons and blood vessels). The fluid, as it circulates, may also distribute chemicals, for example, the opium or morphine-like substances, and sleep factor recently isolated from the brain. Primary control of respiratory activity is believed to be exerted by the H^+ and HCO_3^- concentrations in the CSF. These latter relationships are presented in Figure 10.6.

CLINICAL CONSIDERATIONS

The term HYDROCEPHALUS refers to an increased accumulation of CSF in the brain, with consequent enlargement of the ventricles. If it occurs in an infant or child, the entire brain en-

TABLE 10.1
PLASMA AND CSF COMPARED FOR SOME MAJOR CONSTITUENTS

CONSTITUENT OR PROPERTY	PLASMA	CSF
Protein	6400–8400 mg%	15–40 mg%
Cholesterol	100–150 mg%	0.06–0.20 mg%
Urea	20–40 mg%	5–40 mg%
Glucose	70–120 mg%	40–80 mg%
NaCl	550–630 mg%	720–750 mg%
Magnesium	1–3 mg%	3–4 mg%
Bicarbonate (as vol % of CO_2)	40–60 mg%	40–60 mg%
pH	7.35–7.4	7.35–7.4
Volume	3–4 liters	200 ml
Pressure	0–130 mm Hg	110–175 mm CSF

larges, inasmuch as the cranial bones are not yet fused. An abnormal rate of increase in head circumference in infants may indicate presence of the condition. In adults, with the cranial bones fused, enlargement of the ventricles and increased intracranial pressure, with compression, may occur.

Causes of hydrocephalus include:

CHOROID PLEXUS TUMOR enlarges the plexus(es) and leads to increased fluid formation. A pneumoencephalogram, in which the CSF in the ventricles is replaced by air, usually demonstrates the presence of such a tumor.

DECREASED ABSORPTION of fluid is usually due to failure of the subarachnoid granulations to develop, and continued production raises CSF volume.

BLOCKAGE OF FLOW, by tumor or failure of the ventricular system to develop, accounts for about one-third of the cases. The cerebral aqueduct, because of its small size, is a common site of blockage of flow.

Treatment of the disorder may be surgical, with the insertion of a tube and valve to bypass the obstruction, if such exists (Fig. 10.7), or by choroid plexus removal. The CSF may be de-

FIGURE 10.5
The subarachnoid villi (granulations).

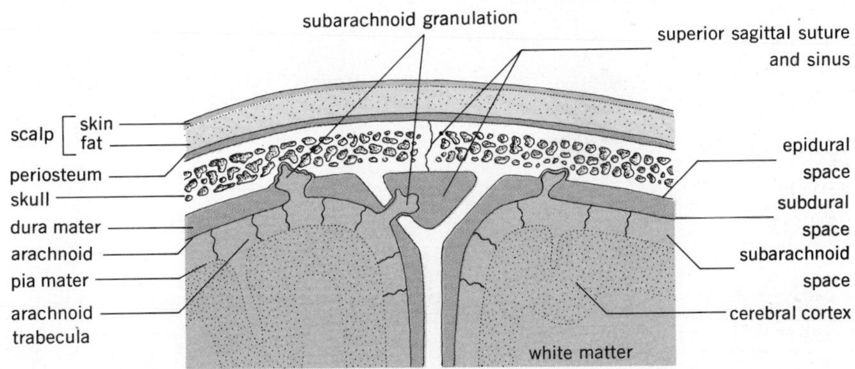

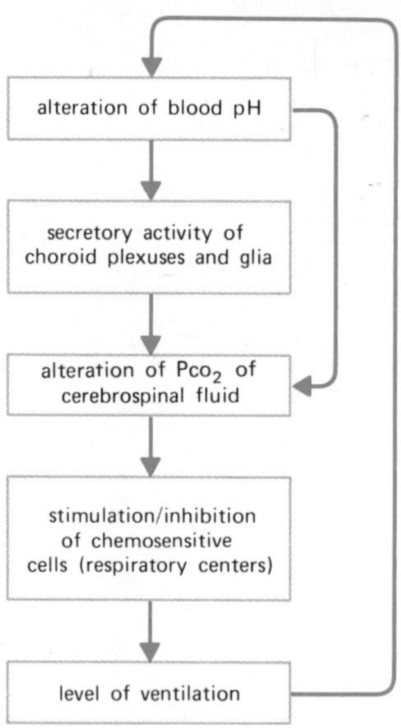

FIGURE 10.6
The relationship of CSF pH to control of ventilation.

livered to the right atrium of the heart or to the peritoneal (abdominal) cavity. Surgical intervention carries certain hazards: plexus removal does not stop secretion of CSF; tubing may become blocked; tissues around the tube may become infected. Pharmacological agents (drugs) avoid such problems, but may not result in adequate treatment if used alone. Acetazolamide is a drug that inhibits CSF formation by 50 percent, and is sometimes used if the disorder is mild.

Homeostasis is again emphasized in our consideration of blood flow and CSF. Brain cells probably require a closer monitoring of their environment than any other body area. This is partly assured by the barrier, and by the integration of other systems into the supply and removal of nutrients and wastes to and from the brain.

FIGURE 10.7
The surgical treatment of hydrocephalus.

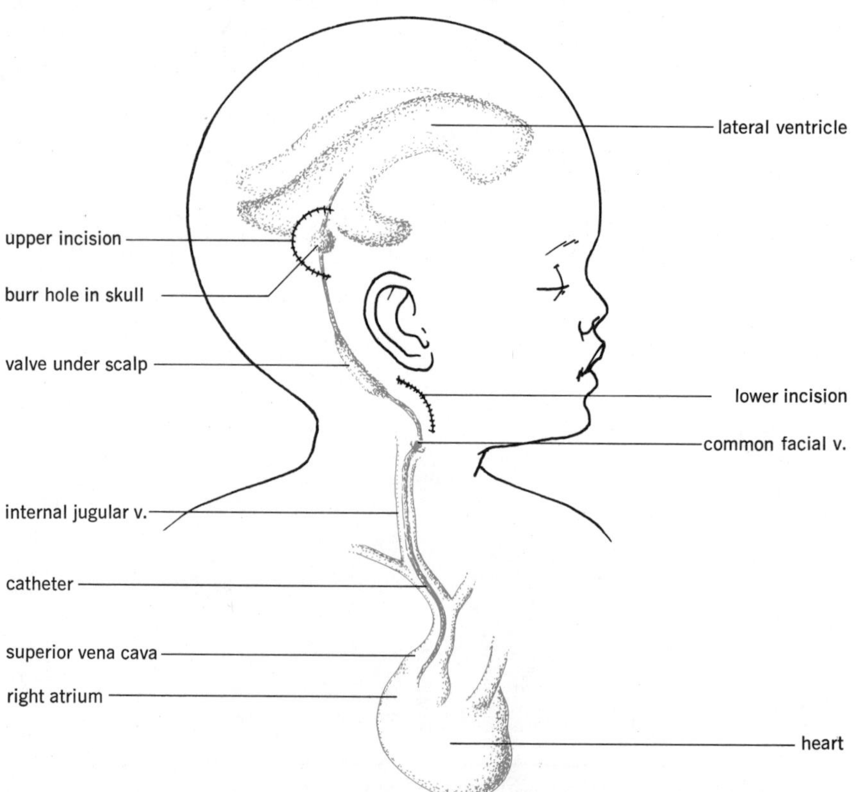

SUMMARY

1. The brain receives blood from the internal carotid arteries and vertebral arteries, and is drained by dural sinuses.
   a. The arterial circle of Willis allows communication between arterial supplies.
2. The spinal cord receives blood from the spinal arteries, and is drained by spinal veins.
3. The blood brain barrier has an anatomical basis, and makes cerebral capillaries less permeable to solutes than capillaries elsewhere in the body.
   a. Capillary pores are lacking, or are fewer and smaller than in other capillaries; cells overlap and there is a continuous basement membrane.
   b. Glia surround capillaries.
4. Vascular disorders of the CNS have several causes.
   a. Vessels may become narrowed or occluded.
   b. Emboli (floating masses) may plug vessels.
   c. Walls may be weakened by disease or congenital failure to develop tissue coats.
   d. All combine to reduce blood flow to the CNS.
5. Symptomatology is similar regardless of cause of diminished blood flow. Some major symptoms are:
   a. Headache
   b. Dizziness and faintness
   c. Unequal reflexes
   d. Muscular weakness
   e. Mental confusion
6. Spaces exist within the CNS, and the CNS is surrounded by membranes or meninges.
   a. There are three meninges: dura mater (protective), arachnoid (surrounds subarachnoid space), and pia mater (vascular).
   b. The brain has ventricles: two lateral in each hemisphere, a third in the diencephalon that connects by an aqueduct to the fourth in the medulla.
7. Cerebrospinal fluid fills the ventricles of the brain and occupies the subarachnoid space.
   a. It is formed by secretion from choroid plexuses and ependymal cells.
   b. It circulates downwards through the ventricles to the subarachnoid space from which it is absorbed by arachnoid granulations and spinal veins.
   c. It cushions the CNS, distributes materials, and is involved in control of respiration (breathing).
   d. Too much CSF in the system may result from plexus tumors, decreased absorption of fluid, or blockage of flow. Hydrocephalus results that may be treated surgically or with drugs.

QUESTIONS

1. What would happen to brain circulation if flow through one internal carotid artery was reduced? What portions of the brain would be most greatly affected?
2. Referring to question 1, is there any way of compensating for reduced carotid flow? Explain.
3. What is the blood-brain barrier? Explain its value. What may be its structural basis?
4. What are some causes of vascular disorders of the CNS? What are the common denominators of cerebrovascular disease?
5. Trace the flow of CSF through the ventricular system to its absorption.
6. In what ways does CSF differ from plasma? What is the significance of such differences?
7. How is CSF formed, and what are its functions?
8. What are some causes for excessive CSF volume? What may happen if volume is increased?

READINGS

Rapoport, Stanley I. *Blood-Brain Barrier in Physiol ogy and Medicine*. Raven, New York, 1976.

11

THE SPINAL CORD

OBJECTIVES

After studying this chapter, the reader should be able to:

☐ Describe the general gross and microscopic appearance of the spinal cord.

☐ Describe the role of the spinal cord and its tracts in impulse conduction to and from the brain.

☐ Describe the role of the spinal cord as a mediator of reflex activity.

☐ Describe the functions of internuncial neurons.

☐ Define what a myotatic reflex is, explain the role of the muscle spindles in such reflexes, and the neurological value of such reflexes.

☐ Explain flexion and crossed extension reflexes, their neural basis, and their value to the body.

☐ Define cord righting reflexes and their value to the body.

☐ Explain what spinal shock is, and the order of return of spinally mediated reflexes after shock.

☐ Explain the genesis of the Brown-Sequard syndrome and what it is.

☐ Account for the results of several infectious and degenerative diseases of the spinal cord.

BASIC STRUCTURE OF THE SPINAL CORD

The spinal cord is the portion of the central nervous system enclosed by the vertebral neural arches. In the adult, the cord measures about 45 cm (18 in) in length and about 2 cm (4/5 inch) at its greatest diameter. Grossly, it presents the features shown in Figure 11.1. A cross section of the cord shows it to have an inner H-shaped mass of GRAY MATTER, consisting primarily of neuron cell bodies and synaptic regions, and an outer area of WHITE MATTER, consisting primarily of myelinated nerve fibers. Each of these gray and white areas is subdivided as shown in Figure 11.2. The dorsal (posterior) gray column receives sensory input from the periphery, while the lateral and ventral (anterior) gray columns contain the cell bodies of motor neurons innervating peripheral effectors (muscle and glands). Internuncial (association) neurons are commonly interposed between the sensory and motor neurons, and impulses from the brain also terminate on these motor neurons. The motor neurons thus serve as a FINAL COMMON PATHWAY for transmitting a variety of influences to peripheral effectors.

THE FUNCTIONS OF THE SPINAL CORD

THE CORD AS A TRANSMITTER OF NERVE IMPULSES

Located within the spinal cord are a number of SPINAL TRACTS (Fig. 11.3). By experiments designed to stimulate or damage these various tracts, it may be demonstrated that they constitute functional regions carrying specific types of information, although there is no visible anatomical difference in their appearance. The tracts are divided into *ascending tracts,* carrying sensory information from the periphery to other levels of the cord and to the brain, and into *descending tracts,* carrying motor impulses to the cord motor neurons. In general, the tracts are named according to their origins and terminations. Table 11.1 summarizes information concerning these tracts.

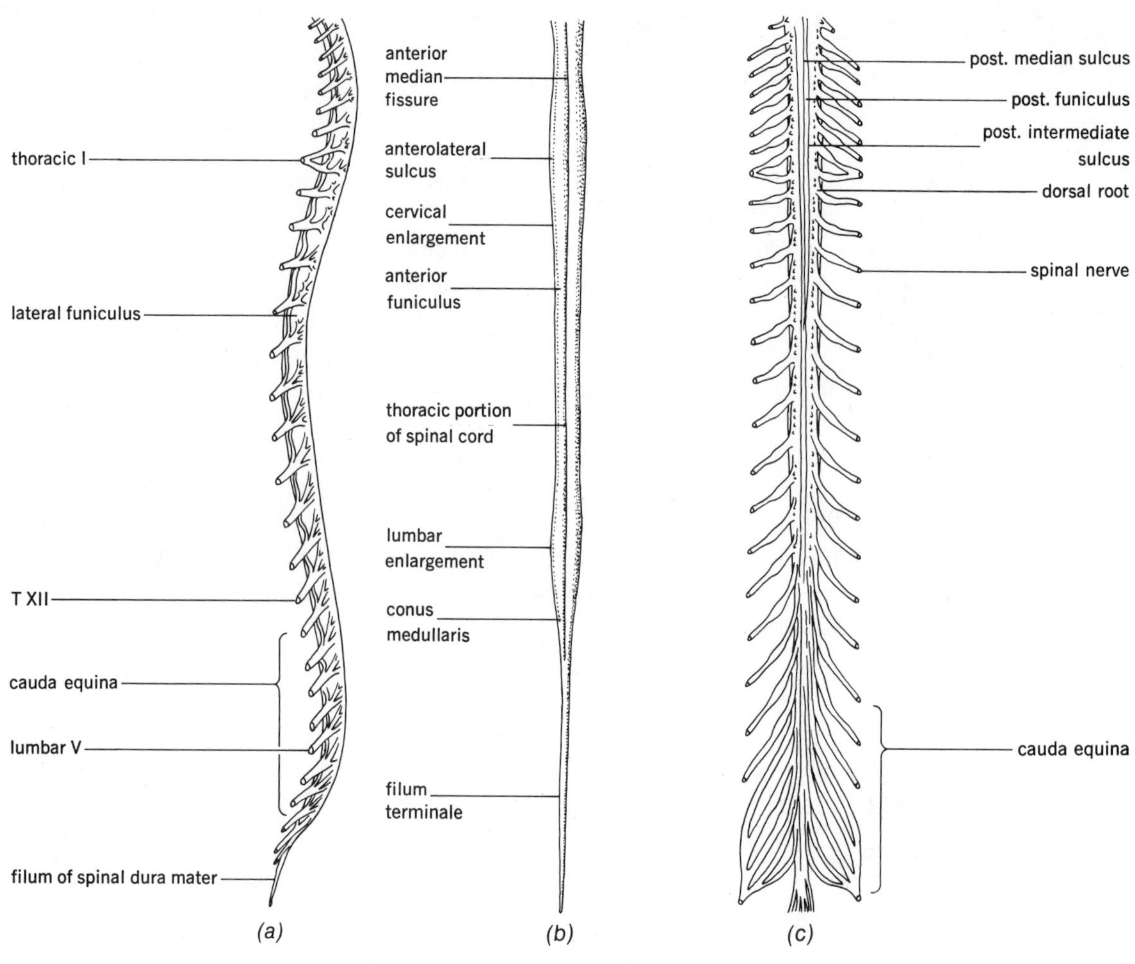

FIGURE 11.1
Three views of the gross anatomy of the spinal cord. (*a*) Lateral view; (*b*) Anterior view; (c) Posterior view.

THE CORD AS A MEDIATOR OF REFLEX ACTIVITY

The internuncial neurons. As stated earlier, internuncial neurons are situated between incoming and outgoing fibers of the spinal cord. They form the bulk of the neurons of the cord, outnumbering motor neurons about 30 to 1. They are smaller than motor neurons (about 15 μm in diameter to 50 μm for motor neurons), and characteristically show "spontaneous" discharge without being stimulated. These cells perform a wide variety of functions that are only now beginning to be understood.

By their physical presence, they permit incoming impulses to travel pathways other than one directly from sensory to motor neurons on the same side of the cord (ipsilateral). Variability of response to sensory input is thus

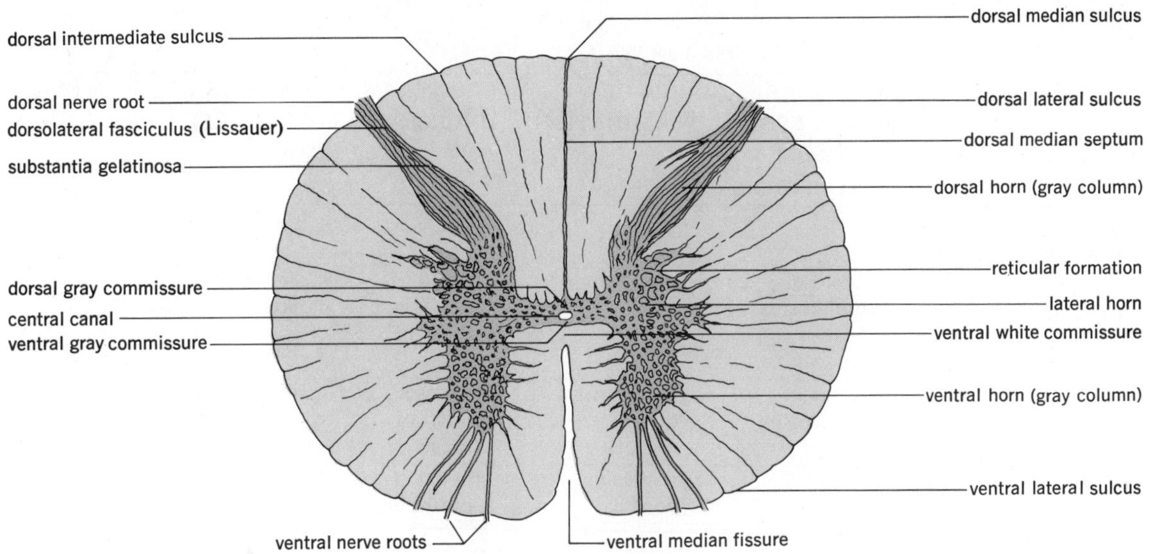

FIGURE 11.2
A cross section of the spinal cord.

FIGURE 11.3
The major spinal tracts or fasciculi. (Color) Sensory or ascending tracts; (black) motor or descending tracts.

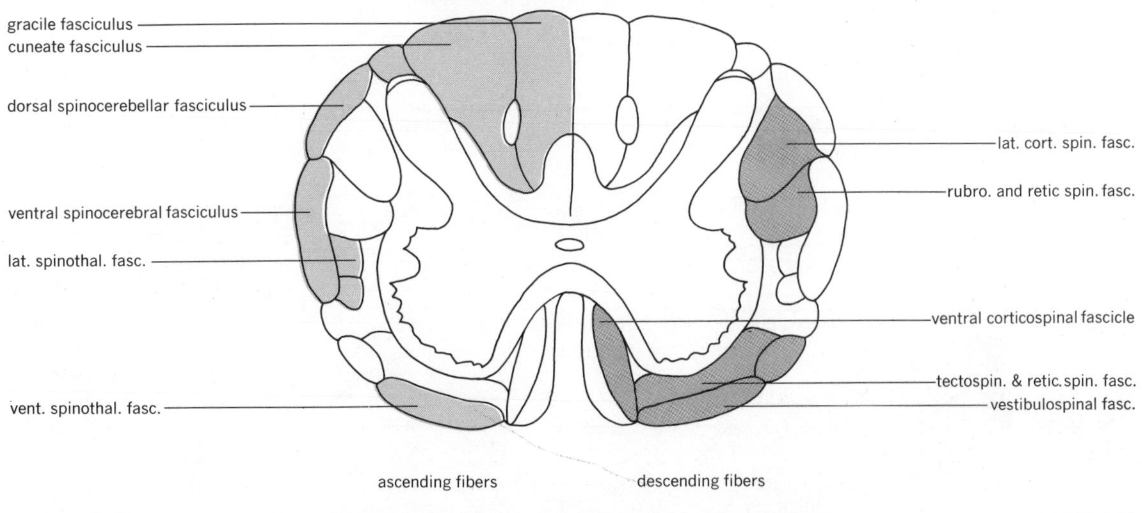

TABLE 11.1
MAJOR SPINAL TRACTS

NAME OF TRACT	AREA OF ORIGIN, OR REGION FROM WHICH TRACT RECEIVES FIBERS	AREA OF TERMINATION	CROSSED OR UNCROSSED	FUNCTION OR IMPULSE(S) CARRIED
Descending				
Corticospinal				
Lateral	Motor areas of cerebral cortex	Synapses in cord	Crossed in medulla, 80% of fibers	Carry voluntary impulses to skeletal muscles
Ventral	Motor areas of cerebral cortex	Synapses in cord	Uncrossed, 20% of fibers	Carry voluntary impulses to skeletal muscles
Rubrospinal	Red nucleus of basal ganglia in cerebrum	Synapses in cord	Crossed in brainstem	Involuntary impulses to skeletal muscles concerned with tone, posture
Reticulospinal	Reticular formation of brainstem	Synapses in cord	Crossed in brainstem	Increases skeletal muscle tone and motor neuron activity
Vestibulospinal	Vestibular nuclei of brainstem	Synapses on motor neurons of cord	Uncrossed	Regulates muscles tone to maintain balance and equilibrium
Ascending				
Spinothalamic				
Lateral	Skin	Thalamus, relays to cerebral cortex	Crossed in cord	Pain and temperature
Ventral	Skin	Thalamus, relays to cerebral cortex	Crossed in cord	Crude touch
Spinocerebellar				
Dorsal	Muscles and tendons	Cerebellum	Uncrossed	Unconscious muscle sense for controlling muscle tone, posture
Ventral	Muscles and tendons	Cerebellum	Uncrossed	Unconscious muscle sense for controlling muscle tone, posture
Gracile	Skin and muscles	Medulla, relays to cerebral cortex	Uncrossed	Touch, pressure, two-point discrimination, conscious muscle sense concerned with appreciation of body position
Cuneate	Skin and muscles	Medulla, relays to cerebral cortex	Uncrossed	Touch, pressure, two-point discrimination, conscious muscle sense concerned with appreciation of body position

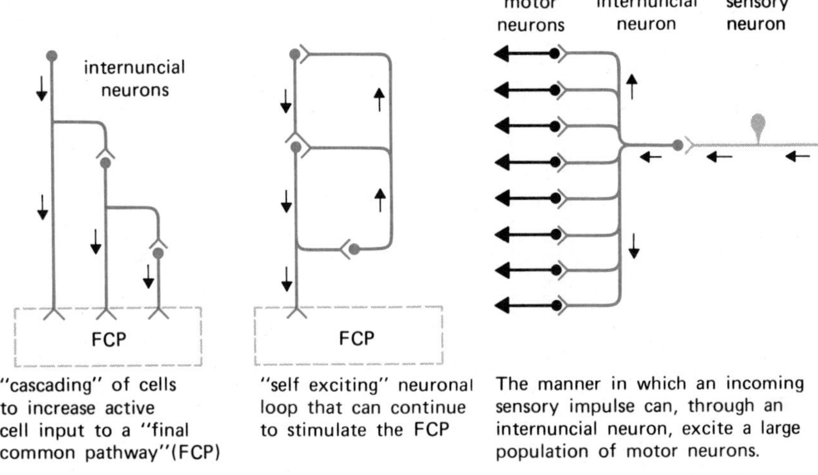

FIGURE 11.4
Mechanisms of action of internuncial neurons.

achieved, by involving the other side of the cord (contralateral), different cord levels, and/or the brain.

They serve as amplifiers of the intensity of incoming impulses by: *cascading of cells or increase in rate of discharge* of individual cells; utilization of neuronal loops to boost signal amplitude; distributing incoming impulses to a wider area of motor neurons, increasing the total amplitude of signals. These devices are shown in Figure 11.4.

They may serve as "valves" or "gates" to pass or block incoming impulses. Some control over impulse passage is thus exerted, thereby determining whether or not a motor neuron will be caused to fire.

They may change the nature of the incoming impulse from excitation to inhibition or vice versa (inversion or conversion). It is unlikely that such changes can occur by secretion of different chemicals from one neuron, so one or more internuncial neurons may be involved.

In summary, internuncial neurons exert a great degree of control over what the nature of the impulse finally delivered to the motor neurons will be.

Myotatic reflexes. The term MYOTATIC REFLEX refers to a reflex contraction of a skeletal muscle elicited by applying a sudden stretch to a muscle or its tendon. The cord is involved in such reflexes as the center in which sensory information is translated into the reflex contraction. The primary receptor for a myotatic reflex is the MUSCLE SPINDLE (Fig. 11.5). Each spindle contains several tiny skeletal muscle fibers known as *intrafusal fibers* (*fusus,* spindle) in contrast to the *extrafusal fibers* of the muscle itself. Two types of intrafusal fibers are distinguished by the arrangement of their nuclei: *nuclear chain fibers* have their nuclei in a row; *nuclear bag fibers* have their nuclei grouped mainly in an enlarged central region of the fiber. Spiral sensory endings of what are called *1a afferent neurons* are wrapped around the central portion of both types of fibers, and the fibers receive what are called *gamma efferent neurons* ($\gamma 1$ for nuclear bag fibers; $\gamma 2$ for nuclear chain fibers). The sensory fibers respond when the fibers are elongated or stretched, and the gamma fibers control the degree of contraction of the intrafusal fibers, thereby determining how much force will be required to

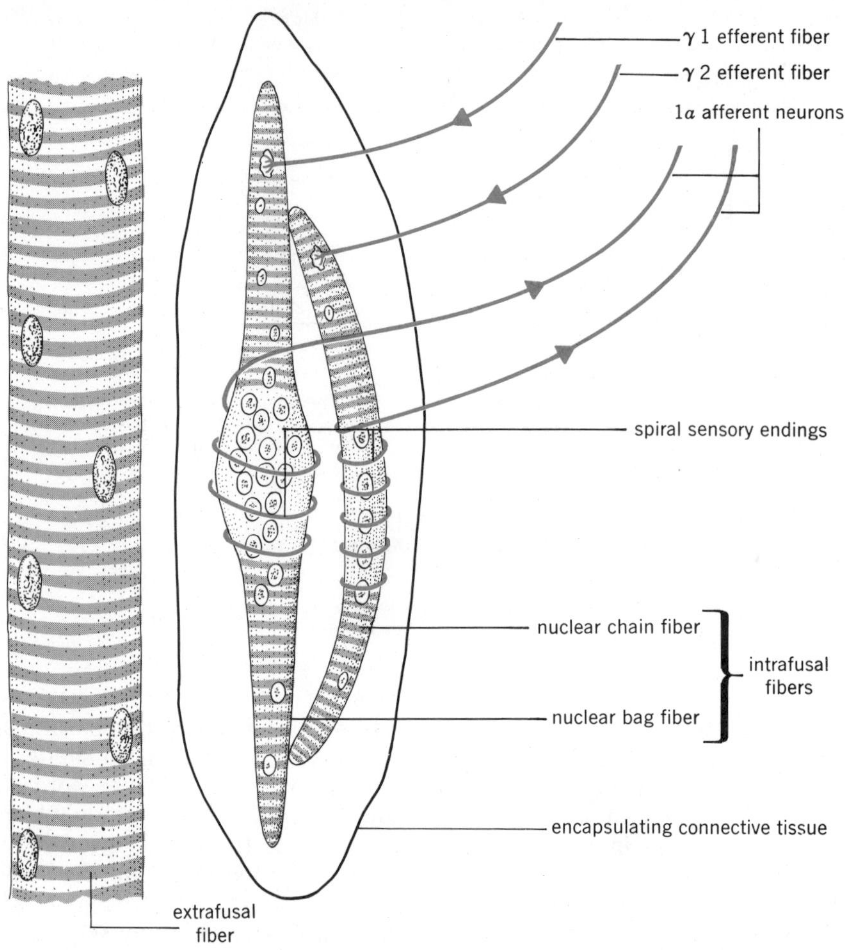

FIGURE 11.5
The structure and innervation of the muscle spindle.

stretch the fiber and cause the sensory nerves to "fire." In general, the greater the contraction of the intrafusal fibers, the less will be the tension required to cause the spindle to "fire." In turn, the cell bodies of the gamma efferent neurons, located in the spinal cord, are influenced by impulses arriving from the cerebellum, basal ganglia, and brain stem. Such signals determine muscle tone and are important in the regulation of posture. The whole system constitutes a servomechanism controlling muscle and the loads it must bear. The general outline of this mechanism is depicted in Figure 11.6.

Golgi tendon organs lie within muscle tendons immediately beyond their attachments to the muscle fibers. The Golgi tendon organ detects tension applied to the tendon by muscle contraction. Signals from the tendon organ are then transmitted into the spinal cord to cause reflex effects in the respective muscle. However, this reflex is entirely inhibitory, the exact opposite to the muscle spindle reflex. The signal from the tendon organ supposedly excites inhibitory interneurons, and these in turn inhibit the alpha motorneurons to the respective muscle.

The myotatic reflexes are often used by the neurologist to assess cord function at different levels, because they are "simple reflexes" that are handled by restricted segments of the spinal cord. Table 11.2 indicates some of the myotatic reflexes used to assess cord function.

Flexion and extension reflexes. These types of reflexes originate from receptors in the skin and deeper tissues that respond to touch, pressure, heat, cold, and tissue damage (pain?). They are primarily protective in nature in that they cause withdrawal of the affected part from a potentially dangerous stimulus. Withdrawal of the stimulated area may upset the equilibrium of the body and require adjustment of other body parts to maintain proper orientation of the body in space. These reflexes are thus more complex than the myotatic reflexes, and usually involve more segments of the cord. They are, nevertheless, mediated by the spinal cord alone.

FLEXOR REFLEXES result when a skin area is stimulated (as by touching a hot object), and the spinal cord distributes impulses to all the flexor muscles of the affected limb. Withdrawal results.

As the stimulated limb is withdrawn from a potentially dangerous stimulus, the limb on the opposite side of the body is extended to support the body or maintain its balance (particularly obvious if one steps on a tack, for example). Because *extension* occurs on the *opposite* side of the body, this type of reflex is called a CROSSED EXTENSION REFLEX. This reflex requires an internuncial neuron or neurons that cross to the opposite side of the cord (Fig. 11.7), and simultaneous inhibition of antagonistic muscles to permit the desired action to occur (reciprocal inhibition). Again, this type of reflex illustrates the importance of the internuncial neurons. Such a system is of great importance in rhythmical movements such as walking, where one limb is moving forward and the other back, as a result of opposite groups of muscles undergoing contraction.

A final example of reflex activity partially mediated by the spinal cord is the CORD RIGHTING REFLEXES. If an animal that has had the brain separated from the cord is laid on its side on a hard surface, the limbs will make movements designed to return the body to an upright position. Such movements are usually uncoordinated and therefore ineffectual, because of the lack of cerebellar involvement. The cord has received sensory information from the skin of the animal and initiates impulses to the muscles to right the body. This illustrates that the cord is involved in such reflexes, but requires the brain to integrate such activity.

It may be indicated that the importance of the spinal cord in reflex activity of muscles decreases as one ascends the phylogenetic scale, with cerebral dominance of all functions becoming more important. A frog can carry on nearly normal motor activity with the cord alone; cats or dogs exhibit independent fore and hind limb movements that are essentially normal but uncoordinated with one another; monkeys show complete incoordination of right to left and fore to hind limb activity. Thus, in humans, the cord remains as a reflex center, but requires higher centers for achieving the greatest benefit from sensory information.

FIGURE 11.6
The neural circuits involved in control of the myotatic reflexes.

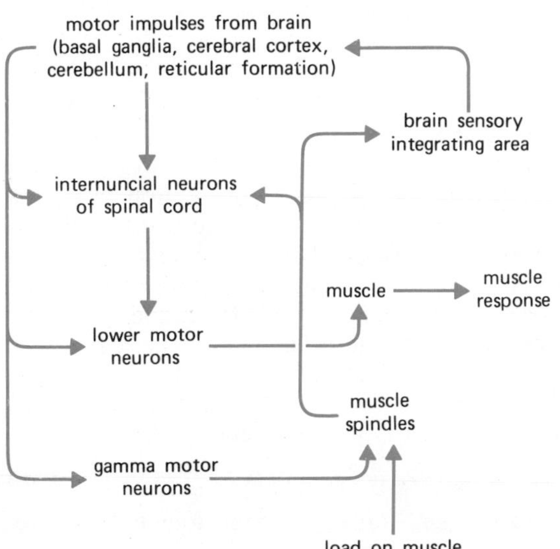

TABLE 11.2
SOME REFLEXES USED TO ASSESS CORD FUNCTION

NAME OF REFLEX	LEVEL OF CORD INVOLVED	HOW ELICITED	INTERPRETATION
Jaw jerk	C3	Tap chin with percussion hammer; jaw closes slightly.	Marked reaction indicates upper neuron lesion.
Biceps jerk	C5.6	Tap biceps tendon with percussion hammer; biceps contracts.	If reflex is absent or greatly exaggerated, cord damage, or damage to sensory or motor nerves is probable.
Triceps jerk	C7.8	Tap triceps tendon with percussion hammer; triceps contracts.	
Abdominal reflexes	T9-L2	Draw key or similar object across abdomen at different levels; muscles contract.	
Knee jerk	L2.3	Tap patellar tendon with percussion hammer; quadriceps contract to extend knee.	
Ankle jerk	L5	Tap Achilles tendon; gastrocnemius contracts to plantar flex foot.	
Plantar flexion	S1	Firmly stroke sole of foot from heel to big toe; toes should flex.	If toes fan and big toe extends (Babinski sign), upper neuron lesion is present.

CLINICAL CONSIDERATIONS

EFFECTS OF VARIOUS TYPES OF LESIONS ON CORD FUNCTION

If the spinal cord is completely cut through (transected), two functional results are evident immediately.

All voluntary movement ceases in those parts of the body served by the cord below the plane of section.

All sensation from the same areas will be lost.

The reason for such losses lies in the interruption of the spinal tracts carrying afferent and efferent impulses.

A third result of section is the development of SPINAL SHOCK, a condition of loss of reflex activity below the level of section. It is believed that the areflexia is due to synaptic depression occasioned by loss of excitatory impulses from the brain. The length of time that the loss of reflexes lasts again depends on the complexity of the animal subject. In the frog, shock lasts only minutes, but in humans, it may last for months. As recovery proceeds, reflexes reappear in a typical sequence.

Myotatic reflexes reappear.

Flexion and crossed extension reflexes reappear.

Visceral reflexes, including bladder evacuating, rectal evacuating, and sexual reflexes reappear. (In humans, erection may occur with tactile stimulation, but ejaculation of seminal fluid is rare).

If the section occurs above the cord levels supplying large numbers of visceral muscular arteries, there may be a profound drop in blood pressure. Some return of vascular tone may occur, but the subject usually demonstrates a postural hypotension (low blood pressure) when passing from lying to sitting positions. This indicates that the reflex mechanisms lead-

FIGURE 11.7
The interrelationships of the flexor reflex, the crossed extension reflex, and the phenomenon of reciprocal inhibition.

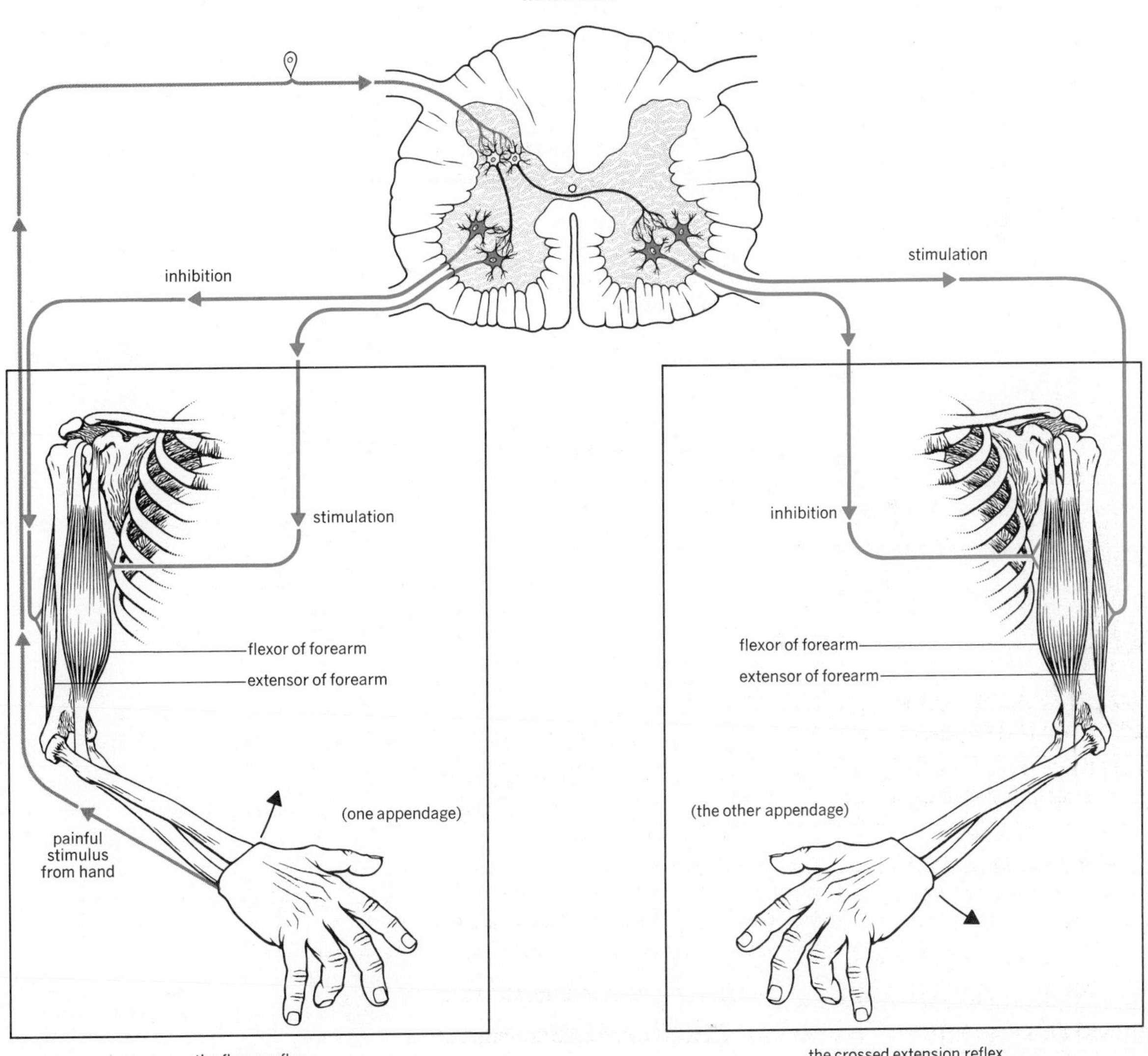

ing to vasoconstriction are not operating effectively to keep the blood pressure elevated. Skin ulcerations (bedsores) are another hazard for a subject with a transected cord, because there is no sensation from the areas distal to the cut and thus there is no awareness of pain or ischemia (lowered blood flow) in the parts being subjected to pressure. Frequent changes of position are thus required.

If the cord is cut only half-way through (hemisection), a characteristic set of symptoms known as the BROWN-SEQUARD SYNDROME will develop. There will be loss of voluntary movement, touch, and pressure sense on the same side as the cut, but loss of pain and thermal sensations on the *opposite* side. The latter effect occurs because the pathways for pain and thermal sensations are crossed in the cord. Figure 11.8 demonstrates the basis for these effects.

FIGURE 11.8
A diagram to illustrate the losses of function occurring with cord hemisection. Hemisection at *A* interrupts descending pathways on the same side (right), touch and pressure pathways from receptors on the same side (right), and pain and thermal pathways from the opposite side (left). Hemisection at *B* has the opposite effects as *A*.

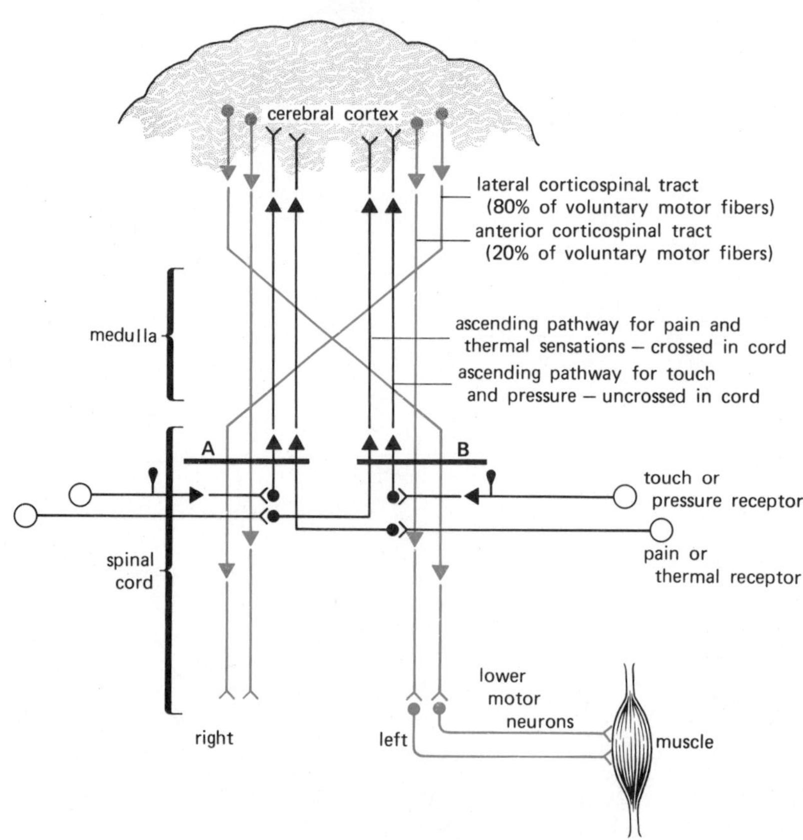

TABES DORSALIS results from syphilitic invasion of the spinal cord. The microorganisms affect the dorsal roots of the spinal nerves and the dorsal funiculi, thereby interrupting pathways from the muscle spindles and the touch-pressure pathways. Pain results from the inflammation, and there is a disturbance of walking. The course of the disease is unpredictable, and response to treatment is variable. It seems wisest to avoid acquiring the organism.

MULTIPLE SCLEROSIS is a disorder characterized by loss of myelin sheaths (demyelination) and development of lesions in the brain and spinal cord. Its cause is unknown, but attempts have been made to relate it to diet, radiation, and heredity, generally without success. Motor function is disturbed and pain and visual complaints are common. Treatment may slow the course of the disease, but cannot halt its ultimate progress.

POLIOMYELITIS, a viral disease, may result in paralysis due to the organism's predilection for the anterior gray columns, wherein are located the motor neurons for skeletal muscles. Paralysis is *flaccid,* that is, the muscle is in a relaxed state. Such paralysis indicates a lower motor neuron (neurons to muscle from cord) lesion, as opposed to an upper motor neuron (from brain to cord) lesion, in which spastic paralysis (muscle in a contracted state) occurs.

By way of summary, the spinal cord of the human may be viewed as a super highway for rapid transmission of sensory information *to* the brain, and for delivery of motor impulses *from* the brain to peripheral effectors. The internuncial neurons give the cord the "power" to alter the nature of incoming signals, and thus control response of motor neurons. Reflex activity, concerned primarily with skeletal muscles, is a basic function of the human cord.

SUMMARY

1. The spinal cord is that part of the CNS lying within the spinal column.
   a. It has several gross features of importance (*see* Fig. 11.1).
   b. A cross section of the cord shows its general internal organization (*see* Fig. 11.2).
   c. Internuncial neurons lie between incoming sensory neurons and outgoing motor neurons.
2. Spinal tracts carry impulses to and from the brain.
   a. Ascending tracts carry sensory information up the cord to the brain.
   b. Descending tracts carry motor information to the cord's internuncial and motor neurons.
3. Internuncial neurons process incoming impulses.
   a. They create diverse pathways for impulses to travel.
   b. They amplify the incoming impulses.
   c. They may block impulses or permit their passage.
   d. They may invert or convert impulses from excitatory to inhibitory or vice versa.
4. Myotatic reflexes are mediated by the spinal cord.
   a. They are reflex muscle contractions depending on muscle spindles in the skeletal muscles.
   b. Muscle spindles are "adjustable" sensory receptors that react to stretching or tension applied to a muscle.
   c. Spindles are responsible for posture and muscle responses to loading.
   d. Myotatic reflexes may be used to assess cord function (*see* Table 11.2).
5. Flexion and crossed extension reflexes withdraw body parts from potentially dangerous stimuli, and maintain body posture.
6. The cord is partly responsible for body righting reflexes, but their coordination and integration requires the brain.
7. Several things result when the cord is transected.
   a. There is loss of voluntary movement below the cut.
   b. There is loss of sensation below the cut.
   c. Spinal shock results. It is a loss of spinal reflex activity, due to temporary synaptic depression. Reflex activity reappears in a certain order.

1) Myotatic reflexes reappear.
2) Flexion and crossed extension reflexes reappear.
3) Visceral reflexes reappear; hypotension occurs that is exaggerated by postural changes.

8. Hemisection of the spinal cord produces the Brown-Sequard syndrome, characterized by:
   a. Loss of touch and pressure, and voluntary movement on the same side as the cut.
   b. Loss of pain and thermal sensations on the opposite side as the cut.
9. Syphilitic infection of the cord (tabes dorsalis) produces loss of sensation; multiple sclerosis is a degenerative demyelinating disease; poliomyelitis is a viral disease causing muscular paralysis.

QUESTIONS

1. How are the neurons and spinal tracts organized within the spinal cord?
2. Name the descending tracts of the spinal cord, indicate their positions on a cross-section diagram of the cord, and indicate the origin, termination, and functions served by each tract you name.
3. Name the ascending tracts of the spinal cord, indicate their positions on a cross-section diagram of the cord, and indicate the origin, termination, functions and whether it is a crossed or uncrossed tract for each that you name.
4. What is an internuncial neuron, and what are some of its functions?
5. Define and give the cord neuronal pathways for flexion and crossed-extension reflexes. What is the value of such reflexes to the body?
6. What is spinal shock? In what order to the spinal reflexes reappear?
7. Compare the effects of transection and hemisection of the cord on body activities.

READINGS

Maugh, Thomas H. II. "Multiple Sclerosis: Genetic Link, Viruses Suspected." *Science 195*:667. 18 Feb 1977.

12

THE PERIPHERAL NERVOUS SYSTEM; SOMATIC AND AUTONOMIC SYSTEMS

OBJECTIVES

THE SPINAL NERVES

ORGANIZATION

FUNCTIONAL CLASSIFICATION OF COMPONENTS

SOMATIC AFFERENTS

SOMATIC EFFERENTS

VISCERAL AFFERENTS

VISCERAL EFFERENTS

SEGMENTAL DISTRIBUTION OF SPINAL NERVES

THE CRANIAL NERVES

THE AUTONOMIC NERVOUS SYSTEM

CONNECTIONS OF THE SYSTEM WITH THE CENTRAL NERVOUS SYSTEM

AFFERENT PATHWAYS

EFFERENT PATHWAYS

AUTONOMIC GANGLIA AND PLEXUSES

DIVISIONS OF THE SYSTEM; STRUCTURE AND FUNCTION

THE PARASYMPATHETIC DIVISION

THE SYMPATHETIC DIVISION

HIGHER AUTONOMIC CENTERS

CLINICAL CONSIDERATIONS

SURGERY AND THE AUTONOMIC SYSTEM

DRUGS AND THE AUTONOMIC SYSTEM

SUMMARY

QUESTIONS

READINGS

OBJECTIVES

After studying this chapter, the reader should be able to:

☐ Describe the origins of the spinal nerves from the cord, and how many come from each part of the cord.

☐ Describe the anatomical components of a spinal nerve.

☐ Describe the four functional components of the spinal nerves, and the areas each serves.

☐ Describe the segmental distribution of the spinal nerves.

☐ Name the 12 pairs of cranial nerves, and give the distribution and functions of each.

☐ Describe the anatomical organization of the autonomic nervous system, the autonomic ganglia, and the two subdivisions of the system.

☐ List and explain the effects of stimulation of the two divisions of the autonomic system.

☐ Describe the role of the higher autonomic centers in the functioning of the system.

☐ Discuss the roles of surgery and drugs in the treatment of autonomic disorders.

THE SPINAL NERVES

ORGANIZATION

The spinal cord gives rise to 31 pairs of spinal nerves, divided into 8 cervical (C), 12 thoracic (T), 5 lumbar (L), 5 sacral (S), and 1 coccygeal (Fig. 12.1). Each pair of spinal nerves exits from the vertebral canal that houses the cord, through foramina (*intevertebral foramina*) lying between the lower parts of the vertebral arches (except the first cervical that exits between skull and first cervical vertebra). The naming and numbering of a particular pair of nerves are done according to the portion of the spinal cord with which the nerves connect, and by the position in the spinal column of the foramen through which the nerve passes. Thus, though the spinal cord substance itself does not reach lower than the second lumbar vertebra, nerves arise from levels of the cord and may pass downwards to exit from their particular intervertebral foramen (*see* Fig. 12.1c).

The term SPINAL NERVE actually refers to the single nerve that passes through its particular intervertebral foramen. It carries both sensory (incoming) and motor (outgoing) impulses. Each member of a pair of spinal nerves (*see* Fig. 12.1, bottom) has a DORSAL (*posterior*) ROOT carrying sensory impulses to the cord from many peripheral body areas (skin, muscles, thoracic and abdominal viscera). The neurons conveying these impulses are unipolar types and their cell bodies are located in the DORSAL ROOT GANGLION found just outside the spinal dural sheath. The incoming dorsal root spreads its constituent fibers over one spinal cord segment (*see* Fig. 12.1a) by a series of rootlets. A VENTRAL (*anterior*) ROOT of a spinal nerve conveys motor impulses from multipolar motor neurons, whose cell bodies are located in the lateral and ventral gray columns, out of the cord. Here too, several rootlets are combined to form the nerve. Peripherally, the spinal nerves send branches (rami) to the muscles and skin, and COMMUNICATING RAMI allow autonomic fibers to separate from the spinal nerve (see Fig. 12.1, bottom).

FIGURE 12.1

(Top) The relationships of the spinal nerves to the spinal cord. (*a*) The cervical region; (*b*) the lumbar region; (*c*) the origin of the 31 pairs of spinal nerves shown on one side only. (Bottom) The connections of one spinal nerve to the cord.

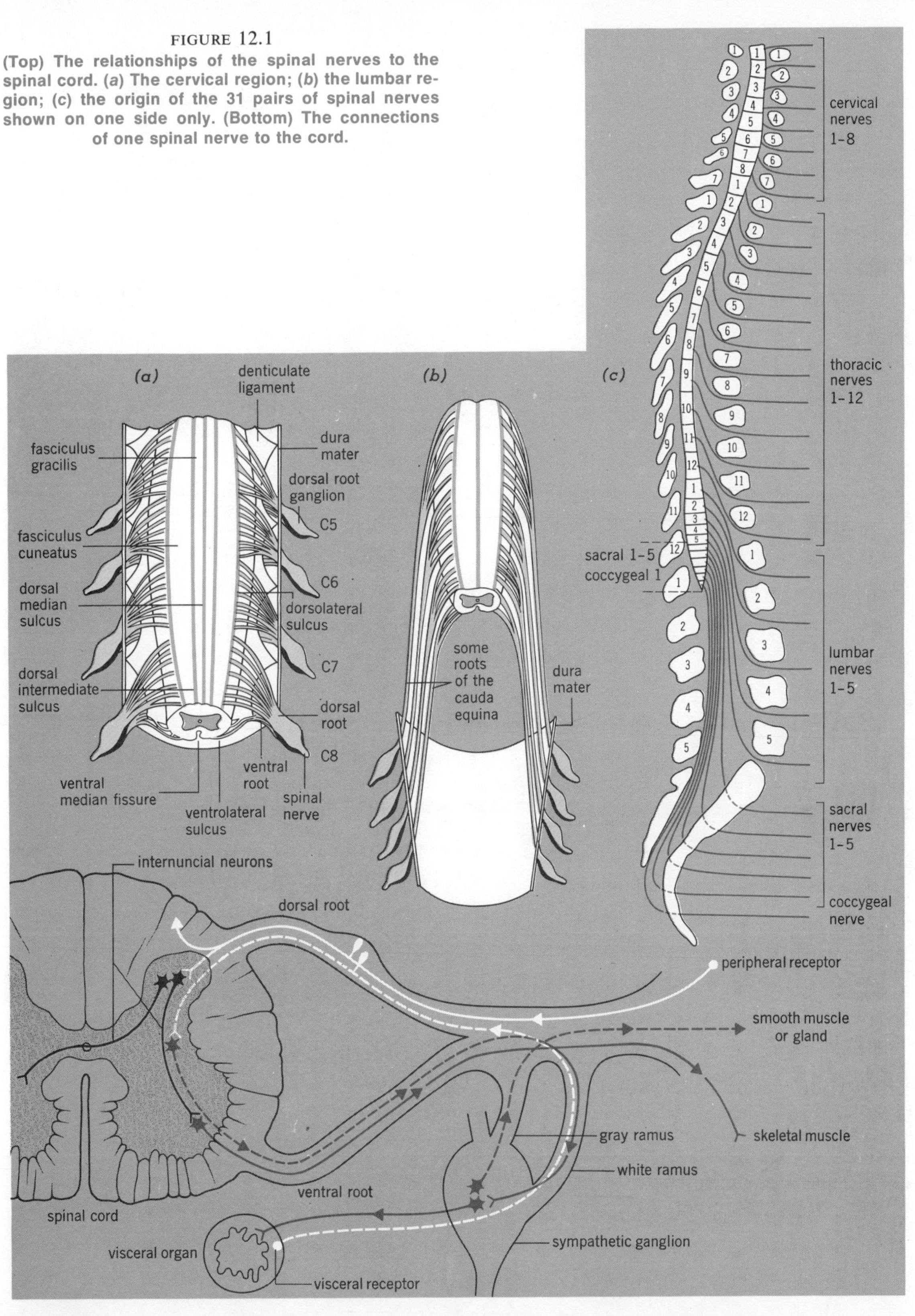

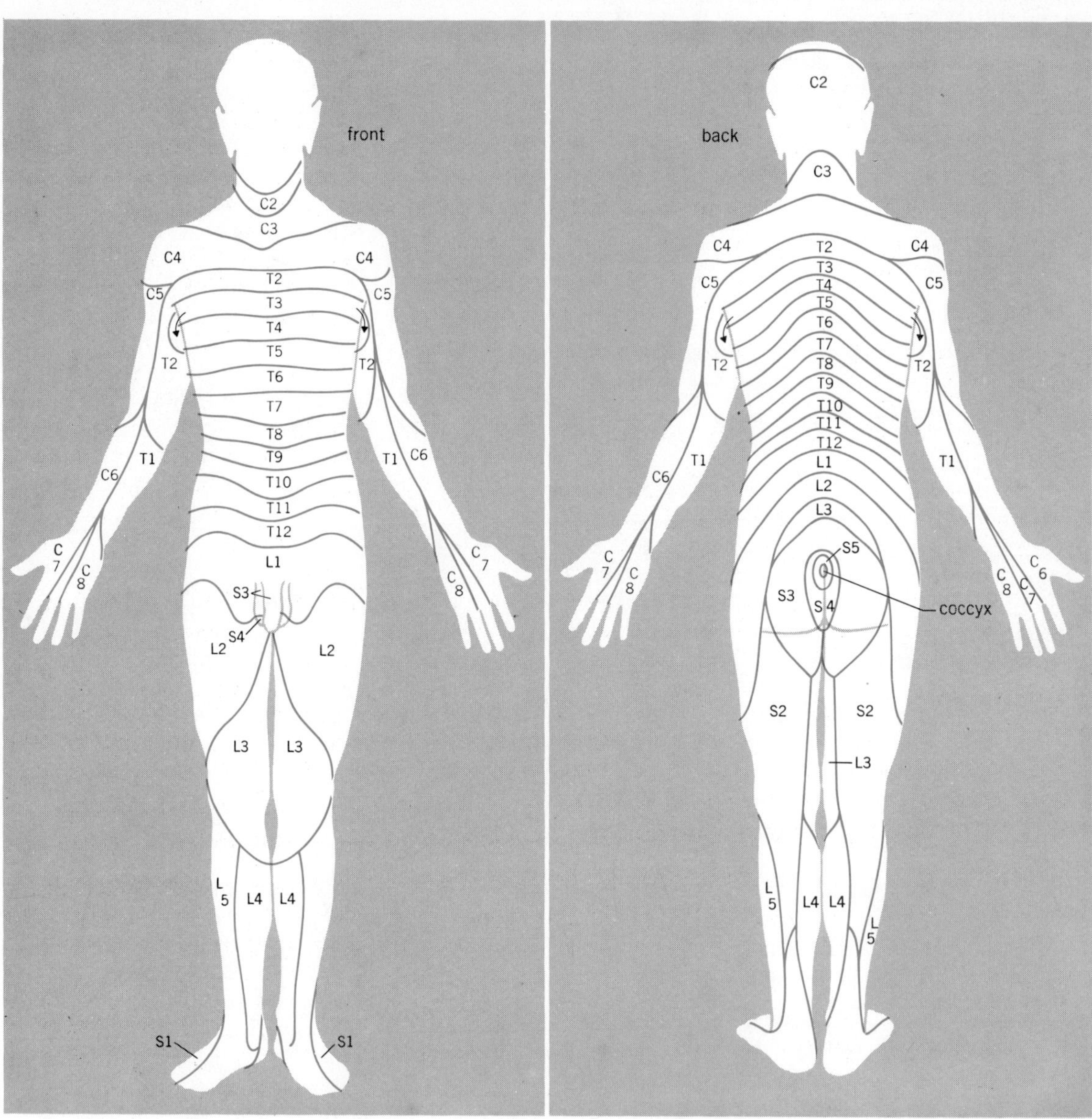

FIGURE 12.2
The segmental distribution of spinal nerves to the body.

FUNCTIONAL CLASSIFICATION OF COMPONENTS

Four functionally distinct types of fibers compose the spinal nerves. They are named SOMATIC AFFERENTS, SOMATIC EFFERENTS, VISCERAL AFFERENTS, and VISCERAL EFFERENTS.

Somatic afferents. Somatic afferent fibers are those carrying sensory impulses from receptors other than those serving sight, hearing, smell, and taste. The *spinal* components of the somatic efferent system serve all but the superior cranial and facial areas of the body in a segmental or regional fashion (Fig. 12.2). The somatic afferents' receptors (whose structure is considered in Chapter 13) are located primarily in the skin, and are activated by alterations in the external environment, such as temperature changes, pressure on the skin, or damage to the body surface. Such receptors, located in the skin and activated by external stimuli, are sometimes called *exteroceptive receptors,* and their sensations of touch, pressure, warmth, cold, and pain may be designated *exteroreceptive sensations*. Also conveyed by spinal nerves as part of the somatic afferent component are sensory fibers arising from receptors in tendons, joints, and skeletal muscles. These receptors are designated *proprioceptive receptors,* and their sensations as *proprioceptive sensations*. Information from these receptors informs us of body position, and direction, rate, and force of movements.

Information of sensory and motor nature is carried within the cord in specific *tracts,* and it would be worthwhile to review the information presented in Table 11.1 relating to tract names, their courses and terminations in the central nervous system.

Somatic efferents. Somatic efferents are the axons of the neurons having their cell bodies in the ventral gray columns of the spinal cord. They are the lower motor neurons of the system supplying skeletal muscles with their activating fibers. They terminate on skeletal muscles by way of neuromuscular junctions (see Chapter 5). It may be recalled that these neurons respond to both "voluntary" and "involuntary" impulses reaching them from several areas of the brain, and that their loss results in a flaccid paralysis.

Visceral afferents. Visceral afferent fibers are derived from receptors located in the walls of abdominal and thoracic blood vessels and organs. Their cell bodies are also in the dorsal root ganglion. These fibers convey information regarding the pressure within the organs, tension in their walls, chemical composition of fluids circulating within the walls, and volume of fluid in a hollow organ. Such receptors may be called *enteroceptive receptors,* and they are activated by changes within the body. They convey *enteroceptive impulses* that are extremely important in maintaining body homeostasis of blood pressure, blood gas composition, and blood flow. These fibers belong to the autonomic nervous system to be described later in this chapter.

Visceral efferents. These fibers also belong to the autonomic nervous system, and are derived from neurons whose cell bodies are located in the lateral gray columns of the spinal cord. These fibers terminate on smooth muscle of the body, and on certain body glands, such as sweat glands.

SEGMENTAL DISTRIBUTION OF SPINAL NERVES

Each pair of spinal nerves supplies a particular region of the body. The development of the muscles and skin and their innervation by the spinal nerves creates a SEGMENTAL DISTRIBUTION of spinal nerves that is retained as development proceeds (Fig. 12.3). A DERMATOME is defined as a skin area supplied by a single spinal nerve's dorsal root. The word implies that the muscles are innervated in a similar segmental pattern. Dermatomes overlap one another by about 30 percent, so that if a given spinal nerve is damaged, total loss in its dermatome will not occur. Knowing the pattern of spinal nerve distribution makes it easier to understand why there is specific area loss of sensation and/or movement if spinal nerves or the cord itself are damaged.

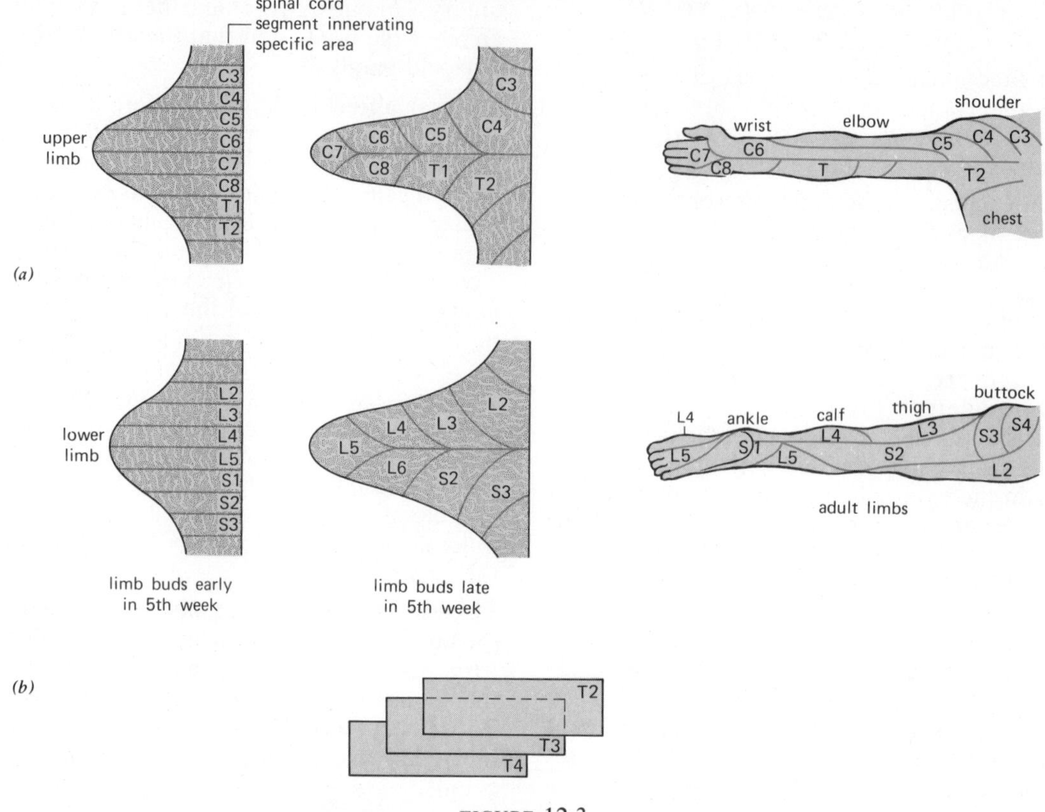

FIGURE 12.3

Diagrams illustrating dermatomal patterns of the limbs. Similar segmentation is shown on the trunk (*see* Fig. 12.2). (*a*) The development of limb dermatomes as the limb buds form the upper and lower appendages (limbs are viewed from the posterior aspect). (*b*) The overlap of innervation of adjacent dermatomes. C = cervical; T = thoracic; L = lumbar; S = sacral. [(*a*) redrawn from Moore.]

THE CRANIAL NERVES

The brain gives rise to 12 pairs of cranial nerves (Fig. 12.4) that supply sensory and motor fibers to structures in the head, neck, and shoulder region; one pair supplies fibers to body viscera. These nerves include some that have somatic components, others have visceral components, and still others serve the organs of special sense (eye, ear, taste buds, olfactory epithelium). Fibers belonging to the supply of special sense organs are sometimes designated as *special* visceral components.

The cranial nerves have NUCLEI within the brainstem that contain the cell bodies for their afferent and efferent nuclei. The positions of these nuclei are illustrated in Figure 12.5. A general description of each nerve follows, and a summary is presented in Table 12.1.

I. *Olfactory nerve*. The receptors for the sense of smell are located in the apices of the nasal cavities and their axons pass through the roof of the cavity to terminate in the olfactory bulb of the brain. The nerve carries only sensory impulses.

II. *Optic nerve*. The receptors for the sense of sight are the rods and cones of the retina. Impulses are carried over the optic nerve to the thalamus. This nerve carries only sensory impulses.

III. *Oculomotor nerve*. The oculomotor nerve is a mixed nerve in that it carries both sensory and motor fibers. The sensory component arises primarily from the retina and conveys impulses related to intensity of light entering the eye. Proprioceptive impulses from the muscles named below are also carried in the nerve. The motor component includes fibers to four of the six extrinsic eye muscles (superior rectus, medial rectus, inferior rectus, inferior oblique) that move the eyeball, the sphincter pupillae of the iris that constricts the pupil of the eye, and the ciliary muscle involved with focusing. The Edinger-Westphal nucleus is that

FIGURE 12.4
A basal view of the brain to illustrate the origins of cranial nerves.

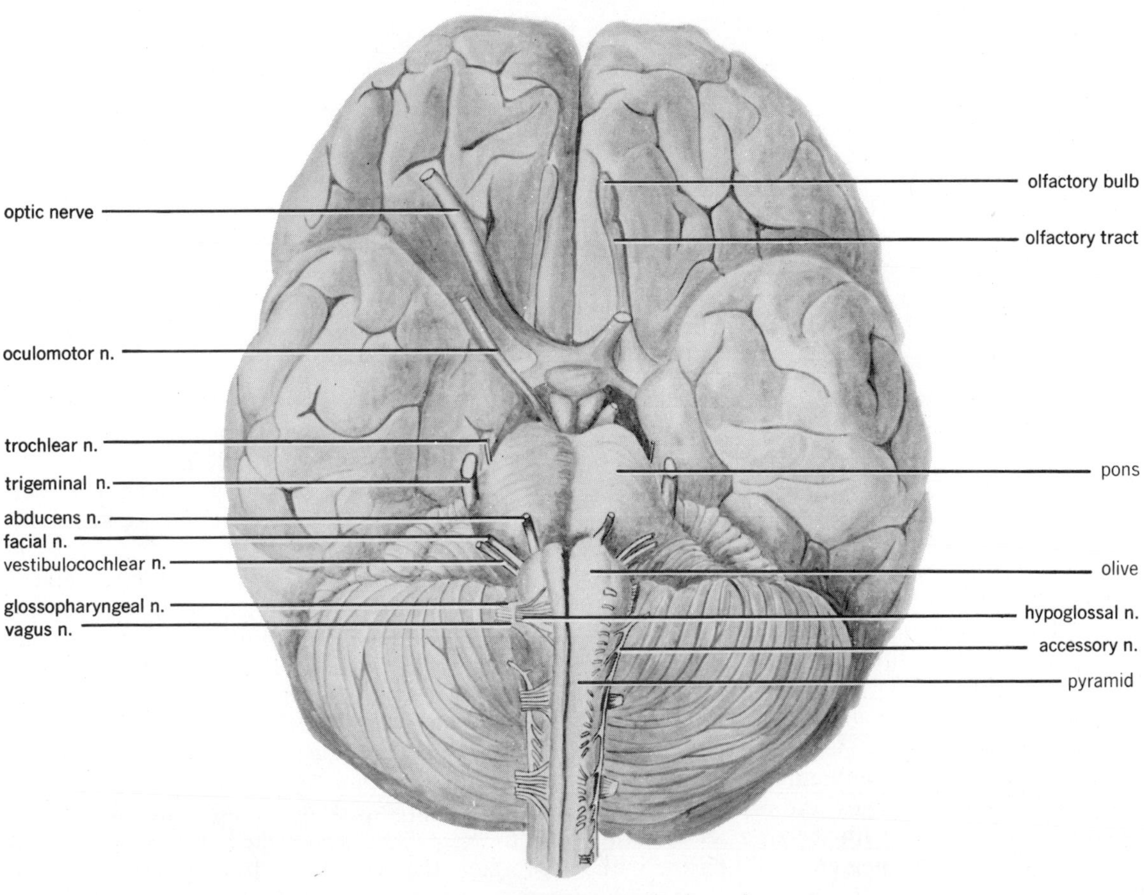

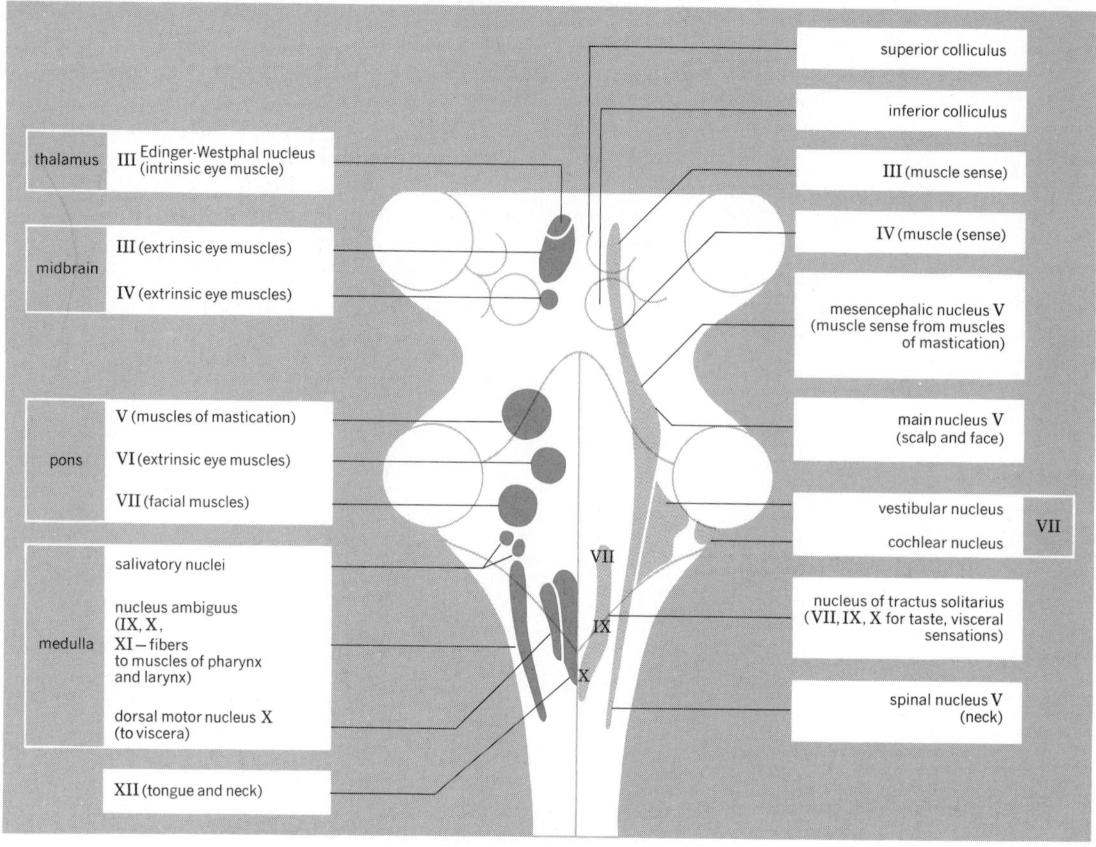

FIGURE 12.5
Diagram illustrating the positions of the cranial nerve nuclei in the brainstem.

part of the oculomotor nucleus concerned with mediating response to light shown in the eye.

IV. *Trochlear nerve*. Also a mixed nerve, the trochlear carries proprioceptive impulses from the superior oblique extrinsic eye muscle, and motor fibers to the same muscle.

V. *Trigeminal nerve*. This nerve is a mixed nerve and is the largest cranial nerve. It is often characterized as "the great sensory nerve of the head." Its sensory division is its largest component, and supplies, by three divisions, the anterior cranium and face. The *ophthalmic nerve* supplies sensory fibers to the area of the cranium above a line extending roughly from the nose through the eyes; the *maxillary nerve* supplies sensory fibers for the area between the above and a line running generally in line with the upper teeth to the ear; the *mandibular nerve* supplies the lower jaw and an area on the lateral side of the skull (Fig. 12.6). The motor component of the nerve supplies certain muscles involved in chewing (masseter, pterygoids, temporalis).

VI. *Abducent (abducens) nerve*. A mixed nerve that carries proprioceptive sensory fibers from, and motor fibers to, the lateral rectus muscle. This muscle is the last of the six extrinsic eye muscles (three nerves thus serve the eyeball muscles: III, IV, and VI).

VII. *Facial nerve*. A mixed nerve in which the motor component predominates, this nerve is often characterized as "the great motor nerve of the face." It supplies the muscles of facial expression that enable us to laugh, frown, pout,

TABLE 12.1
SUMMARY OF THE CRANIAL NERVES

NERVE	COMPOSITION M = MOTOR; S = SENSORY	ORIGIN	CONNECTION WITH BRAIN OR PERIPHERAL DISTRIBUTION	FUNCTION
I. Olfactory	S	Nasal olfactory area	Olfactory bulb	Smell
II. Optic	S	Ganglionic layer of retina	Optic tract	Sight
III. Oculomotor	MS	M—Midbrain	4 of 6 extrinsic eye muscles (superior rectus, medial and inferior rectus, inferior oblique)	Eye movement
		S—Ciliary body of eye	Nucleus of nerve in midbrain	Focusing, pupil changes, muscle sense
IV. Trochlear	MS	M—Midbrain	1 extrinsic eye muscle (superior oblique)	Eye movement
		S—Eye muscle	Nucleus of nerve in midbrain	Muscle sense
V. Trigeminal	MS	M—Pons	Muscles of mastication	Chewing
		S—Scalp and face	Nucleus of nerve in pons	Sensation from head
VI. Abducent	MS	M—Nucleus of nerve in pons	1 extrinsic eye muscle (lateral rectus)	Eye movement
		S—1 extrinsic eye muscle	Nucleus of nerve in pons	Muscle sense
VII. Facial	MS	M—Nucleus of nerve in lower pons	Muscles of facial expression	Facial expression
		S—Tongue (ant. 2/3)	Nucleus of nerve in lower pons	Taste
VIII. Vestibulocochlear (Statoacoustic, acoustic, auditory)	S	Internal ear: balanced organs, cochlea	Vestibular nucleus, cochlear nucleus	Posture, hearing
IX. Glossopharyngeal	MS	M—Nucleus of nerve in lower pons	Muscles of pharynx	Swallowing
		S—Tongue (post 1/3), pharynx	Nucleus of nerve in lower pons	Taste, general sensation
X. Vagus	MS	M—Nucleus of nerve in medulla	Viscera	Visceral muscle movement
		S—Viscera	Nucleus of nerve in medulla	Visceral sensation
XI. Accessory	M	Nucleus of nerve in medulla	Muscles of throat, larynx, soft palate, sternocleidomastoid, trapezius	Swallowing, head movement
XII. Hypoglossal	M	Nucleus of nerve in medulla	Muscles of tongue and infrahyoid area	Speech, swallowing

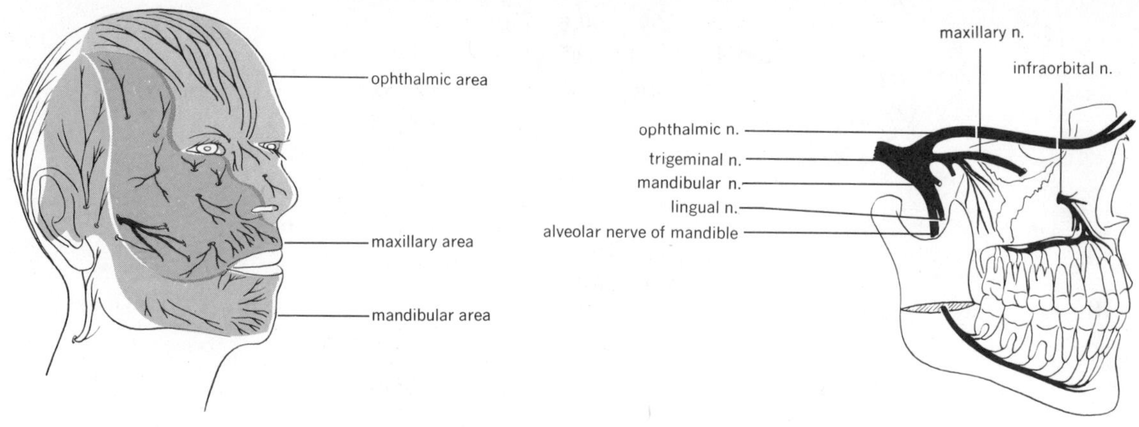

FIGURE 12.6
The distribution of the Vth cranial nerve (trigeminal).

and assume other expressions related to our emotions, and to move the face in speech. The sensory component is derived from taste buds located on the anterior two-thirds of the tongue.

VIII. *Vestibulocochlear (statoacoustic, acoustic, auditory) nerve.* A purely sensory nerve, it carries impulses from the cochlea (*cochlear nerve*) for hearing, and organs of equilibrium (*vestibular nerve*) of the inner ear for postural and equilibrial reactions.

IX. *Glossopharyngeal nerve.* A mixed nerve, the sensory component serves taste buds on the posterior one-third of the tongue, and motor fibers carry impulses to throat muscles involved in swallowing.

X. *Vagus nerve.* Sometimes referred to as the most important cranial nerve of the autonomic nervous system, the vagus nerve carries motor and sensory fibers from many abdominal viscera (Fig. 12.7). The figure shows the motor fibers of the nerve; sensory fibers follow a similar distribution. The sensory component originates in enteroceptive receptors in the organs indicated; the motor component supplies smooth muscle of organ walls, blood vessels, and glands.

XI. *Accessory (spinal accessory) nerve.* A motor nerve, this nerve supplies fibers to throat muscles involved in swallowing, and two muscles of the upper back and neck (trapezius and sternocleidomastoid).

XII. *Hypoglossal nerve.* A motor nerve, this nerve supplies fibers to the muscles of the tongue and upper anterior neck. These muscles are involved in speech and swallowing.

Noting the distribution and functions of the cranial nerves, one might appreciate that it is possible to test their functional integrity by relatively simple methods. Table 12.2 presents methods of testing cranial nerve function, with comments.

THE AUTONOMIC NERVOUS SYSTEM

The autonomic nervous system has been classically defined as that part of the peripheral nervous system supplying motor fibers to smooth and cardiac muscles and glands, operating "automatically" at the reflex and subconscious levels. Such a definition is too limited, for it is obvious that there are fibers of sensory nature that pass from organs to the central nervous system (Fig. 12.8) and that form the afferent limbs of many autonomic reflex arcs. Today, we recognize that there are VISCERAL AFFERENT FIBERS, carrying sensory input from organs, and VISCERAL EFFERENT FIBERS to organs.

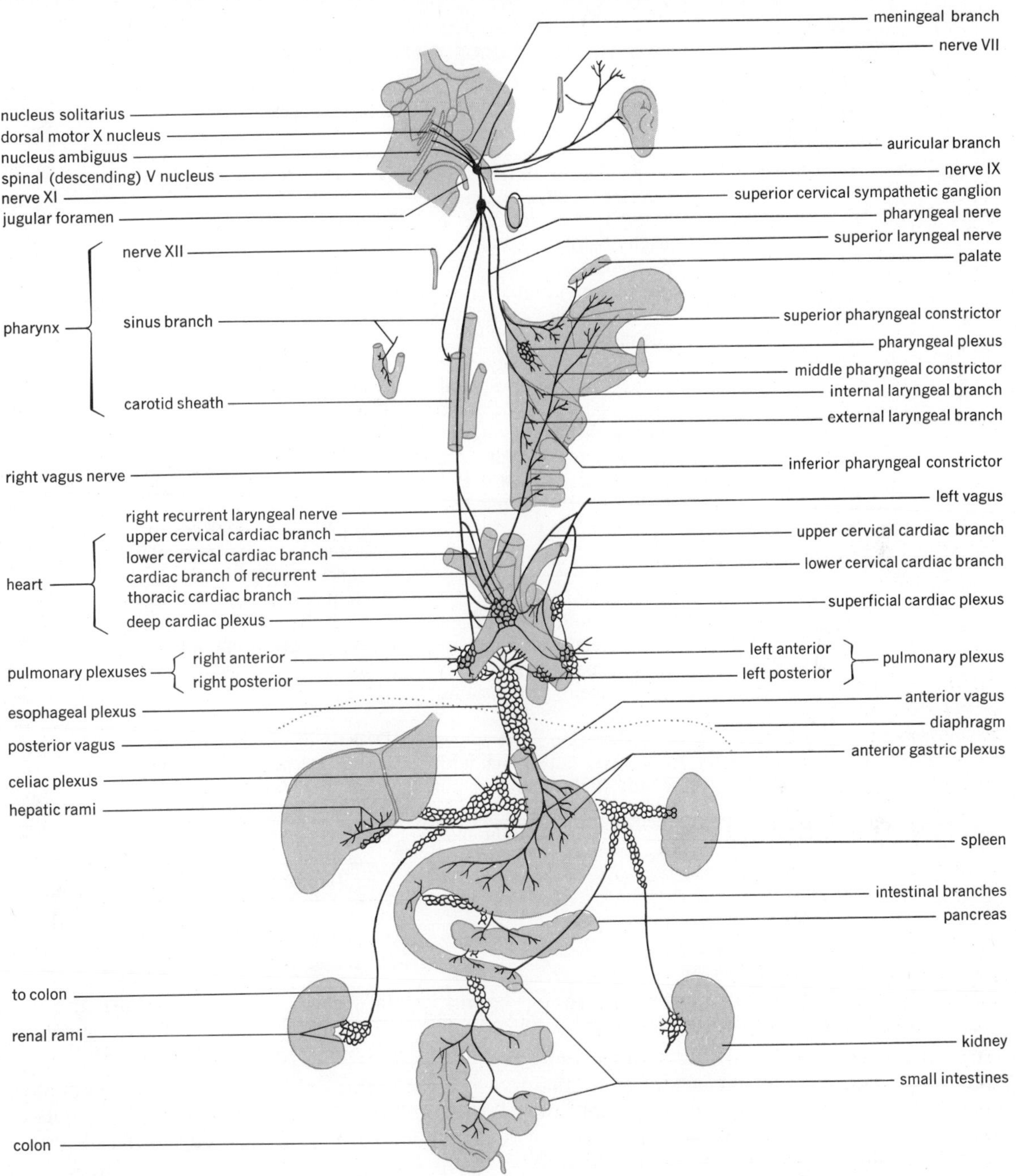

FIGURE 12.7
The distribution of the Xth cranial nerve (vagus).

TABLE 12.2
METHODS OF ASSESSING CRANIAL NERVE FUNCTION

NERVE BEING TESTED	METHOD	COMMENTS
Olfactory	Ask patient to differentiate odors of coffee, bananas, tea, oil of cloves, etc., in each nostril.	Test may be voided if patient has a cold or other nasal irritation.
Optic	Ask patient to read newspaper with each eye; examine fundus with ophthalmoscope.	Perform test with lenses on, if patient wears them.
Oculomotor	Shine light in each eye separately and observe pupillary changes; look for movement by moving finger up, down, right, and left; have subject's eyes follow finger movements.	Look for defects in both eyes at same time. Light in one eye causes other one to constrict to lesser degree.
Trochlear		Requires special devices. If eye movement is normal, nerves are probably all right.
Trigeminal	Motor portion: have patient clench teeth and feel firmness of masseter; have patient open jaw against pressure.	Subject should exhibit firmness and strength in both tests.
	Sensory portion: test sensations over entire face with cotton (light touch) or pin (pain).	Deficits in particular areas indicate problems in one of the three branches.
Abducent	Perform test for lateral movement of eyes.	Must be very careful to observe any deficit in lateral movement.

CONNECTIONS OF THE SYSTEM WITH THE CENTRAL NERVOUS SYSTEM

Afferent pathways. Visceral afferent fibers enter the spinal cord through the posterior (dorsal) roots, in company with somatic afferents from the skin and skeletal muscles. One neuron reaches from the periphery to the central nervous system and its cell bodies are found in the dorsal root ganglia of the spinal nerves. Synapses usually occur in the dorsal gray column. Visceral afferents may also enter in various cranial nerves, and in this case the cell bodies are found with the sensory nuclei of those nerves located in the brainstem.

Efferent pathways. Visceral efferent fibers have their cell bodies in the lateral gray columns of the spinal cord, or in the motor nuclei of certain cranial nerves. Those that leave by way of anterior spinal nerve roots separate from the somatic fibers by way of the WHITE RAMI. They may undergo synapses in ganglia (see below) close to the cord or pass through the ganglia to

TABLE 12.2 (continued)

METHODS OF ASSESSING CRANIAL NERVE FUNCTION

NERVE BEING TESTED	METHOD	COMMENTS
Facial	Have subject wrinkle forehead, scowl, puff out his cheeks, whistle, smile. Note nasolabial fold (groove from nose to corners of mouth).	Nerve supplies all facial muscles. Fold is maintained by facial muscles and may disappear in nerve damage.
Vestibulocochlear (statoacoustic, acoustic, auditory)	Cochlear portion: test auditory acuity with ticking watch; repeat a whispered sentence; use tuning fork. Compare ears. Examine with otoscope.	Ears may not be equal in acuity; note.
	Vestibular portion: not routinely tested.	If subject appears to walk normally and keep his balance, nerve is usually all right.
Glossopharyngeal and Vagus	Note disturbances in swallowing, talking, movements of palate.	
Accessory	Test trapezius by having subject raise shoulder against resistance; have subject try to turn head against resistance.	Strength is index here.
Hypoglossal	Have subject stick out tongue. Push tongue against tongue blade.	Tongue should protrude straight; deviation indicates same side nerve damage. Pushing indicates strength.

synapse in ganglia closer to the organs innervated. If a synapse does occur in the ganglia close to the cord, the nerve fibers usually rejoin the spinal nerve by way of the GRAY RAMI, and are then distributed to organs located within the skin and skeletal muscles (mostly blood vessels and skin glands). Two autonomic neurons usually extend from the CNS to the organ innervated. The one passing from the CNS to a synapse in the periphery is called the PREGANGLIONIC NEURON; the one passing from the synapse to the organ supplied is called the POSTGANGLIONIC NEURON. Transmission from pre- to postganglionic neurons is chemical, utilizing acetylcholine. Postganglionic neurons produce either acetylcholine or norepinephrine at their endings, and this accounts for the different effects exerted on the organs by the two divisions of the autonomic system.

AUTONOMIC GANGLIA AND PLEXUSES

The term GANGLION is usually used to refer to

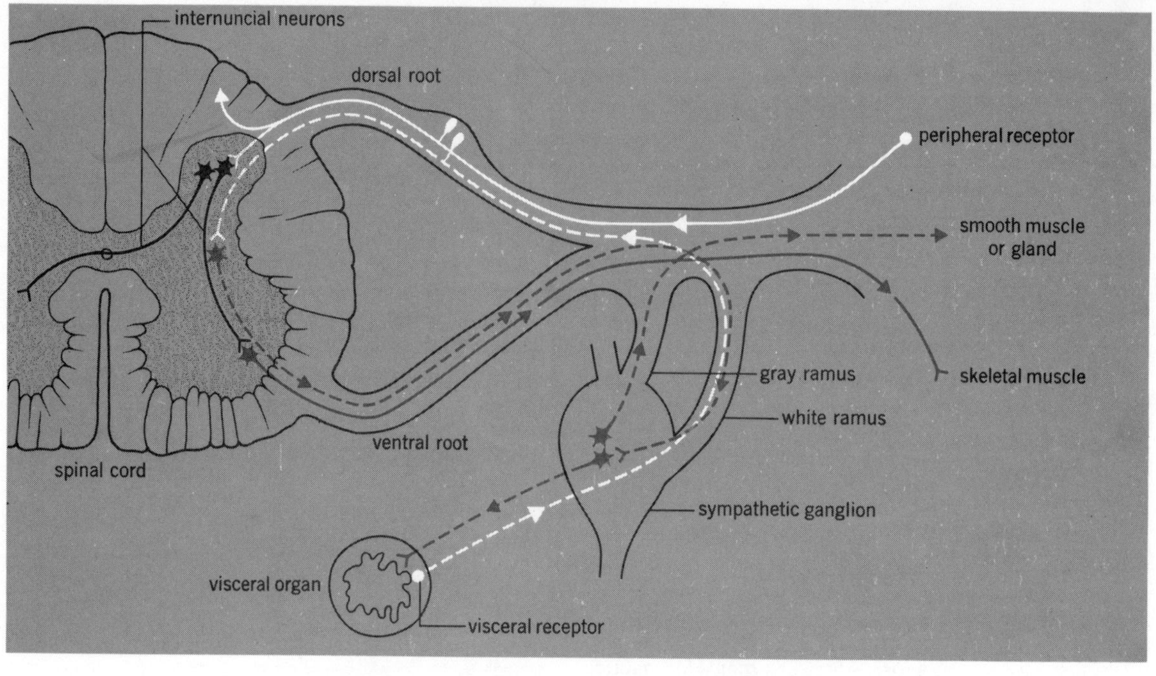

FIGURE 12.8
The connections of sensory and motor neurons to the spinal cord. Dashed lines indicate autonomic fibers.

masses of nervous tissue that contain nerve cell bodies and/or synapses and that are located in the periphery, outside of brain and spinal cord. The term *plexus* usually refers to a network of nerve fibers or blood or lymphatic vessels. In describing the nervous system, the two terms are *sometimes* used interchangeably, with the term ganglion understood to mean a synaptic region.

Synapses between pre- and postganglionic neurons occur in one of three areas in the periphery.

The LATERAL or VERTEBRAL GANGLIA, also called the *sympathetic* ganglia, form paired chains of 22 ganglia that lie alongside the vertebral bodies in the thoracic and abdominal cavities. They receive preganglionic fibers from the thoracic and lumbar spinal nerves.

The COLLATERAL or PREVERTEBRAL GANGLIA form several groups of nerves and synapses in association with visceral organs or large blood vessels. The *cardiac, celiac* (solar), and *mesenteric plexuses* are three important ones located near the heart, spleen, and urinary bladder, respectively. These plexuses receive preganglionic fibers from the cranial, thoracic, lumbar, and sacral portions of the central nervous system.

TERMINAL GANGLIA or plexuses lie close to or within the organ supplied. The nerve plexuses in the wall of the intestine are examples of terminal ganglia. These ganglia receive preganglionic fibers from the cranial and sacral portions of the central nervous system.

DIVISIONS OF THE SYSTEM: STRUCTURE AND FUNCTION

Preganglionic autonomic fibers form three outgoing groups (outflows). The CRANIAL OUT-

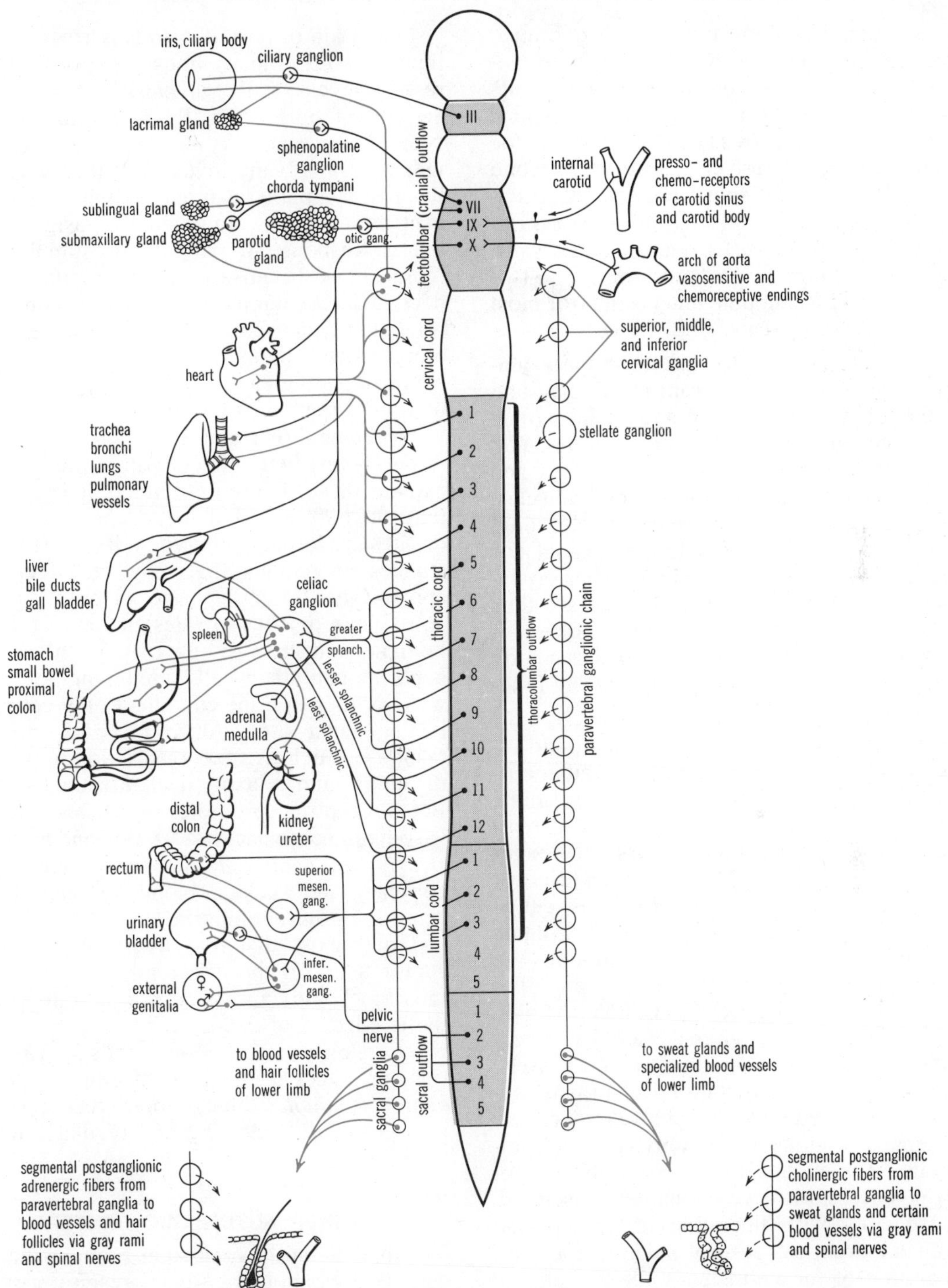

FIGURE 12.9
The autonomic nervous system.

FLOW is composed of the motor fibers of cranial nerves III, VII, IX, and X. The THORACOLUMBAR OUTFLOW consists of the visceral efferents derived from all thoracic and lumbar spinal nerves. The SACRAL OUTFLOW consists of the visceral efferents derived from the second through fourth sacral spinal nerves. Functionally, the fibers are grouped into a PARASYMPATHETIC (*craniosacral*) DIVISION and a SYMPATHETIC (*thoracolumbar*) DIVISION. The two divisions provide a double innervation for most (but not all) body organs (Fig. 12.9).

The parasympathetic division. The parasympathetic division, formed from the cranial and sacral outflows, supplies fibers to all autonomic effector organs except the adrenal medulla, sweat glands, splenic smooth muscle, and blood vessels of skin and the skeletal muscles. The preganglionic fibers are relatively long; they synapse in collateral or terminal ganglia or both. The postganglionic fibers are shorter, secrete acetylcholine at their ends, and are thus called *cholinergic* fibers. The cranial portion of this division tends to protect and conserve body resources and to preserve normal resting functions.

The sympathetic division. The sympathetic division is formed by the thoracic and lumbar outflows. Preganglionic fibers are relatively short; they synapse in vertebral and/or collateral ganglia. The postganglionic fibers are longer and most of them secrete norepinephrine at their endings. These fibers are thus called *adrenergic* (from noradrenalin, another name for norepinephrine). This division supplies the same organs as does the parasympathetic division, creating a dual innervation, and is the only nerve supply to the adrenal medulla, sweat glands, skin and skeletal muscle blood vessels, and spleen. In the latter group of organs, activity is controlled only by increase or decrease in sympathetic stimulation. Activity of the sympathetic division tends to increase utilization of body resources. Massive sympathetic discharge is associated with liberation of epinephrine (adrenalin) from the adrenal medulla and with preparation of the entire body for a "fight-or-flight" posture against the agent causing the great stimulation.

Operation of both divisions is TONIC or continuous, so that many body functions, even at resting levels, reflect a balance between the activity of both parts. Response to the parasympathetic system is more specific, with individual organs being influenced; this is because the parasympathetic preganglionic fibers connect with fewer "individual" postganglionic fibers. Response to sympathetic stimulation is more diffuse because preganglionic fibers connect to larger numbers of postganglionic neurons, and because norepinephrine is destroyed more slowly than acetylcholine. Table 12.3 summarizes the effects on body organs of stimulation of the two divisions.

Norepinephrine and epinephrine cause different sympathetic effects within the body. To explain these differential effects it is assumed that two different kinds of receptor substances or sites are found in effector cells—that is, substances or regions that may be affected by epinephrine, norepinephrine, or both. The two substances or sites are designated ALPHA RECEPTORS and BETA RECEPTORS. In theory, epinephrine can affect cells containing both sites, while norepinephrine can affect only cells containing alpha sites. Alpha receptors promote responses such as vasoconstriction, dilation of the pupil, and relaxation of intestinal muscle. Beta receptors promote responses such as vasodilation, cardiac acceleration and increased strength of heart contraction, vasodilation of coronary and skeletal blood vessels, and relaxation of uterine muscle.

Drugs exist that may compete for the receptor sites on the effector organs, and may thus block the expression of a certain sympathetic effect. For example: *phenoxybenzamine* blocks alpha receptors and leads to vasodilation, an effect useful in controlling hypertension; *propanolol* blocks beta receptors and leads to cardiac slowing and regularization of heart beat.

HIGHER AUTONOMIC CENTERS

The medulla of the brainstem contains the vital centers (vasomotor, cardiac, respiratory) that are involved in controlling blood pressure and breathing. The cerebral cortex contains, in the

TABLE 12.3 (continued on p. 194)

EFFECTS OF AUTONOMIC STIMULATION

ORGAN AFFECTED	PARASYMPATHETIC EFFECTS	SYMPATHETIC EFFECTS
Iris	Contraction of sphincter pupillae; pupil size decreases	Contraction of dilator pupillae; pupil size increases
Ciliary muscle	Contraction; accommodation for near vision	Relaxation; accommodation for distant vision
Lacrimal gland	Secretion	Excessive secretion
Salivary glands	Secretion of watery saliva in copious amounts	Scanty secretion of mucus rich saliva
Respiratory system:		
Conducting division	Contraction of smooth muscle; decreased diameters and volumes	Relaxation of smooth muscle; increased diameter and volumes
Respiratory division	Effect same as on conducting division	Effect same as on conducting division
Blood vessels	Constriction	Dilation
Heart:		
Stroke volume	Decreased	Increased
Stroke rate	Decreased	Increased
Cardiac output and blood pressure	Decreased	Increased
Coronary vessels	Constriction	Dilation
Peripheral blood vessels:		
Skeletal muscle	No innervation	Dilation
Skin	No innervation	Constriction
Visceral organs (except heart and lungs)	Dilation	Constriction

frontal lobes, areas that produce both sympathetic and parasympathetic effects. These regions may be tied in with expression of emotions, such as anger and rage.

The hypothalamus appears to be involved to a great degree in the control of what are sometimes called the vegetative (involuntary) functions of the body. Many of the functions of the hypothalamus discussed in Chapter 7 are autonomic in nature.

CLINICAL CONSIDERATIONS

NEURITIS is a term referring to a syndrome of reflex, sensory, motor, and vasomotor symptoms that occurs as a result of a lesion in a nerve root to or from the central nervous system, or as a result of a lesion of a peripheral nerve. Mechanical damage (trauma), infections (bacteria or viruses), or toxins (poisons) are commonly involved as etiologic agents in neuritis. Pins-and-needles sensations, pain, numbness, muscular weakness, and sweating of the affected body part are some of the more common symptoms that develop.

NEURALGIA is a term referring to pain that develops within the distribution of a peripheral sensory nerve when no apparent cause for the pain can be demonstrated. The pain is commonly described as intense, stabbing, or shooting in quality, and is difficult to relieve. Analgesics usually afford only temporary relief.

SHINGLES (*herpes zoster*) is an acute infection of sensory nerves as they enter the spinal

TABLE 12.3 (continued)
EFFECTS OF AUTONOMIC STIMULATION

ORGAN AFFECTED	PARASYMPATHETIC EFFECTS	SYMPATHETIC EFFECTS
Stomach:		
Wall	Increased motility	Decreased motility
Sphincters	Inhibited	Stimulated
Glands	Secretion stimulated	Secretion inhibited
Intestines:		
Wall	Increased motility	Decreased motility
Sphincters:		
Pyloric, iliocecal	Inhibited	Stimulated
Internal anal	Inhibited	Stimulated
Liver	Promotes glycogenesis; promotes bile secretion	Promotes glycogenolysis; decreases bile secretion
Pancreas (exocrine and endocrine)	Stimulates secretion	Inhibits secretion
Spleen	No innervation	Contraction and emptying of stored blood into circulation
Adrenal medulla	No innervation	Epinephrine secretion
Urinary bladder	Stimulates wall, inhibits sphincter	Inhibits wall, stimulates sphincter
Uterus	Little effect	Inhibits motility of nonpregnant organ; stimulates pregnant organ
Sweat glands	No innervation	Stimulates secretion (produces "cold sweat" when combined with cutaneous vasoconstriction)

cord and is due to a virus. Recovery occurs as antibodies are produced to the virus.

CRANIAL NERVE FUNCTIONS are easily assessed by simple procedures, an important part of a neurological examination, presented in Table 12.2.

Some clinical conditions are life-threatening and are believed to be the result of excessive autonomic activity. One of these is HYPERTENSION (*high blood pressure*). Surgical or chemical intervention designed to relieve the excessive autonomic activity (if this is the cause) is sometimes employed.

SURGERY AND THE AUTONOMIC SYSTEM

Because the effect of sympathetic stimulation on most visceral blood vessels is to constrict them, and since constriction of large numbers of blood vessels raises the blood pressure, removal of the sympathetic effect should lower the blood pressure. A SYMPATHECTOMY, which removes the sympathetic ganglia from the levels of T10-L2, may be performed to alleviate hypertension. If the procedure is effective, blood pressure will fall, but since the ganglia are gone, the individual has lost the ability to raise blood presssure in stress or in changes of position.

VAGOTOMY, or sectioning of the vagal supply to the stomach, is sometimes employed to treat ulcers. The operation diminishes secretion of gastric glands as a result of stress, lowering the amounts of acid and enzymes in the stomach when no food is present.

DRUGS AND THE AUTONOMIC SYSTEM

An alternative to surgical intervention in disorders of autonomic function is to employ drugs that either block or depress the effect of one of the divisions of the system, increase the activity of a division, or mimic the effects of stimulation of a division.

Drugs that mimic effects of stimulation of a given division are said to be *parasympathomimetic* or *sympathomimetic*. Those that prevent synaptic transmission are *ganglionic blocking agents*. Table 12.4 presents a summary of some drugs that affect the autonomic system, with notes as to how they act. Most are employed to alleviate hypertension by causing vasodilation.

A chemical that produces the same effect as cutting of the parasympathetic nerves is termed a *parasympatholytic* drug. Such drugs usually operate by blocking the effect of acetylcholine at myoneural junctions and organs supplied by postganglionic cholinergic neurons. Thus, the cells affected are chiefly smooth muscle, cardiac muscle, and exocrine gland cells. Smooth muscle is usually caused to relax, heart rate increases, secretion is inhibited, and the pupils of the eye dilate. Atropine and scopolamine are examples of parasympatholytic drugs.

TABLE 12.4
DRUGS AND THE AUTONOMIC SYSTEM

DRUG	HOW ACTS	USE	COMMENTS
Reserpine	Depletes norepinephrine from postganglionic endings by increasing release. Causes decrease in heart action.	To alleviate hypertension	Is one of, and the most potent of a series of alkaloids derived from the Rauwolfia plant. Norepinephrine loss results in vasodilation and fall in blood pressure, and decrease in heart action.
Guanethidine	As above, but exerts no effect on heart.	As above	Limited side effects make it a "clinically advantageous drug."
Methyldopa	Lowers brain and heart content of norepinephrine by interfering with synthesis.	As above	May produce toxicity of liver.
Hydralazine	Depress vasoconstrictor center and inhibit sympathetic stimulation.	As above	Many side effects are common.
Veratrum (a series of plant alkaloids)	Slow heart action, stimulate vagus (parasympathetic) activity	As above	Some side effects.
Hexamethonium	Blocks pre- to post-ganglionic transmission. Anticholinergic.	As above	
Acetylcholine	Increase or mimics parasympathetic stimulation. Causes some vasodilation.	As above	Not routinely employed. It is rapidly destroyed and affects many organs other than blood vessels.
Atropine	Inhibits acetylcholine and parasympathetic effects	Eye examinations; before general anaesthesia to dry respiratory secretions.	Sympathomimetic
Pilocarpine	Mimics parasympathetic stimulation.	Treatment of glaucoma.	Parasympathomimetic

Atropine or its analogs are commonly used to dilate the eye for examination.

A *sympatholytic* agent is one that blocks the effect of epinephrine or norepinephrine. Such agents therefore are effective in lowering blood pressure by causing arteriolar dilation, and in stimulating gastrointestinal motility. Ergot and its derivatives are natural sympatholytic agents, and propanolol, azapetine, phenoxybenzamine, and piperoxan are synthetic sympatholytic agents. Because of transient action and often troublesome side effects, such agents are not drugs of choice in the treatment of hypertension.

SUMMARY

1. The peripheral nervous system consists of the spinal nerves, cranial nerves, and their organization into the somatic and autonomic nervous systems.
2. There are 31 pairs of spinal nerves, divided into 8 cervical, 12 thoracic, 5 lumbar, 5 sacral, and 1 coccygeal.
   a. The nerves exit through foramina between vertebrae.
   b. Each nerve has a dorsal (sensory) and ventral (motor) root.
   c. Peripheral rami and communicating rami provide pathways for somatic and autonomic fibers.
3. Four functional types of fibers compose each spinal nerve.
   a. Somatic afferents (sensory) are derived from skin, skeletal muscle, and tendons and joints.
   b. Somatic efferents (motor) supply skeletal muscle.
   c. Visceral afferents (sensory) carry impulses from thoracic and abdominal viscera.
   d. Visceral efferents (motor) supply smooth and cardiac muscle, and glands.
4. Spinal nerves are distributed in segmental fashion to skin and muscles.
5. There are 12 pairs of cranial nerves that supply sensory and motor fibers to structures of the head, neck, and body viscera. (The nerves are named and described as to function; *see* Table 12.1).
6. The autonomic nervous system controls visceral activity and glandular secretion.
7. The efferent autonomic pathway usually contains two neurons and one synapse. The neurons are designated as pre- and postganglionic neurons. Transmission between them is by acetylcholine. Postganglionic fibers produce either acetylcholine or norepinephrine.
8. Autonomic ganglia and plexuses provide areas for synapse of pre- and postganglionic neurons.
9. Two divisions of the autonomic system are recognized.
   a. The parasympathetic (craniosacral) division is composed of cranial and sacral outflows, and conserves body resources.
   b. The sympathetic (thoracolumbar) division consists of thoracic and lumbar outflows and increases utilization of body resources.
   c. Most body viscera and glands receive fibers from both divisions, creating a dual innervation. Sweat glands, the adrenal medulla, spleen, and blood vessels of skin and skeletal muscles receive only sympathetic fibers.
   d. Effects (*see* Table 12.3) of both systems are tonic (continuous) and usually antagonistic.
10. Higher centers in the brainstem, hypothalamus, and cerebral cortex are responsible for autonomic effects.
11. In treatment of hypertension or ulcers, surgical or chemical intervention may be necessary.
    a. Sympathectomy involves removal of sympathetic ganglia to stop vasoconstrictive impulses to visceral blood vessels.
    b. Vagotomy reduces gastric secretion in ulcer formation.
    c. Drugs may block vasoconstrictive effects.

QUESTIONS

1. What are the anatomical and functional components of the spinal nerves?
2. In what ways are the somatic and visceral components of the spinal nerves alike? Different?

3. If each of the four functional components of a spinal nerve were to be cut or blocked with drugs, what would be the pattern of loss on the body? Explain.
4. What cranial nerve(s)
   a. Are involved in eye movement?
   b. Are involved in taste?
   c. Give sensation to the face?
   d. Let us smile, frown, speak?
   e. Control visceral activity?
   f. Are involved in balance?
5. What are the components of the peripheral nervous system that are included in the autonomic nervous system?
6. How are the differential effects of the two parts of the autonomic system explained?
7. What organs receive only sympathetic nerves? How is their function controlled if only one division supplies them?
8. What brain areas serve autonomic functions? Are these functions sympathetic or parasympathetic in nature?
9. Propanolol is described as a "beta-adrenergic blocking agent." What does this mean, and what is the rationale for the use of such agents in control of heart action or blood pressure?

READINGS

Lefkowitz, Robert J. "β-adrenergic Receptors: Recognition and Regulation." *New Eng. J. Med. 295*:323. Aug 5, 1976.

13

SENSORY FUNCTIONS OF THE NERVOUS SYSTEM

OBJECTIVES

After studying this chapter, the reader should be able to:

☐ Define what a receptor is, and explain what the properties common to all receptors are.

☐ Construct a classification of receptors according to their adequate stimulus.

☐ Describe the receptors for touch and pressure, their cord pathways and termination of those pathways in the cerebrum.

☐ Explain how it is possible to discriminate pressure and touch intensity and two-point separation.

☐ Name the receptors for thermal sensations, their cord pathways, and termination of those pathways in the cerebrum.

☐ Account for the "upside-down" representation of sensory function in the cerebrum.

☐ Describe the various types of pain.

☐ Describe the receptors for pain, its cord pathways, and the brain areas for discrimination and localization of pain.

☐ Define what kinesthetic sense is, its receptors, cord pathways, and termination of those pathways in the brain.

☐ Define what a synthetic sense is, name the synthetic senses, and describe how they are produced.

The ability to maintain homeostasis by adjustments of function implies the presence of structures, RECEPTORS, capable of detecting alterations in the internal and/or external environments of the body. Passage of the impulses generated in response to change must be transmitted over SENSORY PATHWAYS (afferent nerves) to the CENTRAL NERVOUS SYSTEM for processing and interpretation. Then, MOTOR PATHWAYS (efferent neurons) will conduct impulses to EFFECTORS for adjustment of a particular process or mechanism. That portion of the system described above dealing with detection, transmission, and interpretation of change constitutes the sensory portion of the nervous system. This portion of the system also enables recognition of location, intensity, and quality of a stimulus and may lead to learning and storage of information.

RECEPTORS

A RECEPTOR is generally considered to be the peripheral ending of an afferent nerve and may be a highly specialized or simple type of ending. An afferent neuron and all of its peripheral and central branches constitutes a SENSORY UNIT that responds to stimuli of appropriate type (a *modality*) delivered within that unit's PERIPHERAL RECEPTIVE FIELD, or area supplied. Different areas of the body may contain different numbers or DENSITIES of a particular type of receptor and therefore SENSITIVITY of the body may vary for that particular stimulus.

BASIC PROPERTIES OF RECEPTORS

Receptors usually respond best, although not exclusively, to one particular type of stimulus (LAW OF ADEQUATE STIMULUS). They thus exhibit some degree of SPECIFICITY to a given type or MODALITY of sensation. Impulses from different receptors are, however, alike in terms of their measurable characteristics, and the different sensations we appreciate are due to the central connections the nerves ultimately make. These principles are contained in the LAW OF SPECIFIC NERVE ENERGIES. (Muller's law). INTENSITY is usually communicated by frequency of discharge or by numbers of active receptors. All receptors appear capable of TRANSDUCING (changing) a particular stimulus into a nerve impulse through permeability changes that lead to the production of a GENERATOR POTENTIAL. This causes a depolarization of the receptor nerve fiber and conduction of impulses. ADAPTATION occurs as frequency of discharge diminishes with continued stimulus of the same strength. Receptors may be of two types in terms of their ability to adapt or accommodate to continued stimulus application. POORLY ADAPTING or tonic receptors either do not accommodate or do so to only a slight degree. They therefore continue to transmit information to the brain as long as they are stimulated. Such receptors are found in muscles, tendons, and joints and convey information as to limb position. Other poorly adapting receptors include those of the inner ear (maculae and cochlea), pain fibers, baroreceptors of the vascular system, and the chemoreceptors of the aortic and carotid bodies. RAPIDLY ADAPTING receptors react strongly while a change is occurring but rapidly decrease or stop reacting when the same strength stimulus is continued. They thus appear to signal *change* in a function. Such receptors include the touch and pressure receptors of the skin and the olfactory cells.

CLASSIFICATION OF RECEPTORS

Receptors may be grouped according to location in the body or by the particular type of stimulus that elicits a nerve impulse. According to location, and from where they receive stimuli, receptors are designated EXTEROCEPTIVE, if located at or near the body surface, ENTEROCEPTIVE, if located within an internal organ other than a muscle or tendon, and PROPRIOCEPTIVE if located in muscle or tendon. A more useful classification, physiologically, is provided by grouping according to adequate stimulus, as indicated in Table 13.1. Grouping by adequate stimulus is the most widely used scheme of classification, and describes the *type* of stimulus to which the receptor responds most effectively.

THE STUDY OF SPECIFIC SENSATIONS; SOMESTHESIA

SOMESTHESIA includes those sensations aroused from receptors other than those for sight, hearing, taste, and smell.

TOUCH AND PRESSURE

Touch and pressure represent different intensities of the same basic stimulus, mechanical distortion of a skin surface. Touch is served by the Meissner's corpuscle, while pressure is served by the Pacinian corpuscle. These receptors are shown in Figure 13.1. Touch may also be served by unspecialized fibers located around hair follicles. These respond when the hair is bent.

Adequate stimulus. The adequate stimuli to create generator potentials in touch and pressure receptors are mechanical forces resulting in distortion of the receptor. The Meissner's corpuscle, lying closer to the skin surface, and consisting of fewer layers of tissue surrounding the nerve terminals, is fired by a smaller distortion of shape than is the Pacinian corpuscle.

Thresholds. Where density of receptors is greatest (e.g., lips, finger tips) sensitivity to a particular stimulus is greatest; each receptor also characteristically appears to have a lower threshold. Thus, both numbers of receptors and characteristics of individual receptors may determine threshold. Discrimination of intensity above threshold strength is nearly linear in proportion to amount of deformation of the skin and depends both on degree of deformation of individual receptors and numbers of receptors involved.

Discrimination of spatial relationships. In areas where receptor density is greatest, it is possible

TABLE 13.1
A CLASSIFICATION OF RECEPTORS BY STIMULUS

ADEQUATE STIMULUS	EXAMPLE(S)
Mechanical	Touch and pressure in skin and tissues.
Pressure, bending, tension	Pressure or stretch receptors (baroreceptors) in circulatory system, alimentary tract
	Equilibrium receptors in inner ear
	Organ of hearing (hair cells of cochlea)
	Kinesthetic receptors in joints, muscles, and tendons
Chemical	Taste
Substances in solution	Smell
Stimulation either by type or concentration of chemical	Carotid and aortic body receptors
	Osmoreceptors in hypothalamus monitoring blood osmotic pressure; postulated in stomach wall
Light	Eye: vision
Thermal change	Heat
	Cold
Extremes of nearly any stimulus	Pain

to discriminate depressions as small as 1 μm in depth, and two points separated by as little as 1.2 mm. Where density is less, ability to perceive two separate points may increase to as much as 60 mm between points. This ability appears to depend almost entirely upon the number of receptors (density) stimulated by the points. This type of discrimination is well illustrated by the blind reading the Braille alphabet. Determination of size and shape of objects as well as quality of surface depends on the sense of touch combined with experience or training. Humans can, for example, determine easily the shapes of squares, spheres, and pyramids by feel alone, providing they have the concept of what these shapes are. They can also detect differences in size of objects if they differ by about 10 percent (Weber-Fechner law). The law accounts for our ability to discriminate differences in stimulus intensity and emphasizes that a change must be of a certain degree to be appreciated.

Conduction pathways for touch and pressure. The afferent (first order) neurons for these sensations enter the spinal cord over the dorsal roots (Fig. 13.2). A synapse may occur in the dorsal column of the spinal cord; if it does, a second (second order) neuron will cross to the opposite side of the cord and ascend through the VENTRAL SPINOTHALAMIC TRACTS to the thalamus for initial processing. If no synapse occurs, the fibers ascend through the dorsal areas of the cord (GRACILE and CUNEATE TRACTS) to a synapse in the brain stem. From this stem synapse, second-order fibers cross to the opposite side of the stem in the MEDIAL LEMNISCUS and proceed to the thalamus. The spinothalamic tracts appear to convey touch information concerned mainly with basic sensation and not with fine localization or discrimination. Fine discrimination of localization, shape, size, and texture is conveyed by the dorsal tracts. Thus, while both areas must be destroyed to completely remove all touch sensations, the greatest impairment occurs with dorsal tract destruction (as in syphilitic destruction in tabes dorsalis). In these tracts, a spatial representation is maintained in terms of body area (Fig. 13.3).

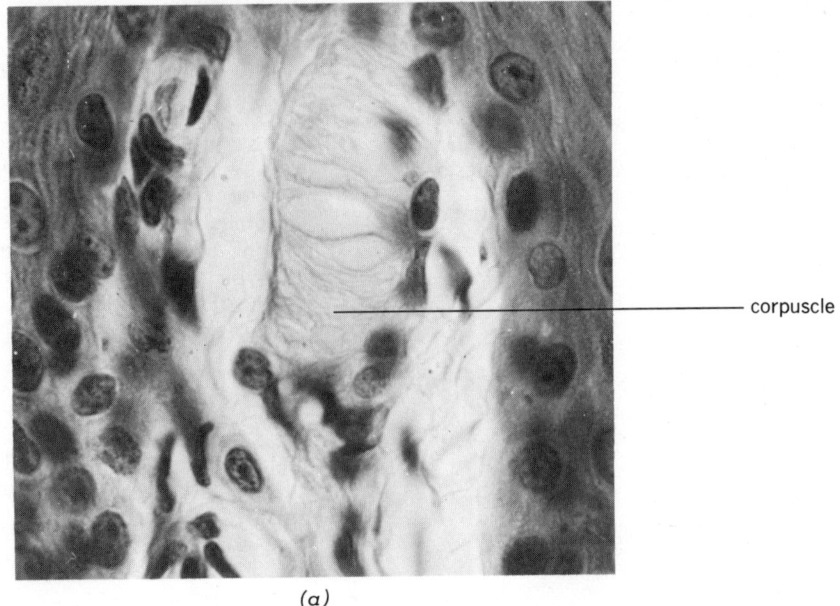

(a)

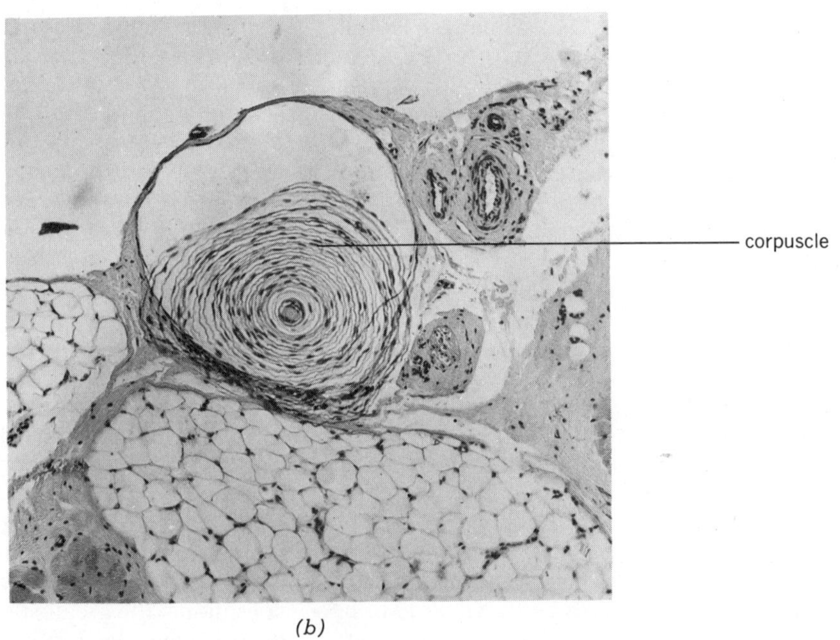

(b)

FIGURE 13.1
The receptors for (*a*) touch (Meissner's corpuscle), and (*b*) pressure (Pacinian corpuscle).

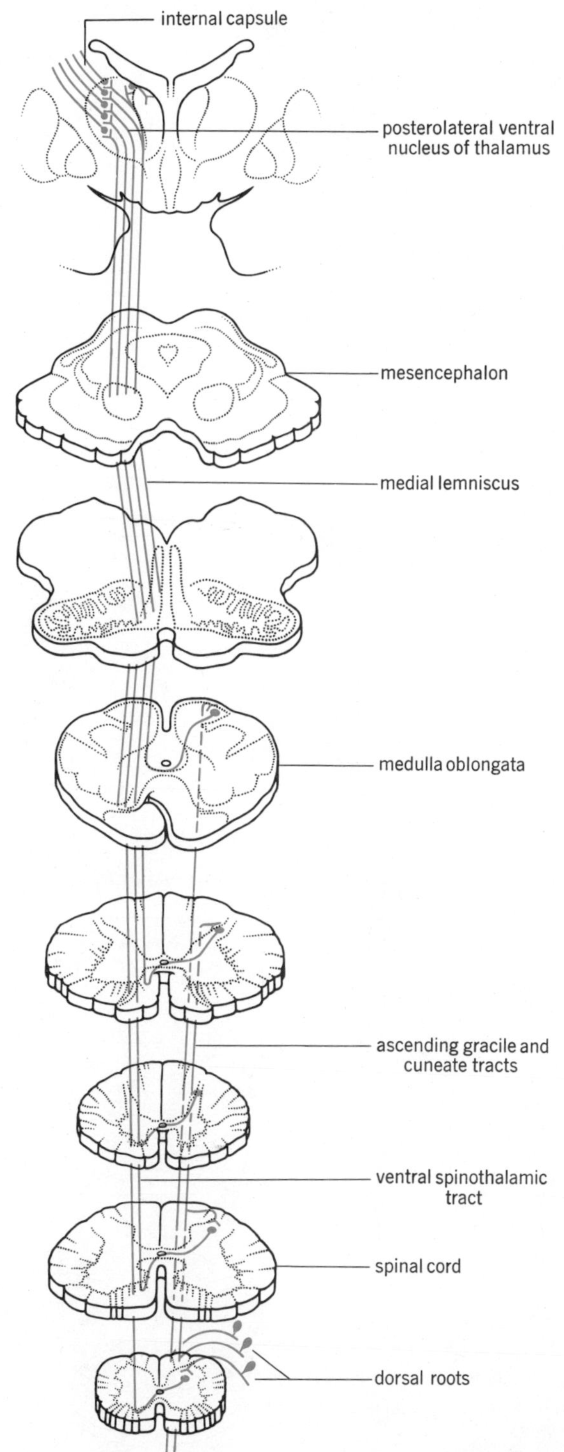

In the THALAMUS, touch and pressure impulses undergo synapses with a third (third order) neuron that distributes impulses to the cerebrum. In the thalamus, the area concerned with tactile sensibility appears to lie in the lateral-ventral complex of nuclei, where a spatial organization is also maintained according to body area (Fig. 13.4). The third-order fibers pass to the sensory areas of the CEREBRUM (areas 3, 1, 2) maintaining a spatial organization that results in an "upside down" representation (Fig. 13.5).

THERMAL SENSATIONS: HEAT AND COLD

Temperature sensations consist of two separate modalities: warmth and cool. The specific receptors, known as Ruffini corpuscles (warmth) and Krauses corpuscles (cold), are located dermally in the skin and therefore are actually monitoring the skin temperature at the depth of their location. Although discrete end organs sensitive to temperature changes are not found in all body areas, distribution of sensitivity appears spotlike, with "cold spots" outnumbering "warm spots" by about 4–10 to 1. Both areas are more numerous on hands and face than elsewhere on the body. Spatial summation occurs according to the area of body exposed to the environment. The rate of change of temperature is also important in determining formation of impulses by temperature receptors; a rate of change of about 0.001°C/sec can suffice to initiate sensation.

Warmth receptors fire at temperatures between 20°C and 45°C, with a peak at 37.5 to 40°C. Cold receptors form impulses between 10°C and 41°C, with a maximum discharge at 15 to 20°C. Interestingly, cold receptors, which normally show little discharge at 45°, begin to discharge again at temperatures of 46 to 50°C. Thus the sensation of cold in stepping suddenly into a hot shower (*paradoxical cold*). Both receptors are actually monitoring temperature differentials in their particular locations and do not sense a change from normal body temperatures (37°C, 98.6°F).

FIGURE 13.2
The cord pathways for touch and pressure (the ventral spinothalamic and gracile and cuneate tracts).

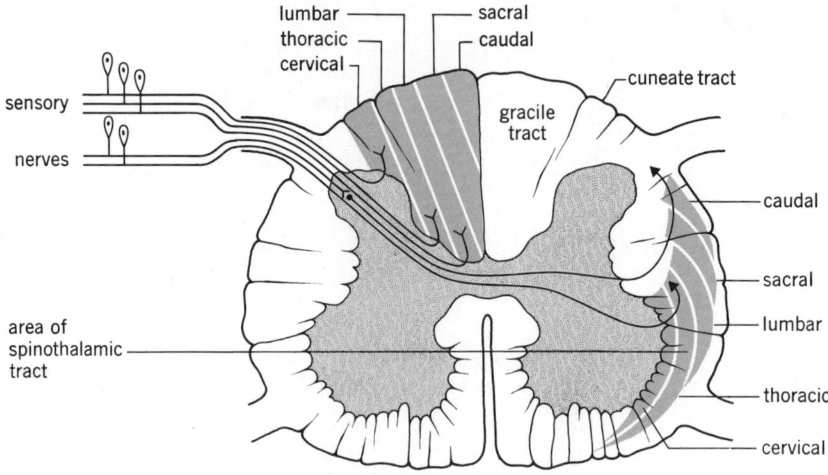

FIGURE 13.3
The spatial representation of touch and pressure in the spinothalamic and gracile and cuneate tracts.

FIGURE 13.4
The right thalamus in frontal section, viewed from the anterior aspect. The diagram shows the locations of the major sensory nuclei, and the spatial representation of the body in the thalamus.

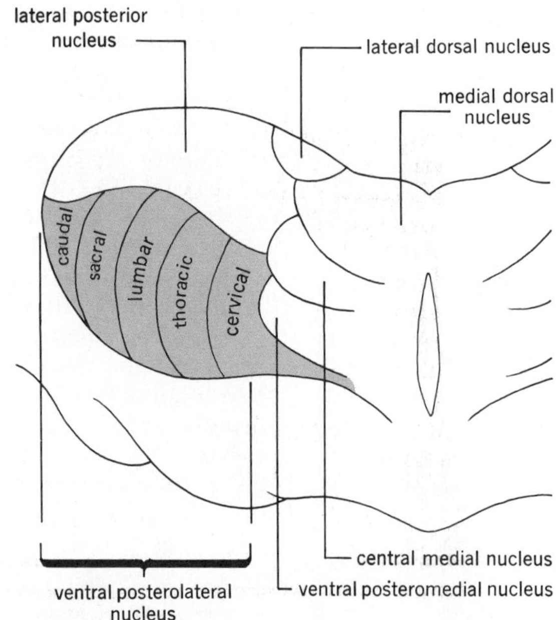

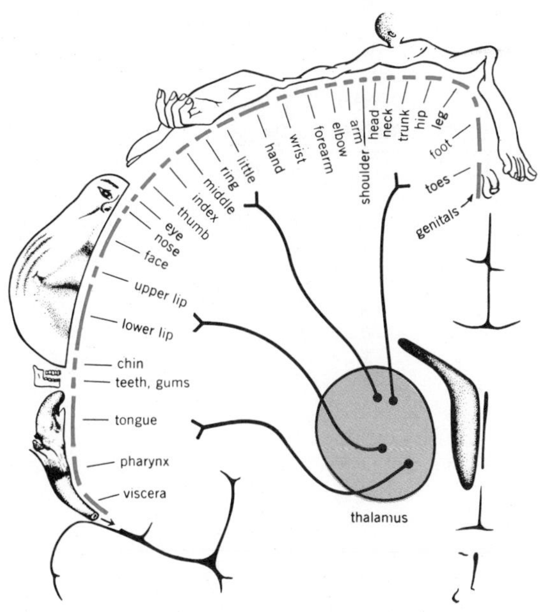

FIGURE 13.5
The distribution of sensory fibers from thalamus to cortex, and the orientation of the sensory homunculus.

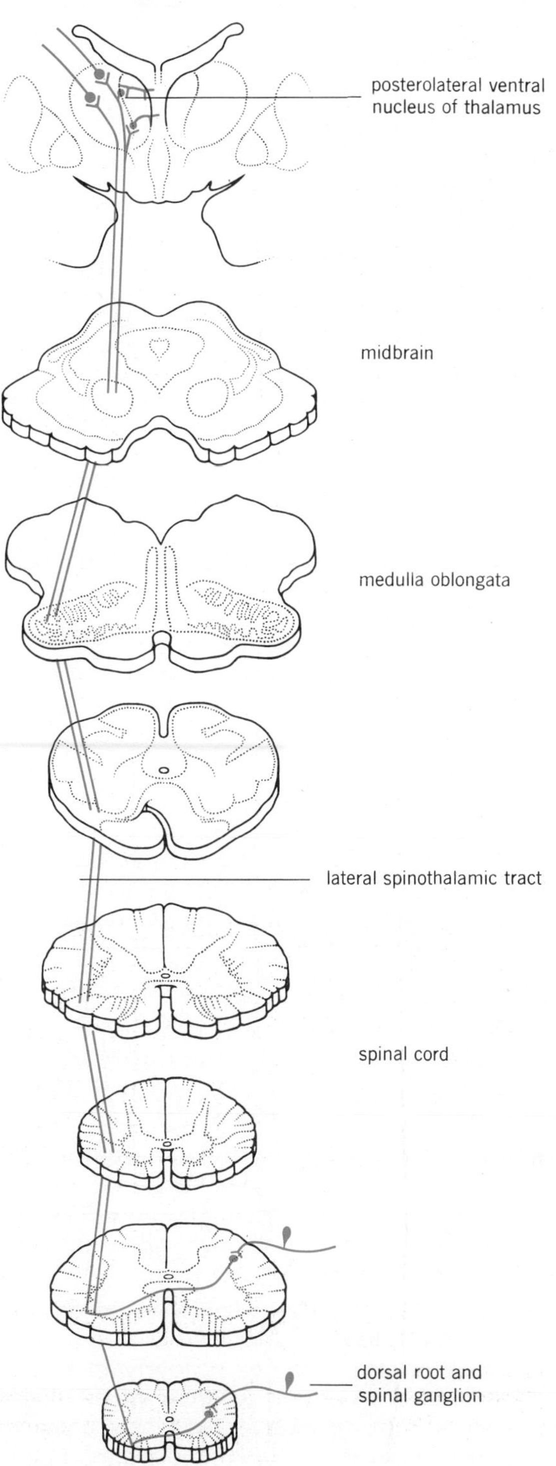

The afferent neurons for thermal sensation enter the cord over the dorsal roots, synapse in the dorsal gray column, and the second-order neurons cross to the opposite side of the cord to enter the LATERAL SPINOTHALAMIC TRACTS (Fig. 13.6). In these tracts, impulses ascend to the THALAMUS where synapse with third order fibers occur. Third-order neurons convey thermal sensations to areas 3,1,2 of the CEREBRAL CORTEX.

Either through cortical to hypothalamic contions and/or thalamic to hypothalamic connections, the thermal information from the periphery affects the heat and cold centers in the hypothalamus and leads to compensatory changes maintaining body temperature nearly constant. These control mechanisms were discussed in connection with the hypothalamus.

FIGURE 13.6
The spinal tract for pain, heat, and cold (the lateral spinothalamic tract).

TABLE 13.2
REFERRED PAIN

TRUE POINT OF ORIGIN OF PAIN	COMMON CAUSE	AREA TO WHICH PAIN IS REFERRED
Heart	Ischemia secondary to infarction	Base of neck, shoulders, pectoral area, arms
Esophagus	Spasm, dilation, chemical (acid) irritation	Pharynx, lower neck, arms, over heart ("heartburn")
Stomach	Inflammation, ulceration, chemical	Below xiphoid (epigastric area)
Gall bladder	Spasm, distention by stones	Epigastric area
Pancreas	Enzymatic destruction, inflammation	Back
Small intestine	Spasm, distention, chemical irritation, inflammation	Around umbilicus
Large intestine	As above	Between umbilicus and pubis
Kidneys and ureters	Stones, spasm of muscles	Directly behind organ or groin and testicles
Bladder	Stones, inflammation, spasm, distention	Directly over organ
Uterus, uterine tubes	Cramps (spasm)	Low abdomen or lower back

PAIN

Pain is a protective sensation for the body in that it indicates overstimulation of any sensory nerve and damage or potential damage to body tissues. It is served by naked nerve endings in skin and organs. Pain is said to occur in three varieties.

"BRIGHT" or PRICKING PAIN is the type experienced when the skin is cut or jabbed with a sharp object. It may be intense but is relatively short lived, and is easily localized.

BURNING PAIN, of the sort experienced when the skin is burned, is slower to develop, lasts longer, and is less accurately localized. It commonly stimulates cardiac and respiratory activity and is difficult to endure.

ACHING PAIN develops when visceral organs are stimulated. It is persistent, often produces feelings of nausea, is poorly localized, and is often referred to areas of the body distant from the actual area where damage is occurring. It is of the utmost importance to the physician, for it may signal life-threatening disorder of vital organs.

There appears to be no particular adequate stimulus for pain; it may be evoked by mechanical, electrical, thermal, or chemical stimuli. All appear to result in permeability changes leading to generation of pain impulses. Adaptation or accommodation does not appear to occur; we may be distracted from a painful stimulus or ignore it, but if we "pay attention" to it, it appears to persist unchanged. ISCHEMIA (lowered blood flow) of an area may, by retaining metabolites or other chemicals in an area, contribute to development of pain.

Afferent neurons enter via the dorsal roots, and synapse in the dorsal gray column. Second-order neurons cross to the opposite LATERAL SPINOTHALAMIC TRACT and ascend to the THALAMUS. Here, it is believed most of the interpretation of pain occurs. Evidence for thalamic interpretation of pain revolves mainly around the fact that removal of the sensory cortex does not eliminate inability to appreciate pain. The cortex may contain areas of pain appreciation and certainly aids in localization of painful stimuli, particularly if touch receptors are stimulated simultaneously with pain fibers.

Referred pain. Sensory impulses from visceral organs and skin often impinge on the same or closely spaced neurons. Intense stimulation of visceral pain neurons may cause irradiation or spread to neurons serving the skin and the sensation "feels as though" it was coming from the skin itself (it is referred to the skin). Visceral pain commonly results from spasm, distention, ischemia, or chemical irritation of an organ. The list presented in Table 13.2 indicates to what body area pain originating in various viscera is referred.

Headache. Perhaps the most common pain complaint is headache. From whatever cause, headache pain is nearly always appreciated as deep, diffuse, and aching in quality; it nearly always causes reflex contractions of cranial muscles that may aggravate the pain. Its causes are diverse.

TENSION applied to blood vessels (arteries and veins) within the skull or to the membranes around the brain causes headache.

DILATION OF THE ARTERIES to the brain, resulting from vasodilation or high blood pressure (increased intracranial pressure) produces pain. Spasm, producing ischemia, may be the cause of migraine headache.

INFLAMMATION of paranasal sinuses or any pain-sensitive structure within the skull may produce headache.

SPASM OF CRANIAL MUSCLES produces so-called "tension headaches" and results most commonly from emotional involvement or fatigue.

EYE DISORDERS, such as spasm of the extraocular muscles or irritation of the conjunctivae, produces pain behind the eyes.

The role of aspirin in the relief of pain (analgesic action) revolves about its ability to depress nervous conduction or raise the threshold of conduction of impulses in or near the hypothalamus.

Disorders of pain perception. PARESTHESIA (abnormal sensation) arises from irritation or damage to peripheral nerves or cord tracts that results in impulses being transmitted to the brain. The brain then localizes the pain in the area from which that nerve impulse would normally have come. HYPERESTHESIA involves a greater than normal sensitivity to pain, that is, a lower pain threshold. The receptor itself may be hypersensitive (primary hyperesthesia) or conduction somewhere along the pathway may be facilitated (secondary hyperesthesia). CORD DAMAGE will produce localized or generalized loss of specific sensations depending on extent, area, and side of damage. Hemisection of the cord produces loss of touch and pressure on the

FIGURE 13.7
The location and general structure of muscle and tendon receptors.

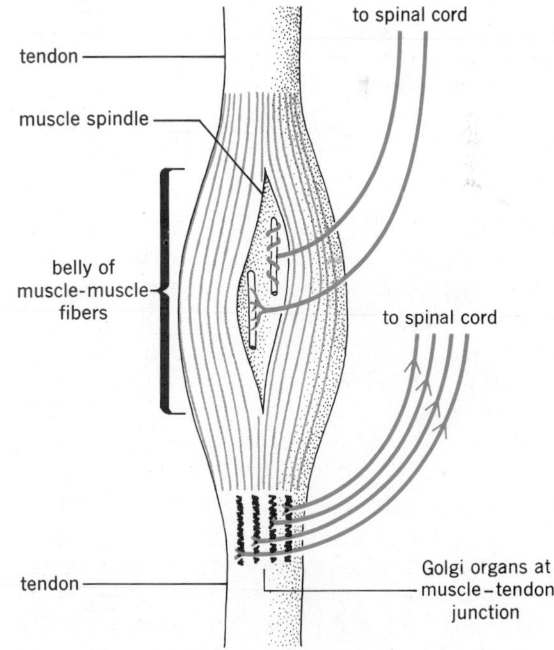

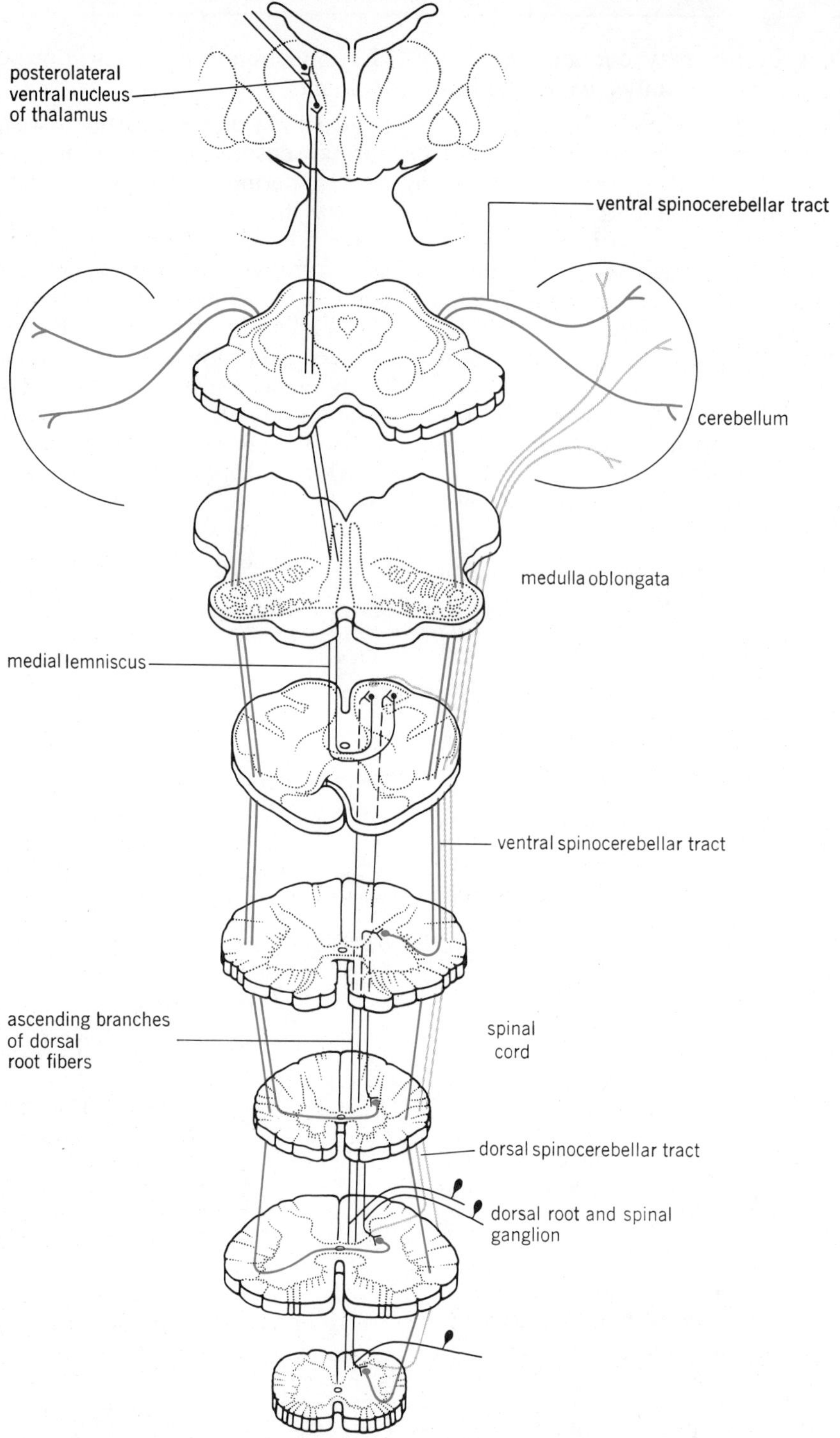

FIGURE 13.8
The spinal tracts for unconscious muscle sense; the spinocerebellar tracts.

same side (ipsilateral) as the damage and loss of pain, heat, and cold on the opposite side (contralateral) as the injury. The THALAMIC SYNDROME arises from ischemia or damage to the posteroventral portion of the thalamus and results in loss of all sensations from the opposite side of the body, loss of ability to control muscular movements precisely (ataxia) and continuous pain, which resists relief by normal methods of control.

KINESTHETIC SENSE

The sense of body position and movement of joints embraces that body sense known as KINESTHESIA. It originates peripherally in sensory organs located in muscles, joints, and tendons (Fig. 13.7). Movement of joints, shortening or stretching of muscles and tendons elicit impulses that pass to the cord and enter the dorsal columns. CONSCIOUS MUSCLE SENSE, or appreciation of body position, is carried primarily in the GRACILE and CUNEATE TRACTS, while impulses related to maintenance of muscle tone, UNCONSCIOUS MUSCLE SENSE, are transmitted primarily over the SPINOCEREBELLAR TRACTS to the CEREBELLUM (Fig. 13.8).

SYNTHETIC SENSES

The body is capable of appreciating such sensations as ITCH, TICKLE, and VIBRATIONAL sense. No specific receptors exist for such sensations; they appear to arise as a result of simultaneous stimulation of one or more of the basic sensations previously described. Itch appears to arise from chemical irritation of pain fibers in the skin, and mucous membranes of the nose. If the stimulus moves across the skin, tickle results. Vibrational sensitivity appears to be due to rhythmic and repetitive stimulation of pressure receptors. The pressure receptors respond maximally to repetitive stimuli at 250 to 300 Hz.

SUMMARY

1. Receptors sense change internally and/or externally to the body and their nerves convey impulses to the central nervous system for interpretation.

2. All receptors show several basic properties.
   a. They respond best to one particular form of change or modality (law of adequate stimulus).
   b. All impulses from different receptors are alike. Differences depend on central connections (law of specific nerve energies).
   c. Intensity of stimulation may be signalled either by increased frequency of receptor discharge or by stimulation of more receptors.
   d. Receptors adapt or slow their rate of impulse discharge with continued stimuli of the same strength.

3. Receptors are classified by their location and their adequate stimulus.
   a. Receptors located at or near body surfaces are exteroceptive; in viscera, enteroceptive; in muscles and tendons, proprioceptive.
   b. By stimulus, receptors may be chemical, pressure, light, thermal, or pain sensitive.

4. Somesthesia refers to touch, pressure, heat, cold, and pain receptors.
   a. Touch and pressure are served by Meissner's and Pacinian corpuscles, pass through ventral spinothalamic, gracile, and cuneate tracts to the thalamus, then to the sensory cortex.
   b. Heat and cold may have specialized corpuscles, are conveyed over the lateral spinothalamic tracts to the thalamus, then to the sensory cortex.
   c. Pain is served by naked nerve endings, is conveyed over lateral spinothalamic tracts and is interpreted in the thalamus. Localization of pain occurs in the sensory cortex. Pain is bright, burning, or aching in character and may be referred to areas other than its actual point of origin.

5. Headache is a type of pain resulting from tension on internal skull structures, increased intracranial pressure, inflammation, cranial muscle spasm, and eye disorders.

6. Paresthesia (abnormal sensation) and hyperesthesia (increased sensitivity) are abnormalities of pain perception.

7. Kinesthesia refers to the sensation of body position and joint movement. It originates from receptors in muscles and tendons.
   a. Conscious muscle sense (of body position) is carried in gracile and cuneate tracts to the cerebrum.
   b. Unconscious muscle sense (muscle tone reflexes) is carried over the spinocerebellar tracts to the cerebellum.
8. Synthetic senses result from simultaneous stimulation of other receptors.
   a. Itch results from chemical irritation of pain fibers.
   b. Tickle results from movement across the skin.
   c. Vibration is due to rhythmic stimulation of pressure receptors.

QUESTIONS

1. What is a receptor? What is a peripheral receptive field?
2. What basic properties do all receptors share?
3. Give examples of chemoreceptors, baroreceptors, and thermoreceptors and describe what the adequate stimulus is for each you list.
4. What is somesthesia?
5. Name the receptor, spinal pathway, and interpretive area for the senses of touch, pain, and unconscious muscle sense.
6. What are the varieties of pain? How do they differ?
7. What are paresthesia and hyperesthesia? How may each be brought about?
8. What are the roles of the thalamus in sensation?
9. What are poorly and rapidly adapting receptors, and what is the significance of two rates of adaptation?

READINGS

Goldstein, Avram. "Opioid Peptides (Endorphins) in Pituitary and Brain." *Science 193:*1081. 17 Sept 1976.

Guillemin, Roger. "Endorphins, Brain Peptides That Act Like Opiates." *New Eng. J. Med. 296:*226. Jan 27, 1977.

Marx, Jean L. "Analgesia: How the Body Inhibits Pain Perception." *Science 195:*471. 4 Feb 1977.

Marx, Jean L. "Neurobiology: Researchers High on Endogenous Opiates." *Science 193:*1227. 24 Sept 1976.

Snyder, Solomon H. "Opiate receptors and Internal Opiates." *Sci. Amer. 236:*44. (Mar) 1977.

Snyder, Solomon H. "Opiate Receptors in the Brain." *New Eng. J. Med. 296:*266. Feb 3, 1977.

14

THE EYE AND VISION

OBJECTIVES

STRUCTURE OF THE EYE

PHYSIOLOGICAL ANATOMY

LIGHT AND LENSES; IMAGE FORMATION

SOME PROPERTIES OF LIGHT AND LENSES

IMAGE FORMATION BY THE EYE

ACCOMMODATION

DISORDERS OF THE EYE

DISORDERS OF IMAGE FORMATION

OTHER DISORDERS OF THE EYE

RETINAL FUNCTION

PHYSIOLOGICAL ANATOMY OF THE RETINA

FUNCTIONS OF THE RODS

FUNCTIONS OF THE CONES

THE VISUAL PATHWAYS

EFFERENT FIBERS

VISUAL ASSOCIATION AREAS

EFFECTS OF LESIONS IN VARIOUS PARTS OF THE VISUAL PATHWAY

OTHER FUNCTIONS OF THE BRAIN IN VISION

SUMMARY

QUESTIONS

READINGS

OBJECTIVES

After studying this chapter, the reader should be able to:

☐ Describe the major parts of the eye, and its associated glands and muscles.

☐ Trace the circulation of aqueous humor within the eye.

☐ State the major differences between composition of aqueous humor and plasma.

☐ Explain how the "strength" of a lens is determined.

☐ Describe in general terms how an image is formed on the retina of the eye.

☐ Account for the common disorders of inage formation.

☐ Explain, in general terms, how the rods and cones function in vision.

☐ Explain the involvement of the retina in visual acuity, fusion of separate images, and in the creation of afterimages.

☐ Trace the visual pathways from eye to brain.

☐ Explain the different types of visual loss occurring with lesions in different parts of the visual pathways.

The eye (Fig. 14.1) is a complex peripheral receptor specialized to respond to quanta of light energy. Additionally, the normal organ is constructed in such a way that it should be able to focus images of the world on the photosensitive portion of the eyeball and create clear and undistorted views of our surroundings.

STRUCTURE OF THE EYE

PHYSIOLOGICAL ANATOMY

The eyeball is an organ approximately 25 mm (about 1 in.) in diameter, located within the ORBIT of the skull. It is protected anteriorly by the EYELIDS that may close reflexly when the eye is stimulated by bright light and during sleep and blinking. The lids may also be closed voluntarily during strong winds and cold or when light is strong. The lids are lined with EYELASHES, about 200 for each eye. The lashes act, when bent, to trigger a blink reflex (as when touched by a foreign object). When the eye is partially closed, some scattering of light by the lashes occurs diminishing the amount entering the pupil. Large SEBACEOUS GLANDS (*Meibomian glands*) lie within the lids and produce the material often found in the eye when we awake.

A two-lobed LACRIMAL GLAND lies on the superior temporal aspect of each eyeball. A number of small ducts empty the lacrimal fluid (tears) onto the surface of the eyeball at the upper outer corner of the lids. Blinking spreads the fluid evenly over the anterior surface of the eyeball. The fluid moistens and cleanses the cornea, lubricates the lids for movement and, through its content of lysozyme (an enzyme), destroys bacteria. Normally, about 1 ml of fluid is produced per day, and it is drained into the nasal cavities by the lacrimal canals, lacrimal sac, and nasolacrimal duct (Fig. 14.2). Production of fluid is increased when the eyeball is irritated or under certain emotional conditions.

Six EXTRINSIC EYE MUSCLES (Fig. 14.3) operate the eyeball to turn it within the orbit. Their names, innervation, and action are presented in Table 14.1. Note the position of each muscle on the eyeball.

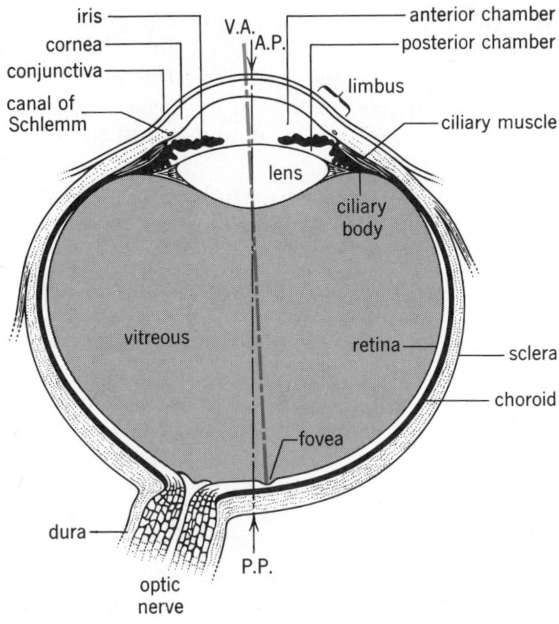

FIGURE 14.1
The right eyeball in horizontal section as viewed from above. A.P. = anterior pole; P.P. = posterior pole (line connecting these is the anatomical axis of the eye); V.A. = visual axis (path of light rays).

The action of the superior oblique occurs because the tendon of the muscle utilizes the wall of the orbit as a pulley to change its direction of pull. The six muscles on each eye are normally coordinated to turn both eyes simultaneously to focus squarely on stationary or moving objects. Failure of both eyes to be directed at the object, due to muscular incoordination, muscles of unequal length, or paralysis, constitutes STRABISMUS.

Three TUNICS or coats surround the eyeball: the outer fibrous tunic includes the SCLERA (white of the eye) and the anterior transparent CORNEA; the middle coat or UVEA includes the anterior CILIARY BODY and the posterior vascular CHOROID; the inner RETINA contains the photosensitive elements.

The liquid AQUEOUS HUMOR, produced by secretion (carbonic anhydrase required) from the processes of the ciliary body, flows forward to fill the anterior and posterior chambers of the eye. From these areas it is drained by the CANAL OF SCHLEMM into the veins of the eyeball. Blockage of the canal raises intraocular pressure and creates GLAUCOMA. The composition of aqueous humor is given in Table 14.2. The gel-like VITREOUS HUMOR is composed of microscopic fibrils in a hyaluronic acid ground substance and fills the eyeball behind the ciliary body. It aids in maintaining the shape of the eye.

The IRIS provides control over the amount of light that enters the eye. Two muscles are present in the iris: the SPHINCTER PUPILLAE (circularly around the pupil) is innervated by parasympathetic nerves and serves to make the pupil smaller (*miosis*); the DILATOR PUPILLAE (radially around the pupil) is innervated by sympathetic nerves and serves to enlarge the pupil (*mydriasis*). Drugs may also cause change of pupil size. *Miotics* constrict and *mydriatics* dilate the pupil. A primary mechanism operating the iris is the PUPILLARY REFLEX, which

FIGURE 14.2
The lacrimal apparatus.

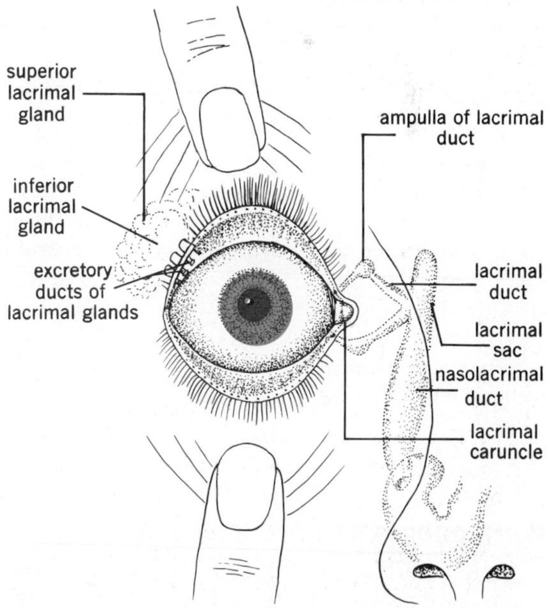

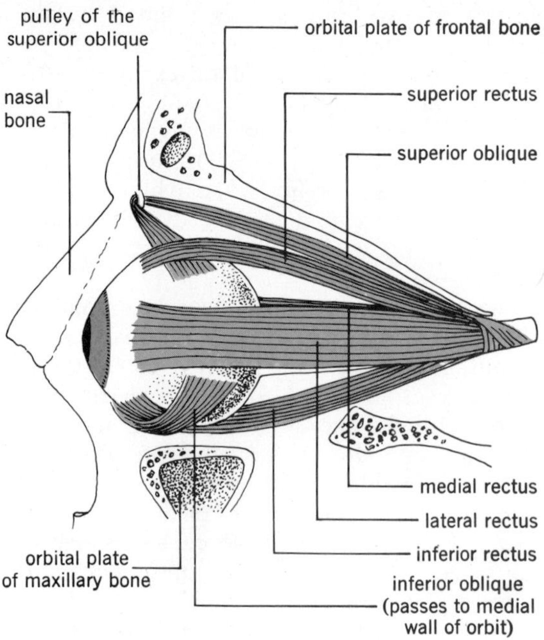

FIGURE 14.3
The six extrinsic muscles of the left eye, viewed from the lateral side. Note how the pulley converts the direction of pull of the superior oblique.

constricts the pupil during strong light stimuli, Afferent nerves pass from the retina to the midbrain, to the nucleus of the oculomotor nerve and over oculomotor fibers to the iris. An ARGYLL-ROBERTSON PUPIL is one that fails to respond to light. The iris also acts to block peripheral light rays that might otherwise enter the edges of the lens and be bent (refracted) to a different degree than rays entering the central portion of the lens. The iris thus prevents SPHERICAL ABERRATION (multiple blurred images from several focal points) and CHROMATIC ABERRATION (multiple colored images from unequal refraction of light waves).

LIGHT AND LENSES; IMAGE FORMATION

SOME PROPERTIES OF LIGHT AND LENSES

Light is a form of energy that behaves like a WAVE composed of individual particles or QUANTA of energy. A continuous spectrum of electromagnetic radiation exists, but only that portion that is visible concerns the eye and vision. The visible portion of the spectrum extends from wavelengths of about 400 to 800 nanometers (nm).

Light may be REFRACTED or bent as it passes from one medium to another (e.g., air to water) or through a system of lenses. A beam of white light may be divided, as in a prism, into its spectral colors ("rainbow"). Only three pure light colors – red, blue, and green – are required to create all other colors.

The power of a lens to refract light is dependent on the substance composing the lens and its curvature. The FOCAL LENGTH of a lens refers to the distance from the lens where a sharp primary image is created. The DIOPTER is used as a unit of measure of the strength of lenses. A convex lens capable of focusing parallel rays of light sharply at 100 cm (1 meter) is said to have a strength of +1 diopter. A convex lens focusing an image nearer than 100 cm has a positive and larger diopter number. For example, a focal distance of 50 cm would give a lens a strength of +2 diopters; 20 cm, +5 diopters. A concave lens is assigned a negative number, and its focal length is extrapolated to a point on the same side as that of the light source. A concave lens focusing to a point 50 cm from the lens would thus have a strength of −2 diopters (Fig. 14.4).

TABLE 14.1
EXTRINSIC EYE MUSCLES

NAME	INNERVATION (CRANIAL NERVE)	EYE TURNS
Lateral rectus	VI Abducent	Laterally
Medial rectus	III Oculomotor	Medially
Superior rectus	III Oculomotor	Superiorly and medially
Inferior rectus	III Oculomotor	Inferiorly and medially
Superior oblique	IV Trochlear	Inferiorly and laterally
Inferior oblique	III Oculomotor	Superiorly and laterally

TABLE 14.2
THE COMPOSITION OF THE AQUEOUS HUMOR COMPARED TO PLASMA (meq/l)

SUBSTANGE	AQUEOUS HUMOR	PLASMA
Na	134.0	140.0
K	4.0	4.2
Mg	1.56	2.0
Ca	2.75	4.75
Cl	104.0	103.0
HCO_3	34.0	27.0
Urea	13.0	15.0
Protein	Trace (albumin)	17.0

IMAGE FORMATION BY THE EYE

Light rays entering the eye are refracted by four structures: the anterior and posterior surfaces of the cornea and the anterior and posterior surfaces of the lens. The anterior surface of the cornea refracts light like a convex lens with a strength of +48.2 diopters; the posterior corneal surface refracts like a concave lens of −5.9 diopters. The lens surfaces are both convex, but the anterior surface has a lesser curvature (+5.0 diopters) than the posterior (+8.3 diopters). The greatest refraction of light thus occurs at the anterior corneal surface. The cornea is however, a fixed lens in the sense that its curvature cannot be changed. It thus initially refracts light to a degree that can then be finely adjusted by the lens that *can* undergo changes in shape during ACCOMMODATION.

Accommodation. A *triad* of changes occurs during accommodation. There are CHANGES IN CURVATURE OF THE LENS of the eye necessary to achieve sharp focusing of a near object's image on the retina, there is PUPILLARY CONSTRICTION, and there is CONVERGENCE (directing both eyes at the subject). The lens is normally pliable and is surrounded by a capsule of elastic connective tissue. It is suspended by the ZONULE (lens ligaments) attached to the

FIGURE 14.4
The refracting power of convex and concave lenses as measured by the diopter.

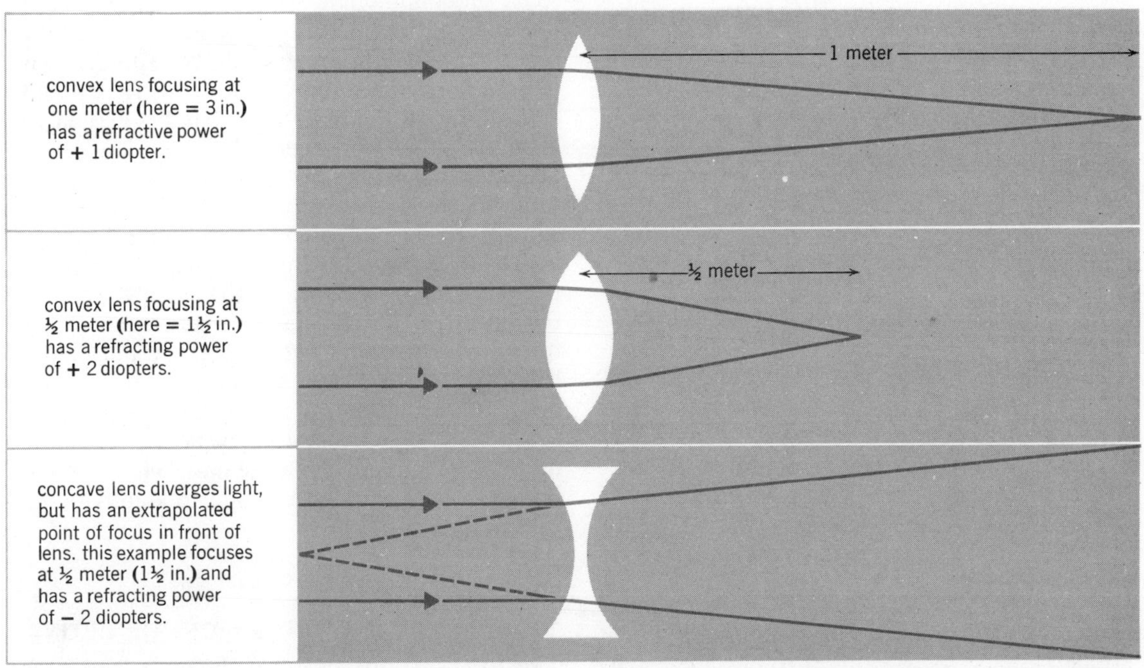

tissue of the CILIARY MUSCLE. The muscle takes its origin anterior to the zonule attachment and inserts upon the choroid. When the muscle contracts the anterior part of the choroid is pulled forward, is stretched, and the tension is relaxed on the zonule. The elastic capsule causes the lens to assume a more rounded shape. Relaxation of the muscle is associated with elastic recoil of the choroid, tension application to the zonule, and flattening of the lens. The ability of the lens to accurately focus images thus depends on alteration of its curvature and consequent change in refractive power. Images that pass through the lens are reversed on the retina relative to their position in the external world.

If the focal lengths of the lens systems of the eye are matched to the length of the eyeball, images will be focused precisely on the retina, and the eye is EMMETROPIC (normal).

Accommodation occurs when objects closer than 20 feet are being viewed. As objects move closer, they will eventually blur because the lens can round up no more. This point is termed the "near point" of vision.

DISORDERS OF THE EYE

Disorders of image formation. If the eyeball is too long for the focal system, images will be focused ahead of the retina, and a blurry image will be formed. Such an eye is MYOPIC (*hypometropic* or *nearsighted*). Myopia is corrected by placement of a concave lens in front of the cornea to diverge entering light rays. If the eyeball is too short for the lens system, images will be focused behind the retina and the eye is HYPERMETROPIC (*farsighted*). This condition is remedied by placement of a convex lens ahead to refract rays more. These relationships are shown in Figure 14.5.

ASTIGMIA or ASTIGMATISM results when the corneal surface does not possess equal curvatures in the lateral and vertical directions. The cornea thus forms two images, one for each curvature. Double vision or blurring of vision usually results.

PRESBYOPIA (*old eye*) results from hardening of the lens with age. It usually assumes a more flattened shape and cannot round up even though the ciliary muscle is still contracting. The subject with presbyopia moves reading matter farther away from the eyes in the attempt to maintain proper focus. Again, convex glass lenses placed ahead of the cornea will correct this problem.

Other disorders of the eye. CATARACT is opacity of the lens or its capsule, or both. Aging may produce cataract as may trauma, infection, or diabetes. The cataract develops in stages beginning with spokelike opacities, proceeding to swelling, shrinkage, and loss of transparency. Removal of the lens (extraction) is the treatment employed when the lens becomes so opaque that it interferes with the pursuit of normal activities. GLAUCOMA is an increased intraocular pressure due to blockage of the canal of Schlemm. Blockage may result from eye infections, congenital malformations, or trauma. Treatment may be attempted by medication (miotics usually) that constrict the pupil and reduce aqueous humor production by carbonic anhydrase inhibition. If medication cannot control the pressure, surgical intervention is indicated, with creation of an outflow canal to be subconjunctival space. The CONJUNCTIVA is a membrane lining the lids and covering the anterior surface of the eyeball. CONJUNCTIVITIS is inflammation of the conjunctiva. TRACHOMA is bilateral viral conjunctivitis and is estimated to affect some 15 percent of the world's population. Untreated, it can lead to blindness through corneal ulceration and destruction. KERATITIS is corneal inflammation usually of bacterial etiology. Corneal transplant may be performed for conditions that badly damage the cornea. The cornea is an avascular structure that derives its nutrients from and places its wastes in the aqueous humor. It is minimally antigenic and does not have to be tissue typed before transplanting. HORDEOLUM (*sty*) involves inflammation of the hair follicles of the lids. It is most commonly due to staphylococcal bacteria.

RETINAL FUNCTION

PHYSIOLOGICAL ANATOMY OF THE RETINA

The RETINA consists of 10 layers of neurons

and cells (Fig. 14.6). The second layer contains the photosensitive elements—the RODS and CONES (Fig. 14.7). Rods and cones are found in all parts of the retina except in two places: the fovea, where cones alone are present; the exit of the optic nerve, or blind spot, where *no* receptors are found. It may be noted that there are three neurons in the retina leading to the optic nerve, a fact that suggests considerable processing of information received from the rods and cones. The presence of two anatomically distinct receptors led to the suggestion (*duplicity theory*) that two different functions were served by the receptors. Subsequently, it has been shown that rods are receptors for low intensity light, and the cones are for higher intensity light vision and color vision.

FIGURE 14.5
(*a*) Emmetropic, (*b*) myopic, and (*c*) hypermetropic eyes. Diagram indicates type of lens required to correct abnormal eyes.

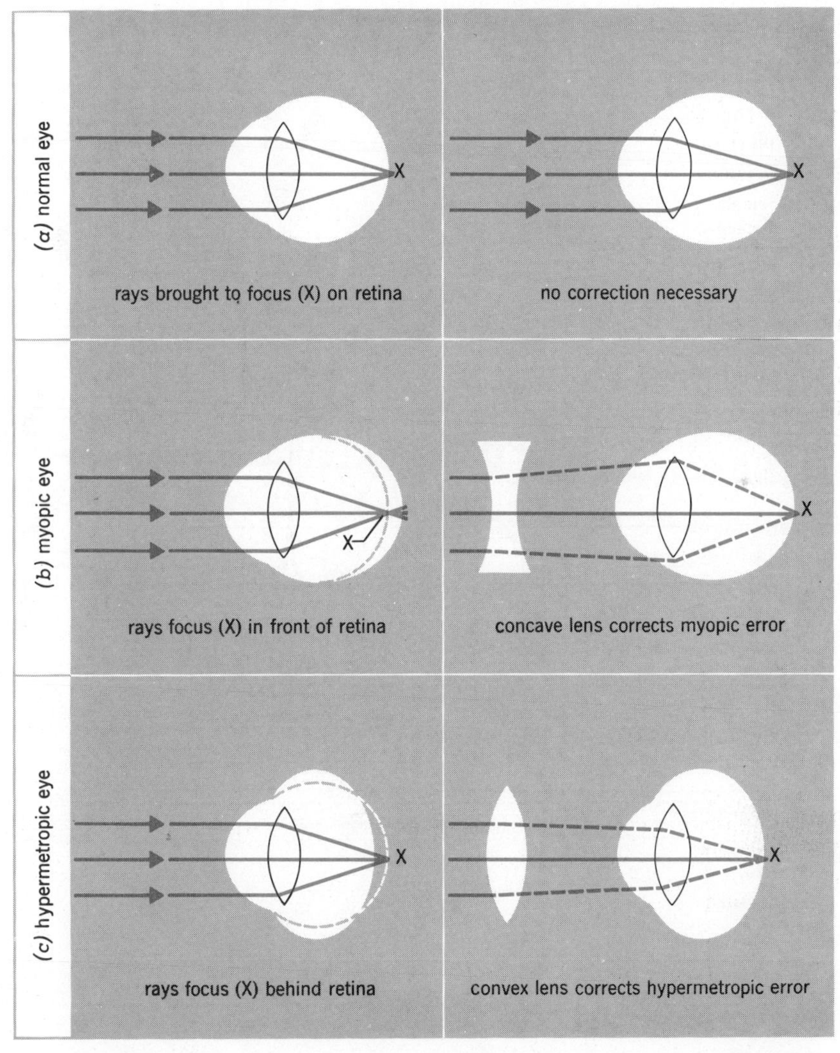

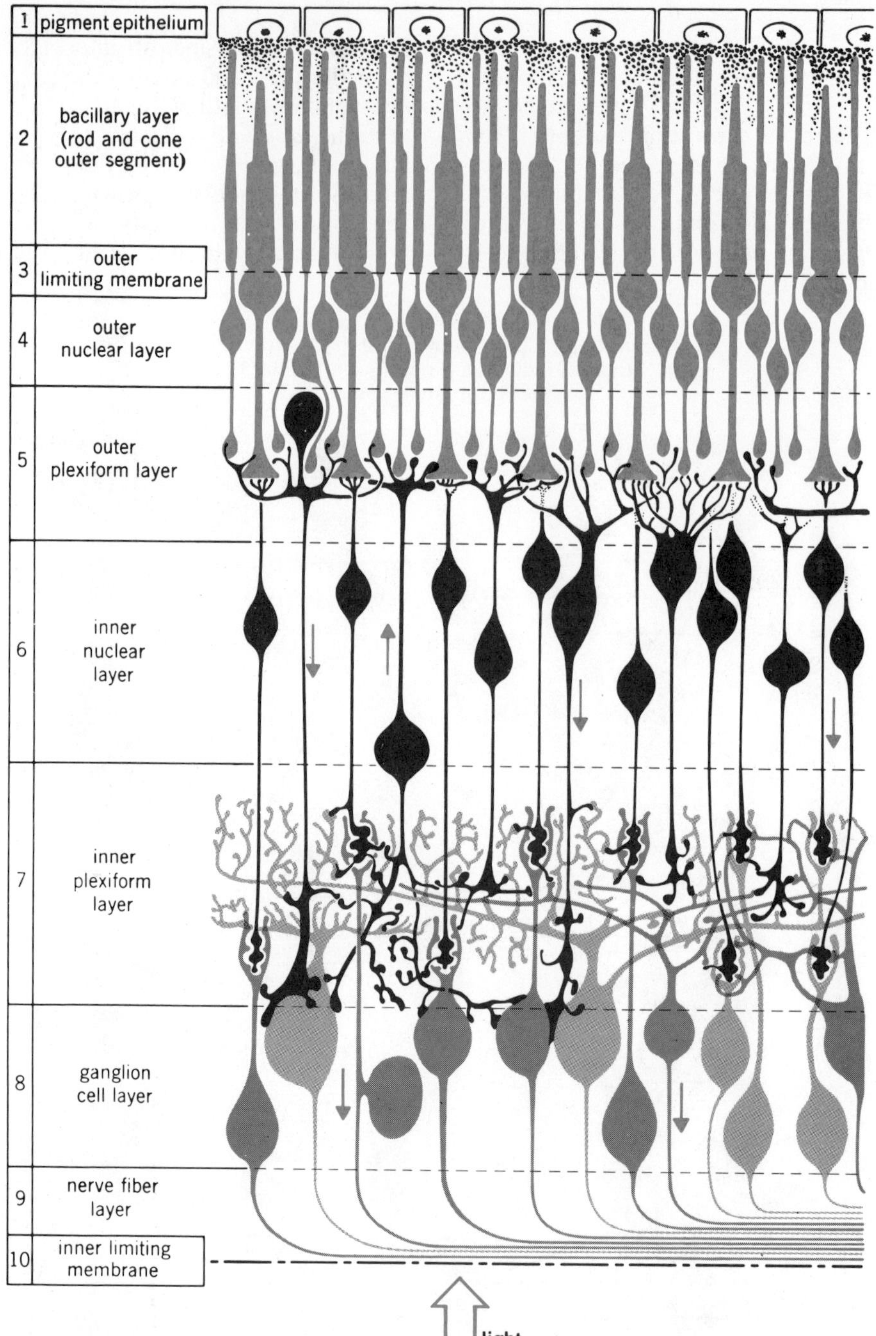

FIGURE 14.6
The organization of the retina. (From *How Cells Communicate* by Bernard Katz. Copyright © September 1961 by Scientific American Inc. All rights reserved.)

FUNCTION OF THE RODS

Rods are receptors for intensities of light less than 0.1 foot candle (the light shed by a candle at 1 foot distance). They are responsible for SCOTOPIC VISION (*night vision*). Rods contain a visual pigment designated as RHODOPSIN. It is a deep purple pigment composed of a derivative of vitamin A designated as RETININE and a protein designated as an OPSIN. Acted upon by light, rhodopsin is changed to LUMIRHODOPSIN, then to METARHODOPSIN. Bleaching (loss of color of the pigment) follows, and the molecule is split into retinine and opsin. The splitting of the molecule is apparently what triggers depolarization of the rod membrane. The disturbance is then transmitted to the rod cell body by a cilium. Regeneration of rhodopsin occurs in the dark by recombination of retinine and opsin. The reactions are summarized below.

FIGURE 14.7
The morphology of the rods and cones.

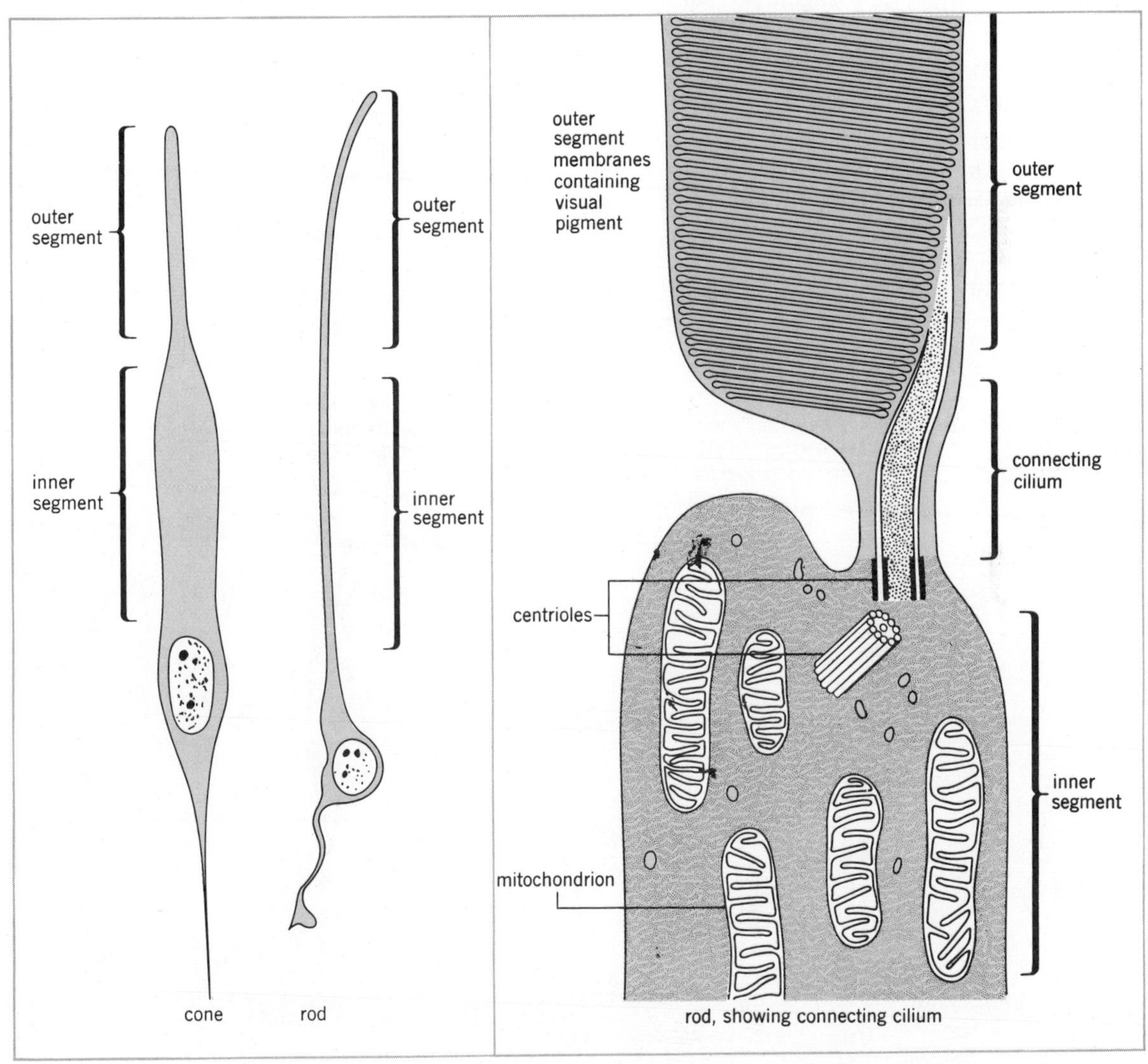

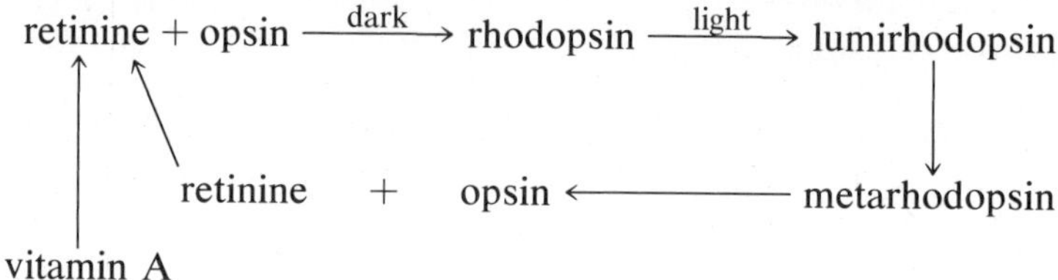

A single quantum of light is apparently effective in splitting a molecule of rhodopsin and leads to the depolarization of the cell. Conduction from the outer segments containing pigment to cell body proper is by a cilium (*see* Fig. 14.7).

FUNCTION OF THE CONES

There appear to be three classes of cones, each with a different visual pigment within it. Each pigment has a different maximum absorption of light of different color, indicating that each responds best to light of given wavelength and therefore color. Absorption maxima for human cones are at 445 nm (blue), 535 nm (green), and 570 nm (red). The absorption curves are shown in Figure 14.8. One notes that there is overlapping of the curves, suggesting that light of a given wavelength does not necessarily stimulate just one cone. Differential stimulation may thus be the basis for discrimination of color, based on "mixing" of red, green, and blue primary colors. It is presumed that light acts on cone pigments in a manner similar to that on rod pigments, and that the production of a retinine and opsin (cone opsin) is requisite to cell depolarization. Amounts of molecular breakdown and thus production of different amounts of substances may result in differential degrees of depolarization of receptors as

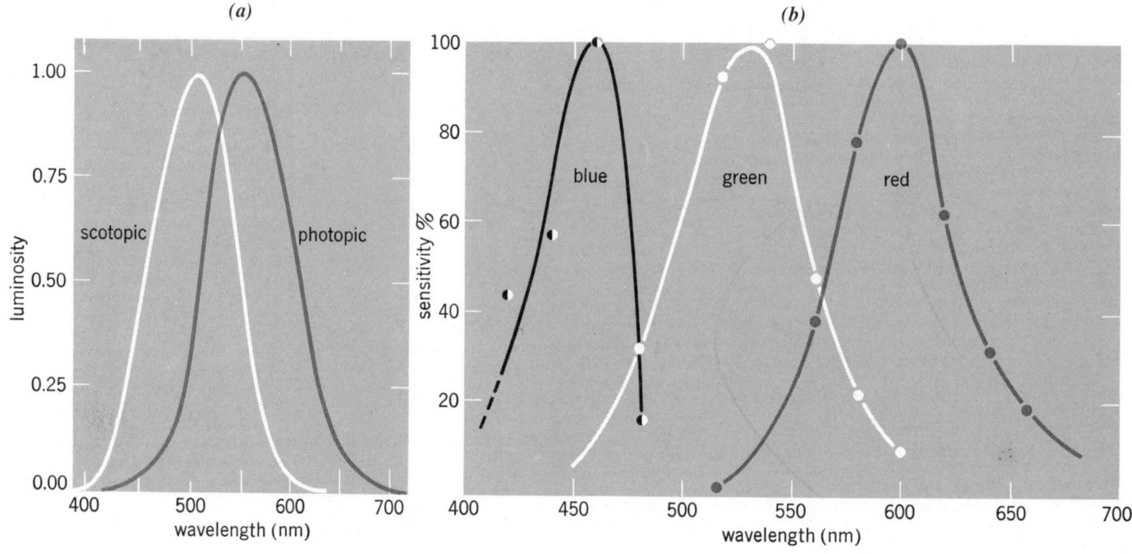

FIGURE 14.8
Absorption curves for: (*a*) rods vs. cones; (*b*) the three types of cones. Data of this type is used to show the existence of two types of receptors (rods and cones), and three types of cones.

the basis for coding of color vision information. With the assumption that there are three separate cones, each with a different pigment, it is easy to account for color blindness. If a cone or its pigment is lacking, the individual will be "blind" or insensitive to that color. An individual is described as having PROTANOPIA, DEUTERANOPIA, or TRITANOPIA according to his lack of red, green, or blue sensitivity respectively. Most color blindness is inherited. Red-green blindness is inherited as a sex-linked recessive. The trait is produced when the genes on one X chromosome are abnormal, while the genes on the other X chromosome are normal. Thus, a female transmits the trait, but a male, receiving the abnormal X chromosome with no other X chromosome to mask the trait, will show the defect.

Electrical discharge from excited rods and cones in the retina may be recorded as an ELECTRORETINOGRAM (*ERG*, Fig. 14.9); a, b, c, and d waves may be identified. The *a wave* is associated with changes occurring in the receptor layer; the *b wave* with changes in the nuclear layer; the *c wave* with changes in the pigment epithelium associated with dark adaptation; and the *d wave* with cone response.

The electrical activity, whatever its nature, is transmitted to the bipolar cells and thence to the ganglion cells. The latter cells appear to be ones that process information from rods and cones. They appear to be cells that generate impulses either to "light-on," "light-off," or both. Coding of information as to duration of light may thus be conveyed.

The retina is responsible for several other visual phenomena.

VISUAL ACUITY, as measured by ability to discriminate two separate lines, depends on the presence of an unlighted cone between two lighted ones. Lines of specified distance and separation are drawn on a card. The card is then removed from the subject by increasing the distance of the card from the subject. As long as the images of the two lines fall on cones separated by an unstimulated cone, the lines will be perceived as separate. The area of greatest acuity lies in the fovea, where only cones are present. Also, packing of cones is quite dense in this region and the diameter of the cones is at the minimum.

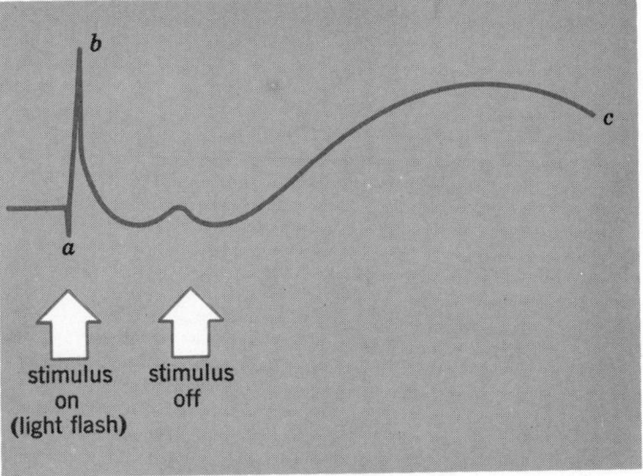

FIGURE 14.9
Mammalian electroretinogram (ERG). The *a* wave is associated with stimulation of the photoreceptors. The *b* wave arises from the inner nuclear layer containing rod-and-cone nuclei. The c wave arises from the pigment epithelium and is associated with metabolic processes of recovery after light stimulation. Note similarity to spike potential.

CRITICAL FUSION FREQUENCY or "persistence of vision" enables resolving of stimuli that are separated in time into a continuous sequence. The best example is the moving picture film consisting of separate pictures that, when run at the proper speed, creates the illusion of continuous motion. Apparently the stimulus delivered to a visual receptor by decomposition of a pigment lasts for a measurable period of time and can be "summated" to a second stimulus.

AFTER IMAGES are views of a scene that persist after stimulation of the retina. They are usually perceived in the complementary color from the original scene and are called *negative after images*. They are explained by bleaching of a particular pigment by light. If stimulated again before regeneration of pigment has occurred, the image will be seen in a color different from that of the original scene. *Positive after images* are perceived in the same shades of light and dark as the original scene but do not usually occur in

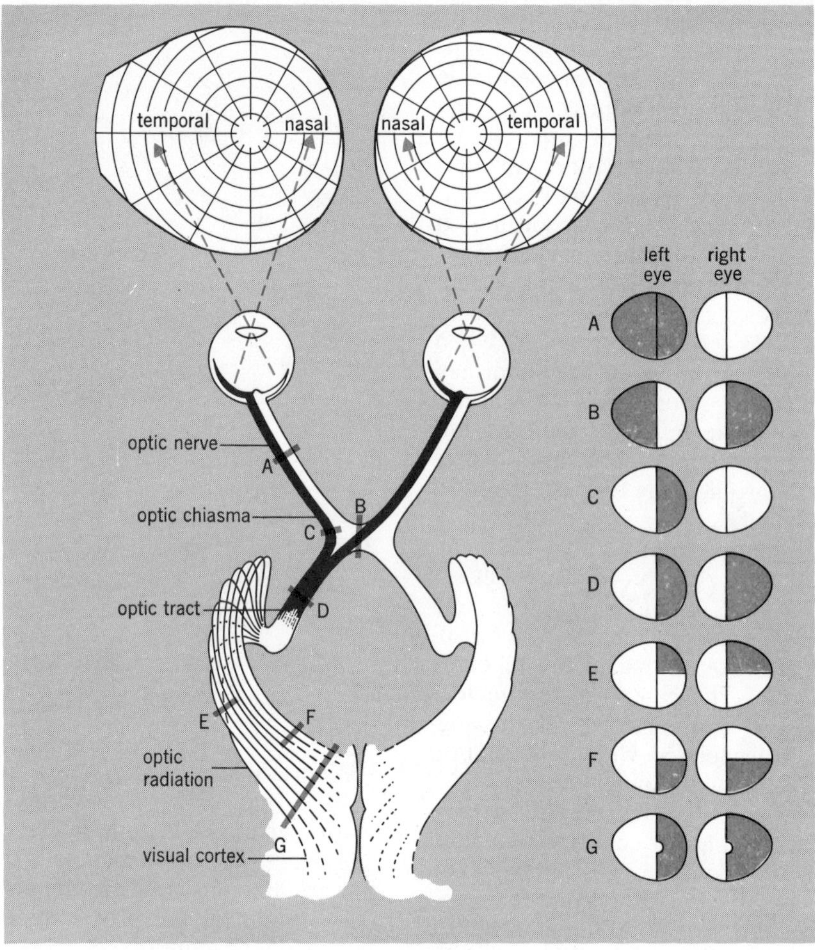

FIGURE 14.10
The central visual pathways with indication of the type of visual defect that will result from lesions in different parts of the system. Shaded areas show deficit of vision.

color. They are explained by persistence of the products of pigment breakdown in the visual receptors.

THE VISUAL PATHWAYS

EFFERENT FIBERS

Impulses originating from the retinal receptors are gathered into the fibers (Fig. 14.10) of the optic nerve and leave the eyeball at the optic disc (an area of blindness due to absence of all receptors). The optic nerve maintains a spatial orientation of fibers from the different areas of the retina. At the OPTIC CHIASMA, the fibers from the nasal portions of each eye undergo a crossing and all fibers proceed to the LATERAL GENICULATE BODY (part of the thalamus) where a synapse occurs. The geniculate body also maintains a spatial organization of fibers. From the lateral geniculate body the GENICULOCALCARINE TRACT conveys impulses to the PRIMARY VISUAL AREAS (17) in the occipital cortex. This area as well maintains a spatial organization.

VISUAL ASSOCIATION AREAS

Areas 18 and 19, lying anterior to 17, are areas where integration, interpretation, and storage of visual information occurs.

EFFECTS OF LESIONS IN VARIOUS PARTS OF THE VISUAL PATHWAY

Figure 14.10 presents the type of visual loss that results from damage in various parts of the system beyond the retina.

OTHER FUNCTIONS OF THE BRAIN IN VISION

The cortex is apparently involved in two other important aspects of vision. BINOCULAR VISION is the result of possessing two eyes that, being separated, view an object from slightly different angles. Integration of the two images produces a SENSE OF DEPTH AND DISTANCE. Inherent in the perception of distance is previous experience and knowledge of sizes of objects "close-up," so that comparison of sizes may be interpreted as distance.

SUMMARY

1. The eye is the organ of vision. It is located in the orbit, is protected by the eyelid, and lubricated by lacrimal fluid. Six eye muscles turn the eye within the orbit.
2. Three tunics surround the eyeball.
   a. The outer fibrous tunic includes the sclera and cornea.
   b. The middle tunic is the uvea and includes the ciliary body and vascular choroid.
   c. The inner retina contains the visual cells, the rods, and cones.
3. Image formation occurs by light entering the cornea, being refracted by the cornea and lens to a focus on the retina.
   a. Focusing occurs by accommodation, which involves change in lens curvature, constriction of the pupil, and movements of the eyeballs.
   b. The iris, by enlarging or constricting the pupil, corrects for spherical and chromatic aberration and controls light entering the eye.
4. Disorders of the eye may involve the cornea, the whole eyeball, the lens, and ocular fluid. Some of the more common disorders are listed below.
   a. Corneal imperfections cause astigmatism.
   b. Eyeballs too long or short for the lens system produce near- and farsightedness, respectively.
   c. Presbyopia is associated with hardening of the lens and the inability to focus near objects.
   d. Cataract is lens opacity.
   e. Glaucoma is increased intraocular pressure as a result of excess aqueous humor.
   f. Conjunctivitis is inflammation of the membrane covering the anterior eyeball.
5. The retina contains rods and cones.
   a. Rods respond to low light intensity and give night vision.
   b. The cones respond to higher intensity of light and give daylight and color vision.
   c. Both types of receptors contain pigments that are broken by light to give stimuli to depolarize the cell.
6. The retina determines several other visual phenomena.
   a. Visual acuity is determined by density of cones in the retina.
   b. Visual persistence in the retina enables fusion of individual events into a continuous sequence.
   c. After images are explained by persistence of products of retinal pigment destruction.
7. The optic nerves, optic tract, and geniculocalcarine tract convey visual impulses to the occipital lobe for processing.
   a. Lesions in each area produce characteristic types of visual loss.
8. The possession of two eyes enables binocular vision and its attendant abilities to perceive depth.

QUESTIONS

1. What structures are involved in image formation by the eye? What does each contribute to the formation of the image?
2. What changes occur during accommodation?
3. Describe the defects present that result in myopia. How is the condition corrected?
4. What evidence exists to suggest the presence of two types of receptors in the retina?
5. Describe the changes that occur in rhodopsin during stimulation by light.
6. How may color blindness be explained in terms of current theories of color vision?
7. Trace the visual pathways.
8. Compare the effects on vision of cutting the left optic nerve and the left optic tract. Explain any differences.

READINGS

Glickstein, Mitchell, and Alan R. Gibson. "Visual Cells in the Pons of the Brain." *Sci. Amer. 235*:90. (Nov) 1976.

Johansson, Gunnar. "Visual Motion Perception." *Sci. Amer. 232*:76. (June) 1975.

Ross, John. "The Resources of Binocular Perception." *Sci. Amer. 234*:80. (Mar) 1976.

Rushton, W. A. H. "Visual Pigments and Color Blindness." *Sci. Amer. 232*:64. (Mar) 1975.

15

HEARING, EQUILIBRIUM, TASTE, AND SMELL

OBJECTIVES

After studying this chapter, the reader should be able to:

- ☐ Describe the general structure of the ear.
- ☐ Describe what sound waves are, and what determines their pitch, intensity, and quality.
- ☐ Describe the functions of the outer ear.
- ☐ Describe the functions of the middle ear.
- ☐ Describe the methods by which the organ of Corti is stimulated.
- ☐ Indicate where low and high frequency sounds are determined in the cochlea, and why frequency determination is primarily a cochlear function.
- ☐ Describe the auditory pathways, and what each part contributes to the process of hearing.
- ☐ List and explain the genesis of the several types of deafness.
- ☐ Describe the equilibrial organs of the inner ear, and how they function.
- ☐ Describe the receptors for taste, the four basic taste sensations, and what the adequate stimulus for each sensation is.
- ☐ Describe the receptors for smell, the seven basic odors, and the relationship of taste and smell.
- ☐ Describe the pathways for taste and smell, and where they terminate in the brain.

The four senses described in this chapter are those of the ear, nose, and tongue. The ear confers the ability to hear and also contains receptors sensitive to movement and head position. Taste and smell complement each other for the appreciation of the fine qualities of our cuisine and may afford protection against toxic substances.

THE EAR AND HEARING

The AUDITORY MECHANISM is subdivided into three portions, the outer, middle, and inner ears (Fig. 15.1). The system gathers and directs sound waves, transforms vibrations in air to vibrations in fluid, and serves to determine pitch and intensity of sound.

THE NATURE OF SOUND WAVES

Sound waves are vibrations of air molecules. A wave produces alternate compressions and decompressions of air molecules and transmits itself like a ripple in a pond. A sound has three important properties: PITCH, INTENSITY, and QUALITY (*timbre*). Pitch (or frequency) of the sound is determined by the cycles (or vibrations) per second (Fig. 15.2). The closer the peaks are together, the higher the pitch of the sound. Intensity refers to the loudness of the sound and is a function of the height of the wave. Intensity is measured in DECIBELS. A decibel is a relative measure of sound intensity and is defined as the logarithm of the power or energy of a sound compared to a standard reference point, or $10 \log_{10} E/E_R$.

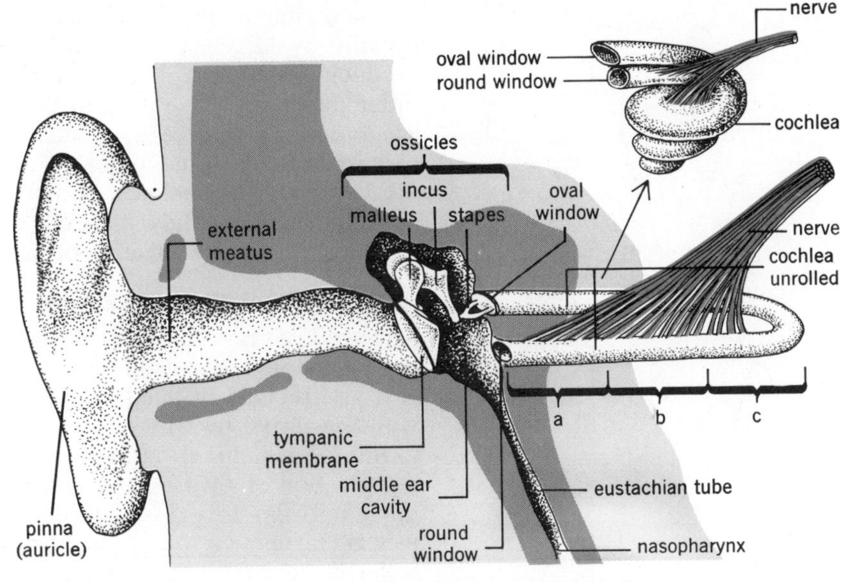

FIGURE 15.1
The ear. *a, b,* and c indicate the parts of the cochlea (1st, 2nd, and 3rd turns).

E = energy of signal
E_R = energy of reference point

The reference is taken as the energy required to make a 1000 Hz pure tone barely audible and corresponds to an energy of 0.0002 dynes per square centimeter (0 decibels). Humans can hear intensities from 0 to about 140 decibels. Beyond about 100 decibels, sound becomes painful and damaging to the ear. Because the decibel scale is logarithmic, each 10 decibel change reflects a tenfold change in power. The ear is thus subjected to a 10^{13} change in intensity that it can appreciate. The decibel ratings of some common sounds are presented in Table 15.1.

The quality of a sound is due to the presence in the vibrating object of overtones or harmonics. The latter are higher tones that faintly accompany the fundamental tone produced by a musical instrument. A sound of given pitch sounds differently when played by a piano, a violin, or a reed instrument.

THE OUTER EAR

The outer ear consists of the AURICLE (*pinna*) and the EXTERNAL AUDITORY TUBE. The auricle, in humans, serves little sound-gathering function. In lower animals, the auricle is more cuplike and may be directed by muscles toward

TABLE 15.1
DECIBEL RATINGS OF SOUNDS

SOUND	RATING—DECIBELS
Absolute silence	0
Watch ticking	20
Residential street, no traffic	40
Stream	50
Automobile at 30 ft	60
Conversation at 3 ft	70
Loud radio	80
Truck at 15 ft	90
Car horn at 15 ft	100
Pneumatic hammer at 3 ft	120
Propeller airplane at 15 ft	130
Jet aircraft at takeoff	150+

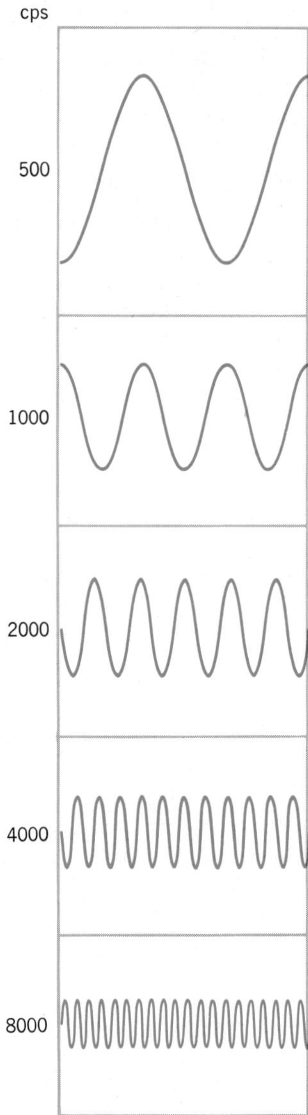

FIGURE 15.2
The relationships of frequency (cycles per second or Hz) of a sound wave to pitch. The greater the frequency, the higher the pitch.

the source of a sound in order to obtain the greatest benefit from the sound. The external auditory canal is S-shaped, courses medially and inferiorly, and directs sound waves to the tympanic membrane. The canal is guarded externally by large hairs and is lined by glands producing cerumen (ear wax), a brownish, bitter material. These devices tend to discourage entry into the canal of dust, large flying objects (insects), and other contaminants. If the cerumen accumulates, it may harden and partially or completely plug the canal with a resulting decrease of hearing acuity.

THE MIDDLE EAR

The middle ear consists of the middle ear cavity and its contents. The MIDDLE EAR CAVITY, an irregular air-filled space, communicates with the throat by way of the PHARYNGOTYMPANIC (*Eustachian*) TUBE. The tube allows equalization of air pressure on the two sides of the ear drum to prevent rupture. The TYMPANIC MEMBRANE (*ear drum*) forms the lateral wall of the cavity. Sound waves striking the membrane cause it to vibrate according to the frequency and intensity of the sound. Three small ear bones (ossicles), the MALLEUS (*hammer*), INCUS (*anvil*), and STAPES (*stirrup*) form a system of levers across the cavity to transmit vibrations of the tympanic membrane to the fluid-filled cochlea. The handle of the malleus is attached to the ear drum, while the footplate of the stapes fits into the OVAL WINDOW of the cochlea. A basic problem arises from the fact that the ear drum vibrates in air, which requires less energy to move it, while the structures of the cochlea are suspended in fluid and require more energy to set them in vibration. A mismatch in impedance (force per unit area required to vibrate) is created in the ratio of about 130:1. The ossicles, acting as a lever system, coupled with the relative areas of the drum and oval window reduces the impedance mismatch to about 4.5:1. (Leverage change = factor of 1.5; areas of drum and oval window 0.55 sq cm to 0.032 sq cm = factor of 17.5.) In short, the pressure of the wave is increased by a factor of about 29 in traversing the ossicles, sufficient to create shock waves in the cochlear fluid. The ossicles also reduce the degree of movement of the oval window as compared to the drum to prevent rupture of the window by movement of the stapes. The tendons of two small muscles, the TENSOR TYMPANI and STAPEDIUS, lie within the middle ear cavity. The tensor tympani attaches to the handle of the malleus and the stapedius to the stapes. They are innervated by the trigeminal and facial nerves respectively. The muscles form the effectors of a reflex arc where the inner ear and auditory nerve form the afferent limb. Their contraction, occurring after a time lag of about 15 msec, limits ossicle movement. The most potent stimulus to this AUDITORY REFLEX is a loud sound. The reflex thus serves a protective function, protecting the inner ear structures from damage by high intensity sound waves. Though transmission of sound waves to the cochlea normally occurs via the ossicles, BONE CONDUCTION may also serve to create shock waves in the cochlear fluid or cause the cochlea to vibrate within the fluid. This phenomenon is utilized in the use of certain types of hearing aids.

THE INNER EAR

The inner ear (Fig. 15.3) consists of the OSSEOUS LABYRINTH or the channels within the bony substance of the temporal bone and the MEMBRANOUS LABYRINTH or living tissues of the inner ear that line the channels. The inner ear contains the ORGAN OF CORTI, the organ of hearing in the cochlea, and MACULAE and SEMICIRCULAR CANALS, organs involved in maintenance of posture and equilibrium.

The cochlea. The channels of the cochlea are fluid filled, with the footplate of the stapes acting on the fluid column of the scala vestibuli. Inward movements of the stapes compress the fluid and result in vibration of the BASILAR MEMBRANE (Fig. 15.4). The basilar membrane consists of some 25,000 to 30,000 strands of tissue that lengthen as the membrane follows the $2\frac{1}{2}$ turns of the cochlea. It is believed that while the entire basilar membrane moves in response to shock waves set up in the cochlear fluid, certain portions of it vibrate to a greater degree and give the cochlea the ability to discriminate pitch. The basis of this ability is embodied in the PLACE THEORY or selective-vibration theory of cochlear function. A string (or fiber of the basilar membrane) will vibrate according to its tension and its mass. Experiments

FIGURE 15.3
The structures of the inner ear.

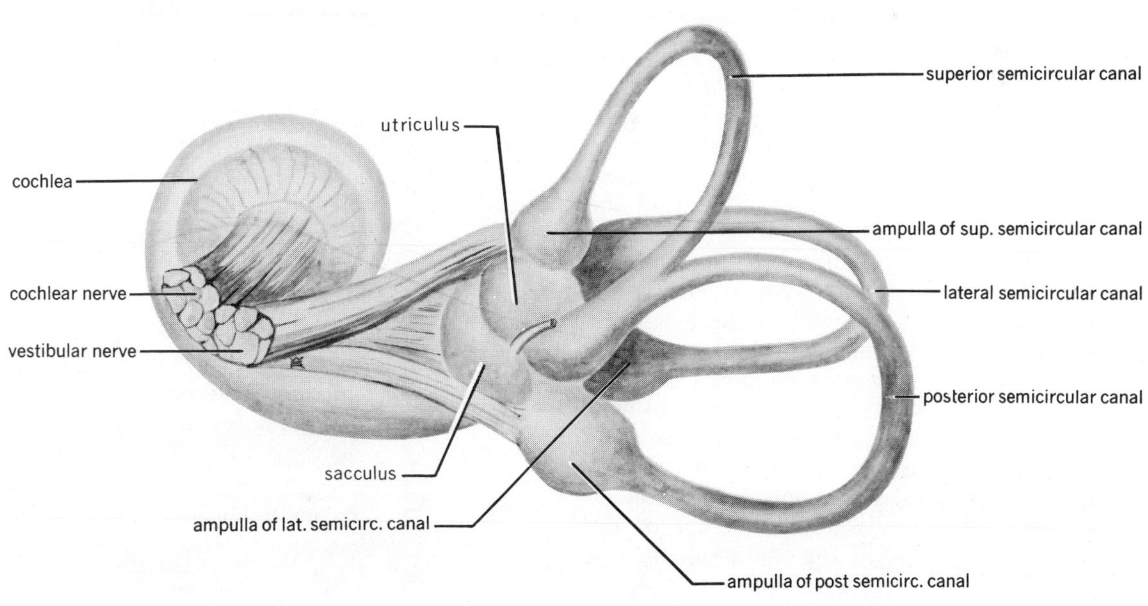

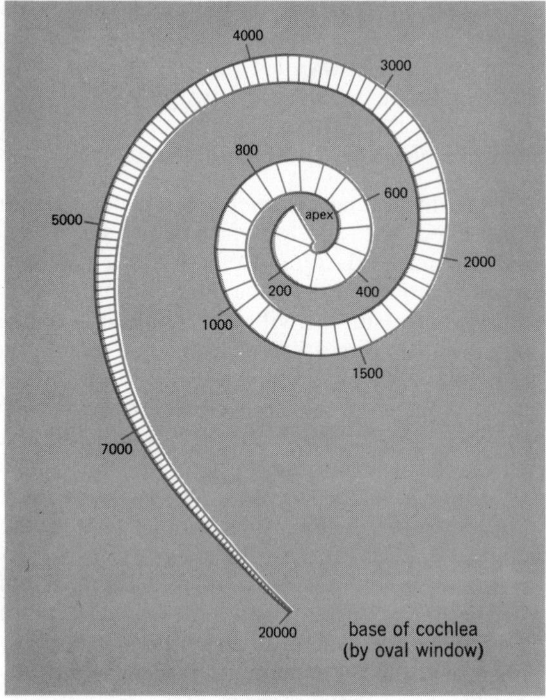

FIGURE 15.4
Pitch analysis by the basilar membrane of the cochlea. Numbers indicate cycles per second (Hz). Membrane averages 32 mm long, 0.04 mm wide at base, 0.5 mm wide at apex.

have shown that the basilar fibers closest to the oval window are under the greatest tension. Lying closest to the oval window, the mass of fluid is least. Towards the apex of the cochlea, tension is less on the basilar fibers and the amount of fluid is greater, resulting in a greater mass to be moved. Sound waves entering the cochlea cause a maximum sympathetic vibration of those fibers "tuned" to the particular frequency of the sound. High-frequency sounds (high pitch) have a greater frequency and are discriminated by fibers closer to the oval window. Low-frequency sounds (low pitch) are discriminated towards the apex of the cochlea. Once discriminated, sound waves are reduced and cancelled by outward movement of the ROUND WINDOW and by the communication of the two scali at the HELICOTREMA.

Transduction of motion of the basilar membrane into nerve impulses is thought to be accomplished through the HAIR CELLS of the organ of Corti (Fig. 15.5). Movement is translated into bending of the hair cells' processes, and depolarization occurs. The exact mechanism by which depolarization occurs is unknown, but probably involves ionic shifts secondary to mechanical stimulation.

A step-by-step summary of the events occurring from the vibration of the eardrum to the cochlea is presented below.

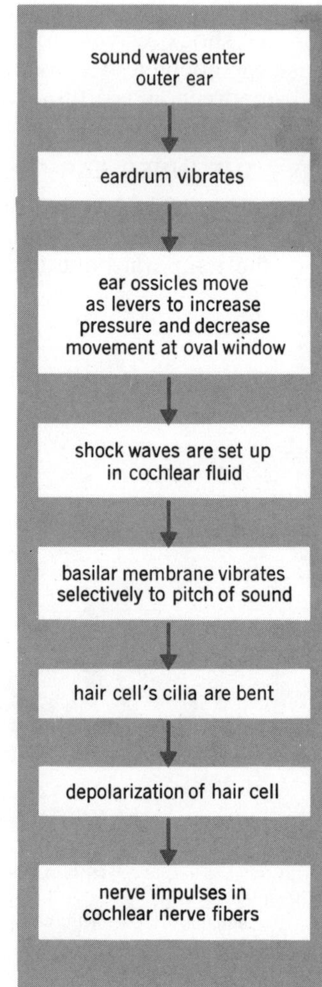

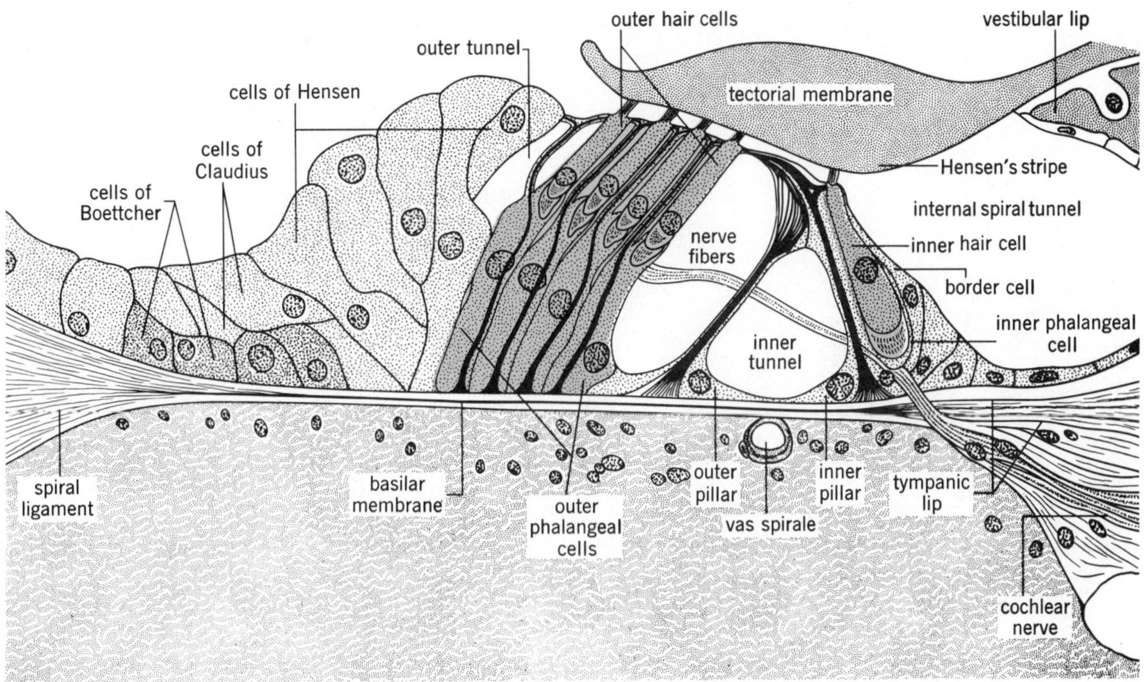

FIGURE 15.5
The organ of Corti; the sensory elements are the hair cells.

AUDITORY PATHWAYS

Nerve fibers innervating the hair cells represent dendrites of neurons whose cell bodies lie in the spiral ganglion of the cochlea (Fig. 15.6). The axons form the fibers of the auditory portion of the VIIITH cranial nerve (commonly called the AUDITORY NERVE). Some 30,000 fibers comprise the nerve, a figure that accords closely enough with basilar fibers to lead to the suspicion that each lateral row of hair cells has its own common neuron. These neurons are the first order neurons in the auditory pathway. They exhibit a random SPONTANEOUS DISCHARGE of 100 Hz or less, show tuning (response to particular sound frequencies), show INHIBITION when tones close to that to which the unit is responding are applied, and CODE IMPULSES by frequency up to 2000 Hz (cell forms impulses at some multiple of the basic frequency). The neurons are also sensitive to anoxia, cold, and drugs.

The first-order neurons proceed to the brain stem where they synapse within the COCHLEAR NUCLEI. These nuclei are located in the lower pons and upper medulla and consist of about a dozen different masses of cell bodies in three main groups; the dorsal, anteroventral, and posteroventral. The incoming fibers branch to achieve a complete representation of the cochlea in each of the three areas. The cochlear nuclei respond to sound intensity by discrete neural activity, which enlarges as the intensity and the sound increases. "Processing" of the input in terms of frequency and intensity occurs here. In the brain stem are other areas serving auditory functions: the OLIVARY BODY sends auditory connections to the motor system and may be involved in reflex responses; the INFERIOR COLLICULUS is involved in motor responses to auditory stimuli; the MEDIAL GENICULATE BODY (actually part of the thalamus) is a major relay station on the way to the auditory cortex.

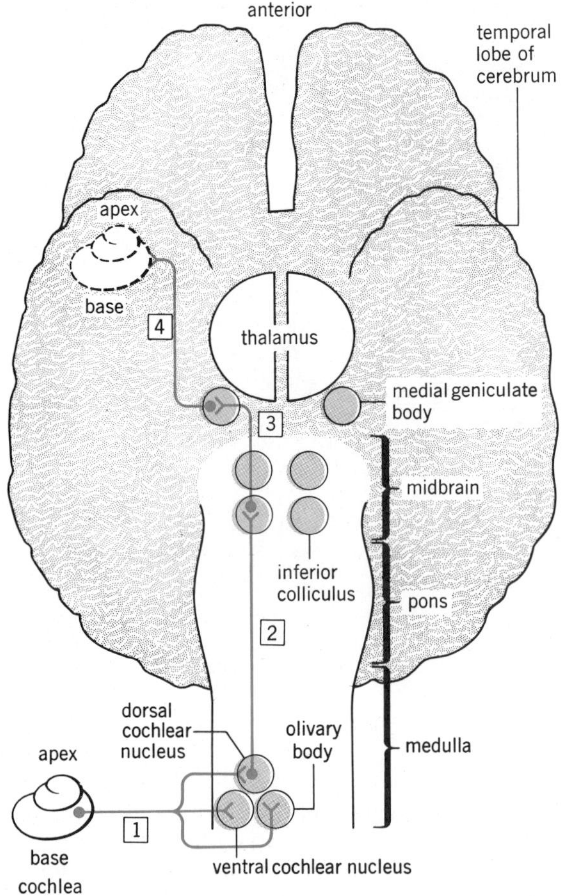

FIGURE 15.6
Diagram of the auditory pathways. Note that four neurons convey impulses to the cortex.

The AUDITORY CORTEX lies in the temporal lobe, in areas 41 and 42 of Brodman. A topographical localization of sound is maintained within the area, with low frequency sound found anteriorly and high frequency posteriorly. Both cochleas are represented in both auditory cortices. Surrounding the auditory cortex are the AUDITORY ASSOCIATION AREAS. Within these areas lie the ability to integrate, remember, and analyze various types of sound input. For example, INTENSITY DISCRIMINATION is hampered by damage to the association areas; DISCRIMINATION OF TONAL PATTERN (sequence of notes) requires these areas; LOCALIZATION of sounds also requires the association areas. Localization implies the ability to determine the source of a sound. A basic requisite for accurate localization is that the sound strikes one ear more strongly than the other; finally, the ability to COMPREHEND the spoken word requires normal function of the association areas.

DISORDERS OF HEARING AND INTERPRETATION

DEAFNESS is defined as loss of ability to hear and may be partial or complete. According to where the problem lies, several types of deafness are described.

TRANSMISSION (*conduction*) DEAFNESS occurs when the sound waves are prevented from reaching the eardrum or by failure of the ossicles to transmit the sound waves to the cochlea. OBSTRUCTION of the external auditory tube by wax or foreign objects, INFLAMMATION of the middle ear (otitis media), FUSION of the ossicles, or FIXATION of the stapes in the oval window (*otosclerosis*) may all result in loss of hearing.

NERVE (*sensory-neural*) DEAFNESS results from damage to the cochlea and/or auditory nerves and/or brain by mechanical or chemical (drug) agents. Some occupations (e.g., foundry workers, aircraft mechanics) are associated with continued high-intensity noise exposure. According to the frequency of the sound, selective degeneration of the organ of Corti may occur, destroying the ability to hear sound of particular frequencies. Lesions in the eighth cranial nerve typically result in loss of hearing in one ear only. Lesions in the brain stem produce loss in both ears, as do lesions of the auditory cortex. If a lesion involves the auditory association areas, various types of aphasia may be produced.

Mixtures of the two types may also occur, as in congenital deafness (born deaf). Several tests for auditory function are routinely employed.

Tests for determination of transmission include the WEBER TEST. This involves placing the handle of a vibrating tuning fork in the midline of the forehead. If the defect is in the outer or middle ears, the sound will be heard in the deaf ear (bone conduction); with a defect elsewhere, the sound is heard in the good ear.

Tests for acuity and pitch discrimination use an audiometer, a device producing sounds of variable pitch and intensity. A written record, the AUDIOGRAM (Fig. 15.7) is produced. Normal individuals hear pitches from about 50–25,000 Hz, and intensities from 0–120 decibels.

Test for determination of central processing of information include word comprehension tests.

THE INNER EAR AND EQUILIBRIUM

The inner ear includes organs for determination of body position and changes of acceleration, leading to maintenance of posture and equilibrium. These are the SEMICIRCULAR CANALS and the MACULAE (Fig. 15.8).

The semicircular canals are three fluid-filled channels in each ear. An enlarged AMPULLA contains a CRISTA. The crista is bent by movement of the fluid as changes in acceleration and direction of motion occur. The three canals are placed in three mutually perpendicular planes to one another so that any head motion will cause excitation of one or more cristae. The MACULAE are primarily ORGANS OF POSITION, not movement, and are activated by gravity or acceleration "pulling" on the OTOLITHS ("ear stones") of the organ. This force bends the cilia of the maculae. The maculae are located in the utricles and saccules of the inner ear. Those of the utricles are horizontal when the head is in the erect posture. They are stimulated by tilting the head or by linear acceleration. The maculae of the sacculus are oblique in erect posture, and are particularly sensitive to vibration. Activation of either of these receptors results in reflex responses (*labyrinthine reflexes*) tending to maintain the head properly in space. Labyrinthine reflexes are categorized into two groups.

ACCELERATORY REFLEXES, arising from stimulation of the semicircular canals, produce response of eye, neck, trunk, and appendages to "starting," "stopping," or "turning" motions. Nystagmus, a horizontal, vertical, or rotary movement of the eyes, occurs on angular acceleration. Limb responses are typically those of extension on the side away from an angular acceleration and toward the force if linear. Thus, a person leans into a curve, forwards on acceleration and back on deceleration.

POSITIONAL REFLEXES, arising from stimulation of the maculae, involve righting reflexes and notification of head position. The pathway for communication of information from the canals and maculae is via the VESTIBULAR PORTION (*vestibular nerve*) of the VIIIth cranial nerve. The fibers pass to brain stem VESTIBULAR NUCLEI, located in the lower lateral pons, and thence to the CEREBELLUM, where appropriate muscular response is provided.

FIGURE 15.7
Audiogram in normal hearing. Hearing range extends from about 20 Hz to about 20,000 Hz and acuity tends to be inversely proportional to frequency.

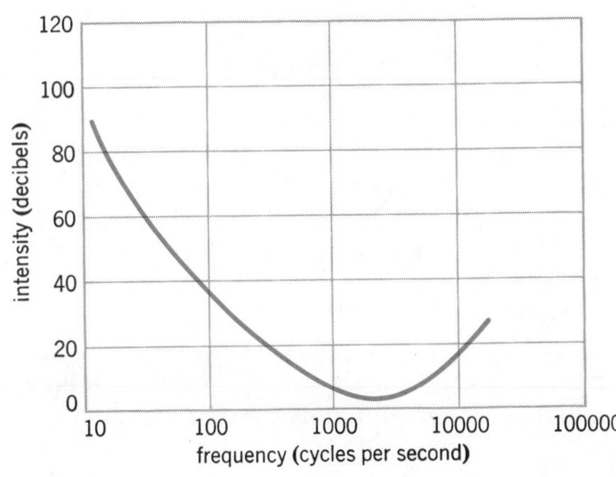

DISORDERS OF AND TESTING OF LABYRINTHINE FUNCTION

Though not strictly a disorder, MOTION SICK-

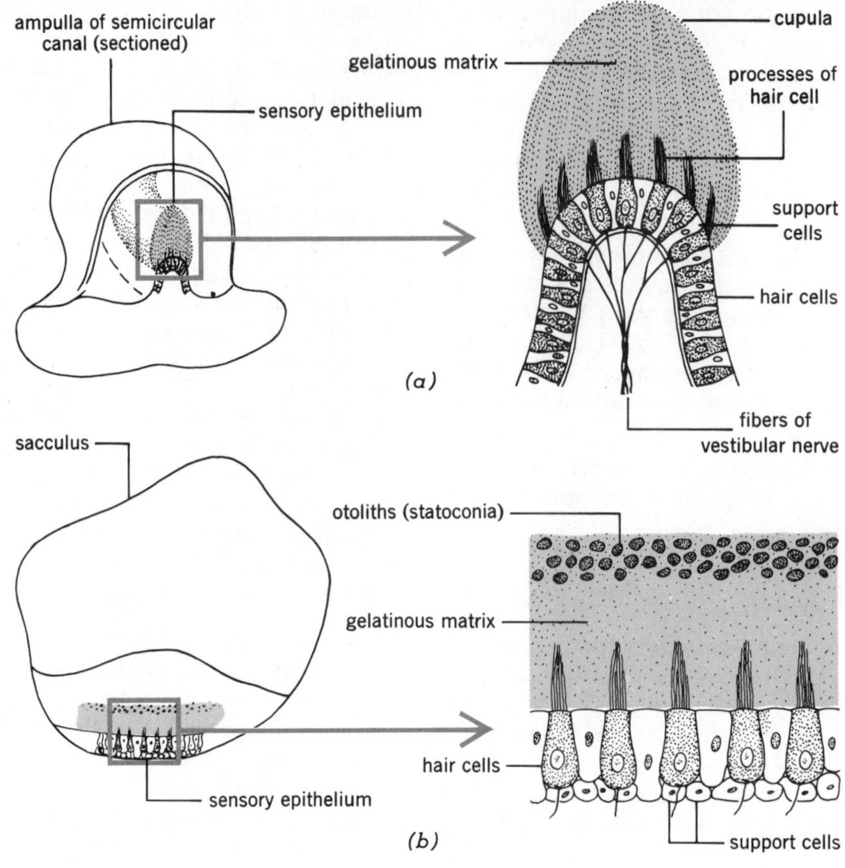

FIGURE 15.8
The balance and equilibrium structures of the inner ear. (*a*) Crista ampullaris of semicircular canal. (*b*) Macula of the sacculus or utriculus.

NESS involves repetitive changes in angular and linear acceleration affecting the receptors. Nausea and vomiting may result. INFECTIONS of the vestibular nerve may result in loss of equilibrium as may TRAUMA that injures the nerve. Indeed, any basic loss of equilibrium *without* change of muscle tone is indicative of altered vestibular function. Testing function typically involves spinning the subject in a Barany chair with the head in various positions to excite the different canals. If nystagmus, dizziness, and loss of equilibrium are *not* produced, a lesion is presumed to exist somewhere in the system.

GUSTATION (TASTE) AND OLFACTION (SMELL)

Taste and smell are chemical senses in that the adequate stimuli for the receptors are chemicals in solution in water or air. The senses reinforce one another; as anyone with a cold knows, one's food becomes "tasteless." Taste may protect against the intake of toxic substances such as alkaloids and, as a component of appetite, plays a role in food selection. It has also been established that animals deficient in certain nutritional elements may exhibit an increased interest in those foods containing the missing elements and that taste provides the sensory clue to proper selection of foods. Smell provides humans with the ability to detect odors that may represent something harmful to them (e.g., spoiled food, insecticides, and gases).

TASTE

The receptors for the sense of taste are the TASTE BUDS (Fig. 15.9) located primarily on the tongue. A few buds may be found on the soft palate, epiglottis, and pharynx. According to the stimulus that causes the maximum response by the taste buds, four basic types of buds and thus four basic taste sensations are described. SOUR tastes are produced by hydrogen ion (acids). Strong acids, because of their greater liberation of hydrogen ion produce a stronger taste than weak acids. The SALTY taste is due to cations of ionized salts, for example, Na^+, NH_4^+, Ca^{++}, Li^+, K^+. Sodium compounds and ammonium compounds are most effective in stimulating salty taste. The SWEET taste requires the presence of compounds with hydroxyl (—OH) groups in the molecule. Many substances meet this criterion, including sugars, alcohol, amino acids, ketones, and lead salts. Of the list of sweet-tasting substances, sugars are second to saccharine (600 times as sweet as glucose) and chloroform (40 times as sweet as glucose) in terms of sweetness. The BITTER taste is produced by alkaloids and long-chain organic molecules. Thus, strychnine, quinine, caffeine, and nicotine all taste bitter.

The buds serving the four basic taste sensations do not have the same threshold for stimulation. The sensation with the lowest threshold (greatest sensitivity) is bitter, followed by sour, salt, and sweet.

Although the manner in which the substance to be tasted causes depolarization of the taste cells is unknown, it has been suggested that the material must fit a depression on the microvilli of the taste cells in the manner of a lock and key. A "fit" is followed by depolarization of the membrane and generation of a nerve impulse.

Buds located on the anterior two-thirds of the tongue pass impulses to the Vth cranial nerve (trigeminal) and thence to the VIIth cranial

FIGURE 15.9
The taste buds.

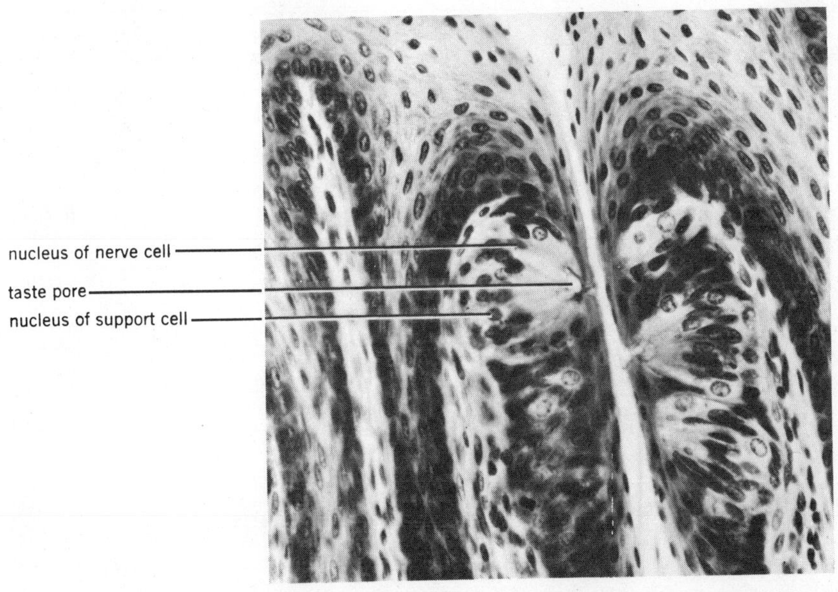

nerve (facial). Buds on the posterior one-third of the tongue send impulses over branches of the IXth cranial nerve (glossopharyngeal). Buds on palate, pharynx, and epiglottis are innervated by the Xth cranial nerve (vagus). The afferent nerves pass to the tractus solitarius in the brainstem. From here impulses pass to the thalamus and from there to the parietal cortex. Fibers also pass to the salivary nuclei and vagus nuclei to serve the salivary reflex and the cephalic phase of gastric secretion. The taste pathways are presented in Figure 23.3.

SMELL

The receptors for the sense of smell are located in the OLFACTORY EPITHELIUM located in the apices of the nasal cavities (Fig. 15.10). The classification of odors is highly subjective and is based on the assumption that molecular shape of the substance and depressions on the microvilli must achieve a match before a depolarization can occur. Knowledge of molecular shape can lead to prediction of what the material will smell like. According to this theory, there are seven basic molecular shapes, and thus seven basic odors. The odors are described as camphoraceous, musky, floral, ethereal, pungent, putrid, and pepperminty.

Thresholds of smell are lower than those for taste. On the average, smell is 25,000 times more sensitive than taste. Concentrations of substances only 10 to 50 times above threshold

FIGURE 15.10
The olfactory epithelium (*a*) light microscope view; (*b*) as seen under the electron microscope.

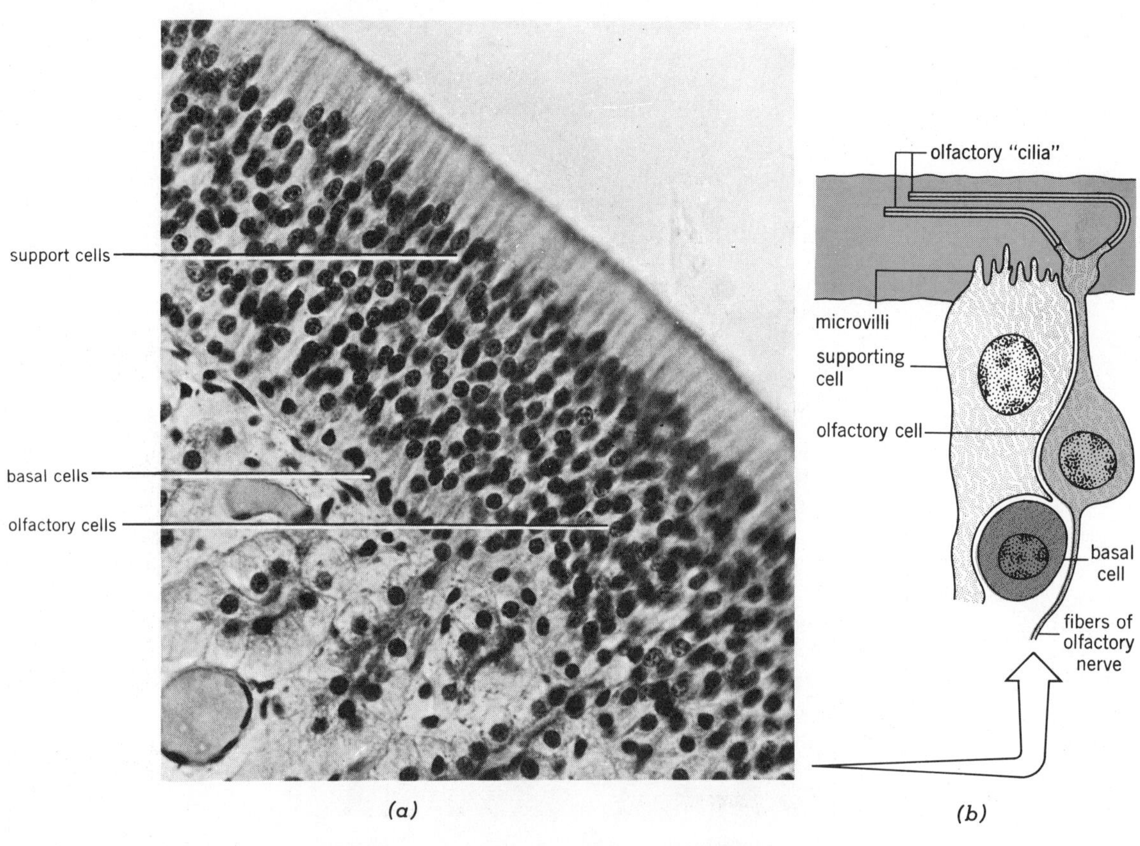

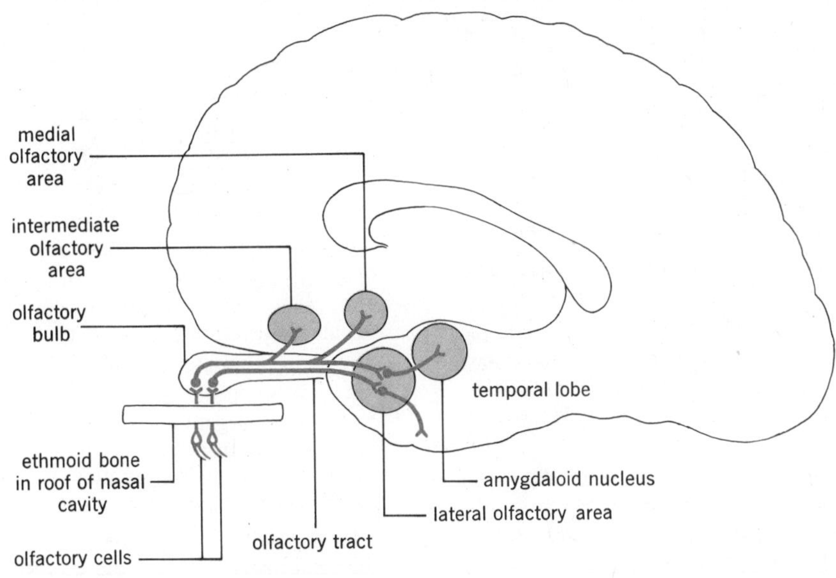

FIGURE 15.11
The basic neural pathways for the sense of smell.

levels trigger maximum firing by olfactory cells. In other words, relatively low concentrations of materials result in maximum response. This suggests that the olfactory receptors are designed for qualitative rather than quantitative detection of odors.

The neural pathways for smell (Fig. 15.11) include the axons of the OLFACTORY CELLS that pass through the roof of the nasal cavity to synapse with mitral cells in the OLFACTORY BULB of the brain. Fibers from the mitral cells form the OLFACTORY TRACT that leads to the OLFACTORY AREAS of the brain. Three olfactory areas are described. A LATERAL OLFACTORY AREA includes a portion of the temporal lobe and a part of the amygdaloid nuclei. This area is termed the *primary olfactory area* because it is the principal area for olfactory awareness. The INTERMEDIATE OLFACTORY AREA lies at the bases of the olfactory tracts in the lower frontal cortex. It appears to be primarily a region for communication and interconnection between the different parts of the olfactory system. A MEDIAL OLFACTORY AREA lies on the medial aspect of the frontal lobe beneath the anterior portion of the corpus callosum. This area appears to be involved with the limbic system in expression of emotion. In rats, a "pleasure center" has been shown to be located here, one that, if implanted with an electrode connected to a self-stimulating device, results in stimulation for its pleasure even to the exclusion of eating.

The olfactory regions send fibers to autonomic centers that can produce reflex behavior in response to olfactory stimuli. Salivation, sexual behavior, and emotional expression are examples of reflex behavior induced by olfactory cues.

DISORDERS of the olfactory system are rare but have considerable importance when present. Interference with the sense of smell may be produced by a tumor pressing on or invading the olfactory tract or bulb. Temporal-lobe lesions may produce olfactory hallucinations, usually of disagreeable odor. Such disagreeable odors often precede the development of an epileptic seizure.

SUMMARY

1. Sound waves are vibrations of air molecules. Sound has three qualities.
   a. Pitch depends on frequency of the sound vibrations.
   b. Intensity refers to the loudness of a sound, and is measured by the decibel.
   c. Timbre refers to the quality of the sound.
2. The outer ear gathers and directs sound waves to the ear drum.
3. The middle ear contains a series of ear bones (ossicles) that reduce the amplitude and increase the force of eardrum movements. This is required, because the organ of hearing floats in fluid.
4. The organ of hearing is contained within the cochlea. It is known as the organ of Corti.
   a. The organ contains hair cells that convert movement into electrical impulses.
   b. The organ determines pitch by selective vibration according to frequency of the incoming sound wave.
5. The auditory pathways conduct impulses from the cochlea to the brainstem cochlear nuclei, then to the auditory area in the temporal lobe of the cerebrum.
6. Disorders of hearing may result from plugging of the outer ear, inability of the ossicles to move, or by damage to the nervous structures of cochlea and/or auditory pathways.
7. The semicircular canals and maculae of the inner ear respond to movement and position. Muscular responses to maintain posture and equilibrium result.
8. Taste is a chemical sense, served by the taste buds. Four tastes exist.
   a. Sour is triggered by H^+.
   b. Sweet is triggered by OH.
   c. Salty is triggered by metallic ions.
   d. Bitter is triggered by alkaloids.
9. Taste pathways include the VIIth and IXth cranial nerves and the parietal lobe.
10. Smell is a chemical sense. Seven subjective senses exist.
    a. The olfactory pathways lead from the nasal cavities to the temporal lobes and frontal lobes.

QUESTIONS

1. How is the mismatch in impedance between the tympanic membrane and the oval window reduced?
2. What functions are served by the structures located in or communicating with the middle ear?
3. How does the cochlea determine pitch?
4. Describe the structure and function of the cristae ampullaris.
5. What are the basic taste sensations and what are the adequate stimuli for each?
6. What are some basic odors? Compare sensitivity of the olfactory organ to that of taste.
7. What is transmission deafness and how can it occur?
8. Why is it that when you have a "cold," your food may become "tasteless"?
9. Trace the auditory pathways, indicating what type of processing of auditory impulses occurs in each area.
10. How are macula and crista ampullaris different or similar in structure and function?

READINGS

DuBois, Grant E., Guy A. Crosby, and Patrick Saffron. "Nonnutritive Sweeteners: Taste-Structure Relationships for Some New Simple Dihydrochalcones." *Science 195:*397. 28 Jan 1977.

Goldbert, J. M., and Cesar Fernandez. "Vestibular Mechanisms." *Ann Rev. Physiol.* Vol 37. Palo Alto, Cal. 1975.

Oster, Gerald. "Auditory Beats in the Brain." *Sci. Amer. 229:*94. (Oct) 1973.

THE INTERNAL ENVIRONMENT: BODY FLUIDS

OBJECTIVES

After studying this chapter, the reader should be able to:

- ☐ Explain why land animals require an "internal environment."
- ☐ Define what is meant by total body water.
- ☐ Describe the two major subdivisions of the total body water: the extracellular and intracellular compartments.
- ☐ List the subdivisions of the extracellular compartment and the percent of the total water each constitutes.
- ☐ Explain how one can measure or calculate the volumes of the various fluid compartments.
- ☐ List the differences in composition of the various compartments.
- ☐ Explain how fluid and electrolytes are transferred between compartments.
- ☐ Show how water and solutes are taken into the body, and how they are absorbed into the extracellular compartment.
- ☐ Explain how fluids and electrolytes are lost from the body, and the amounts lost by each route.
- ☐ Explain the mechanisms involved in fluid and electrolyte defense by the body.
- ☐ Explain how excessive intake or loss of fluids and electrolytes can alter extra- and intracellular fluid volumes.
- ☐ Explain the genesis of dehydration and edema.

Life is said to have originated in the sea. The sea provided rich stores of nutrients for the organisms living therein, and it also supplied a means of carrying away from the organisms their often toxic products of activity. As organisms colonized the land, they carried with them a fluid in their bodies to serve the same functions as were served before—provision of nutrients and removal of wastes. In humans, this "sea" constitutes the body fluids that form the INTERNAL ENVIRONMENT for the body cells. Wherever there is a living cell in the human body, there will be fluid within and around it, and the composition and characteristics of the fluid must be narrowly regulated to ensure survival of the living units. This regulation requires constant expenditure of energy on the part of living cells.

FLUID COMPARTMENTS

TOTAL BODY WATER

In the organism as a whole, and within its various tissues and organs, are found membrane-separated fluid compartments. Collectively, they form the TOTAL BODY WATER that amounts to about 55 to 60 percent of the adult body weight. Individual variation exists due to age (older persons have less fluid volume), and amount of adipose (fat) tissue in the body (more fat means relatively less water, because fat tissue is nearly water-free). Thus, body water is often expressed as a percentage of lean (fat-free) body mass; it is relatively constant at about 72 percent of the lean body weight. Within this total body water are found the major solute cations (sodium, potassium, hydrogen, calcium, and magnesium), and anions (chloride, bicarbonate, and protein) in a dissolved or suspended state. By active processes, such as pinocytosis and active transport, and by the passive processes of diffusion, filtration, and dialysis, the solutes are distributed between the various subdivisions of the body water. Solute distribution then largely determines water distribution by the process of osmosis.

Within the organism and its parts, the body fluids may be divided into two major compartments: EXTRACELLULAR FLUID (ECF) lying outside of the cells, and INTRACELLULAR FLUID (ICF) within cells. The boundaries between these compartments are the cell membranes. Extracellular fluid has several additional subdivisions, each separated from the other by membranes. The relationships of these several compartments, and their percentage of total body water are shown, for the adult, in Figure 16.1.

EXTRACELLULAR FLUID

EXTRACELLULAR FLUID accounts for about 37.5 percent of the body fluids. It represents the internal environment land animals have incorporated, and may be subdivided as follows.

The BLOOD PLASMA, or liquid portion of the substance contained within the cardiovascular system, amounts to about 7 PERCENT of the body water. It is circulated by the action of the heart, and forms the major source of fluid and solutes for all other compartments.

INTERSTITIAL (*tissue*) FLUID forms the compartment outside of the blood vessels that bathes the cells. LYMPH is interstitial fluid that has entered lymphatic vessels of the body. These fluids move slowly through the intercellular spaces and lymph vessels, but are in equilibrium with the plasma for major inorganic cations and anions. Together, these two fluids form about 18 PERCENT of the body water.

FLUIDS OF DENSE CONNECTIVE TISSUE (as in skin, ligaments and tendons) AND BONE are actually a part of the interstitial fluid, but their water exchanges so very slowly that it *behaves* as though it was a separate compartment. These fluids account for about 10 PERCENT of the body water.

TRANSCELLULAR FLUIDS are those extracellular fluids separated from other compartments by *epithelial membranes*. Included here are cerebrospinal fluid, the ocular fluids (aqueous and vitreous humors), synovial fluid (in joints) pleural fluid (around the lungs), peri-

FIGURE 16.1
Relationships of body water compartments. Numbers represent the percent that each compartment comprises of total body water in an adult.

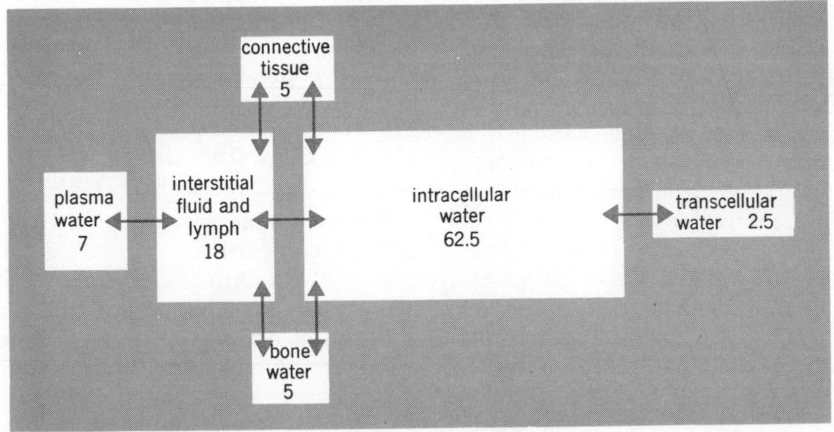

TABLE 16.1
COMPARTMENTS OF THE BODY WATER

COMPARTMENT	APPROXIMATE % OF BODY WEIGHT					APPROXIMATE % OF BODY FLUID				
	PREMATURE	NEWBORN	1 YR.	2 YR.	OVER 18 YR.	PREMATURE	NEWBORN	1 YR.	2 YR.	OVER 18 YR.
Extracellular fluid	**50**	**40**	**30**	**24**	**21.4**	**56**	**53**	**47**	**40**	**37.5**
Plasma	5	5	5	5	5	6	7	7	7	7
Interstitial fluid (including connective tissue and bone)	45	35	25	19	15	50	46	40	33	28
Transcellular fluid	—	—	—	—	1.4	—	—	—	—	2.5
Intracellular fluid	**40**	**35**	**34**	**36**	**38.6**	**44**	**47**	**53**	**60**	**62.5**
Total body water	**90**	**75**	**64**	**60**	**60**	**100**	**100**	**100**	**100**	**100**

toneal fluid (in the abdominal cavity), pericardial fluid (around the heart), and the fluids within the hollow organs of the digestive, respiratory, and urogenital systems. Because these fluids are produced by secretory (active) processes, their composition is characteristic for the specific fluid, and is often different from that of the rest of the ECF. Collectively, transcellular fluids account for about 2.5 PERCENT of the body water.

INTRACELLULAR FLUID

INTRACELLULAR FLUID is fluid within living cells and accounts for about 62.5 PERCENT of the body water. It differs in composition even within the different regions of one cell type because of the many organelles that are present within a cell. Each organelle tends to have an environment within it that ensures its own optimal operation. In considering the fluid balances of the body, one ignores these differences, and views the ICF as a single compartment.

It may again be emphasized that body water content is age-dependent. Table 16.1 summarizes some changes that occur in the various body fluid compartments according to age. One may note that ECF tends to decrease as one ages, because of the accumulation of tissue with maturation. Also, body water is not equally distributed within various tissues. Table 16.2 indicates water content of various organs and tissues of the adult. In general, the more active the tissue is, the higher its water content.

COMPOSITION OF THE VARIOUS COMPARTMENTS

Methods of measurement. The volume of a given fluid compartment cannot be measured directly. DILUTION TECHNIQUES for estimating compartment volume utilize the principle that injection or infusion of a chemical substance will eventually result in a uniform distribution of that chemical in the available fluid volume. If one knows the concentration or quantity of the substance originally injected, and the final concentration after distribution has occurred, the volume of distribution may be calculated by the formula

$$V = \frac{Q}{C}$$

where

V = the volume of distribution
Q = quantity of substance injected
C = final concentration in a sample of the fluid compartment

For example, let us presume we have a beaker full of water, and we desire to determine the volume of the water without pouring it into a measuring cup. Add 25 mg of dye (Q) to the fluid in the beaker, mix and analyze for dye concentration; it is 0.05 mg per ml (C). Then

$$V = \frac{25 \text{ mg}}{0.05 \text{ mg/ml}} = 500 \text{ ml}$$

Employed in the body, this technique depends on finding a substance that will largely confine itself to the compartment being measured. The substance should also not be metabolized by body cells. There are no such ideal substances, so that corrections must be made, particularly if the substance is excreted in the urine. Thus, the equation

$$\text{Volume of distribution} = \frac{\text{Quantity administered} - \text{Quantity excreted}}{\text{Concentration of sample}}$$

is utilized. Lastly, it must be emphasized that figures obtained for fluid compartment volumes are approximate because of leakage or transport from one compartment to another, and only three compartments may be measured; the others must be calculated. Table 16.3 summarizes some of the materials used to measure fluid compartment volumes and the data obtained.

Ionic composition. In general, the subdivisions of the ECF (except transcellular fluids) are nearly identical in composition of electrolytes and small molecules, and differ mainly in protein content. Sodium, bicarbonate, and chloride are the chief electrolytes of ECF. Plasma contains the plasma proteins that pass with difficulty through capillary walls and thus interstitial fluid is low in protein concentration.

Intracellular fluid contains potassium and magnesium as the principal electrolyte cations, and organic phosphate and protein as the major

TABLE 16.2
DISTRIBUTION OF WATER IN VARIOUS TISSUES AND ORGANS

TISSUE/ORGAN	PERCENT WATER	PERCENT BODY WEIGHT	L. IN 70 KG MAN
Skin	72.0	18.0	9.07
Muscle	75.7	41.7	22.10
Skeleton	31.0	15.9	3.45
Brain	74.8	2.0	1.05
Liver	68.3	2.3	1.10
Heart	79.2	0.5	0.28
Lungs	79.0	0.7	0.39
Kidneys	82.7	0.4	0.23
Spleen	75.8	0.2	0.11
Blood	83.0	7.7	4.47
Intestine	74.5	1.8	0.94
Adipose	10.0	9.0	0.63
Total body	62.0	100.0	43.40

TABLE 16.3
MEASUREMENT OF BODY WATER COMPARTMENT VOLUMES

COMPARTMENT	MEASURED OR CALCULATED	SUBSTANCE(S) EMPLOYED	VOLUME IN 70 KG MAN (LITERS)
Total water	Measured	Heavy water	46
ECF	Measured	Polysaccharides Disaccharides (sucrose) Radioactive sodium	17
Plasma	Measured	Radioactive serum albumin	3
Interstitial fluid	Calculated (ECF minus plasma)		14
ICF	Calculated (total minus ECF)		29

anions. Figure 16.2 and Table 16.4 show the distribution of cations and anions in the three major body fluid compartments. The total cation concentration equals the total anion concentration in each compartment so that the fluids are electrically neutral. Additionally, each compartment is in osmotic equilibrium with the other (each contains the same *concentration* of osmotically active particles) so that *total body water is determined by the total quantity of solutes in the body*. If the osmolarity of any one compartment should change, an osmotic fluid shift will occur. These statements should underline the importance of the

TABLE 16.4
COMPOSITION OF EXTRACELLULAR AND INTRACELLULAR FLUID COMPARTMENTS
(values in meq/l unless otherwise indicated)

CONSTITUENTS AND PROPERTIES	EXTRACELLULAR FLUID		INTRACELLULAR FLUID
	PLASMA	INTERSTITIAL FLUID	
Sodium	142	145	10
Potassium	4	4	160
Calcium	5	5	2
Magnesium	2	2	26
Chloride	101	114	3
Sulfate	1	1	20
Bicarbonate	27	31	10
Phosphate	2	2	100
Organic acids	6	7	—
Proteins	16	1	65
Glucose (av)	90 mg%	90 mg%	0–20 mg%
Lipids (av)	0.5 g%	—	—
pH	7.4	7.4	6.7

homeostatic mechanisms responsible for maintenance of the total solute and water balance of the body.

EXCHANGES OF FLUID AND ELECTROLYTES BETWEEN COMPARTMENTS

Water may be considered to be freely exchangeable between all body fluid compartments. Its passage across membranes is determined primarily by passive forces. OSMOSIS, controlled by differences of solute concentrations on the two sides of a membrane, and FILTRATION, occurring as a result of pressure of the blood, are the processes involved in water movement.

Solutes may move by ACTIVE processes, or as a result of DIFFUSION.

FIGURE 16.2

A comparison of the constituents of the three major compartments of the body water. (Reproduced with permission, from J. L. Gamble, Jr., CHEMICAL ANATOMY, PHYSIOLOGY, AND PATHOLOGY OF EXTRACELLULAR FLUID, 6th ed., Harvard University Press, Boston, 1954.)

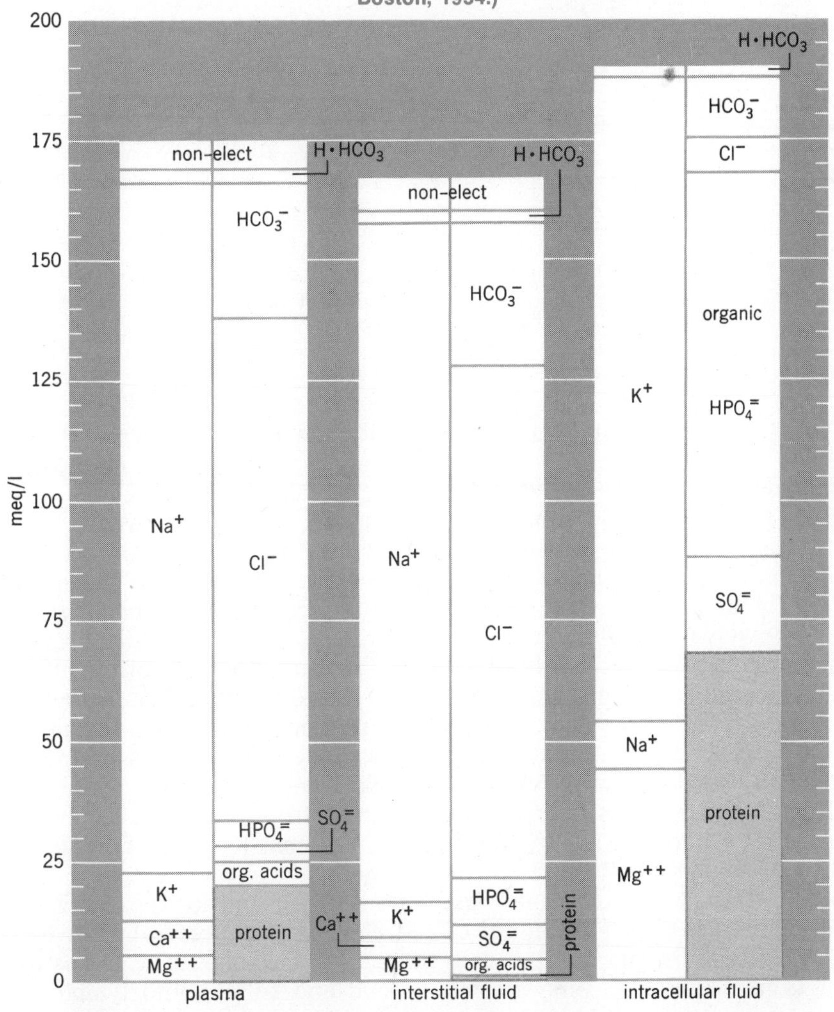

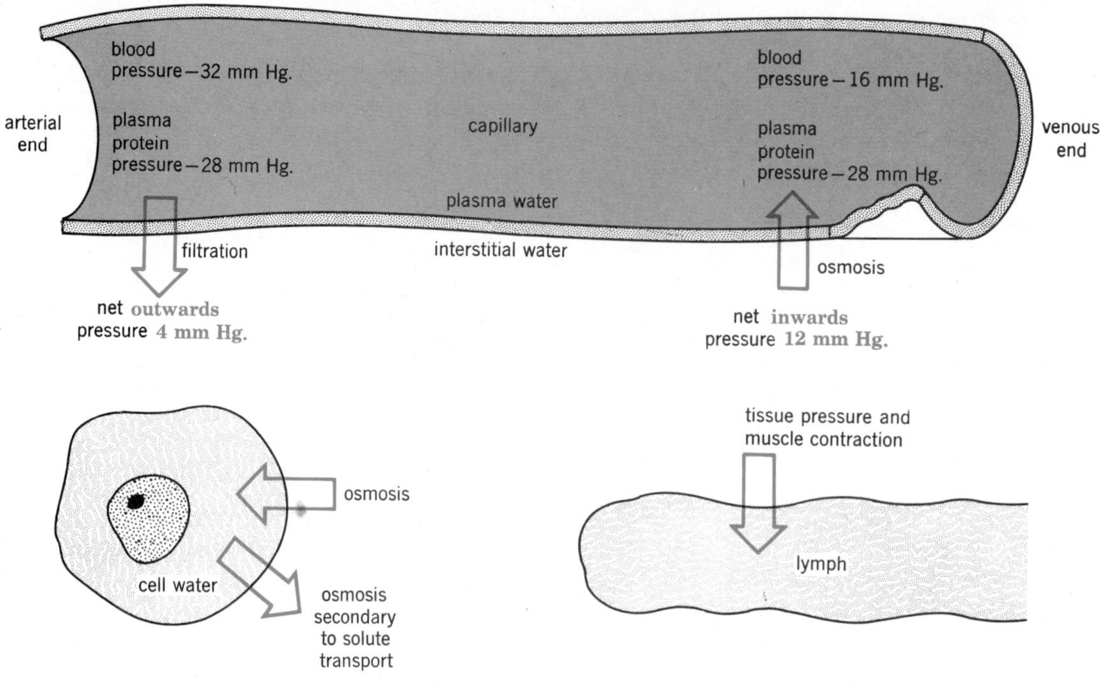

FIGURE 16.3
Some determinants of water movement between body water compartments.

PLASMA AND INTERSTITIAL FLUID

Distribution of fluid and solutes between the plasma and interstitial fluid is determined by the hydrostatic (blood) pressure and osmotic forces operating in the blood capillaries. Figure 16.3 shows the forces involved. The arterial end of a capillary receives blood under a pressure of about 32 mm Hg. Due primarily to the plasma proteins (the albumins and globulins), and to the operation of the Donnan-Gibbs equilibrium, the plasma at the arterial end of the capillary has an osmotic pressure of about 28 mm Hg. Because blood pressure exceeeds osmotic pressure by 4 mm Hg, a filtration of fluid and diffusible solutes (bulk flow) occurs *out of* the capillary. Thus, water, inorganic ions, and small molecules such as glucose and urea pass to the interstitial fluid and assume concentrations of near equality between plasma and interstitial fluid. A small leakage of plasma proteins also occurs from blood capillaries. As the blood passes through the capillaries, it loses pressure until, at the venous end of the capillary, the blood pressure is less than the osmotic pressure of the plasma, and osmotic flow of water occurs *into* the capillary. Solutes may be swept along as the water osmoses (bulk flow).

About 90 percent of the filtered water is returned to the blood capillaries by the bulk flow described above. The remainder enters lymphatic vessels in the tissues carrying with it the small amounts of plasma proteins that have leaked from the blood capillaries, and filtered solutes that have not been taken up by cells. The lymphatics eventually empty their contained fluid and solutes back into the blood vascular system, preserving its osmotic pressure and fluid volume. The forces causing *entry* of fluid and solutes into lymph vessels are promoted by muscular contractions (Figure 16.4) that alternately compress and decompress the vessels. The same mechanism influences the *movement* of lymph. Also aiding movement of materials into lymphatic capillaries is a 2 to 5 mm Hg tissue pressure (interstitial fluid pressure) causing flow into the lymphatics, where pressure is 0.2 to 1.2 mm Hg.

INTERSTITIAL FLUID AND INTRACELLULAR FLUID

Most animal cell membranes are considered to be freely permeable to water. The force responsible for water movement from interstitial to intracellular compartments is osmosis. As stated earlier, although the *types* of solutes differ markedly between ECF and ICF, *total numbers* of osmotically active particles are the same in the two areas. Thus, there will normally be *no net shifts* of fluids into or out of cells. Cells, however, contain a much greater concentration of nondiffusible materials that can act osmotically, and to ensure that no net entry of water occurs into cells, they actively pump out solutes from the cell interior. Sodium ion appears to be one substance entering the cell by diffusion that is actively removed from the cell interior. Thus, osmolarity is equilibrated inside and outside the cell and no net water flow ensues in one direction or the other.

FIGURE 16.4
The relationships of lymphatic vessels to skeletal muscles and fascial planes of the body. The anchoring filaments tend to expand the vessel; muscle contraction tends to compress the vessels (↓). Lymph tends to move to the right (→) and is prevented from backflowing by the valve. A "bellows" or "pumping" action thus draws tissue fluid into the vessels and moves it onwards.

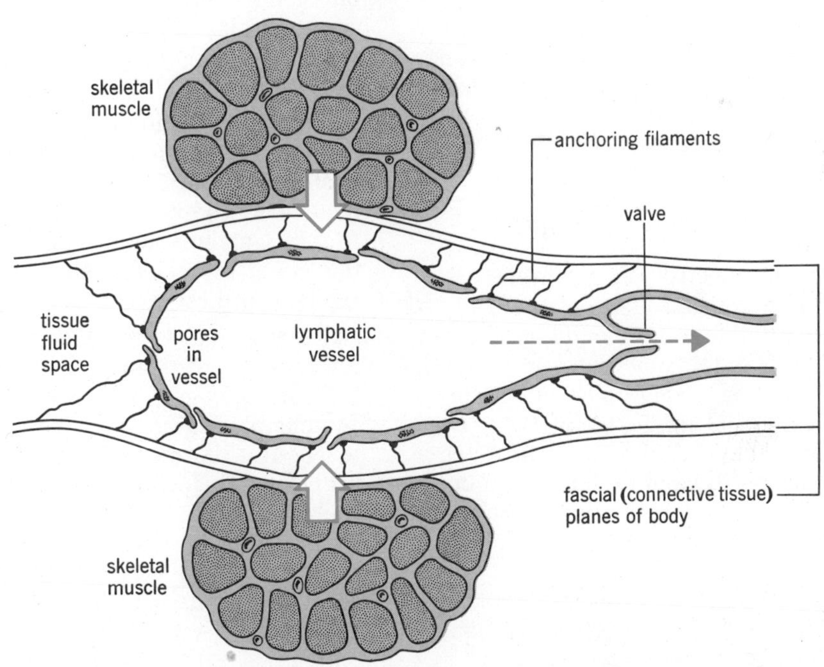

EXCHANGES OF FLUID AND ELECTROLYTES BETWEEN THE EXTRACELLULAR COMPARTMENT AND THE EXTERNAL ENVIRONMENT

FLUID AND ELECTROLYTE INTAKE

The fact that the total body water volume and electrolyte composition tend to remain remarkably constant on a day-to-day basis reflects the operation of a balance between intake and loss of these substances. Addition of an average of about 2500 ml of fluid per day to the total body water volume occurs. Of this total, about 2300 ml is INGESTED in food and drink, with some 200 ml added by metabolic pathways that produce water. Ingestion is voluntarily adjustable, while metabolic water production is fairly constant on a normal diet.

The THIRST MECHANISM is activated when cells in the lateral preoptic area of the hypothalamus are stimulated by any condition raising the osmolarity of blood reaching the area. Thus, excessive salt intake or retention, heavy sweating, a loss of body fluids as in burns, all cause the development of a greater solute concentration in the bloodstream. Also, salivary secretion diminishes and a drying of the mouth cavity results. These effects combine to create a sensation of thirst that normally stimulates fluid intake.

FLUID AND ELECTROLYTE ABSORPTION

Ingested fluids and electrolytes enter the alimentary tract. The stomach absorbs practically no water or electrolytes. The major area for absorption of both is the small intestine, with the colon contributing a small part of the absorption of water (400 ml/day) and Na^+ (60 meq/day). Active removal of both electrolytes (Na^+, Cl^-, K^+ primarily) and the end products of digestive activity (simple sugars, amino acids) from the intestine creates an osmotic gradient that causes water movement from this area. The rate of solute removal, particularly Na^+ and K^+, is controlled by certain adrenal cortical hormones (e.g., aldosterone).

Absorbed water and electrolytes of inorganic nature pass into the interstitial compartment surrounding the intestinal epithelial cells, and from there to the plasma.

FLUID AND ELECTROLYTE LOSS

Fluid and electrolyte loss from the body as a whole occurs through two general types of routes.

OBLIGATORY LOSS, or loss that cannot be controlled occurs through the *lungs, perspiration, feces,* and *evaporation from the mouth.*

Through the LUNGS, about 300 ml of water is lost per day as air saturated with water vapor is exhaled. (One can see the condensed water as the "breath" if one exhales on a cold day.) The loss through the lungs is of water alone.

PERSPIRATION is normally of the *insensible* type, in which there is a continual secretion of fluid that immediately evaporates from the skin surface. About 500 ml of water is lost per day in this fashion. *Sensible* perspiration is the heavy sweating that occurs when working hard, and this type may carry out sodium chloride with it. Sensible perspiration may reach several liters per day in volume.

FECAL EXCRETION OF WATER and MOUTH EVAPORATION amounts to about 200 ml per day. Fecal excretion carries out of the body each day about 3 meq of Na^+ and Cl^-, 10 meq of K^+, 45 meq of Ca^{++}, and 20 meq of Mg^{++}. Calcium and magnesium losses by this route represent a significant portion of the loss of these two ions.

FACULTATIVE LOSS of fluids and electrolytes refers to controllable (within limits) loss, with the *kidneys* as the organs involved here. In general, fluid loss via the URINE amounts to about 1500 ml per day, and is inversely proportional to losses via the obligatory routes. Solute loss consists primarily of NaCl and urea. The kidney thus acts to stabilize the osmolarity and volume of the ECF by adjusting the solute and water losses via the urine independently of one another.

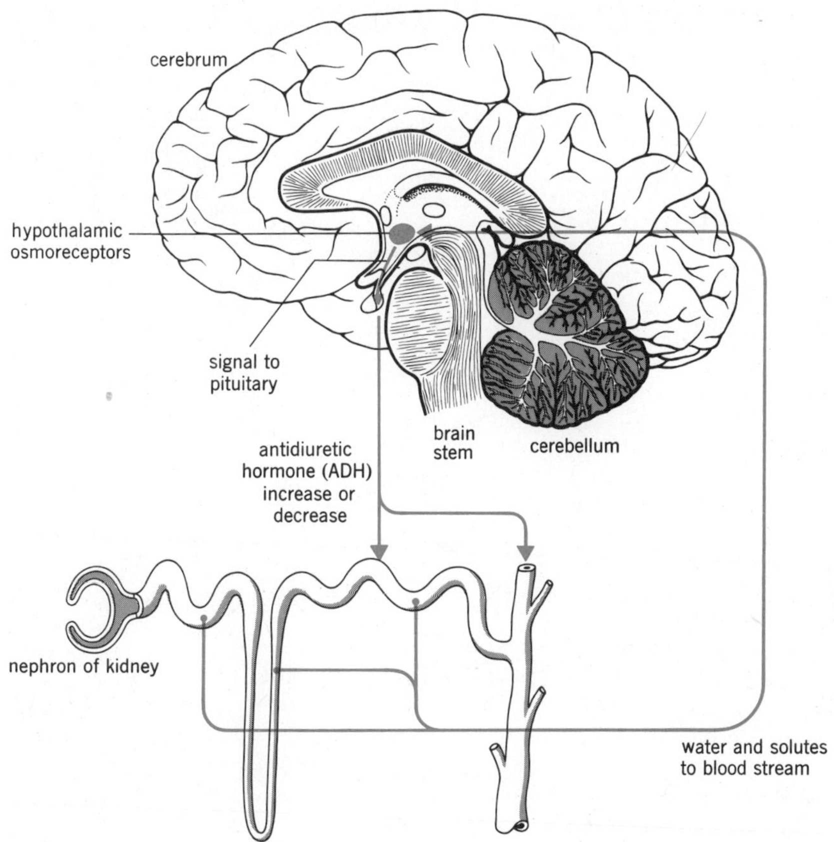

FIGURE 16.5
The ADH mechanism for control of ECF fluid volume and osmotic pressure.

REGULATION OF VOLUME AND OSMOLARITY OF THE EXTRACELLULAR FLUID

REGULATION OF VOLUME of the ECF is accomplished by controlling both Na^+ and water excretion. Control is achieved by two hormone dependent systems.

The hypothalamus contains cells that act as osmoreceptors, that is, they swell or shrink according to the osmolarity of the blood reaching them (Fig. 16.5). The same cells are neurosecretory, and produce a hormone known as ANTIDIURETIC HORMONE (*ADH*), that is passed over nerve fibers to be stored in the posterior lobe of the pituitary gland. The stimulus for release of ADH appears to be nerve impulses passing to the posterior lobe. Also influencing the hypothalamic cells are impulses passing to it over nerves from volume receptors sensitive to the "fullness of the bloodstream" located in the left atrium of the heart, and possibly the carotid arteries and aorta. If blood volume is decreased, or if the blood reaching the hypothalamus has a higher than normal osmolarity, ADH secretion is increased and more water is permitted to pass from the kidney tubules back into the circulation. This effect increases blood

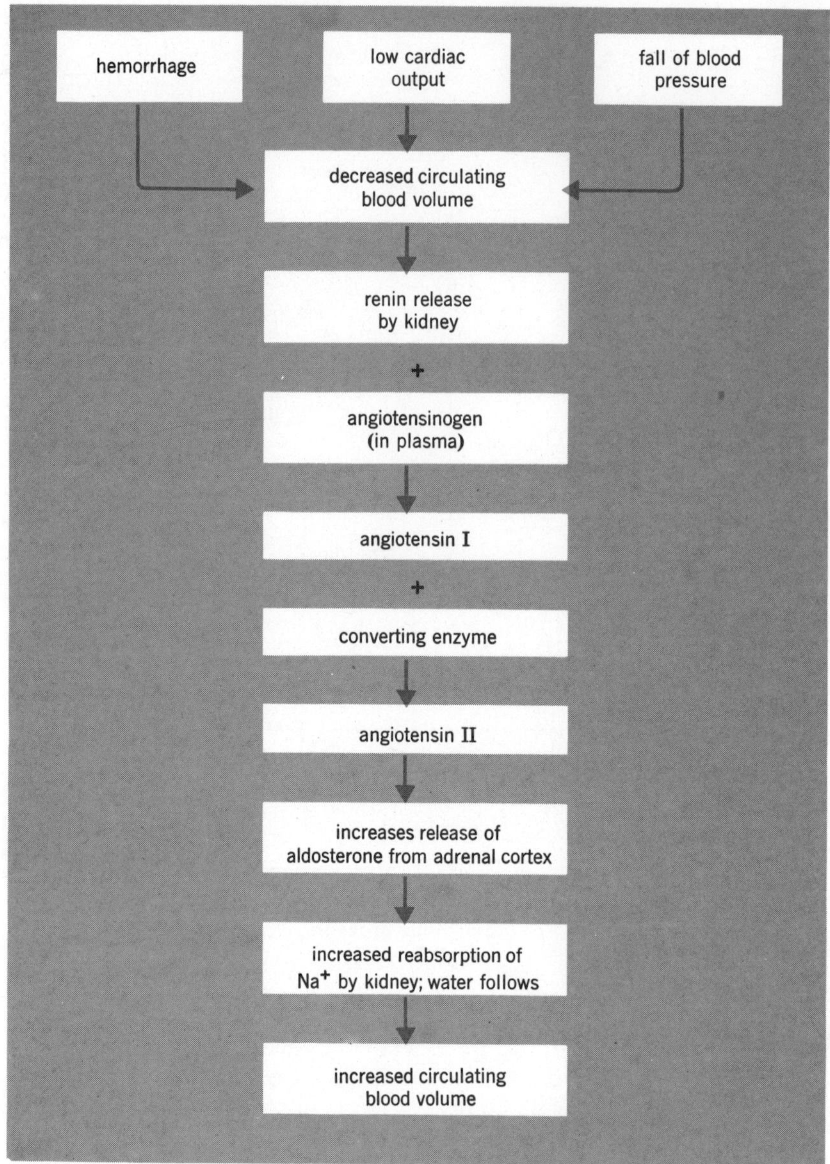

FIGURE 16.6
The aldosterone mechanism for regulation of ECF volume.

volume and dilutes the more concentrated blood. The reverse series of reactions tends to decrease the blood volume and concentrate a "too dilute" blood.

The mechanism of ADH action on permitting water to be passed out of the kidney tubules is believed to be the effect of ADH to increase the synthesis of a compound called cyclic adenosine monophosphate (c-AMP). ADH is believed to bind to cell receptors in the distal and collecting tubules. The binding causes an enzyme, adenylcyclase, to increase the production of c-AMP from ATP. c-AMP then causes an increase in the size and/or number of pores in the cell membranes lining the tubule, and more water passes out of the tubules. Remember that water movement is passive according to an osmotic gradient. The kidney mechanism responsible for creation of this osmotic gradient is called the *countercurrent multiplier,* and its operation is detailed in Chapter 25.

REGULATION OF VOLUME and OSMOLARITY of the ECF occurs through the action of ALDOSTERONE, a hormone of the adrenal cortex that increases renal reabsorption of Na^+. Figure 16.6 outlines the series of events in the production and action of aldosterone. Any condition causing a lower circulating blood volume, or a decreased blood flow or oxygen level to the kidneys, causes the kidneys to release an enzymelike substance known as RENIN. The source of the renin is the *juxtaglomerular apparatus* (Fig. 16.7), a structure involving the blood vessels and tubules of the kidney. Renin then converts a plasma precursor called ANGIOTENSINOGEN to ANGIOTENSIN I. The latter compound is converted to ANGIOTENSIN II by *converting enzyme* in the plasma. One action of angiotensin II is to stimulate release of aldosterone from the adrenal cortex, and aldosterone then increases active reabsorption of Na^+ from the kidney tubules. The reabsorbed Na^+ passes into the bloodstream. As positively charged Na^+ is reabsorbed, anions follow, and water osmotically follows these solutes. Thus, both the solute and water content of the bloodstream is increased, and volume is restored.

FIGURE 16.7
A representation of the juxtaglomerular apparatus. (*a*) Diagram of kidney blood vessels and nephron showing the location of the juxtaglomerular apparatus. (*b*) The juxtaglomerular apparatus responsible for renin secretion. It consists of two portions: granular cells of the afferent arteriole (juxtaglomerular cells), and the large cells of the distal tubula (macula densa).

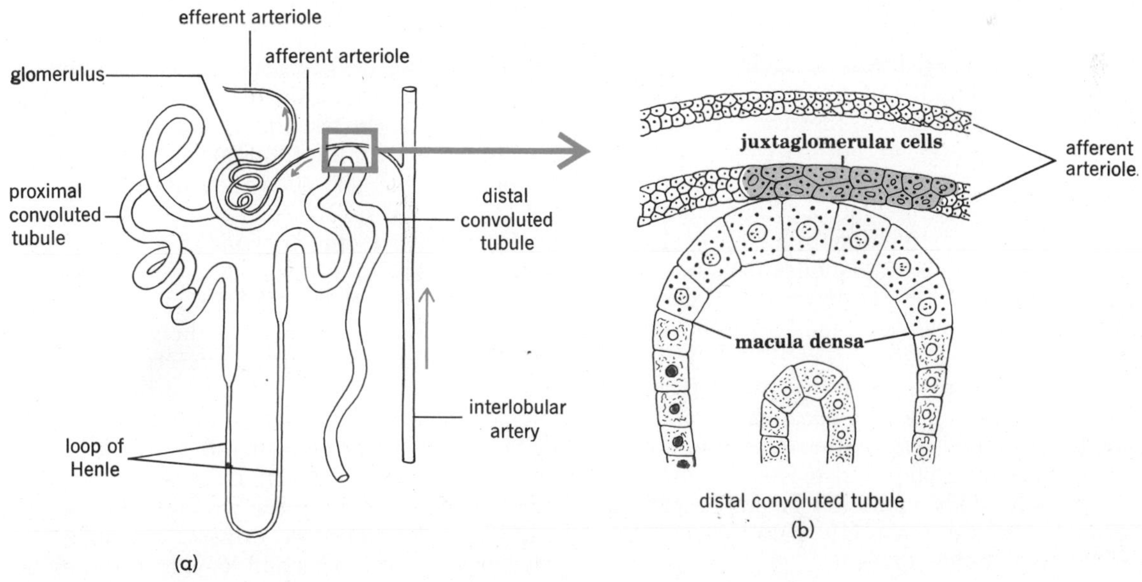

Some investigators suggest the presence of a NATRIURETIC HORMONE that is released when either blood volume or ECF volume has been expanded. It causes increased excretion of Na^+ into the urine (by preventing its reabsorption) and water follows. The site of production of the hormone is not known; its release is believed to be due to nerve impulses arriving at the site of production from volume receptors postulated to occur in the heart, thorax, liver, or brain.

This section of the chapter has indicated that regulation of fluid and solute intake and output, osmolarity, and volume of ECF is essential to the normal operation of cells. Regulation is accomplished primarily by regulation of ECF sodium concentrations, and this function is handled primarily by the kidney. The relationships between ECF osmolarity and volume and the hormones controlling Na^+ levels forms a self-regulating homeostatic control device that maintains ECF fluid volume nearly constant on a day-to-day basis.

CLINICAL CONSIDERATIONS

HYDRATION; "WATER INTOXICATION"

HYDRATION is a term referring to the results of excessive water intake, decreased loss of water, or increased reabsorption of water from the kidney because of ADH administration. In such a case, the excess water is evenly distributed in both the ECF and ICF compartments, causing an increased water volume, with dilution of solutes in both areas. Excessive water intake may produce the syndrome of water intoxication in which cellular function is disturbed by the dilution of cellular electrolytes. Disorientation, convulsions, and coma may result, as well as gastrointestinal dysfunction, muscular weakness, and abnormal cardiac rhythms.

A condition resembling hydration results if solute loss, as in a severe diarrhea, is relatively greater than fluid loss by the same route. ECF becomes diluted and fluid may enter the cells osmotically. This condition is sometimes referred to by the apparently paradoxical name of "hypotonic dehydration."

DEHYDRATION

This term refers mainly to loss of water from the cells. Excessive losses of fluid or excessive intake of solutes renders the ECF hypertonic to the ICF and water moves out of the cells. This condition is sometimes called "hypertonic dehydration" because the ECF becomes hypertonic to the cells. In the case of excessive water loss, both ECF and ICF volumes will decrease, and concentration of solutes will increase. Excess solute intake (e.g., NaCl) will cause increase of solute concentration in the ECF that will draw water from the cells. Thus, ECF volume increases as ICF volume decreases, and the solute increase in the ECF is equalized. Figure 16.8 depicts these changes.

ISOTONIC EXPANSION AND CONTRACTION OF FLUID VOLUMES

If a solution of isotonic NaCl is infused intravenously into a human, the total solute and water concentrations of the ECF will remain unchanged. There will be no net shift of ICF volume because no osmotic gradient has been created by the infusion. However, ECF volume will increase by the volume of fluid infused. Expansion of the interstitial compartment will occur as fluid and electrolytes are filtered into it from the bloodstream and EDEMA of the tissues will occur. If fluid and sodium are lost from the ECF in equivalent amounts, the volume of the ECF is reduced with no change in ICF. This condition occurs in hemorrhage and extensive burns and is sometimes called "isotonic dehydration." Figure 16.8 also depicts the changes occurring during isotonic changes in the ECF.

Edema as a clinical condition has several causes. CARDIAC EDEMA arises as a result of failure of a ventricle to eject a normal amount of blood (heart failure), and there will be accumulation of blood in the vessels that empty into the atrium ahead of the ventricle. For example, right ventricular failure leads to increased blood volume in the systemic veins, venous pressure rises and opposes the osmotic return of water into the venous end of the capillaries. PERIPHERAL EDEMA, evidenced by swelling of the ankles and legs, and accumula-

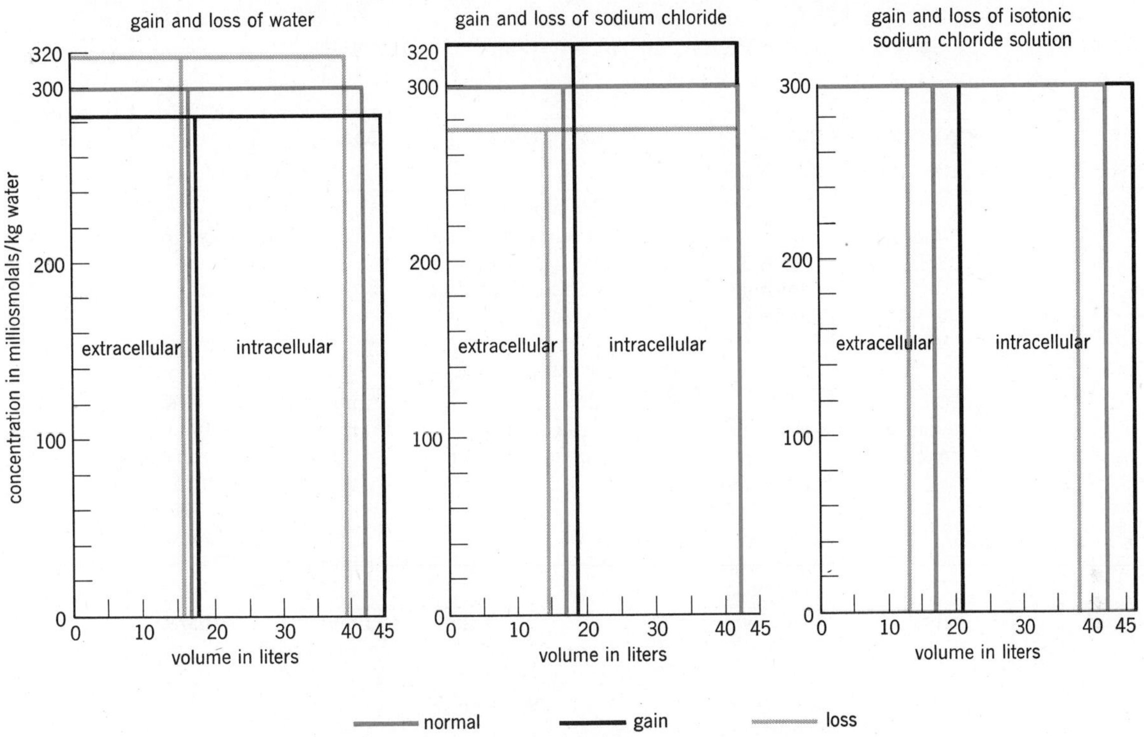

FIGURE 16.8
Examples of clinical states characterized by changes in volume and osmolar concentration of extracellular and intracellular compartments. ***Ordinate,*** **Concentration in mOsm per kg water.** ***Abscissa,*** **Volume of the compartment in liters. Colored lines represent the initial normal state, black lines the final state after** ***addition*** **of fluid or solute, and gray lines the final state after** ***loss*** **of water or solute. (After Darrow and Yannet,** ***J. Clin. Invest.,*** **1935,** ***14,*** **266–275.)**

tion of fluid in the abdomen (*ascites*), will occur. Failure of the left ventricle raises pulmonary blood volume and PULMONARY EDEMA will result. RENAL EDEMA occurs if there is loss of plasma proteins through a damaged or diseased kidney. The osmotic force drawing fluid from interstitial compartment to the plasma is decreased and the interstitial volume increases. NUTRITIONAL EDEMA causes edema by the same mechanism as does renal damage, that is, plasma proteins are decreased. The reason here, however, is due to lack of protein in the diet, and the consequent deficiency of amino acids for liver plasma protein synthesis. If lymph flow is occluded, as in parasitic blockage of flow (elephantiasis), or by a tumor, *local edema* of the body part drained by the blocked vessels will occur. SALT RETENTION, or excessive salt intake, is associated with the retention of enough water to keep the fluids isotonic. The excessive fluid volume may cause an INCREASED FILTRATION of fluid from capillaries, causing increase of interstitial volume.

The interrelationships between solute concentration, fluid absorption, and filtration and osmotic forces are extremely important. Keep in mind that water follows solutes, and filtration is a function of fluid volume.

TABLE 16.5

A SUMMARY OF SYMPTOMS OF COMMON FLUID AND ELECTROLYTE DEFICITS AND EXCESSES

CONDITION	HISTORY/LAB FINDINGS	SYMPTOMS	COMMENTS
ECF volume deficit	Decreased intake of fluid and electrolytes Vomiting Diarrhea Rise of hematocrit*	Dry skin and mucous membranes Oliguria (scanty urine production) Weight loss (5–15%) Lassitude	Appropriate response is isotonic infusion
ECF volume excess	Excessive isotonic IV infusion Kidney disease Congestive heart failure Hematocrit decreases	Edema Rapid weight gain	Appropriate response is to increase fluid output as by diuretic administration, or, give hypertonic NaCl and monitor ECF osmolarity
Sodium deficit	Heavy sweating with water intake Adrenal disease Hyponatremia (< 137 meq/L)	Mental confusion Convulsions Abdominal cramps Diarrhea	Give NaCl, perhaps as salt pills Raise ECF sodium levels
Sodium excess	Excessive isotonic IV infusion Watery diarrhea Inadequate water intake Hypernatremia (> 147 meq/L)	Mania/convulsions Dry mucous membranes Oliguria	Administer hypotonic fluid/water by mouth
Potassium deficit	Diarrhea Vomiting Burns Starvation Low sodium diet Hypokalemia (< 4 meq/L)	Flabby muscles Anorexia Cardiac arrhythmias	Administer K^+ as citrate or KCl
Potassium excess	Burns Severe trauma Kidney disease Adrenal disease Hyperkalemia (> 5.6 meq/L)	Oliguria → anuria Diarrhea Cardiac arrhythmias	Administer 10% calcium gluconate for heart Hemodialysis

* The hematocrit measures the relative volumes of blood cells and plasma. If fluid volume decreases, the number of cells increases in relative fashion and vice versa.

The clinical importance of regulating the solute and water concentrations of the fluid compartments is obvious. Deficits or excesses of fluid, sodium, and potassium are among the most commonly encountered disorders, and are associated with symptoms characteristic of each. Table 16.5 summarizes some of these conditions.

SUMMARY

1. The body fluids form an internal environment for body cells that supplies their necessary nutrients and carries off their products of metabolism.
2. Total body water amounts to 55 to 60 percent of the adult body weight.
   a. Age increase results in total water decrease, as a percent of total body weight.
   b. Increase of adipose tissue results in a relatively lower percent of water in the body.
   c. The body water contains anions and cations that determine the water distribution of the body.
   d. Active and passive processes determine the solute distribution.
   e. Total body water is subdivided into extracellular fluid (ECF) outside cells, and intracellular fluid (ICF) inside cells. Those subdivisions are called fluid compartments, since they are membrane-separated.
3. Extracellular fluid accounts for about 37.5 percent of the total body fluid. It has several subdivisions or compartments.
   a. Plasma—7 percent of body fluid; fluid in blood vessels and heart.
   b. Interstitial (tissue fluid) and lymph—fluid outside of blood vessels and cells; 18 percent of body fluid.
   c. Fluids of dense connective tissue and bone—10 percent of body fluid.
   d. Transcellular fluids—fluids of hollow organs and body cavities; 2.5 percent of total body fluid.
4. Intracellular fluid—within cells; 62.5 percent of body water.
5. Volumes of several of the body fluid compartments are measured indirectly by the dilution method; others are calculated.
   a. A substance is injected, distributes itself in a fluid volume, and is analyzed for its concentration.
   b. A typical adult has a total body water volume of 46 L, ECF volume of 17 L, of which plasma volume is 3 L, interstitial volume is 14 L, and ICF volume is 29 L.
6. ECF contains Na^+, Cl^-, and HCO_3^- as the dominant electrolytes.
   a. Plasma has a higher protein content than interstitial fluid.
   b. ECF osmolarity is the primary determinant of ICF and total body water volume.
7. ICF contains K^+, Mg^{++}, organic phosphate, and protein as the primary electrolytes.
8. Both passive and active processes determine fluid and electrolyte exchange between compartments.
   a. Filtration causes movement of materials out of the plasma; osmosis returns 90 percent of filtered water to the plasma compartment from the interstitial area.
   b. Lymphatic vessels return water and protein to the bloodstream.
   c. Cells control their exchanges of water by controlling their active removal of solutes.
9. Water intake and production is about 2.5 L/day.
   a. Food and drink account for most of the addition to fluid volume. Solutes enter by this route.
   b. Thirst plays a role in adjusting fluid intake; it depends on blood osmolarity and salivary secretion.
   c. Ingested fluids are absorbed passively from small and large intestines secondarily to active solute absorption.
10. Water and electrolyte loss normally balances intake.
    a. Fluid is lost via lungs, perspiration, feces (solutes too), and evaporation from the mouth. These represent fixed or uncontrollable routes of loss.
    b. Controllable losses of fluids and electrolytes are handled by the kidney.
11. ECF volume regulation is handled by an osmotic-hypothalamic volume-receptor-antidiuretic hormone sequence that regulates fluid loss via the kidney.
12. Regulation of ECF osmolarity is handled primarily by a kidney renin-angiotensin-aldosterone mechanism that regulates sodium reabsorption and therefore water reabsorption.

13. Hydration refers to excessive volume of, first, the ECF, and then the ICF. Cellular function suffers as dilution occurs.
14. Dehydration refers to ICF loss of water, and results when ECF osmolarity increases.
15. Isotonic fluid ingestion or infusion increases ECF fluid compartment volume, and interstitial volume as filtration of water and solutes occurs. There will be no net water shifts between interstitial and ICF, since no change in solute and water concentrations of ECF has occurred.
16. The effects of excessive intake of fluid or of solute loss in production of edema are explored, as are the effects of cardiac, kidney, and nutritional disorders.

QUESTIONS

1. What are the three largest body water compartments? Give the approximate percent of body weight for each in the adult.
2. Which of the three largest compartments do the kidneys regulate and, thus, ultimately regulate all of the fluid compartments of the body?
3. Compare the water requirements of a newborn baby, a year-old-child, and an adult. Explain the differences.
4. What are the sources and routes of loss of body water?
5. What forces are responsible for water movement into and out of each fluid compartment?
6. Compare the compositions of intracellular and extracellular fluid for the predominant ions.
7. How is the volume of the ECF regulated?
8. What is the ADH mechanism for control of fluid osmotic pressure?
9. What is the effect on body fluid volume of a high aldosterone level?
10. Describe the operation of the renin-angiotensin mechanism and its effects on body fluid volumes.
11. Explain what would happen to ECF and ICF fluid volumes and solute concentrations in the following situations:
    a. IV infusion of 2 liters of isotonic saline.
    b. IV infusion of 2 liters of 2 percent saline solution.
    c. Loss of 2 liters of blood by hemorrhage.

READINGS

Burke, S. R. *The Composition and Function of Body Fluids*. Mosby. St. Louis, 1972.

Caldwell, Peter R. B., Beatrice C. Seegal, and Konrad C. Hsu. "Angiotensin-Converting Enzyme; Vascular Endothelial Localization." *Science 191:*1050. 12 March 1976.

Christensen, H. N. *Body Fluids and Their Neutrality*. Oxford University Press. New York, 1963.

Gamble, J. L., Jr. *Chemical Anatomy, Physiology, and Pathology of Extracellular Fluid.* Harvard University Press. Boston, 1954.

Kuntz, I. D., and A. Zipp. "Water in Biological Systems." *New Eng. J. Med. 297:*262, Aug. 4, 1977.

Maxwell, Morton H., and Charles R. Kleeman (Eds). *Clinical Disorders of Fluid and Electrolyte Metabolism*. 2nd ed. McGraw-Hill. New York, 1972.

Peart, W. S. "Renin-Angiotensin System." *New Eng. J. Med. 292:*302. Feb 6, 1975.

Weldy, N. J. *Body Fluids and Electrolytes*. Mosby. St. Louis, 1972.

17

ACID–BASE BALANCE OF THE BODY

OBJECTIVES

After studying this chapter, the reader should be able to:

☐ Explain the importance of regulating hydrogen ion concentration in the body fluids.

☐ Define what acids and bases are and explain why one substance may be a stronger acid or base than another.

☐ Explain the difference between titratable acidity and free acidity.

☐ Explain where hydrogen ion comes from in the body.

☐ Describe the various methods the body utilizes to regulate hydrogen ion accumulation in the body.

☐ Explain the importance of a buffering system in acid-base balance.

☐ Explain what the alkali reserve is.

☐ Define acidosis and alkalosis.

☐ List the four major types of acid-base disturbance, and show how each comes about.

The previous chapter has indicated the importance and necessity of regulating the solute and fluid constituents of the extracellular fluid. One of the most precisely regulated constituents of that fluid is the hydrogen ion (H^+). The term acid-base balance actually refers to the regulation of H^+ concentration in the ECF.

Hydrogen ion must be regulated for several reasons. HYDROGEN ION INFLUENCES THE ACTIVITY AND STRUCTURE OF PROTEIN MOLECULES, many of which are enzymes. Each enzyme operates best in a particular concentration of H^+ and changes in H^+ concentration (expressed as $[H^+]$) will result in abnormal patterns of enzyme action. HYDROGEN ION INFLUENCES THE DISTRIBUTION OF OTHER IONS between ECF and ICF. Hydrogen ion concentration influences the activity of drugs, ions, and hormones, in many cases affecting the binding of these substances to plasma proteins. This in turn may determine the availability of these substances for metabolic reactions.

Concentration of hydrogen ions is often expressed by the symbol pH. pH is defined as the negative logarithm (or logarithm of the reciprocal) of the hydrogen ion concentration,

$$\text{pH} = -\log_{10} [H^+] \text{ or } \text{pH} = \log_{10} \frac{1}{[H^+]}$$

and it represents the concentration of free H^+ in the body fluids.

By definition, hydrogen ion concentration is inversely related to pH (the greater the $[H^+]$, the lower the pH), and for each unit of pH change (as from 2 to 3), there is a tenfold change in $[H^+]$. The pH scale ranges from 0 to 14, or from strongly acid (weakly basic) solutions to weakly acid (strongly basic) solutions. At a pH of 7, a solution is neither acidic nor basic (alkaline), but is said to be neutral, since concentrations of H^+ and alkaline materials are equal. Figure 17.1 shows some values of pH in a variety of body fluids, secretions, and excretions. Values shown for plasma, interstitial fluid, and cerebrospinal fluid represent the values of ECF, and pH of these fluids is the one that must be closely regulated as they are the cellular environment. Values varying greatly from those of ECF are secretions of

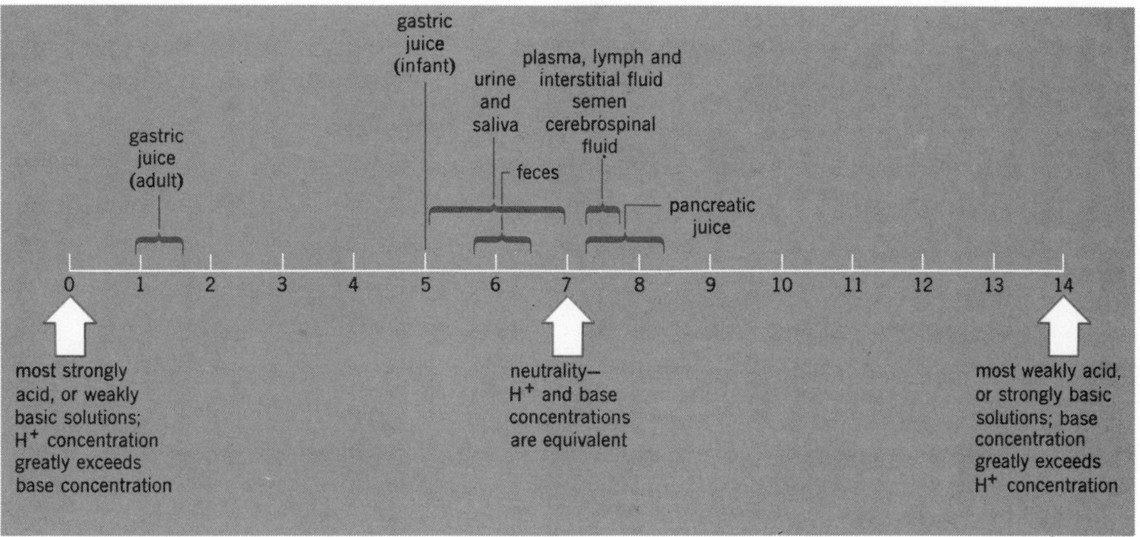

FIGURE 17.1
The range of pH values in some body fluids.

special cells and usually require protective devices in those organs in which such strong solutions are found.

ACIDS AND BASES

DEFINITIONS

ACIDS are materials that, in water, are capable of losing or donating hydrogen ions (*dissociation*) as free ions; BASES are materials that either release (dissociate) hydroxyl groups (OH^-) in solution or that accept free H^+. The combination of an acid and a base in equal ratios produces a solution that is neutral. The degree of dissociation of an acid determines how many H^+ it will release in solution and therefore the pH of the solution. STRONG ACIDS have a high degree of dissociation, liberating much H^+. Hydrochloric acid, HCl, dissociates nearly 100 percent in solution and is a strong acid:

$$HCl \longrightarrow H^+ + Cl^-$$

WEAK ACIDS have low degrees of dissociation and give up fewer H^+ in solution; they "prefer" to remain in the undissociated state as exemplified by acetic acid (the length of the arrows indicates the direction of the predominating chemical reaction):

$$HAc \rightleftharpoons H^+ + Ac^- \text{ (Ac = acetate ion)}$$

Obviously, it would be advantageous as far as controlling pH is concerned if the body acids were predominantly weak, so that fewer free H^+ would be present.

STRONG BASES bind H^+ readily, as do sodium or potassium hydroxide; they also release large quantities of OH^- in solution. WEAK BASES bind H^+ less strongly or release fewer OH^- in solution; bicarbonate ion (HCO_3^-) is a weak base. An acid and the base (anion) it forms upon dissociation are termed a CONJUGATE (joined) ACID-BASE PAIR. For example, HCl—Cl^-, H_2CO_3-HCO_3, and HAc-Ac^- are such pairs. Often, the anion is present in a solution due to the dissociation of a salt of the anion. For example, sodium bicarbonate dissociates into Na^+ and HCO_3^-; the resultant solution is basic (alkaline) due to the presence of HCO_3^-, and not Na^+, for Na^+ cannot accept H^+.

Some materials can either accept *or* donate H^+, and are said to be AMPHOTERIC. In this case, pH of the solution around the molecule will determine whether it accepts or donates H^+:

hydrogen acceptor (base)

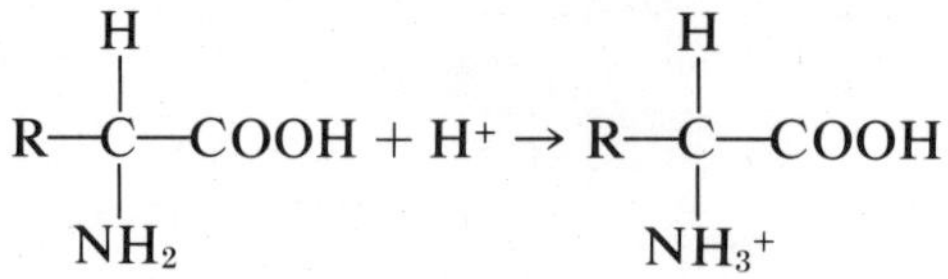

$$R-\underset{NH_2}{\overset{H}{\underset{|}{\overset{|}{C}}}}-COOH + H^+ \rightarrow R-\underset{NH_3^+}{\overset{H}{\underset{|}{\overset{|}{C}}}}-COOH$$

hydrogen donator (acid)

$$R-\underset{NH_2}{\overset{H}{\underset{|}{\overset{|}{C}}}}-COOH \rightarrow R-\underset{NH_2}{\overset{H}{\underset{|}{\overset{|}{C}}}}-COO^- + H^+$$

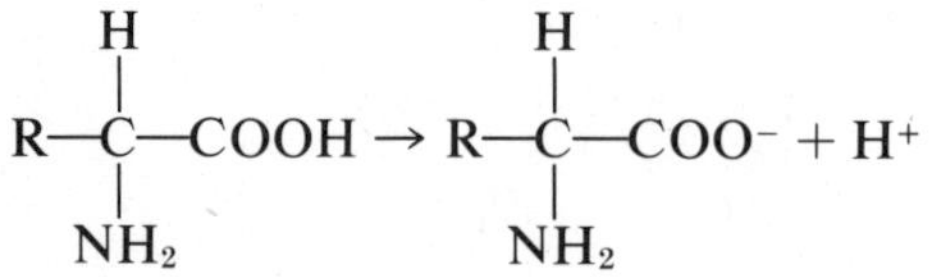

NEUTRALIZATION

This term refers to the combination of any base with an acid, where H^+ and base combine on a one-to-one basis so that base and H^+ are equal in the combined and free forms. Such a solution is thus neutral (neither acid nor basic) in reaction and has a pH of 7.0.

TITRATABLE ACIDITY

An *equivalent* of an acid or base is the molecular weight of the acid or base (in grams) divided by the number of *replaceable* (bound or free) hydrogen or base anions in the molecule. One equivalent of acid will neutralize one equivalent of base. To measure the amount of titratable acid in a solution, one adds basic solution until the acid is neutralized as evidenced by the attainment of a pH of 7.0. Titratable acidity measures *potential* acidity, and includes both free and bound H^+.

FREE OR IONIC ACIDITY

These terms refer to the *concentration* of *free* H^+ in a solution, that is, those already dissociated. (pH expresses the free $[H^+]$). It differs from titratable acid in that what is being referred to here is *only* those H^+ *already dissociated,* and does not include those that are bound but replaceable in a molecule. Thus, an acid having a greater degree of dissociation, and which liberates much H^+ will be a stronger acid than one liberating fewer H^+, but which might require the same amount of base to neutralize it. For example, 10 ml of 0.1 normal sodium hydroxide will neutralize 10 ml of 0.1 normal HCl; 10 ml of 0.1 normal sodium hydroxide will also neutralize 10 ml of 0.1 normal acetic acid. Both acids have the same titratable acidity, but HCl is a strong acid liberating much H^+ in solution.

SOURCES OF H^+ IN THE BODY

Among the sources that contribute to the increase of H^+ in the adult body fluids are the following.

As cells metabolize substances, some 300 liters of CARBON DIOXIDE are produced each day. Carbon dioxide reacts with water to form carbonic acid according to the equation

$$CO_2 + H_2O \rightarrow H_2CO_3$$

Carbonic acid is a weak acid, but so much of it is formed that when it dissociates,

$$H_2CO_3 \rightleftharpoons H^+ + HCO_3^-$$

some 13,000 milliequivalents of H^+ are added to the body fluids daily. These reactions form the primary source of H^+ for body fluids.

Metabolism of foodstuffs also produces what are called FIXED ACIDS. These include sulfuric, phosphoric, and hydrochloric acids, produced as the result of the metabolism of sulfur-containing amino acids and phosphoproteins and phospholipids. Yet another source of H^+ is the PRODUCTION OF ORGANIC ACIDS during such metabolic schemes as glycolysis and the Krebs cycle. Examples include pyruvic, lactic, glutaric, and other acids. These acids tend to dissociate H^+ from their carboxyl groups, adding to the free H^+ of the body fluids. Lastly, INGESTION OF ACID-PRODUCING substances adds H^+ to the body. Salicylates (aspirin) and ammonium compounds liberate H^+ in solution.

DEFENSE OF ACID-BASE BALANCE

BUFFERING

The primary method for removing free H^+ is by the process of BUFFERING. Buffering occurs in a solution in the presence of a weak acid and its corresponding conjugate base. The conjugate base is the anion of a completely ionized salt having the same anion as that of the weak acid. Such a combination, it may be remembered, is referred to as a *conjugate acid-base pair*, also known as a buffer pair or buffer system. If H^+ is added to a solution containing such a pair, the excess H^+ combines with the anion provided by the salt, and a weak acid is formed that ties up much of the free H^+. The pH of a solution in which a buffering system is functional is then dependent on the concentrations of the conjugate base anion (A^-) and related weak acid (HA). The *ratio* of $[A^-]$ to [HA], as given in the equation $pH = pK + \log_{10} \frac{[A^-]}{[HA]}$, is an extremely important factor in acid-base balance. The pK ($-\log_{10}$ K) is a numerical constant for the weak acid of a particular buffering system (K represents the degree of ionization of the acid.).

Specific buffer systems in the body include:

CARBONIC ACID AND SODIUM BICARBONATE.

$$\underset{\text{(weak acid)}}{H_2CO_3} \rightleftharpoons H^+ + HCO_3^-$$

$$NaHCO_3 \longrightarrow Na^+ + \underset{\text{(base)}}{HCO_3^-}$$

then,

$$H^+ + HCO_3^- \rightleftharpoons H_2CO_3$$

The body normally maintains a ratio of 20 molecules of $NaHCO_3$ to 1 molecule of H_2CO_3, regardless of the actual concentrations of the two substances. Thus, a large reserve of HCO_3^- is available to combine with H^+. This system is the primary buffer system of the ECF. (In this system, pK = 6.1, and log 20/1 = 1.3, yielding a pH of 7.4, the normal blood pH.)

HYDROGEN PROTEINATE, AND SODIUM OR POTASSIUM PROTEINATE.

Proteins are found in the ECF, primarily within the plasma, and within cells. Sodium is an ECF ion, while potassium is an ICF ion; thus, the cation varies according to location. Amino acids in either area dissociate or gain H^+ on their carboxyl (COOH) groups. The system operates as

$$\underset{\text{(weak acid)}}{\text{H proteinate}} \rightleftharpoons H^+ + \text{proteinate}^-$$

$$\underset{\text{(base)}}{\text{Na (or K) proteinate}} \longrightarrow Na^+ \text{ or } K^+ + \text{proteinate}^-$$

then,

$$H^+ + \text{proteinate}^- \rightleftharpoons \text{H proteinate}$$

PHOSPHATE SYSTEMS.

These are chiefly intracellular: as may be recalled, organic phosphate is a primary anion inside cells. They operate according to the equation

$$\underset{\text{(weak acid)}}{KH_2PO_4} \rightleftharpoons H^+ + KHPO_4^-$$

$$K_2HPO_4 \longrightarrow K^+ + \underset{\text{(base)}}{KHPO_4^-}$$

then,

$$H^+ + KHPO_4^- \rightleftharpoons KH_2PO_4$$

DILUTION

Some hydrogen ion is not buffered and remains as free H^+ in the body fluids. As it is produced, it is diluted by the fluid volume. Although this does not remove the ion, it tends to prevent local accumulation that might interfere with metabolic reactions.

LUNG ELIMINATION OF CARBON DIOXIDE

The weak acid, carbonic acid, formed when carbon dioxide reacts with water is said to be a *volatile acid*. This is true because the CO_2 that forms the acid is a gas, and can be eliminated via the lungs *if* the CO_2 can be recovered from carbonic acid. The use of the lungs as a means of defending acid-base balance revolves around their ability to eliminate CO_2 as it is formed by metabolic processes, and thus to prevent or decrease the reaction of CO_2 and water. Therefore, fewer free H^+ are dissociated from the carbonic acid.

The cells lining the capillaries of the lungs contain an enzyme called CARBONIC ANHYDRASE. The enzyme catalyzes the reaction

$$H_2CO_3 \longrightarrow CO_2 + H_2O$$

As H_2CO_3 is broken down into CO_2 and H_2O, and as the CO_2 is exhaled, more H^+ and HCO_3^- combine to form carbonic acid, lowering $[H^+]$ in the ECF,

$$H^+ + HCO_3^- \longrightarrow H_2CO_3 \longrightarrow \underset{\text{(exhaled)}}{CO_2} + H_2O$$

This method is one that is very quick to lower H^+ in the fluids, since elimination of CO_2 is a direct function of rate and depth of breathing, and because rate and depth of breathing are stimulated by rise of H^+ in the bloodstream. Thus a self-regulating mechanism is created.

KIDNEY SECRETION OF HYDROGEN ION

An additional mechanism for adjustment of $[H^+]$ in body fluids revolves about the kidneys' ability to actively secrete H^+ from the blood into the tubules of each kidney's nephrons. This process requires a longer time to bring ECF pH back to normal than does lung elimination of CO_2. Nevertheless, it can, over a period of several days, compensate or restore pH to normal. It operates regardless of lung elimination of CO_2, and if lung disease exists that limits elimination of CO_2, this mechanism becomes extremely important.

THE ALKALI (ALKALINE) RESERVE

The ALKALI RESERVE, or CO_2 COMBINING POWER is the amount of base in the plasma, principally in the form of bicarbonates, that is available for the neutralization of fixed acids such as hydrochloric acid or lactic acid. More specifically, it refers to the concentration of bicarbonate ion in plasma that has been equilibrated with a gas mixture containing carbon dioxide at a partial pressure (P_{CO_2}) of 40 mm Hg. The latter figure represents an average P_{CO_2} for adult plasma. Total CO_2 content in millimoles per liter is determined in the equilibrated sample, and 1.3 millimoles per liter is subtracted from the total, because at a P_{CO_2} of 40 mm Hg, 1.2 millimoles of CO_2 per liter is dissolved (not as bicarbonate) in the plasma. The magnitude of the reserve gives an indication of the degree of an alkalosis or acidosis that may exist in an acid-base disturbance, since changes from the normal adult value of 24 to 31 meq HCO_3^- per liter reflects (inversely) the concentration of H^+ that has been neutralized. Use of the reserve as the *only* indication of degree of pH disturbance should be discouraged, as determinations of pH and blood carbon dioxide levels are more meaningful.

Although the various mechanisms causing maintenance of normal H^+ levels in the body fluids have been considered separately, it must be appreciated that all cooperate to maintain normal $[H^+]$ in the body fluids. Buffering may be considered as sort of a temporary storage form of H^+ until lung and kidney activity can actually eliminate the H^+, or its source, from the body. Normal pH cannot be maintained unless all components of the system are operating properly. Figure 17.2 presents in pictorial form the mechanisms acting to maintain normal ECF pH within the range of 7.4 ± 0.04, and emphasizes that pH disturbance represents an *imbalance* between H^+ and base substances in the body.

ACID-BASE DISTURBANCES

Blood is the fluid most commonly analyzed to discover if the mechanisms regulating body $[H^+]$ are operating properly. The term ACIDEMIA is used to describe a condition in which arterial blood has a pH of less than 7.36. ACIDOSIS defines the processes that caused the acidemia. ALKALEMIA refers to an arterial blood pH greater than 7.44. ALKALOSIS describes the mechanisms leading to the development of alkalemia.

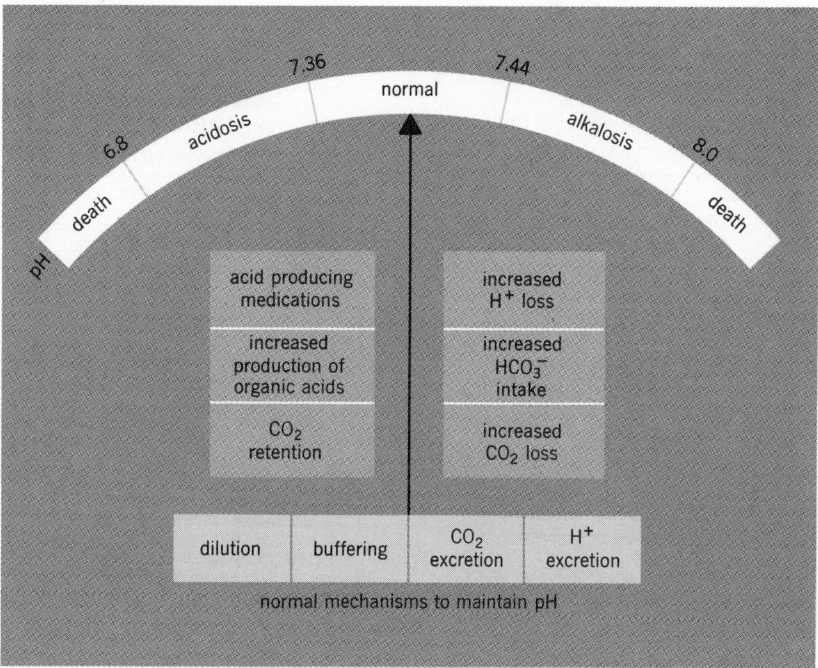

FIGURE 17.2
Factors determining body fluid pH.

The clinical diagnosis of a pH disturbance requires three quantities: arterial blood pH, concentration of carbon dioxide in arterial blood, and arterial bicarbonate concentration. Venous blood contains a higher CO_2 level than arterial blood because it is draining the tissues that produce CO_2. Its pH is *normally* lower than that of arterial blood (7.35 to 7.4).

Disturbances of acid-base balance have respiratory or nonrespiratory (metabolic) causes, and are generally divided into four types: respiratory acidosis, metabolic acidosis, respiratory alkalosis, and metabolic alkalosis.

RESPIRATORY ACIDOSIS

RESPIRATORY ACIDOSIS occurs because of abnormal retention of CO_2. Increased amounts of H_2CO_3 are present, and pH is decreased because the HCO_3^-/H_2CO_3 ratio is smaller. The condition results because of failure of CO_2 to diffuse from blood to lungs or because of failure of the lungs to eliminate CO_2 to the atmosphere. Thus, respiratory paralysis, loss of elastic tissue in the lungs, collapse or kinking of small bronchioles of the lungs (the latter two are common in emphysema), or filling of the alveoli with fluid (as in pneumonia and pulmonary edema), all result in CO_2 accumulation in the bloodstream. If the lung route for CO_2 elimination is decreased, the kidneys secrete more H^+ and reabsorb more HCO_3^-, raising the bicarbonate/carbonic acid ratio, and tending to compensate for the decreased pH. Remember, however, that kidney compensation is a slower process than lung elimination of CO_2, so that active addition of HCO_3^- (as by the intravenous route) may be required if the acidosis is severe.

METABOLIC ACIDOSIS

This condition results from accumulation of excessive fixed acids as a result of abnormal or exaggerated metabolic processes, or as a result of excessive loss of base (HCO_3^-). In diabetes mellitus, increased quantities of ketoacids are

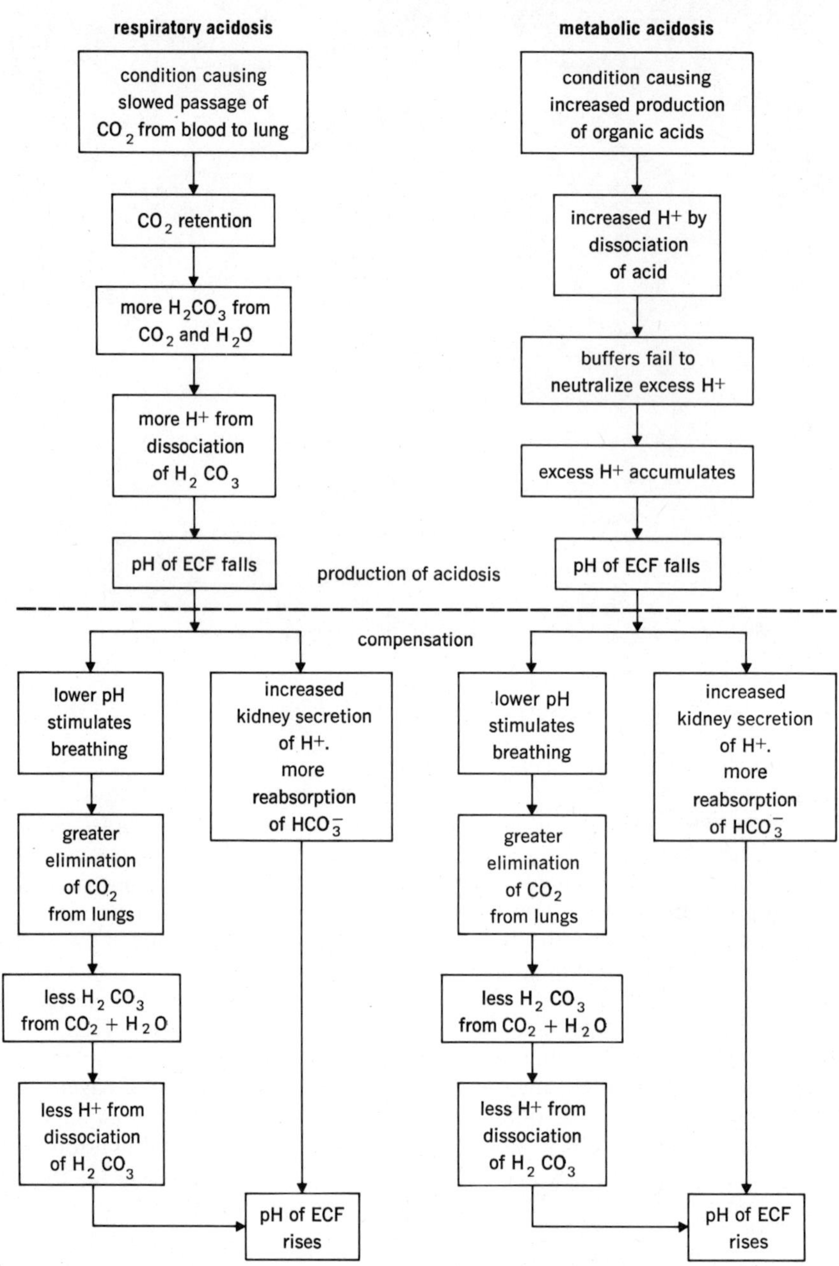

FIGURE 17.3
Production and compensation of respiratory and metabolic acidosis.

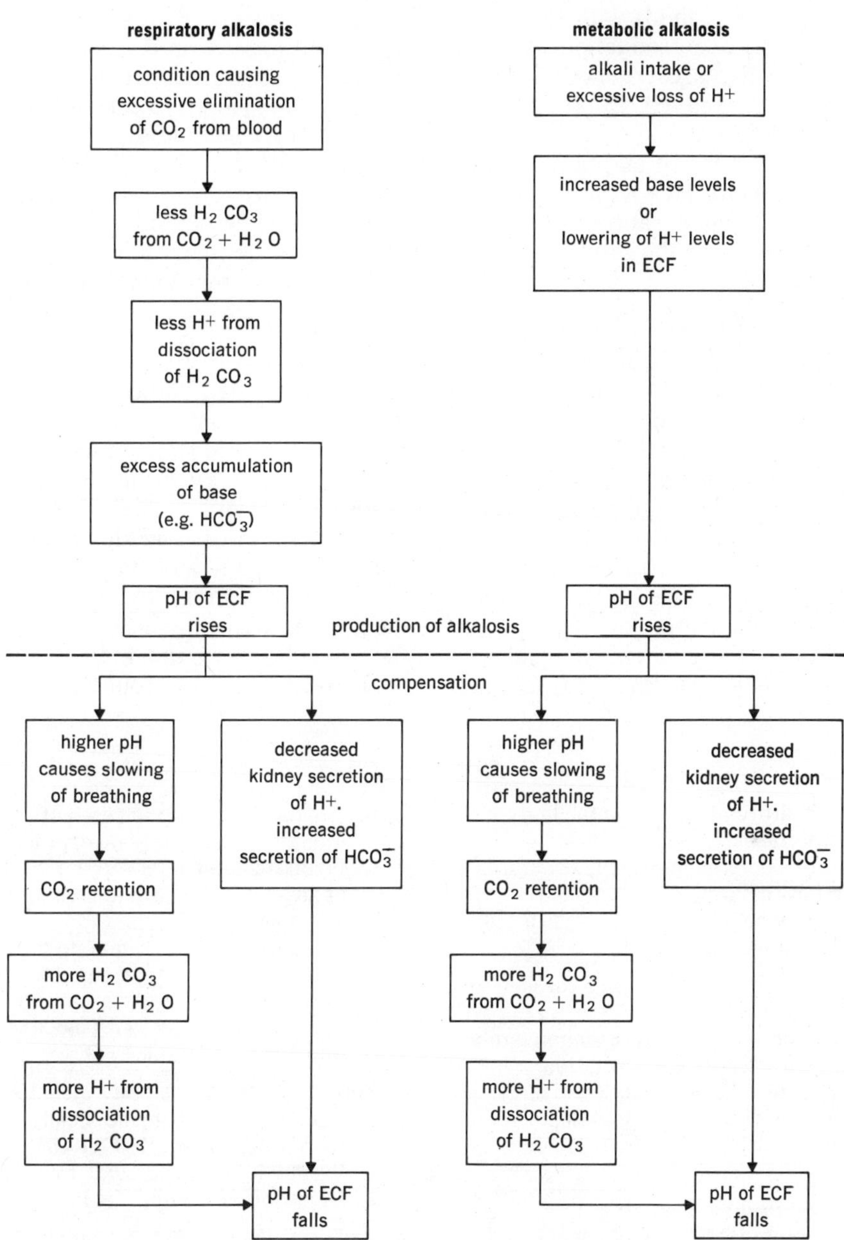

FIGURE 17.4
Production and compensation of respiratory and metabolic alkalosis.

produced that dissociate their H^+; in a severe diarrhea, loss of intestinal fluids with their contained HCO_3^- occurs; overproduction of acids utilizes much of the alkali reserve. All of these combine to lower plasma HCO_3^- relative to H_2CO_3, and the pH falls.

The kidneys again secrete additional quantities of H^+ and reabsorb HCO_3^-, and the condition moves toward compensation.

Figure 17.3 summarizes the production and compensation of acidosis.

RESPIRATORY ALKALOSIS

If retention of CO_2 causes respiratory acidosis, excessive elimination of CO_2 represents the cause of respiratory alkalosis. Any condition that results in alveolar hyperventilation will "blow off" excessive amounts of CO_2. Carbonic acid levels in the blood decrease, base is present in excess, and the pH rises. Voluntary hyperventilation, anxiety, fever, encephalitis, and salicylate poisoning are some common causes of excessive elimination of CO_2 through increased breathing. The higher pH results in a slowing of breathing, and less CO_2 is eliminated; at the same time, kidney excretion of H^+ is decreased. Bicarbonate reabsorption is decreased (excretion is also increased), and an alkaline urine is produced during compensation.

Common symptoms reported during respiratory alkalosis include dizziness, tingling of the hands and/or feet, and muscular spasms. These symptoms are due to a diminished cerebral blood flow as vessels constrict in the "absence" of CO_2, and to increased muscular excitability secondary to a more alkaline pH.

METABOLIC ALKALOSIS

This condition results from excessive loss of H^+, or from the addition of base to the body fluids. Severe vomiting, that results in loss of H^+ via the gastric juice, or excessive consumption of antacids by those suffering from "indigestion" or ulcers, cause a base excess relative to acids; pH rises. The kidney increases its excretion of HCO_3^- and the condition moves toward compensation. Figure 17.4 summarizes the genesis and compensation of alkalosis. Further details as to the kidney involvement in acid-base balance are found in Chapter 25.

SUMMARY

1. Regulation of hydrogen ion (H^+) in the body is of paramount importance.
   a. Hydrogen ions determine protein structure.
   b. Hydrogen ions determine enzyme activity.
   c. Hydrogen ions help determine ionic distribution across cell membranes.
2. The symbol pH is used to express hydrogen ion concentration.
   a. A change of one pH unit represents a tenfold change in hydrogen ion concentration.
   b. A pH of 7 represents neutrality (acid and base components are equal); a pH less than 7 represents acidity; one greater than 7, alkalinity.
3. Definitions:
   a. Acids give up or dissociate H^+.
   b. Bases give up or dissociate OH^- or accept H^+.
   c. Strong acids give up much H^+.
   d. Weak acids give up less H^+.
   e. Strong bases readily bind H^+.
   f. Weak bases bind H^+ less readily.
   g. A conjugate acid-base pair consists of an acid and the anion (base) it dissociates.
   h. Amphoteric substances act both as acids (donators) and bases (acceptors).
   i. Neutralization refers to a 1 to 1 combination of an acid and a base to give a neutral (pH 7.0) solution.
   j. Titratable acidity refers to all H^+ (free and potentially free) that can be titrated with a base.
   k. Free acidity refers to dissociated H^+ in a solution. This determines pH.
4. Sources of H^+ in the body include:
   a. The reaction of metabolically produced CO_2 with water to form H_2CO_3 that dissociates H^+.
   b. Production of fixed acids by metabolism.
   c. Ingestion of acid-producing substances.
5. Defense against H^+ accumulation involves several mechanisms.
   a. Buffering utilizes conjugate bases to combine with H^+ to form a weak acid. Bicarbonate ion, proteins, phosphate ions are major bases. The equation $pH = pK + \log \frac{[A^-]}{[HA]}$ is used to calcu-

late pH in the presence of a buffer system.
   b. Dilution removes local concentrations of H^+.
   c. Lung elimination of CO_2 removes the major *source* of H^+.
   d. Kidney secretion of H^+.
6. The alkali reserve is base available in excess of normal body needs to handle H^+.
7. Acid-base disturbances reflect abnormalities in the mechanisms adjusting hydrogen ion concentrations in the body fluids.
   a. Acidemia means a blood $pH < 7.36$.
   b. Acidosis defines the processes causing acidemia.
   c. Alkalemia means a blood $pH > 7.44$.
   d. Alkalosis defines the processes causing alkalemia.
8. Four primary types of acid-base disturbances are described.
   a. Respiratory acidosis implies CO_2 retention due to defects in ventilation or CO_2 diffusion.
   b. Metabolic acidosis implies excessive fixed acids in the bloodstream or a base deficit.
   c. Respiratory alkalosis implies excessive elimination of CO_2.
   d. Metabolic alkalosis implies excessive loss of H^+ or base intake.
   e. The lungs and kidneys work together to achieve compensation of acid-base disorders.

QUESTIONS

1. Why is defense of acid-base balance so important to the body?
2. What is implied regarding hydrogen ion concentration in a shift from a pH of 6 to a pH of 8?
3. Which of the following represent a substance acting as an acid and which represent a substance acting as a base? Explain your choices.
   a. $H_2CO_3 \rightarrow H^+ + HCO_3^-$
   b. $HCO_3^- + H^+ \rightarrow H_2CO_3$
   c. $NH_3 + H^+ \rightarrow NH_4^+$
   d. $H^+ + OH^- \rightarrow H_2O$
   e. $CH_3—CHNH_2—COOH \rightarrow CH_3 + CHNH_2—COO^- + H^+$
   f. $CH_3—CHNH_2—COOH + H^+ \rightarrow CH_3—CHNH_3^+—COOH$
4. Explain the difference between titratable acid and free acid.
5. Name the two major sources of daily addition of H^+ to the body fluids.
6. What are the three major methods of maintaining H^+ concentrations within normal limits?
7. If an individual with emphysema was said to have a respiratory acidosis, what signs or symptoms would that individual most likely exhibit, and what would laboratory tests show concerning the blood gases and alkali reserve?

READINGS

Davenport, H. W. *The ABC of Acid-Base Chemistry*. 5th ed. University of Chicago Press. Chicago, 1969.

O'Brien, Donough. "The Management of Fluid and Electrolyte Problems in Childhood." *Pediatric Basics*. No. 17. Gerber Products Co. Fremont, Mich., 1977, pp 10–14.

Pitts, R. F. *Physiology of the Kidney and Body Fluids*. 3rd ed. Year Book Publishers. Chicago, 1974.

Tuller, M. A. *Acid-Base Homeostasis and Its Disorders*. Med. Exam. Publishing. New York, 1971.

Winters, R. W., K. Engel, and R. B. Dell. *Acid-Base Physiology in Medicine*. London Co. Westlake, Ohio, 1969.

18

THE BLOOD AND LYMPH

OBJECTIVES

THE BLOOD

A GENERAL DESCRIPTION OF PROPERTIES AND FUNCTIONS

COMPONENTS OF THE BLOOD

PLASMA

FORMED ELEMENTS

HEMOSTASIS

COAGULATION

ANTICOAGULATION

DISORDERS OF CLOTTING

LYMPH

FORMATION, COMPOSITION, AND FLOW

LYMPH NODES

SUMMARY

QUESTIONS

READINGS

OBJECTIVES

After studying this chapter, the reader should be able to:

☐ List the volumes and percents that plasma and formed elements constitute of the blood.

☐ Give the functions of the blood, with specific examples of each function listed.

☐ Describe the significance of the erythrocyte sedimentation rate in determining blood characteristics.

☐ List the major components of the plasma and their functions.

☐ List the various types of plasma proteins and their functions.

☐ Describe the sources of the plasma components.

☐ List the formed elements of the blood, their morphology, life history, and functions.

☐ Describe hemoglobin regarding production, function, and fate.

☐ Describe the role of iron in oxygen transport and outline its metabolism.

☐ Describe what anemia is and some of its causes.

☐ Define hemostasis, and explain the reactions that occur to prevent blood loss.

☐ Explain how blood may be rendered incapable of coagulating.

☐ Discuss the formation, composition, and flow of lymph.

☐ Describe the structure and function of lymph nodes.

THE BLOOD

A GENERAL DESCRIPTION OF PROPERTIES AND FUNCTIONS

Blood is classified as a connective tissue with a complex liquid intercellular material, the PLASMA, in which cells or cell-like structures are suspended, the FORMED ELEMENTS. Furthermore, plasma forms one of the extracellular fluid compartments, one that despite its small volume (3L) is one of the most dynamic in terms of turnover of constituents. Blood comprises about 7 percent of the body weight, and, in the "standard" 70 kg (154 lb) male adult, amounts to 5 to 6 liters in volume. The adult female, because of a usually smaller body size, averages 4.5 to 5.5 liters in volume. Plasma accounts for 55 to 57 percent of the blood volume, the formed elements 43 to 45 percent. Blood is easily obtained from superficial veins of the body, and because it circulates to and from the vicinity of the body cells, its composition reflects cellular activity. Therefore, analysis of the blood may give a good idea of the status of body function.

Functions of the blood may be considered to center around two main activities: TRANSPORT and REGULATION OF HOMEOSTASIS.

Transport. Because of its liquid intercellular material, the blood can DISSOLVE and/or SUSPEND many materials, and carry them to and from the cells as the blood is caused to circulate by heart action.

Amino acids, lipids, carbohydrates, minerals, vitamins, and enzymes are transported from their absorption sites in the gastrointestinal tract to the cells.

Red blood cells bind oxygen picked up in the lungs and carry it to the cells; carbon dioxide is transported from the cells to appropriate organs for excretion or elimination.

Heat is carried from the cells to the skin and lungs for elimination.

Hormones are carried from their sites of production to body cells that respond to them.

Excess body water is carried to appropriate organs of excretion.

Regulation of homeostasis. Regulation of homeostasis is concerned with such activities as regulation of the volume of the interstitial compartment, regulation of body fluid pH, regulation of body temperature, protection against infection, and protection against blood loss.

Regulation of the interstitial compartment volume. Fluids in the interstitial compartment are derived from FILTRATION occurring in the blood capillaries, and the plasma protein content causes the OSMOTIC RETURN of interstitial fluid to the capillaries. Thus, passive exchange of water and solutes between plasma and tissue fluid depends, in part, on the blood composition.

Regulation of pH. Blood contains bicarbonate, phosphate, protein, and hemoglobin BUFFERS that resist changes in pH. The blood also transports acids and bases to appropriate organs of excretion, an aid to pH regulation.

Regulation of body temperature. Plasma water ABSORBS MUCH HEAT of metabolic activity with relatively small changes in its own temperature. It then carries this heat to appropriate organs for elimination.

Protection against infection. The blood contains chemicals known as ANTIBODIES that can neutralize foreign chemicals that enter the body, and several of the white cells of the bloodstream are good PHAGOCYTES. Such cells can engulf and in many cases digest microorganisms. Phagocytes also remove dead cells and aid in the healing process after injury.

Protection against blood loss. Loss of blood (hemorrhage) may occur if blood vessels are damaged. The body cannot tolerate extensive losses of blood without harmful consequences. As part of the mechanisms preventing blood loss, the blood COAGULATES or clots. These reactions involve both the plasma and formed elements.

COMPONENTS OF THE BLOOD

By drawing a sample of blood and adding an anticoagulant, then allowing it to settle in a tube or hastening settling by centrifugation, whole blood may be separated into two major fractions: formed elements and plasma.

Settling or centrifuged to the bottom of the tube will be the FORMED ELEMENTS. The rate at which the erythrocytes (red cells) settle out of whole blood without centrifugation is measured by the SEDIMENTATION RATE. The cells first tend to form stacks, like a pile of coins, called *rouleaux*. These masses *settle out rapidly,* and then the mass may be packed. Rate of sedimentation depends on a number of factors including: shape of cells (abnormally shaped cells do not form rouleaux, and the rate is slower); concentration of plasma proteins (increased amounts facilitate rouleaux formation); presence of infections, toxemia, anemia (all increase sedimentation rate). Although the test does not diagnose a particular disease, it may be used to monitor recovery from disease or to monitor the efficacy of treatment. Normal sedimentation rate is about 2 mm per hour for newborns, 3 to 7 mm per hour for men (mean 3.7), and 3 to 15 mm per hour for women (mean 9.6). Leucocytes and platelets sediment more slowly and form a "buffy coat" or layer on top of the red cells. Centrifugation packs the cells and the red cells normally form a column constituting 43 to 45 percent of the volume of the whole blood (apparent hematocrit). A PACKED CELL VOLUME (*PCV*) or HEMATOCRIT (*Hct*) refers to the percent volume of whole blood occupied by packed red cells. The use of radioactively labeled plasma indicates that about 4 percent of the plasma remains in the packed red cell layer. To obtain a *corrected Hct,* the value of the *apparent* Hct should be multiplied by 0.96. The leucocytes and platelets form about 1 percent of the blood volume.

The straw-colored fluid remaining after the cells have settled or been centrifuged out is the *plasma*. It is an extremely complex substance, normally constituting 55 to 57 percent of the volume of the whole blood.

PLASMA

Some 90 percent of the plasma consists of WATER. INORGANIC CONSTITUENTS, including Na^+, K^+, Cl^-, HCO_3^-, Ca^{++}, and others, compose about 1 percent of the plasma. Na^+ and Cl^- are most plentiful; as may be recalled, plasma is an extracellular fluid. The inorganic constituents are essential for certain osmotic properties of the plasma, and are involved in buffering and maintenance of proper excitability for cells.

The PLASMA PROTEINS constitute 7 to 9 percent of the plasma and are substances characteristic for the plasma. They are large enough that they pass through blood capillary membranes with difficulty and tend to remain within the bloodstream. Here they are primarily responsible for osmotic return of filtered water to the capillaries from the interstitial fluid. Other functions served by the proteins include:

Contributing to the viscosity of the plasma.

Creating a suspension stability in the blood that aids in maintaining dispersion of materials.

Serving as a reserve of amino acids (not usually used for metabolism but available).

Serving as buffers.

By relatively simple techniques such as adding certain concentrations of salts to the plasma, three main fractions of proteins may be separated. They are designated as ALBUMINS, GLOBULINS, and FIBRINOGEN.

Albumins are the most plentiful (55 to 64 percent) of the proteins, and are present to the extent of 4 to 5 gm per 100 ml of blood. They are also the smallest (molecular weights 69,000 to 70,000) of the proteins. They contribute most of the osmotic pressure of the plasma (*oncotic pressure*) that "pulls" water into capillaries, and serve to bind many substances for transport through the plasma [barbiturates, thyroxin (partly on albumins)].

Globulins constitute about 2 percent of the plasma proteins and are present to the extent of 2 to 3 gm per 100 ml of blood. Their molecular weights range from 150,000 to 900,000. Three fractions may be separated in the globulins by electrophoresis, in which the plasma is subjected to an electric current and the proteins migrate at different rates according to their size and electrical charge.

ALPHA GLOBULINS have molecular weights of 150,000 to 160,000. They serve the general functions of the proteins and also bind many substances for transport including lipids, thyroxin (*TBG*—thyroxin binding globulin), copper (on ceruloplasmin), and cortisol. Angiotensinogen is an α-globulin previously described in connection with control of aldosterone secretion.

BETA GLOBULINS have molecular weights of 160,000-200,000. They also serve the general functions of the plasma proteins and bind certain substances such as iron and cholesterol. Several of the clotting factors, including prothrombin, are members of the alpha and beta groups.

GAMMA GLOBULINS have molecular weights of 150,000 to 900,000 and this group contains the immunoglobulins (*Ig*) or antibodies. By further procedures, such as ultracentrifugation, five types of Ig have been isolated. They are designated as: IgA, IgM, IgG, IgD, and IgE. Each arises as a result of a particular type of antigenic challenge, and some are found in a variety of secretions. As detailed in the following chapter, Ig's are one of several mechanisms giving the body protection against chemical challenges.

Fibrinogen is a soluble plasma protein with a molecular weight of about 200,000 and it occurs in the plasma to the extent of 0.15 to 0.3 gm per 100 ml. It is converted to insoluble *fibrin* as the blood coagulates.

Table 18.1 summarizes these facts and others about the plasma proteins.

Concentrations of plasma proteins vary little in persons with good health. Albumins and globulins normally maintain about a 2:1 ratio; this ratio is designated as the A/G ratio. Levels of proteins decrease during starvation, liver damage, and renal disease. A primary sign of lowered plasma protein levels is the develop-

MAJOR CONSTITUENTS OF PLASMA

COMPONENT	AMOUNT	COMMENTS
Water	90% of plasma	Dissolves, suspends, causes ionization, carries heat
Electrolytes (meq/l)	About 1%	Create osmotic pressure, buffer, irritability in all tissues
Sodium	136–145	
Potassium	3.5–5.0	
Calcium	4.5–5.0	
Magnesium	1.5–2.5	
Chloride	100–106	
Bicarbonate	26–28	
Phosphate	2	
Protein	17	
Other	6	
Nitrogenous substances (mg/100 ml blood)		Nutrients or wastes
Nonprotein nitrogen (NPN)	33	
Blood urea nitrogen (BUN)	8–25	
Creatinine	0.7–1.5	
Amino acids	0.13–3.0	Amount depends on each acid, 26 have been found in plasma
Plasma proteins	6–8% of plasma	As a group, contribute to blood viscosity, reserve of amino acids, clotting, antibodies
Albumins	4.5%, 4–5 g%	Oncotic pressure (plasma colloidal osmotic pressure)

ment of *edema,* as filtered plasma water is not returned to the bloodstream and remains in the interstitial compartment.

Water and electrolytes of the plasma are derived almost entirely by absorption from the alimentary tract. Amino acids produced by digestion of proteins are also absorbed from the gut, as are simple sugars resulting from digestion of carbohydrates. Such substances are also delivered to the bloodstream by body cells. In starvation, protein intake is decreased and fewer amino acids are available for synthesis into plasma proteins. This accounts for the decrease of plasma proteins during starvation. Albumins, alpha and beta globulins, and fibrinogen are produced by the liver. Gamma globulins are contributed by plasma cells, and by the disintegration of white cells in the bloodstream. Waste materials such as CO_2 are contributed by all body cells as they carry on cellular metabolism. Urea is derived from liver cells as they metabolize amino acids.

TABLE 18.1 (continued)

COMPONENT	AMOUNT	COMMENTS
Globulins	2%	
Alpha	510–980 mg%	Serve general functions of plasma proteins
Beta	550–1010 mg%	
Gamma	1200–2150 mg%	Antibodies
A	140–260 mg%	Most common in secretions, general lytic
M	70–130 mg%	First to appear
G	700–1450 mg%	Natural and acquired antibodies
D	300 mg%	Function unknown
E	100 mg%	Reagins, incomplete antibodies as those in allergies
Fibrinogen	0.3%; 0.15–0.30 g%	Clotting
Lipids (mg/100 ml blood)		
Fatty acids	190–420	Usually in transit or fuels
Cholesterol	150–280	
Triglycerides	About 20	
Glucose (mg/100 ml blood)	60–110	Preferred fuel source for cell activity
Enzymes	A wide variety acting on all categories of materials. Some increase with pathological conditions in the body, and are thus aids in diagnosis. Lactic dehydrogenase (LDH) and serum glutamic-oxaloacetic transaminase (SGOT) increase in heart damage	
Vitamins		
A, D, tocopherol, thiamine, riboflavin, B_6, nicotinic acid, B_{12}, folic acid, C, and others		

FORMED ELEMENTS

Three categories of formed elements are found in normal human peripheral blood. Their morphology is presented in Figure 18.1. The categories are, ERYTHROCYTES (red blood cells or corpuscles), LEUCOCYTES (white blood cells), and PLATELETS or thrombocytes.

Erythrocytes. Mature erythrocytes, as they occur in the bloodstream, are biconcave non-nucleated discs averaging 8.5 μm in diameter, 2 μm thick at their edges, and 1 μm thick in their center. This shape provides the maximum surface area for the volume of the cell, and therefore the greatest possible diffusion surface for passage of gases. Although mature erythrocytes lack a nucleus, they consume glucose, oxygen, and ATP, and liberate carbon dioxide. The metabolism is directed primarily toward fueling active transport systems that maintain ionic homeostasis between the cell and the plasma.

TABLE 18.2

NUMBERS OF ERYTHROCYTES BY AGES

AGE	MEAN, MILLION PER MM^3	COMMENTS
Newborn	6.1	Life *in utero* is life at a low oxygen concentration, thus more red cells are present.
1 day	5.6	At birth, a loss of red cells occurs with breathing of higher O_2 levels. An increase in number occurs at puberty.
3 weeks	4.9	Males have more muscle than females, and thus have a greater oxygen demand, hence more cells
2 months	4.5	
1 year	4.6	
10 years	4.8	
Adult male	5.4	
Adult female	4.8	

Numbers. Erythrocytes are the most numerous of the formed elements, averaging in the adult about 5×10^6 cells per mm^3. Numbers are higher in the adult male (range 4.5 to 6.0×10^6 cells/mm^3) than in the adult female (range 4.3 to 5.5×10^6 cells/mm^3). The apparent reasons for this difference reside in the greater mass of muscular tissue in the male that requires a greater oxygen supply, gonadal hormones that stimulate production, and a greater oxygen utilization that also increases production. Numbers are also age dependent, as shown in Table 18.2.

Production. During intrauterine life, the yolk sac, spleen, liver, lymph nodes, and red bone marrow serve as sites of erythrocyte production. At or shortly before birth, all areas except the red bone marrow in such areas as the sternum, skull bones, vertebrae, ribs, and ends of long bones cease production of red cells. The cells originate from a stem cell common to all formed elements, and pass through a series of stages (Figure 18.2) to form the mature cell. HEMOGLOBIN, a respiratory pigment capable of combining with both oxygen and carbon dioxide, accumulates during the latter stages of development, reaching maximum concentration just before loss of the nucleus.

Life history. Mature cells pass from the bone marrow at rates estimated to be about 3.5 million per kg of body weight per day. They circulate for 90 to 120 days, carrying out their primary functions of oxygen and carbon dioxide transport, and then "die." Aged erythrocytes undergo PHAGOCYTOSIS by cells of the *reticuloendothelial system* located primarily in the liver, spleen, and bone marrow. How these phagocytes "recognize" old cells is not known; as the cells age, they may undergo subtle chemical changes that enable them to be differentiated from younger cells. Also, cells containing abnormal types of hemoglobin are selectively phagocytosed, further strengthening the theory that chemical differences are responsible for phagocytosis of particular cells.

TABLE 18.3

REQUIREMENTS FOR ERYTHROCYTE PRODUCTION

SUBSTANCE	DESCRIPTION AND/OR USE
Lipid	Cholesterol and phospholipids; incorporated in membrane and stroma
Protein	Incorporated into cell membrane
Iron	Incorporated into hemoglobin
Amino acids	Incorporated into hemoglobin
Erythropoietin	A glycoprotein, it is released from the kidney with hypoxia, hemorrhage, and excessive androgen secretion and stimulates production of erythrocytes
Vitamin B_{12}	Used in the formation of DNA in nuclear maturation
Intrinsic factor	A mucopolysaccharide produced by the stomach. It combines with vitamin B_{12} and insures absorption of the vitamin from the gut
Pyridoxin	Increases the rate of cell division
Copper	Catalyst for hemoglobin formation
Cobalt	Aids in synthesis of hemoglobin
Folic Acid	Promotes DNA synthesis

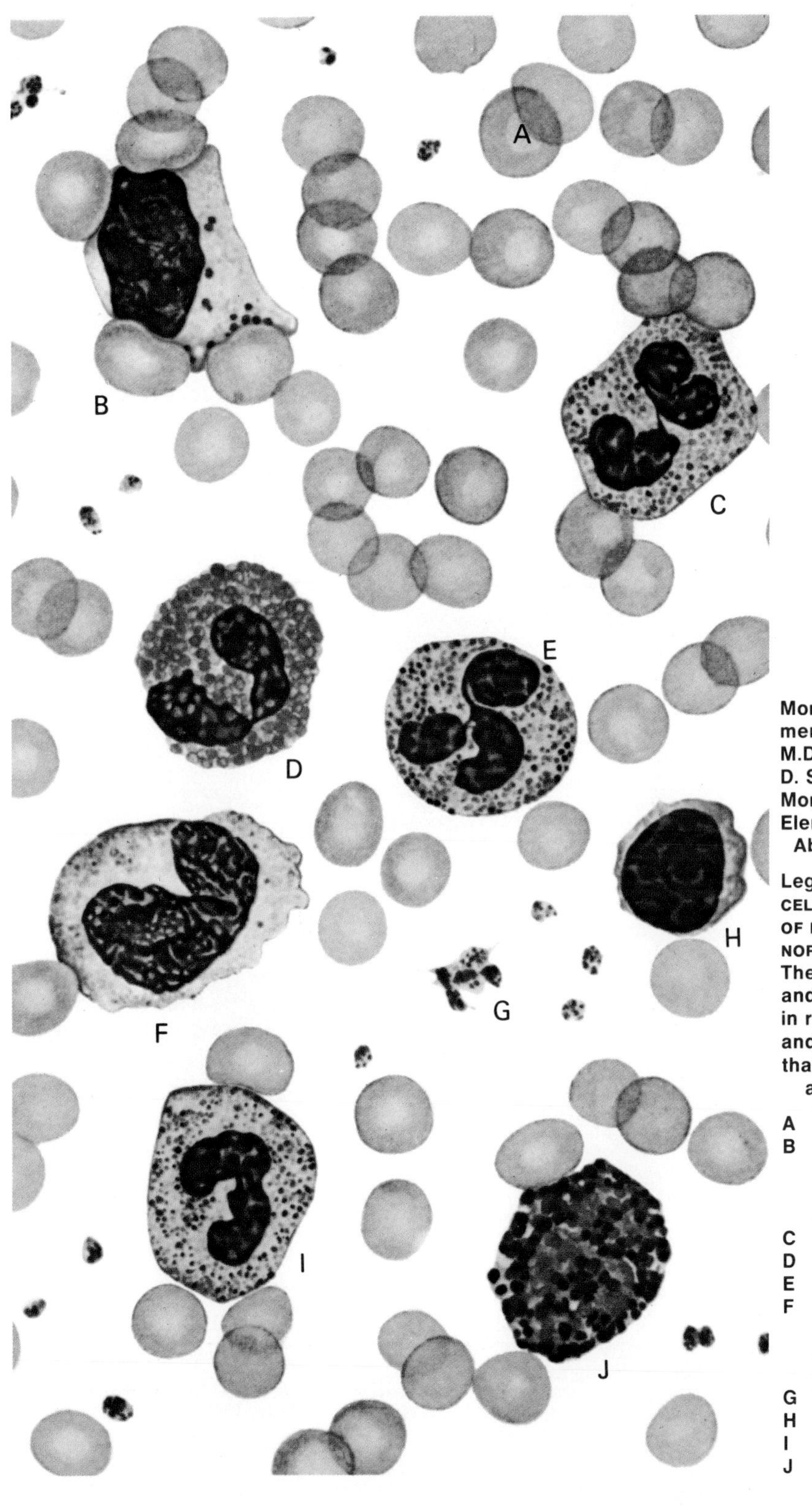

FIGURE 18.1
Morphology of blood elements. Courtesy L. W. Diggs, M.D. From L. W. Diggs, D. Sturm, A. Bell, "The Morphology of Human Blood Elements," 3rd edition, Abbott Laboratories, 1954.

Legend key.

CELL TYPES FOUND IN SMEARS OF PERIPHERAL BLOOD FROM NORMAL INDIVIDUALS.

The arrangement is arbitrary and the number of leukocytes in relation to erythrocytes and thrombocytes is greater than would occur in an actual microscopic field.

- **A** Erythrocytes
- **B** Large lymphocyte with azurophilic granules and deeply indented by adjacent erythrocytes
- **C** Neutrophil, segmented
- **D** Eosinophil
- **E** Neutrophil, segmented
- **F** Monocyte with blue gray cytoplasm, coarse linear chromatin and blunt pseudopods
- **G** Thrombocytes
- **H** Lymphocyte
- **I** Neutrophilic band
- **J** Basophil

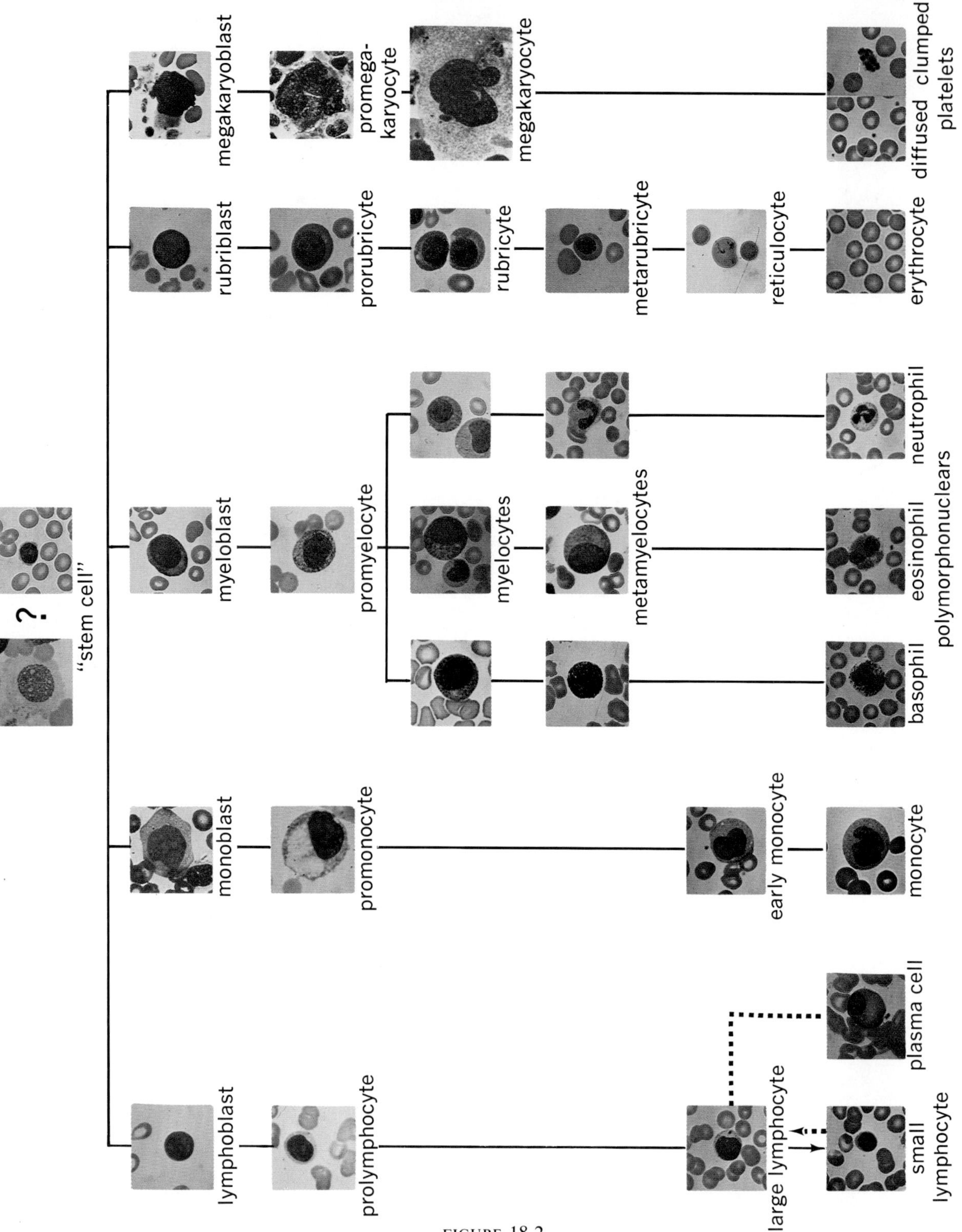

FIGURE 18.2
THE MATURATION OF BLOOD CELLS.
The exact morphology of the stem cell is still open to question. Wright stain. 1200×. (Courtesy of American Society of Hematology National Slide Bank and Health Sciences Learning Resources Center, University of Washington. Used with permission.)

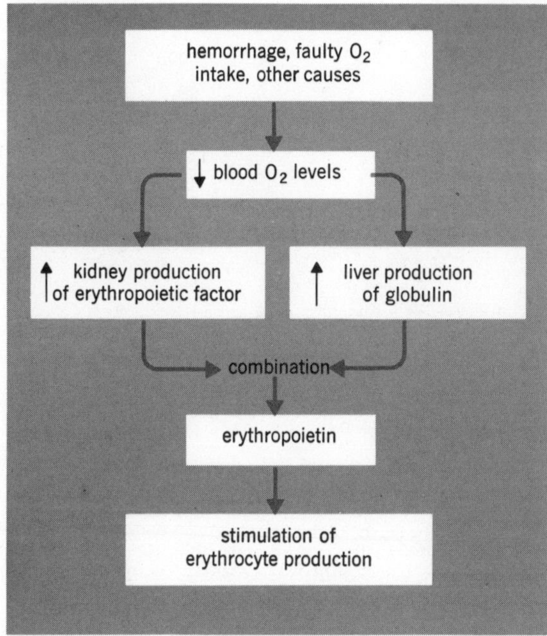

FIGURE 18.3
The mechanism of erythropoietin secretion effect.

The phagocytes largely recycle the erythrocyte components for reuse by body cells in general, and in the synthesis of new hemoglobin.

Requirements for the production of erythrocytes. Many substances are required for the production of erythrocytes. The more important ones and their roles are presented in Table 18.3. ERYTHROPOIETIN (Figure 18.3) is a substance that stimulates erythrocyte production. The kidney, in response to *hypoxemia* (lowered arterial oxygen levels), produces *renal erythropoietic factor (REF)*. REF acts on a plasma globulin precursor to form erythropoietin that in turn stimulates mitosis of marrow stem cells and their maturation to erythrocytes. Erythropoietin can achieve an increase in the circulating erythrocyte mass of about 2 percent, usually sufficient to elevate the oxygen-carrying capacity of the blood to normal levels.

Hemoglobin. Oxygen and some carbon dioxide transport depends on the presence of the respiratory pigment HEMOGLOBIN in the erythrocytes. The substance increases the ability of the blood to carry oxygen by 60 fold (plasma carries 1/3 ml O_2/100 ml; whole blood 20 ml/100 ml). Thirteen to sixteen (13 to 16) gm of hemoglobin per 100 ml of blood represents a normal amount.

Hemoglobin has a molecular weight of about 67,000 and is composed of pigmented HEME molecules attached to four polypeptide chains that constitute the GLOBIN portion of the molecule. Normal adult hemoglobin, designated hemoglobin (*Hgb*) A, contains two different pairs of polypeptide chains called *alpha* and *beta chains*. Each has a particular sequence of amino acids in it that is genetically determined. Fetuses have a type of hemoglobin designated Hgb F. In this type of hemoglobin the alpha chains are the same as in Hgb A, but the beta chains are replaced by two *gamma chains*. The latter differ from beta chains in the amino acid sequences in the globin. Both hemoglobin A and F are normal hemoglobins. Hgb F has a greater affinity for oxygen than does Hgb A, an advantage to the fetus living in a low-oxygen environment. Any genetic mutation that leads to an alteration in the amino acid sequences (such as a substitution or deletion), will lead to the production of an abnormal hemoglobin molecule. Hgb S, the characteristic hemoglobin of sickle cell anemia, has had a valine substituted for a glutamic acid in position six of the beta chains as the *only* alteration. This is sufficient to cause the Hgb S to precipitate into spindle-shaped aggregations under conditions of low blood oxygen levels. Sickled cells are more liable to hemolysis and phagacytosis. The change is reversible by raising blood oxygen levels. Sickle cell sufferers may thus exhibit anemia, signs of hypoxemia, jaundice, and weakness.

There are other abnormal hemoglobins; at least 50 have been discovered. Each is characterized by alterations of the globin chains, and gives the molecule different properties. Not all of these alterations prevent normal oxygen transport, and most individuals lead normal though sometimes restricted lives.

Synthesis of hemoglobin. As implied above,

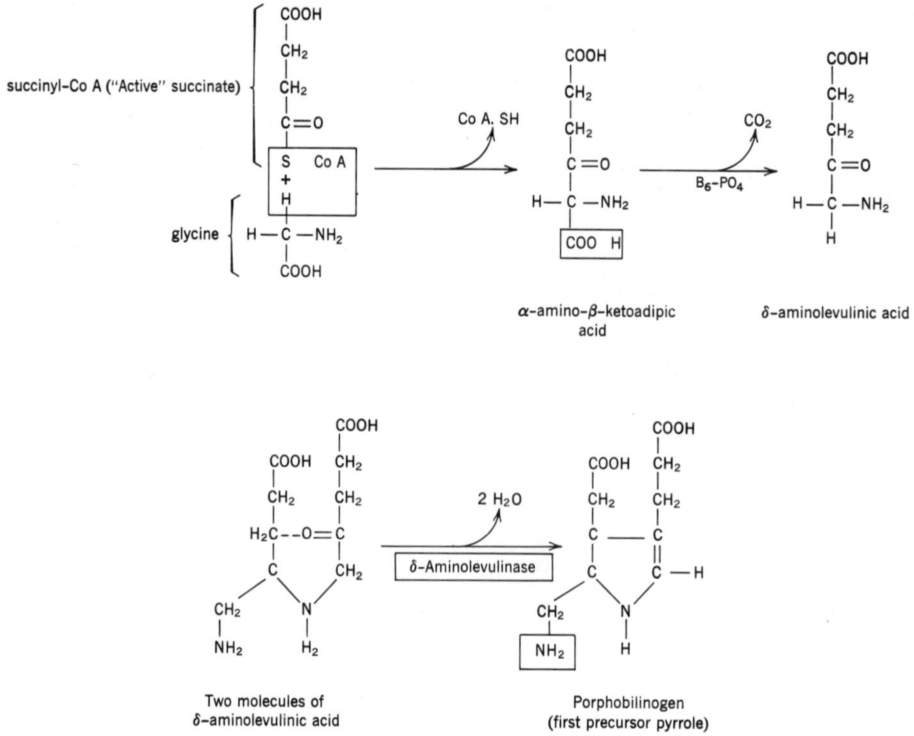

succinyl-Co A ("Active" succinate)

glycine

Co A. SH

CO_2

B_6-PO_4

α-amino-β-ketoadipic acid

δ-aminolevulinic acid

2 H_2O

δ-Aminolevulinase

Two molecules of δ-aminolevulinic acid

Porphobilinogen (first precursor pyrrole)

FIGURE 18.4
The steps in the synthesis of hemoglobin.

synthesis of hemoglobin requires the synthesis of two components, the *heme* and the *globin*.

Heme synthesis (Fig. 18.4) begins with simple materials and forms a ring-like structure called a *pyrrole*. Four pyrroles are combined into a *porphyrin,* to which iron is added to form heme. A critical enzyme in the sequence is an isomerase; deficiency or abnormality in this enzyme will lead to the production of an abnormal heme that will be excreted.

Globin synthesis has been discussed in a preceding section of this chapter.

Relationship of iron to the synthesis. Iron is essential to the transport of oxygen. Iron is absorbed from the first part of the small intestine by active transport mechanisms and is absorbed three times more rapidly if it is in the ferrous (Fe^{++}) state, as opposed to the ferric (Fe^{+++}) state. Amounts of iron required to sustain iron homeostasis range from 0.5 mg/day for an adult male, to as much as 2.0 mg/day for a menstruating female. Iron absorbed by intestinal epithelial cells or released by erythrocyte destruction is transferred to the plasma where it combines with a beta-globulin named TRANSFERRIN (*siderophilin*). The liver stores about 60 percent of the body's iron, receiving it from the transferrin. In the liver, iron combines with a cellular protein designated APOFERRITIN, to form FERRITIN, the storage form. Ferritin gives up iron as required for body use. Figure 18.5 summarizes the metabolism of iron.

Formation of pigments during heme destruction. As reticuloendothelial phagocytes destroy the heme fraction of the hemoglobin molecule, after the normal lifespan of the erythrocytes, BILIVERDIN is formed. Biliverdin undergoes a reduction to BILIRUBIN, which is released into the plasma. It combines with plasma albumins and is transported to the liver. It is

uroporphyrinogen III

6H

auto-oxidation in light

uroporphyrin III

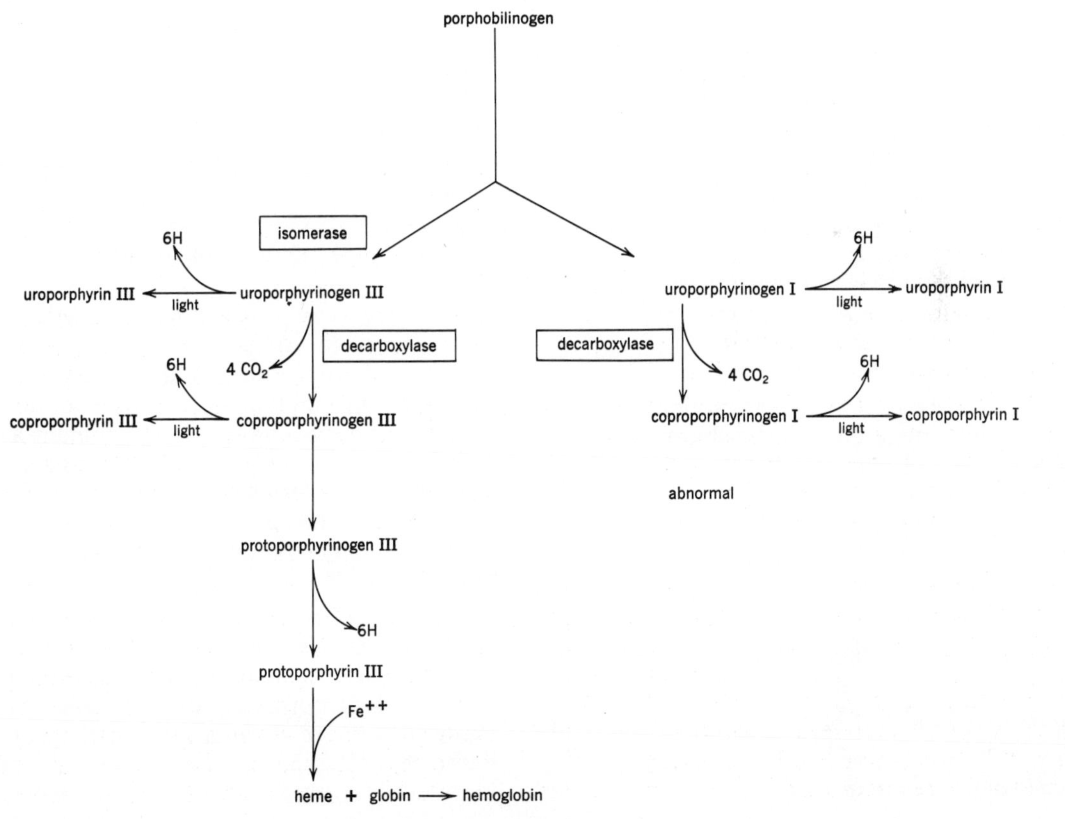

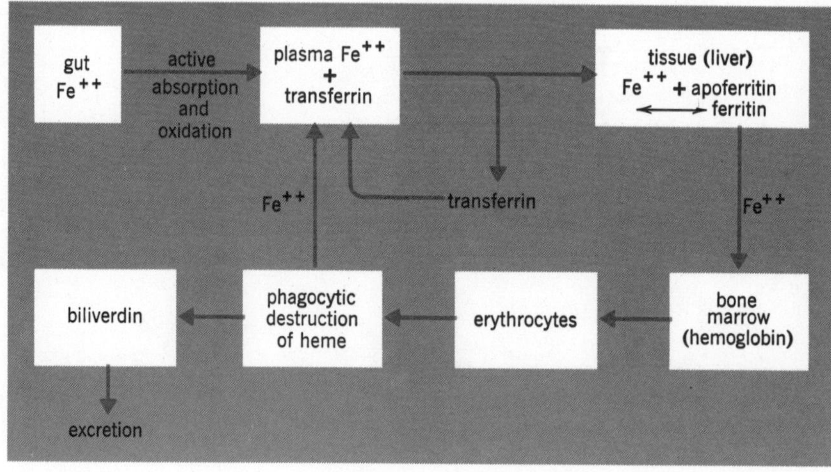

FIGURE 18.5
The metabolism of iron in the body.

released from the albumins and is absorbed by liver cells. It next undergoes *conjugation* with glucuronic acid to form a BILE SALT; this conjugation makes the bilirubin water soluble. Excreted into the small intestine via the bile, bilirubin then is acted on by bacteria and is converted to UROBILINOGEN. Some of the urobilinogen remains in the intestine and imparts the orange-brown color to the feces; the remainder is absorbed by the intestine into the bloodstream, and is eventually excreted by the kidney as the amber coloring matter of the urine. These relationships are summarized in Figure 18.6.

Clinical tests for hepatic function often include a determination of plasma bilirubin levels. When total plasma bilirubin is persistently greater than 2 mg/100 ml, the sclerae (whites of the eyes), mucous membranes, and skin may become yellowed, and JAUNDICE or ICTERUS is said to have developed. Among the causes of elevated bilirubin levels are:

Excessive destruction of red blood cells (malaria, sickle cell anemia, hemolytic disease).

Defective uptake of bilirubin by the liver (as in infectious hepatitis).

Obstruction of the bile duct carrying bile to the small intestine (gall stones), that causes excessive entry of the pigment into the bloodstream.

A temporary or *physiological jaundice* occurs shortly after birth, as the excessive numbers of red cells are brought to normal levels by phagocytosis and hemolysis. The infant is now in a high oxygen environment and does not need as many oxygen-carrying units as it did *in utero*.

Other functions served by the erythrocytes. Hemoglobin (a conjugated protein) is a good BUFFERING compound, and the erythrocytes also contain potassium bicarbonate for buffering purposes. The units also contain an enzyme, carbonic anhydrase, that speeds the conversion of carbonic acid to carbon dioxide and water. Transport of CO_2 across membranes is thus speeded by the action of the enzyme that ensures a high diffusion gradient.

Abnormalities in erythrocyte formation include the processes that result in abnormal shape of cells (*poikilocytosis*), abnormal size of cells (*anisocytosis*), or numbers of cells (*polycythemia* or *anemia*). Hemorrhage, heavy exercise, and exposure may result in a release into the bloodstream of immature erythrocytes that are larger than normal and that may contain nuclei or fragments of nuclei. Table 18.4 summarizes some of the common types and causes of ANEMIA. POLYCYTHEMIA (increased numbers of red cells to greater than 6 million/mm^3) may be a response to decreased oxygen content in the air that is breathed. This is caused by release of erythropoietin and is called a *physiological polycythemia*. *Pathological polycythemia* results from tumorlike conditions in the bone marrow that vastly overproduce erythrocytes. The blood flows very sluggishly because of the great increase in viscosity resulting from the presence of the increased number of cells. Extraction of oxygen from the hemoglobin is thus greater, and the person typically shows a bluish color to the skin (*cyanosis*).

FIGURE 18.6
The metabolism of the heme portion of the hemoglobin molecule.

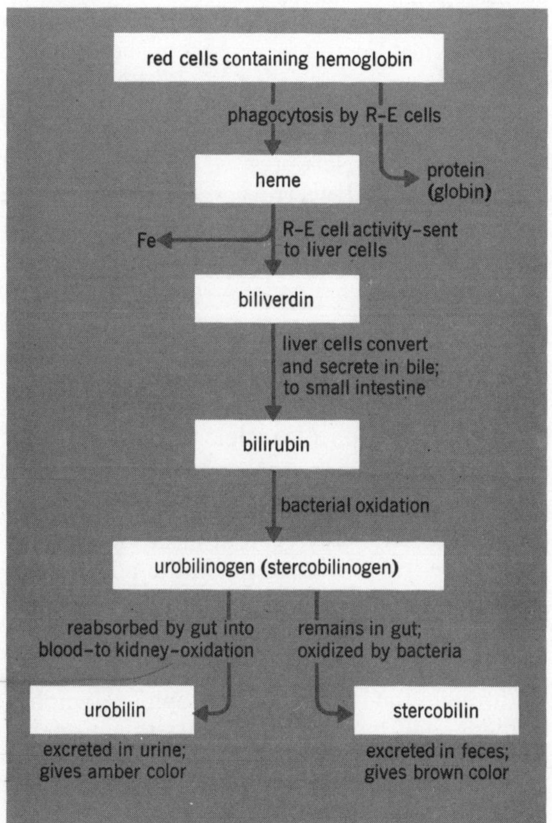

Leucocytes (white blood cells). The leucocytes are a heterogeneous population of nucleated cells, lacking in hemoglobin. There are five varieties, originating from two different body areas. The morphology, characteristics, and origins of these cells are described in Table 18.5 and are shown in Figures 18.1 and 18.2.

Numbers. Leucocytes are the least numerous of the formed elements; numbers may vary physiologically with age and state of the body. Some values of the cells according to age are presented in Table 18.6. Leucocytes are counted in the DIFFERENTIAL COUNT. In this procedure, 100 stained leucocytes are counted on a slide and numbers of each type are recorded. The final number in each category are then their percent of total leucocyte number.

Life history. GRANULAR (*myeloid* or *bone marrow origin*) LEUCOCYTES include neutrophils, eosinophils, and basophils. They are also known as granulocytes or polymorphonuclear (*PMN*) leucocytes and are shed into the bloodstream and remain functional for 12 hours to three days. When liberated into the bloodstream, about 75 percent possess two or three lobes in their nucleus. As they age, the number of lobes increases. The classification of myeloid leucocytes may be tabulated in series on a page with the younger (fewer lobes) leucocytes on the left and the older leucocytes (more lobes) to the right. The term "*shift to the left*" implies a greater proportion of younger forms in the bloodstream, and a "*shift to the right*" implies a greater proportion of older cells. The relative proportions may suggest the presence of abnormal marrow activity. NONGRANULAR (*lymphoid*) LEUCOCYTES, that is, lymphocytes and monocytes, have been estimated to remain functional for 100 to 300 days under normal body conditions. The normal production of both categories of cells appears to depend primarily on the levels of *adrenal steroid hormones* and whether or not *infection* is present in the body. Steroid hormones increase the production of neutrophils, while products of infec-

TABLE 18.4

SOME COMMON TYPES OF ANEMIA

	TYPE OF ANEMIA	CAUSES	CHARACTERISTICS	SYMPTOMS	TREATMENT
Loss Increased	*Hemorrhagic*			Shock	
	Acute	Trauma Stomach ulcers Bleeding from wounds	Cells normal		Transfusion
	Chronic (iron deficiency)	Stomach ulcers Excessive menses	Cells small (microcytes), deficient in hemoglobin content	None or Fatigue	Iron administration
	Hemolytic	Defective cells Destruction by parasites, toxins, antibodies	Young cells (reticulocytes) very prominent, serum haptoglobin[a] reduced. Morphology of cells may be abnormal when detected	None or Fatigue	Dependent on cause
Formation Decreased	*Deficiency states*				
	Folic acid	Nutritional deficiency	Cells large (macrocytes), with normal hemoglobin content	None or Fatigue	Folic acid
	Vitamin B_{12}	Lack of intrinsic factor in stomach (pernicious anemia)	Cells large with normal hemoglobin content		Administration of vitamin B_{12}
	Hypoplastic or *aplastic*	Radiation Chemicals Medications	Bone marrow hypoplastic	None or Fatigue	Transfusion Androgens Cortisone Remove cause

[a] A protein strongly binding hemoglobin in plasma.

tion tend to increase both neutrophils and lymphocytes. Additionally, a *leucocyte mobilizing factor* is released by injured cells. The cells are lost from the circulation by migration through the walls of the alimentary tract, by destruction (pus formation) during infections, or by disintegration in the bloodstream. *Mononucleosis,* a debilitating infectious disease thought to be viral in origin, results in an increase of abnormal mononuclear cells (lymphocytes and monocytes) in the bloodstream, fatigue, fever, and in the appearance of an abnormal antibody in the bloodstream (Paul-Bunnel antibody).

Requirements for production. Materials required for leucocyte production are the same as for cells in general, but especially important is folic acid. This substance is required for the normal production of nucleic acid as the cells form.

Functions of the leucocytes. All leucocytes possess, to some degree, four basic PROPERTIES that relate to their functions in the body.

AMEBOID MOTION. By the production of cytoplasmic extensions or pseudopods, leucocytes are capable of independent movement through the tissues. Neutrophils, lymphocytes, and monocytes possess this property to the greatest degree.

CHEMOTAXIS. The cells are attracted to (*positive chemotaxis*) or repelled from (*negative*

TABLE 18.5
SUMMARY OF FORMED ELEMENTS

ELEMENT	NORMAL NUMBERS	ORIGIN (AREA OF)	DIAMETER (μM)	MORPHOLOGY	FUNCTION(S)
Erythrocytes	4.5–5.5 million /mm	Myeloid (marrow)	8.5 (fresh) 7.5 (dry smear)	Biconcave, non-nucleated disc; flexible	Transports O_2 and CO_2 by presence of hemoglobin; buffering
Leucocytes	6000–9000 /mm^3		9–25		
Neutrophil	60–70% of total	Myeloid	12–14	Lobed nucleus, fine heterophilic specific granule	Phagocytosis of particles, wound healing. Granules contain peroxidases for destruction of microorganisms
Eosinophil	2–4% of total	Myeloid	12	Lobed nucleus; large, shiny red or yellow specific granules	Detoxification of foreign proteins? Granules contain peroxidases, oxidases, trypsin, phosphatases. Numbers increase in autoimmune states, allergy, and in parasitic infection (schistosomiasis, trichinosis, strongyloidiasis)
Basophil	0.15% of total	Myeloid	9	Obscure nucleus; large, dull, purple specific granules	Control viscosity of connective tissue ground substance? Granules contain heparin (liquefies ground substance) serotonin (vasoconstrictor), histamine (vasodilator)
Lymphocyte	20–25% of total	Lymphoid			
Small			9	Nearly round nucleus filling cell, cytoplasm clear staining	Phagocytosis of particles, globulin production
Large			12–14	Nucleus nearly round, more cytoplasm	
Monocyte	3–8% of total	Lymphoid	20–25	Nucleus kidney of horseshoe shape, cytoplasm looks dirty	Phagocytosis, globulin production
Platelets (thrombocytes)	250,000–350,000 /mm^3	Myeloid	2–4	Chromomere and hyalomere	Clotting

TABLE 18.6
NUMBERS OF LEUCOCYTES BY AGES

AGE	NUMBER OF CELLS (per mm^3)
Newborn	10,000–45,000
3 days–4 years	5,000–24,000
4–7 years	6,000–15,000
8–18 years	4,500–15,000
Adult	5,000– 9,000

chemotaxis) areas of injury or inflammation. A specific polypeptide (*leucotaxine*), nucleic acids, and positively charged particles are strong positive chemotaxic agents. Negatively charged particles repel most leucocytes.

PHAGOCYTOSIS. The ability to engulf and digest or kill bacteria and products of cell death is best developed in the neutrophils, lymphocytes, and monocytes. The latter two cells may also migrate into the tissues and become macrophages, larger cells with an increased capacity for phagocytosis.

DIAPEDESIS. The ability to pass through the walls of capillaries to reach an area of inflammation or infection.

The FUNCTIONS of the cells thus relate to defense against bacteria, foreign particulate matter, and in the removal of the debris resulting from injury. Functions are shown in Table 18.5.

Leukemia. Leukemia is regarded as a neoplastic disease of the blood-forming tissues that causes unrestricted production of mature leucocytes and their precursor cells. Numbers may reach 250,000 to 500,000/mm^3. Unlimited production of white cells depresses erythrocyte and platelet formation and may lead to anemia and bleeding tendency.

Causes of leukemia are not definitely known, although viruses and radiation have been shown to cause the disease. The disease occurs in acute and chronic forms. Acute leukemia is sudden in onset, causes anemia, bleeding is usually present, and response to treatment is usually poor. The chronic form occurs more often in adults and may be myelocytic (leucocytes of marrow origin are increased) or lymphoid (leucocytes of lymphoid origin are increased). General weakness, fatigue, and enlargement of the spleen are common symptoms.

Four types of leukemia are generally distinguished.

Acute lymphoblastic leukemia (*ALL*) is a disease primarily of children, and is characterized by increased numbers of immature lymphocytes. Anemia, due to decreased production of red cells is found in 90 percent of patients with ALL.

Acute myeloblastic leukemia (*AML*) can occur at any age, and is characterized by increase in all of the granulocytes (neutrophil, eosinophil, basophil) and by increased numbers of immature granulocytes.

Chronic lymphocytic leukemia (*CLL*) occurs most commonly in the middle aged and elderly, and is characterized by greatly increased numbers of small, mature lymphocytes. Anemia is milder than in the acute leukemias.

Chronic myelocytic leukemia (*CML*) occurs more commonly in young adults and is characterized by increased numbers of all granular leucocytes.

Radiation and chemotherapy are the measures most frequently employed to treat leukemia. The acute forms of the disease are more difficult to arrest than the chronic forms; remissions of as much as 15 years have been achieved with adequate therapy that has been instituted early.

HODGKIN'S DISEASE is a chronic lymphoma of unknown cause, characterized by enlargement of the lymph nodes, spleen, and liver. White cell numbers do not increase greatly in this disorder, but a giant cell with many-lobed or multi-nuclei (Reed-Sternberg cells) appear in the nodes. There is no certain cure for the disease, but radiation and chemotherapy are employed to lengthen life.

The conditions described above are examples of *pathological leucocytoses. Physiological leucocytoses* (of nonpathologic cause and usually transient) occur in newborn infants, during pregnancy, in emotional disturbances, menstruation, and in severe exercise.

Platelets. PLATELETS or *thrombocytes* (see Fig. 18.1) originate in the bone marrow from a giant cell known as a *megakaryocyte*. The units are 2 to 4 μm in diameter, and consist of a granular structure (chromomere) surrounded by a light-colored area (hyalomere). They are not nucleated and carry one of several materials essential to the clotting of the blood. Their numbers are normally 250,000 to 300,000 per mm^3 of blood.

HEMOSTASIS

This term refers to the reactions that occur to prevent or minimize loss of blood from the blood vessels if they are injured or rupture. Three types of reactions take place in hemostasis:

There is NARROWING OF THE MUSCULAR BLOOD VESSELS (*vasoconstriction*) in the area of vascular damage. This narrowing results in smooth muscle spasm as a direct product of mechanical trauma, to sympathetic reflexes triggered primarily by the pain associated with the trauma, and by the release of vasoconstrictive chemicals, chiefly serotonin, and by the cellular damage that occurs. This reaction reduces blood flow to the injured area, and facilitates the accumulation of factors required for coagulation to occur.

A PLATELET PLUG IS FORMED at the site of the torn vessels. Platelets are rather sticky, and when they contact the exposed collagen (a connective tissue protein) at the site of injury, they adhere to the torn surfaces. They release ADP (adenosine diphosphate) as they adhere and this speeds the aggregation and breakdown of more platelets. This plug can seal small ruptures in vessel walls, and is the primary mechanism that operates to seal the microscopic tears in capillary walls that are a normal part of everyday life.

The blood COAGULATES or CLOTS by changing from a liquid to a semisolid condition (a sol-gel change).

Coagulation. The series of reactions that culminate in a sol-gel change in blood consistency are termed COAGULATION. In order for coagulation to occur, at least 12 factors or chemical substances must be available. These are presented in Table 18.7. In simplest terms, coagulation involves the conversion of the soluble plasma protein FIBRINOGEN into a netlike mass of insoluble FIBRIN protein strands. The latter forms a mat across the torn blood vessels, and subsequently entangles red cells, white cells, and platelets to form a clot. Two "systems" promoting clotting are recognized as occurring in the body, and the reactions of coagulation occur in three phases. The EXTRINSIC SYSTEM is activated when cells are damaged and release a TISSUE FACTOR. This factor reacts with factor VII in the presence of Ca^{++}, and activates factor X. Factor X then reacts with factor V, a phospholipid from platelets, and in the presence of Ca^{++}, develops an enzymelike activity that converts prothrombin to thrombin. Thrombin then activates factor XIII, which in the presence of Ca^{++}, changes fibrinogen to fibrin.

In the INTRINSIC SYSTEM, a more complicated series of reactions occurs. Activating factors, including contact of the blood with a foreign surface (resulting from injury) are produced, and activate factor XII. In turn factor XI changes inactive factor IX to an active state. Active IX then changes factor VIII to an active state and this activates factor X. From this point onwards, the extrinsic and intrinsic schemes are the same. The reactions described are depicted in Figure 18.7.

The clot formed by these reactions undergoes *syneresis* or clot retraction as a result of platelet release of a substance known as *thrombosthenin*. It resembles actomyosin, the contractile proteins of muscle. The clot "shrinks" and forms an even tighter block across the injured vessels. As tissue repair proceeds, the clot is gradually dissolved. To ensure this, an activator changes a plasma component named PLASMINOGEN (*profibrinolysin*) to PLASMIN (*fibrinolysin*) that dissolves the clot. Neutrophils and reticuloendothelial cells then phagocytize the products of clot dissolution.

Anticoagulation. Any procedure or chemical

TABLE 18.7

FACTORS DEFINITELY IMPLICATED IN BLOOD COAGULATION

INTERNATIONAL COMMITTEE DESIGNATION	SYNONYMS	ORIGIN	LOCATION
Factor I	Fibrinogen	Liver	A plasma protein
Factor II	Prothrombin	Liver	A plasma protein
Factor III	Thromboplastin	By series of reactions in blood; also found, as such, in cells	Produced in the clotting process or released into fluids by injured cells
Factor IV	Calcium	Food and drink; from bones	As Ca^{++} in plasma
Factor V	Labile factor (accelerator globulin)	Liver	Plasma protein
Factor VI*			
Factor VII	SPCA (serum prothrombin conversion accelerator)	Liver	Plasma
Factor VIII	AHF (antihemophilic factor) AHG (antihemophilic globulin)	Liver	Plasma
Factor IX	PTC (plasma thromboplastin component) Christmas Factor	Liver	Plasma
Factor X	Stuart-Prower factor; develops full factor III power	Liver	Plasma
Factor XI	PTA (plasma thromboplastin antecedent)	Liver	Plasma
Factor XII	Hageman factor; contact factor; initiates reaction	?	Plasma
Factor XIII	Fibrin stabilizing factor; renders fibrin insoluble in urea. (Laki-Lorand factor)	?	Plasma
Platelet Factor	Cephalin	Marrow	Platelets

* No longer considered a separate entity; considered to be identical to Factor V.

that renders the blood incapable of coagulation is an ANTICOAGULANT. If one of the necessary materials in the coagulation process is removed, the process will not go to completion. Of the various materials present, Ca^{++} is probably the easiest to remove. Mixing withdrawn blood with sodium or ammonium oxalate or citrate results in an exchange of Na^{+} or NH_4^{+} for Ca^{++}. The blood has been DECALCIFIED. HEPARIN is an organic anticoagulant that may be injected into the body and acts to decrease the thromboplastin generation. It also is antithromboplastic and interferes with the reactions of Phase II. DICOUMAROL, a product first isolated from spoiled sweet clover, interferes with liver synthesis of prothrombin, and Factors V and VII. Its effect is not fully developed until 36 to 48 hours after intake but lasts longer than the other organic anticoagulants.

Disorders of clotting. The hemophilias are genetically determined conditions that result in failure of the blood to clot. HEMOPHILIA A, or

FIGURE 18.7

Schematic representation of blood coagulation. The formation of fibrin may be considered to occur in three stages: (1) formation of a prothrombin converting factor through intrinsic or extrinsic mechanisms; (2) conversion of prothrombin, and enzymatically inactive plasma globulin, to thrombin, an enzymatically active procoagulant; and (3) conversion of fibrinogen, a soluble plasma protein, into fibrin, an insoluble plasma protein, through the action of thrombin. (Modified from O. D. Ratnoff and D. Bennet. *Science* 179 (4080):1291, 1973. Copyright © 1973 by the American Association for the Advancement of Science.)

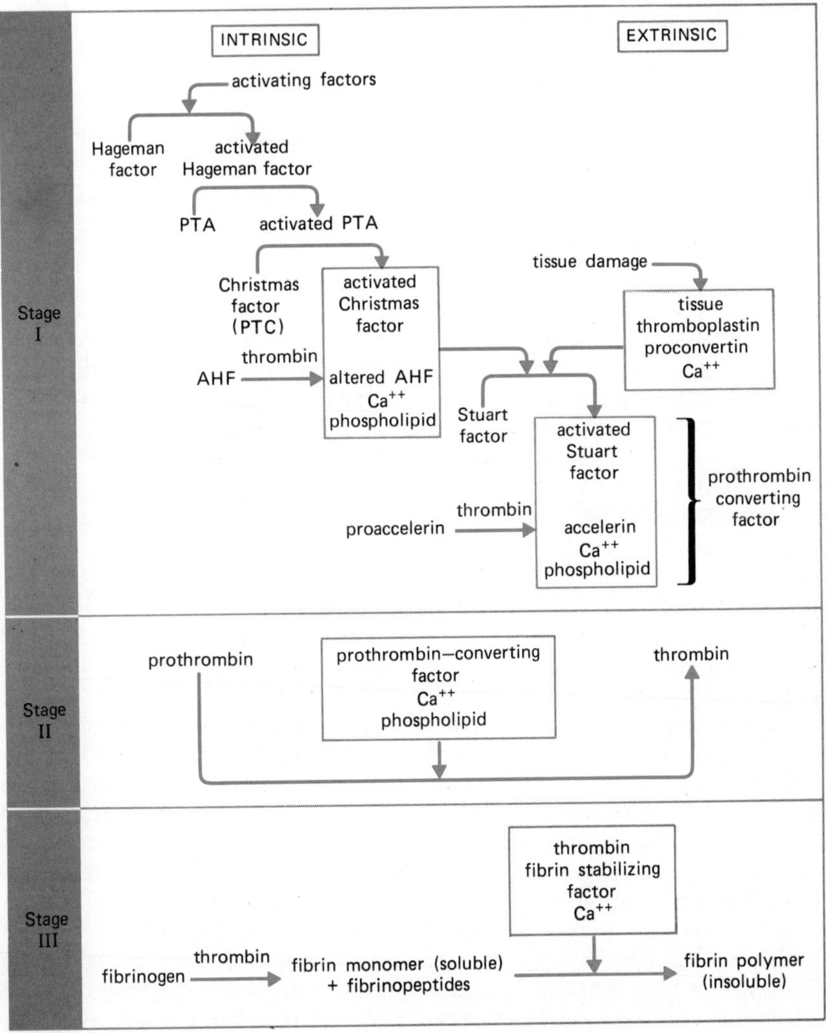

TABLE 18.8
A COMPARISON OF BLOOD AND LYMPH FOR VARIOUS COMPONENTS (VALUES IN MG PERCENT UNLESS OTHERWISE NOTED)

CONSTITUENT	LYMPH (THORACIC DUCT)	PLASMA
Calcium	7.7	9.5
Chloride	335	335
Inorganic phosphate	3.9	4.0
Potassium	18.3	19.1
Sodium	290	290
Protein (total) g %	2.8–3.6	6.5
Albumin g %	1.6–2.4	3.5
Globulin g %	1.2	2.5
Fibrinogen g %	0.2	0.4
Amino acids	2.4	2.4
Nonprotein nitrogen	23.4	25.7
Urea	22	24
Glucose	136	135
Lipid g %	0.2–7.3 (intestinal)	0.3–0.6
Cholesterol	75	120
Leucocytes (lymph cells)/mm^3	3500–75,000	1500–1850

classical hemophilia, is mediated by a *sex-linked recessive gene* and results in failure to synthesize AHF (factor VIII). HEMOPHILIA B, or *Christmas disease,* is also transmitted by a *sex-linked recessive gene* and results in failure of PTC (factor IX) synthesis. HEMOPHILIA C is transmitted by an *autosomal dominant gene* and results in failure of PTA (factor XI) synthesis. Afibrinogenemia, or, more correctly, HYPOFIBRINOGENEMIA (deficiency of fibrinogen in the blood), may be either congenital or acquired following severe liver damage.

Blood may spontaneously clot within the vessels to give an intravascular clot or THROMBUS. Slow blood flow and the presence of roughened surfaces on a blood vessel (as in atherosclerosis) appear to have a role in intravascular clotting. A thrombus may be quite serious if thrombosis of a vital vessel occurs; for example, thrombosis of coronary or cerebral vessels. The term EMBOLUS may refer to a floating clot that has detached from its site of formation. The danger here is that the clot may lodge in a vital vessel and result in *embolism.* Clots floating from the lower appendages appear particularly dangerous because they have a good chance of passing through the right side of the heart and lodging in the lungs to create *pulmonary embolism.*

LYMPH

FORMATION, COMPOSITION, AND FLOW

Tissue or interstitial fluid is formed by the filtration of materials from blood capillaries under the hydrostatic pressure of the heart action, and by fluid secretion from body cells. All components of the blood except red cells and platelets may normally be found in tissue fluid. Upon entering the lymph vessels, the fluid is properly termed lymph. Concentration of all diffusible substances in lymph is nearly the same as that of plasma, while protein concentration is always less than that of plasma and is variable according to where the sample is taken. The reason for lower protein concentration of lymph protein is because filtration of protein is slower than that of the other substances. Some representative figures for various components in blood and lymph are presented in Table 18.8.

ENTRY OF TISSUE FLUID, and any particles or materials within it, into lymphatic vessels is facilitated by several factors.

Lymphatics tend to run in the fibrous fascial planes of the body (Fig. 18.8). Relaxation of the tissues around the vessels tends to expand them and to draw fluid into the vessels. Contraction of muscle compresses the vessels, moving the fluid onwards.

Permeability of lymphatics is greater than that of blood capillaries allowing easy penetration of protein and other large molecules. Particulate matter may enter by phagocytosis or pass directly through the walls as in wounds. The exact method of entry is unknown.

Classically, it has been taught that pressure gradients decrease from blood capillary (15 to 0.7 mm Hg.) to interstitial space (2.2 to 0.2 mm Hg.) lymphatic (1.2 to 0.2 mm Hg.). Hydrostatic pressures thus force fluid into the lymphatics. Increase of venous pressure will increase lymph flow by providing more fluid for the lymphatics. Recently, it has been demonstrated that interstitial fluid pressure may be negative (−7 mm Hg.). In this case, expansion and compression of the tissue "pumps" fluid into the lymph vessels.

FLOW OF LYMPH through the vessels varies, depending primarily on the level of muscular

FIGURE 18.8
The relationships of lymphatic vessels to skeletal muscles and fascial planes of the body. The anchoring filaments tend to expand the vessel; muscle contraction tends to compress the vessels (↓). Lymph tends to move to the right (→) and is prevented from backflowing by the valve. A "bellows" or "pumping" action thus draws tissue fluid into the vessels and moves it onwards.

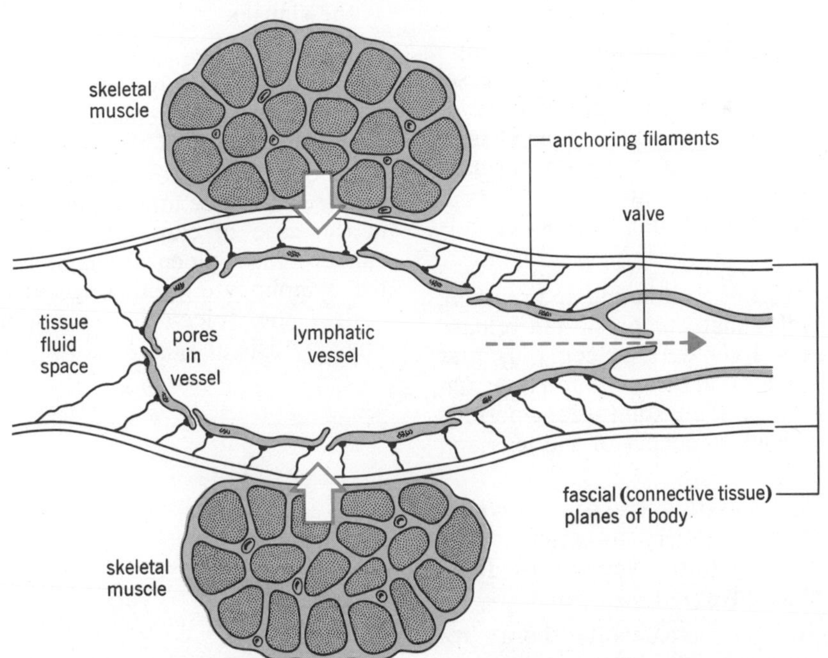

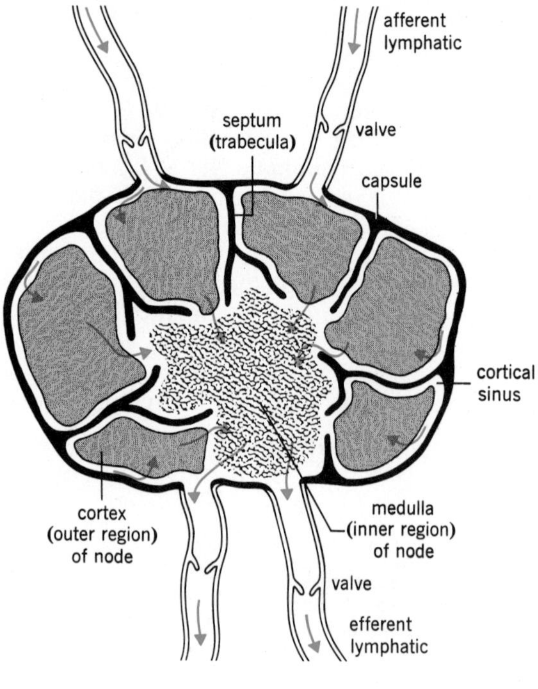

FIGURE 18.9
A diagram of a lymph node.

activity in the body. A typical amount of lymph drained from a medium-sized lymphatic vessel of a human would be about 1.5 ml/min at moderate activity levels. Direction of flow is controlled by valves that permit only a movement of fluid towards the subclavian veins.

Flow of lymph through the vessels depends on several factors.

MASSAGING ACTION of contracting and relaxing skeletal muscles that compress and release the lymphatics. Even during sleep, muscular movements occur that insure lymph flow. On the other hand, standing still for long periods of time may lead to edema of the legs and feet.

A HYDROSTATIC PRESSURE GRADIENT, which decreases from periphery to entry of the ducts into the subclavian veins, provides a means of causing lymph flow.

INSPIRATORY ACTIVITY expands the thorax, reduces the pressure on the ducts, and they expand. Like a suction pump, lymph is drawn into the expanded vessel from more peripheral areas, both above and below the thorax.

The discussion so far has dealt with the function of the lymph in returning excess fluid and filtered protein to the blood-vascular system. Other functions served are associated with defense and leucocyte production.

THE LYMPH NODES

The lymph nodes (Fig. 18.9) are ovoid or rounded structures varying in size from 1 to 25 mm in diameter. They are placed in the course of most medium-sized lymph vessels and thus have incoming (*afferent*) and outgoing (*efferent*) lymphatics. Fibrous tissue surrounds the node and an internal STROMA of reticular connective tissue forms the basic outline of the nodes. Lymph moves through the nodes by way of LYMPH SINUSES. The node contains two types of cells:

PRIMITIVE (*reticular*) CELLS that give rise to lymphocytes and/or plasma cells.

MACROPHAGES or phagocytes, members of the reticuloendothelial system, line the walls of the sinuses.

Lymphocytes are shed into the efferent lymph flow and ultimately enter the bloodstream. Plasma cells may remain in the node or may seed other areas of the body. They represent areas of antibody production. The macrophages engulf particulate matter entering the node and, in the case of carcinoma, trap metastasizing cancer cells. Spleen, thymus, and tonsils share the lymphocyte producing function with the nodes; the spleen is an important area of phagocytosis of aged erythrocytes.

SUMMARY

1. Blood is a connective tissue with two major components.
   a. The liquid portion is the plasma.
   b. Formed elements are cells or cell-like units.
   c. Blood has a volume of about 5 L in the adult, or 7 percent of the body weight.
2. The blood has two major functions.
   a. It transports: nutrients, oxygen, heat, wastes, hormones and excess water.
   b. It is involved in regulation of homeostasis of water content, pH, temperature, and protects against foreign chemicals, microorganisms, and blood loss.
3. The blood has a suspension stability.
   a. Cells may settle out of it or be centrifuged.
   b. The sedimentation rate reflects the stability of the blood.
4. Plasma is the liquid portion of the blood.
   a. It normally constitutes about 55 percent of the blood volume.
   b. It contains about 90 percent water, 1 percent inorganic substances, and 7 to 9 percent proteins.
   c. Plasma proteins include albumins, globulins, fibrinogen. Albumins contribute most of the colloidal osmotic pressure of the plasma; globulins are of several types and are important in immune reactions; fibrinogen is a material required for blood clotting.
   d. Plasma constituents are derived by absorption from the gut (water, salts, amino acids, simple sugars, fats) from synthesis in the liver (proteins) and from endocrines (hormones).
5. The formed elements are of three types.
   a. Erythrocytes (red blood cells) are biconcave discs, are the most numerous, and carry oxygen and carbon dioxide. They are formed in bone marrow, last 3 to 4 months in the bloodstream, and are phagocytosed when old; their components are recycled. Their production is controlled by availability of building blocks and chemical factors.
      1) Hemoglobin is a pigment in the erythrocytes that carries oxygen.
      2) There are several normal and abnormal hemoglobins. Type is genetically determined.
      3) Iron is essential to heme formation; its metabolism is described (see Fig. 18.5).
      4) Anemia refers to low hemoglobin concentrations.
   b. Leucocytes (white blood cells) are of five types.
      1) Granular leucocytes originate in bone marrow and include neutrophils, basophils, and eosinophils. Neutrophils are good phagocytes; the other two are mainly in transit and will become tissue cells.
      2) Nongranular leucocytes are produced primarily in lymph organs. Lymphocytes and monocytes are both good phagocytes.
      3) White cells protect and aid in wound healing.
      4) Production is controlled mainly by adrenal steroids and products of infection and inflammation.
   c. Platelets originate in bone marrow, are not nucleated, and carry one of the factors necessary for blood clotting.
6. Hemostasis refers to reactions that prevent loss of blood from blood vessels.
   a. Vascular responses (vasoconstriction) reduce blood flow to an injured area.
   b. Platelets form a plug across a rupture.
   c. Coagulation of the blood occurs; fibrin is formed.
      1) An extrinsic system provides fibrin starting with injured cells.
      2) An intrinsic system utilizes substances entirely within the bloodstream to produce fibrin.
      3) A clot, once formed, shrinks, and then is dissolved as healing proceeds.
   d. Anticoagulants remove one or more of the materials necessary for clotting, or prevent their production.
   e. Hemophilias are genetically determined disorders of clotting that result from failure to produce one or more of the materials essential for clotting.
7. Lymph
   a. Is formed by plasma filtration.
   b. Is the fluid that enters lymph vessels from the interstitial compartment.
   c. Is moved by muscular contractions and breathing movements.
   d. Carries tissue water and proteins back to the bloodstream.
8. Lymph nodes
   a. Produce nongranular white cells.
   b. Contain macrophages that cleanse the lymph.

QUESTIONS

1. In what ways is the blood an important vehicle for transport of materials?
2. Name five homeostatic functions in which the blood is involved.
3. What are the two major components of the blood? What percent of total blood volume does each constitute? What percent of the body weight?
4. What factors are involved in determining the sedimentation rate of the blood?
5. What are the major constituents of the plasma? Give at least one function for each constituent.
6. Where are the plasma proteins produced, what are the three main protein categories, and what function(s) does each category serve?
7. Where are erythrocytes produced before and after birth, and what functions do they serve?
8. Describe briefly the life history of a red cell.
9. What might happen, and why, if there was a substitution of an amino acid in the normal chains of adult hemoglobin?
10. Describe the metabolism of iron in the body.
11. What is urobilinogen? Where does it come from and how is it eliminated from the body?
12. What is the relationship of the liver to bilirubin formation and excretion?
13. What is anemia, and what are some of its causes?
14. What are the basic differences between the extrinsic and intrinsic schemes of blood coagulation?
15. What materials constitute the end products of the three phases of the clotting scheme?
16. In what ways does lymph differ from plasma? How do those differences arise?
17. What factors are involved in
    a. Entry of lymph into lymph vessels?
    b. Movement of lymph through lymph vessels?
18. What functions do lymph nodes serve?

READINGS

Erslev, A. J. *Pathophysiology of Blood.* Saunders. Philadelphia, 1975.

Peschle, Cesare, and Mario Condorelli. "Biogenesis of Erythropoietin: Evidence for Pro-Erythropoietin in a Subcellular Fraction of Kidney." *Science 190:*910. 28 Nov 1975.

Tullis, James L. *Clot.* Thomas. Springfield, Ill., 1976.

Weiss, H. J. "Platelets: Physiology and Abnormality of Function." *New Eng. J. Med. 293:*531. Sept 11, 1975. *293:*580. Sept. 18, 1975.

BODY DEFENSES AGAINST DISEASE; MECHANISMS OF PROTECTION, ANTIGEN-ANTIBODY REACTIONS, IMMUNITY, BLOOD GROUPS

OBJECTIVES

After studying this chapter, the reader should be able to:

☐ Outline the various mechanisms the body employs to protect itself against disease.

☐ Discuss the mechanisms by which the skin and external body surfaces protect themselves.

☐ Describe how internal body organs are protected.

☐ List the mechanisms of protection that are available within the blood and lymph streams.

☐ Give the purposes of the inflammatory response and the steps in the response.

☐ Describe how wound healing takes place.

☐ Define what the immune system is and what its functions are.

☐ Characterize antigens and antibodies.

☐ Define what an antigen-antibody reaction is, and give its purpose(s).

☐ Describe the roles of complement and properdin in the antigen-antibody reaction.

☐ Discuss current theories of how antibodies are produced.

☐ Explain the origin and functions of T- and B- lymphocytes.

☐ List some diseases against which immunity may be acquired, and methods of achieving that immunity.

☐ Explain the relationships of interferon and transfer factor to the development of immunity.

☐ Explain what allergy is and what the characteristics of the several types of allergic responses are.

☐ Explain what the problems of transplantation are on an immunological basis.

☐ Define what autoimmunity is and how it arises.

☐ Explain the genesis of immunodeficiency.

☐ Explain the origins of the ABO and Rh blood groups.

☐ Discuss the basis of antigen-antibody reactions that may occur with blood transfusion or pregnancy.

☐ Explain how blood types are determined.

☐ Outline the considerations involved in transfusion of blood or other fluids.

☐ List some possible transfusion reactions.

THE VARIOUS MECHANISMS OF PROTECTION

We are constantly faced, from without and from within, by factors that have a great potential to harm us. Radiations of various sorts assault our body from the sun and space in general; microorganisms on our external body surfaces and within several body systems that open on that surface are, in many cases, pathogenic; foreign substances enter the body in our food and drink or in the air we breathe; toxic substances are produced by normal metabolic activity and must be detoxified or eliminated before they reach critical levels.

The various devices the body utilizes to protect itself constitute its total defense capability, which is, as we shall see, considerable. This chapter indicates the scope or range of body defenses, and concentrates primarily on those dealing with immunity. Other mechanisms are considered in greater detail in appropriate chapters: airborne materials and defense against them are described in the respiratory chapter; protection against ingested pathogens and toxic products of metabolism is treated in the digestive system chapter, elimination of toxic substances is explained in the urinary system chapter.

Finally, because the blood groups represent examples of antigen-antibody systems, they are considered after the reader has an understanding of what antigens and antibodies are, and how they interact.

THE EXTERNAL BODY SURFACES

The skin is considered to be the largest organ of the body. It accounts for about 7 percent of body weight, and has a surface area of about 1.75 m^2 (about 3000 $in.^2$) in the average adult. It serves a variety of functions that may properly be considered protective.

Whole skin offers a MECHANICAL BARRIER not only to the entry of living organisms *into* the body, but also to the loss of essential body constituents *from* the body. One has only to have dealt with a burn victim to appreciate the potential for infection, dehydration, and loss of vital inorganic and organic constituents such a patient presents. The basis of the protection against entry of organisms resides in the KERATINIZED SURFACE LAYERS of cells of the skin, and in a SURFACE FILM that covers the skin. KERATIN is a fibrous protein that accumulates in the cells of the epidermis as they move from deeper layers of the skin toward the surface. Keratin is tough, insoluble in water, and resistant to the action of weak acids, weak alkalis, and most proteolytic (protein digesting) enzymes. The SURFACE FILM of the skin is composed of water, lipids, amino acids, and polypeptides derived from the secretions of the sweat and sebaceous glands and from the breakdown of the keratinized surface cells. It has a pH of 4 to 6.8 and forms an antiseptic layer that retards the growth of bacteria and fungi on the skin surface. The water and lipid content tends to moisten the skin and keep it from drying and cracking, since cracks would permit the entry of microorganisms into the body. CERUMEN is the brownish waxlike material secreted by the glands of the external ear. It also discourages the growth of microorganisms and is bitter to the taste, a fact that tends to discourage the entry of insects into the outer ear. TEARS contain *lysozyme* (muramidase), a bacteriostatic enzyme that protects the anterior surface of the eyeball.

At the junction in the skin of its nonkeratinized and keratinized layers is a double electrical layer of H ions externally and OH ions internally. This layer tends to limit the passage of charged substances in either direction through the skin. Excess body heat is largely lost via the skin; heat radiated from cutaneous blood vessels is utilized to evaporate body surface water. Lastly, mechanical attrition of the skin results in a thickening of the skin, further providing a pad or cushion (e.g., calluses) against wear-and-tear. Melanocytes, pigment-containing cells in the skin, provide variable amounts of pigment that can absorb solar radiation. Increase in their number accounts for tanning of the skin on exposure to sunlight.

THE INTERNAL BODY ORGANS

Organs that open upon the external body surfaces, such as the mouth, anal canal, nasal cavities, vagina, and urethra are protected by MUCOUS MEMBRANES. Mucus, produced by glands or cells in these organs tends to trap and remove potentially harmful substances. The vagina also contains much glycogen in its surface cells. The glycogen is broken down by the normal vaginal flora to organic acids that create an acid pH (about 5) that retards microorganism growth.

SALIVA has a pH of 5.8 to 7.1, an acid reaction that tends to retard microorganism growth as well as providing cleansing of the mouth surfaces. It is not uncommon for infection of the salivary glands to occur in dehydrated or feverish individuals as the mouth dries with diminished salivary secretion.

HYDROCHLORIC ACID, produced by the stomach, is a very effective bacteriolytic agent. pH in the stomach averages about 1.5, a strongly acidic environment.

Normal URINE is acidic and antibacterial, and moves away from the kidney so as to "wash" materials toward the external aspect of the body. The kidney itself, in the sense that it can excrete toxic materials (e.g. urea), forms an additional means of protection against accumulation of toxins.

The LIVER, by its ability to DETOXIFY or render harmless certain chemicals produced in the body or ingested, and by its many PHAGOCYTIC MACROPHAGES, offers further protection against chemicals and microorganisms.

The BLOOD, with its content of phagocytic cells and antibodies, forms a final line of defense against invaders.

INFLAMMATION AND WOUND HEALING

If the skin or a mucous membrane is penetrated by some agent, the area around the point of penetration typically becomes swollen, tender, and warm to the touch, reddened (hyperemic), and painful. These symptoms and responses are part of a response to injury known as the INFLAMMATORY REACTION. The purpose of such a reaction is twofold:

To destroy or neutralize the agent responsible for the reaction.

To facilitate repair of the injury.

The response is essentially the same regardless of whether the agent is a chemical, a radiation, a microorganism, or mechanical. Several steps typically follow the application of the injurious agent to the body.

There is a transient VASOCONSTRICTION of the blood vessels in the injured area that tends to localize the agent. The basis of such a constriction is in direct stimulation of the vessels themselves, in nervous reflexes associated with the pain of the injury, and in the release of serotonin by injured cells.

As the products of cell destruction reach the vessels, they DILATE. Histamine is a chemical, also released by damaged cells, that causes vasodilation. The area thus receives a greater blood flow (*hyperemia*), and more leucocytes are brought to the area to combat the invader and hasten repair. The area also becomes reddened and warmer as the result of the hyperemia.

LYSOSOME RELEASE from damaged cells provides enzymes to digest microorganisms and products of cell death, clearing the way for repair.

VENULES DILATE because of relaxation of their smooth muscle and CAPILLARIES DILATE and become more permeable due to relaxation of precapillary sphincters, allowing a greater volume of blood to enter the vessels. An *inflammatory exudate* is produced. "Clotting" of this exudate may "wall off" the injured area and slow the passage of noxious substances from the area to the bloodstream. The accumulation of exudate may also result in *swelling* of the injured area.

LEUCOCYTES MARGINATE (line up on or adhere) inside the capillaries in the injured area, and then they pass through (*diapedesis*) the vessel walls to join the battle against the agent(s).

PHAGOCYTOSIS of agents and dead cells sets the stage for replacement of lost cells by mitosis of remaining cells, or for healing by scar tissue formation.

How long the inflammatory process takes,

and its eventual outcome are influenced by several factors.

The *nature and intensity of the irritating agent.* If the invading agent is bacterial, it may produce an enzyme (hyaluronidase) that liquefies the gellike intercellular material and facilitates the spread of the agent. Recovery will obviously be slower if a larger body area is involved. Agents that cause deep or extensive injury will result in a longer healing process.

The *duration of the irritation.* Any given agent, if it lasts a long time, will cause more extensive destruction. BURNS, either from thermal, chemical, or radiation sources, may be classed as first, second, or third degree in severity. In a FIRST-DEGREE BURN, only the epidermis is involved and reddening with no blistering occurs. In a SECOND-DEGREE BURN, dermis and deeper tissues are involved but some epidermal remnants are always present in the burned area. There is blistering. In a THIRD-DEGREE BURN, bones, muscles, and tendons are usually involved and no epidermal remnants are present in the burned area. Ulceration is common. In general, the deeper the burn, the greater the protein, fluid, and electrolyte loss, and the greater the danger of (and from) bacterial contamination.

The *tissue affected.* In general, the more vascular the tissue, the faster the recovery. Tissues of younger people recover more quickly than older ones, presumably because of greater rates of mitosis and replenishment of cells and intercellular substance.

The *effectiveness of the defensive response.* Implied here is the ability of the white blood cells to get to the injured area, the effectiveness of phagocytosis, the ability to wall off the infection, and the ability to produce antibodies.

WOUND HEALING is said to occur by a three-step process, and serves to repair damage in organs that cannot repair themselves by regeneration of cells like the original ones. (Skin and liver of humans appear to be the primary organs that can regenerate themselves.) Using a cutting instrument on the skin as the example of trauma, the first step in would healing is an initial flow of blood into the gap created by the cutting instrument. This flow fills the space, and the clot formed unites the cut edges of the wound. As the clot dries and forms a scab, the inflammatory process occurs as the second step. Lastly, invasion by leucocytes occurs, and removal of debris begins; at the same time scar tissue (collagenous fibers produced by *fibroblasts,* connective tissue cells) is formed, and mitosis of epidermal cells occurs to create a surface like the original one. When this surface is nearly complete, the scab sloughs off. Figure 19.1 depicts these events on a day-to-day basis. The healing process is essentially the same regardless of the tissue involved, with the exception that epidermis usually forms a surface like the "old" one, while other tissues merely heal by scar tissue formation.

IMMUNE REACTIONS

The IMMUNE SYSTEM is defined as a diffuse set of organs containing about 10^{12} antibody molecules, and weighing about two pounds. More specifically, the organs involved include the thymus gland, lymph nodes, the spleen, the tonsils, all other aggregates of lymphoid tissue in the walls of body organs, the lymphocytes of blood and lymph, and all of the individual lymphocytes and plasma cells dispersed throughout the epithelial and connective tissues of the body. Its function, stated in general terms, is to PROTECT THE BODY FROM MACROMOLECULES that may enter it on viruses, microorganisms, or cells (as in an organ or tissue transplant), or which may arise from within the body. Central to this mechanism of protection is the ability of the immune system to distinguish between foreign macromolecules and those that are a natural part of the individual's body. The year 1955 is often taken as the time when "modern immunology" began, for it was in this year that the Salk polio vaccine became available. This marked a fusion of theoretical immunology and applied immunology that has opened the gateway to effective conquest of infectious disease.

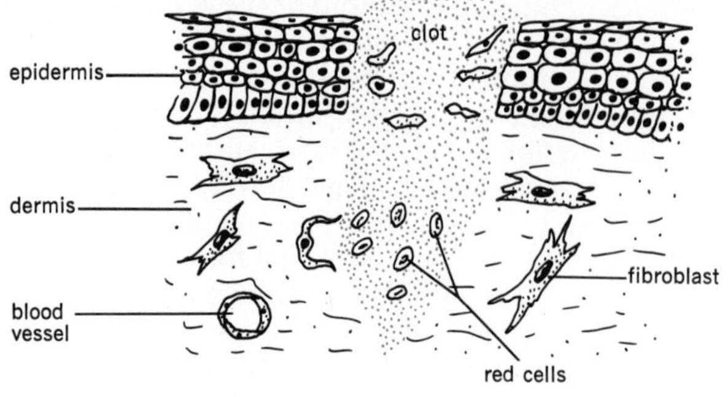

1. Gap fills with blood from damaged vessels

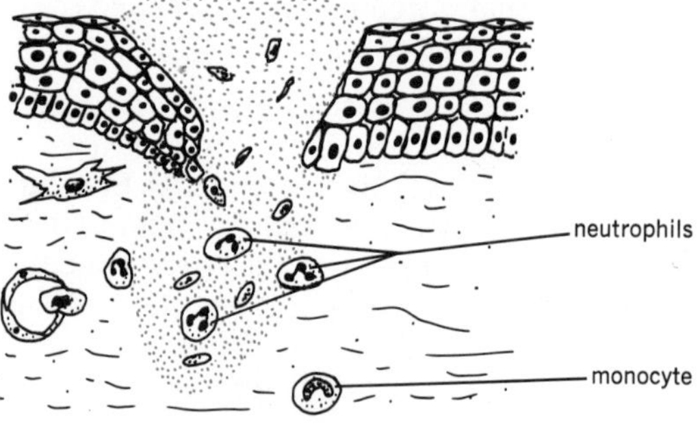

2. 1 day later, neutrophils enter wound and begin to clean up debris. Epidermal cells undergo mitosis and enter wound

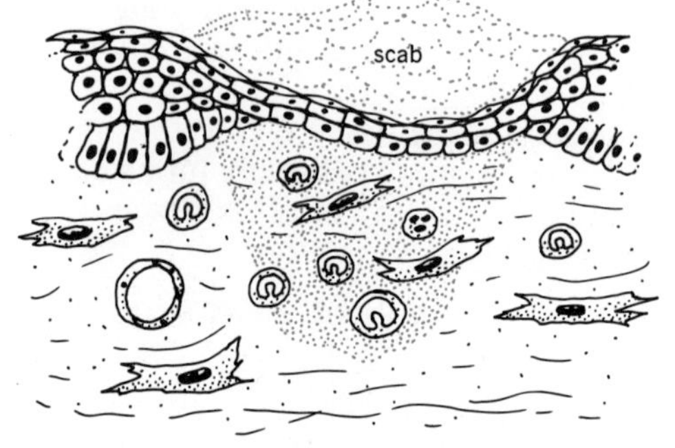

3. 2 days later, monocytes and fibroblasts enter wound. Debris is phagocytosed, new dermal fibers are formed, epidermal layer continuous across wound

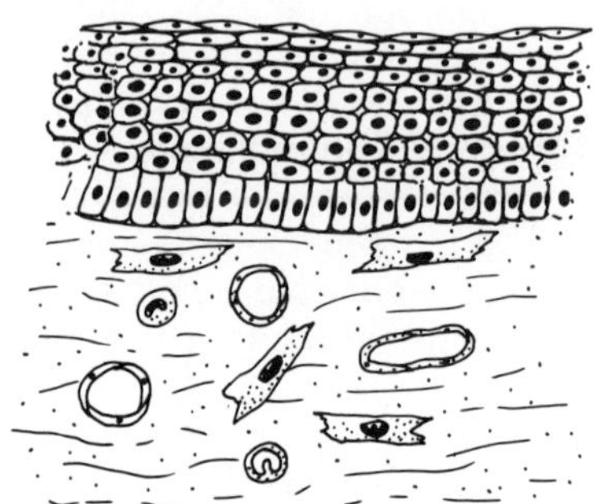

4. 7 days later, scab is gone, new epidermal layer formed, wound vascularized, new dermal material formed

FIGURE 19.1
The basic stages in wound healing. (Redrawn from "Wound Healing," by Russel Ross. Copyright © June, 1969, by Scientific American, Inc. All rights reserved.)

ANTIGENS AND ANTIBODIES AND THEIR INTERACTIONS

DEFINITIONS AND CHARACTERISTICS

An ANTIGEN is typically described as a large molecule (a *macromolecule*) with a molecular weight of 10,000 or more. It may be part of a virus, bacterium, or a foreign tissue cell, or some part of such structures. Antigens may also arise from chemical changes that occur within our own bodies, changes that, for example, alter the sequences of amino acids in a protein of the body. Chemically, they may be proteins, polysaccharides, nucleic acids, or combinations of these.

Each antigen has a complicated shape and somewhere on its surface is a sequence of 3 to 10 amino acids or other molecules that constitutes that molecule's EPITOPE or *antigenic determinant.* It is the epitope that an antibody "recognizes" and that leads to a reaction between the two. An immune system is capable of recognizing epitopes on any protein or other antigen produced by any of the millions of

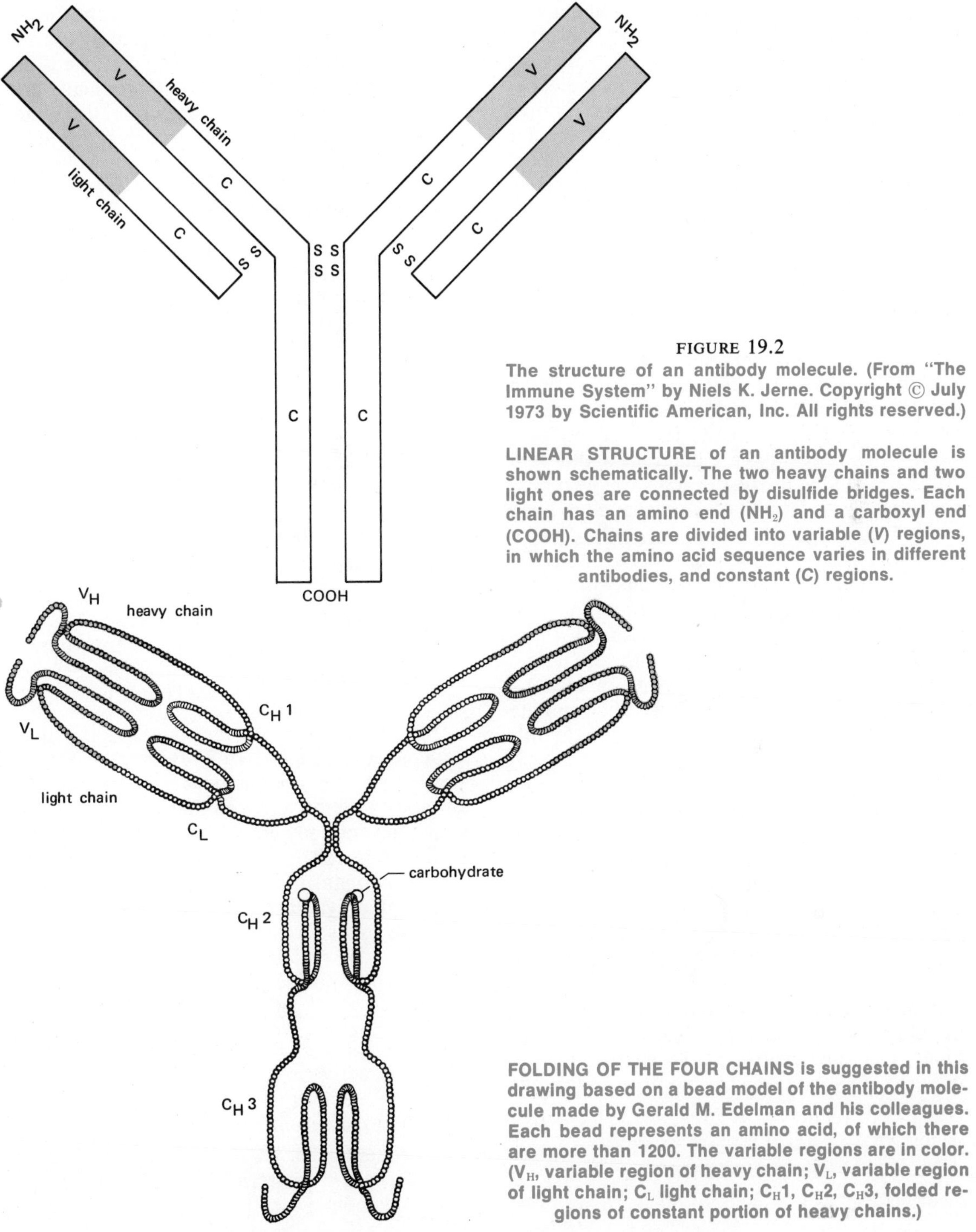

FIGURE 19.2

The structure of an antibody molecule. (From "The Immune System" by Niels K. Jerne. Copyright © July 1973 by Scientific American, Inc. All rights reserved.)

LINEAR STRUCTURE of an antibody molecule is shown schematically. The two heavy chains and two light ones are connected by disulfide bridges. Each chain has an amino end (NH_2) and a carboxyl end (COOH). Chains are divided into variable (*V*) regions, in which the amino acid sequence varies in different antibodies, and constant (*C*) regions.

FOLDING OF THE FOUR CHAINS is suggested in this drawing based on a bead model of the antibody molecule made by Gerald M. Edelman and his colleagues. Each bead represents an amino acid, of which there are more than 1200. The variable regions are in color. (V_H, variable region of heavy chain; V_L, variable region of light chain; C_L light chain; C_H1, C_H2, C_H3, folded regions of constant portion of heavy chains.)

plants, animals, or microorganisms, but can distinguish foreign epitopes from those that compose its own body.

An ANTIBODY is itself a protein molecule, belonging to that group of body proteins known as the gamma globulins. It has a molecular weight of 150,000 to 900,000 and is produced by PLASMA CELLS in response to challenge by a particular antigen. An antibody is made up of four polypeptide chains (Fig. 19.2), two identical *light chains* and two identical *heavy chains.* A light chain contains about 214 amino acids and a heavy chain about twice as many. In most of the molecule, the amino acid sequences are the same in different antibodies, but in both the light and heavy chains, extending over an area of about 50 amino acids, there are what are termed VARIABLE REGIONS. In these regions, amino acids sequences are different in different antibodies, and at the tip of each variable region is a COMBINING SITE that has a shape enabling it to recognize a particular antigen's epitope and make the antibody stick to the molecule having that epitope. Inherent in this discussion is the possibly startling consideration that if at each of the 50 positions of both chains there was a choice of having just two different amino acids, there would be a possibility of forming 2^{100} different molecules. In short, the immune can potentially produce *at least* that many different antibody molecules!

Interaction between an antibody and the particular antigen that called forth its production constitutes the ANTIGEN-ANTIBODY REACTION. The reaction is designed to neutralize or remove the antigen as a threat to the body and may take a variety of visible or nonvisible forms, including: *neutralization,* in which the effect of the antigen is removed; *precipitation,* in which an aggregate is formed, settles out, and is phagocytosed; *agglutination,* in which cells are clumped together in masses; or *lysis,* in which cells are destroyed. In each case, the antigen's epitope attaches to the antibody's combining site.

THE COMPLEMENT SYSTEM AND PROPERDIN

In order for lysis of cells to occur, a series of serum factors collectively called COMPLEMENT must be present. There are 11 proteins in the complement system and, when activated, enzyme activity is developed that eventually creates holes in the outer surfaces of cells. Water and ions pour into the cell through these holes and the cell swells and ruptures.

More specifically, the steps in the complement reaction are as follows:

An antibody molecule of the IgM or IgG series has sites on it (combining sites) that enable it to bind to epitopes on antigens. The antibody also has binding sites on it for complement C1q, the largest (molecular weight 400,000) of the complement molecules.

C1q binds to sites on the antibody.

Molecules of another member of the system, designated C1s, are activated, and they activate the C4, C2, and C3 molecules, and it is believed that the latter three combine to form yet another enzyme.

This new enzyme cleaves C5, and C5 combines C6 and C7, the entire assembly binding to sites on the cell membrane. The cell membrane is rendered more permeable to salts, water, and other small molecules that flow into it. The cell swells and bursts.

The generalized scheme of these reactions is depicted in Figure 19.3. It may be emphasized that the complement system *requires* antibody as a marker, and that the antigen with which the antibody combines is generally on the surface of a cell.

The PROPERDIN PATHWAY is an alternative means of activating C5 and may not, according to some investigators, require antibody to initiate the lytic reaction. It has thus been termed a more nonspecific type of protective mechanism. In this pathway the following steps occur:

Enzymes activating C3 and C5 are produced by the reaction of bacterial polysaccharides with blood serum precursors.

The properdin enzymes then assemble on the surface of a bacterial cell and activate the complement attack sequence. It may thus be that the properdin pathway can be activated by antibodies other than those of the IgM–IgG categories, or that no antibodies at all are required. Only the antigens, represented by the bacterial polysaccharides, may be required to initiate cell destruction.

THEORIES OF ANTIBODY PRODUCTION

Two general categories of theories have been advanced to explain how antibodies are produced against the literally millions of antigens the body may encounter during its lifetime.

FIGURE 19.3

An outline of complement activation leading to lysis of cells. (Redrawn from *Scientific American*, "The Complement System" by M. M. Mayer, Nov. 1973.)

Legend
(a) C1q binds to IgG antibody. When bound, it becomes enzymatically active.
(b) C4 approaches, is split and part of it (C4b) binds to adjacent cell membrane.
(c) Bound C4b is shown; C2 contacts activated C1 complex, and is split; part of it (C2a) binds to C4b.
(d) C3 and C5 are split by activated C1, part of their molecules attach to complex developing on cell membrane.
(e) Other molecules are split and join membrane complex.
(f) A hole (shaded area) develops in membrane; ions flow, water enters, cell swells and breaks.

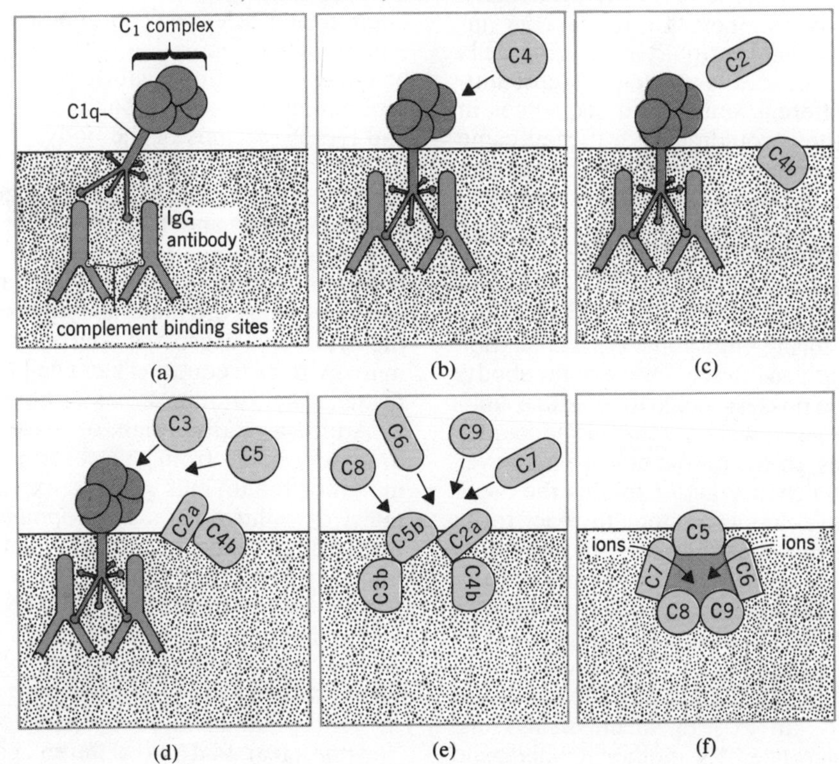

The INSTRUCTIVE (*template*) THEORY suggests that the antigen acts as a form around which a "standard" unfolded gamma globulin molecule could be molded to give a certain shape that was a mirror image of the antigen at a certain point along its course. The shape achieved using the antigen as a template would then be retained by chemical bonding within the folds. The antibody would then separate and possess a binding site for other antigen molecules of the type that initially caused it to fold in a particular configuration. According to this theory, antibody activity is not present until the antigen enters the body and causes the globulin chain to fold into a particular configuration. This theory also does not allow for duplication of particular antibodies, but only the "standard" molecule.

The SELECTIVE (*modified clonal*) THEORY, on the other hand, suggests that the capacity to synthesize different antibodies is already present in the genetic apparatus, and that contact with an antigen "switches on" the production of that antibody. Support for this theory includes the facts that: antibody-producing cells *do not* have demonstrable antigen content; antibodies have different amino acid sequences in them (which they wouldn't have if they came from a standard molecule); and, mutations can produce a loss of ability to synthesize *particular* antibodies. Furthermore, the selective theory postulates *a separate cell* or group of cells (*a clone*) for the production of each antibody. These clones have arisen by evolution and mutation during embryonic cell replication before any antigens have entered the body. Thus, we may possess cells to produce antibodies to antigens we may never encounter during our lives, or to combat any we may ever encounter. This theory also explains the usual inability of the immune system to react to its own body chemicals as they are produced during development as the result of clone destruction by an antigen-antibody reaction as new chemicals appear. An alternative hypothesis to explain "self-tolerance" is the recently advanced idea that chemicals may be produced that "shut off" an existing mechanism. Thus, the normal tolerance of a mother for her child *in utero* (who contains chemicals foreign to *her* body coded by the father's genes) may be explained.

The "true" explanation of how antibodies are produced will only be unraveled by continuing research on this important question. Genetic bases seem to have the edge at the moment, as will be further discussed in the section on immunodeficiency later in this chapter.

THE IMMUNE RESPONSE

DEVELOPMENT OF THE IMMUNE SYSTEM

Two distinct but interdependent systems of immunity protect the body from antigenic challenge. One is designated the CELL-MEDIATED IMMUNE RESPONSE that combats fungi, initial invasion by viruses, and initiates the rejection of foreign tissues. In this type of response "sensitized" lymphocytes are produced that have antibody-like molecules on their surfaces, and the whole cell must contact an antigen. The other is HUMORAL IMMUNITY, achieved by production and release of antibodies by modified lymphocytes known as *plasma cells,* which provides protection against bacteria and virus reinfection. These antibodies enter the blood and lymph streams of the body.

The basis for such division of labor lies in the development of two different cell lines during embryonic and fetal life. These cell lines arise from a common stem cell, but acquire different potentials as development proceeds. Both cell lines originate as lymphocytes within the lymphoid organs of the body and in bone marrow from a common stem cell that gives rise to them and to other types of blood cells.

A clue as to the origin of these two types of immunity came from the observations that removal of the thymus gland in experimental animals, or failure of its development in humans resulted in defective development of the immune system. Such conditions resulted in defective cell mediated immunity, but did not always change humoral immunity. Later it was shown that removal of a lymphoid organ in birds, known as the bursa of Fabricius, resulted in depression of antibody production. It thus became clear that two different populations of lymphocytes were produced by the body, and that each mediated different types of immune response. These two types of lymphoid cells were subsequently designated as T-CELLS (from thymus) and B-CELLS (from the bursa, or its mammalian equivalent, possibly the tonsils, appendix, bone marrow, or lymph nodes). The cells can be distinguished from one another by certain surface markers similar to the antigens of the blood groups.

Stem cells giving rise to T-cells actually originate in the yolk sac early in embryonic development, and then migrate to the developing thymus where they undergo rapid proliferation. As the cells mature, they are "endowed" with T-cell characteristics either by contact with the epithelial cells that form the structural framework of the gland, or by hormones (e.g., thymosin) produced by such cells. T-cells are involved in cell-mediated immunity. If confronted with an antigen, T-cells are activated, they enlarge, divide, and produce and release molecules called *lymphokines*. The latter participate in the destruction of the foreign antigen, and in the attraction of macrophages to a battlefield. T-cells may also release chemicals (also included in lymphokines) that stimulate B-cell production of antibodies. T-cells leave the thymus by way of the bloodstream and pass through various body organs, perhaps encountering other antigens. If such a contact is made, the T-cell will be stimulated to divide and gives rise to a clone of cells responsive to that antigen. They next enter the lymphatic vessels and are returned to the bloodstream for further circulation throughout the body. This T-cell "circulation" increases the probability that they will encounter foreign antigens or malignant cells.

B-cells originate in the bursa or its mammalian equivalents, and differentiate into plasma cells—cells capable of synthesizing antibodies and releasing them into the bloodstream or retaining them on the cell. Such cells are the basis of the body's humoral immunity. The stem cells are thought to migrate to bursal-type areas from the yolk sac and are there induced to become B-cells. Thymic hormone may be necessary to assure the maturation of B-cells to plasma cells or to stimulate antibody production. As the cells mature, they produce either IgM (first), IgG (second) or IgA (third). It appears that one cell can produce all three Ig's, as long as it remains in the bursa but becomes committed to the production of one type of Ig on leaving the bursal areas. Antigens induce the maturation of B-cells either by direct interaction with antibodies on the B-cell surface, or through the intermediary of T-cells via macrophages.

The relationships of T- and B-cells are depicted in Figure 19.4.

The end result of these cellular differentiations is to give the body the ability to respond, by some type of reaction, to any type of foreign substance that may enter or be produced within the body.

TYPES OF IMMUNITY AND DURATION OF PROTECTION

The term immunity is often understood to refer to the PROTECTION achieved against second exposures to certain antigens or the microorganisms that carry them after having initially contacted the antigen. Any method of achieving immunity obviously requires antibody or cell-mediated responses. Those antibodies that remain on or in cells last longer than those that are freed into the bloodstream and may give lifelong immunity, for they avoid "wear-and-tear" of circulation and possible metabolism or excretion. Humoral or circulating antibodies to diseases may require periodic "boosters" of antigen to retain adequate blood levels.

Some of the methods by which immunity may be achieved include the following:

The use of LIVE, ATTENUATED (weakened) ORGANISMS that retain the power to induce antibody formation but that do not usually cause clinical disease (e.g., polio, measles).

The use of SIMILAR ORGANISMS to provoke the production of antibodies similar to or identical with those occasioned by another organism (e.g., cowpox for smallpox).

The use of KILLED ORGANISMS that cause no

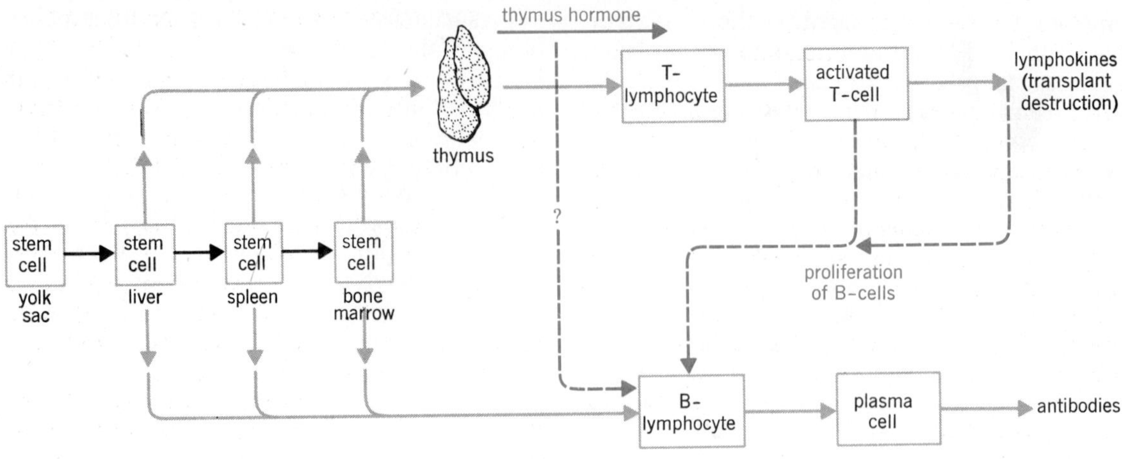

FIGURE 19.4
Development and interrelationships of B- and T- cells. (see text for explanation).

disease but induce antibody formation (e.g., polio, typhoid).

The use of TOXOIDS, which are toxins (harmful chemicals) that have been modified to remove their toxic effect but not their ability to stimulate the production of antibodies (e.g., diphtheria, tetanus).

The injection of pooled HUMAN GAMMA GLOBULIN, which may contain antibodies to a variety of antigens.

The use of ANTITOXINS, antibodies to specific toxins or organisms (e.g., antitoxins to rattlesnake venom).

There are several "types" of immunity.

TISSUE *(cellular)* IMMUNITY. Some antibodies, once produced, remain within the plasma cell. Such antibodies may be retained for long periods of time and may confer lifelong immunity in some cases; for example, measles.

HUMORAL IMMUNITY. Other antibodies, rather than remaining in the plasma cells, are secreted into the bloodstream. Here they are subject to destruction or elimination, and thus do not usually confer immunity for extended periods of time. Examples include immunity to diphtheria and tetanus, which both require periodic "boosters" to restimulate antibody production.

ACTIVE IMMUNITY. This results when the individual contracts a disease or is given a mild case of it by vaccination. He or she responds by production of an antibody.

PASSIVE IMMUNITY. This occurs when an individual is given an antibody from another person or animal by injection. It gives immediate protection, such as its use in rattlesnake bite or exposure to the tetanus organism.

PASSIVE-ACTIVE IMMUNITY. An injection of antibody combined with sufficient antigen to trigger antibody production. It provides immediate protection and provokes long-range protection.

ACTIVE-PASSIVE IMMUNITY. Antibodies are given at the time of exposure to reduce the severity of the disease, and contracting the disease itself confers immunity. An example is giving gamma globulin injections to a child who is incubating measles.

VACCINATION is a method widely used today to induce immunity to a variety of diseases.

Injection or the introduction into the body of dead antigenic bacteria, attenuated live organisms, or some of their chemical products is utilized to trigger antibody production with resultant immunity. Table 19.1 shows some of the diseases against which we may be protected by vaccination.

In the development of active immunity, a specific time period must elapse between the arrival of the antigen and the production of sufficient antibody to have an immunizing effect. It usually requires one to two weeks after the first presentation of an antigen for significant quantities of antibodies to appear. On a second exposure to the same antigen, antibodies appear within days, reach higher concentrations, and last longer than on a first exposure. This "secondary response" is the basis for boosters and maintains immunity to specific diseases.

INTERFERON AND TRANSFER FACTOR

Additional protection against viral infection is provided by a nonspecific antiviral agent known as INTERFERON. The contact of any virus with a cell results in cellular production of the material that is then excreted from the cell. Interferon prevents the biosynthesis of new virus particles, possibly by blocking cellular m-RNA synthesis of the needed proteins or by blocking transcription of m-RNA itself that would lead to production of viral RNA. Uninfected cells are also protected by interferon, perhaps by preventing viral attachment to cell membranes. It would appear that interferon could be used as an agent to treat nonspecific viral diseases, such as the "common cold" or "flu." Things that have prevented such use include difficulty in purifying it, difficulty in obtaining quantities sufficient to make it generally available for use,

TABLE 19.1

SOME DISEASES WITH METHODS AND DURATION OF PROTECTION

DISEASE	METHOD USED	DURATION OF PROTECTION
Measles Rubella Rubeola	Attenuated organism	Expected to be life[a]
Poliomyelitis	Attenuated or killed organism	After a series of injections or oral administrations, boosters are required about every 10 years
Pertussis (whooping cough)	Killed organism	About 10 years
Smallpox	Similar organism toxoid	3–10 years
Diphtheria	Toxoid	About 10 years
Tetanus	Toxoid Antitoxin (for exposure)	To 10 years after first booster or if penetrating injury of skin
Mumps	Attenuated organism	Expected to be life[a]
Typhoid	Killed organism	3 years (minimum)
Influenza	Attenuated organism	1 year for a given strain

[a] Vaccines have been available too short a time to assess their protection for life; they presently produce a lasting effect.

and the fact that animal interferon is not effective in humans (species specificity).

TRANSFER FACTOR is an antibody-like material with a molecular weight of less than 10,000 whose structure has not been elucidated. It may be isolated from the white cells of individuals who have acquired resistance to a particular disease and, when given to victims of that same disease, may cause dramatic amelioration or "recovery" from the disease. Cytomegalovirus (a virus with similarities to the herpes virus that causes "fever blisters" and certain venereal diseases) that can cause a type of encephalitis with brain damage, fungus infections, leprosy organisms, and Valley Fever (coccidioidomycosis) have all been shown to respond favorably to transfer factor administration. Trials of the material on multiple sclerosis, rheumatoid arthritis, hepatitis, and certain types of cancers have yielded promising though sometimes equivocal results. Characterization and further research are needed to clarify the roles of transfer factor in the body and as a possible therapeutic agent.

ALLERGY

Allergies represent an incomplete or abnormal reaction to an antigen. They are typically chronic and occur in individuals who have a hereditary predisposition to develop allergies. An antigen may trigger incomplete or partial antibody production, not enough to confer immunity. Also usually associated with allergy is some type of reaction that results in tissue damage and the liberation of histamine. To HISTAMINE RELEASE may be attributed many of the symptoms of allergy such as watery nasal discharge, hives, and edema. It also accounts for the waxing and waning nature of the symptoms and the shift from one organ or system to another. The use of ANTIHISTAMINES in relief of allergic symptoms is based on this consideration. Three types of allergies are usually recognized.

PRECIPITIN ALLERGIES. These occur when enough antibody is produced to react with an antigen and cause precipitation of the antigen-antibody complex. Because most of these reactions occur between a tissue antibody and the antigen, severe tissue damage results and may cause the following conditions.

Anaphylactic shock. Very low blood pressures, weak heart action, edema, and circulatory collapse are the chief symptoms of anaphylactic shock. They are thought to be due to release of cellular substances (histamine, serotonin, bradykinin) when the antigen combines with tissue-fixed (cellular) antibodies. The condition develops very rapidly and gives little time for effective treatment. The "penicillin reaction" exhibited by some individuals is an example of an anaphylactic reaction.

Anaphylaxis may be passively transferred in the serum from the allergic individual to a normal person as in a transfusion. After a time lapse, contact by the normal individual with the antigen responsible for the shock in the allergic person may cause shock in the normal individual.

Serum sickness. Similar in symptoms to anaphylactic shock, this develops more slowly (one to two weeks) and is more amenable to treatment. It results when serum from a foreign animal is injected into a human.

Arthus reaction. A local inflammatory reaction appears at the site of contact with the antigenic substance.

DELAYED-REACTION ALLERGIES. Exemplified by the reactions of the tuberculin skin test, and the reactions to poison oak and ivy, these allergies cause initial sensitization to an antigen, but a second exposure is required to get a reaction.

ATOPIC ALLERGIES. These include hay fever, bronchial asthma, and dermatitis. They result from production of incomplete antibodies (reagins or IgE). Atopic allergies differ from most allergies in that they are inherited, the antibody is deposited in cutaneous (skin) tissues and may enter the bloodstream, and edema is the primary reaction that appears. Symptoms include bronchiolar muscle spasm, pruritus (severe itching), irritation of the nasal mucosa with nasal discharge, and urticaria (hives).

Allergic respiratory disease represents one of the most common types of allergies in children. Allergic rhinitis (nasal discharge), allergic bronchitis, and bronchial asthma are the most common. The interference with breathing in asthma is due to bronchiolar muscle spasm. Increased mucus secretion and edema may exaggerate the blockage of air movement.

Many types of allergies, particularly atopic allergies, may be treated by giving the individual regular doses of the antigen into the skin in an attempt to build up antibody production in the process of *desensitization.* Specific IgG or blocking antibodies (nonallergic, not tissue fixed) are increased, and symptoms are relieved by removing the allergen before it can combine with the tissue-fixed antibody.

TRANSPLANTATION AND TISSUE REJECTION

The replacement of defective or worn out organs of the human by animal or other human organs and thus to prolong life remains one of the methods available for extending the human lifespan. The major obstacle to such heroic measures has been, and continues to be, the antigenic nature of the tissues transplanted between different individuals. Previous sections of this chapter have indicated that cells carry surface molecular markers that specify individuality even more precisely than do our fingerprints. Only in the case of identical twins, who arise from a single egg, is there genetic programming to avoid a triggering of an immune response when tissues are introduced from one body into another. The events that ensue when a tissue or cell that is "not self" is introduced into a body in which it did not originate constitute the REJECTION PHENOMENON, and usually culminate in destruction of the introduced substance or organ.

The rejection phenomenon is common to all classes of vertebrates and may represent a mechanism primarily designed to prevent parasitism, and reduce the spread of malignant disease and its transference by sexual means.

Accordingly, if transplantation is to become a "routine" procedure, methods of reducing the antigenicity of the tissues of the donor or of reducing the immune response of the recipient must be achieved. Some of the methods employed are:

X-RAY OR GAMMA RADIATION kills the cells responsible for antibody production. Radiation is, however, indiscriminate in terms of cell destruction and may damage bone marrow and lymphoid organs and lower resistance to disease to a point where ordinary stressors may prove fatal to the recipient.

TISSUE MATCHING, utilizing lymphocytes in a type of cross match, determines how close the antigens of the donor match those of the prospective recipient. The closer the match, the less treatment will be required to combat the rejection.

DRUG THERAPY, using cortisonelike materials, is utilized to reduce inflammatory reactions arising from the transplant.

ANTILYMPHOCYTE SERUM, prepared by injecting human lymphocytes into horses, is used to react with human lymphocytes in the recipient. If the cells are destroyed, they cannot become plasma cells and thus produce antibodies.

LYMPH, collected from the lymph ducts, may be TREATED TO REMOVE WHITE CELLS from the fluid, and the fluid is returned to the recipient. The lymphocytes are thus not available as potential antibody producers.

DEVELOPMENT OF TOLERANCE TO FOREIGN ORGAN ANTIGENS through the injection of small doses of these antigens over long periods of time may reduce the necessity for any external therapy and is one of the most promising avenues of investigation currently in vogue.

CULTURING OF TISSUES BEFORE TRANSPLANTING has recently been shown to result in a higher degree of "takes" by transplanted tissue. Skin, adrenals, and other organs have been cultured. It has been suggested that the tissues revert to an embryonic state on culturing, and are not antigenic, thus they do not stimulate antibody production.

Transplants that are AUTOLOGOUS, that is, from one part of a person's body to another or from one identical twin to another, are genetically identical and do not stimulate antibody production. HOMOLOGOUS transplants, from one individual to another member of the same species, are less successful because of rejection. HETEROLOGOUS transplants, from another animal to a human are least successful.

AUTOIMMUNITY

The production of antibody and an ensuing reaction between that antibody and *one of a body's own components acting as an antigen* is the basis for AUTOIMMUNITY ("reaction to self"). Substances comprising the normal body constituents may be altered chemically as a result of disease processes and become an effective antigen, or there may be a genetic or acquired fault in the immune system that causes it to fail to recognize molecules as "self."

Today, it is useful to categorize autoimmune disorders into four basic types.

MONOCLONAL, in which only a certain type of cell produces antigen and it (or they) become the "target" of the antigen-antibody reaction.

INFILTRATIVE, in which specific organs are invaded by T- and/or B-cells and a reaction occurs.

Those that may involve as yet obscure but more deepseated ABNORMALITY of the immune system, such as abnormal T- or B-cells.

Those involving COMBINATIONS of the other categories.

Monoclonal autoimmune disease is exemplified by *idiopathic thrombocytopenia* (low blood platelet levels), and by *thyrotoxicosis* (a disease of the thyroid).

In idiopathic (a disease without obvious cause) thrombocytopenia, about 85 percent of the cases are associated with the appearance of a platelet autoantibody in the bloodstream that destroys platelets and their precursor cells in the bone marrow. The cause of the appearance of the antibody is not known.

In thyrotoxicosis, a gamma globulin designated LATS (*l*ong *a*cting *t*hyroid *s*timulator) is produced, possibly as a consequence of thyroid disease. It acts to chronically stimulate thyroid synthesis and release of thyroxin, its hormone.

In infiltrative disease, such as Hashimoto's thyroiditis there is extensive infiltration of the thyroid gland by (primarily) B-cells, and an antibody to thyroid tissue appears in the bloodstream. The thyroid may be extensively damaged by the reaction between its cells and the antibody.

Systemic lupus erythematosus (SLE) represents a chronic, inflammatory, autoimmune disorder in which abnormal lymphocytes may be produced. It affects the skin, joints, kidneys, nervous system, and other tissues. Antibodies to nuclear components—particularly DNA—are present, and it has been suggested that the disorder requires T-cells that are highly resistant to destruction and that stimulate B-cells to produce high amounts of antibodies.

In any event, autoimmune processes result in destruction of essential body structures with often toxic and possibly fatal results.

IMMUNODEFICIENCY

IMMUNODEFICIENCY is a term referring to genetic or drug-induced deficiencies in the function of the immune system. It results in inadequate response to antigenic challenge with resulting infection by pathogens to which most of us can normally acquire resistance.

Causes of such deficiency would seem obvious:

B- or T-cells are abnormal and cannot carry out their functions properly.

Abnormal or missing Ig's result from mutations in the genetic apparatus.

Among the examples of immunodeficiency may be described the following.

A (or *hypo*) *gammaglobulinemia* refers to lack or low blood levels of gamma globulins. Failure of B-cells to transform into plasma cells, or failure of the thymus to develop T-cells or produce its hormone may account for the loss of immunity. These disorders represent genetically determined abnormalities. Acquired disorders may result from use of drugs in transplant and cancer cases. Such drugs inhibit the cell divisions of cancer cells, and,

TABLE 19.2
ANTIGEN-BLOOD GROUP RELATIONSHIP AND FREQUENCY

ANTIGEN PRESENT	BLOOD GROUP (SAME AS ANTIGEN)	FREQUENCY OF POPULATION % WHITE	% BLACK
A	A	40.8	27.2
B	B	10	19.8
AB	AB	3.7	7.2
Neither A nor B	O	45.5	45.8

indeed, any rapidly dividing cells such as lymphocytes that are responding to the antigenic challenge provided by cancer cells.

Several cases of children born with severe immunodeficiency have recently been described. Isolation in "germ-free" quarters may be required to prevent contact with common pathogens that could have fatal effects. Use of transfer factor and embryonic thymus transplant has, in some cases, reduced the severity of this disorder.

THE BLOOD GROUPS

On the surface of the erythrocytes are found genetically determined chemical substances that act as antigens. Because they are substances unique to the bloodstream, these materials are designated as ISOANTIGENS, ISOAGGLUTINOGENS, or AGGLUTINOGENS. Two main categories of substances are recognized: one group is designated as the ABO group, the other as the Rh system.

THE ABO GROUP

Two basic isoantigens are involved, designated **A** and **B.** With two antigens, four possible combinations may exist. The cell may have one or the other, both, or neither antigen on its surface. The particular antigen present determines the blood group. The relationships of antigens to blood group and frequency of occurrence are shown in Table 19.2.

Genetically, **A** and **B** inheritance is dominant to **O,** and each is the result of two allelic genes (on chromosomes, occupying the same locus, and controlling the same characteristics). Therefore, the following genotypes could exist.

BLOOD GROUP	GENOTYPE
A	AA, Ai
B	BB, Bi
AB	AB
O	ii

where

A = groupA } codominant and dominant to group O
B = group B }
AB = group AB
i = group O – recessive to both A and B

The plasma contains ANTIBODIES *(isoantibodies, isoagglutinins,* or *agglutinins)* that correspond to the antigens described above. They are designated by the terms **a,** *Anti-A,* or *alpha* and **b,** *Anti-B,* or *beta.* Agglutinins are not present at birth. They are gamma globulins whose production is believed to occur between two and eight months of age in response to **A** and/or **B** antigens taken in foods, particularly meats. In the blood of any one given individual, *the antibody present is the reciprocal of the antigen.* Recall the clonal hypothesis, which states that the cells capable of forming corresponding antibodies to the blood antigens are destroyed embryonically, avoiding a reaction as new antigens are produced. The setup of both antigen and antibodies in the various blood groups would thus be as follows:

BLOOD GROUP	DOMINANT ANTIGEN(S)	ANTIBODY
A	A	b
B	B	a
AB	AB	none
O	none	ab

The importance of the blood groups becomes apparent when one considers the necessity of transfusing blood between one individual and another. It must be remembered that the *mixing of red cells having a given antigen with plasma containing a corresponding antibody will result in an antigen-antibody reaction.* The bloods must "match" or correspond as far as the antigens are concerned if such a reaction is to be avoided. The bloods may be TYPED and CROSS MATCHED to determine the compatibility or "likeness" of blood given during transfusion. Typing for the **ABO** system involves placing pure agglutinin (**a** or **b,** available commercially) on a microscope slide, mixing with it the unknown blood, and watching for a visible antigen-antibody reaction. It is important to remember that when typing the blood one looks for a reaction, while in the actual transfusion one avoids the possibility of a reaction. When inspecting the slide of blood and antibody, one should ask: What had to be present on the erythrocytes of the unknown blood in order to give an antigen-antibody reaction with a known antibody? The following chart summarizes the possible combinations that may result (plus indicates a reaction; minus indicates no reaction).

ANTIBODY		GROUP AND ANTIGEN PRESENT IS
a	b	
+	−	A
−	+	B
+	+	AB
−	−	O

In a cross match, donor's erythrocytes are mixed with recipient's plasma ("major side"), and recipient's erythrocytes are mixed with donor's plasma ("minor side"). Because the antigens are carried on the erythrocytes, and there may be thousands of antigen molecules on one red cell, if there is a likelihood of a reaction, it will most probably occur on the major side. Also, if small quantities of blood are transfused, antibodies in the donor's plasma are usually diluted by the recipient's blood volume.

The O type blood has been called a "universal donor" because there are no antigens on the donor's erythrocytes, and therefore a major side reaction cannot occur. However, O type blood has both antibodies, and a minor side reaction is possible. Similarly, AB individuals are called "universal recipients," because their plasma has no antibodies to react with AB antigens. Again, a major side reaction won't occur, but a minor side one can.

THE Rh SYSTEM

This system is composed of three allelic genes for each type. The dominant genes are designated conventionally as **CDE,** and the recessives as **cde.** Eight different genotypes are possible and are typed into two categories as **Rh** POSITIVE and **Rh** NEGATIVE. In general, if the individual has a large **D** gene, he will be **Rh** POSITIVE (about 85 percent of Caucasian Americans, 88 percent of black Americans); lack of a **D** gene results in **Rh** NEGATIVE (about 15 percent of Caucasian Americans, 12 percent of black Americans). For example:

$$\left.\begin{array}{l}\text{CDE}\\ \text{cDE}\\ \text{cDe}\\ \text{CDe}\end{array}\right\}\ \text{Rh}^{+}\qquad \left.\begin{array}{l}\text{Cde}\\ \text{CdE}\\ \text{cdE}\\ \text{cde}\end{array}\right\}\ \text{Rh}^{-}$$

An individual who is **Rh** POSITIVE has the antigen on the erythrocyte and has no corresponding antibody in the plasma. An individual who is **Rh** NEGATIVE has no antigen on the cell and, similarly, has no antibody in the plasma. However, the **Rh** NEGATIVE person has never had the clones challenged and still possesses the ability to form antibody if positive cells ever enter the body. There are two general sets of circumstances under which **Rh** POSITIVE cells might enter the body of an **Rh** NEGATIVE individual.

The first is TRANSFUSION. Blood is typed for the Rh factor in the same manner as for ABO factors, that is, by placing serum with Rh antibody on a slide, adding blood to it, and watching for a reaction. A reaction means Rh positive and the antigen is present; no reaction means Rh negative, and the antigen is not carried by the blood typed. If the person typing the blood makes an interpretive error, positive cells could be transfused into a negative person.

The second situation involves a possible reaction between an Rh positive child *in utero* and an Rh negative mother. The inheritance of Rh factor depends on two genes, with Rh positive dominant to Rh negative. The child will inherit one gene for the condition from each parent. If the father is Rh negative, there is no possibility of a child–mother difficulty due to Rh factors.

The following examples will illustrate the conditions under which a reaction between child and mother might occur.

GENOTYPE OF			
FATHER	MOTHER	CHILD (one gene from each parent)	INTERPRETATION
RhRh	RhRh	RhRh	Child and mother are Rh^+. No reaction possible.
Rhrh	Rhrh	RhRh or Rhrh or rhrh	Child is either Rh^+ (3 chances in 4) or Rh^- (1 chance in 4) relative to a positive mother. No reaction is possible between child and mother.
Rhrh or RhRh	rhrh	Rhrh or rhrh	Child is Rh^+, mother is Rh^-. 50:50 possibility of reaction exists.

Normally, the blood supplies of the mother and child are separated by membranes in the placenta. If the membranes become permeable to the child's red cells, positive cells could enter the blood of the negative mother and cause her body to produce antibodies to the child's cells. The antibodies may then make their way back into the child through the placental membranes where an antigen-antibody reaction may occur. On the other hand, the antibodies may remain in the mother's bloodstream to possibly affect a future Rh positive child, by passing across the placenta.

ERYTHROBLASTOSIS FETALIS or HEMOLYTIC DISEASE OF THE NEWBORN *(HDN)*, is a hemolytic disorder that results in the child when antibodies pass from mother to child during its sojourn *in utero*. If there is evidence that the child is developing erythroblastosis, and is considerably less than 32 weeks of age, it may be too immature and anemic to survive a Cesarean section (surgical removal through the mother's abdomen) and an exchange transfusion (see below). In these circumstances, it may be given fetal transfusions of group O, Rh negative blood injected into the peritoneal cavity to sustain life until delivery and specific supportive care is possible (at about 32 weeks). If suspected severe erythroblastosis is confirmed after the child is born, an exchange transfusion may be performed. Group O, Rh negative or group specific (A, B, AB) Rh negative blood is used. Mild cases may require no treatment or only a transfusion of a small quantity of red blood cells that matches the child's.

Research has shown that Rh antibody concentrations usually rise in the bloodstream of the Rh negative mother *after* the birth of her Rh positive child (about 72 hours). This result is interpreted to mean that it is at placental separation *(afterbirth)* that fetal blood may enter the maternal circulation, presumably through wounds created by placental separation. To desensitize the mother, that is, to remove these antigens, massive doses of ANTI D GAMMA GLOBULIN (RhoGAM*, Rho-Immune†) may be given within 72 hours after each childbirth. The antibody reacts with fetal Rh antigens present in the maternal circulation to form a complex that is excreted through the kidney. The mother's blood is thus cleared of antigens, and the chance that a baby will be born with erythroblastosis is reduced.

An Rh positive mother is no threat to an Rh negative child *in utero*. Even if the mother were to pass antigens to the child, the immaturity of its immune system and the dilution of fetally produced antibodies by the mother's great blood volume (seven to eight times that of the child) results in concentrations of antibody too low to be a threat.

OTHER BLOOD-GROUP SUBSTANCES

Genetic variants or mutants of common factors have been demonstrated in certain *family* groups. Their ANTIGENS have been designated **M, N, S, s, P,** KELL, DUFFY, LEWIS, KIDD, DIEGO, LUTHERAN, and others. Although reactions involving these antigens are rarely seen, transfusion of such a blood into a recipient may cause *sensitization* in the recipient. That is, the recipient will produce antibodies to the antigen, and these antibodies will remain for long periods of time in the bloodstream.

The **Hr** *(cde)* SYSTEM is another system of antigens and is supposedly the RECIPROCAL OF THE RH SYSTEM. If Rh antigens are absent, Hr antigens are thought to be present. Several Hr factors have been demonstrated in this system, the most important of which are **Hr** *(c)*, **Hr**′ *(d)*, and **Hr**″ *(e)*.

* Ortho Chemical Co.
† Lederle Laboratories

TRANSFUSION

In any transfusion, the two bloods involved are typed and CROSS-MATCHED. As stated earlier, the latter procedure tells if the bloods are compatible for all factors, not just the ABO and Rh factors. In this test, erythrocytes from the donor are mixed with plasma of the recipient *("major side")* and plasma of the donor is mixed with the erythrocytes of the recipient *("minor side")*. Evidence of an antigen-antibody reaction in either mixture is sufficient grounds for rejecting the donor's blood as suitable for transfusion.

Transfusion reactions. Among the events that may occur in the case of transfusion of incompatible bloods are the following:

HEMOLYSIS OF RED CELLS, with consequent anemia.

KIDNEY FAILURE as a result of chemicals released by the antigen-antibody reaction and resultant renal vasoconstriction.

PLUGGING OF KIDNEY TUBULES by hemoglobin released by the lysis of erythrocytes.

FEVER as a result of toxins released by the antigen-antibody reaction.

Transfusion fluids. BLOOD is obviously the ideal transfusion fluid, inasmuch as it supplies all needed blood components. Unfrozen blood may not be stored for more than two to three weeks, because the cellular elements begin to disintegrate. Methods that involve quick freezing indicate the possibility of storage for several years. PLASMA (blood with formed elements removed) is used to aid in maintenance of blood volume. It contains all the organic materials of the blood, including antibodies and protein, and thus it must be type specific to avoid any antigen-antibody reaction. Plasma aids in maintaining blood volume because its proteins draw water osmotically into the capillaries from the tissues. Plasma may be stored indefinitely. SOLUTIONS OF COLLOIDS or *"plasma expanders,"* such as albumin and dextran, may be transfused to maintain blood volume. These are artificial solutions and also draw water from tissue spaces into the bloodstream. They thus increase the volume of the plasma at the expense of the interstitial fluid. Dextran, a synthetic polysaccharide, is the colloid most often used as a plasma expander. It consists of aggregates of molecules from 10,000 to 100,000 molecular weight. The small aggregates filter through the kidney; the larger ones remain in the vessels and are metabolized or broken down over the course of several days. The substance thus exerts a relatively long-lived effect and is not antigenic (as albumin could be). The PACKED CELL VOLUME *(hematocrit)* is monitored to determine the need for more or less expander to be added to the bloodstream. SOLUTIONS OF CRYSTALLOIDS, such as NaCl, may be employed to temporarily increase the blood volume. Their effect is only temporary because the particles are too small to remain in the capillaries, and they are filtered rapidly into the tissues. Thus, the use of crystalloids may be associated with the development of *edema.*

SUMMARY

1. The body possesses a variety of mechanisms designed to protect it.
   a. The skin presents a mechanical barrier to the entry of foreign substances; an acidic surface film acts as an antiseptic to retard microorganism growth; ionic layers prevent the passage of charged substances; tears wash and cleanse the eye.
   b. Mucous membranes protect the systems opening on the body surfaces.
   c. Saliva, hydrochloric acid, and the acid urine protect the mouth, stomach, and urinary systems.
   d. The blood contains cells and chemicals that protect.
2. Penetration of a body surface causes an inflammatory reaction. It neutralizes agents causing the reaction and speeds healing. It has several steps:
   a. Vasoconstriction
   b. Vasodilation
   c. Lysosome release
   d. Formation of an inflammatory exudate
   e. Entry of leucocytes to the injured area.
   f. Phagocytosis of debris and healing is begun.
3. The duration of the inflammatory process is determined by the nature and length of the insult and effectiveness of body response.
4. Wound healing occurs by cell replacement (skin and liver primarily) or by formation of scar tissue.
5. The immune system consists of lymphocytes and antibody molecules.
   a. It protects the body against macromolecules.
   b. It can usually distinguish between molecules that are foreign and those that are "self."
6. Antigens are large protein, polysaccharide, or nucleic acid molecules that cause the body to produce a response to themselves.
   a. They have epitopes that form a binding point for antibodies.
7. Antibodies are gamma globulins reacting specifically with particular antigens.
   a. They have a specific form.
   b. They have combining sites for antigens and other molecules.
8. An antigen-antibody reaction occurs between the epitope of an antigen and the combining site of its corresponding antibody.
   a. The reaction may neutralize the antigen, precipitate it, or cause cell lysis (breakdown).
9. Two systems other than antigens and antibodies operate to cause cell lysis.
   a. The complement system is a series of enzymes that produces holes in the cell membrane. It requires antibody binding of complement.
   b. The properdin-enzyme system operates in the same way, but responds without antibody binding.
   c. Both systems increase cell membrane permeability to water and ions, the cell swells and bursts.
10. Two general theories of antibody production have been advanced.
    a. The instructive theory has the antigen acting as a "template" to mold gamma globulin molecules to specific shapes. One molecule is required for variable antibody production.
    b. The selective theory suggests that capability to produce millions of antibodies already exists in the body through evolution and

mutation. An antigen "turns on" capability of production.

11. The "immune system" (see 5 above) supplies cell-mediated immunity, in which cells with antibodies on their surfaces react with an antigen, and humoral immunity, in which cells produce and release antibodies into body fluids. Each handles particular types of antigenic challenge.
12. T-cells arise in the thymus and mediate cellular immunity. B-cells arise in other lymphatic tissue and bone marrow and produce circulating antibodies.
13. Immunity to various diseases may be conferred by the immune response.
    a. Active immunity usually involves contracting the disease or is acquired by injection of antigen.
    b. Passive immunity is given by antibody administration.
14. Interferon is an antiviral substance that inhibits viral replication and entry into healthy cells. Transfer factor is an antibody-like substance that can aid in recovery from certain diseases.
15. Allergy is an incomplete or abnormal response to an antigen. It does not confer immunity.
    a. Precipitin allergies cause shock and collapse.
    b. Delayed-reaction allergies require a second exposure to an antigen to develop a reaction.
    c. Atopic allergies are genetic and are common (hay-fever, asthma, dermatitis).
16. Transplantation involves introduction of foreign antigens into a person, and causes a rejection response. A variety of measures are employed to reduce antigenicity of the transplant or the individual's immune response to the antigen.
17. Autoimmunity arises when "self" is not recognized and internal antigen-antibody reactions occur that destroy tissue. Examples are presented.
18. Immunodeficiency refers to failure to produce antibodies. T- and B-cell response is inadequate.
19. The blood groups are categorized by particular antigen-antibody systems that are genetically determined. Antigens are on erythrocytes; antibodies in the plasma.
    a. The ABO group consists of two antigens *(A* and *B),* and two antibodies *(a* and *b),* usually present in reciprocal relationship. For example, Ab, Ba, AB⁻, and Oab.
    b. The Rh group consists of several genotypes that test out Rh positive (antigen present) or Rh negative (antigen absent). Rh negative individuals can produce anti-Rh antibodies.
    c. Transfusion of blood requires compatibility of antigens and antibodies to avoid a reaction.
    d. Hemolytic disease can result if incompatibility during blood exchange occurs. For example, transfusion or pregnancy.
20. Replacement of lost blood may be made by transfusion of whole blood, plasma, colloids, or crystalloids. The choice employed depends on the effect to be achieved and for how long.

QUESTIONS

1. What mechanism(s) exist(s) to protect the body against the potential effects of each of the following?
    a. Excess body fluid volume.
    b. Microorganisms on the external body surface.
    c. Bacterial or viral invasion of the body.
2. How is the inflammatory process of value to the body in combating a harmful agent? What are the steps in this process?
3. What constitutes the immune system, and what are its functions?
4. What are the characteristics of antigens and antibodies, and how do they relate to one another?
5. How is the complement system related to an antigen-antibody reaction? What does the activated system do to a cell?
6. Contrast the two major theories of antibody production.
7. What are the jobs of T- and B-lymphocytes? From where do each originate?
8. Contrast active and passive immunity as to how each is achieved.
9. How do allergies differ from immunity? What type of a reaction is a "penicillin reaction" and why is it so threatening to the body?
10. What considerations must be kept in mind if a person was to need a kidney transplant? How can the rejection phenomenon be reduced?
11. What determines a person's blood type?

12. What are some of the reactions that can occur if bloods are mismatched for transfusion and then transfused?

13. In Rh disease between a child *in utero* and its mother, how can the severity of the reaction to that child and to subsequent children be reduced?

READINGS

Allison, A. C., and J. Ferluga. "How Lymphocytes Kill Tumor Cells." *New Eng. J. Med. 295:*165. July 15, 1976.

Bach, Fritz H., and Jon I. van Rood. "The Major Histocompatibility Complex." *New Eng. J. Med. 295:*806. Oct 7, 1976. *295:*872. Oct 13, 1976. *295:*927. Oct 21, 1976.

Beaven, Michael A. "Histamine." *New Eng. J. Med. 294:*30. Jan. 1, 1976. *294:*320. Feb 5, 1976.

Beer, Alan E., and Rupert E. Billingham. "The Embryo as a Transplant." *Sci. Amer. 230:*36. 1974.

Brand, Ann, Douglas G. Gilmour, and Gideon Goldstein. "Lymphocyte-Differentiating Hormone of Bursa of Fabricius." *Science 193:*319. 23 July 1976.

Burke, Derek C. "The Status of Interferon." *Sci. Amer. 236:*42. (Apr) 1977.

Cooper, Max D., and Alexander R. Lawton III. "The Development of the Immune System." *Sci. Amer. 231:*58. 1974.

Dannenbert, Arthur J. "Macrophages in Inflammation and Infection." *New Eng. J. Med. 293:*489. Sept 4, 1975.

Franklin, Edward C. "Some Impacts of Clinical Investigation on Immunology." *New Eng. J. Med. 294:*531. Mar 4, 1976.

Greenberg, Harry B., et al. "Human Leukocyte Interferon and Hepatitis B Virus Infection." *New Eng. J. Med. 295:*517. Sept 2, 1976.

Marx, Jean L. "Immunology: Role of Immune Response Genes." *Science 191:*277. 23 Jan 1976.

Maugh, Thomas H. "Chemotherapy: Antiviral Agents Come of Age." *Science 192:*128. 9 April 1976.

Mayer, Manfred M. "The Complement System." *Sci. Amer. 229:*54. Nov 1973.

Raff, Martin C. "Cell-surface Immunology." *Sci. Amer. 234:*30. May 1976.

Rocklin, Ross E., et al. "Absence of an Immunologic Blocking Factor in Women Who Chronically Abort." *New Eng. J. Med. 295:*1209. Nov 25, 1976.

Roitt, Ivan M. *Essential Immunology.* Blackwell Scientific Publications. Oxford. 1974.

Scientific American-Readings. *Immunology.* Freeman. San Francisco, 1976.

Sela, Ilan, Sidney E. Grossberg, I. James Sedneak, and Alan H. Mehler. "Discharge of Aminoacyl-Viral RNA by a Factor from Interferon-Treated Cells." *Science 194:*527. 29 Oct 1976.

Shin, Hyun S., Gary R. Pasternack, James S. Economou, Robert J. Johnson, and Michael L. Hayden. "Immunotherapy of Cancer with Antibody." *Science 194:*327. 15 Oct 1976.

Solomon, Alan. "Cellular Regulation of Humoral Immunity." *New Eng. J. Med. 293:*928. Oct 30, 1975.

Strob, Ranier, Ross I. Prentice, and E. Donnall Thomas. "Marrow Transplantation for Aplastic Anemia: Factors Associated with Rejection." *New Eng. J. Med. 296:*61. Jan 13, 1977.

Tong, Myron J., J. Scott Nystrom, Allan G. Redeker, and G. June Marshall. "Failure of Transfer-Factor Therapy in Chronic Active Hepatitis B." *New Eng. J. Med. 295:*209. July 22, 1976.

Volpe, E. Peter, and James B. Turpen. "Thymus' Central Role in the Immune System of the Frog." *Science 190:*1101. 12 Dec 1975.

Wara, D. W., et al. "Thymosin Activity in Patients with Cellular Immunodeficiency." *New Eng. J. Med. 292:*70. Jan 9, 1975.

THE HEART

OBJECTIVES

THE HEART: ANATOMY AND BASIC PHYSIOLOGY

THE STRUCTURE OF THE HEART

THE NODAL TISSUE

CARDIAC MUSCLE

THE HEART AS A PUMP

CARDIAC OUTPUT

FACTORS CONTROLLING ALTERATIONS IN CARDIAC OUTPUT

CARDIAC RESERVE

MEASUREMENT OF CARDIAC OUTPUT

ENERGY SOURCES FOR CARDIAC ACTIVITY; THE CORONARY CIRCULATION

ALTERATIONS IN RATE AND RHYTHM OF THE HEARTBEAT

ABNORMAL RHYTHMS OF THE HEART

THE CARDIAC CYCLE

TIMING

THE ELECTROCARDIOGRAM

PRESSURE AND VOLUME CHANGES

HEART SOUNDS

SUMMARY

QUESTIONS

READINGS

OBJECTIVES

After studying this chapter, the reader should be able to:

- ☐ Describe the role of the heart in blood circulation.
- ☐ Explain what is meant by a double circulation.
- ☐ Briefly describe the chambers and valves of the heart.
- ☐ List the three types of functionally important tissue composing the heart.
- ☐ Describe the role of the nodal tissue in causing the heart beat.
- ☐ Explain the basis of spontaneous depolarization of the nodal tissue.
- ☐ Trace the pathway of depolarization within the heart.
- ☐ Describe the functional anatomy of cardiac muscle.
- ☐ List the physiological properties of cardiac muscle and relate them to the tasks the heart carries out.
- ☐ Explain the genesis of extrasystoles.
- ☐ Define cardiac output and its components.
- ☐ List the factors controlling cardiac output, and which component of the output is most greatly affected by each factor.
- ☐ Explain the effects of ionic changes and drugs on heart action.
- ☐ Explain the nervous reflexes that control heart rate.
- ☐ Define cardiac reserve.
- ☐ Explain what heart failure is.
- ☐ Show how cardiac output may be measured.
- ☐ Explain what substances are necessary for continued heart activity and how they are delivered to the muscle.
- ☐ Describe some abnormal rhythms of the heart and how they are brought about.
- ☐ Define what a cardiac cycle is, and list the events occurring within it.
- ☐ Explain the timing of contraction and relaxation in a cycle.
- ☐ Diagram and explain the parts of a normal electrocardiogram.
- ☐ Explain and relate the pressures that occur during heart activity to valve operation and blood movement.
- ☐ List and explain the genesis of the heart sounds.

Chapter 18 indicated that the blood is an efficient medium for transport of materials through the body because it is circulated. Circulation implies a source of energy that *causes* blood movement. Blood circulation occurs because of a rhythmical pumping action of the organ known as the HEART. The blood is circulated through BLOOD VESSELS, that at all times contain the blood. This creates what is termed a *closed circulation.* In completing a circuit

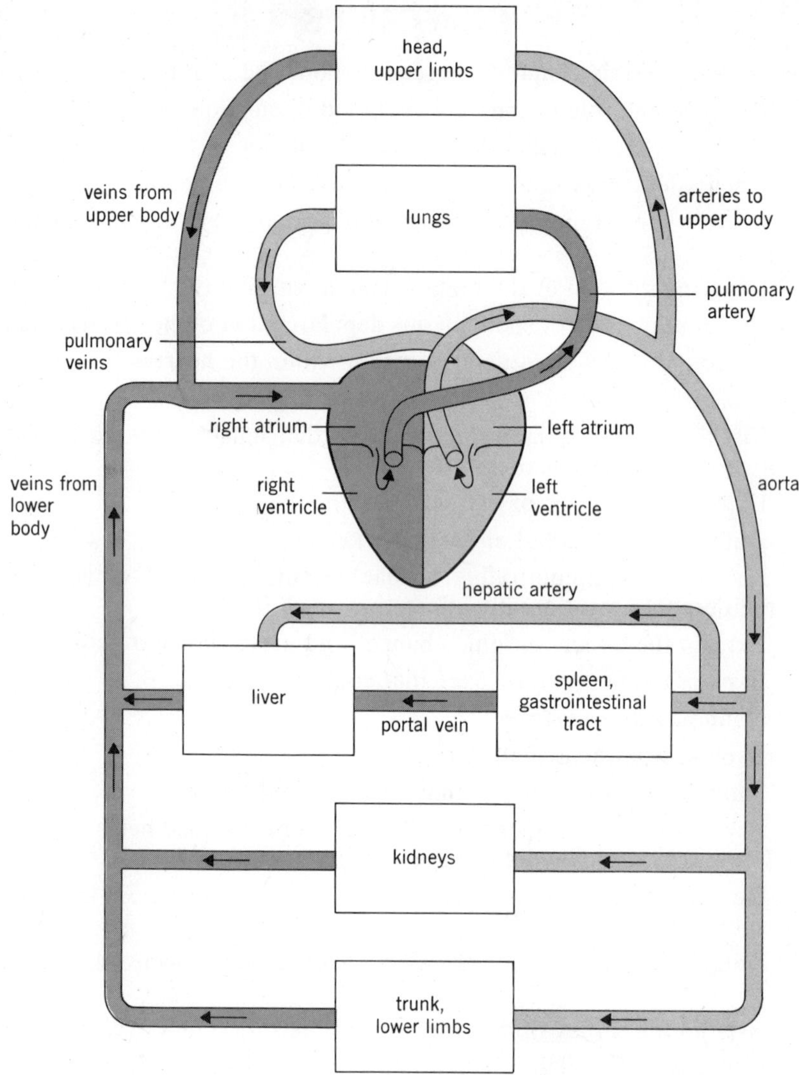

FIGURE 20.1
A schematic representation of the circulatory system.

through the body, the blood will pass twice through the heart, and through two circulations. The SYSTEMIC CIRCULATION begins with the left ventricle of the heart and carries blood to the body generally; it returns blood to the right atrium of the heart. The PULMONARY CIRCULATION begins with the right ventricle, pumps blood to the lungs for oxygenation and carbon dioxide removal, then carries blood to the left atrium. The entire system functions in a coordinated manner to ensure adequate blood pressure and flow to active tissues according to their needs at a given time. This coordination involves inherent controls within the heart and vessels themselves and extrinsic chemical and nervous influences reaching them from other body areas.

The general plan of the circulatory system is presented in Figure 20.1.

THE HEART: ANATOMY AND BASIC PHYSIOLOGY

THE STRUCTURE OF THE HEART

The heart is a hollow muscular organ acting as the pump that creates pressure to circulate the blood through the body. It is a four-chambered organ, divided into right and left halves. Each half is composed of an upper chamber or ATRIUM that *receives* blood from a particular region of the body. The lower VENTRICLES act as *pumping* chambers to circulate the blood through pulmonary and systemic circulations (*see* Fig. 20.1). Valves, placed between the atria and ventricles and in the bases of the pulmonary artery and aorta that leave the right

FIGURE 20.2
Diagram of the heart and its major blood vessels. The heart is viewed from the front.

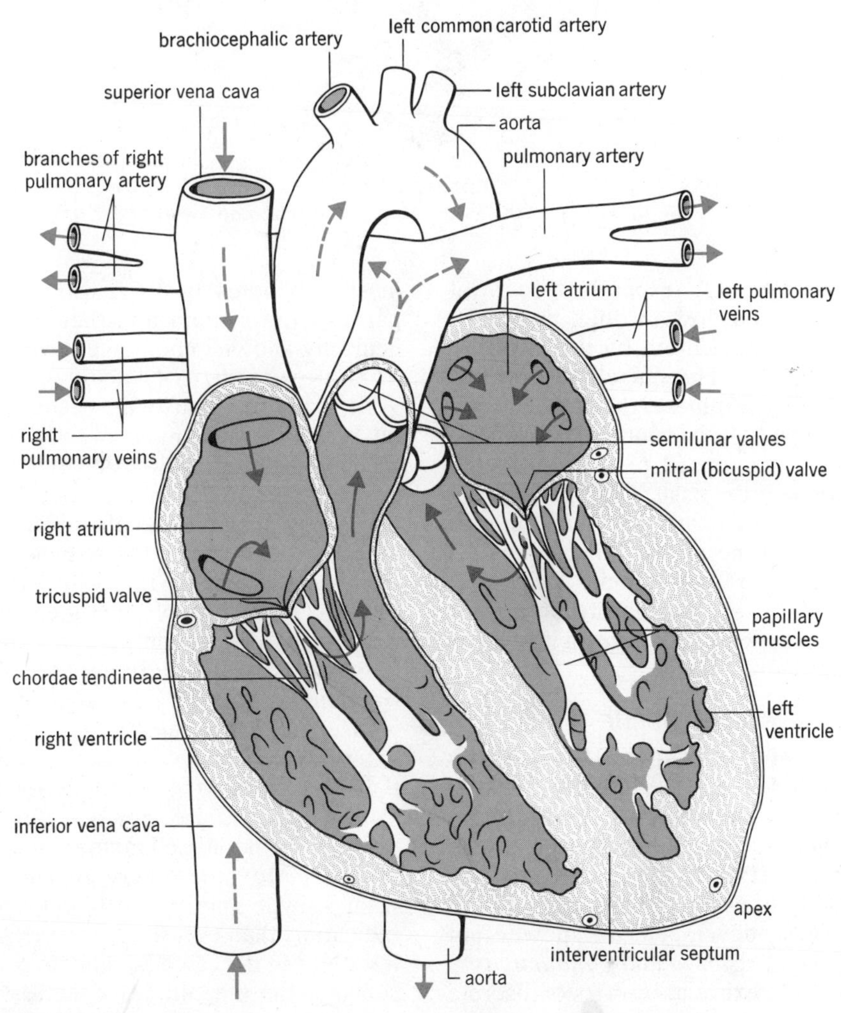

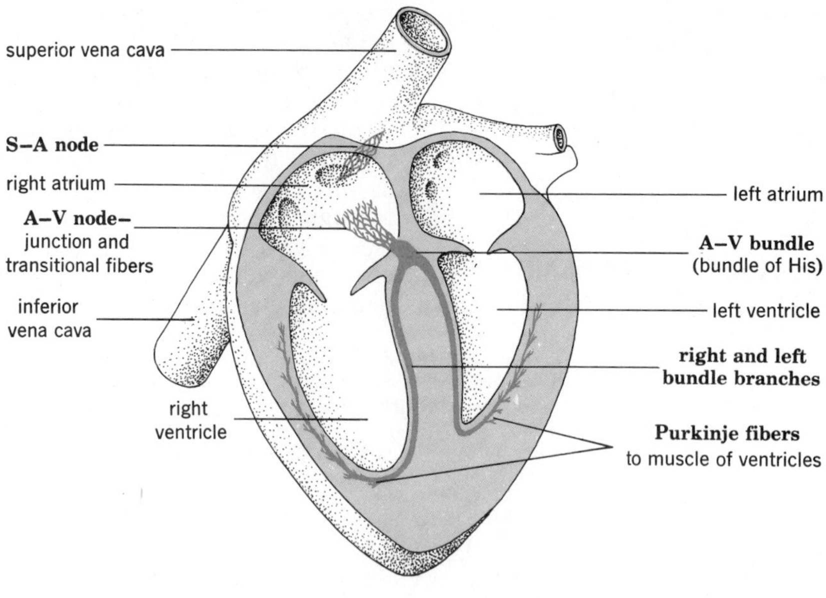

FIGURE 20.3
The locations of the nodal tissue in the human heart.

ventricle and left ventricle respectively, control the direction of blood flow through the heart. Actual pumping is carried out by the CARDIAC MUSCLE of the heart. The muscle constitutes about half of the 300 gram weight of the normal adult human heart. The remaining half of the heart's weight is composed of noncontractile nodal and connective tissue composing the *fibrous skeleton* of the heart. The skeleton provides attachment for the cardiac muscle bundles that course in several directions within the heart, and provides supporting structures for the valves. The general structure of the heart is presented in Figure 20.2.

THE NODAL TISSUE

The heart is capable of beating independently of any outside influence. It possesses what is termed AUTORHYTHMICITY. This ability is conferred on the heart by specialized tissue known as NODAL TISSUE. This tissue is derived from cardiac muscle but has largely lost its ability to contract and has developed the powers of *spontaneous depolarization* and *conductivity*.

The nodal tissue exists as a series of discrete masses, bundles, or fibers within the heart. The parts of the system and their locations in the heart are shown in boldface type in Figure 20.3. The SINOATRIAL *(SA)* NODE acts as the PACEMAKER for the heartbeat, because it spontaneously depolarizes most rapidly of all parts of the nodal tissue. Its basic rhythm is about 80 depolarizations per minute, but this rate can be altered by nervous and chemical factors. Spontaneous depolarization is believed to result from a mechanism that allows "leakage" of sodium ions through the membranes of the nodal cells. The membrane potential of nodal tissue cells is originally about −55 to −60 mv (*remember:* the negative sign indicates a cell whose interior is electrically negative to the exterior; also, the values are considerably less than the 70 to 90 mv values for nerve cells). Potassium ions are slowly diffusing outwards through the nodal cell membranes, as they do in all cells. Movement here appears to be somewhat slower than in nerve cells, for example, indicating that potassium permeability is a bit less than in nerve cells. This is part of the reason that the magnitude of the resting potential is 55 to 60 mv. Permeability to sodium ion in nodal cell membranes is greater than that in nerve cells, and an inwards "leakage" of sodium ions occurs. The negativity of the interior portion thus decreases with time as the sodium ions enter, and the transmembrane potential falls. At some point, the threshold for depolarization is reached and the fibers depolarize. Sodium permeability is greatly increased and a reversal potential occurs. Then, as the repolarization process proceeds, permeability to K^+ increases and that to Na^+ decreases, the sodium is pumped out, and K^+ diffuses outwards again. Leakage of Na^+ again commences and the cycle is repeated. Again, it may be noted that this series of events may occur in any part of the nodal system, but that it occurs most rapidly in the SA node, which consequently dominates the rest of the system.

The action potentials produced by the SA node are transmitted to the muscle cells of the atria and pass through these cells depolarizing them and causing their contraction. Speed of conduction of the impulse through atrial muscle is about 0.3 m per second.

The distance from the SA node to the ATRIOVENTRICULAR *(AV)* NODE (the next part of the nodal system) is shorter than from SA node to the farthest part of the left atrium. Thus, it becomes necessary to temporarily "hold up" the further passage of impulses through the nodal tissue until the atria have completed their contraction. This delaying tactic is carried out by small, slow conducting *junctional* and *transitional fibers* within the AV node. About 0.1 sec is consumed in traversing these fibers, sufficient to allow the wave of depolarization to reach all parts of the atria. Once into the AV BUNDLE, impulse transmission is quite rapid over the bundle, its BRANCHES, and the PURKINJE FIBERS; about 2 to 4 m per second. The object here is to pass the wave of depolarization rapidly to the large mass of ventricular muscle for as nearly a simultaneous contraction as possible.

Only about 0.06 seconds elapses between entry of the wave into the AV bundle, and excitation of the farthest reaches of the ventricular muscle. Only the layer or two of cardiac muscle cells closest to the Purkinje fibers is depolarized directly by the nodal tissue. The last step is for the depolarization to spread through the ventricular muscle mass and cause its contraction. Spread of the wave through the ventricular muscle occurs at a speed of 0.3 to 0.4 m per second, and tends to follow the orientation of muscle bundles within the ventricles. If the question has arisen as to why the impulse generated and passed to the atrial muscle cannot directly pass from here to the ventricular muscle, the answer is because the cardiac skeleton is so oriented as to separate the atrial and ventricular muscle masses by nonconducting tissue. Thus, the nodal system remains the only *normal* route by which the ventricles may

TABLE 20.1
SOME CHARACTERISTICS OF NODAL TISSUE AND CARDIAC MUSCLE

AREA	VELOCITY OF CONDUCTION (M/SEC)	INHERENT RATE OF DISCHARGE (IMPULSES/MIN)
S-A node	0.05	70–80 (pacemaker)
Atrial muscle	0.3	—
A-V node	0.05–0.1	40–60
A-V bundle	2–4	35–40
Purkinje fibers	2–4	15–40
Ventricular muscle	0.3	—

become stimulated. Table 20.1 summarizes some facts related to conduction in the nodal tissue and cardiac muscle.

CARDIAC MUSCLE

The cardiac muscle forming the contractile tissue of the heart (Fig. 20.4) is a type of STRIATED (striped) muscle. Its fibers branch and are separated from neighboring cells by INTERCALATED DISCS that occur at the I bands of the muscle. The discs represent areas where membranes of adjacent cells ends approach one another; a gap of 150 to 200 Å usually separates the cells, but at intervals along the disc, the membranes approach one another as closely as 20 Å. These areas of close approach of membranes are believed to have low electrical resistance that allows a wave of depolarization to pass unhindered from one cell to another. Thus, stimulation of one part of the cardiac muscle affects all parts, and the entire mass *behaves* as though it was a single cell. This behavior is referred to as a FUNCTIONAL SYNCYTIUM (*syn,* together + *cyte,* cell). This type of behavior is essential if near-simultaneous contraction of a chamber is to occur, with development of maximum pressure on the blood.

Cardiac muscle has larger T-tubule systems (occurring only at Z lines), and a simpler sarcoplasmic reticulum than does skeletal muscle. Triads do not seem to be present, as the SR cisternae do not form large expansions next to the T-tubules. Apparently, no part of a cardiac muscle fiber is more than 2 to 3 μm from an external surface or a T-tubule, a fact of importance in excitation-contraction coupling and nutrition of the fibers.

The contractile mechanism for cardiac muscle is the same as that described for skeletal muscle (Chapter 5). Myofibrils are fewer in cardiac than in skeletal muscle, which may be partly responsible for the somewhat slower contraction time cardiac muscle shows.

Cardiac muscle follows the ALL-OR-NONE-PRINCIPLE, whereby an adequate stimulus is followed by a maximal contractile response, the strength depending on the conditions present at the time. For example, epinephrine (adrenalin) increases the strength of the contractile response, as do norepinephrine, serotonin, and digitalis. Temperature rise tends to increase strength, while fall decreases it. Sodium, potassium, and calcium ions must be present in normal amounts and ratios one to another or strength of contraction may decrease. Effects of these and other substances on heart action are considered in greater detail in later sections.

Cardiac muscle, like skeletal muscle, exhibits a length–tension relationship. This relationship is embodied in the LAW-OF-THE-HEART *(Frank-Starling law).* According to the law, as cardiac fibers are elongated, strength developed by contraction will increase, to the point where the tension resulting from elongation is so great that it begins to cause mechanical damage to the muscle. In the heart *in situ* (in place in the body), elongation of fibers occurs when the ventricles fill with blood. Greater filling, as occurs during exercise, causes a stronger contraction that empties the greater quantity of blood from the ventricles. The term END DIASTOLIC VOLUME refers to the amount of blood in the ventricles just before contraction (systole) occurs, and it is this volume that determines stretch on the muscle and the strength of its contraction.

After depolarization, cardiac muscle fibers repolarize more slowly than skeletal muscle. Cardiac muscle passes through an ABSOLUTE REFRACTORY PERIOD, during which no stimulus, no matter how strong, will cause a second response. This is followed by a RELATIVE REFRACTORY PERIOD, during which the muscle may respond to a stronger-than-normal stimulus. The absolute refractory period extends partway into the relaxation phase of muscle activity and ensures that the muscle will partially relax and therefore *allow some filling* of the ventricular cavities before the next contraction occurs. Therefore *some* blood will be ejected with each beat to maintain a circulation. The long refractory period also assures that the muscle cannot normally be tetanized.

After the relative refractory period has ended, another adequate stimulus occurring before the next SA node stimulation arrives may cause a VENTRICULAR EXTRASYSTOLE (additional contraction). It is less strong than a

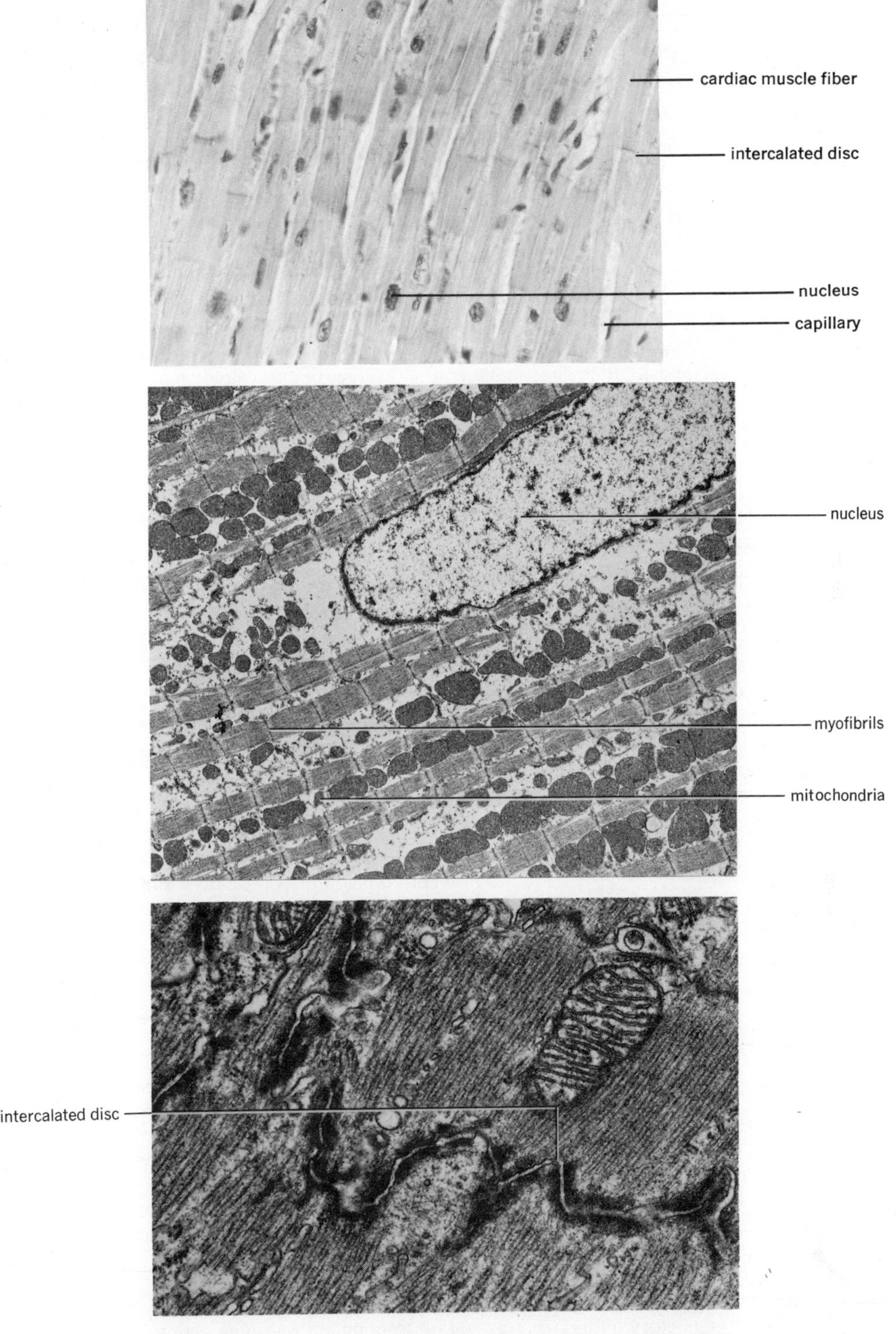

FIGURE 20.4
The histology of cardiac muscle.

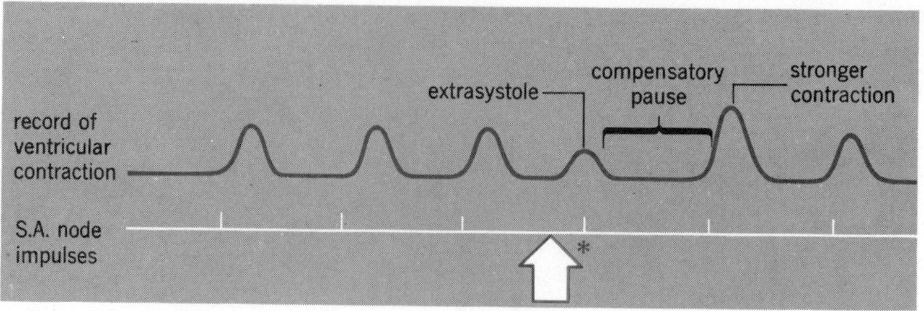

FIGURE 20.5
The genesis of an extrasystole and the compensatory pause. Stimulus applied at ↑ causes extrasystole; next normal SA node impulse arriving at * finds muscle refractory from extrasystole. Muscle does not respond again until SA node impulse arrives at normal time.

TABLE 20.2
A SUMMARY OF PROPERTIES OF NODAL TISSUE AND CARDIAC MUSCLE

TISSUE	PROPERTY	COMMENTS
Nodal	Autorhythmicity	Allows a basic rate of beat (70–80/min) to be established. Not dependent on outside factors
	Conductivity	Fastest in A-V bundle onwards to excite ventricular muscle nearly simultaneously. Slowest in A-V node
Cardiac Muscle	Syncytial arrangement of fibers	Allows rapid spread of depolarization through a mass of muscle. Gives near simultaneous contraction for greatest pressure development
	Follows the all-or-none law	Maximal contraction occurs if stimulus is strong enough to cause depolarization. Strength varies according to environment and mechanical factors (tension)
	Follows the law of the heart	Increased tension results in stronger contraction. Allows variation in amount of blood pumped. Adjustability
	Has long refractory period	No tetanus. Some filling before contraction to maintain pumping action

normal systole, because the ventricle has not filled to its usual capacity and the stretch applied to the muscle fibers is less than normal. Thus, the law-of-the-heart applies. The next normal SA node impulse arrives and finds the muscle refractive from the extrasystole. Therefore, there is a skipped beat (a COMPENSATORY PAUSE) that continues until the next SA node impulse arrives. During this lengthened time, the ventricles fill to a greater degree than normal, and the ensuing contraction will be stronger due to the law-of-the-heart. These relationships are diagrammed in Figure 20.5.

The properties of the nodal tissue and cardiac muscle described in the preceding sections may be said to give the organ as a whole an INHERENT RHYTHMICAL BEAT, whose strength can be ADJUSTED TO NEED, and which normally provide for CONTINUAL CIRCULATION of blood through the body. The properties of this muscle type might well be compared to the properties of skeletal and smooth muscle presented in Chapter 5, to see how each type of muscle is suited for the tasks it carries out in the body. Some properties of the nodal and cardiac tissues are summarized in Table 20.2.

THE HEART AS A PUMP

CARDIAC OUTPUT

When the cardiac muscle contracts, pressure is created on the blood within the ventricles, and blood is moved from those chambers into the pulmonary and systemic circulations. The volume of blood leaving *each* ventricle during one minute's time is termed the CARDIAC OUTPUT. Cardiac output *(CO)* is the product of two other factors, the rate of heart beat, or STROKE RATE *(SR)* and the volume of blood ejected from a ventricle per beat, or STROKE VOLUME *(SV)*. Stated as a mathematical formula

$$CO(\text{ml/min}) = SR\ (\text{beats/min}) \times SV\ (\text{ml/beat})$$

As one can see, beats cancel in the equation, leaving *CO* to be expressed as ml/min.

Over any period of time, the normal heart will eject the same amount of blood from each ventricle, although minor variations may exist on a beat-to-beat basis.

Stroke rate is easily determined by resting the fingers on a superficial artery in the body and counting the pulsations for a given period of time (convert to a *minute* basis if counted for less than a minute). Stroke volume is calculated by finding the difference between the *end systolic volume* (the volume of blood remaining in the heart at the end of contraction or systole) and the *end diastolic volume* (the volume of blood in the heart at the end of relaxation or diastole). For example, if end systolic volume is 50 ml, and end diastolic volume is 110 ml, then the stroke volume is 60 ml (110 − 50). Note that it is not necessary for the heart to empty its chambers with each beat. This provides a reserve that can be emptied more completely if conditions require. This indicates that end systolic volume is variable, but end diastolic volume is less so because it depends on the basic size of a ventricle and the small degree to which the ventricle can enlarge as it fills with blood.

A normal resting cardiac output is calculated as lying between 4.2 and 5.6 liters per minute

$$(SR = 70\text{–}80\ \text{beats/min};\ SV = 60\text{–}70\ \text{ml/beat}).$$

During exercise, heart rates may increase to 120 beats/min, and volume to 120 ml/beat; thus, an active cardiac output may reach over 14 liters/min. Increase of SR beyond about 140 beats/min decreases the time allowed for ventricular filling and is associated with a fall of cardiac output. So, the output does not show a linear relationship to heart rate.

FACTORS CONTROLLING CARDIAC OUTPUT

Obviously, there are two quantities, stroke rate and stroke volume, that can be varied together or separately to alter cardiac output. Thus, we shall be concerned with those factors controlling each of these quantities. Taken in their entirety, the mechanisms involved in controlling rate and volume involve the following.

Venous inflow. This term refers to the volume of blood returned to the right atrium from the body generally. It basically represents the volume of blood that is available to enter the ventricles (end diastolic volume) and be emptied from those chambers at the next contraction

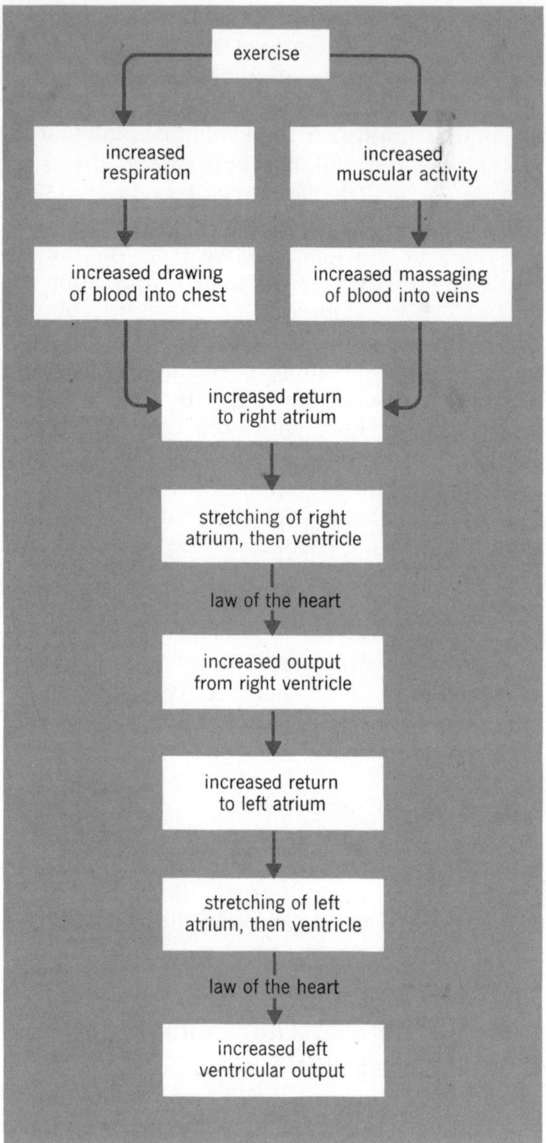

FIGURE 20.6
The effects of exercise on heart action.

(stroke volume). These interrelationships are shown in Figure 20.6, using exercise as the factor increasing venous return.

Peripheral resistance. The muscular arteries of the body have circularly disposed smooth muscle around them. Contraction of the muscle diminishes the size of the vessel *(vasoconstriction)* while relaxation increases it *(vasodilation)*. Constriction reduces blood flow to tissues and organs and causes more blood to remain in the vessels between heart and tissues. This is said to increase peripheral resistance and requires a higher output and pressure on the part of the heart to maintain flow through the narrowed vessels. The ventricles decrease end systolic volume by a stronger contraction. These relationships are shown in Figure 20.7.

Temperature. Though of little physiological importance in the human, exercise is associated with the metabolism of increased quantities of fuels, with increased production of heat. The heat is absorbed by the bloodstream and raises its temperature slightly. The reactions of the SA node are accelerated slightly by elevated temperature, and heart rate increases by perhaps three to four beats per minute. The rapid transfusion of large volumes of blood that is still cold (from storage in a "blood bank") has been associated with an increased incidence of heart block, indicating that transmission of impulses is affected by a fall of temperature of the blood.

Chemicals. Several IONS, including Na, K, and Ca, are essential to normal heart action.

Sodium. The chief function of extracellular sodium ion is to maintain osmotic relationships and excitability. The mechanisms keeping Na^+ concentrations within normal limits are very efficient (*see* Chapter 16) and thus little effect is seen in the intact animal with changes in ECF sodium concentrations. On an isolated heart, the effects of alterations in sodium concentration may be more easily demonstrated. *Increase* of Na^+ concentration is without effect until levels are raised to 2 to 3 times the normal ones. Hyperpolarization of the nodal and cardiac tissue then occurs, and HEART RATE DECREASES as the tissue is less easily depolarized (i.e., it takes *longer* for the sodium leakage to cause the transmembrane potential to decrease to threshold levels). If Na^+ concentrations are *decreased* to 10 to 20 percent of normal, rate of depolarization of

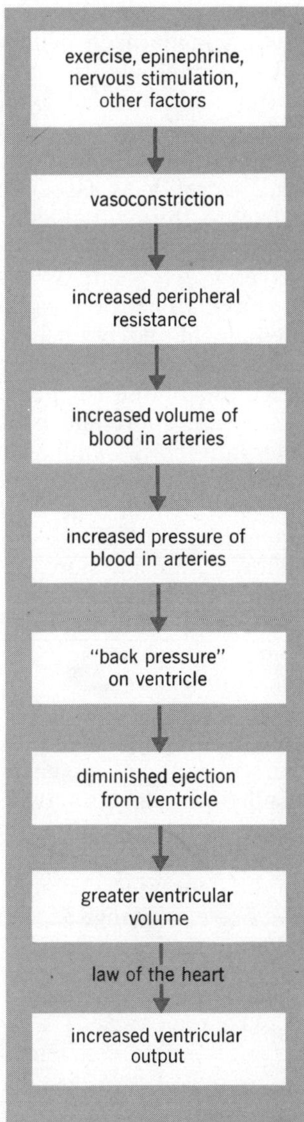

FIGURE 20.7
The effects of vasoconstriction upon heart action.

fibers is also slowed, and the HEART RATE DECREASES. Less sodium is available to pass through the membrane, and it again takes longer for threshold levels to be reached.

Potassium. Increases of K^+ in ECF to 2 to 3 times normal values decrease conduction velocity through the nodal tissue and causes HEART BLOCK, ARRHYTHMIAS, and FIBRILLATION (rates to 300 to 400 beats/min). This may be associated with hypopolarization and a greater ease of depolarization, with reaction to stimuli that are usually not effective. Ectopic foci (*see* below) develop more easily with elevated potassium levels. *Decrease* of potassium concentrations in ECF causes hyperpolarization and a DECREASED STROKE RATE.

Calcium. Increases in ECF calcium concentrations SLOW HEART RATE as hyperpolarization occurs. Also, STROKE VOLUME IS INCREASED, presumably due to a stronger contraction as more calcium is made available for the contraction process. Prolonged elevation of ECF calcium levels creates a state resembling tetany, with failure of the muscle to relax; this is called "calcium rigor." *Decrease* of ECF calcium levels causes hypopolarization and increased excitability. Ectopic foci and fibrillation may result.

Effects of substances other than ions on heart action are quite varied.

Catecholamines, including epinephrine, norepinephrine, and serotonin, act to INCREASE both RATE OF HEART BEAT (chronotropic effect) and strength of beat (inotropic effect) RAISING STROKE VOLUME. Rate increases arise from augmentation of spontaneous depolarization rates, possibly by altering membrane permeability to Na^+. Inotropic effects may occur by increasing Ca^{++} influx into the cells.

Digitalis is a "cardiac glycoside" (Fig. 20.8) that also exerts an inotropic effect in the manner described above. It is used to increase the efficiency of cardiac contraction. It also increases the rate of spontaneous depolarization, decreases conduction velocity of cardiac muscle, and shortens the refractory period of the muscle. It is a toxic substance, and side effects of its use include nausea, vomiting, diarrhea, headache, and increased incidence of arrhythmia and heart

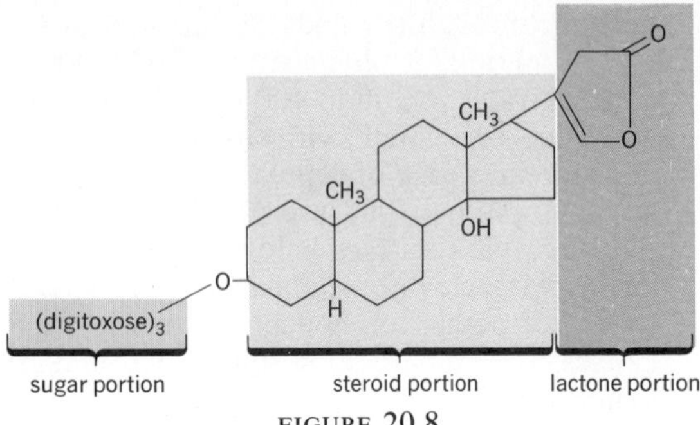

FIGURE 20.8
The chemical structure of digitalis.

block. The line between a therapeutic and toxic dose is very fine and subject to individual variation and its effects should be closely monitored in each individual.

Atropine, caffeine, and *camphor* all INCREASE HEART RATE; *amylnitrite, nitroglycerine, acetylcholine,* and small quantities of *alcohol* all DECREASE HEART RATE.

Oxygen levels of the blood, unless extremely low or prolonged, have little direct effect on heart function. If low and prolonged, *hypoxemia* interferes with aerobic metabolism of the muscle and causes weakened contractions and FALL OF STROKE VOLUME. *Excessive carbon dioxide* levels exert an anesthetic-like effect and cause depression of SA and AV node function. STROKE RATE DECREASES. *Fall of pH* has the same effect as raising of CO_2 levels; *rise of pH* is associated with hyperexcitability and thus STROKE RATE INCREASES.

Nervous factors. Nervous factors provide the finest degree of control of stroke rate and therefore of cardiac output. Nerves have virtually no effect on stroke volume.

The nerves to the heart are derived from both divisions of the autonomic nervous system (Fig. 20.9).

Sympathetic nerves. The sympathetic supply forms what are termed the CARDIAC NERVES, and they arise from the first five segments of the thoracic spinal cord. Their activity is in turn controlled by nerves from the medulla. The preganglionic fibers synapse in the sympathetic ganglia, from which the postganglionic fibers innervate the SA and AV nodes, and atrial and ventricular musculature. These nerves are adrenergic, secreting norepinephrine at their ends in the heart. Because norepinephrine secretion is associated with an increased heart rate, the sympathetic nerves are designed as CARDIOACCELERATOR NERVES.

The *parasympathetic* supply is carried in the VAGUS NERVE. The preganglionic fibers leave the medulla of the brainstem and pass to the *cardiac plexus* located between the aorta and the bifurcation of the trachea. The postganglionic fibers are short and pass from the plexus to innervate the SA and AV nodes and atrial musculature. These nerves are cholinergic, secreting acetylcholine at their terminations. Acetylcholine creates a hyperpolarized state and slows transmission of impulses through the nodal tissue, and thus heart rate decreases. Because of this effect on heart rate, the vagal fibers to the heart are designated as CARDIOINHIBITORY NERVES.

Both sets of nerves are TONICALLY ACTIVE, that is, *continually* conducting impulses. Therefore, the heart rate at any given moment reflects the result of a balance between stimulation and inhibition. The situation is analagous to driving an automobile with both the accelerator and brake depressed. To alter the speed of the vehicle, one can change accelerator pressure, *or* brake pressure, or *do both together.* The latter option is the one followed in controlling heart rate, for it makes for a smoother and more rapid adjustment of cardiac output to body needs.

The medullary neurons from which the vagus nerves to the heart arise, or in the case of the sympathetic, those influencing the thoracic preganglionic fibers, constitute the CARDIOINHIBITORY and CARDIOACCELERATOR CENTERS. These centers display no spontaneous or automatic activity of their own, but serve as the coordinating areas for sensory input from many parts of the body. Thus, cardiac reflexes (reflex arcs) are created (Fig. 20.10 and Table 20.3) that finely tune heart rate to body need.

The sources of afferent (sensory) input to

these centers are varied. Some of the more important are presented below.

BARORECEPTORS respond to alterations in pressure within the walls of the organs in which they are located. The most concentrated and functionally important baroreceptive regions for cardiac control lie in the two CAROTID SINUSES, located in the bifurcation of the common carotid arteries of the neck, and in the AORTIC ARCH. The sinuses are supplied by branches of the *glossopharyngeal nerve* (IX cranial nerve) while the aorta is innervated by

FIGURE 20.9

A diagram showing the arrangement of the autonomic nerves to the heart. Only one side of each set is represented for clarity (T = thoracic).

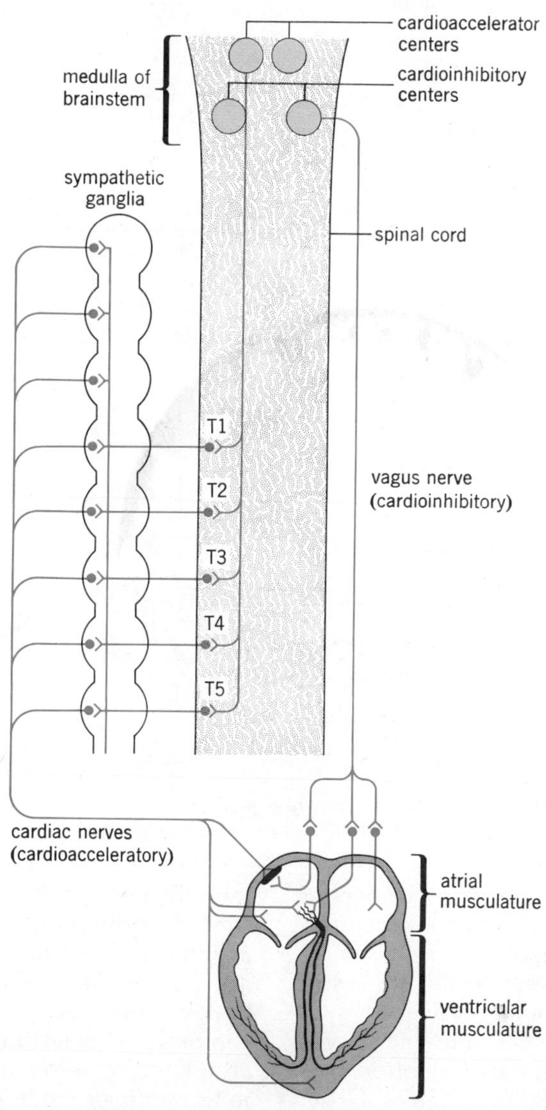

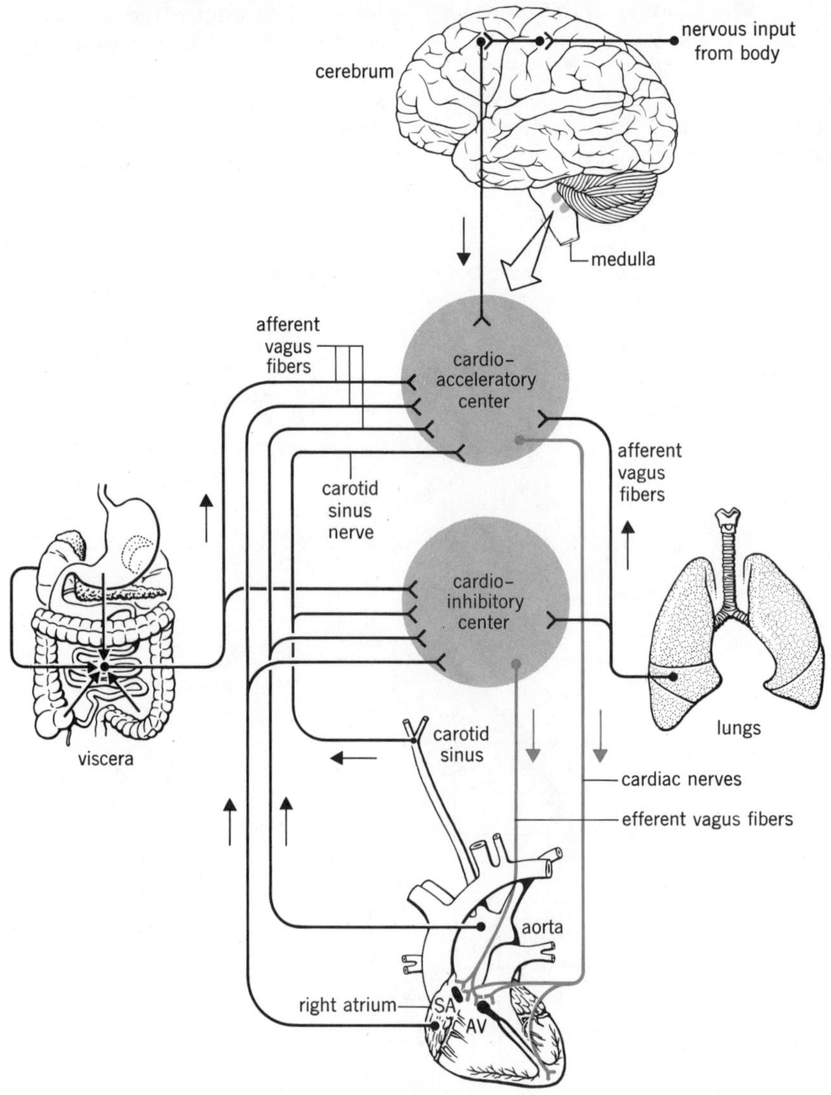

FIGURE 20.10
Cardiac reflex pathways.

the *aortic nerves* that join the sensory fibers of the *vagus nerves*. In general, these receptors respond to the stretch imposed upon them by an increase of blood pressure within the vessel and to the rate of change of that pressure, by an increased rate of firing. The afferent nerves terminate on both the cardioaccelerator centers (CAC) and cardioinhibitory centers (CIC); the effect of increased blood pressure on the CAC is inhibitory, and on the CIC, the effect is stimulatory. Thus, the "foot is taken off the accelerator and pressure is applied to the brake." The STROKE RATE DECREASES. Receptors similar to those of the sinus and aortic arch have been found in the pulmonary artery, in the walls of the left atrium, the pericardium, lungs, and alimentary tract. They also cause a slowing of heart rate by the mechanism described above.

CHEMORECEPTORS sensitive to blood oxygen and carbon dioxide levels are also found in the carotid sinus and aorta. Although their primary effect is on rate and depth of breathing, these receptors also influence heart rate. In general, the receptors increase their frequency of "firing" as blood oxygen levels decrease, as blood carbon dioxide levels increase, or as blood flow decreases in either area. The effect of mild changes in gas levels on heart rate is to speed it, by inhibition of the CIC, and stimulation of the CAC. Also, there is peripheral vasoconstriction, and the increased rate and increased peripheral resistance elevate the blood pressure. Excessive CO_2 anesthetizes the cardiac muscle and rate will fall.

Any stimulus that is *very strong* on any somatic peripheral nerve usually results in pain. Such stimulation excites the CAC and inhibits the CIC, increasing heart rate and raising cardiac output and blood pressure. This pathway may involve the cerebrum to interpret the pain, and shows the effects of brain involvement on heart rate.

In diving animals, and in the human as well, submersion of the head triggers a slowing of heart rate and vasoconstriction in all body areas except the heart, brain, and lungs. The source of the afferent input is not known, but may be in the nostrils.

This discussion has emphasized several points that may now be summarized.

Cardiac output may be altered by changing either stroke rate or stroke volume or both (Fig. 20.11).

Stroke volume is controlled primarily by the end diastolic volume and the operation of the Frank-Starling law that relates filling to strength of contraction.

Stroke rate is controlled primarily by nervous and chemical factors.

Table 20.4 summarizes the control mechanisms involved in cardiac output.

CARDIAC RESERVE

As the foregoing discussion has indicated, the heart has a great capacity to increase its output, so as to supply changing body requirements for blood. Table 20.5 presents data relating to distribution of the cardiac output and output at

TABLE 20.3
CARDIAC REFLEXES

NAME OF REFLEX	INPUT FROM	STIMULUS TRIGGERING	*EFFECT ON CIC[a]	*EFFECT ON CAC	*EFFECT ON SR[a]	*EFFECT ON SV
Aortic depressor	Baroreceptor (sensitive stretch) in aorta	Stretch of aorta	+	—	↓	0
Carotid sinus	Baroreceptor in carotid sinus	Stretch of sinus	+	—	↓	0
—	Baroreceptors in pulmonary arteries	Stretch of arteries	+	—	↓	0
Bainbridge reflex (existence disputed)	Vena cava or right atrium	Increased filling of atrium (stretching)	—	0/+	↑	0
—	Cerebrum—anger/fear	Stress, thoughts	—	+	↑	0
—	Strong stimulation of any nerve	Pain	—	+	↑	↑

* + (stimulate); — (inhibit); 0 (no change); ↑ (increase); ↓ (decrease).
[a] CIC = cardioinhibitory center; SR = stroke rate
CAC = cardioacceleratory center; SV = stroke volume

TABLE 20.4

MECHANISMS CONTROLLING CARDIAC OUTPUT

MECHANISM	COMPONENT(S) OF CO AFFECTED MOST	COMMENTS
Increased venous return to right atrium	Stroke volume	Brought about by law of the heart (increased tension causes stronger contraction and greater emptying)
Increased peripheral resistance	Stroke volume	As above
Temperature	Stroke rate	Effect on nodal tissue primarily
Chemicals		
Epinephrine	Both	Increases rate and strength
CO_2	Stroke volume	Greater filling causes more tension and stronger contraction
pH	Stroke volume	May act in same manner as CO_2
Acetylcholine	Both	These chemicals are liberated at nerve endings and cause change in rate. Acetylcholine slows, and norepinephrine speeds heart rate
Norepinephrine	Both	
Nervous reflexes	Stroke rate	Allows input from any body area; most necessary for continual adjustment to body requirements

TABLE 20.5

SOME FIGURES SHOWING DISTRIBUTION OF THE CARDIAC OUTPUT TO VARIOUS BODY AREAS AND TOTAL CARDIAC OUTPUT OF A NORMAL ADULT DURING VARIOUS ACTIVITY LEVELS

	ACTIVITY LEVEL AND BLOOD FLOW (ml/min)			
AREA SUPPLIED	REST	LIGHT	STRENUOUS	MAXIMAL
Visceral organs (gut, liver, pancreas, lungs)	1400	1100	600	300
Kidneys	1100	900	600	250
Brain	750	750	750	750
Heart (coronary)	250	350	750	1000
Skeletal muscles	1200	4500	12500	22000
Skin	500	1500	1900	600
Other	600	400	400	100
Total cardiac output	5800	9500	17500	25000

Note that blood is diverted away from "nonessential" areas (gut) to skeletal muscles and heart as activity increases. Note also that brain supply remains constant.

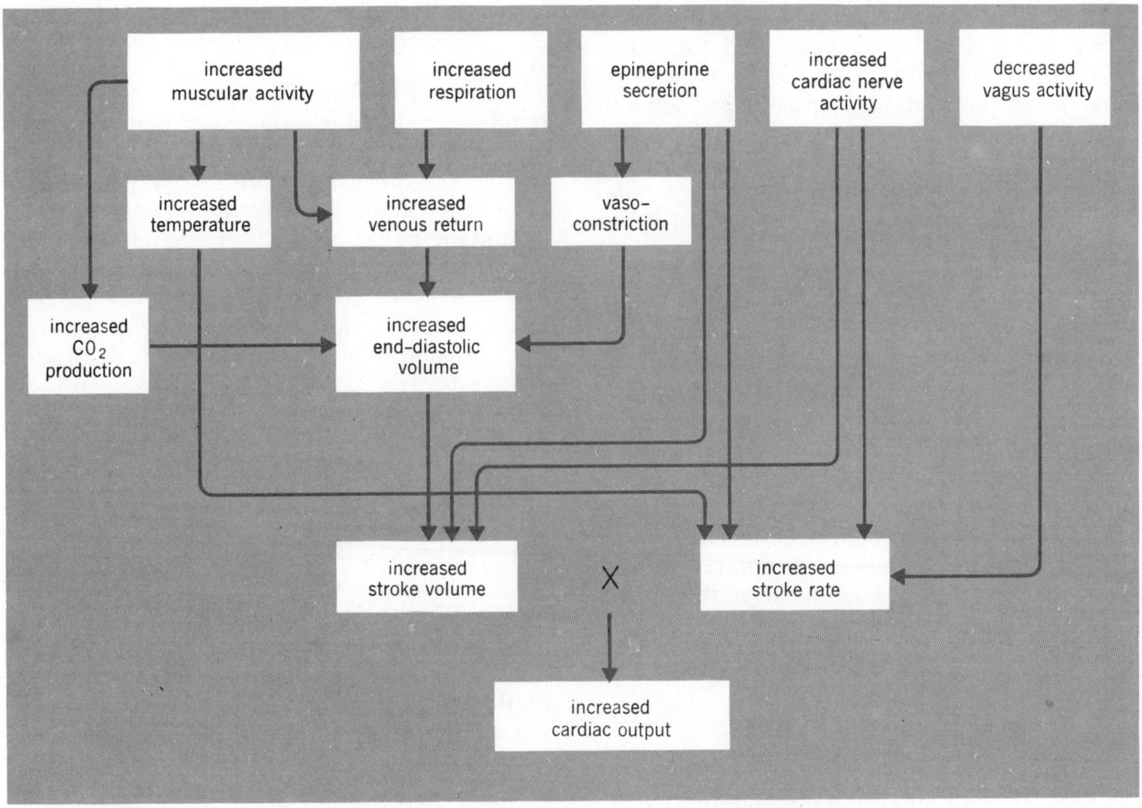

increased muscular activity
increased respiration
epinephrine secretion
increased cardiac nerve activity
decreased vagus activity
increased temperature
increased venous return
vaso-constriction
increased CO_2 production
increased end-diastolic volume
increased stroke volume
X
increased stroke rate
increased cardiac output

FIGURE 20.11
Some factors responsible for increase in cardiac output.

FIGURE 20.12
A ventricular function curve. Note that output increases according to input to a maximum value.

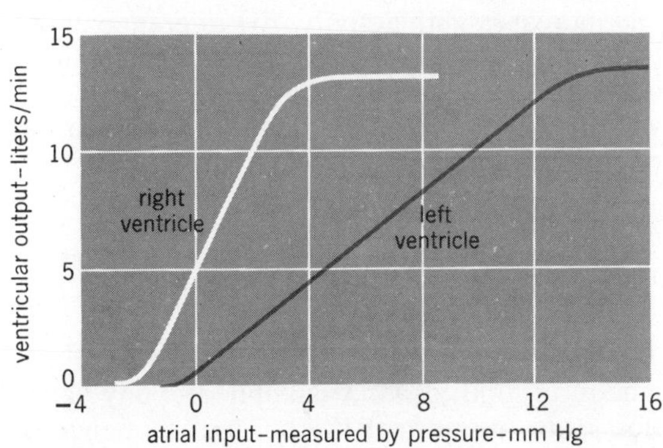

various activity levels. This ability constitutes the CARDIAC RESERVE. The magnitude of the response may be indicated by a *ventricular function curve* (Fig. 20.12). The curve relates inflow to the ventricles as the major determinant of output. Note that output increases to a maximum value imposed by the anatomical characteristics of the heart, primarily the volume of its chambers.

Inability of the cardiac output to keep pace with body demands for nutrients and waste removal constitutes HEART FAILURE. This occurs when resting cardiac output falls to the level of 2 to 2.5 liters per minute. Tissues are starved, suffocated, submerged, and poisoned by failure of the kidney to excrete wastes. One ventricle or both may fail. If the right ventricle fails, blood accumulates in the veins and capillaries returning blood to the heart, and venous pressure rises. This commonly results in edema of appendages and in the abdomen (ascites). Left ventricular failure usually causes pulmonary edema as fluid accumulates in the lungs, and low systemic blood pressure as left ventricular output declines.

MEASUREMENT OF CARDIAC OUTPUT

It is difficult to measure cardiac output accurately. Flowmeters may be installed in the course of a blood vessel, but this requires surgical procedures. The Fick method and dilution techniques are often employed and do not require surgery.

The FICK METHOD compares the oxygen or carbon dioxide content of femoral vein and femoral artery blood, and, knowing the rate of O_2 absorption or CO_2 elimination per minute by the lungs, calculates cardiac output by the equation

$$\text{Cardiac output} = \frac{O_2 \text{ absorbed or } CO_2 \text{ eliminated (ml/min)}}{\text{AV } O_2 \text{ or } CO_2 \text{ difference (ml/L)}}$$

For example:

$$\begin{aligned}
O_2 \text{ intake} &= 250 \text{ ml/min} \\
\text{Arterial } O_2 &= 19 \text{ ml } O_2/100 \text{ ml blood} \\
\text{Venous } O_2 &= 14.5 \text{ ml } O_2/100 \text{ ml blood} \\
\text{AV difference} &= 4.5 \text{ ml } O_2/100 \text{ ml blood} \\
&\quad \text{or } 45 \text{ ml } O_2/\text{L blood} \\
\text{Cardiac output} &= \frac{250 \text{ ml } O_2/\text{min}}{45 \text{ ml } O_2/\text{L blood}} \\
&= 5.5 \text{ L/min}
\end{aligned}$$

Dilution methods require injection of a known quantity of a substance into the circulation, followed by a determination of the dilution of the substance during a specified period of time. A curve is obtained, the area of which may be calculated and related to cardiac output. The rapidity of dilution, and the slope and height of the curve depend on the volume of blood circulated (cardiac output) in a measured span of time.

ENERGY SOURCES FOR CARDIAC ACTIVITY; THE CORONARY CIRCULATION

The myocardium (cardiac muscle) of the heart requires continual supplies of nutrients and oxygen if its activity is to continue normally. A pair of CORONARY ARTERIES arise just behind the cusps of the aortic valve. These vessels, containing blood just oxygenated in the lungs, give many branches that eventually form capillary beds in the myocardium. About 70 percent of the coronary flow is returned by CORONARY VEINS that empty into the right atrium through the CORONARY SINUS. The remaining 30 percent of the flow passes directly into the ventricular cavities from the capillary beds.

During the period of ventricular systole, flow through the coronary vessels is nearly zero for two reasons: contraction squeezes shut the vessels between the cardiac cells; the openings of the coronary arteries in the aorta are sealed by the flaps of the open aortic valve. Flow begins when the muscle relaxes and is thus intermittent, and the muscle is even more dependent on flow during this time. Flow is about 250 ml per minute at rest, and can rise to 1000 ml/min with strenuous exercise.

Carbohydrates account for about 35 percent of the energy supply of the human heart. Materials utilized that are carbohydrates or the products of carbohydrate metabolism include glucose, pyruvic acid, and lactic acid. About 60 percent of the energy supply is obtained from the metabolism of fatty acids. Ketone body and amino acid metabolism account for the remaining 5 percent. Oxygen consumption at rest is about 9 ml per 100 g of heart tissue per minute, and increases with activity. pH tolerance of the heart is quite wide, with beats being maintained between pH 5 and 10. The pH tolerance and wide use of nutrients as fuels contribute to the continued action of this vital organ.

ALTERATIONS IN RATE AND RHYTHM OF THE HEARTBEAT

ABNORMAL RHYTHMS OF THE HEART

The term ARRHYTHMIA is applied to any variation from the normal rhythm and sequence of excitation of the heart. Arrhythmias result from the following factors:

Alterations in rate of SA node activity.

Interference with conduction.

Presence of ectopic foci.

Various combinations of the above.

Alterations in rate of SA node activity. BRADYCARDIA (*decreased rate*), or TACHYCARDIA (*increased rate*), of SA node activity, and thus heart rate, is normally associated with breathing and sleep. Heart rate increases on inspiration, decreases on expiration, and decreases due to increased vagal stimulation during sleep.

Abnormal alterations may also be associated with drug intake. Some drugs that increase heart rate are atropine, caffeine, epinephrine, and camphor. Some that decrease rate are amylnitrite, nitroglycerine, and alcohol in small quantities.

Interference with conduction. HEART BLOCK results from damage to various parts of the nodal conducting system, or from excessive vagal stimulation, and results in an obstruction to the passage of electrical impulses from atria to ventricles. Three degrees of block are recognized.

FIRST-DEGREE BLOCK causes a pause or delay in transmission of depolarization, and thus slows ventricular rate without separation of atrial and ventricular beating rates.

SECOND-DEGREE BLOCK (*partial* or *incomplete block*) results in some sinoatrial impulses being missed by the ventricle. A slight separation of atrial and ventricular rhythm results (e.g., 75 beats/min of the atria, with 60 beats/min of the ventricles).

THIRD-DEGREE BLOCK (*complete block*) results in complete separation of atrial and ventricular beat, with each area controlled by that part of the nodal system closest to the muscular tissue. Thus, the atria may still be responding to sinoatrial stimulation (rate 80 beats/min) while the ventricles are responding to atrioventricular node or bundle stimulation (rate 40 beats/min). If cyanosis, faintness, and convulsions are seen, insufficient cardiac output is indicated. Third-degree block may require implantation of an ARTIFICIAL PACEMAKER if the rate falls to a point where adequate cardiac output cannot be maintained.

The pacemaker provides a source of stimulations that is delivered to the heart by way of long wires or leads. The leads may be inserted through a vein (*transvenous route*) into the right ventricle for temporary or permanent pacing or surgically inserted in the ventricular wall (*transpericardial route*) for permanent pacing. The latter procedure involves opening the chest. The pacer may provide one of three types of pacing.

SET OR ASYNCHRONOUS PACING. This provides a constant rate of stimulation regardless of sinoatrial activity or alterations in muscular activity.

SYNCHRONOUS PACING. The pacer fires on detection of sinoatrial node activity and delivers ventricular stimulations that vary with SA node activity. This type of pacing thus varies with muscular activity levels.

DEMAND PACING. The pacer is set to fire only when it detects a rate of sinoatrial node discharge that is less than some preset value. Thus, this type only stimulates the ventricles when the heart itself fails to achieve an adequate rate of beat.

Some of the hazards associated with pacemaker installation include infections, thrombosis, power failure, and lead breakage. Isotope-powered pacemakers, and those that may be recharged from outside the body without removal of the pacemaker may improve the reliability of these instruments.

Presence of ectopic foci. *Flutter,* with coordinated heart rates of 200 to 300 beats per minute, and *fibrillation,* with incoordinated beats, are abnormal rhythms of the heartbeat. In most cases, they are due to areas of the heart, other than nodal tissue, assuming pacemaker activity. Such areas compete with the SA node to stimulate the muscle. The muscle may be divided into several separately contracting areas, each controlled by its own pacemaker. These

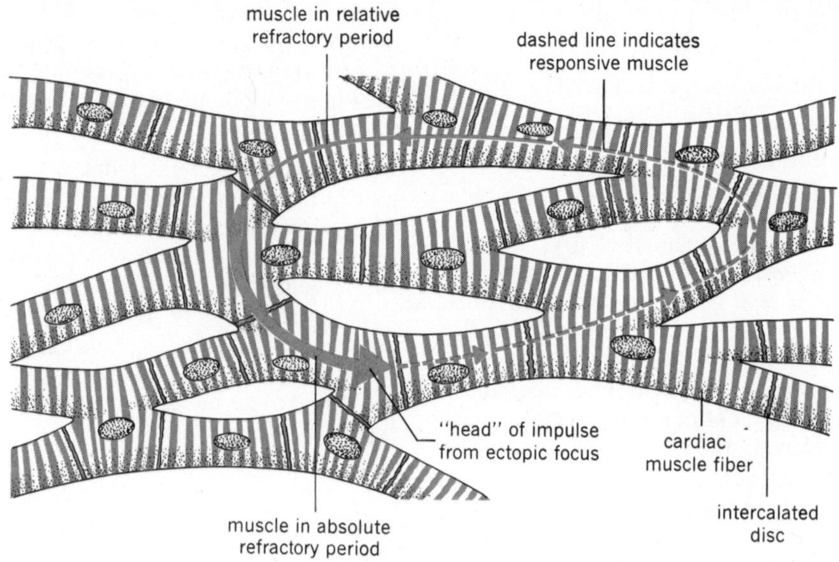

FIGURE 20.13
The development of a circular conduction pathway (circus movement) in a large mass of cardiac muscle. A prime requisite for the development of such a pathway is that it be long enough to ensure that the muscle be past the refractory periods and capable of responding to the stimulus as it traverses the pathway.

abnormal areas of depolarization are the ECTOPIC FOCI. The development of CIRCULAR repetitive areas of depolarization constitute "*circus movements*" (Fig. 20.13). Different areas of the muscle contract at different times and coordination is lost. In most cases, throwing the heart into a simultaneous state of depolarization by the application of external shock (*defibrillation*) usually reestablishes the coordination necessary for efficient circulation of blood.

THE CARDIAC CYCLE

A cardiac cycle is defined as one complete series of events during heart activity. Because electrical activity precedes all other events in the cycle, the beginning of a cycle is usually taken as the initiation of a wave of depolarization by the SA node. Within the framework of a single cycle, three important events occur.

The generation of recordable ELECTRICAL DISTURBANCES as the depolarization wave spreads through the heart.

Contraction of cardiac muscle and the generation of PRESSURE and VOLUME CHANGES within the heart.

The GENERATION OF SOUNDS through valve action and blood flow as the heart muscle contracts.

TIMING (FIG. 20.14)

In a preceding section, it was stated that the rate of heartbeat varies according to many factors. A typical resting rate may be considered to be about 70 beats/minute. At such a rate, one complete cycle takes about 0.8 sec. ATRIAL SYSTOLE (*contraction*), requires about 0.1 sec with ATRIAL DIASTOLE (*relaxation*) occupying about 0.1 sec, and ATRIAL DIASTASIS (*rest*) occupying 0.6 sec. Ventricular systole takes about

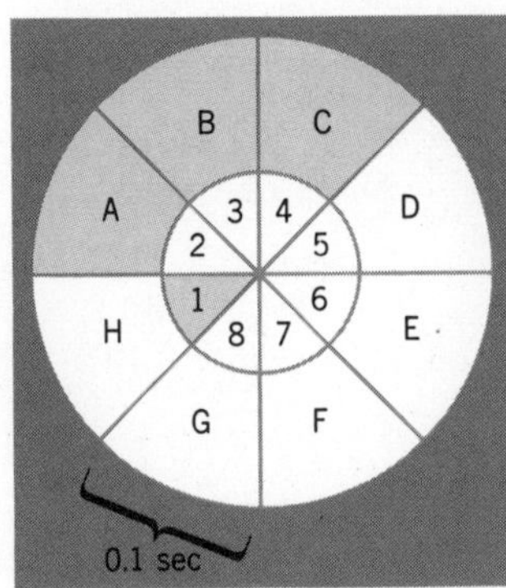

FIGURE 20.14
The timing of a cardiac cycle. Rate of beating is 70 to 72 beats/minute. 1 = atrial systole; 2 = atrial diastole; 3–8 = atrial diastasis. A–C = ventricular systole; D = ventricular diastole; E–H = ventricular diastasis.

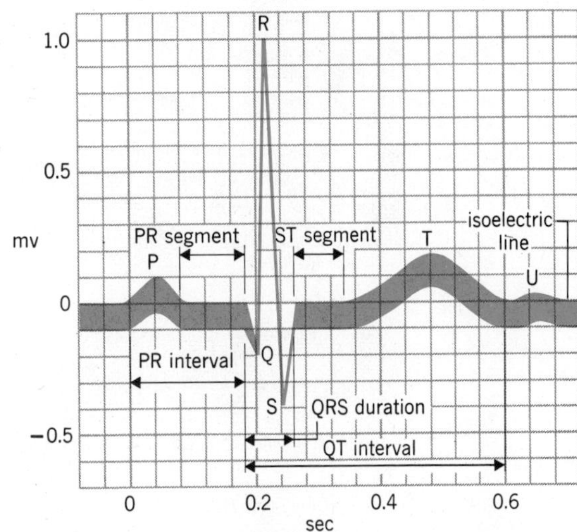

FIGURE 20.15
The normal electrocardiogram. *U* waves are not present in every ECG.

0.2 sec, diastole about 0.2 sec, and diastasis 0.4 sec.

THE ELECTROCARDIOGRAM

As the depolarization wave initiated by the SA node sweeps over the heart, active tissue becomes electrically negative to inactive areas. This generates a cardiac electrical field that may be recorded as an ELECTROCARDIOGRAM (*ECG* or *EKG*). A normal ECG is shown in Figure 20.15. The various deflections are designated by letters; between letters are a variety of intervals or complexes. The deflections, intervals, and complexes are summarized with an explanation of their causes in Table 20.6.

TABLE 20.6
PHASES OF THE ELECTROCARDIOGRAM

ITEM (DESIGNATION)	APPROXIMATE TIME (SEC)	CAUSES OR EVENTS OCCURRING
P wave	0.1	Excitation of the atria
PQRS interval	0.12–0.2	Wave travels through A-V node, bundle, and Purkinje fibers
QRS complex	0.08	Excitation of the ventricles
T wave	0.1	Repolarization of the ventricles
QT interval	0.35	Time required for complete excitation and recovery of the ventricles
S-T segment	0.1	Represents time for complete excitation of ventricles
U wave	0.1	Thought to be due to slow repolarization of papillary muscles. Appears with hypokalemia. Not present in all ECGs

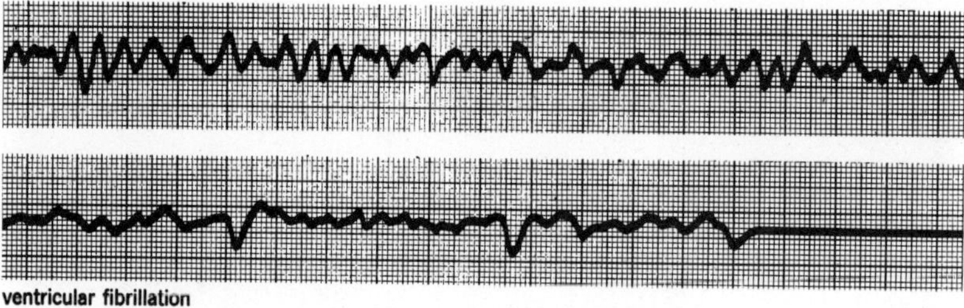

ventricular fibrillation

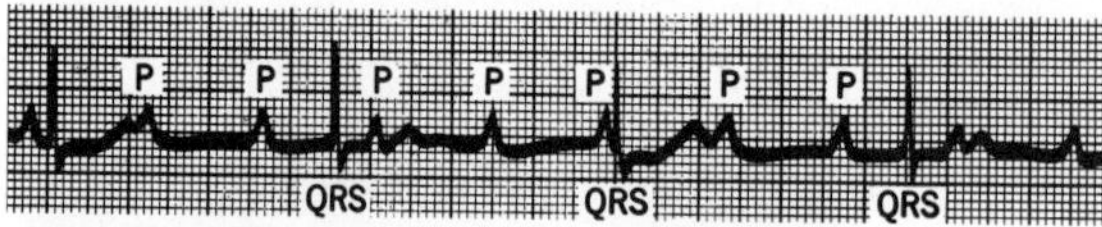

complete heart block (atrial rate, 107; ventricular rate, 43)

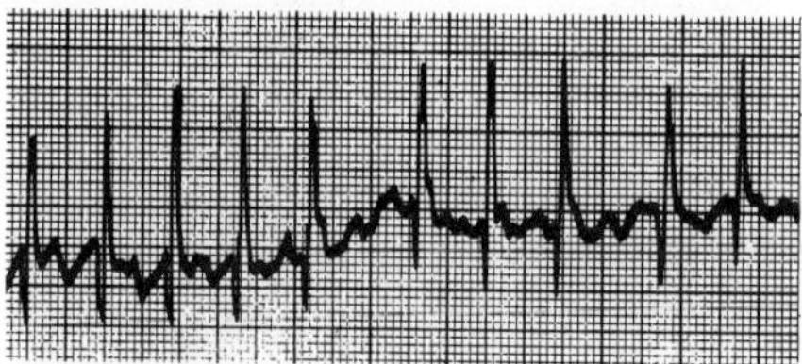

atrial fibrillation

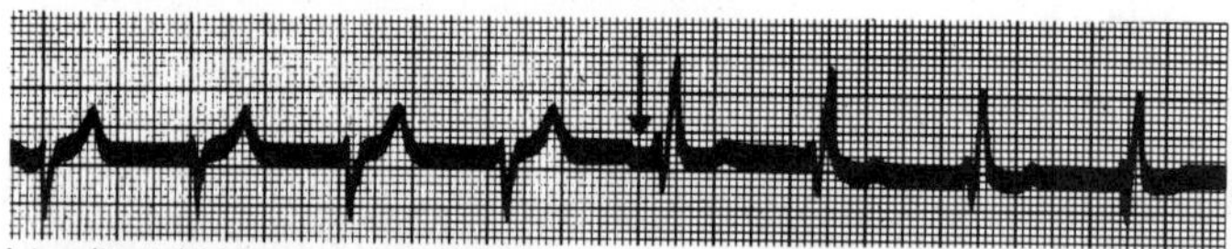

intermittent right bundle branch block

FIGURE 20.16
Some abnormal electrocardiograms in various conditions.

The value of the ECG lies in the fact that any alterations of conduction and status of the nodal or cardiac tissue must and will be reflected as a deviation from the normal ECG. Alterations in polarity or voltage are seen in the vertical plane; alterations in timing are seen horizontally. The analysis of ECGs has proceeded to the point that definitive alterations are characteristic of specific disorders of the heart (Fig. 20.16). The ECG of a damaged heart may be followed as the organ recovers, and this gives clues as to the rapidity and completeness of the recovery process.

In recording an ECG in a human subject, electrodes may be attached to any part of the body. Conventionally, the so-called STANDARD LEADS are attached to the two wrists and to the left ankle. This type of placement creates a triangle within which the heart is enclosed. *Lead I* lies between the right and left wrists. *Lead II* lies between the right wrist and left ankle, and *Lead III* between the left wrist and left ankle (Fig. 20.17). The electrical deflections in these standard leads are shown in Figure 20.18. The height of the deflections in these leads is determined by the amount of tissue depolarizing and by the status of the muscle and nodal tissue.

Nine ADDITIONAL LEADS may be employed to assess the function of particular areas of the myocardium. Six positions on the chest designated V_1 to V_6 (Fig. 20.17) may be employed.

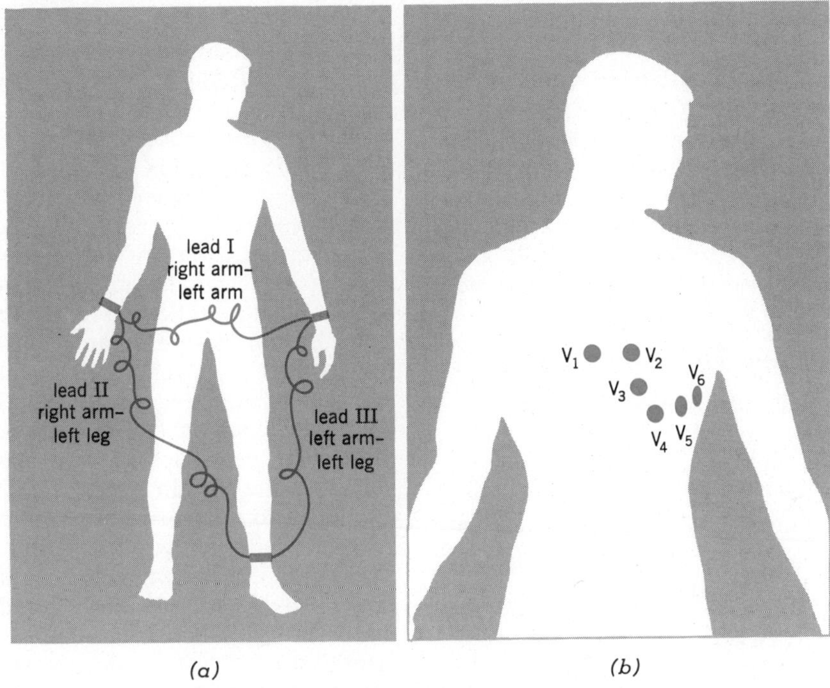

FIGURE 20.17
(a) The three standard electrocardiographic leads, (shown not connected to the machine), and (b) the placement of the chest leads.

FIGURE 20.18
A normal 12-lead electrocardiogram. Direction of deflections and height vary in each lead. (From PHYSIOLOGY OF MAN, 4th ed., by L. Langley, Van Nostrand Reinhold Co., 1971.)

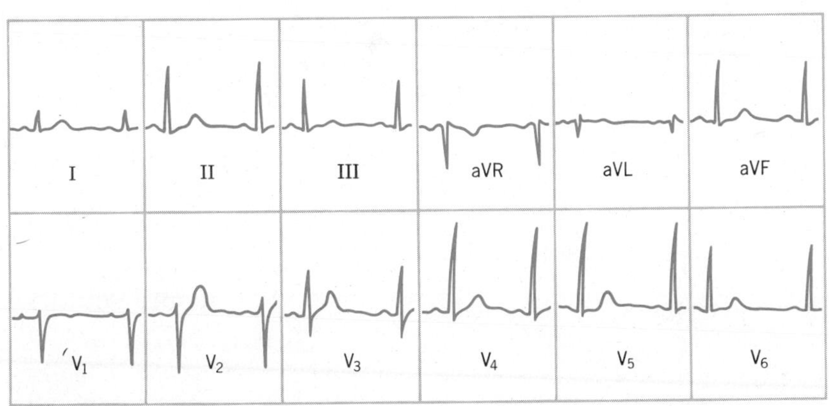

Wiring the three standard leads so that the positive lead is alternately on the left ankle, right arm, or left arm, produces three leads designated respectively, *aVF, aVR,* and *aVL.* This type of wiring "looks at the heart" from the particular lead that is positive, and gives an electrical picture of that surface of the organ. Normal records from chest leads and the aV leads are shown in Figure 20.18. Note that height and direction of the various deflections are not the same in all leads.

PRESSURE AND VOLUME CHANGES

Pressure changes occur within the atria, ventricles, and arteries leaving the ventricles (aorta and pulmonary artery). Significant volume changes occur only in the ventricles as they contract and relax. These changes and their correlation with ECG and heart sounds are shown in Figure 20.19.

Atrial contraction commences immediately after the P wave on the ECG. The right atrial pressure rises from about 3 mm Hg DIASTOLIC PRESSURE, to a SYSTOLIC PRESSURE of 7 mm Hg; left atrial pressure rises from 3 mm Hg diastolic to about 10 mm Hg systolic. This force imparted to the blood by atrial contraction is of minimal importance to ventricular filling. Most of the filling of the ventricle occurs when the ventricle relaxes and creates a lowered pressure that "draws" blood into the expanding chamber. Filling continues during diastasis and terminates with initiation of ventricular systole. The atria thus become receiving and storage chambers for blood returning from lungs and body. During atrial contraction, the tricuspid and mitral valves are open because the ventricular pressure is lower than that of the atria; blood flows from atrium to ventricle. Atrial excitation is followed by the QRS segment of the ECG and ventricular excitation; the ventricles contract. Ventricular pressure begins to rise. It rises above that of the atria, and, at this point, the tricuspid and mitral valves are firmly closed (first heart sound). There is no exit of blood from the ventricles, since they have not generated sufficient pressure to open the arterial valves, closed from the preceding cycle. The ventricles thus enter a period of ISOMETRIC CONTRACTION. The right ventricle must achieve a pressure of about 18 mm Hg to open the pulmonary valve; the left ventricle must create about 80 mm Hg pressure to open the aortic valve. The thicker wall of the left ventricle renders the ventricle capable of generating this higher pressure. At the point where ventricular pressure exceeds arterial pressure, the valves of the arteries are opened and blood is rapidly EJECTED from the ventricles. Peak systolic pressures of about 30 mm and 130 mm Hg are created by the right ventricle and the left ventricle respectively. A momentum is imparted to the blood that causes continued ejection but at a slower rate, even though the ventricles are now starting to relax and pressure within them is falling. Pressure continues to decline within the ventricles until pressure falls below that within the arteries. The blood attempts to return to the ventricles and, in so doing, closes the arterial valves (second heart sound). An incissura (notch) in the arterial pressure curve occurs as the blood rebounds from the closed valve and sets up a secondary shock wave in the artery and pressure rises. The ventricles are again closed chambers, and no blood is moving into them; a period of ISOMETRIC RELAXATION is entered into. Fall of pressure in the ventricles continues until it drops below that of the atria. At this point, the tricuspid and mitral valves open and ventricular filling begins. Note that movement of blood, and valve action, is secondary to pressure changes.

The volume of the ventricles during diastasis is normally 100 to 150 ml per chamber. Under resting conditions, about two-thirds of the contained blood is ejected as the ventricles contract. This leaves a reserve of blood in the ventricle that may be ejected by a more forcible contraction. This ability to increase the amount of blood per beat is an important factor in changing the cardiac output to meet varying body demands.

HEART SOUNDS

Two distinct sounds are normally heard through a stethoscope as the heart works. The FIRST HEART SOUND occurs early in ventricular

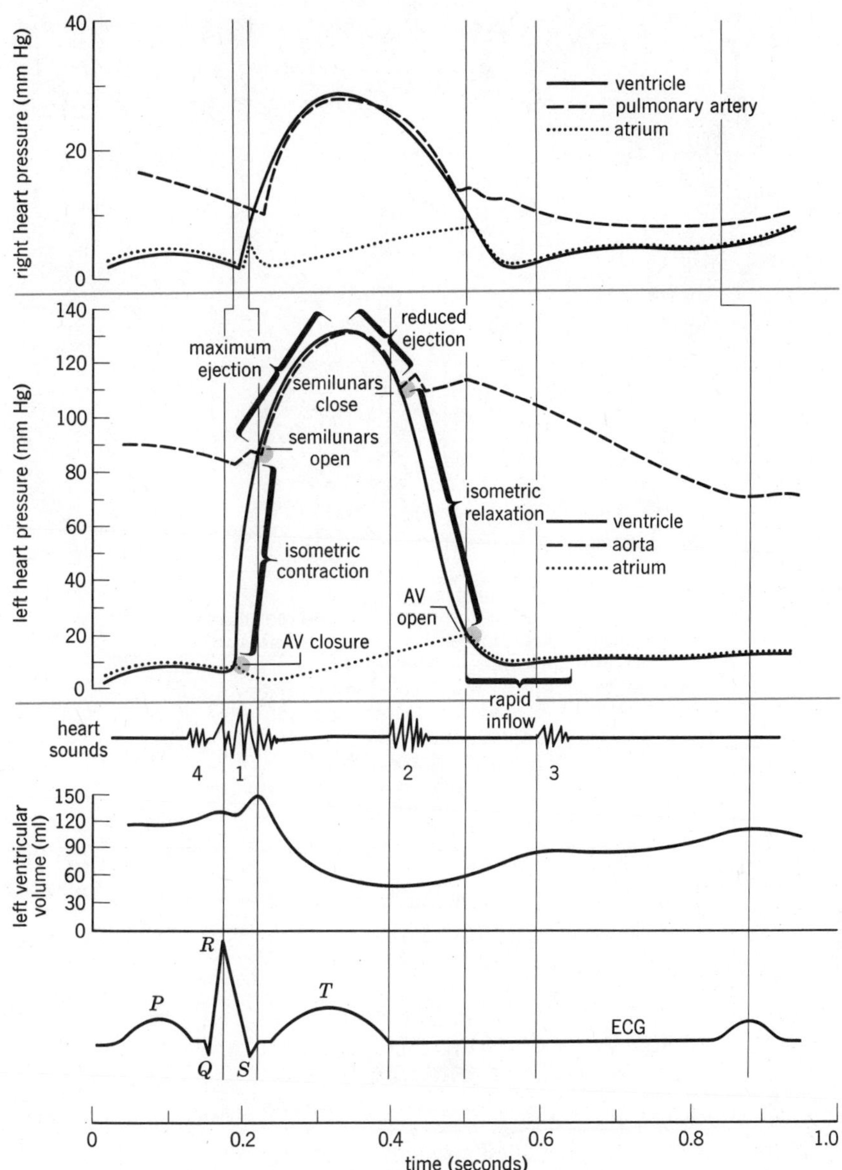

FIGURE 20.19
Correlation of events in a cardiac cycle.

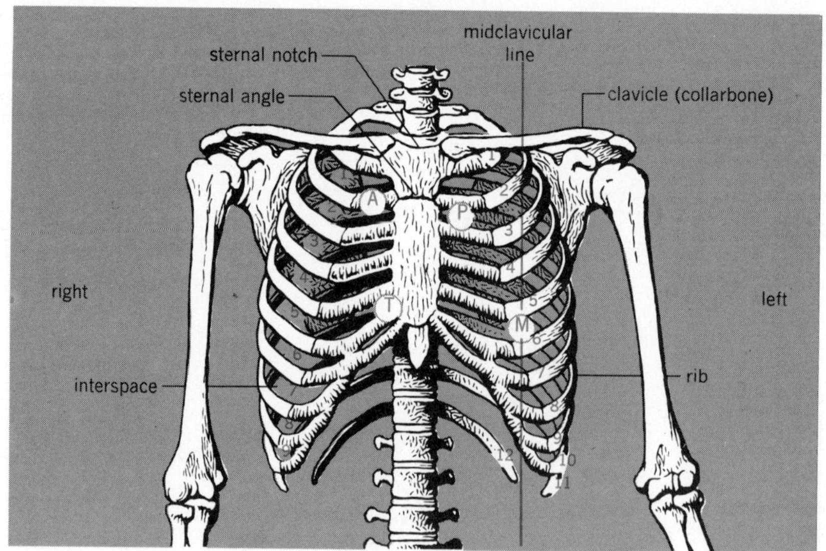

FIGURE 20.20
Chest areas where adult valve sounds may be heard most clearly. Interspaces are numbered according to the rib that lies above. The second rib is the one easiest to locate because it attaches at the sternal angle about two inches below the sternal notch. A = aortic valve; P = pulmonary valve; T = tricuspid valve; M = mitral (bicuspid) valve.

TABLE 20.7
A SUMMARY OF EVENTS OCCURRING DURING THE CARDIAC CYCLE

EVENT	CAUSE(S)	COMMENTS
ECG (electrocardiogram)	Depolarization of nodal tissue and cardiac muscle	Characteristic pattern for normal and abnormal states. Has diagnostic value
Pressure changes	Contraction of cardiac muscle	Pressure changes eject blood from a chamber and operate the valves
Two sounds normally heard	1. Closure of valves 2. Turbulence of blood flow 3. Muscle and valve vibrations	First sound due to all three causes. Second sound due only to valve closure.

contraction, is rather drawn out, and is of low frequency (30 to 45 Hz). Three factors contribute to its production.

VIBRATION OF THE ATRIOVENTRICULAR VALVES during and after closure.

TURBULENCE in the stream of blood as it is ejected through the arterial openings.

VIBRATIONS IN THE MUSCLE of the ventricles as they contract.

The SECOND HEART SOUND occurs as ventricular pressure begins to fall from peak values. It is shorter in duration than the first sound and is higher pitched (50 to 70 Hz). CLOSURE OF THE ARTERIAL VALVES causes this sound.

With amplification, a third and fourth heart sound (Fig. 20.19) may be detected. The third heart sound occurs during or after atrial contraction and is attributed to the tensing of ventricular walls due to the rush of blood from atrium to ventricle. The fourth heart sound may be heard during atrial contraction and is attributed to the muscular contraction of these chambers.

Intensity of the sounds depends on the force with which the heart is beating, becoming louder with stronger activity. Splitting of the sounds may occur, particularly the second sound, since the valves do not close synchronously (together). The sites where the sounds created by the action of each valve may be best appreciated are shown in Figure 20.20.

MURMURS may occur due to regurgitation (backflow) of blood when valves fail to close completely. *Functional murmurs* are no threat to the body because they usually occur only during strenuous exercise and are outgrown. *Resting murmurs* are a cause for concern, for they usually reflect the presence or result of an abnormal process in the body. Narrowing (*stenosis*) of an orifice or artery may also produce modification of a sound. The heart sounds thus have some diagnostic value to the physician. The major events of the cardiac cycle are shown in Figure 20.19 and are summarized in Table 20.7.

SUMMARY

1. The heart creates a pressure to circulate blood through the body's blood vessels.
   a. The right ventricle and vessels to and from the lungs constitute the pulmonary circulation.
   b. The left ventricle and the vessels to and from the body generally constitute the systemic circulation.
2. The heart is a four chambered hollow organ with valves to control direction of blood flow.
   a. It consists of cardiac muscle for contraction.
   b. It has nodal tissue to initiate contraction.
   c. It has a fibrous connective tissue skeleton.
3. Nodal tissue originates and distributes depolarization that causes cardiac muscle contraction.
   a. The SA node acts as the pacemaker of the heart.
   b. AV node, bundle, bundle branches, and Purkinje fibers distribute depolarization to the cardiac muscle.
   c. Nodal tissue exhibits spontaneous depolarization, explained by sodium leakage into the fibers.
   d. The nodal tissue assures a nearly simultaneous stimulation of the ventricular muscle.
4. Cardiac muscle contracts because of nodal tissue stimulation and creates pressure to circulate the blood.
   a. It is a variety of striated muscle.
   b. It exists in a functional syncytium; stimulation to any part is distributed to all parts of the muscle.
   c. The muscle follows the all-or-none principle, giving maximal contraction whose strength may vary according to the environment of the muscle at the time of contraction.
   d. Greater tension on the muscle results in a stronger contraction (law of the heart).
   e. The muscle has relatively long refractory periods that aid in its ability to pump blood.
   f. Extrasystoles may occur due to stimuli arriving at the muscle other than through the nodal tissue.
5. Cardiac output is mls of blood pumped per minute by one ventricle.

a. It is the product of rate of beat and mls ejected per beat.
b. It can be varied by changing rate of beat and amount of blood ejected per beat.

6. Factors controlling rate of beat include
   a. Temperature; increase raises rate and vice-versa.
   b. Chemicals; certain changes in the ionic environment alter rate; catecholamines and digitalis increase rate, as do atropine and caffeine; nitroglycerine and *excessive* CO_2 levels decrease rate.
7. Factors controlling volume of ejection include
   a. Venous inflow to the atria.
   b. Pressure in the system against which the heart ejects blood.
   c. Calcium: increased levels result in a stronger contraction that empties the ventricles more completely.
   d. Catecholamines increase volume of ejection; hypoxemia decreases ejection.
8. Nervous influences are most pronounced on heart rate.
   a. Cardioaccelerator (sympathetic) nerves increase rate.
   b. Cardioinhibitory nerves (parasympathetic) decrease rate.
   c. The nerves are derived from centers that are reflexly controlled; sensory input from blood vessels and body viscera generally results in lowered heart rate; pain and cerebral activity generally increase rate.
9. The cardiac reserve refers to the ability of cardiac output to be raised to meet body demands.
   a. Heart failure occurs when output cannot keep pace with demand.
10. Cardiac output may be measured by flowmeters, arteriovenous gas differences, or by dilution techniques.
11. Blood for sustaining heart activity is delivered via the coronary circulation.
    a. Fats, carbohydrates, and amino acids are the sources of energy for cardiac activity.
    b. Oxygen supply is essential for normal activity.
    c. pH tolerance is quite wide.
12. Abnormal rhythms of the heart may result from
    a. Alterations of SA node activity secondary to drug intake or ionic imbalances.
    b. Conduction problems through the nodal tissue; several degrees of heart block may result.
    c. Ectopic foci or areas of altered excitability may cause fibrillation.
13. A cardiac cycle is one complete series of events in cardiac activity.
    a. At resting rates of beat, a cycle usually takes about 0.8 second.
    b. An electrical record of activity constitutes an electrocardiogram.
       1) It may be taken from leads placed in different body areas.
       2) It has several parts and phases that may be interpreted to reflect heart status.
       3) Abnormal electrocardiograms are associated with abnormalities of the heart's functioning.
    c. Pressure changes occur in the heart's chambers and in the arteries leaving those chambers that move the blood and operate the valves.
    d. Heart sounds occur as valves close and muscle contraction occurs. Abnormalities of the sounds are usually associated with valve malfunctioning.

QUESTIONS

1. Compare the functions of the various portions of the vascular system. How is the heart related to these vessels?
2. What devices are available to the heart to enable increase in volume of blood pumped when the demand for increased circulation occurs?
3. Explain the role of the nodal tissue in establishing the rhythmical nature of the heartbeat.
4. In what ways is cardiac muscle especially suited to its tasks of circulation of the blood and adjustment of amount of blood pumped?
5. Describe the role of nerves in alteration of heart rate.
6. Discuss the role of cardiac reflexes in controlling blood pressure.
7. Suppose that a kidney malfunction results in potassium retention and calcium loss. What would be expected in terms of derangement of heart action?
8. What is the physiological justification for the use of epinephrine and digitalis as stimulants to

heart action? Are there any cautions associated with the use of these chemicals?

9. What defect is implied in "heart block"? Speculate as to the effect of the various degrees of block on maintenance of adequate cardiac output.
10. What is the value of an electrocardiogram? How does the recording indicate changes in conduction times? In strength of depolarization (and therefore condition of the cardiac muscle)?
11. Discuss the pressure changes occurring within the ventricles that lead to ejection of blood and filling of the ventricles.
12. What are the causes of the first and second heart sounds? How are modifications of these sounds related to valvular defects?
13. How does an increased return of blood to the right atrium stimulate cardiac output? How is the increased venous return brought about?

READINGS

Benditt, Earl P. "The Origin of Atherosclerosis." Sci. Amer. *236*:74. (Feb) 1977.

Blackburn, Henry. "Contrasting Professional Views on Atherosclerosis and Coronary Disease." *New Eng. J. Med. 292*:105. Jan 9, 1975.

Braunwald, Eugene. "Current Concepts: Determinants of Cardiac Function." *New Eng. J. Med. 296*:86. Jan 13, 1977.

Cannon, Paul J. "The Kidney in Heart Failure." *New Eng. J. Med. 296*:26. Jan 6, 1977.

Kolata, Gina Bari. "Coronary Bypass Surgery: Debate Over Its Benefits." *Science 194*:1263. 17 Dec 1976.

Kolata, Gina Bari. "Detection of Heart Disease: Promising New Methods." *Science 194*:1029. 3 Dec 1976.

Kolata, Gina Bari, and Jean L. Marx. "Epidemiology of Heart Disease: Searches for Causes." *Science 194*:509. 29 Oct 1976.

Marx, Jean L. "After the Heart Attack: Limiting the Damage." *Science 194*:1147. 10 Dec 1976.

Marx, Jean L. "Cardiovascular Disease and the Forms It Takes." *Science 194*:511. 29 Oct 1976.

Marx, Jean L. "Sudden Death: Strategies for Prevention." *Science 195*:39. 7 Jan 1977.

21

THE ROLE OF THE BLOOD VESSELS IN CIRCULATION

OBJECTIVES

After studying this chapter, the reader should be able to:

☐ Describe the role of blood vessels in control of blood pressure, flow, and volume adjustment in the cardiovascular system.

☐ Describe the functional anatomy of the arteries, capillaries, and veins.

☐ Summarize the factors leading to creation of blood pressure.

☐ Summarize the "laws" governing the flow, pressure, and velocity of fluid flow in tubes.

☐ Apply these laws to the same parameters concerning the flow of blood through blood vessels.

☐ Define systolic, diastolic, pulse, and mean pressure as applied to the vascular system.

☐ Define the role of arteries in control of blood pressure and describe the characteristics of blood pressure, flow, and velocity in arteries.

☐ Define what the microcirculation is, and how pressure, flow, and velocity ensure the exchange role of this part of the vascular system.

☐ Explain the role of the venous circulation in the body.

☐ Describe the factors ensuring blood flow through veins.

☐ Describe how blood vessel size is controlled, and relate this to maintenance of pressure homeostasis.

☐ Describe the effects of postural changes, exercise, and temperature and chemical changes on the circulation.

☐ Describe the special characteristics of circulation in several special areas of the body, including the lungs, brain, skin, liver, and skeletal muscles.

☐ Define what shock is, list the various types, and describe what happens to the circulation in each type.

☐ Describe the compensatory reactions the body exhibits in shock.

☐ Describe the various types of hypertension, some possible causes, and the aims of treatment.

The blood vessels of the body contain the blood and allow a pressure to be maintained within those vessels. If the heart simply pumped the blood into an open cavity (as it does in some invertebrates) pressure would be lost. The vessels also direct the blood flow to specific organs so as to ensure their adequate nutrition and removal of wastes. In other words, a *directed* flow of blood is established by a vascular system. The vessels also permit direction of blood flow to be altered—that is, when a tissue becomes active, that area will receive a greater flow, at the expense of course, of an area not needing as much flow. Changes in pressure are often associated with redirection of flow. A vascular system further permits, with heart action, the regulation of blood pressure.

The discussion to follow is directed primarily toward the systemic circulation, with special areas discussed separately.

PHYSIOLOGICAL ANATOMY OF THE BLOOD VESSELS

ARTERIES are defined as vessels carrying blood *away from the heart,* regardless of the level of oxygenation of that blood. The largest of the arteries, such as the pulmonary artery and the aorta, are termed ELASTIC ARTERIES, because the greater thickness of their walls is composed of *elastic connective tissue* membranes. These vessels are stretched as the ventricles eject blood into them. This stretching action tends to cushion the rise of systolic pressure that would occur if the vessels were rigid. As the ventricles relax, the elastic recoil of these vessels pushes the blood onwards through the arteries and maintains a continual movement of blood. Medium-sized and small arteries (arterioles) are formed by branching from the large arteries, and they contain many layers of *smooth muscle* in their walls. They are thus named MUSCULAR ARTERIES. As the muscle contracts or relaxes, the total cross-sectional area of the vessels is altered, resulting in alterations of peripheral resistance, blood pressure, and blood flow. Though smaller in diameter than the elastic arteries, muscular arteries (especially the arterioles) are very numerous, so that the *total* cross-sectional area of the vessels increases. CAPILLARIES are smaller and still more numerous, presenting even more total cross-sectional area to the blood flow. They form net-

FIGURE 21.1
The histology of blood vessels.

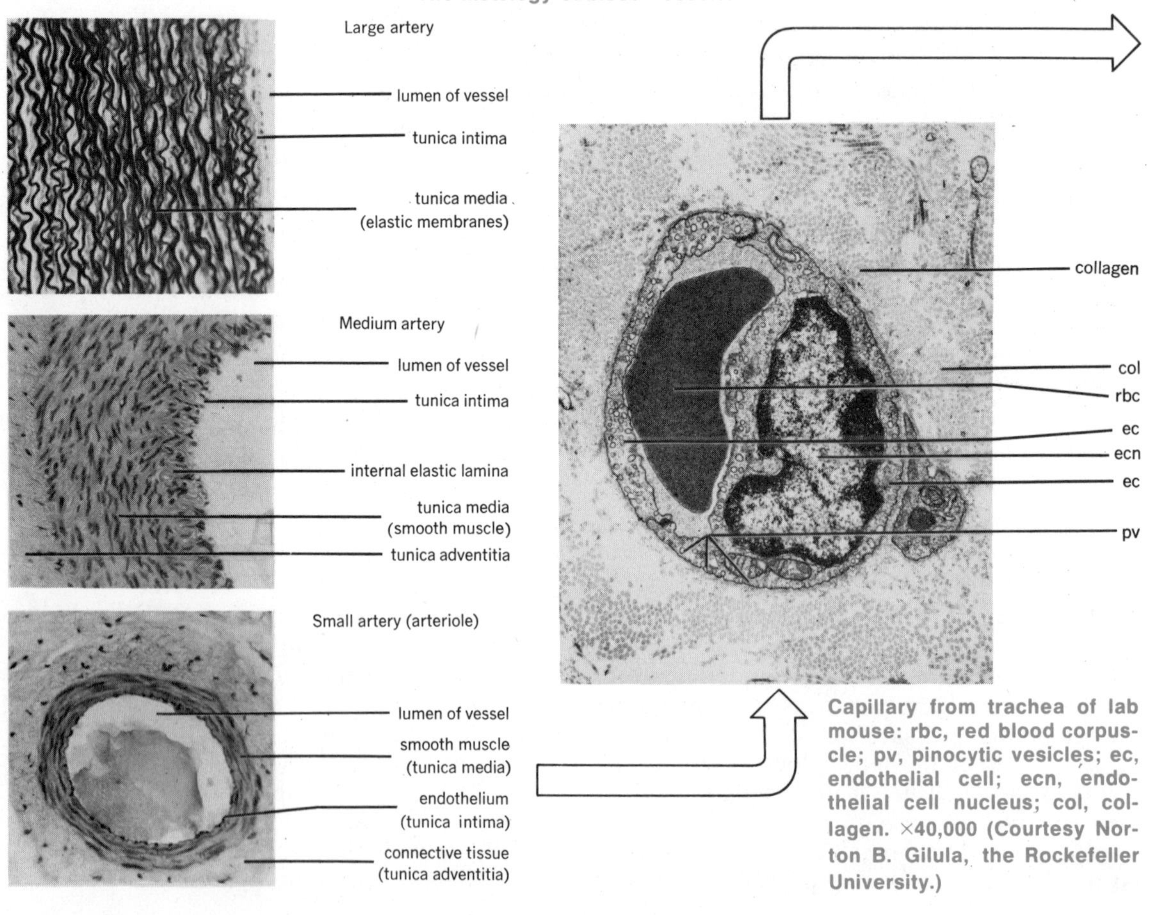

Capillary from trachea of lab mouse: rbc, red blood corpuscle; pv, pinocytic vesicles; ec, endothelial cell; ecn, endothelial cell nucleus; col, collagen. ×40,000 (Courtesy Norton B. Gilula, the Rockefeller University.)

works or *beds* within body tissues and organs, and their walls are extremely thin. Capillaries serve as the *area of exchange* of materials between the blood and interstitial fluid compartment. The latter surrounds cells and thus a pathway is established between blood and cells. A capillary and its supplying and draining vessels form a "microcirculatory unit" that forms a basic homeostatic unit of the vascular system. Small veins or VENULES drain the capillary beds. They have *collagenous connective tissue* as the major component of their walls, a nonelastic type of connective tissue. They are somewhat permeable to water and small solute molecules, and play a role in exchange of materials. The larger venules and MEDIUM-SIZED VEINS have some smooth muscle in their walls, and can, to a degree, alter their size and thus change the volume of the venous circulation. Veins are mainly volume rather than pressure vessels. The LARGE VEINS, such as the *venae cavae* that carry blood to the right atrium of the heart, have a connective tissue wall and serve as volume and conducting vessels. The major anatomical features of each type of vessel is presented in Figure 21.1.

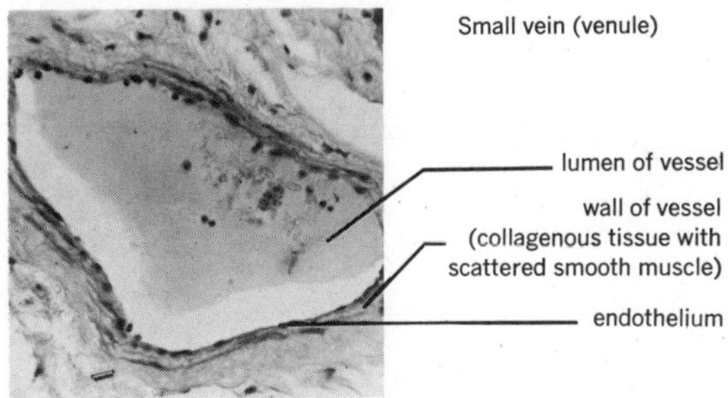

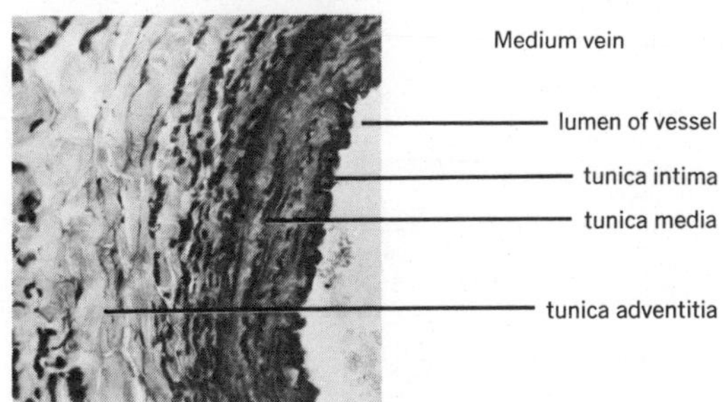

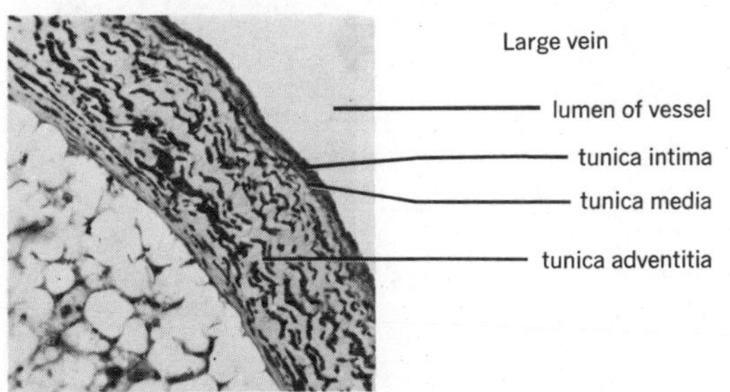

The entire vascular system may be viewed in the manner of two trees placed with their branches together (Fig. 21.2). It should be noted again that the smaller vessels are more numerous and thus present a greater surface and cross-sectional area to the flow of blood. These factors primarily determine pressure and flow in the vessel system.

PRINCIPLES GOVERNING PRESSURE AND FLOW OF FLUIDS IN TUBES

The laws of hydrodynamics that govern the flow of fluids through rigid tubes may be adapted to the flow of blood through the blood vessels. Although the analogy is not perfect, because the vessels are *not* rigid and can change their size either passively or actively, the principles will give a basis for interpreting why characteristics of pressure and flow change in the different parts of the vascular system.

First, let us review the five factors involved in creation and maintenance of pressure in a system of tubes.

CARDIAC OUTPUT (the pumping action of the heart) determines the volume of blood pumped per minute into a system of tubes having a given capacity at a given time. This action creates a basic pressure and flow that depends on the amount pumped and the volume of the vessels. If output goes up, and the volume of the vessels remains constant, pressure and flow must rise. If output stays the same, but volume increases, pressure must fall. These examples suggest that pressure is basically determined by the force

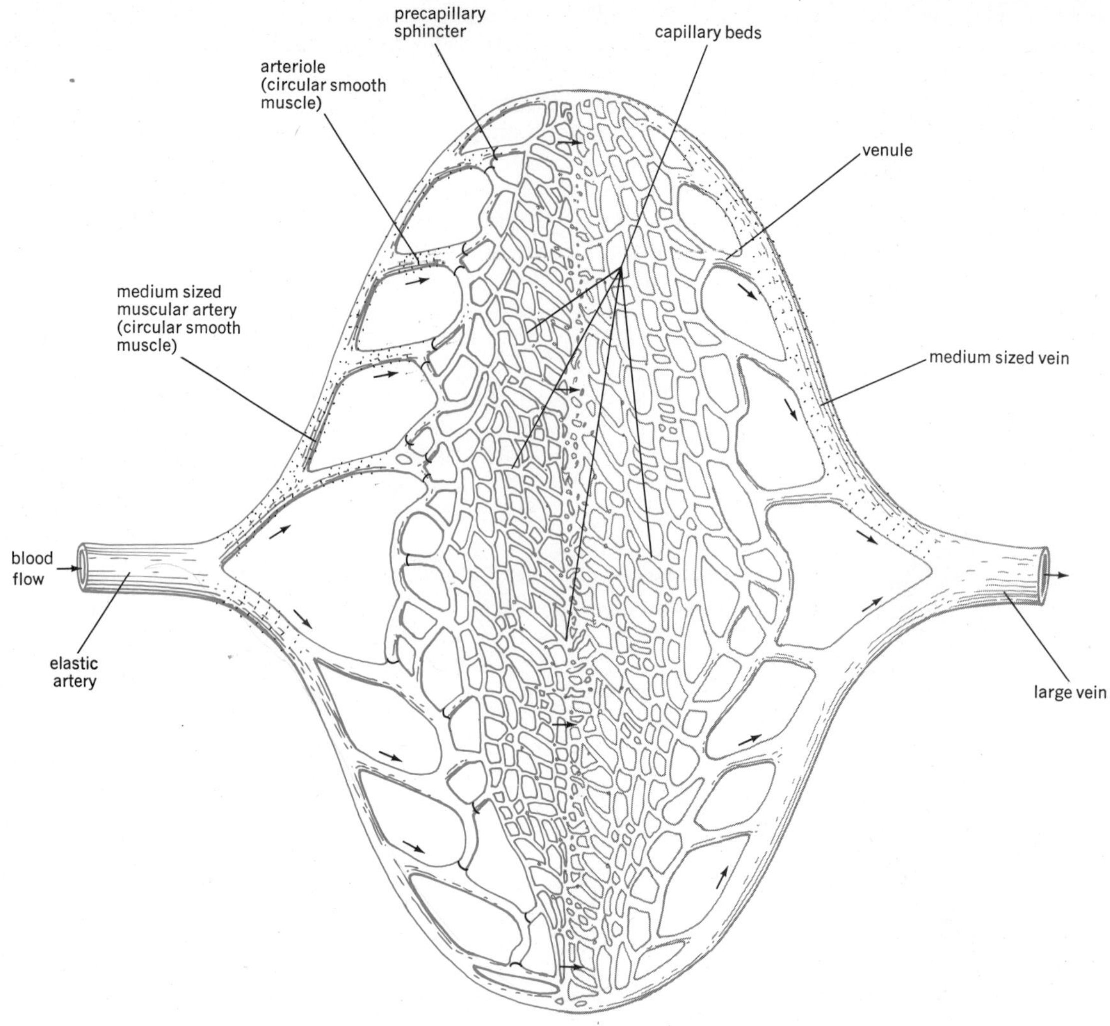

FIGURE 21.2
A representation of the relationships of the "vascular tree."

being exerted on the vessel walls, and that this in turn is a function of volume of fluid and volume of the container.

PERIPHERAL RESISTANCE refers to the difficulty encountered by the blood in passing through the small arteries of the vascular system. Remember that it is these vessels that can actively change their size (smooth muscle in their walls). Thus, narrowing of the vessels increases peripheral resistance and raises pressure in the vessels proximal to the narrowing.

Cardiac output and peripheral resistance are obviously subject to control, and it is primarily alterations in these two factors that govern pressure and flow in the circulation. The remaining three factors are not generally subject to control.

The ELASTICITY OF THE LARGE ARTERIES basically determines the height of the systolic pressure when the ventricles contract. If the tubes become less elastic, as often occurs with aging, systolic pressure rises, and the recoil pushing blood onwards is lost.

The CIRCULATING VOLUME OF BLOOD determines pressure in a system of given capacity. Conditions such as salt retention are associated with fluid retention as well and pressure rises.

The VISCOSITY OF THE BLOOD determines its resistance to flow. A "thicker" fluid flows less easily than a "thinner" one, and obviously requires a greater application of force to circulate it. Polycythemia and hemoconcentration are two conditions that result in an increased viscosity of the blood.

The principles governing pressure and flow in tubes may be stated as follows.

Resistance to flow is directly proportional to the length and inversely proportional to the total cross-sectional area presented by the tubes. This may be interpreted to mean that the longer a vessel of a given size is, the greater the friction presented to the fluid flow and a pressure drop will occur. Also, as vessels branch and become smaller and more numerous, the *total* surface area of the vessels increases; again a pressure drop occurs, as frictional surface is increased.

Flow is directly proportional to the fourth power of the radius of the tube. Thus, if a tube has its radius doubled, flow does not double, but increases by a factor of 16 (2^4). Also, *total cross-sectional area increases* in a nonarithmetic fashion, bringing great flow changes as more numerous and smaller vessels are formed.

Pressure within the vessels is directly proportional to cardiac output and inversely proportional to the total cross-sectional area of the vessels. Pressure should thus be highest closest to the heart and should decrease as the blood traverses the increasing number of smaller vessels in the periphery.

Velocity of flow is directly proportional to pressure and inversely proportional to total cross-sectional area of the vessels. Flow should thus be rapid in arteries where pressure is high and surface is low, should be slower where there are more vessels (e.g., capillary beds), and should increase again, as in veins, where area is less, even though pressure has fallen.

Pressure, flow, and resistance are related by the equation

$$\text{Flow} = \frac{\text{Pressure}}{\text{Resistance}}$$

which may be solved for any one of the three components, given the other two.

Resistance to flow is directly proportional to the viscosity of the blood. Poiseuille's law takes the viscosity into account as a determinant of blood flow and states

$$\text{Flow} = \frac{\text{Pressure} \times \text{Radius}^4}{\text{Length} \times \text{Viscosity}}$$

Figure 21.3 is a graphic representation of the interrelationships of cross-sectional area, pres-

FIGURE 21.3
Relationships of cross-sectional area, blood pressure, velocity, and blood flow in the several areas of the vasculature.

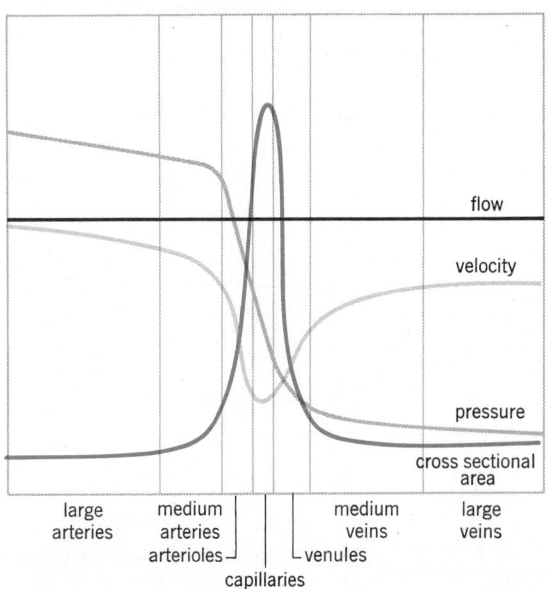

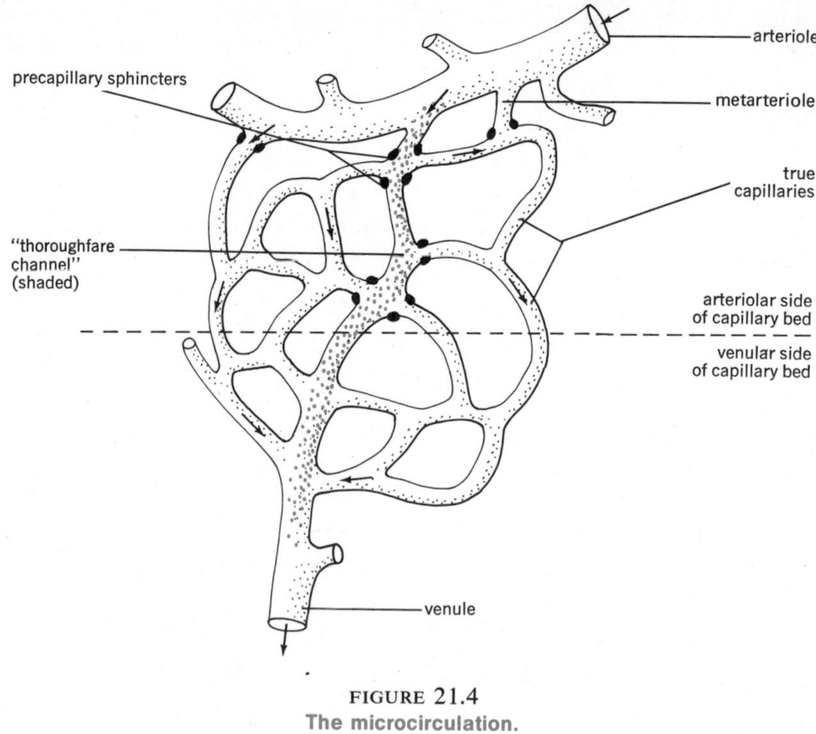

FIGURE 21.4
The microcirculation.

sure, flow, and velocity in the several regions of the vascular system. Applying the principles stated above leads to the conclusion that cross-sectional area is the primary determinant of the other factors. In obese persons, each pound of fat is said to add 1 mile of capillaries to the circulation. Thus, the heart must work harder to circulate blood through the resistance imposed by these capillaries. Venous and lymphatic outflow from a given organ must equal inflow on a volume/time basis, else fluid would accumulate in some part of the system.

Let us now look at the various parts of the vascular system and detail the relationships of these factors.

PRESSURE AND FLOW IN THE SYSTEMIC CIRCULATION

PRESSURE AND FLOW IN ARTERIES

Blood is pumped into the aorta from the left ventricle at a SYSTOLIC PRESSURE (highest pressure generated by contraction) of about 125 mm Hg (range 100 to 150 mm Hg). As the ventricle relaxes, pressure falls to a DIASTOLIC PRESSURE (lowest pressure determined by peripheral resistance and elastic recoil) of about 75 mm Hg (range 60 to 90 mm Hg). The PULSE PRESSURE, or *difference* between systolic and diastolic pressures is about 50 mm Hg. The MEAN ARTERIAL BLOOD PRESSURE, represented by

$$\frac{\text{(Systolic pressure + 2 diastolic pressure)}}{3}$$

is about 100 mm Hg. Resistance to flow in the aorta is slight because cross-sectional area is low (2.5 cm^2), so velocity is high (100 to 140 cm/sec), and flow is great (5 to 6 L/min.). At the end of the aorta in the abdomen, cross-sectional area has not changed appreciably, but some 25 cm of vessel has been traversed, so that mean pressure has fallen to about 95 mm Hg. Blood next flows into the muscular arteries that present some 60 cm^2 of surface area. Mean pressure thus drops to about 40 mm Hg at the beginning of the capillary beds. The systemic arteries contain, on the average, about 20 percent of the total blood volume.

PRESSURE AND FLOW IN CAPILLARIES

Blood next enters the MICROCIRCULATION. This is defined as an *arteriole,* its branches known as *metarterioles,* the *capillaries* served by the metarterioles, and the *venules* draining the capillary beds. The anatomy of the microcirculation is depicted in Figure 21.4. It may be noted that there may be vessels passing directly from arterioles to venules without supplying capillary beds; these vessels form *arteriovenous shunts.* Also, metarterioles contain a few smooth muscle cells in their walls and can act as *sphincters* ("valves") that control how much of a capillary bed will contain blood. Thus, the amount of blood in a given capillary is subject to control. (The skin is a primary example in which blood flow is controlled for temperature homeostasis.)

At the arteriolar ends of capillaries, pressure is about 40 mm Hg, and decreases to about 16 mm Hg at the venous ends of the vessels (Fig. 21.5). A higher "entrance pressure" into the capillary ensures filtration of materials to the interstitial compartment; a lower "exit pressure" allows osmotic return of water and bulk flow of solutes to occur because of plasma oncotic pressure (conferred by plasma proteins). Velocity of capillary flow is slowest here, about 0.3 mm/sec because cross-sectional area is greatest (2500 cm²). Capillaries contain about 5 percent of the blood volume.

Starlings "law of the capillaries" emphasizes some of the factors responsible for governing exchange of materials between plasma and interstitial compartments. The relationship is summarized as

Capillary hydrostatic (blood) pressure + Interstitial fluid osmotic pressure (filtration force) = Blood osmotic pressure + Interstitial fluid hydrostatic pressure (absorption force); or, CHP + IFOP = BOP + IFHP.

If the forces balance, no net shifts of fluid will occur between blood and tissue fluid. Blood will gain fluid if CHP + IFOP < BOP + IFHP,

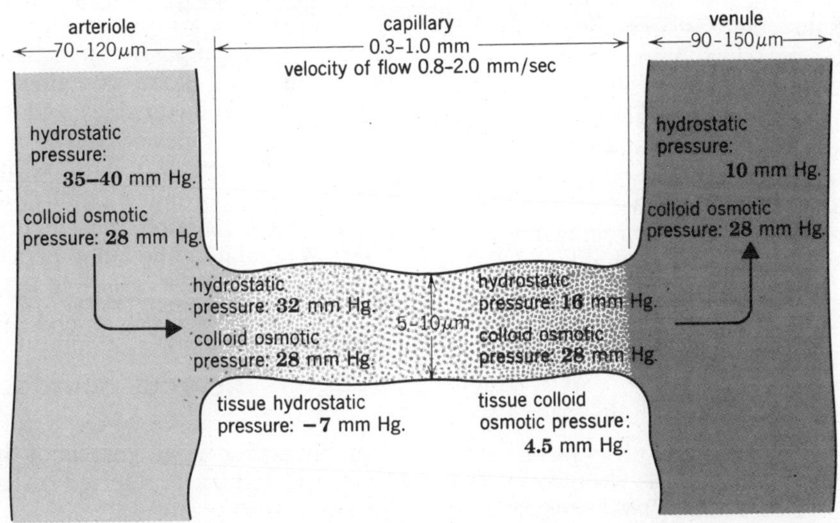

FIGURE 21.5
Capillary dynamics. Pressures favor filtration of fluids from arteriolar end of capillary and osmotic return at the venous end. Flow is slow because of the large number of capillaries and their large surface area.

and looses fluid if CHP + IFOP > BOP + IFHP. In general, the former situation operates at the venous end of the capillary, the latter at the arteriolar end. These relationships set the basis for bulk flow. *Diffusion* accounts for the passage of gases. This situation represents an average for capillaries in general, but does not always apply to individual capillaries under different conditions, since some may not be filled with blood, or may exist in a tissue area where conditions are different from the model.

Capillaries are also active in *endocytosis* and *exocytosis*, providing an additional means of exchanging materials across their walls.

Lastly, the permeability of capillaries depends on their size. The "capillary pores," hypothesized to exist in their walls, may be stretched by an increased amount of blood in the vessel, and thus exchange of materials may be speeded if a tissue becomes active (as muscle). This provides an "automatic" mechanism to supply active areas with more nutrients and to remove the wastes of activity.

PRESSURE AND FLOW IN VEINS

As veins join to form the fewer and larger vessels beyond the capillary beds, total cross-sectional area decreases from the 2500 cm^2 of the capillaries to about 325 cm^2 in all the veins of the body. Pressure has largely been lost in traversing the arterioles and capillary beds, and continues to decline as the length factor operates, because there is no pump in this area to elevate the pressure. Venous pressure decreases from about 10 mm Hg in the tissue venules, to 0 mm Hg at the right atrium. Velocity increases as cross-sectional area decreases, and reaches a value of about 60 cm/second in the venae cavae. Veins contain about 60 percent of the blood volume. The remaining blood volume is contained within the heart (6%), arteries (20%), capillaries (5%), and pulmonary vessels (9%).

Blood flows rapidly through the veins, although pressure is low in these vessels. Among the factors aiding flow through the veins are the following (Fig. 21.6).

The MASSAGING ACTION OF SKELETAL MUSCLES (the "venous pump"). As muscles contract, they not only become shorter, but thicker. Veins located on or between muscles are alternately compressed and decompressed. The blood is "squeezed" and is moved toward the heart as valves in the veins close to prevent backflow.

BREATHING MOVEMENTS, particularly inspiration, increase the volume of the chest cavity, and decrease the pressure therein. The venae cavae are expanded and blood is "drawn" into the larger and lower pressure area of the veins. This increases flow into the right atrium of the heart.

GRAVITY operates to speed return to the heart through those veins above the heart, when the body is in the standing posture.

VENTRICULAR RELAXATION draws blood from the right and left atria, which is then replaced by blood from the venae cavae and pulmonary veins.

The RESIDUAL PRESSURE (*vis a tergo*) of the blood in the veins after passing through the capillary beds aids flow. Pressure decreases to the right atrium, and blood only flows from higher to lower pressure areas.

CONTROL OF SIZE OF BLOOD VESSELS

In previous sections it has been stated that blood vessels may passively or actively change their size. Elastic arteries and capillaries enlarge or diminish passively in size according to the volume of blood contained within their cavities. Muscular arteries and veins respond actively to two types of stimuli: those delivered by way of nerves, and those depending on chemical stimuli and the nature of the muscle itself. These reactions are of primary importance in regulating the blood pressure to ensure adequate flow to the cells, and to keep the pressure from reaching heights that can damage the vessels or tissues they supply.

NERVOUS CONTROL

Two types of nerve fibers may be traced to most muscular blood vessels of the body. Collectively, they are referred to as the VASOMOTOR NERVES.

VASOCONSTRICTOR FIBERS, derived from the sympathetic division of the autonomic nervous system, tonically stimulate contraction of smooth muscle in the walls of vessels, narrowing them and increasing both peripheral resistance and blood pressure. Their effect is mediated by the production at the nerve endings of NOREPINEPHRINE (*noradrenalin*), and the fibers are thus called ADRENERGIC

VASODILATOR FIBERS, derived from the sympathetic and parasympathetic divisions of the autonomic nervous system, inhibit contraction of vascular smooth muscle and allow an increase of vessel diameter with a consequent decrease of peripheral resistance and blood pressure. In skeletal muscle, *sympathetic* cholinergic fibers cause dilation of their muscular vessels. The effect of these nerves is mediated by the production of ACETYLCHOLINE (*ACh*) at the nerve endings, and these fibers are called CHOLINERGIC.

Vasoconstrictor fibers are more numerous and more widely distributed than vasodilator fibers. It should be kept in mind that while vasoconstriction is brought about by stimulation of vasoconstrictor nerves, vasodilation may occur either by active vasodilation through nerves or by inhibition of vasoconstrictor nerves. Thus, it appears that vasodilator fibers are of lesser importance to the body in carrying out normal control of cardiovascular function. Thus, when the sympathetic nervous system discharges impulses, vessels other than in heart and skeletal muscles are constricted, while those to the heart and skeletal muscles may be dilated. Dilation is due to inhibition of tonic

FIGURE 21.6
The skeletal and respiratory pumps. (*a*) The skeletal muscle pump. Relaxed muscle (left) allows blood to flow through open valve. Contracted muscles (right) become shorter and thicker, compressing the vein. Blood attempts to go away from the heart, closing the valve. Blood is thus massaged toward the heart. (*b*) The respiratory pump. Inspiration increases thoracic volume, decreases intrathoracic pressure, and expands the thin-walled venae cavae. Blood is drawn into the larger vessels and speeds the flow through the venous system, increasing blood flow to the right atrium (RA).

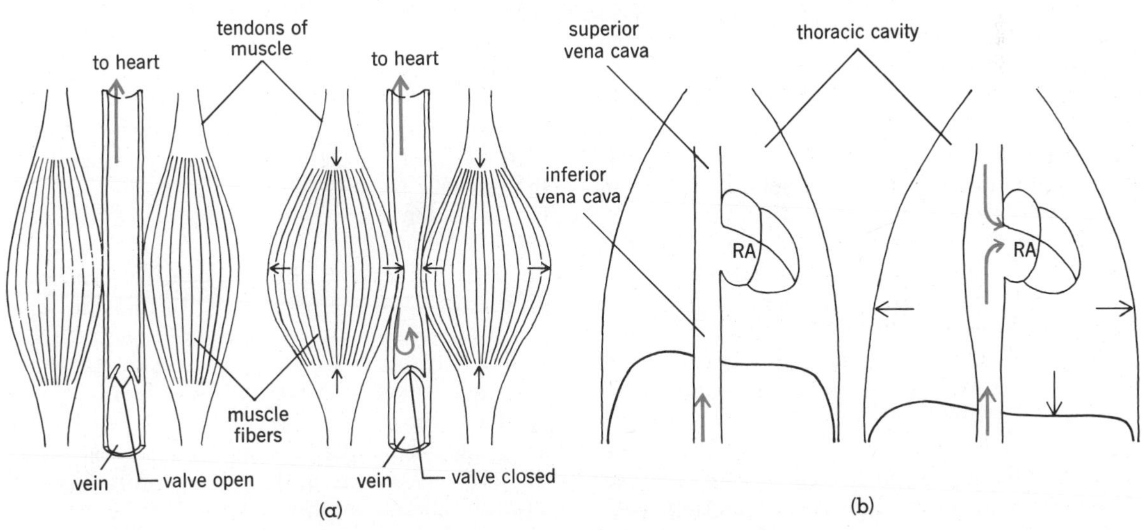

vasoconstrictor activity. The blood will take the path of least resistance, and a redistribution of blood to the muscles must then occur.

Vasomotor nerves originate from VASOMOTOR CENTERS located in the medulla of the brain stem. Separate vasoconstrictor and vasodilator areas appear to exist. These centers appear to be influenced by impulses arriving from several body areas.

PRESSURE SENSITIVE RECEPTORS (*baroreceptors*) are located in the AORTIC ARCH, SUBCLAVIAN and CAROTID ARTERIES, and LUNGS. They include the same receptors controlling heart rate discussed in Chapter 20. These receptors discharge when pressure in the designated areas rises above some critical value. The baroreceptors send impulses to the medulla, and INHIBITION of vasoconstrictor activity and heart rate occurs. Large numbers of blood vessels dilate, cardiac output decreases, and blood pressure is reduced.

BARORECEPTORS, located in the right ATRIUM, serve a pressor reflex and cause an acceleration of heart rate by stimulation of the cardioaccelerator centers. Connections between these centers and the vasoconstrictor centers give a slight vasoconstriction.

The MIDBRAIN appears to have a relay station in it serving the sympathetic vasodilator reflex to skeletal muscle.

The HYPOTHALAMUS contains areas concerned with emotional behavior and temperature regulation. A "heat loss area" is located in the anterior portion of the hypothalamus, and causes, among its other effects, a dilation of cutaneous blood vessels, constriction of visceral vessels, and perspiring. Another area, in the posterior portion of the hypothalamus, brings about essentially the opposite reactions to those described above. These areas then, through effects exerted on blood vessels, determine the degree of heat radiated through the skin, and can affect blood pressure by the change in vessel diameter. The emotional involvement of the hypothalamus is reflected in flushing or blanching of the skin in association with anger, fright, or embarrassment. Little effect is seen on blood pressure since relatively few peripheral vessels are involved.

The CEREBRAL CORTEX, if stimulated in the motor region, results in increased heart rate and vasoconstriction.

These vasomotor reflexes are summarized in Figure 21.7.

The changes in vessel size just described occur primarily in the arterial side of the circulation but may also affect veins, which usually respond in the same manner as the arteries. Decrease in the diameter of veins results in changes in two parameters (a particular measure of physiological response) of circulatory dynamics.

The VOLUME OF THE VENOUS RETURN TO THE RIGHT ATRIUM IS INCREASED. Constriction of veins increases venous return by raising venous pressure. Constriction also avoids an increase in the 60 percent of blood normally present in the veins and minimizes "pooling."

The PRESSURE IN THE CAPILLARIES IS RAISED. Venous constriction increases capillary pressures, and may increase fluid filtration in the capillaries. Edema may result if the blood pressure in the capillaries exceeds the oncotic pressure of the plasma.

NONNERVOUS CONTROL

Nonnervous control of blood vessels is exerted mainly by chemical, thermal, and mechanical factors.

EPINEPHRINE, produced by the adrenal medulla, and NOREPINEPHRINE, produced by the adrenal medulla and nerve endings on certain effectors, are released in physiologically significant quantities when there is stress applied to the body. The body vessels respond differentially to the chemicals. In the skin, kidney, and spleen, the effect of both chemicals is vasoconstrictive. In heart, liver, skeletal muscle, and lungs, a vasodilating effect of both agents is seen.

CARBON DIOXIDE excess causes local vasodilation. It would appear the contraction of smooth muscle is inhibited by CO_2 excess.

Direct effects of HEAT and COLD. Application of either heat or cold to denervated skin produces a vasodilation. Damage to the vessels or the release of histamine by the stimulus is thought to account for this effect.

The production of BRADYKININ, a polypeptide, results from the action of enzymes released from certain gland cells on tissue fluid proteins. It is a potent vasodilator.

SMOOTH MUSCLE responds to stretching by a CONTRACTION. Sudden increases in pressure within a muscular vessel thus cause vasoconstriction, followed by a slow dilation. This response probably protects the vessels from overdistention.

These various control mechanisms result in blood flow being maintained to active tissues, and normally protect the vessels from high pressures. Heart action and vessel changes compliment one another to achieve the desired change of blood pressure.

FIGURE 21.7
Vasomotor reflex pathways.

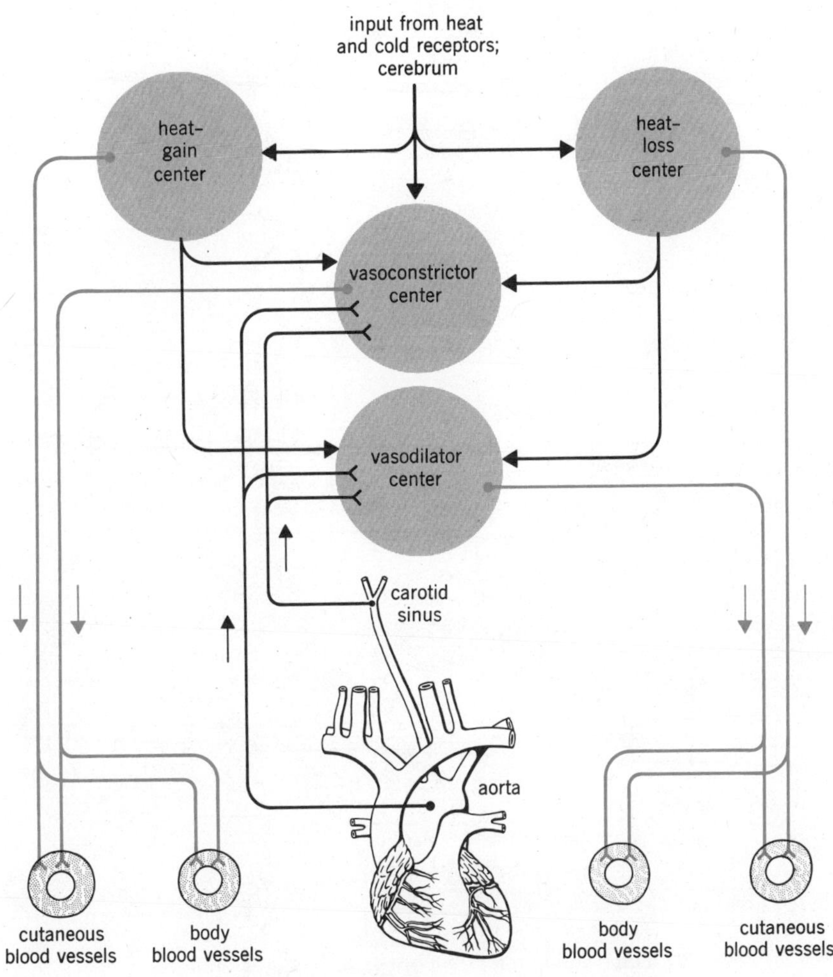

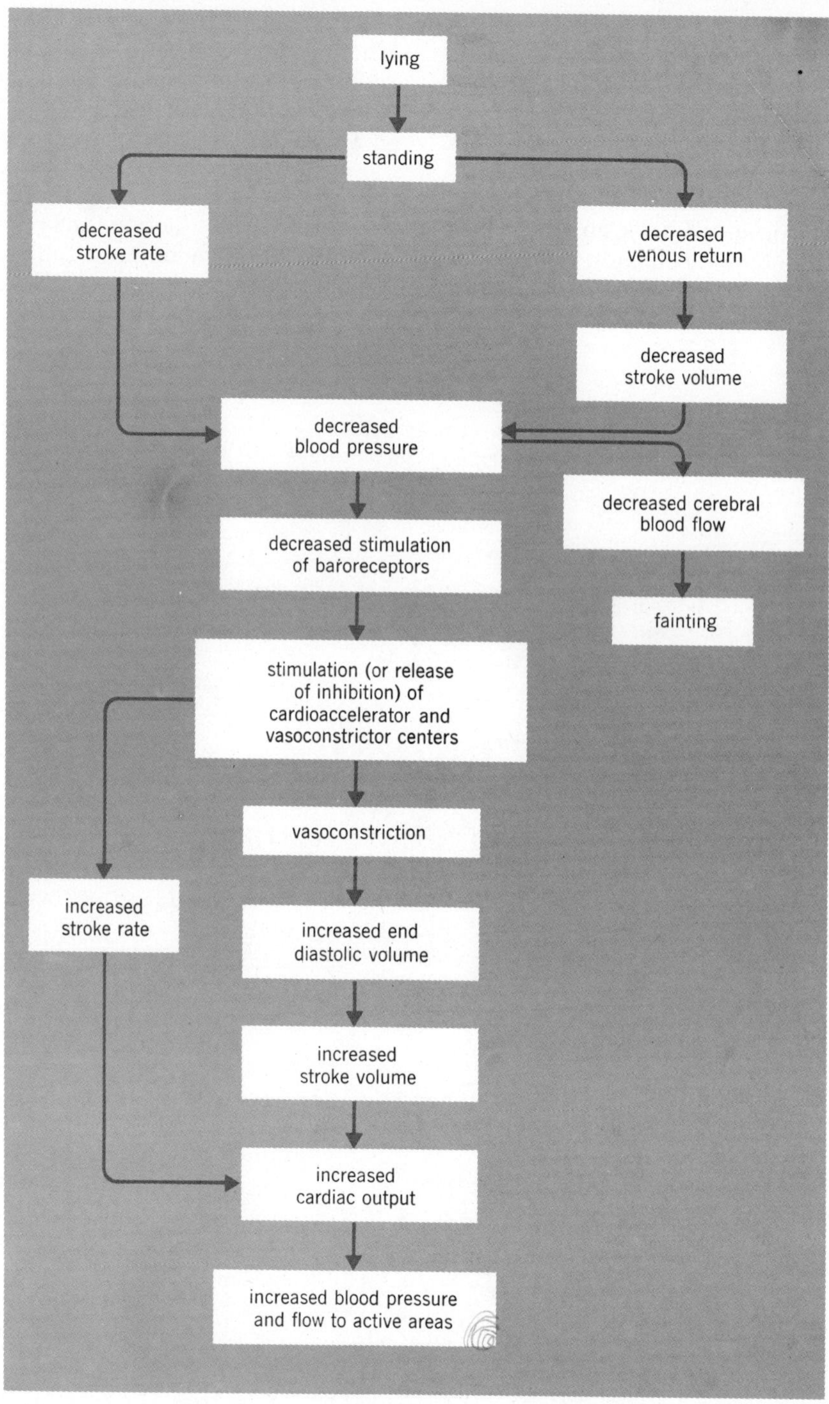

FIGURE 21.8
Vascular and cardiac responses to postural changes.

COORDINATION OF HEART ACTION AND VESSEL RESPONSE IN VARIOUS STATES

Normal systemic circulation responses illustrate the coordinated operation of some of the factors just discussed. Pressure and flow change normally in the body according to changes in the following four basic factors:

POSTURE. In passing from the lying to standing position, the force of gravity tends to cause pooling of blood in veins below the heart. Fall of venous return and blood pressure results. Compensatory nervous mechanisms that operate to counteract these changes are:

An increase in *contraction of skeletal muscles* compressing the veins.

Venous constriction, to maintain volume of the venous circulation, and to insure adequate venous return.

A *decrease in cardiac output* (about 20%) that lowers the amount of blood pumped into the circulation. Rate is 5 to 10 beats higher in the erect position, but stroke volume is lowered 10 to 50 percent. Shifts of fluid volume from upper to lower parts of the body as gravity operates reduces cardiac filling and stroke volume decreases, at least temporarily.

Failure of these compensatory mechanisms to maintain blood flow may lead to postural hypotension, failure of the brain to be adequately perfused, and fainting. The sequence of events occurring in response to postural change is shown in Figure 21.8.

EXERCISE. A redistribution of blood to active muscles and away from the viscera is the outstanding circulatory adjustment in addition to a rise of cardiac output. This redistribution is achieved by vasodilation of the vessels to skeletal muscles, and constriction of visceral vessels. A 20-fold increase of the blood supply to active skeletal muscle may be achieved, with cardiac output increasing only 5-fold. These changes are summarized in Figure 21.9.

TEMPERATURE. Changes in the blood flow through the skin account for the pronounced pressure effects of exposure to a hot environment. Due to cutaneous vasodilation, diastolic pressure may fall 40 mm Hg or more, with rise of cardiac output of only 10 to 20 percent. This may give rise to feelings of fatigue and drowsiness. Exposure to cold causes cutaneous vasoconstriction and an increase in cardiac output to maintain circulation in the face of an increased peripheral resistance.

CARBON DIOXIDE. Increases in CO_2 levels of the tissues, during exercise, for example, is associated with localized dilation of arterioles and increased flow. The capillaries dilate passively with increased arteriolar flow. Independent of nerves, this effect automatically insures an increased blood flow to an active tissue.

CIRCULATION IN SPECIAL REGIONS

PULMONARY CIRCULATION

The pulmonary circulation, compared to the systemic circulation, is a LOW PRESSURE, LOW RESISTANCE circulation. The vessels are shorter, larger, and thinner walled than those of the systemic circulation, and consequently may be perfused at pressures about one-fourth those of the systemic circulation. The vessels have less muscle in their walls as well. Fall of pressure across the pulmonary capillary beds is only about 10 mm Hg. The pulmonary vessels normally contain about 9 percent of the blood volume or about 450 ml in the 70 kg adult. About four seconds is required to traverse the pulmonary circulation.

Blood reaching the lungs through the pulmonary arteries not only is oxygenated in the lungs but nourishes the terminal portions of the respiratory system (respiratory bronchioles, alveolar ducts, and sacs). The bronchi and bronchioles are nourished by the bronchial arteries derived from the aorta.

Vessels in this system are provided with vasomotor nerves, but these nerves have more effect on the bronchial than on the pulmonary vessels. Reflex dilation of vessels has been

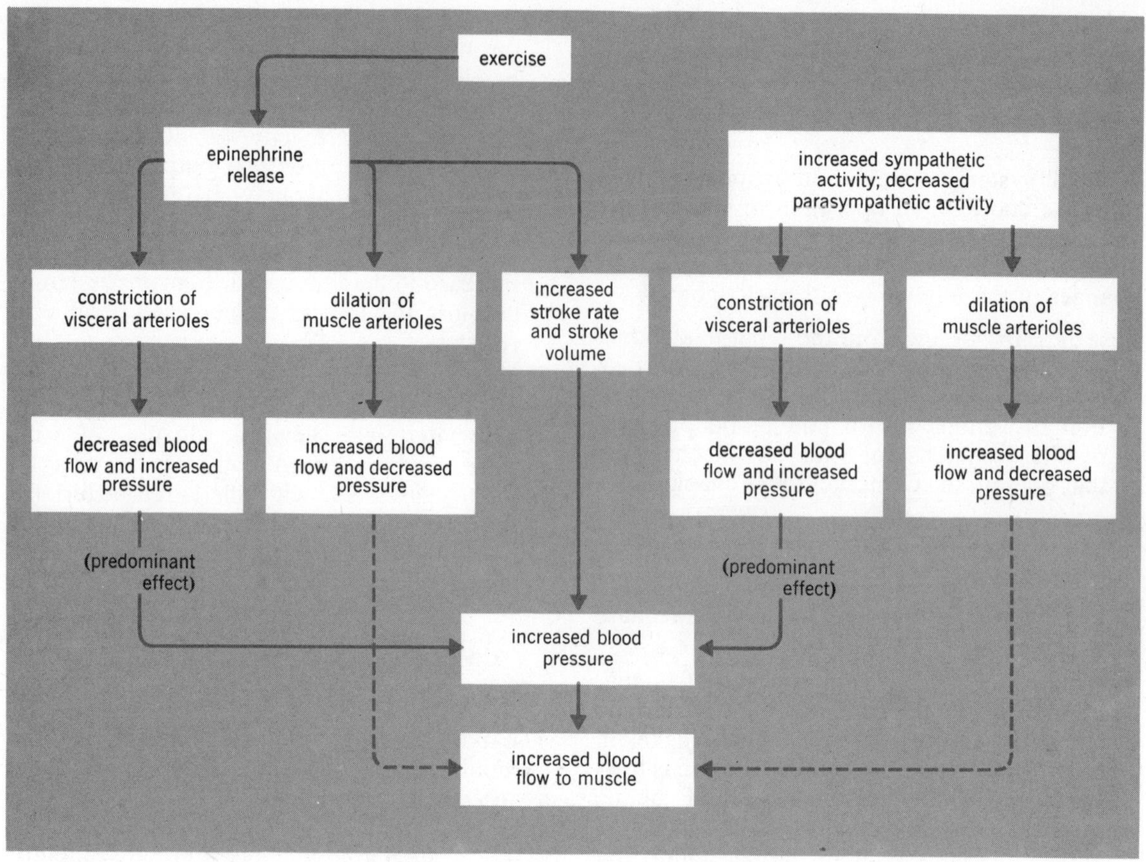

FIGURE 21.9
Vascular and cardiac responses to exercise.

demonstrated through stimulation of baroreceptors. Epinephrine, norepinephrine, histamine, and decrease of systemic arterial P_{O_2} to 80 percent of normal cause pulmonary vasoconstriction; increased CO_2 causes vasodilation.

Pulmonary blood flow is also influenced by breathing. Inspiration draws more blood into the right atrium and results in increased right ventricular output. Expiration tends to result in the opposite effect. Thus, perfusion of the lungs tends to vary by a small amount during breathing cycles.

CEREBRAL CIRCULATION

The striking thing about the cerebral circulation is its relative constancy in the face of rather wide variations in pH, systemic pressure, and chemical influences. Because the brain is enclosed in a set of rigid containers (dura, skull), the volume of the cranial contents cannot vary without damage to the brain itself; thus, large variations in amount of blood entering the skull cannot be tolerated. Average flow is about 750 ml/min, with the cerebrum receiving about 14 percent of the left ventricular output and consuming about 18 percent of the O_2 supplied by the lungs. Control is exerted on the circulation by intrinsic and extrinsic mechanisms.

Intrinsic control (*autoregulation*). Increased CO_2 concentration results in vasodilation and this speeds the removal of CO_2. Oxygen decrease has little effect until levels about 50 percent of normal are reached. At low oxygen concentration, dilation occurs. Both increase and decrease of pH cause constriction of cerebral arterioles.

Extrinsic control. This is exerted mainly through the carotid sinus reflex that ensures that blood will perfuse the brain under adequate but not excessive pressure.

CORONARY CIRCULATION

The vessels to the heart are derived from the aorta just beyond the aortic valve. A right and left artery supply the respective sides of the heart in most individuals. Four to 5 percent of left ventricular output goes to the myocardium via the coronary arteries.

Pressure in the system is high, as high initially as the aortic pressure. Flow is at or near zero through the vessels during ventricular systole due to the compression of myocardial capillaries during heart contraction. During this time, blood is moving through all other body vessels. A period of relaxation and rest between contractions thus allows decompression of the coronary vessels and also assures filling of the ventricles.

Fall in O_2 tension, rise of CO_2, or decreased pH enormously increase flow (200 to 300%) by causing coronary vasodilation. Many nerves are found in the organ but seem to have minimal effect on the coronary vessels. Because heart rate is nervously controlled and results in change of pressure, and because pressure determines the basic rate of perfusion of the coronary circulation, no direct control is exerted or needed on the vessels themselves.

CUTANEOUS CIRCULATION

The anatomical arrangement of the skin vessels (Fig. 21.10) favors the increase or decrease of heat radiation from the organ. The capillary loops basically provide a means of allowing more blood into the skin for heat radiation. Shunting the blood through arteriovenous anastomoses to bypass the capillaries retains heat and elevates the blood pressure. Regula-

FIGURE 21.10
The pattern of blood vessels supplying the skin.

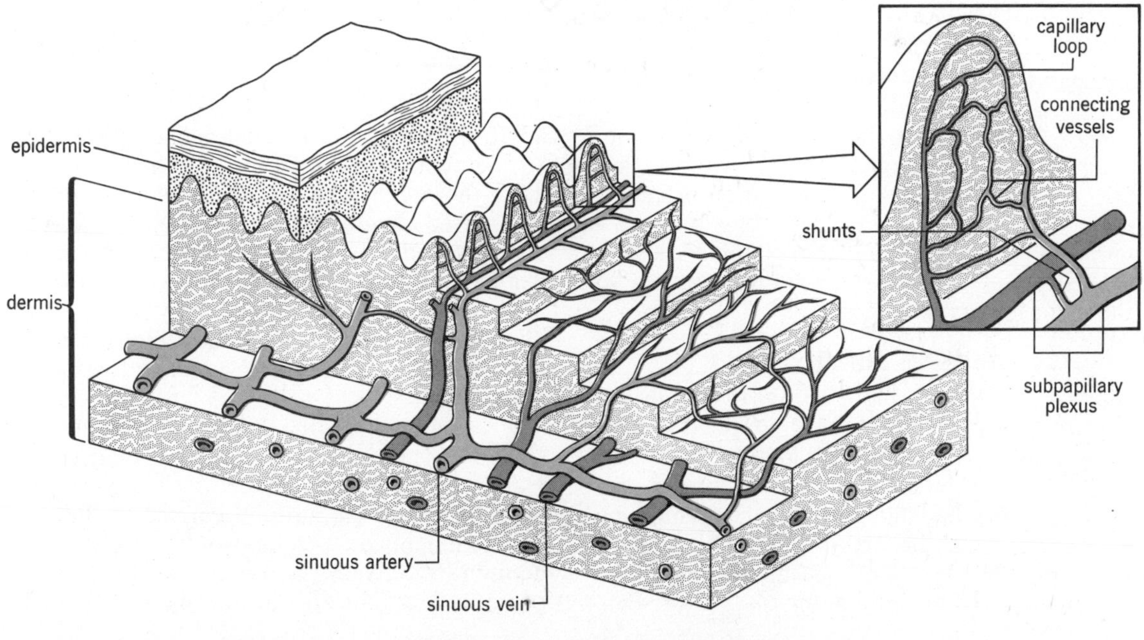

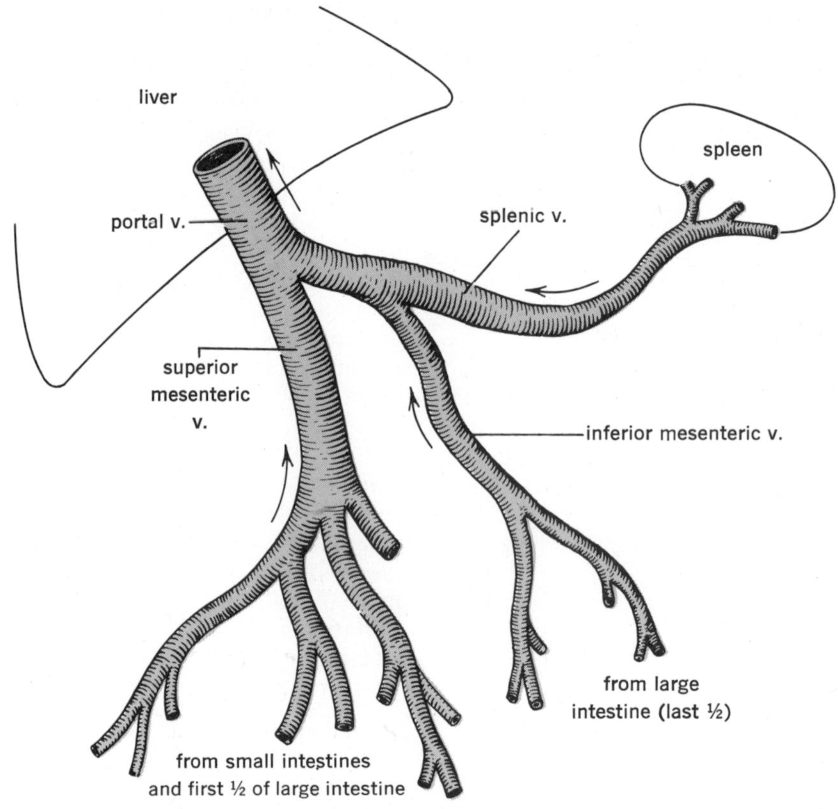

FIGURE 21.11
The hepatic portal circulation.

tion of flow is mainly nervous by way of tonic, sympathetic, adrenergic vasoconstrictors that are inhibited to produce vasodilation on application of heat. The only other factor of importance in controlling caliber of cutaneous vessels is the release of histamine upon damage to the skin; it causes dilation.

HEPATIC CIRCULATION

The liver receives blood from two sources. Oxygenated (arterial) blood is supplied via the hepatic artery, derived ultimately from the aorta. Venous blood rich in nutrients is provided via the hepatic portal vein, formed by vessels draining the organs of the alimentary tract (Fig. 21.11). Combined blood flow is approximately 1400 ml/min in the 70 kg male adult, with about 20 percent being arterial flow, 80 percent venous flow. The hepatic veins drain the organ. The volume of blood in transit or "stored" within the liver causes important changes in the dynamics of the circulation. For example, during heart failure, pooling of blood in the organ appears to occur by active vasodilation and may thus remove blood volume and ease the load on an already weakened organ.

Pressure within the highly permeable liver sinusoids is quite low, usually less than 10 mm Hg. Thus, the high concentration of solutes absorbed from the gut and carried to the liver (nutrients) inside the sinusoids is allowed to equilibrate outside the vessels and little or no osmotic flow of water occurs into the vessels.

SPLENIC CIRCULATION

The spleen, like the liver, appears to serve as an organ for blood storage (Fig. 21.12). About 250 ml of blood can be retained within the organ. Sphincterlike arrangements of tissue are found on the vessels entering the splenic sinusoids. Contraction of these "sphincters" may prevent blood from entering the sinusoids, or, after blood *has* entered the sinusoids, closure will prevent "washing out" of blood elements. Retention within the sinusoids may aid in the phagocytosis of aged erythrocytes and may provide a store of erythrocytes for injection into the bloodstream during contraction of the organ under stress.

SKELETAL MUSCLE CIRCULATION

Blood flow through skeletal muscle nourishes the muscle during states of rest and activity. Because the mass of skeletal muscle is so great, the vessels of the muscles play an important role in adjustments of circulatory dynamics. The vessels appear to be provided with sympathetic adrenergic constrictor fibers and some sympathetic cholinergic dilator fibers. The predominant control lies with the constrictors, and diameter of the vessels is altered by changes in the degree of vasoconstrictor tone. Reflexes that impinge on the medullary vasomotor centers thus bring about changes in caliber that are important to control of blood pressure. Elevated blood pressures cause reflex dilation of skeletal vessels and vice versa.

FIGURE 21.12
Scheme of the splenic circulation.

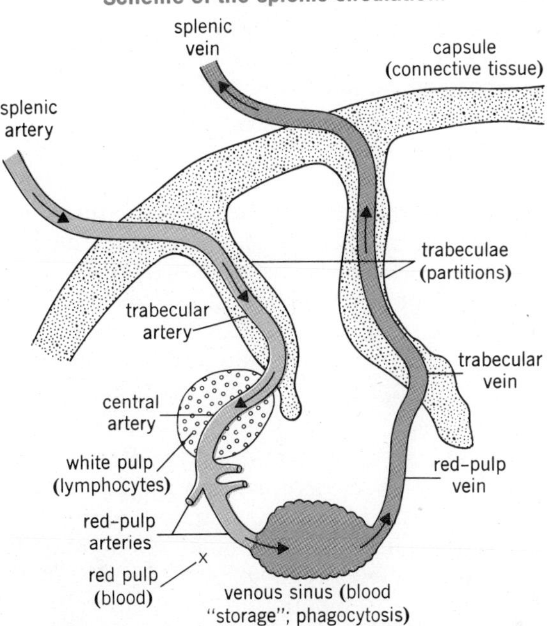

SHOCK

SYMPTOMS OF SHOCK

SHOCK is a condition characterized by the development of certain CHARACTERISTIC SIGNS. Among these are:

Pallor

Cold extremities

Decreased body temperature

Depression of the central nervous system

Rapid, weak pulse

Low blood pressure

Reduced cardiac output

Regardless of the cause of the symptoms, the most threatening developments are LOW CARDIAC OUTPUT and LOW BLOOD PRESSURE. Inadequate cerebral and renal circulation may result. In order to overcome the effects of shock, the body exhibits a number of compensatory mechanisms that attempt to restore blood flow and pressure to within normal limits. This will be considered under the specific types of shock described below.

CLASSIFICATION OF SHOCK

Shock is usually classified on the basis of the cause.

HEMORRHAGIC SHOCK, results from loss of blood or plasma. This is also termed *hypovolemic shock* since circulating blood volume is reduced.

CARDIOGENIC SHOCK, resulting from disturbances in heart action, is characterized by low output and low blood pressure.

VASCULAR SHOCK, resulting from alterations in

blood vessel size from nervous, hormonal, or chemical influences, alters blood pressure.

TRAUMATIC SHOCK, resulting from trauma to muscle and other tissues as in automobile accidents.

More than one type of shock may occur in a given individual at a time, depending on the cause.

Hemorrhagic shock. Internal or external bleeding through ruptured or damaged vessels may result in rapid or slow losses according to location. Among the responses the body makes to blood loss are:

Prevention of continued loss by coagulation and vasoconstriction.

Restoration of lost volume by shifts of fluids from interstitial to plasma compartments; by increased intake of fluids (thirst); and by increased secretion of ADH and aldosterone.

Decrease in volume of the vascular system is achieved by nerve impulses from the sympathetic nervous system that constrict arteries and veins and by epinephrine secretion, which has the same effect. Not only is vascular volume reduced but pressure is elevated, unless blood loss is severe. These reflexes are initiated through pressure sensitive receptors in the body that are not stimulated when pressure falls. The normal effect of impulses from such receptors is to inhibit the activity of the vasoconstrictor center. Thus, if no impulses are present, the vasoconstrictor center is released from inhibition, and muscular vessels are diminished in diameter.

Heart rate is stimulated by epinephrine and sympathetic discharge. Cardiac output is raised.

Shunting of blood from areas such as the skin and kidney to the brain and heart. Failure of the organs from which blood is removed may occur.

A loss of about 5 percent of the blood volume is required before changes in circulatory dynamics can be detected, so efficient are the compensatory reactions. Compensation may be achieved if loss does not exceed 10 to 15 percent. Above 15 percent, compensation becomes more and more difficult, and death will result with a loss of 25 to 30 percent. Changes and compensatory mechanisms involved in hemorrhage are depicted in Figure 21.13.

BURN SHOCK resembles hemorrhagic shock with several important differences.

Plasma rather than whole blood *is lost*. Leakage into the tissues produces edema and hemoconcentration.

Burned tissue releases toxins and is extremely subject to infection.

Toxins may influence the operation of other organs. For example, toxins may cause vasodilation and increase venous and capillary permeability. Renal blood flow may decrease to the point where filtration ceases and kidney tubules are damaged.

Cardiogenic shock results when the action of the heart is disturbed. The heart may fail to fill properly, as when the pericardial sac fills with fluid (cardiac tamponade); it may lose pumping efficiency when there is an infarction (blockage of coronary supply and tissue death); there may be decreased emptying. The common denominators in this disorder are decreased cardiac output and raised venous pressure. Cyanosis is common as circulation slows.

Vascular shock is *neurogenic* if, as a result of sympathetic discharge, vasodilation of skeletal muscle diverts blood from the brain. Pooling of blood in veins occurs, retarding venous return. It may be *humoral* if toxins that are released during tissue damage, such as histamine or other chemicals, are produced in quantities sufficient to cause widespread vasodilation and consequent fall in blood pressure.

Traumatic shock always involves damage to large masses of skeletal muscle and results in blood loss at the site of injury. Response is basically the same as for hemorrhagic shock.

A final word is perhaps appropriate regarding shock. Compensatory reactions do exist to combat the effects of shock. Most involve sacrificing one body function to aid another. These

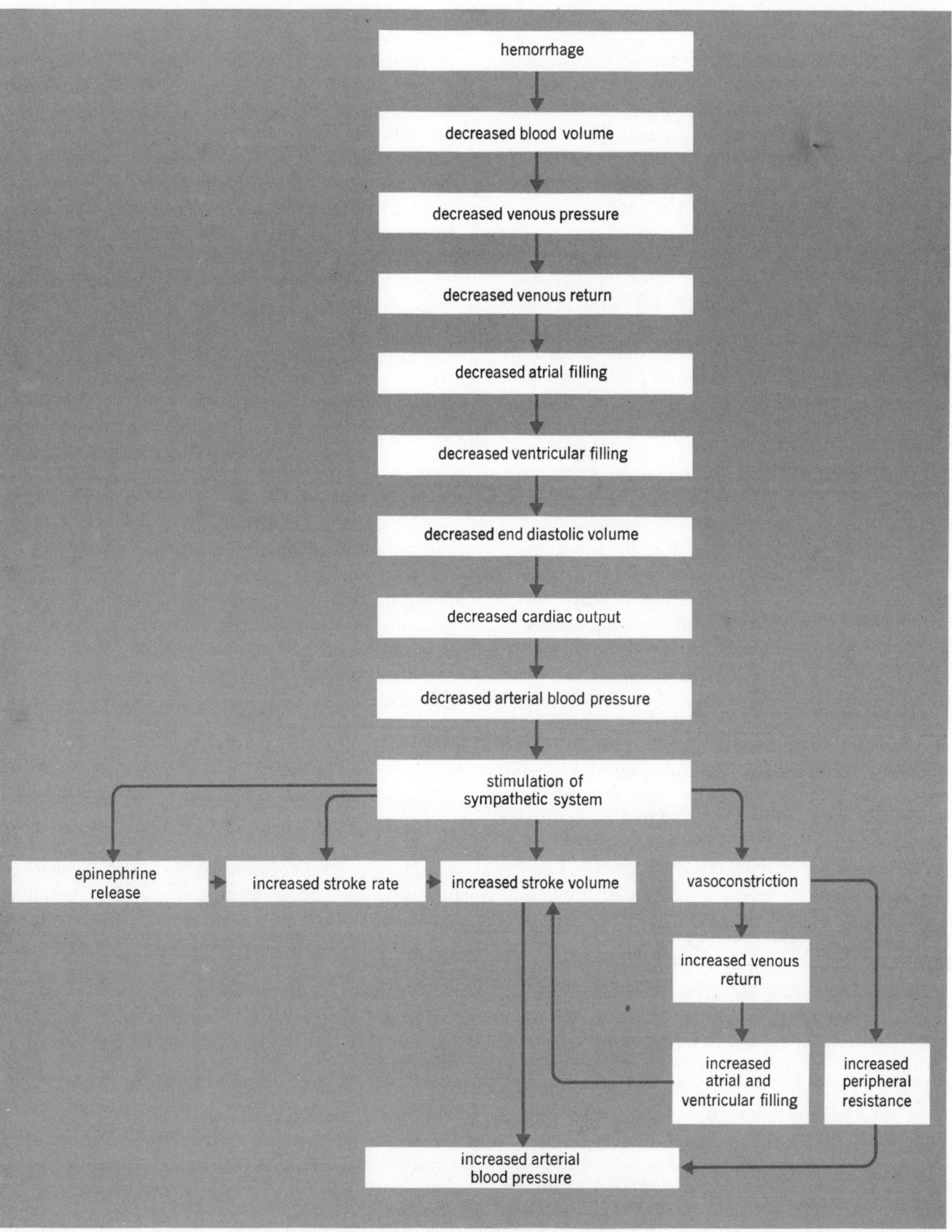

FIGURE 21.13
Some factors involved in genesis of hemorrhagic shock and its compensation.

mechanisms may therefore set up vicious circles that aggravate the shock. For example, diversion of blood from the kidney to the brain may result in retention of toxins that aggravate the condition. Restoration of blood volume, pressure and heart action therefore become the primary aims of shock treatment.

HYPERTENSION

Systolic pressures of 140 to 150 mm Hg and diastolic pressures of 90 to 100 mm Hg are generally regarded as the upper limits of normal. Sustained elevation of systolic pressure, diastolic pressure, or both, above these limits, is termed HYPERTENSION or high blood pressure. Sustained elevation of the systolic pressure alone is termed SYSTOLIC HYPERTENSION; elevation of diastolic pressure alone is termed DIASTOLIC HYPERTENSION. If a cause of the hypertension can be determined, the hypertension is designated as SECONDARY HYPERTENSION; that is, it occurs secondary to some other demonstrable disorder. If no specific cause for the hypertension can be discerned, the hypertension is designated as ESSENTIAL or PRIMARY HYPERTENSION.

Among the causes of secondary systolic hypertension are increased cardiac output, and rigidity of the walls (arteriosclerosis) of the aorta and main arteries. If systolic pressure rises to 200 mm Hg or above, a danger of rupture of the blood vessels exists, particularly in the brain [cerebrovascular accidents (CVA), strokes].

Diastolic hypertension appears to be a particularly dangerous condition and results in vascular damage that may affect the operation of the organs that the affected vessels serve. Vessels serving the kidneys, liver, pancreas, brain, and retina appear to be particularly prone to damage. High pressures in these vessels leads to sclerotic (hardening and thickening) changes and reduction in diameter of the vessels. Blood flow diminishes and the supplied area undergoes degenerative changes secondary to deprivation of nutrients. A common denominator present in diastolic hypertension, regardless of cause, is generalized peripheral vasoconstriction. Investigation of the causes of the vasoconstriction has led to the development of several hypotheses to explain the vascular spasm.

Goldblatt demonstrated many years ago that partial or complete occlusion of the renal arteries, which causes *renal ischemia,* resulted in the production of an enzyme designated RENIN. The source of renin is believed to be the JUXTAGLOMERULAR APPARATUS of the kidney (Fig. 21.14). The stimulus to renin release is diminished oxygen supply and/or pressure to the kidney. Renin acts on a plasma globulin produced by the liver named ANGIOTENSINOGEN, and converts it to ANGIOTENSIN I. A plasma converting enzyme then changes angiotensin I to ANGIOTENSIN II, the most powerful vasoconstrictor known. If plasma renin levels may be shown to be elevated, hypertension of this type is designated RENAL HYPERTENSION. Angiotensin II also stimulates adrenal cortical production of ALDOSTERONE, a hormone that increases kidney reabsorption of sodium. Sodium retention is associated with retention of sufficient water to maintain the blood isotonic with tissue cells. Thus, blood volume increases, and blood pressure rises. As time passes, blood renin levels may be shown to diminish, without corresponding fall in blood pressure. The hypertension may thus be said to have become *irreversible.*

As an explanation of irreversibility, it has been suggested that a renal hormonal mechanism, whose function is to cause arteriolar dilation, has been interfered with. Chemical compounds named PROSTAGLANDINS, which are derivatives of unsaturated fatty acids, have been shown to be produced by the kidney, lungs, brain, seminal vesicles, spleen, adrenal, liver, stomach, small intestine, and other organs. Among the effects of prostaglandins is a potent vasodilating action on peripheral arterioles. The chemicals are theorized to be diminished in production in renal ischemia and thus fail to counteract the vasoconstriction brought about by renin and sympathetic activity. A sudden onset of hypertension, onset before age 20 or after age 50, a history of renal trauma, and lack of a hypertensive history in the immediate

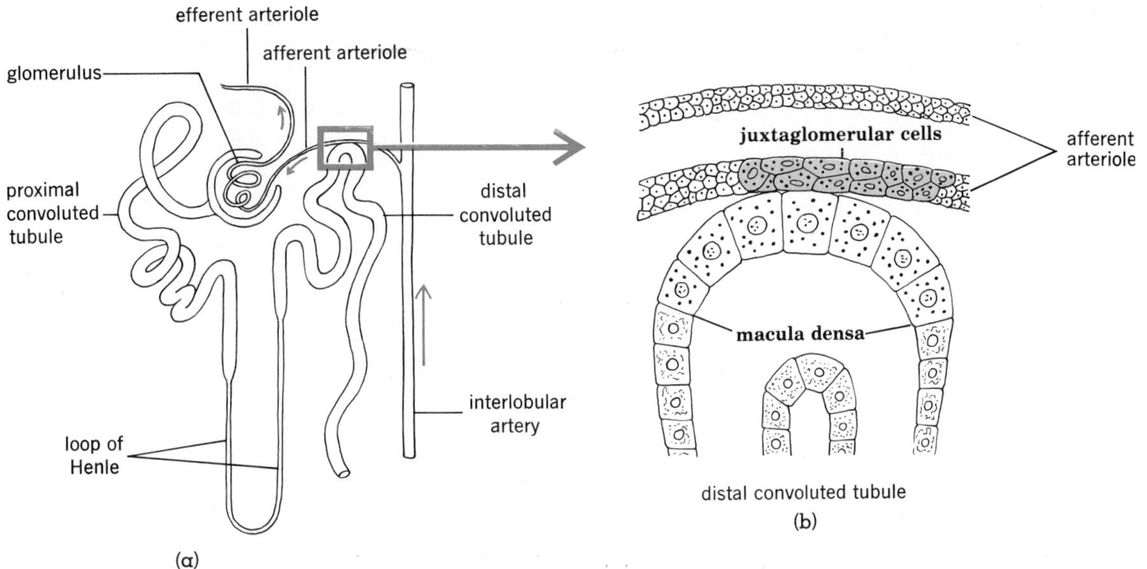

FIGURE 21.14
A representation of the juxtaglomerular apparatus. (a) Diagram of kidney blood vessels and nephron showing the location of the juxtaglomerular apparatus. (b) The juxtaglomerular apparatus responsible for renin secretion. It consists of two portions: granular cells of the afferent arteriole (juxtaglomerular cells), and the large cells of the distal tubule (macula densa).

family, suggest failure of an antihypertensive mechanism to operate.

Another theory to account for peripheral vasoconstriction is that it is due to NEUROGENIC causes. According to this theory, excessive activity in the sympathetic nervous system leads to vasoconstriction, or the reflex mechanisms, which inhibit the activity of the vasoconstrictor center, are not operating properly. Excessive sympathetic activity may accompany anxiety, stress, or psychological disorders. In some cases of hypertension, the baroreceptors in the carotid sinus, aorta, lungs, and elsewhere in the vascular system become insensitive or have a reduced sensitivity to the pressure within the vessels in whose walls they are located. Thus, increased pressure does not result in reflex inhibition of vasoconstrictor activity. Drugs that block the effect of sympathetic nerves may be employed to treat neurogenic hypertension. If this proves ineffective, the sympathetic ganglia from the tenth thoracic to the first or second lumbar may be removed. Such a surgical procedure is termed a SYMPATHECTOMY and deprives blood vessels of the abdomen and lower limbs of vasoconstrictor impulses. They dilate, lowering blood pressure. A disadvantage of this procedure is that the vessels cannot undergo constriction, and blood may tend to pool in the lower parts of the body as position tends toward standing.

EXCESSIVE INTAKE OF SODIUM, as sodium chloride (common table salt), requires retention of sufficient water to render the blood isotonic to tissue cells. In some hypertensive individuals, simple restriction of salt intake will aid treatment since the kidney will excrete the excess water. Diuretics have the effect of promoting both sodium and water excretion. The mechanism of production of hypertension by excessive sodium intake is

increased Na intake
↓
elevated blood Na levels
↓
increased H_2O retention
↓
increased blood volume
↓
increased blood pressure

Reduction of blood pressure upon restriction of sodium intake is by

decreased Na intake
↓
decreased blood Na levels
↓
hypotonic plasma
↓
decreased ADH production
↓
increased excretion of H_2O
↓
decreased blood volume
↓
decreased blood pressure

EXCESSIVE PRODUCTION OF ADRENAL STEROIDS, or the use of steroids in the treatment of inflammation (e.g., rheumatoid arthritis), is associated with sodium and water retention. Again, increased blood volume raises blood pressure.

Prognosis in hypertension does not depend on the height to which the pressure rises but on the appearance of complications. In turn, appearance of complications depends on interactions between factors of diet, genetics, and stress.

HYPOTENSION

Hypotension refers to chronically low blood pressure, usually of a degree that normal organ perfusion cannot be maintained. IDIOPATHIC HYPOTENSION has no known causes, but is associated with sympathetic insufficiency, degenerative changes in brain motor nuclei, and evidence of failure to synthesize norepinephrine. These factors are thought to combine to cause low cardiac output, and peripheral vasodilation, lowering blood pressure. HEART FAILURE, due to myocardial infarction, results also in low cardiac output and hypotension. HYPOVOLEMIA (reduced blood volume) such as might follow hemorrhage, reduces blood pressure. The transient POSTURAL HYPOTENSION that follows assumption of erect posture has been described in a previous section.

Treatment is aimed at restoring sympathetic activity, as by epinephrine infusion, increasing contractility of the heart, or restoring blood volume (transfusion).

SUMMARY

1. The blood vessels serve as containers for the cardiac output, and maintain and adjust blood flow and pressure to the body tissues and organs.
2. There are several categories of blood vessels.
   a. Arteries carry blood away from the heart. Large elastic arteries receive the ventricular discharge and recoil to move blood onwards through the vasculature. Muscular arteries can change size to control pressure and flow.
   b. Capillaries are very thin-walled and serve as the area of exchange of materials between blood and cells. The microcirculation serves as the functional unit of this part of the vascular system.
   c. Venules and veins carry blood to the heart. They serve mainly as volume vessels.
3. Five factors combine to create and maintain pressure in the cardiovascular system.
   a. Cardiac output creates a basic pressure.
   b. Peripheral resistance adjusts flow and pressure.
   c. Elasticity of the large arteries cushions the "shock" of ventricular ejection.
   d. Blood volume determines pressure when related to the capacity of the vessels.
   e. Blood viscosity determines its resistance to flow.

4. Several principles govern pressure, flow, and velocity of flow of vessels.
   a. Total cross-sectional area is inversely related to resistance, directly related to flow, inversely proportional to pressure, and inversely proportional to velocity.
   b. Resistance is directly proportional to viscosity.
   c. These principles are always observed as the blood circulates. They determine the characteristics of circulation in arteries, capillaries, and veins.

5. Some characteristics of circulation in various vessels are:
   a. Pressure and velocity is highest in large arteries.
   b. Cross-sectional area is lowest in arteries.
   c. Cross-sectional area increases greatly as blood passes through arterioles and capillary beds. Pressure drop is greatest here, and velocity is lowest (to allow time for exchange).
   d. Pressure is low in veins, but velocity increases as cross-sectional area decreases.
   e. Flow through veins is ensured by skeletal muscle contraction, breathing, gravity, ventricular relaxation, and residual pressure.

6. Change in size of blood vessels may occur passively or actively.
   a. Elastic arteries and capillaries change passively according to the volume of blood within them.
   b. Muscular vessels are served by nerves that provide for active change in size. These are termed vasomotor nerves.
      1) Vasoconstrictor nerves are sympathetic, and are controlled by vasoconstrictor centers in the medulla. These centers are in turn influenced by reflexes that originate in peripheral receptors.
      2) Vasodilator nerves are sympathetic and parasympathetic, and are controlled by vasomotor centers in the medulla. These are also influenced by reflexes originating in peripheral receptors.
      3) Vasomotor and cardiac responses are integrated to achieve change in blood pressure.
      4) Different vessels may be caused to either dilate or constrict at the same time to cause redistribution of blood to active areas.
   c. Chemical factors provide a nonnervous and more local method of altering blood flow.

7. Vascular and cardiac changes are described as a result of postural changes, exercise, temperature changes, and states of excessive CO_2 production.
   a. Postural changes are generally associated with vasoconstriction as more erect posture is assumed. This maintains blood pressure to the head.
   b. Exercise causes a redistribution of blood to skeletal muscles and away from the viscera as a result of differential changes in vessel size.
   c. Temperature rise is associated with an increased flow to the skin.
   d. CO_2 accumulation causes local dilation.

8. There are special characteristics of blood flow in particular body regions.
   a. The pulmonary circulation is a low-pressure, low resistance circulation. Blood vessel size in this circulation is determined mainly by chemical factors.
   b. The cerebral circulation is relatively constant. Flow is maintained by autoregulatory devices (mainly chemical) and carotid sinus reflexes.
   c. The coronary circulation is intermittent, high in pressure, and vessel size depends mainly on chemical factors.
   d. Skin circulation is designed for temperature regulation, with capillary loops and arteriovenous shunts to permit filling or nonfilling of the capillary beds.
   e. The hepatic circulation consists of separate arterial and venous supplies for oxygen and nutrients.
   f. The splenic circulation is designed for "storage" of blood and phagocytosis.
   g. Skeletal muscle circulation is designed to permit great increases of flow when the muscles become active. Control is nervous via change in level of activity of vasoconstrictor nerves.

9. Shock is characterized by fall of cardiac output and vasodilation; these cause low blood pressure. There are several types of shock.
   a. Hemorrhagic shock causes loss of blood volume.
   b. Cardiogenic shock results from low cardiac output.

c. Vascular shock results from generalized vasodilation secondary to nervous or chemical factors.
d. Traumatic shock is the result of mechanical damage to the body. Reaction resembles that of 9a above.

10. Shock is compensated by various body mechanisms, including
    a. Coagulation of blood to seal vessel damage.
    b. Shifts of fluid from interstitial to plasma compartments.
    c. Vasoconstriction.
    d. Stimulation of heart action by epinephrine release.
    e. Shunting of blood from nonvital to vital areas by differential vasomotor changes.

11. Hypertension refers to high blood pressure. It is dangerous because of the potential for mechanical damage to the cardiovascular system.
    a. Systolic hypertension refers to persistent elevation of systolic pressure. Causes include high cardiac output and arteriosclerosis.
    b. Diastolic hypertension refers to persistent elevation of diastolic pressure. It is usually due to generalized peripheral vasoconstriction.
    c. Secondary hypertension results secondarily to some other disorder such as renal disease or excessive sympathetic nerve discharge.
    d. Essential or primary hypertension has no demonstrable cause.
    e. Among the causes advanced for hypertension are:
        1) Excessive renin production.
        2) Failure to secrete prostaglandins.
        3) Vascular spasm due to sympathetic nerve discharge.
        4) Salt retention.

12. Hypotension is low blood pressure. It may be due to sympathetic insufficiency, weak heart action, or loss of blood volume.

QUESTIONS

1. It has been said that the circulatory system exists to serve the capillaries. Do you agree or disagree? Defend your choice.
2. What ensures that time will be allowed for exchange of materials to occur in the capillary beds?
3. Why may it be said that total cross-sectional area of the blood vessels is the determining factor for blood flow and pressure in the vessels?
4. Why is blood flow most rapid in the aorta? Cite the hydrodynamic laws that apply.
5. Compare the factors that cause blood flow through arteries with those causing flow through veins.
6. What might an elevated venous pressure indicate as far as heart action and venous volume are concerned?
7. How is size of muscular vessels regulated to ensure homeostasis of blood pressure?
8. What is a "microcirculatory unit," and what is its physiological significance?
9. In what ways does the pulmonary circulation differ from the systemic circulation? The cerebral circulation?
10. What are the common denominators in shock, regardless of cause? What are some of the compensatory changes that occur to combat shock?
11. What is hypertension? What are some of its dangers?
12. List some causes for hypertension and hypotension, and advance a treatment, based on physiological grounds, for each cause you list.

READINGS

Friedman, Richard, and Junichi Iwai. "Genetic Predisposition and Stress-Induced Hypertension." *Science 193:*161. 9 July 1976.

Marx, Jean L. "Hypertension: A Complex Disease with Complex Causes." *Science 194:*821. 19 Nov 1976.

Vatner, Stephen F., and Eugene Braunwald. "Cardiovascular Control Mechanisms in the Conscious State." *New Eng. J. Med. 293:*970. Nov 6, 1975.

22

THE RESPIRATORY SYSTEM

OBJECTIVES

After studying this chapter, the reader should be able to:

☐ Explain the general roles of a respiratory system in the economy of the body.

☐ List the several steps in the processes by which oxygen is delivered to the tissues and carbon dioxide removed from them.

☐ Name the organs composing the conducting and respiratory divisions of the respiratory systems.

☐ Give the functions of the conducting and respiratory divisions.

☐ Explain the steps that occur to cause air to be drawn into, and driven out of, the lungs.

☐ List some common diseases of the respiratory system that result in alterations of ventilatory mechanics, and cite the parameters that are altered in each disease.

☐ Quote values for oxygen and carbon dioxide in the atmosphere, lungs, blood, and tissues, and relate these to the direction each gas will diffuse.

☐ Explain the laws that govern volume, pressure, and rate of diffusion of gases.

☐ Explain the various volumes and capacities of the lung.

☐ Describe some tests used to assess pulmonary function.

☐ Explain which way, and why, oxygen and carbon dioxide will pass between lung and blood, and blood and tissues, and the factors that govern oxygen uptake by the blood.

☐ Show the various ways that oxygen and carbon dioxide are transported in the bloodstream.

☐ Explain the factors that determine the extent to which oxygen and carbon dioxide are carried by the bloodstream.

☐ Describe the importance of the lungs in regulation of the body's acid-base balance.

☐ Explain the several mechanisms by which the body is protected from potentially harmful inhaled materials.

☐ Locate the respiratory centers in the brain and explain the contribution of each to normal respiratory rhythm.

☐ Explain the contributions of central and peripheral regulatory mechanisms to the maintenance of normal breathing patterns.

☐ Explain the differences between the respiratory and circulatory systems of the fetus and newborn, and the changes that occur at birth.

☐ Define some disorders or abnormalities of respiration and how they are caused.

☐ Explain the effects of changes in gas pressure on the body physiology.

Active body cells require large and continual supplies of oxygen to meet their metabolic requirements. Metabolism itself produces large quantities of carbon dioxide that must be eliminated lest the body's acid-base balance be upset. The major functions of a respiratory system center about its ability to provide a large surface area through which oxygen may be acquired and through which carbon dioxide may be eliminated. Because passage of both gases occurs by diffusion, devices to maintain high pressure gradients of the two gases at the diffusing surface must be provided. Thus, the muscles of respiration, and the circulatory system work closely with the organs of respiration to ensure that diffusion gradients are maintained. The respiratory system must also provide protective devices to guard against the entry into the body of noxious materials in the inhaled air and to warm and moisten the air breathed in. Finally, the system must be controlled so as to match ventilation to the varying levels of body activity.

It may be said that all of these activities are interrelated, but for purposes of understanding each activity more clearly, several phases may be listed as though they occurred separately.

PULMONARY VENTILATION involves bringing air into contact with the diffusing surface, and emptying the respiratory organs of accumulated carbon dioxide. Thus, ventilation is usually divided into INSPIRATORY (*intake*) and EXPIRATORY (*output*) phases.

EXTERNAL and INTERNAL RESPIRATION refer to the processes by which gases are (respectively) exchanged between the respiratory organ and the bloodstream, and the bloodstream and body cells.

TRANSPORT OF GASES to and from body cells involves the bloodstream.

CONTROL OF RESPIRATION involves nervous and chemical factors to adjust demand to acquisition and elimination of gases.

THE PHYSIOLOGICAL ANATOMY OF THE RESPIRATORY SYSTEM (FIG. 22.1)

The nasal cavities, pharynx, larynx, trachea, bronchi, and bronchioles (to the terminal bronchioles) are designated as the CONDUCTING DIVISION of the respiratory system. These organs serve to transport or CONDUCT air, and to HUMIDIFY, CLEANSE, and WARM the incoming air. Their walls are too thick to permit any significant gas exchange with surrounding capillary networks. Thus, air within these tubes is called DEAD AIR (nonexchangeable) and the volume of the tubes containing the air is the ANATOMICAL DEAD SPACE. The RESPIRATORY DIVISION of the system is characterized by the presence in the organs of thin-walled ALVEOLI or *air sacs* that are thin enough walled to permit gas diffusion between their cavities and surrounding capillary networks. Not all alveoli necessarily are equally ventilated during resting breathing, and there may be very low blood flow through the capillary networks of some alveoli. The air within such nonfunctional alveoli constitutes a PHYSIOLOGICAL DEAD SPACE, so named because little gas exchange is occurring in those areas. In the normal adult person, the anatomical dead space amounts to about 150 ml, and all alveoli are functional to some extent. In certain states, such as asthma and emphysema, physiological dead space may reach 1 to 2 liters, and result in hypoxygenation of the blood and/or carbon dioxide retention.

The conducting division provides warming of inhaled air by heat radiation from the blood vessels in the walls of the tract. Moistening of incoming air (humidifying) occurs by evaporation of water from the mucus covering of the organs, by the production of serous (watery) secretions from glands in the walls of the organs, and by loss of fluids through the epithelial linings from extracellular fluid compartments. Moistening of the incoming air is essential to prevent drying of the diffusing membranes of the air sacs; diffusion will not occur through dry membranes. Incoming air is also cleansed by the conducting division. The epithelium is covered by a mucus coating secreted by glands in the epithelium and in the walls of the organs, and the mucus layer is moved toward the throat by cilia located on the epithelial cells. The combination of mucus coating and ciliary activity is called the MUCOCILIARY ESCALATOR. Immuno-

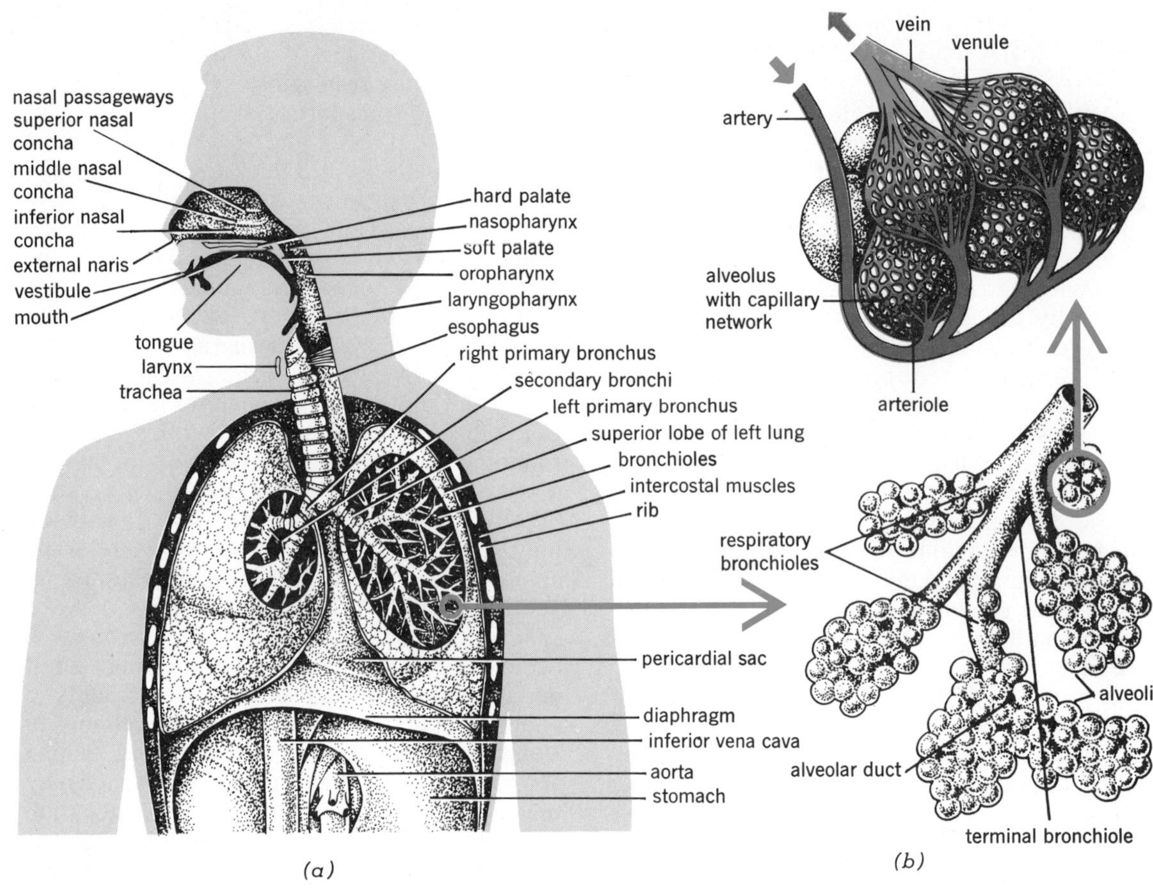

FIGURE 22.1
A representation of the (a) conducting and (b) respiratory divisions of the respiratory system.

globulins (antibodies), particularly immunoglobulin A (*IgA*), are secreted by the glands in the walls of the conducting division and offer protection against foreign antigens.

THE MECHANICS AND PHYSICS OF VENTILATION

Pulmonary ventilation is defined as the INSPIRATION (breathing in) and EXPIRATION (breathing out) of the air of the lungs. Of any given amount of air breathed in at rest, about 30 percent of it remains in the conducting division, is not available for exchange with the blood, and constitutes DEAD SPACE VENTILATION. The remaining 70 percent reaches the alveoli, is available for exchange with the blood, and constitutes ALVEOLAR VENTILATION.

RESPIRATORY MOVEMENTS

Inspiration. The muscles of inspiration, the diaphragm and external intercostal muscles, are caused to contract by nerve impulses reaching them over the phrenic nerves and intercostal nerves respectively. During inspiration, the diaphragm descends, causing an increase in the vertical dimension and volume of the chest (thoracic) cavity (belly or abdominal breathing). Simultaneously, contraction of the external intercostal muscles, and certain neck and chest muscles (sternocleidomastoid, pectoralis minor, scalenes) causes an upward and outward movement of the ribs (costal or chest breathing). This movement enlarges the side-to-side and front-to-back dimensions, and thus the volume of the chest cavity is increased. During normal breathing (*eupnea*), about one-third of the increase in volume of the chest cavity is contributed by the action of the diaphragm and two-thirds by the intercostals, neck, and chest muscles. The "standard (70 kg) man" achieves a total volume increase in the chest cavity of about 500 ml during normal inspiration. The alterations described so far are diagrammed in Figure 22.2.

FIGURE 22.2
Changes in thoracic dimensions on expiration and inspiration.

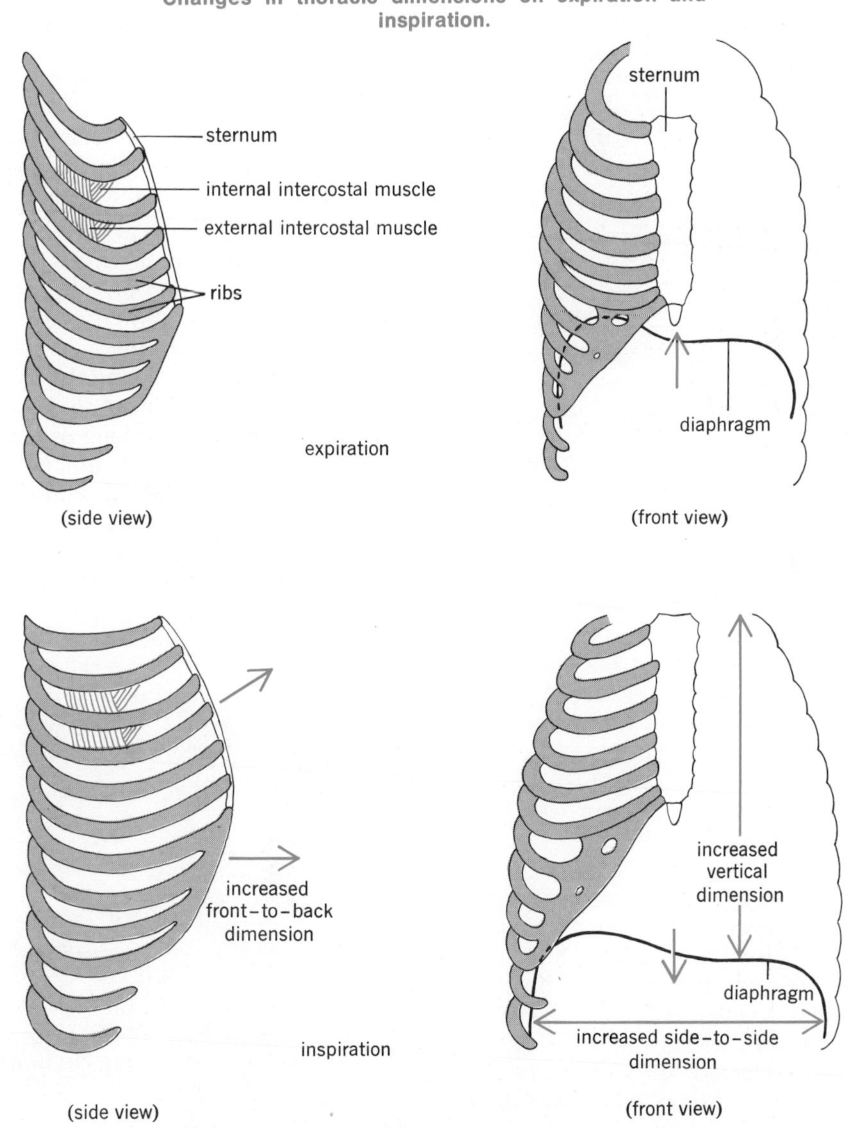

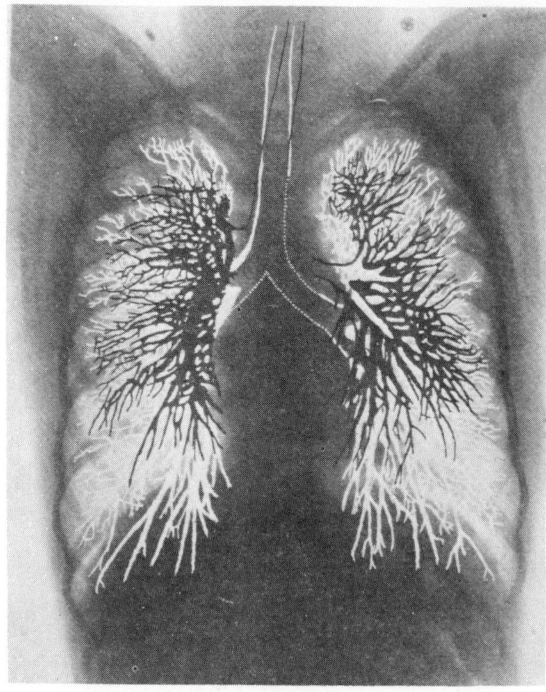

FIGURE 22.3
Changes in bronchi during breathing: inspiration (white) and expiration (black). (Courtesy New York Academy of Medicine. From Macklin, *American Journal of Anatomy* 35:303, 1925.)

As the volume of the chest cavity is increased by the muscular action, there are alterations in the pressure within the chest (thoracic) cavity. The pressure within this cavity is designated the INTRATHORACIC PRESSURE and is normally less than atmospheric pressure. Conventionally, a pressure less than atmospheric is designated as a negative pressure. Measurements of intrathoracic pressure, at rest, show it to be about 757 mm Hg, or −3 mm Hg below atmospheric pressure (760 mm Hg). During eupneic inspiration, the pressure falls to 755 mm Hg, or −5 mm Hg. The lungs cohere to the walls of the thorax and to the diaphragm because of fluid films between the lung and chest wall surfaces. As the size of the thorax increases, the lungs follow, and their size is increased. The pressure within the lungs is designated the INTRAPULMONIC PRESSURE, and this pressure falls as the size of the lungs is expanded. Since the lungs communicate with the atmosphere by way of the respiratory tubes, a fall of intrapulmonic pressure causes air to enter the lungs to equalize the pressure. Normally equal to atmospheric pressure at rest, the intrapulmonic pressure decreases to about 757 mm Hg (−3 mm Hg) during eupneic inspiration. Air flow into the lungs is initially rapid, since the pressure differential is greatest at this time. As air moves into the lungs, the pressure differential disappears, and the intrapulmonic pressure returns to atmospheric at the end of inspiration. Inspiration also stretches the elastic tissue of the lung. It may be noted that lung expansion is not uniform in all directions (Fig. 22.3). The greatest expansion occurs in the lower and outer portions of the organ; apical portions of the lungs may thus contain a larger number of poorly aerated alveoli.

Expiration. Expiration at rest is an entirely passive process, requiring no muscular activity to bring it about. Relaxation of the diaphragm and external intercostal muscles allows return of thoracic size to normal, and the elastic tissue of the lung, stretched during inspiration, recoils. The elastic recoil of the lung tissue raises intrapulmonic pressure to about 763 mm Hg (+3 mm Hg), and air is forced out of the lungs; intrathoracic pressure returns to its normal value of 757 mm Hg. These pressure alterations are shown in Figure 22.4. Expiration may become

FIGURE 22.4
Fluctuations in intrathoracic and intrapulmonic pressures during breathing.

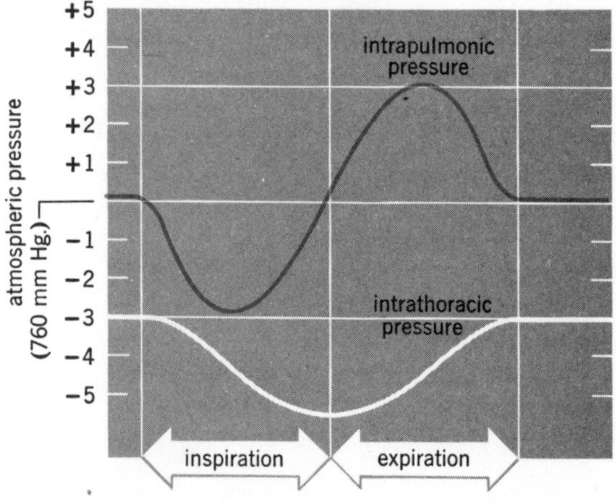

the pressure exerted by a bubble in a liquid is given by:

$$P = \frac{2T}{r} \quad \begin{array}{l} P = \text{pressure} \\ T = \text{surface tension} \\ r = \text{radius} \end{array}$$

the smaller the radius, the greater is the pressure. If two bubbles are connected as in the following diagram; the one having the smaller radius (*b*) will tend to empty into the larger (*a*).

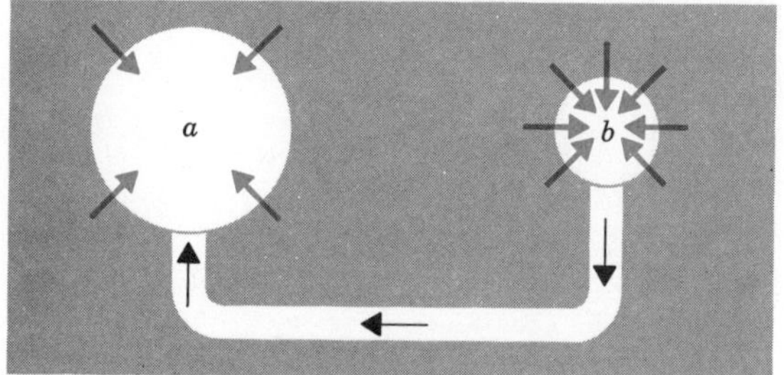

FIGURE 22.5
The forces tending to cause alveolar collapse, with alveoli acting as bubbles.

active under conditions that require rapid and forceful breathing. Two mechanisms aid forced expulsion of air from the lungs; stimulation of the internal intercostal muscles actively depresses the ribs; contraction of abdominal muscles forces the abdominal viscera against the diaphragm, hastening its return to the elevated position. Active expiration may also occur in emphysema, where elastic tissue in the lungs has been reduced.

SURFACE TENSION OF THE ALVEOLI; SURFACTANT

Alveoli behave like bubbles of gas in the fluids of the lung; that is, they have a tendency to collapse, depending on the surface tension of the bubble. Surface tension is caused by the attraction between molecules at the fluid-air surface, and tends to cause a bubble to assume the smallest diameter for its volume. If the molecules at the fluid-air surface are water molecules, the surface forces are quite high, and the tendency is for the surface tension to pull the bubble into a collapsed condition, that is, to assume the smallest size. Additionally, if two bubbles are connected to one another, the smaller bubble, having higher surface tension forces, tends to empty into the larger bubble (Fig. 22.5). In the lung, where alveoli are arranged side-by-side, the tendency of one alveolus to empty into another cannot be tolerated because this will lead to the collapse of alveoli. Solving the problem of the tendency of alveoli to collapse depends on the presence on the alveolar surface of a material other than water. A phospholipid, designated SURFACE ACTIVE AGENT or SURFACTANT, has been demonstrated on the alveolar surface. It is produced by a special type of alveolar cell that forms part of the lining of the alveoli. It has the remarkable property of lowering alveolar surface tension 7–14 times that expected from an air-water junction. This means that, as the alveoli become smaller (as on expiration), surface tension does not increase, and the tendency to collapse is minimized; similarly, the tendency for one alveolus to empty into an adjacent one is minimized. The action of the surfactant in the alveoli may thus be compared to springs that are stretched as the alveoli are expanded on inspiration but that have less tension as the alveoli become smaller on expiration. If the phospholipid is reduced or missing, alveolar collapse may occur, or the alveoli may be very difficult to inflate if collapsed. In infants, a RESPIRATORY DISTRESS SYNDROME (*RDS*) known as HYALINE MEMBRANE DISEASE (*HMD*) is attributed to insufficient amounts of surfactant at the alveolar surface. Alveolar collapse occurs, with reduction of diffusion surface, and the result is the development of labored and difficult breathing.

EXCHANGE OF AIR

VOLUMES AND CAPACITIES

The act of breathing results in the rhythmic exchange of volumes of air. Several lung volumes and capacities are designated.

An instrument known as a SPIROMETER is used to measure lung volumes. It consists basically of a piston in a cylinder type of arrangement, except that the cylinder is usually water filled, and the piston is caused to move by air being exhaled into the machine.

Lung volumes. The term TIDAL VOLUME (*TV*)

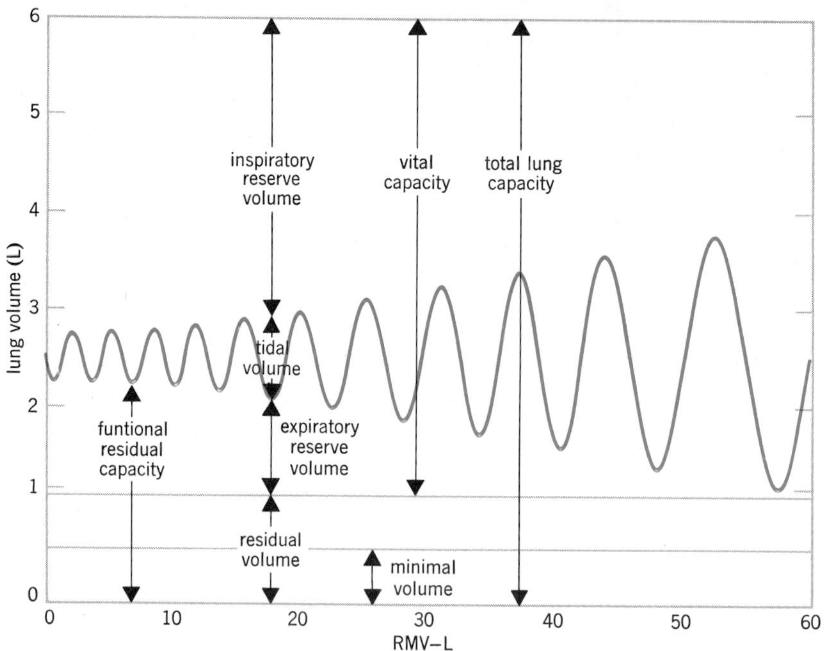

FIGURE 22.6
Subdivisions of lung air according to respiratory minute volume (RMV). As tidal volume increases with increased depth of breathing, inspiratory and expiratory reserve volumes decrease.

refers to the amount of air entering or leaving the lungs during normal inspiration or expiration. In the "standard man," tidal volume averages 500 ml. INSPIRATORY RESERVE VOLUME (*IRV*) refers to the amount of air that may be inhaled above tidal volume (*TV*) by a maximum inspiratory effort. This averages about 3000 ml. EXPIRATORY RESERVE VOLUME (*ERV*) refers to the amount of air which may be forcibly exhaled after a normal expiration has expelled the tidal volume (*TV*). This averages 1100 ml. RESIDUAL VOLUME refers to the amount of air remaining in the lungs after a maximum expiratory effort. Since the alveoli do not collapse on exhalation, some air always remains in the alveoli. It averages about 1200 ml. If the intrathoracic pressure is equalized with that of he atmosphere (as in *pneumothorax,* where a ommunication is created between the intraoracic space and the outside), the lungs will collapse like a deflated balloon. Even so, air remains trapped in collapsed alveoli and in the tissue spaces. This air is called MINIMAL VOLUME and averages about 600 ml. The relationships of these volumes is presented in Figure 22.6.

Lung capacities. Adding the lung volumes in various combinations gives several lung capacities. The VITAL CAPACITY (*VC*) is the sum of tidal, inspiratory and expiratory reserve volumes. This capacity represents the total exchangeable air volume of the lungs. The FUNCTIONAL RESIDUAL CAPACITY (*FRC*) is the sum of expiratory reserve and residual volumes. This represents the air remaining within the lungs after a normal expiration and is responsible for gas exchange with the blood between inspirations. TOTAL LUNG CAPACITY (*TLC*) is the sum of vital capacity and residual volume. It gives an idea of the size of the lungs.

The adult human normally breathes 14 to 16 times per minute. The rate of breathing multiplied by the tidal volume gives the volume per minute of air supplied to the respiratory system. This quantity is termed the RESPIRATORY MINUTE VOLUME (*RMV*). At typical tidal volumes (500 ml), and 15 breaths per minute, about 7500 ml/min of air is drawn into the lungs. Not all of this air is available for exchange; about 150 ml per breath remains in the anatomical dead space. With increased activity, there is an increased demand for air, and both rate and depth of breathing increase. A MAXIMUM VENTILATORY CAPACITY is set by the vital capacity and the ability of the respiratory muscles to continue rapid rates of contraction despite development of fatigue. Normally, the maximum ventilatory capacity is about 25 times the resting RMV and indicates the reserve capacity inherent within the system.

THE WORK OF BREATHING

Exchange of air is usually achieved with a minimum amount of effort. EUPNEA is a term referring to normal, quiet breathing. In HYPERVENTILATION, or an increase in rate and depth of breathing, the work involved may increase to where DYSPNEA or labored breathing results. Dyspnea may also imply respiratory distress resulting from inadequate alveolar ventilation. The work of breathing is largely determined by airway resistance, lung compliance, and lung elastance.

AIRWAY RESISTANCE refers to the ease with which air flows through the tubular structures of the respiratory system. Ease of flow depends on the length of the tubes and their diameter. A smaller tube requires more work to cause air flow through it, much in the same fashion as it takes more "pull" to draw fluid through a small straw than through a larger one. Resistance is also created if alveoli have not emptied properly, and the next inhalation finds air trying to enter alveoli already filled with air. This latter situation occurs in asthma and in emphysema. Thus, any condition that narrows the tubes or causes incomplete emptying of alveoli will require more effort to maintain the same level of flow against the increased resistance.

LUNG COMPLIANCE* refers to the ability of the alveoli and lung tissue to be expanded on inspiration. Normally the lungs are easily expanded, due to their content of easily stretched elastic tissue. If the elastic tissue is replaced by fibrous nonelastic tissue (as in some infections and in emphysema) the work required to expand the lung will be greatly increased, and labored breathing usually results (compliance is reduced).

LUNG ELASTANCE refers to the ability of the elastic tissues of the lung to recoil during expiration and force air out of the lungs. If elastic tissue is diminished, the ability of the lungs to recoil would be decreased, and increased abdominal muscle activity would be required to aid in emptying the lungs.

Figure 22.7 indicates the effects of some of these factors on the work of breathing, with work indicated by oxygen consumption.

MEASUREMENT OF PULMONARY FUNCTION

Several tests may be performed to assess an individual's level of pulmonary function or to discover if a derangement of function is present.

The MAXIMUM VOLUNTARY VENTILATION (*MVV*) is measured by having the subject breathe as deeply and rapidly as possible into a spirometer for 15 sec and converting the volume of exchange to liters per minute. Normal values should be between 120 to 200 liters/min. This test gives an indication of the status of the respiratory muscles and the ease of air flow through the system.

The FORCED EXPIRATORY VOLUME, TIMED (FEV_t) is determined by having the subject inhale as deeply as possible, then exhale as rapidly as possible into a spirometer that not only measures volume but the time required to exhale. If the lungs are normal, 68 percent of the vital capacity should be emptied in the first 0.5 sec, 77 percent in 0.75 sec, 84 percent in

* Compliance may be defined as a ratio between pressure increase (ΔP) and change in volume (ΔV). Expressed in a formula: $C(\text{compliance}) = \frac{\Delta V}{\Delta P}$. A normal value is 0.2L/cm water pressure.

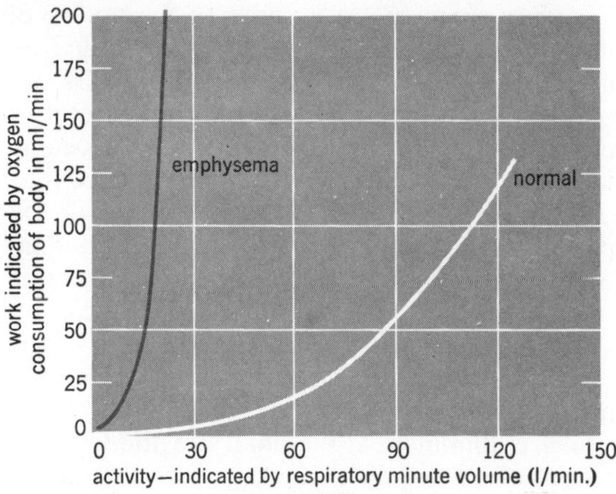

FIGURE 22.7
The oxygen consumption of a normal individual, and one whose compliance, elasticity, and airway resistance are altered by emphysema.

1.0 sec, 94 percent in 2 sec, and 97 percent in 3 sec. Times lengthened beyond these values suggest decreased elastance or increased expiratory airway resistance.

A MAXIMUM EXPIRATORY FLOW RATE (*MEFR*) may be measured by a flow meter, with the subject exhaling forcibly through the device. Values of 350 to 500 liters/min are considered normal. This test is related primarily to airway resistance.

ALTERATIONS OF VENTILATORY MECHANICS IN DISEASE

Pulmonary disorders constitute a major threat to health in 10 to 15 percent of adult Americans beyond the age of 40. Coal miners, exposed for many years to coal dust, smokers, and those living in heavily polluted environments suffer a variety of disorders in lung mechanics. Several major disorders warrant mention, with some consideration of what is changed with reference to ventilation mechanics in these disorders.

"BLACK LUNG" is the common term for *pneumoconiosis* of coal workers. A variety of occupations are associated with exposure to air containing various dusts. In coal mines, inhalation of airborne coal dust results in large amounts of dust accumulating in macrophages in the alveolar spaces, or penetrating the lung tissue itself to accumulate as thick sheaths around the respiratory bronchioles. In the alveoli, dust laden macrophages reduce gas diffusion through alveolar walls; in the respiratory bronchioles, collapse or narrowing of the tubes results, and the sufferer must work harder to draw air through the narrowed tubes. Dyspnea results, and air exchange may be reduced to the point where even mild activity cannot be carried out. Similar changes occur in *asbestiosis,* associated with the inhalation of asbestos fibers. Asbestiosis is additionally associated with inflammation of the lung tissue where the fibers have penetrated the walls of the alveoli, and *edema* and *fibrosis* may further reduce the elastic properties of the lung and increase the work of breathing.

ASTHMA is designated as a *bronchiospastic disease,* in which the smooth muscle, particularly of the terminal bronchioles, undergoes a contraction that effectively prevents either inflation or deflation of alveoli. The causes of the muscle spasm may include histamine release, dusts acting as allergens, exposure to cold air, atmospheric pollutants, or psychologic stress. Clinically, there is dyspnea, with difficult expiration predominating, coughing and wheezing. Relief of the spasm becomes the primary aim of treatment, and may be achieved by inhalation of muscle relaxing chemicals (ephedrine, isoproterenol) or by removal of the patient from the irritating environment.

EMPHYSEMA is a *chronic obstructive pulmonary disease* (*COPD*) characterized by loss of septae between alveoli, collapse and kinking of terminal and respiratory bronchioles, loss of lung elasticity, and production of a thick viscous mucus. Diffusion surface is lost, there is difficulty inflating and deflating the lungs, and dyspnea may develop on exertion. "Pink puffers" is a term referring to emphysematous individuals who are normal as to most lung mechanics, but who, on exertion, develop breathlessness. Their mucous membranes show a normal pink coloration indicating little disturbance of blood oxygenation. "Blue bloaters" exhibit chronic bronchial irritation and narrowing associated with cyanosis at rest, and per-

TABLE 22.1
COMPOSITION OF EARTH'S ATMOSPHERE

GAS	PERCENT IN DRY AIR
Nitrogen	78.09
Oxygen	20.94
Carbon dioxide	0.03
Rare gases (argon, neon, helium, xenon)	0.94

sistent edema of the lower extremities. The latter usually indicates an associated congestive heart failure. Such individuals show increased airway resistance, and abnormal expiratory flow rates.

THE COMPOSITION OF RESPIRED AIR

Humans breathe an atmosphere that (ideally, and ignoring pollutants) contains the proportions of gases shown in Table 22.1. About 4/5 of the mixture consists of nitrogen, and about 1/5 consists of oxygen. Carbon dioxide is in low concentration. In reporting concentrations of gases in the body, it is conventional to use the term PARTIAL PRESSURE. This is a figure, expressed in mm Hg, obtained by multiplying the percent of a gas in a mixture times the total pressure of the mixture. The resulting value expresses the contribution of that particular gas to the total pressure as though it was present alone. The value is usually represented as P_{O_2}, P_{CO_2}, and so on, with the figure in mm Hg following the symbol.

As atmospheric air is inhaled, it is mixed with the air remaining in the respiratory system from the last exhalation, which contains lower pressures of oxygen and higher pressures of carbon dioxide than the atmospheric air. The partial pressures of atmospheric nitrogen and oxygen are thus "diluted" or decreased, and the partial pressure of carbon dioxide is increased. This mixture of air may be sampled in the trachea. If one breathes out forcibly, expelling with a maximal effort the last part of the expiratory reserve volume, a sample of the gas expelled with the last effort is presumed to reflect the composition of alveolar air. This air contains less oxygen than that within the trachea, because oxygen has been diffusing from the lungs into the bloodstream. Ventilation normally maintains P_{O_2} in alveolar air at about 100 mm Hg. P_{CO_2} is about 40 mm Hg in the alveoli, and the gas is saturated with water vapor at a P_{H_2O} of 47 mm Hg. Blood leaving the lungs ("arterial" blood) will have a P_{O_2} nearly equal to that of the alveoli, while P_{CO_2} will be the same as that of the alveoli. (The reasons for this are explained in the next section.) The tissue and cellular area of the body has the lowest P_{O_2}, since the gas is being used in cellular metabolism; P_{CO_2} is highest in this area, reflecting its production as metabolism proceeds and its diffusion from the blood. Blood leaving the tissues ("venous" blood), has had about 30 percent of its original oxygen content

TABLE 22.2
PARTIAL PRESSURES IN mm Hg. OF GASES IN VARIOUS AREAS OF THE BODY

GAS	ATMOSPHERE	TRACHEA; MIXTURE OF "IN" AND "OUT" AIR	ALVEOLI	ARTERIAL BLOOD	VENOUS BLOOD	TISSUES
Nitrogen	596	564	573	573	573	573
Oxygen	158	149	100	95	40	40
Carbon dioxide	0.3	0.3	40	40	46	46
Water vapor	5.7 av.	47	47	47	47	47
Total	760	760	760	755	706	706

removed in the tissue area, and has picked up carbon dioxide for transport to the lungs. The values of these gases in various areas of the body is presented in Table 22.2.

The Table indicates that Po_2 decreases from atmosphere to alveoli to blood to tissues. Because passage of oxygen occurs by diffusion—a passive process proceeding from "high-to-low"—it should become apparent that oxygen must move from "outside to inside" the body. Conversely, Pco_2 decreases from tissues to blood to lung to atmosphere; therefore, the diffusion gradient for carbon dioxide is from "inside to outside" the body.

THE GAS CONTENT OF THE BLOOD

The behavior of gases in or outside of the body is governed by a series of "laws." These laws are cited below for purposes of review, because they are related to the exchange of gases between lung and blood, and blood and tissues.

At the same temperature and volume, equal numbers of gas molecules will exert the same pressure (*Avogadro's hypothesis*). Thus, pressure and therefore diffusion depends only on numbers of molecules, provided that temperature and volume remain essentially constant. In the body, diffusion of a given gas is therefore most rapid when its diffusion gradients are high due to adequate ventilation.

In a mixture of gases, each gas exerts a pressure independent of all other gases (*Dalton's law*). Total pressure is thus the sum of individual pressures, and each gas will diffuse according to its own pressure gradient (partial pressure). One thus may consider each gas as behaving as though it alone were present.

When a gas is compressed, its pressure increases proportionately to its volume decrease (*Boyle's law*). Any decrease of volume must therefore increase gas pressure and speed its diffusion or movement.

If the volume of a gas is kept constant, its pressure is proportional to temperature (*Charles' law*). Higher temperatures increase kinetic motion of gases, more collisions per unit time occur and pressure increases. Thus, rate of diffusion increases to a more or less constant value at body temperature.

The solution of a gas in a liquid is directly proportional to the pressure of gas to which the liquid is exposed (*Henry's law*). Thus, more gas dissolves in a liquid if its partial pressure is greater at the air-liquid interface.

EXCHANGE OF GASES BETWEEN LUNG AND BLOOD

Oxygen. The diffusion of gases from lung to blood or vice versa, occurs through six media. Using as the example the diffusion of oxygen from lung to blood, the media would be: the alveolar membrane, the basement lamina (or membrane) beneath the alveolar cells, a thin interstitial fluid layer, the capillary endothelial membrane, the plasma of the bloodstream, and the membrane of the erythrocyte. The thickness of the membranous portions of this diffusing surface does not exceed 0.7 μm (Fig. 22.8). The longest distance that oxygen must diffuse through is the liquid of the plasma. Oxygen is not very soluble in the plasma, and thus the blood in the lungs does not achieve a full equilibrium with oxygen in the alveoli (*see* Table 22.2). Also, the blood spends only a second or two in the pulmonary capillaries, so time for equilibrium to be achieved is not allowed. An additional factor determining overall oxygenation of the blood is the fact that all alveoli are not perfused to the same degree with blood, and not all are ventilated equally. A good index of pulmonary gas exchange is the ratio of alveolar ventilation to cardiac output. In normal male adults, this ratio is about 0.85, indicating that at rest, about 85 percent of maximum diffusion capacity is obtained. Exercise causes increases toward 100 percent as pulmonary blood flow increases and more alveoli are ventilated, but will not reach 100 percent because both cardiac output and speed of blood flow increases and ventilation increases as well. Blood leaving the lungs is thus about 98 percent oxygenated during high levels of activity.

Carbon dioxide. Venous blood returning from

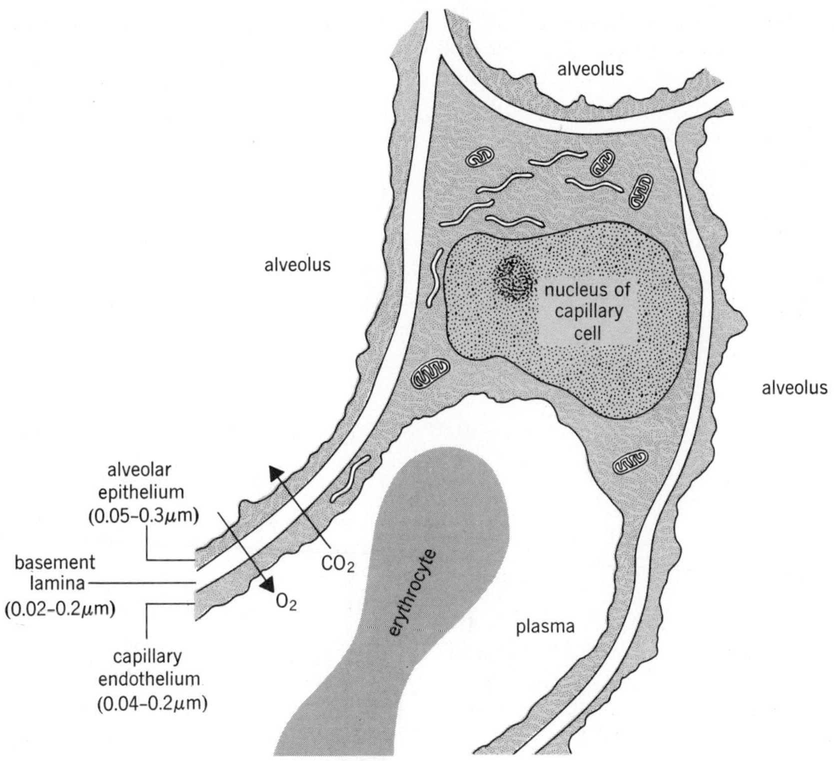

FIGURE 22.8
The alveolar-capillary membrane drawn from an electron micrograph at 20,000×.

the tissues contains much carbon dioxide in three forms: attached to hemoglobin, in the form of bicarbonate ion (HCO_3^-), formed largely in the erythrocytes, and dissolved in the plasma. Diffusion of hemoglobin-bound and dissolved CO_2 proceeds according to diffusion gradients. Release of CO_2 from HCO_3^- is aided by the enzyme CARBONIC ANHYDRASE. This enzyme catalyzes the reaction

$$\underset{\text{carbonic acid}}{H_2CO_3} \xrightarrow[\text{anhydrase}]{\text{carbonic}} CO_2 + H_2O$$

To replace the H_2CO_3, HCO_3^- combines with H^+ in the red cell and plasma

$$HCO_3^- + H^+ \longrightarrow H_2CO_3$$

Thus, release of CO_2 is very rapid, and an equilibrium is achieved between blood and lung. This device also serves as a means of regulating the plasma acid-base balance by removing CO_2 that came from H_2CO_3.

EXCHANGE OF GASES BETWEEN BLOOD AND TISSUES

Oxygen. Oxygen diffuses from blood to cells at resting levels of activity to the extent that about 30 percent of the oxygen is removed from the erythrocyte hemoglobin. A strong diffusion gradient is ensured by the continual use of oxygen by the cells as they metabolize materials. Extraction of oxygen from the blood increases with activity, as in skeletal muscle, and the oxygen in the bloodstream may be reduced to about 20 percent of its original value.

Carbon dioxide. Continual production of carbon dioxide and high levels in the tissues (50 mm Hg at resting levels) ensures a constant diffusion of the gas into the bloodstream. Carbon dioxide then reacts with water to form carbonic acid (primarily in the erythrocytes), combines with erythrocyte hemoglobin, or remains dissolved as CO_2 in the body fluids.

Gas passage through capillary walls occurs through a membrane about 1.0 μm in thickness. Thus, gas passage is rapid, the rate of passage being determined by slopes of the diffusion gradients.

Table 22.3 presents data reflecting the daily exchange of oxygen and carbon dioxide in the body as a whole and in selected tissues.

TRANSPORT OF GASES BY THE BLOOD

OXYGEN TRANSPORT

For oxygen, physical solution of the gas in the water of the plasma is inadequate to supply the demands of the body cells. By physical solution, only 0.3 ml of O_2 will dissolve in 100 ml of plasma. Yet, normal blood may be shown to contain 20 ml of O_2 per 100 ml. The difference in oxygen carrying capacity is due to the presence of hemoglobin in the erythrocytes, with each gram of hemoglobin carrying 1.34 ml of oxygen. Thus, 98.5 percent of the oxygen transported in the bloodstream is carried on hemoglobin as a compound called OXYHEMOGLOBIN. Oxygen combines with the iron atoms of the four heme molecules found in a molecule of hemoglobin. Because oxygen combines as a molecule (O_2) with the hemoglobin, each hemoglobin molecule can carry eight atoms of oxygen. The four atoms of iron in a hemoglobin molecule do not become oxygenated or deoxygenated simultaneously or at the same rate. Oxygen is taken on in steps, with each step having its own equilibrium constant (K). Thus, oxygen uptake may be represented as: (hemoglobin = *Hgb*.)

$$\mathrm{Hgb_4} + \mathrm{O_2} \underset{}{\overset{K_1}{\rightleftharpoons}} \mathrm{Hgb_4O_2}$$

$$\mathrm{Hgb_4O_2} + \mathrm{O_2} \overset{K_2}{\rightleftharpoons} \mathrm{Hgb_4O_4}$$

$$\mathrm{Hgb_4O_4} + \mathrm{O_2} \overset{K_3}{\rightleftharpoons} \mathrm{Hgb_4O_6}$$

$$\mathrm{Hgb_4O_6} + \mathrm{O_2} \overset{K_4}{\rightleftharpoons} \mathrm{Hgb_4O_8}$$

Thus, some molecules of hemoglobin take up or give off oxygen more readily than others.

Several factors determine how much oxygen will be taken up by the hemoglobin.

P_{O_2}. If more oxygen is available in the alveoli for diffusion into the blood, uptake by hemoglobin will be increased. The relationship between P_{O_2} and percent saturation of hemoglobin is not a linear one (Fig. 22.9), a fact that has considerable physiological importance. The curve that may be constructed relating P_{O_2} to percent saturation of hemoglobin with oxygen is termed an OXYGEN DISSOCIATION CURVE.

pH. The lower the acidity of the plasma, the less oxygen will be taken up by the hemoglobin, or, the more will be driven off it. This

TABLE 22.3
OXYGEN CONSUMPTION AND CARBON DIOXIDE PRODUCTION PER DAY IN SEVERAL TISSUES AND ORGANS

TISSUE/ORGAN	MASS (kg)	O_2 CONSUMED PER DAY (L.)	CO_2 PRODUCED PER DAY (L.)
Brain	1.4	67	67
Heart	0.3	42	36
Kidney	0.3	25	20
Digestive (all)	2.6	73	51
Skeletal muscle	31.0	71	57
Skin	3.6	17	14
Others	24.0	72	58
Whole body	63.0	368	294

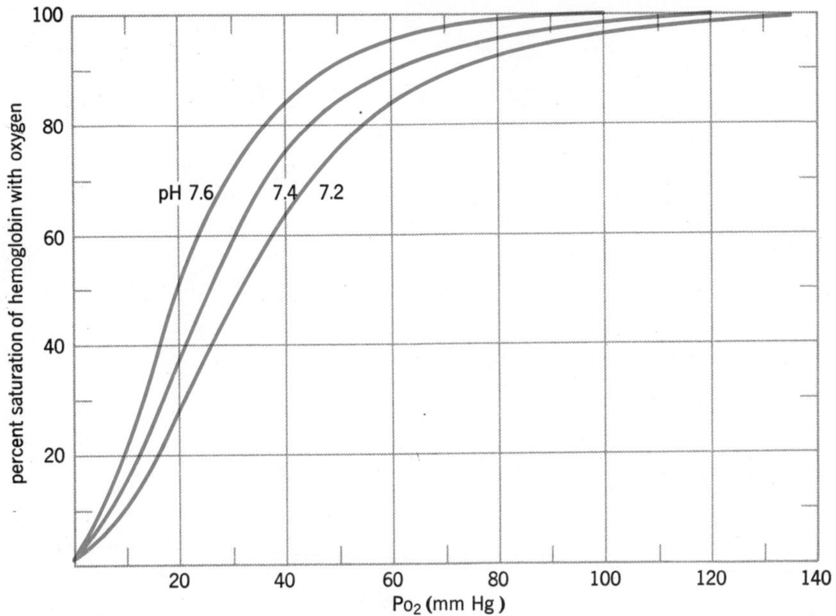

FIGURE 22.9
An oxygen dissociation curve for adult hemoglobin.

effect is designated the BOHR EFFECT, or a shift of the oxygen dissociation curve to the right. The main factor governing pH of the plasma is the amount of CO_2 in the bloodstream, for CO_2 reacts with water to form carbonic acid that then dissociates hydrogen ions to cause a fall of pH. Thus, increased tissue activity that produces more CO_2 ensures more O_2 being driven off the hemoglobin.

Temperature. Increased temperatures decrease the amount of O_2 the hemoglobin can hold (Fig. 22.10). Increased temperature in active tissues thus drives more oxygen from the hemoglobin.

DPG levels. DPG stands for the molecule 2,3 diphosphoglycerate. It is produced within the red cells themselves from the nucleoside inosine, along with pyruvic acid (from glycolysis) and inorganic phosphate. It is normally present in erythrocytes in amounts equal in molecular concentration to the hemoglobin itself. In turn, DPG concentration is influenced by several factors:

The amount of reduced hemoglobin (Hgb without oxygen on it). More reduced hemoglobin causes an increased production of DPG.

Hypoxemia. Exposure to low environmental O_2 levels results in low blood oxygen levels (hypoxemia). DPG synthesis is increased.

pH. If pH falls, DPG levels are increased.

CO_2 concentration. More CO_2 (and thus a fall of pH) causes increased DPG synthesis.

Temperature. A rise of temperature causes increased DPG synthesis.

DPG binds in the center of the hemoglobin molecule to the beta chains, cross-linking them and decreasing the gap between these chains (there is a change in the structure of the molecule). The cross-linking also creates additional chemical bonds that must be broken for oxygenation to occur. Thus, the affinity for oxygen is reduced by the binding of DPG. What this means is that binding of DPG causes a release of O_2 from the hemoglobin, or prevents its ready uptake. If DPG is low, as it is in fetal blood, O_2 is easily bound to hemoglobin and

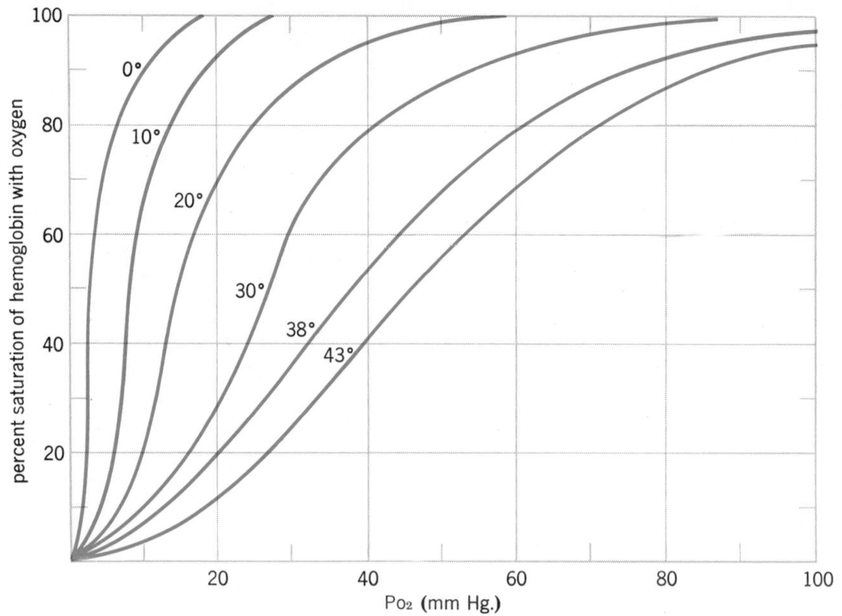

FIGURE 22.10
Oxygen dissociation curves for adult hemoglobin showing the effect of temperature (°C) on hemoglobin saturation.

tends to remain on the pigment, and is not as readily released to the tissues.

Further study of the oxygen dissociation curve shows it to have an *S* or sigmoid shape. This shape indicates that large changes in P_{O_2} in the range from 40 to 100 mm Hg have relatively little effect on the amount of oxygen carried by the hemoglobin. Therefore, the blood contains much O_2 in the ranges of P_{O_2} to which the body is normally exposed.

CARBON DIOXIDE TRANSPORT

Carbon dioxide is carried in several ways by the blood.

About 64 percent is carried in the form of BICARBONATE ION according to the equation

$$CO_2 + H_2O \rightarrow \underset{\text{(carbonic acid)}}{H_2CO_3} \rightarrow \underset{\text{(hydrogen ion)}}{H^+} + \underset{\text{(bicarbonate ion)}}{HCO_3^-}$$

Part of the bicarbonate ion is formed within the plasma by the above reaction; part is formed within the erythrocyte, and HCO_3^- concentration increases within the red cell until it begins to diffuse out of the cell into the plasma. To maintain electrical neutrality, plasma chloride ion moves into the cell (the *chloride shift*).

Up to about 27 percent of the carbon dioxide is carried attached to the globin portion of the hemoglobin molecule, in the form of CARBAMINOHEMOGLOBIN.

About 9 percent is carried in PHYSICAL SOLUTION in the plasma.

Hydrogen ion produced from the dissociation of carbonic acid is buffered by hemoglobin and by the plasma buffering systems to resist changes in pH. Figure 22.11 presents a summary of the chemical reactions occurring at the tissues and lungs as gas exchange proceeds. An explanation of these changes follows.

At the tissues, carbon dioxide dissolves in plasma water or enters the erythrocytes to react with cellular water. In either case, carbonic acid is formed, which dissociates into hydrogen ion and bicarbonate ion. The hydrogen ion produced in the plasma is buffered by

FIGURE 22.11
The chemical reactions occurring as gases diffuse between cells, blood, and lungs. (*a*) Gas exchange between blood and tissues. (*b*) Gas exchange between blood and lungs.

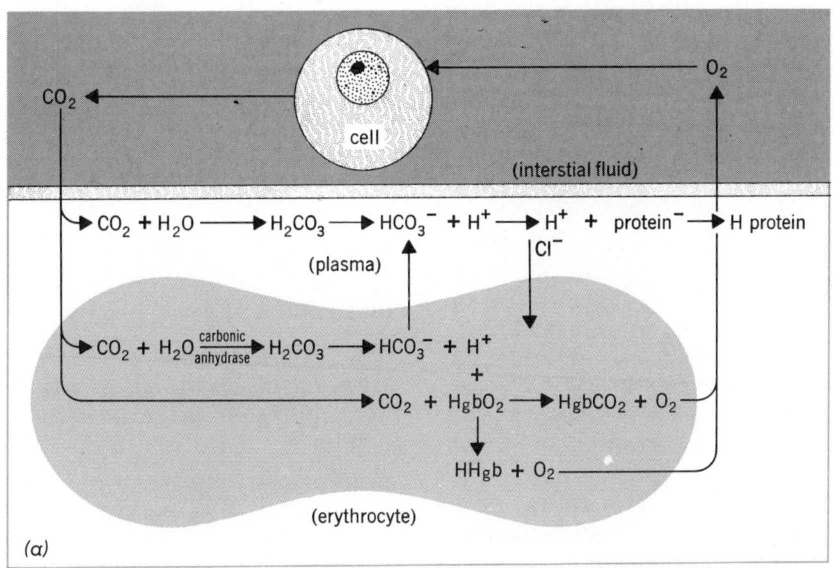

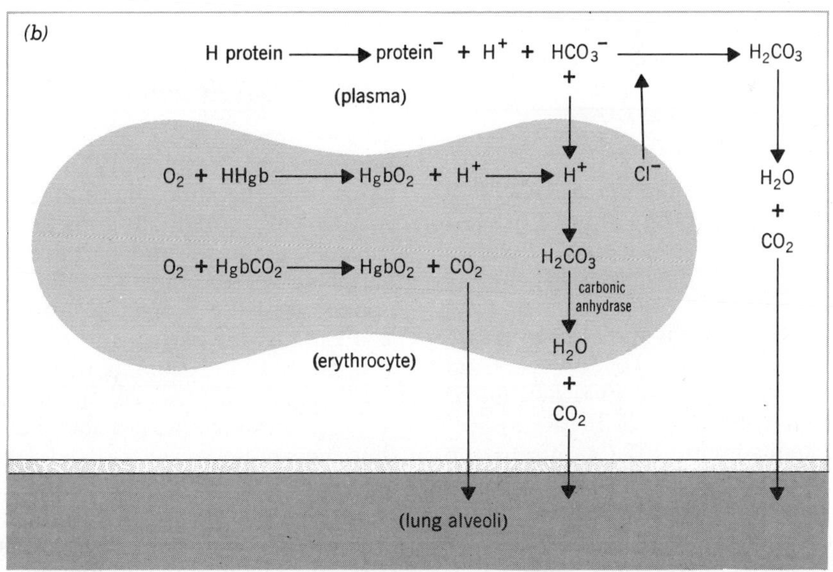

reacting with plasma proteins; the hydrogen ion produced within the erythrocytes does not pass through the cell membrane and is buffered by hemoglobin. Bicarbonate ion accumulates within the erythrocytes and soon begins to diffuse out of the cell into the plasma. A loss of negative charge results, and, to return electrical neutrality, chloride ion moves from plasma into the erythrocytes (CHLORIDE SHIFT). Some carbon dioxide displaces O_2 from the hemoglobin, and the excess H^+ in the erythrocyte drops the pH and also drives O_2 from the hemoglobin.

At the lungs, the action of carbonic anhydrase liberates CO_2 from erythrocyte carbonic acid. To reform carbonic acid requires the hydrogen ion from within the cell and bicarbonate ion from the plasma. As bicarbonate ion moves into the erythrocyte, an excess of negative charges accumulates; and chloride ion moves back into the plasma. The high oxygen levels in this region cause displacement of CO_2 from the hemoglobin and the formation of oxyhemoglobin results. Hydrogen ion is released from its combination with protein as bicarbonate enters the erythrocyte, and the protein is made available to buffer more H^+.

These processes illustrate the interrelationships between gas transport and the buffering systems of the body.

THE ROLE OF THE LUNG IN ACID-BASE BALANCE

The lung, through its elimination of CO_2, is a prime regulator of the pH of blood and tissues. The role of the lungs in regulation of the body's acid-base balance (*see* also Chapter 17) centers around the facts that (1) elimination of CO_2 prevents the reaction of CO_2 and H_2O to create H ion through the dissociation of carbonic acid and (2) any increase in blood level of CO_2 or fall of pH stimulates the act of breathing, removing increased amounts of CO_2 from the alveoli and thus increasing the diffusion gradient for CO_2 from the blood to lungs. This ultimately reduces blood P_{CO_2}. We thus have the operation of a self-regulating mechanism that adjusts ventilation to blood CO_2 level.

OTHER FUNCTIONS OF THE LUNGS

The lungs have traditionally been regarded only as areas for gas exchange between air and bloodstream. Their functions other than gas exchange have recently been summarized, and include:

Secretion of surfactant by alveolar "pneumonocytes."

Production of:

- *Kinins,* compounds that cause vasodilation
- *Kallikreins,* compounds that convert inactive kininogens to kinins
- *Kininases,* compounds that destroy the kinins.

Inactivation of gastric hormones (e.g., gastrin).

Conversion of angiotensin I to angiotensin II.

Metabolizing serotonin so that it does not continue to exert vasoconstrictive effects on the body.

Synthesis and release of prostaglandins.

Metabolism of insulin.

Production of enzyme inactivators (e.g., α-antitrypsin that prevents trypsin from digesting the connective tissue of the lungs).

PROTECTIVE MECHANISMS OF THE RESPIRATORY SYSTEM

Because the respiratory system transports gases from the atmosphere to the lungs, a wide variety of potentially dangerous microorganisms, antigens, and particulate matter may be inhaled. To aid in the understanding of the operation of the protective mechanisms, recall that the greater part of the respiratory system is lined with a mucus coating secreted by mucous glands in the subepithelial connective tissue and the epithelial goblet cells. Also recall that the epithelium is ciliated. There are lymphatics, alveolar macrophages, and reflex mechanisms that aid in dislodging or combatting foreign materials in the system. Several specific mechanisms may be cited as providing protection.

The mucociliary escalator. The presence of a sticky mucus layer on the epithelium of the system allows trapping of particles as incoming air contacts the surface. Coordinated ciliary activity moves the mucus layer posteriorly in the nasal cavities to the throat where the mass is usually swallowed. A similar mechanism operates from the terminal bronchioles upward through the conducting division. Continuous cleansing of the system above the terminal bronchioles is provided. In normal individuals, the rate of mucus production and removal by ciliary action is nicely balanced, and mucus rarely comes to our conscious attention. One of the demonstrated effects of tobacco smoke is to slow the ciliary action and change the nature of the mucus, usually making it more viscous. Such mucus is harder to move and may result in retention of potentially harmful substances.

Alveolar macrophages. Since both cilia and goblet cells are not found beyond the terminal bronchioles, a cleansing mechanism is provided for the respiratory division. Phagocytic cells (alveolar macrophages) are found in the alveoli of this division. These cells engulf and destroy bacteria, particles of dust, foreign antigens, and other harmful substances. Their numbers are partially dependent on the levels of contamination of inhaled air, increasing as the load of pollutants increases. Their phagocytic ability may also vary, becoming less as the amounts of chemical contaminants increase. These cells thus provide an extremely important line of defense.

Filtering. The presence of the large hairs around the nostrils tends to restrict the entry of large objects into the system.

Secretion of immune globulins. The presence of immunoglobulin A (*Ig A*) has been demonstrated in the secretions of the glands in all parts of the respiratory tree. The substance is a nonspecific antibody whose production is apparently triggered by a wide variety of antigenic challenges. The amount produced is directly proportional to the degree of antigenic challenge and forms an important defense mechanism against foreign chemicals.

Lymphatics. Lymphatics are present in somewhat greater numbers in the respiratory and digestive systems than elsewhere in the body. The combination of lymphocytes in the mucous membranes, and lymphatic vessels carrying lymph to the lymph nodes, aids removal of matter that has entered the tissues themselves.

Reflex protective mechanisms include sneezing and coughing. The impulses responsible for initiating a sneeze originate from irritation of trigeminal nerve endings in the nasal cavities, while those initiating a cough are the result of irritation of glossopharyngeal and vagus endings in the pharynx, larynx, trachea, bronchi, and alveoli. Both reflexes involve a deep inspiration brought about by stimulation of the inspiratory centers. A violent expiration, aided by powerful contractions of the abdominal muscles, produces air flow velocities that may approach the speed of sound. In a sneeze, the uvula and soft palate are positioned to allow air to escape both through the nose and mouth; in a cough, the uvula blocks the nasal cavities and the full force of the expiration is directed through the mouth.

CONTROL OF BREATHING

CENTRAL INFLUENCES

Located within or adjacent to the medulla and pons are groupings of nerve cells that constitute the RESPIRATORY CENTERS (Fig. 22.12). Within these areas are paired regions designated as the *inspiratory centers,* and that, when stimulated, lead to a maximal inspiratory effort. Dorsal to these areas are paired regions that, if stimulated, initiate expiration or partially inhibit existing inspiratory effort. These regions are designated as the *expiratory centers.* In the pons are paired regions that also tend to inhibit inspiratory activity with the major effect of inhibiting the drive provided by the medullary reticular formation. These areas are designated the *pneumotaxic centers.* In the reticular formation there are paired areas that provide a stimulation or "drive" to the inspiratory centers and are designated as the *apneustic centers.* The *nuclei of the vagus nerves* in the medulla receive impulses from the lungs and other body

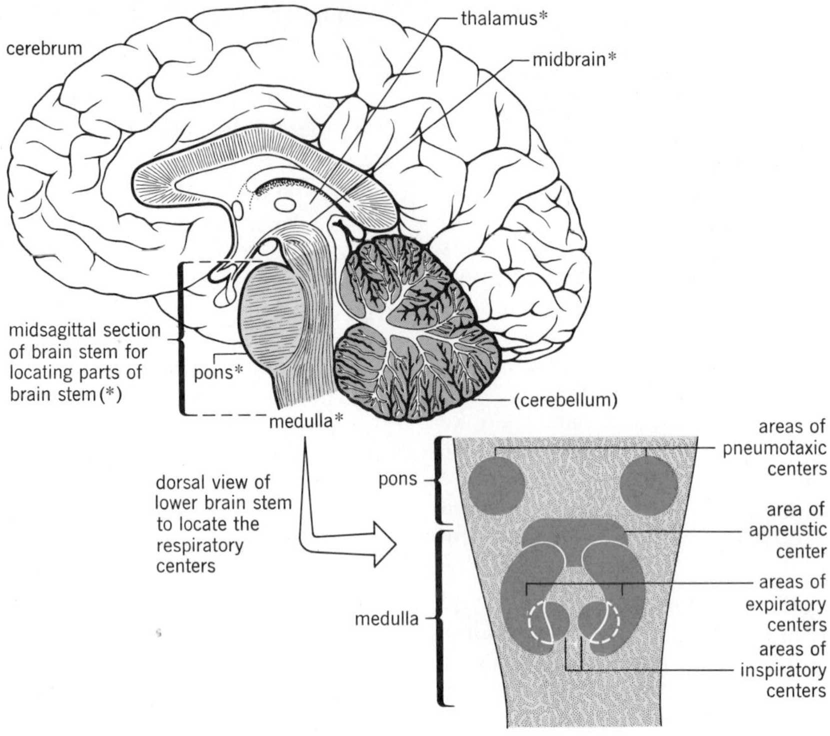

FIGURE 22.12
The location of the respiratory centers.

viscera and contribute to inhibition of inspiration. Lastly, an unnamed center in the region of the *fourth ventricle* of the brain monitors the pH of the cerebrospinal fluid and stimulates breathing if pH of the fluid decreases.

Integration of the activity of these centers may be described as follows.

The inspiratory centers have an inherent rhythmicity and send impulses via the phrenic nerves to the diaphragm, and intercostal nerves to the external intercostal muscles. These impulses cause muscular contraction drawing air into the lungs.

As lung tissue is stretched, stretch receptors send impulses to the vagal nuclei that in turn exert an inhibitory effect on the inspiratory centers. Simultaneously, the same impulses are directed to the expiratory centers. These are activated and provide an additional inhibitory effect on the inspiratory centers.

Impulses from the inspiratory centers reach the pneumotaxic centers; they in turn inhibit the activity of the apneustic centers, and the "drive" to the inspiratory centers is reduced.

All of these influences combine to interrupt inspiratory activity, the muscles relax, and expiration of air occurs. These events are depicted in Figure 22.13.

Other central influences on breathing include connections between the LIMBIC SYSTEM (*see* Chapter 9) and the respiratory centers to allow for involuntary modification of breathing during emotional states.

Thus, while the basic desire to breathe is involuntary and establishes a basic rhythm to breathing, that rhythm may be modified by a variety of factors.

THE STIMULUS TO BREATHING

Mention was made in an earlier section that the inspiratory centers exhibited a basic inherent rhythm. What is the source of that rhythm?

The most potent *stimulus* to breathing is provided by increasing P_{CO_2} in inspired air. For example, an increase of 10 percent in inspired CO_2 will cause an 8 fold increase in ventilation. This procedure causes a rise of blood P_{CO_2} and an apparent stimulus to the inspiratory centers. There are two mechanisms advanced to explain this effect.

FIGURE 22.13
The integration of central breathing control mechanisms (+, stimulation; −, inhibition)

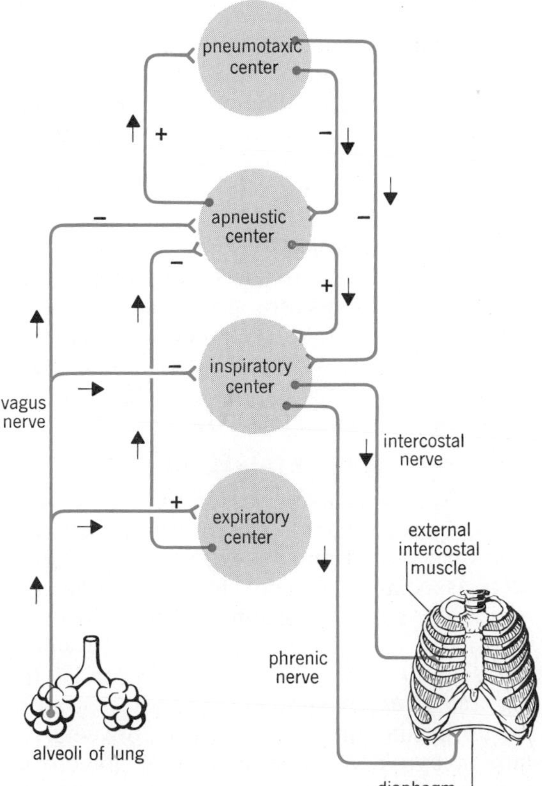

CO_2 exerts a direct effect on the inspiratory centers, or on central chemoreceptors located in the 4th ventricle, that then provide a stimulus to the inspiratory centers.

CO_2 reacts with water in the cerebrospinal fluid or cells of the inspiratory center to create carbonic acid that then dissociates *hydrogen ion.* The H^+ then stimulates breathing by causing a fall of pH that triggers the central chemoreceptors or acts directly on the cells of the inspiratory center.

Of less importance to breathing is the effect of lowered P_{O_2} on central chemoreceptors or on the inspiratory center itself. Fall of P_{O_2} in the bloodstream to about one-half normal values stimulates breathing. Because of the shape of the oxygen dissociation curve, extreme decreases would have to occur in P_{O_2} of the atmosphere or there would have to be some block to normal ventilation to cause P_{O_2} of the blood to fall to such low levels. Thus, hypoxemia, unless severe, is not as potent a stimulus to breathing as is a rise of CO_2.

PERIPHERAL INFLUENCES

The REFLEX MECHANISMS that alter basic respiratory patterns include those arising from *baroreceptors* located in the body viscera, and *chemoreceptors* that monitor CO_2, O_2, and H^+ levels in several body organs.

MECHANICAL STIMULATION (*stretching*) of baroreceptors located in the *lungs, aorta,* and *carotid sinuses* normally provides for the most important modifications of breathing.

The HERING-BREUER REFLEX is initiated when inspiration inflates the lungs. The receptors are located in the respiratory division of the system, the part that is most greatly stretched by expansion with air. The frequency of discharge of these baroreceptors increases as the tissues are stretched and are carried via vagal afferents to the brainstem. The vagal nuclei are excited and provide inhibitory impulses to the inspiratory and apneustic centers, and stimulation to the expiratory centers. Inspiration is interrupted.

Baroreceptors in the aorta and carotid sinuses pass impulses over the vagus and glossopharyngeal nerves to the brainstem. Stimulation of these receptors brings about *apnea* (cessation of breathing) through inhibition of inspiratory center activity, and also, through connections to cardiac and vasomotor centers, causes fall of blood pressure (vasodilation) and slowing of heart rate (cardiac output falls). These are the same baroreceptors discussed in control of heart rate and blood pressure in previous chapters. Thus, respiratory and cardiovascular activity is integrated to provide less O_2 for less activity.

Within the aorta and carotid sinus are chemoreceptors that monitor blood P_{CO_2}, P_{O_2}, and hydrogen ion concentration. They are known as the AORTIC BODY and CAROTID BODIES. Cells called *glomus cells* in these bodies respond to lowered P_{O_2}, elevated P_{CO_2}, or decreased pH by an increased rate of electrical discharge. The impulses are directed centrally where they produce acceleration of breathing and rise of blood pressure through vasoconstriction and this in turn ultimately causes an increased cardiac output. As with the central chemoreceptors, a rise of P_{CO_2} or decrease of pH is a more potent stimulus than is lowered P_{O_2}. Acetylcholine has been shown to be increased in the bodies by any procedure that causes hypoxia in the cells, or that raises CO_2, and may be the substance by which the stimulatory effects of the glomus cells are produced. Chemoreceptive reflex mechanisms provide the body with two important capabilities:

- A very fast reacting control mechanism, since the receptors are monitoring P_{O_2} and P_{CO_2} of blood coming from the lungs.
- A system that can, on its own, provide stimulation to the entire neural mechanism of breathing that will ensure protection against subnormal P_{O_2} of the bloodstream (providing, of course, that there is adequate atmospheric oxygen to be inspired).

Reflex mechanisms involved in control of breathing are shown in Figure 22.14.

THE FETAL RESPIRATORY SYSTEM

The fetus does not, of course, receive its oxygen or eliminate its carbon dioxide through its lungs, but utilizes the placenta as its respiratory organ. The assumption of air breathing at birth provides one of the most dramatic events in all of nature. The associated cardiovascular changes, though not as perceptible, are nonetheless as essential to survival as the respiratory changes.

THE PLACENTA AS A "LUNG"

The human placenta provides membranes of about 3.5 μm in thickness (the lung, 0.7 μm or less), a total surface area of about 12 m^2 (lung of the adult, 50 m^2) and contains a blood volume of about 45 ml (adult lung, 75 to 100 ml). The greater thickness of the diffusing surface slows gas diffusion, but the circulation time is slower, somewhat offsetting the slowed diffusion. The exchange surfaces, known as *placental villi,* float in pools of maternal blood known as the *intervillous spaces,* and the maternal blood acts as the reservoir of nutrients and oxygen the fetus requires, as well as serving as the route for elimination of fetal wastes. As far as gases are concerned, P_{O_2} in the intervillous spaces is about 44 mm Hg. Fetal blood, reaching the placenta through the umbilical arteries has a P_{O_2} of about 18 mm Hg. Oxygen diffuses until fetal P_{O_2} reaches about 30 mm Hg, drawing the maternal P_{O_2} to about 40 mm Hg in the uterine veins. Carbon dioxide concentrations in fetal umbilical artery blood are about 48 mm Hg (maternal blood, 40 to 45 mm Hg), thus CO_2 diffuses into the intervillous spaces. These relationships are presented in Figure 22.15. Oxygen uptake is aided by the presence of fetal hemoglobin (Hgb F) and low levels of DPG, so that fetal blood exhibits a greater affinity for oxygen at low P_{O_2}.

THE FETAL LUNG

The fetal lung itself passes through several stages of development. The GLANDULAR STAGE occurs between 1 and 16 weeks of life *in utero* and is associated with development of the conducting division of the lungs. The CANALICULAR STAGE occurs between 16 and 24 weeks and is associated with vascularization of the lung and the development of the respiratory division of the lungs. From 24 to 26 weeks onward, progressive development of alveoli occurs in what is termed the ALVEOLAR STAGE.

FIGURE 22.14

The basic reflex mechanisms affecting breathing. The stimulus affecting the receptor is shown in parentheses beneath the receptor name. A plus sign (+) indicates stimulation at the center; a negative sign (–) indicates inhibition.

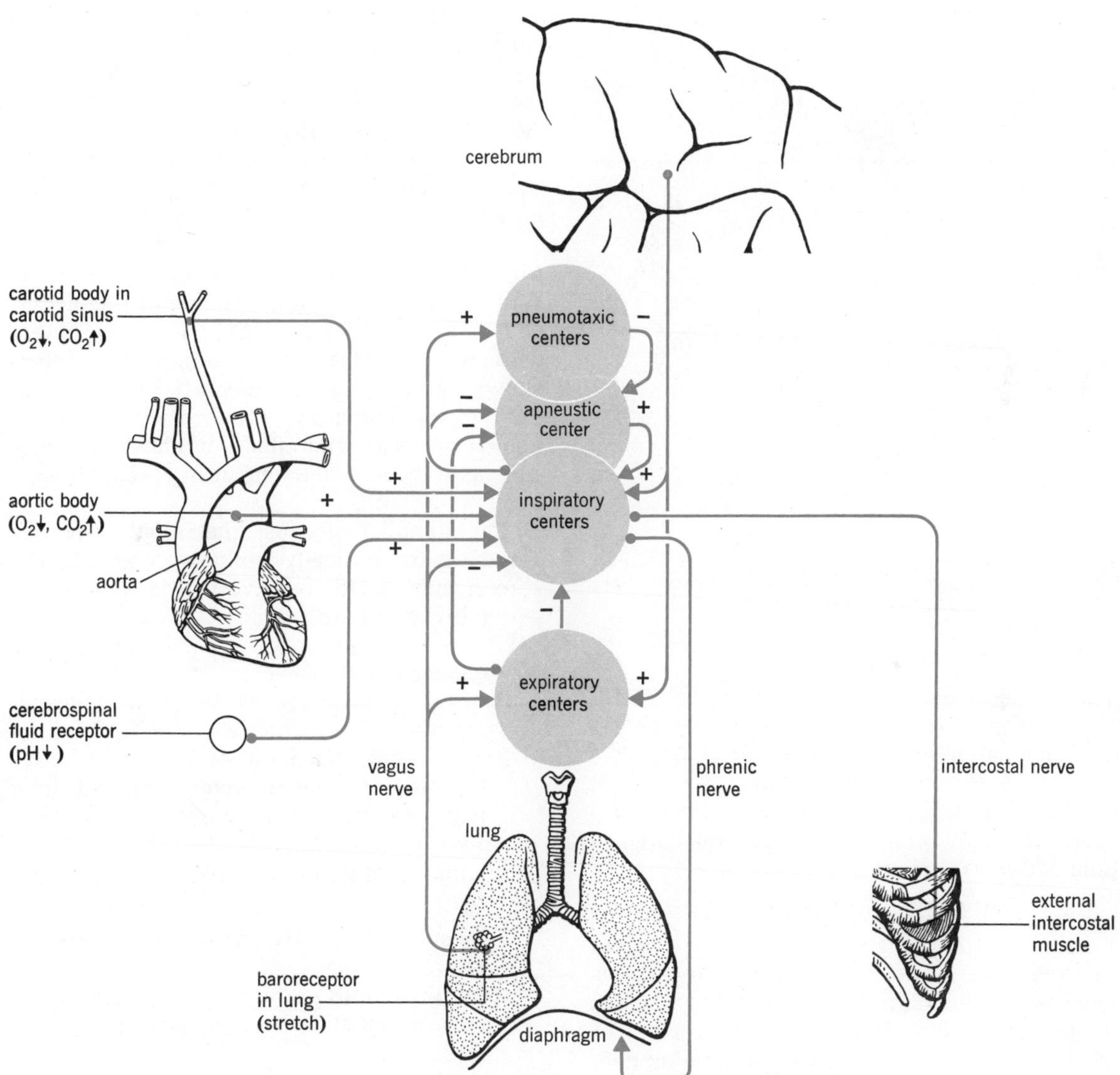

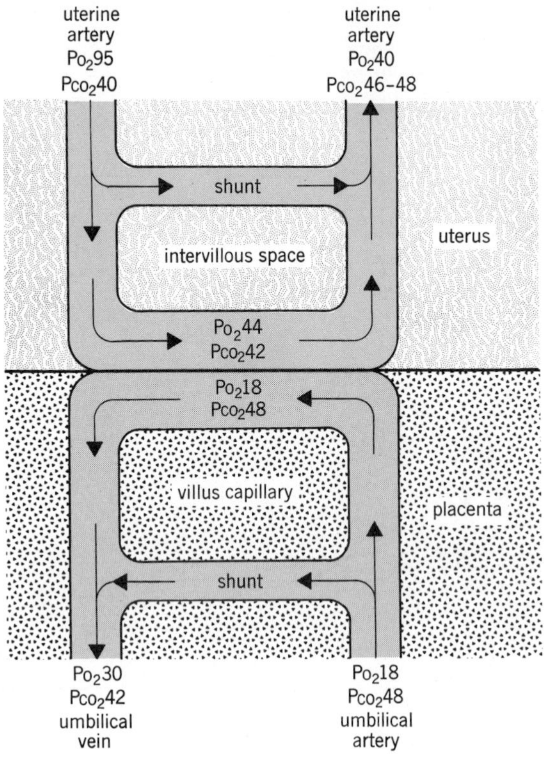

FIGURE 22.15
Oxygen and carbon dioxide tensions in the placenta and uterus.

These stages are presented in Figure 22.16. Until sufficient alveoli are formed to permit gas exchange, an infant born prematurely has little chance of survival. Additionally, the system contains amniotic fluid that the fetus has inhaled and that reduces gas diffusion between lung and bloodstream.

THE FETAL CIRCULATION

Fetal circulation (Figure 22.17) is constructed to bypass the lungs. The *placental circulation,* with its umbilical arteries and vein, the *foramen ovale* between right and left atria, and the *ductus arteriosus* between the pulmonary artery and aorta, effectively shunt the blood past the lungs.

CHANGES AT BIRTH

Birth by the normal route (through the vagina) provides a compression of the chest that is believed to express up to 10 ml of amniotic fluid from the nostrils, mouth, and trachea, so as to partially clear the system. Delivery of the chest allows it to expand and air is drawn into the lungs to partially fill them. Viscosity of the lung fluid, stiffness (compliance) of the lung tissue, and less surfactant than in the adult lung all conspire to resist initial inflation of the alveoli. Before the first true breath occurs, the infant may force additional quantities of air into the system by what is termed "frog breathing," that is, air in the throat is pushed by pharyngeal contraction into the lungs. Next comes diaphragmatic contraction and the generation of a much larger fall in intrathoracic pressure than was generated by decompression of the chest. With this first true breath, filling of the alveoli and their expansion is usually assured. Infants delivered by Caesarean section have been shown to have greater difficulty in expanding the lungs, presumably because they have been denied the beneficial effects of chest compression and recoil occurring during vaginal passage.

The stimuli that cause the initial active inspiration are basically unknown. Among the stimuli believed to be involved in stimulating active breathing are:

Nonspecific

- Position (upside down to drain fluid by gravity)
- Pain ("spank the baby" – stroking is better because the incompletely inflated lung may be collapsed if the chest is struck)
- Auditory and visual stimuli
- Excitation of pulmonary baroreceptors

Specific

- Exposure to a colder environment than that *in utero*
- Acidotic condition (high P_{CO_2}) in the bloodstream that stimulates the inspiratory centers
- Stimulation of inspiratory centers through chemoreceptors, by low P_{O_2} in the bloodstream

In any event, there is no reason to believe

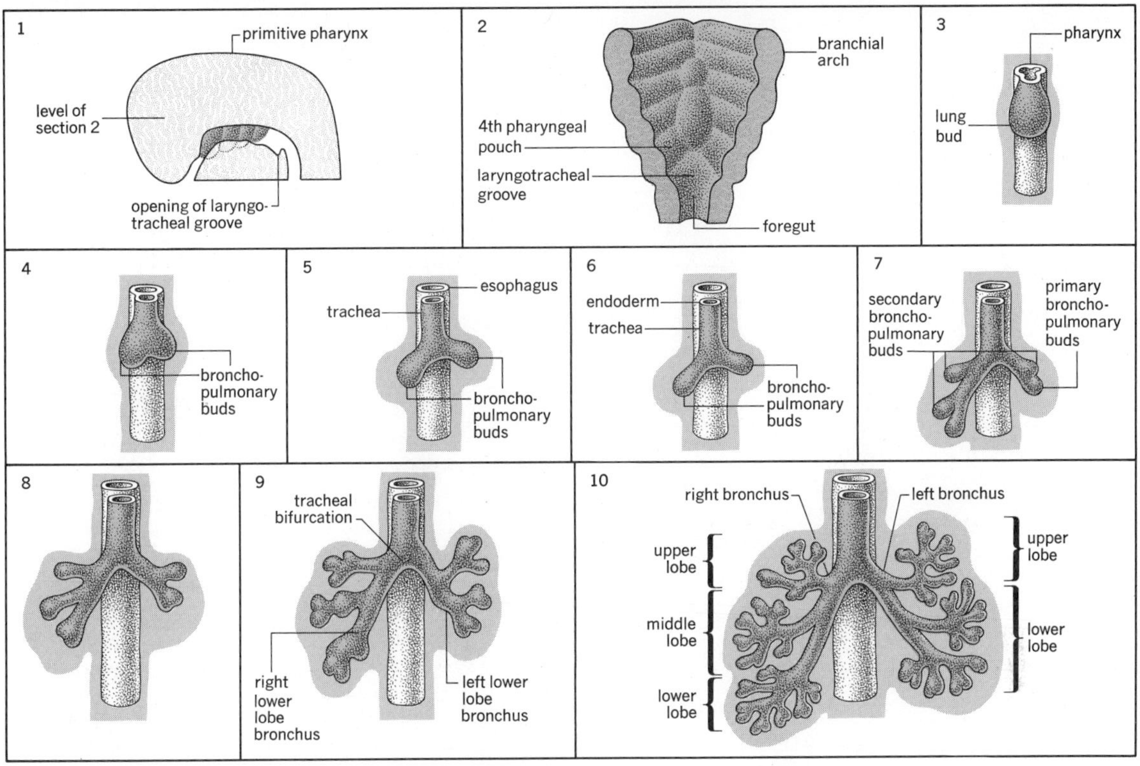

FIGURE 22.16

The development of the lower respiratory system. Numbers 1 and 2, 3½ weeks; 3–6, 4 weeks; 7 and 8, 5 weeks; 9, 6 weeks; 10, 8 weeks (*above*). The development of the alveoli at 24 and 26 weeks in the newborn (*below*).

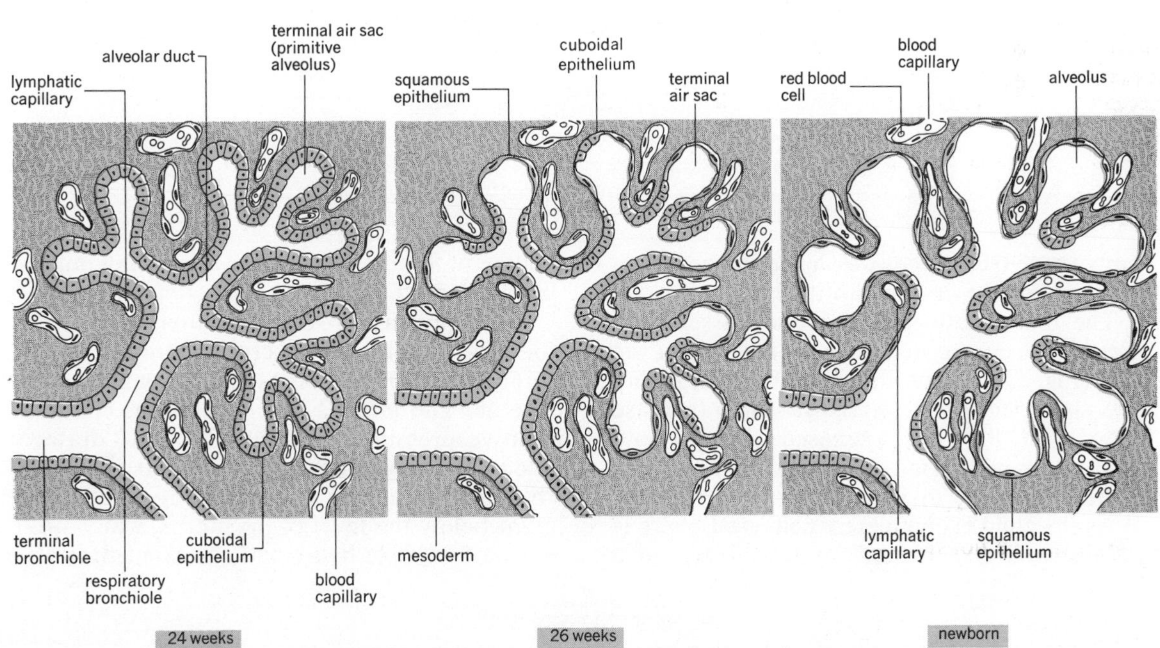

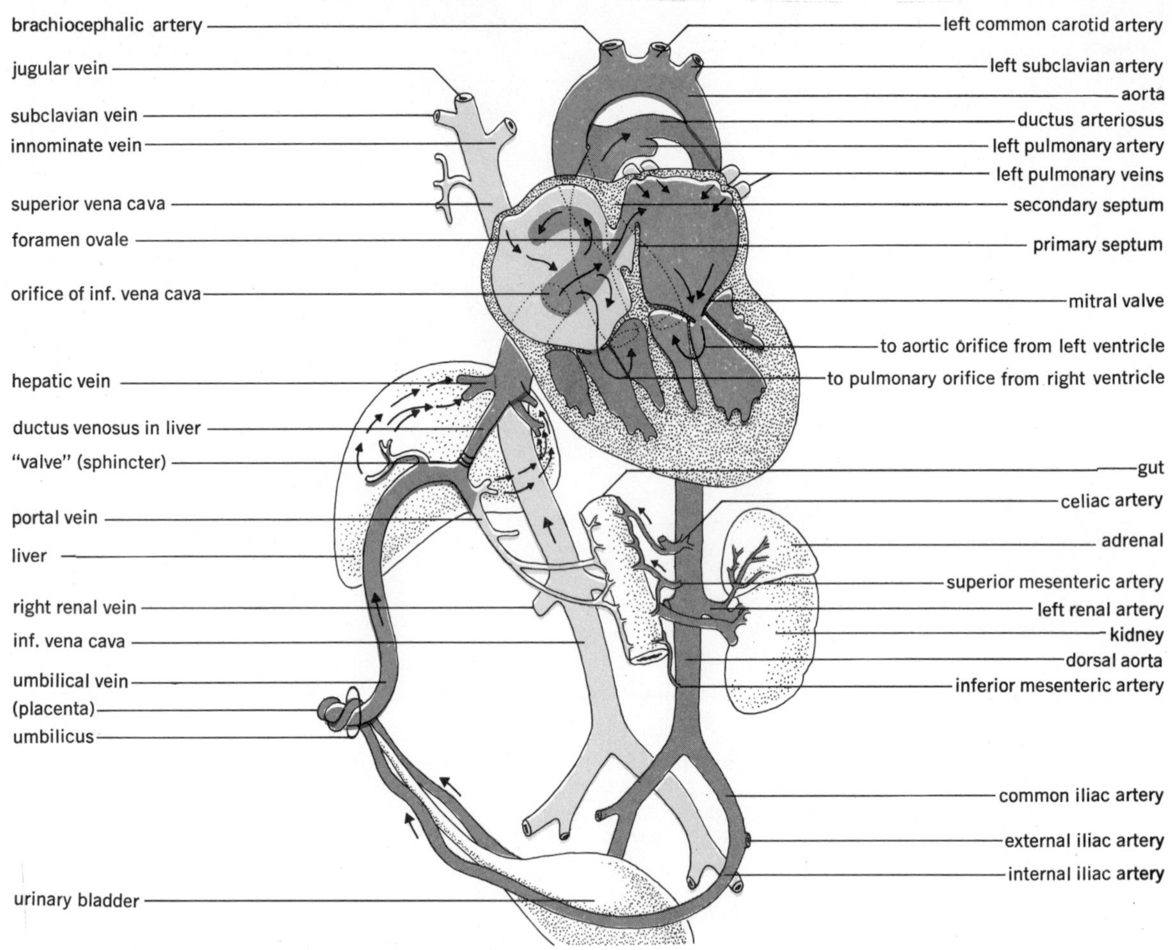

FIGURE 22.17
The plan of the fetal circulation. Colors indicate state of oxygenation (solid color, highest; lighter color, lowest; gray, intermediate).

that one of these stimuli is more important as the cause of the first breath than another.

The first breath sets in motion a series of events that ensure continued expansion of the lung (Fig. 22.18) and its perfusion with blood. The end result of these changes is an increase in alveolar P_{O_2} and increased vascular and lymph flow with removal of fluid that remains in the respiratory division.

As the child is completely delivered, there is a "placental transfusion" of fetal blood in a volume of 80 to 100 ml, *if* the cord is not immediately tied off.

The "transfusion" occurs as uterine contractions that expel the fetus create a pressure head (to 60 mm Hg) that forces blood from placenta to fetus. The infant's first inspirations create a negative pressure (−30 mm Hg) that "draws" blood from the umbilical veins and placenta. The transfusion may be speeded by holding the infant below the level of the uterus, which provides a favorable hydrostatic gradient. It is obvious that an infant whose cord is clamped 5 seconds after delivery (a common practice) will receive little or none of the fetal placental blood. Erasmus Darwin summarized the value of this transfusion by stating:

> Another thing injurious to the child is the tying and cutting of the naval string too soon, which should always be left until the child has not only repeatedly breathed but till all pulsations in the cord cease. As otherwise the child is much weaker than it ought to be, a part of the blood being left in the placenta which ought to have been left in the child.

About three minutes appears to allow transfusion of fetal placental blood to occur. The higher blood pressure occasioned by the injection of fetal placental blood into the infant is believed to aid the elimination of H^+ through the kidneys, to increase the perfusion of the lungs, to speed gas diffusion across the lung surfaces, and to accelerate the release of surfactant from the alveolar walls. It is thus generally accepted that the transfusion represents a beneficial event as far as the infant is concerned.

Clamping and cutting of the umbilical cord obviously removes one of the shunts of the fetal circulation. The umbilical arteries close proximal to the naval in response to the higher P_{O_2} occasioned by infant respirations. The ductus arteriosus also responds to higher P_{O_2} and normally begins closing at about 15 minutes after birth; closure is complete 6 to 12 hours after birth. The foramen ovale closes *functionally* within minutes after birth as pressures rise in the left atrium of the heart. Eventual "growing together" of the opposed flaps of tissue of the foramen ensures a permanent closure of the shunt.

Functioning of the respiratory control mechanisms ensuring a rhythmical pattern of breathing is immature at birth, and the infant is more

FIGURE 22.18

Some effects of birth and first breaths on the lung and circulation.

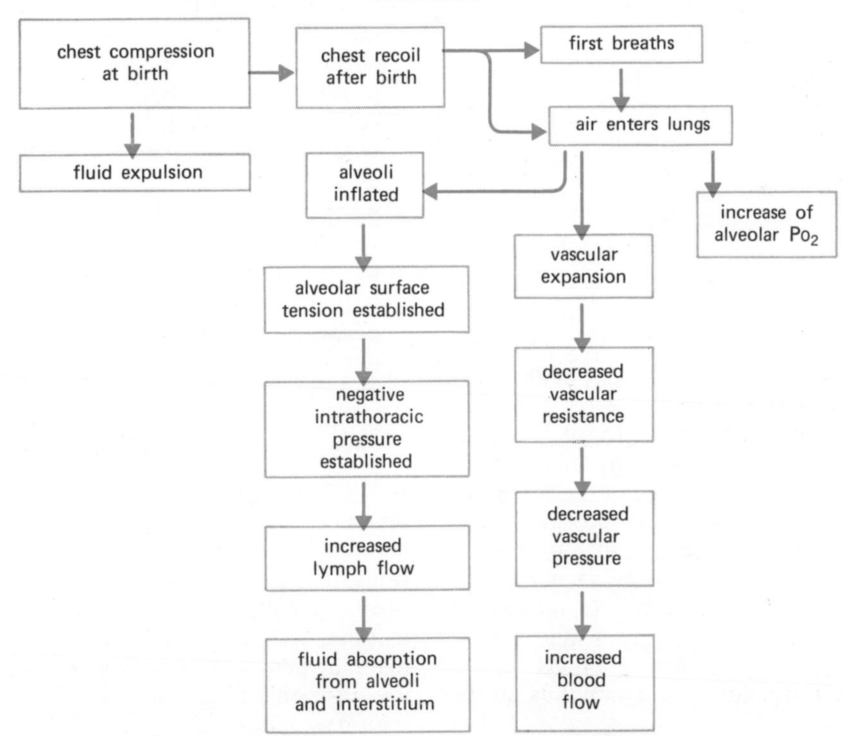

susceptible to apnea (cessation of breathing) and periodic (Cheyne-Stokes) breathing than an older child. The respiratory centers are less sensitive to blood Po_2 and Pco_2, chemoreceptors are less sensitive, and the stiffness of lung tissue may diminish the effect of the Hering-Breuer reflex. The *sudden-infant-death syndrome* (*SIDS*), characterized by death of an infant due to respiratory arrest, may (according to one of many theories of cause of SIDS) have its origin in this immaturity of the normal breathing regulatory mechanisms.

Further discussion of fetal and neonatal respiratory and circulatory function may be obtained by reference to the end-of-chapter readings.

ABNORMALITIES OF RESPIRATION

RESPIRATORY INSUFFICIENCY

This condition develops when the respiratory system cannot meet tissue demands for oxygen supply and/or carbon dioxide removal and is often the result of inadequate alveolar ventilation. Inadequate alveolar ventilation may result from a variety of conditions.

Narcosis (depression of the central nervous system). Drugs that depress sensitivity of the respiratory centers to CO_2 and H^+ result in CO_2 retention, fall of pH, and decreased Po_2 in arterial blood. Such drugs are termed *depressants*. Fortunately, such drugs also usually lead to decreased demand for oxygen, and thus tissue O_2 levels are not appreciably decreased relative to demands. Respiratory insufficiency due to narcosis therefore has the better prognosis for eventual recovery.

Respiratory disease. Emphysema, pneumonia, and atelectasis (alveolar collapse) typically reduce the diffusing surface, increase the distance for gas diffusion, prevent normal expansion and/or recoil of the lungs or combinations of these. The end results will be elevated Pco_2 and depressed Po_2 of the blood. Since tissue demand for O_2 and CO_2 production is usually unchanged or elevated (as with the increased metabolism associated with fever) mild to severe degrees of asphyxia may result. Hypoxia and acidosis at the cellular level may thus prove fatal.

Hypoxia. Regardless of cause, continued hypoxia (low arterial Po_2) results in the establishment of "vicious circles" that result in increasing degrees of cellular damage. A decrease in blood pressure, decreased cardiac output, vasodilation, release of toxins due to cell damage, and other factors combine to ultimately cause irreversible damage to respiratory neurons. Death is the usual result.

Apnea. This term refers to cessation of breathing, and it is usually understood to mean temporary interruption only. Several mechanisms are advanced to explain the development of apnea.

A REDUCED STIMULUS TO A NORMAL SET OF RESPIRATORY CENTERS is one factor. For example, voluntary hyperventilation (increased rate *and* depth of breathing) reduces blood Pco_2 by "blowing off" alveolar CO_2 and allows greater diffusion from the blood. Until blood Pco_2 returns to normal, the desire to breathe is absent.

A second factor involves INHIBITION OF DISCHARGE OF THE CENTERS as in prolonging the Hering-Breuer reflex. For example, holding the breath at the end of inspiration prevents deflation of the lungs; apnea will result until blood Pco_2 rises to the point where breathing need overrides the reflex inhibition.

A third factor is ABNORMAL RESPIRATORY CENTERS that require a higher than normal Pco_2 to stimulate them. In periodic breathing (Cheyne-Stokes respiration) periods of hyperpnea (increased depth of breathing) alternate with periods of apnea. The hyperpnea reduces alveolar and blood Pco_2, reduced stimulus to respiration is present, and apnea ensues until metabolism restores Pco_2 to higher than normal levels.

Dyspnea. This implies labored breathing and may be associated with a variety of disease entities. It is a symptom rather than a disorder itself. Acidosis, poliomyelitis, emphysema, asthma, pneumonia, atelectasis, cardiac failure, anemia, hemorrhage, fever, and use of stimulants are only a few of the conditions associated with dyspnea.

Polypnea. Polypnea (*tachypnea*), or acceleration of rate of breathing without accompanying increase in depth, occurs during fever (late), pain, and hypoxia. If depth is no greater than dead space volume (150 ml in normal adult), no alveolar ventilation will occur since all the inspired air remains in tubes that are too thick walled to allow diffusion. Thus, hypoxia may be a result, as well as being a cause.

RESUSCITATION

Resuscitation of individuals who have suffered

FIGURE 22.19
Mouth to mouth resuscitation. (Courtesy American Red Cross.)

WHEN BREATHING STOPS SECONDS COUNT SAVE A LIFE BY ARTIFICIAL RESPIRATION

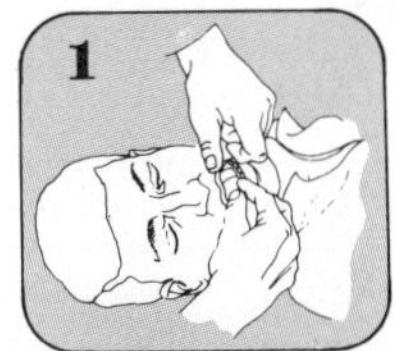

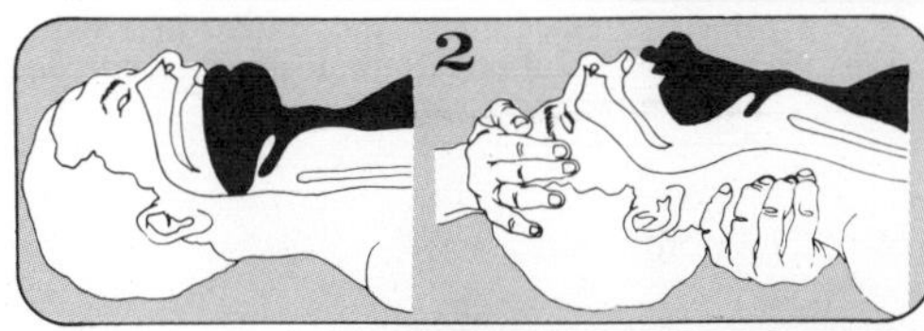

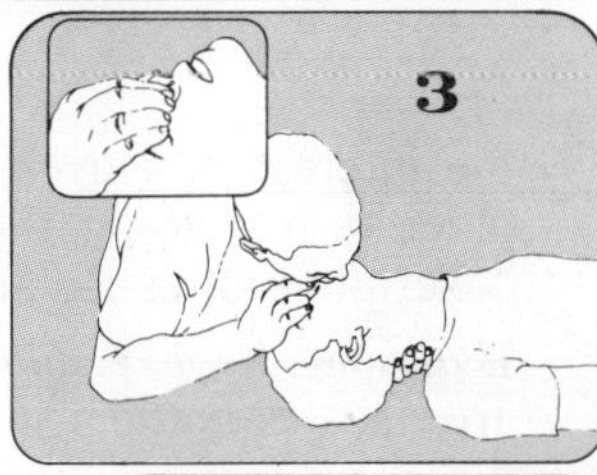

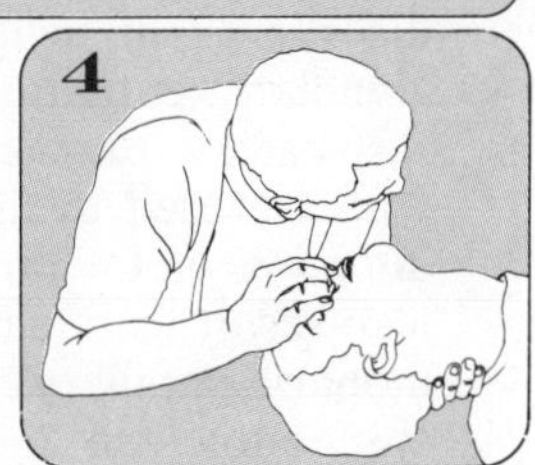

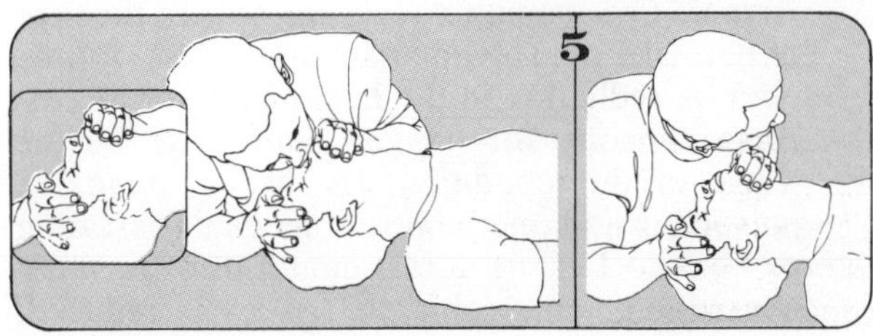

MOUTH-TO-MOUTH METHOD

1. If foreign matter is visible in the mouth, wipe it out quickly with your fingers, wrapped in a cloth, if possible.
2. Tilt the victim's head backward so that his chin is pointing upward. This is accomplished by placing one hand under the victim's neck and lifting, while the other hand is placed on his forehead and pressing. This procedure should provide an open airway by moving the tongue away from the back of the throat.
3. Maintain the backward head-tilt position and, to prevent leakage of air, pinch the victim's nostrils with the fingers of the hand that is pressing on the forehead.

   Open your mouth wide; take a deep breath; and seal your mouth tightly around the victim's mouth with a wide-open circle and blow into his mouth. If the airway is clear, only moderate resistance to the blowing effort is felt.

   If you are not getting air exchange, check to see if there is a foreign body in the back of the mouth obstructing the air passages. Reposition the head and resume the blowing effort.
4. Watch the victim's chest, and when you see it rise, stop inflation, raise your mouth, turn your head to the side, and listen for exhalation. Watch the chest to see that it falls.

   When his exhalation is finished, repeat the blowing cycle. Volume is important. You should start at a high rate and then provide at least one breath every 5 seconds for adults (or 12 per minute).

   When mouth-to-mouth and/or mouth-to-nose resuscitation is administered to small children or infants, the backward head-tilt should not be as extensive as that for adults or large children.

   The mouth and nose of the infant or small child should be sealed by your mouth. Blow into the mouth and/or nose every 3 seconds (or 20 breaths per minute) with less pressure and volume than for adults, the amount determined by the size of the child.

   If vomiting occurs, quickly turn the victim on his side, wipe out the mouth, and then reposition him.

MOUTH-TO-NOSE METHOD

5. For the mouth-to-nose method, maintain the backward head-tilt position by placing the heel of the hand on the forehead. Use the other hand to close the mouth. Blow into the victim's nose. On the exhalation phase, open the victim's mouth to allow air to escape.

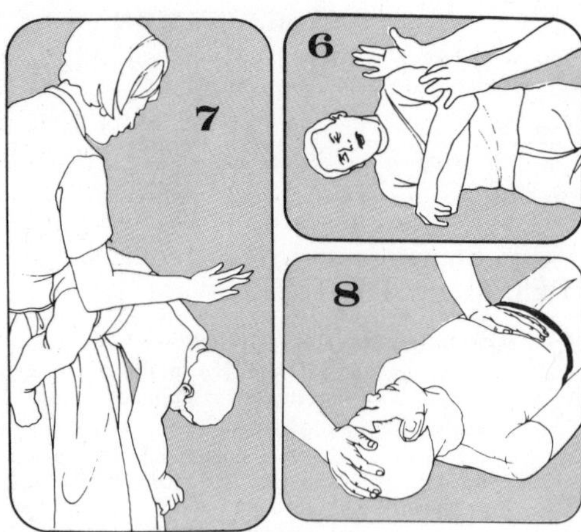

RELATED INFORMATION

6. If a foreign body is prohibiting ventilation, as a last resort, turn the victim on his side and administer sharp blows between the shoulder blades to jar the material free.
7. A child may be suspended momentarily by the ankles or turned upside down over one arm and given two or three sharp pats between the shoulder blades. Clear the mouth again, reposition, and repeat the blowing effort.
8. Air may be blown into the victim's stomach, particularly when the air passage is obstructed or the inflation pressure is excessive. Although inflation of the stomach is not dangerous, it may make lung ventilation more difficult and increase the likelihood of vomiting. When the victim's stomach is bulging, always turn the victim's head to one side and be prepared to clear his mouth before pressing your hand briefly over the stomach. This will force air out of the stomach but may cause vomiting.

When a victim is revived, keep him as quiet as possible until he is breathing regularly. Keep him from becoming chilled and otherwise treat him for shock. Continue artificial respiration until the victim begins to breathe for himself or a physician pronounces him dead or he appears to be dead beyond any doubt.

Because respiratory and other disturbances may develop as an aftermath, a doctor's care is necessary during the recovery period.

THE AMERICAN NATIONAL RED CROSS

Fig. 22.19 continued

respiratory insufficiency or failure requires ventilation of the alveoli to restore arterial P_{O_2} and assurance of sufficient heart action to deliver the blood to the tissues. MOUTH-TO-MOUTH RESUSCITATION (Fig. 22.19) provides a nonmechanical means of lung ventilation. RESPIRATORS of a positive pressure type (Fig. 22.20) force air intermittently into the lungs, and expiration occurs by elastic recoil. Negative pressure devices (Fig. 22.21) decompress and compress the environment around the patient and "breathe for him."

HYPOXIA

In a preceding section, the term "hypoxia" was used. The word, strictly defined, refers to lowered oxygen tension at the cellular level. Insufficient oxygen at the cellular level may be secondary to diminished levels of oxygen in the blood, known as HYPOXEMIA. Types of hypoxemias and hypoxias are presented below with some causes listed (synonomy is presented in the interest of correlation between newer and older terminology; the newer terms are more descriptive of cause and characteristics).

Hypotonic hypoxemia (*anoxic anoxia*). Lowered arterial P_{O_2} is the primary characteristic of this disorder. It is usually the result of inadequate alveolar ventilation due to obstruction of a respiratory passageway, exposure to altitude with lowered available oxygen, or diminished diffusion of O_2 through alveolar walls. It may also result from shunting of blood past the pulmonary capillary beds (as in failure of fetal shunts to close).

Isotonic hypoxemia (*anemic anoxia*). In the face of normal P_{O_2} in alveolar air, arterial P_{O_2} is lowered. A reduction in the hemoglobin or its carrying capacity for oxygen appears to be the cause of this condition. In carbon monoxide poisoning and methemoglobinemia (oxidation of Fe^{++} to Fe^{+++}), total amount of oxygen delivered on hemoglobin to tissues is reduced.

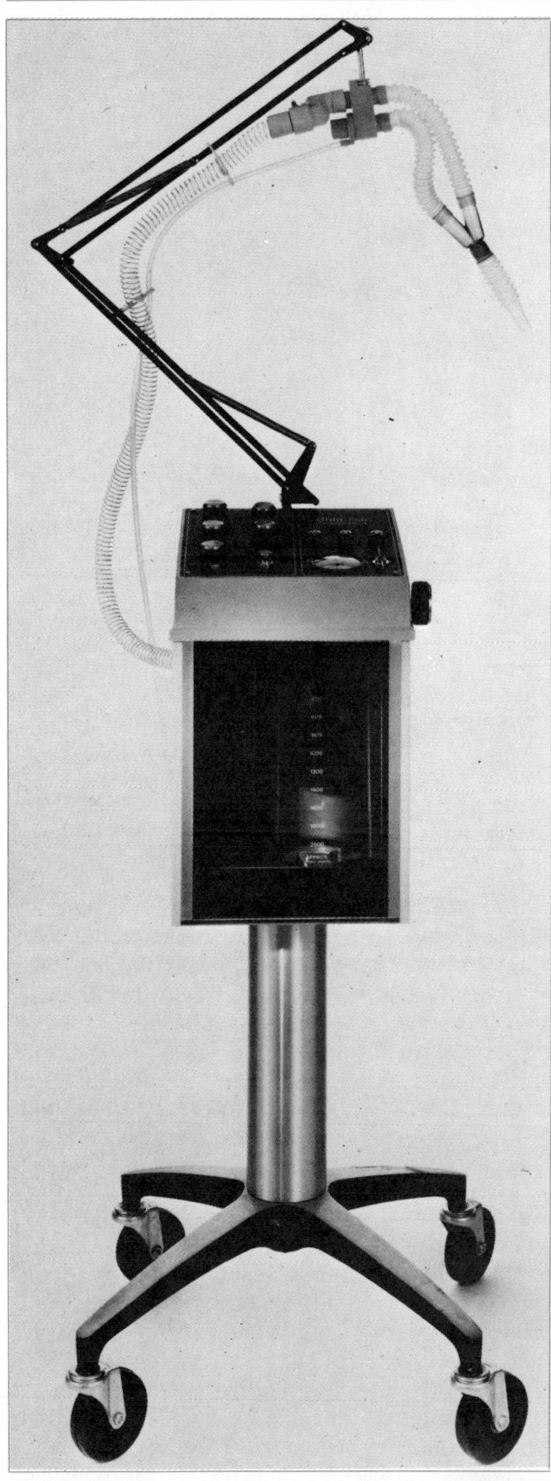

Less oxygen combines with ferric (Fe^{+++}) iron. Also, snake venoms or chemicals that destroy erythrocytes may reduce the oxygen carrying capacity.

Hypokinetic hypoxia (*stagnant anoxia*). Basically a circulatory problem, this condition results when blood flow is impeded in some way. Hemoglobin concentrations, alveolar ventilation and diffusion are all within normal limits. Cardiac failure, embolism or thrombosis, hypotension, and acidosis of metabolic origin may lead to this condition.

Overutilization hypoxia. In strenuous exercise, convulsions, or where there are narrowed vessels, the demand for oxygen may exceed the supply. Arterial O_2 concentration decreases, and CO_2 rises. O_2 debt results (a temporary shortage of O_2 for oxidation of metabolites).

Histotoxic hypoxia. Cellular poisoning, as with cyanide, sulfides, or excessive O_2 levels, causes inability of cells to utilize oxygen. Cell death typically results. All other factors (ventilation, diffusion, transport) are normal.

HYPOXEMIA

Decreased P_{O_2}, or hypoxemia, is a common result of exposing oneself to increasing altitude above sea level. The diminishing availability of oxygen in the atmosphere may lead to insufficient oxygenation of arterial blood. Effects on the body are determined primarily by the rate at which the decreased oxygen supply is encountered. In short, if exposure is gradual, the possibility for the body to adjust or acclimatize to the lowered oxygen pressure exists. Figure 22.22 shows some of the relationships between altitude and oxygen pressure. At about 18,000 feet, pure oxygen must be breathed if adequate arterial P_{O_2} is to be maintained. At about 44,000 feet, *pressure* must be added to the pure oxygen, as total pressure of H_2O and CO_2 in the lungs will equal the environmental pressure and no O_2 will be found in the lungs.

FIGURE 22.20
A positive pressure respirator. It forces air under pressure into the lungs. (Courtesy Ohio Medical Products, Division of Airco, Inc.)

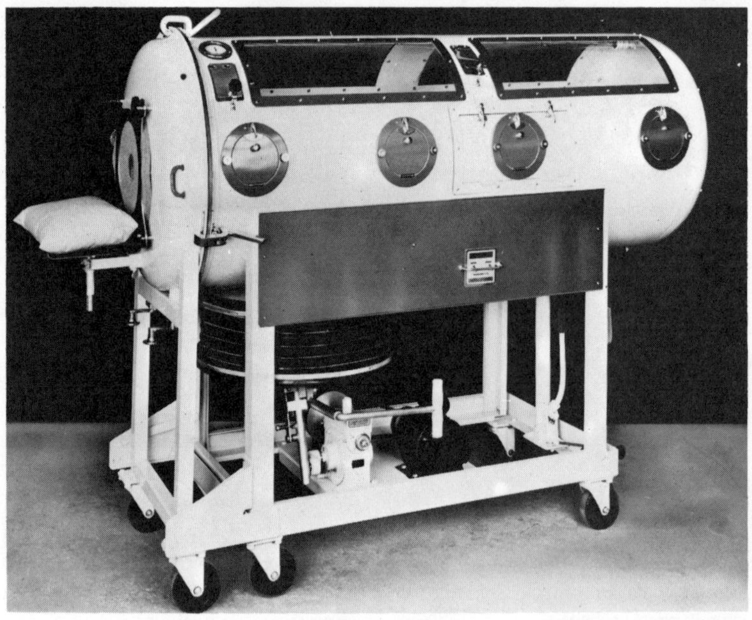

FIGURE 22.21
A negative pressure respirator. It operates by lowering pressure around the chest. The chest expands, and air enters the lungs. (Courtesy Warren E. Collins, Inc.)

Acclimatization. Adjustment of the body to altitude occurs through changes that ensure continued supply of oxygen to cells in the face of decreased oxygen in the environment. Several changes occur during acclimatization.

INCREASED RATE AND DEPTH OF BREATHING. Relatively greater concentrations of CO_2 stimulate respiration, increasing RMV, and elimination of CO_2.

MOVEMENT OF O_2 DISSOCIATION CURVE TO THE LEFT. Increase in pH, caused by ridding the body of more CO_2, causes a left shift; this also results in a greater affinity for O_2 at lower ambient P_{O_2}.

INCREASE IN RED CELLS. Lowered P_{O_2} in inspired air and resultant hypoxemia causes release of erythropoietin. Stimulation of red cell production results. More cells per volume of blood increases oxygen carrying capacity.

INDIVIDUAL TOLERANCE to hypoxia is variable. Probably dependent on circulatory adjustments, some individuals retain their performance better than others in the face of oxygen deprivation.

Adjustments such as those described require time to complete. A *resident* at high altitude can thus expect to adjust better than one who enters and leaves such environments rapidly (e.g., aviators). Time of acclimatization varies and seems to depend upon altitude and individual response. Complete acclimatization to altitudes of 15,000 to 18,000 feet may take as long as five to nine weeks.

GAS PRESSURE AND PHYSIOLOGY

Humans' entry into the inner space (undersea) and outer space environments has created physiological problems associated with supply of oxygen, removal of carbon dioxide, and accumulation of nitrogen and carbon monoxide. In some cases, artificial environments must be created to sustain life, environments that differ in gas composition and pressure from that to which humans are normally adjusted.

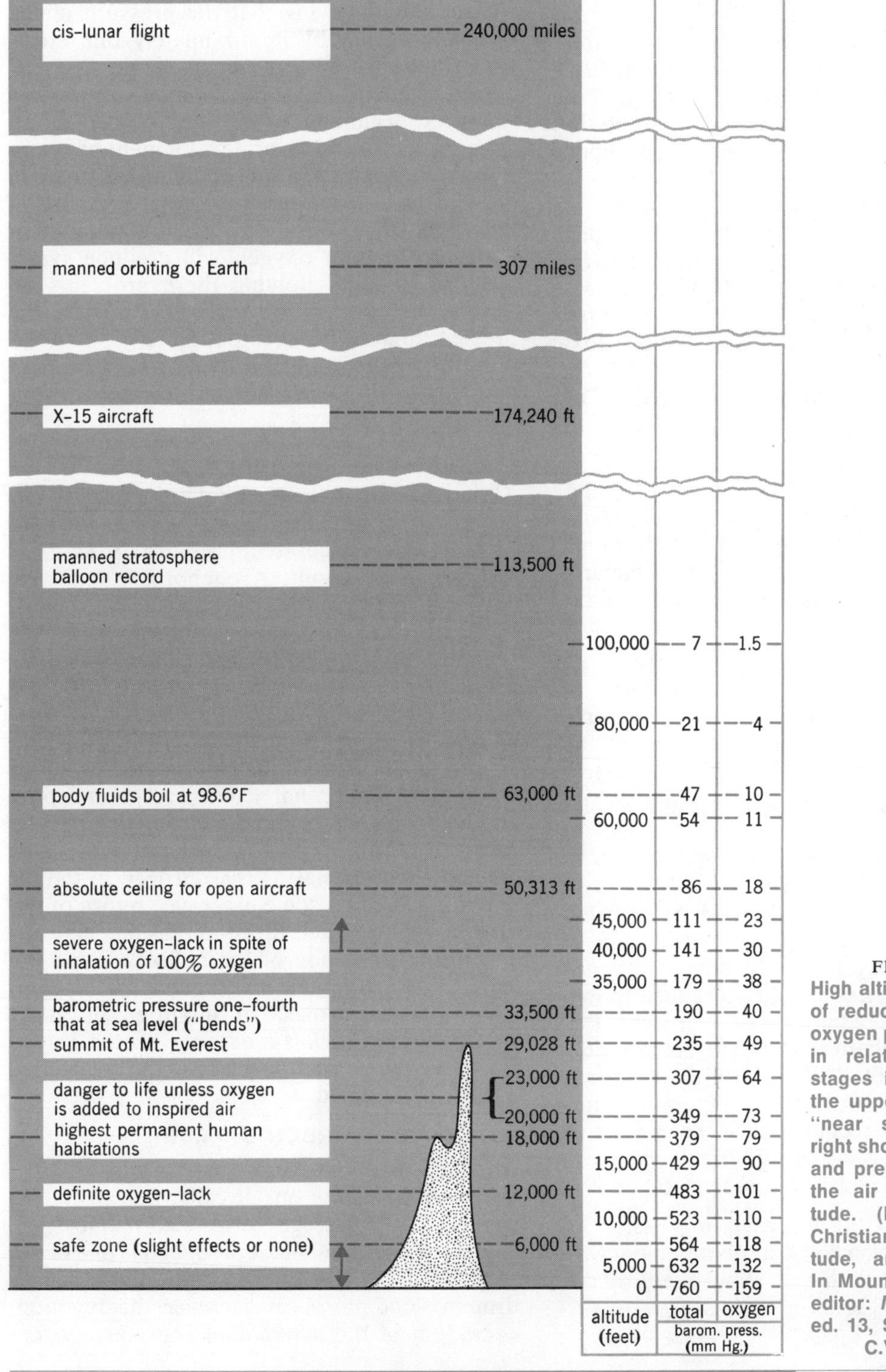

FIGURE 22.22
High altitude. Critical levels of reduced barometric and oxygen pressure are shown in relation to significant stages in the conquest of the upper atmosphere and "near space." Scales at right show total air pressure and pressure of oxygen in the air at increasing altitude. (From Lambertsen, Christian J.: Anoxia, altitude, and acclimatization. In Mountcastle, Vernon B., editor: *Medical Physiology*, ed. 13, St. Louis, 1974, The C.V. Mosby Co.)

OXYGEN

Exposure to oxygen at high partial pressures for periods of time exceeding 12 hours may lead to the development of oxygen toxicity. The mechanism of the toxicity is believed to be the result of the action of excessive oxygen upon intracellular metabolic processes, particularly those reactions concerned with transfer of hydrogen. A number of dehydrogenases that depend for their activity on sulfhydryl (*SH*) groups may be put permanently into the oxidized state as disulfide (*SS*) and be rendered incapable of hydrogen transport. Oxygen toxicity is, at least in the earlier stages, reversible. Later, irreversible damage may result. The SYMPTOMS OF OXYGEN TOXICITY are seen in many body areas.

THE LUNGS. Irritation of the linings of the respiratory system leads to coughing, bronchopneumonia (inflammation of terminal bronchioles and alveoli), loss of surfactant, pulmonary edema, and other pathological changes.

THE CENTRAL NERVOUS SYSTEM. At pressures of 2 atmospheres* or more, toxic effects of oxygen on neural function include the production of nausea, dizziness, tingling of the hands, convulsions, and unconsciousness.

THE EYE. In the adult, vasoconstriction results that may interfere with the blood supply to the retina. Loss of peripheral vision leads to "gun-barrel vision," where one has a visual field similar to that produced by peering down a gun barrel. Ultimately, damage to retinal cells may result. In infants, RETROLENTAL FIBROPLASIA (*RLF*) has been shown to result from exposure to increased Po_2 (as in an incubator). Extreme constriction of retinal vessels leads to cell death by asphyxia, detachment of the retina, and blindness. It is thus recommended that a limit of 40 percent oxygen be observed when exposure will exceed 12 hours in duration.

Implications of the foregoing discussion as to the use of pure oxygen in diving should be obvious. Starting with one atmosphere at the surface and knowing that the pressure of the ocean increases by approximately one atmosphere for each 33 feet increase in depth, it becomes clear that toxic effects may be expected to occur beginning at 33 feet, or two atmospheres. The rapidity of development of symptoms will depend not only on depth but on time. At 100 feet, for example, a total pressure of four atmospheres will exist; a few minutes of exposure to pure oxygen will produce symptoms of toxicity. Solving these problems requires reduction of oxygen pressure. At the present limit (600 feet depth and about 20 atmospheres) of human's diving, an atmosphere containing about 10 percent oxygen is employed.

* One atmosphere is equal to 760 mm Hg or approximately 15 lb/sq in. (psi).

NITROGEN

Although inert in the metabolic sense, nitrogen exerts important mechanical effects on the body. Under excessive pressure, NITROGEN NARCOSIS may result. A euphoria is followed by impairment of higher nervous system functions, unconsciousness, and interference with synaptic function. The mechanism involves many effects; a recent theory suggests that the formation of gas hydrate "micro-crystals" stabilizes, or renders incapable of change, the side chains of nerve cell chemicals and results in diminution of neuronal activity. At great depths another problem is encountered, that of the entry of nitrogen into the tissues in greater amount than normal. If the nitrogen in the air is not replaced by some other gas, return of the diver to the surface (decompression) may result in the formation of gas bubbles in the tissues and circulation (the "bends"). Helium, which is often used in mixtures of gases breathed when diving at extreme depths, leaves the tissues very rapidly and thus the bends may usually be avoided.

CARBON DIOXIDE

At high partial pressures, carbon dioxide produces toxic effects on all cells, due to its extremely rapid passage across cell membranes. The most common effect produced is one of depression of cellular activity. Vasodilation (with drop in blood pressure), decreased heart action, excitation of the sympathetic nervous system and adrenal medulla, are signs of CO_2 toxicity. The mechanism of production of these effects is either by increased amounts of molecular CO_2 or increased $[H^+]$ on cell processes.

CARBON MONOXIDE (CO)

A normal product of cellular metabolism, CO is produced at rates that are too low to result in toxic effects. At higher levels, the major effects are produced by the fact that CO binds more strongly to hemoglobin than oxygen. Oxygen is thus displaced from the pigment; hypoxemia and hypoxia result. Symptoms of CO poisoning include severe frontal headache, fainting, collapse, and the imparting of a cherry red color to arterial blood as carboxyhemoglobin (HgbCO) is formed.

In space travel, where it is necessary to create artificial environments to sustain astronauts, the problems become those associated with supply of O_2, and removal of toxic gases (CO_2, CO). "Scrubbers," which chemically remove CO_2 and CO from the atmosphere are easily constructed and are capable of keeping the concentrations of these gases at levels of 0.2 to 0.3 percent. The United States has employed a 100 percent oxygen atmosphere at ⅓ atmosphere (5 psi) to insure adequate oxygenation. On return to the earth's atmosphere, there are no recompression problems, and the excess oxygen is metabolized. Thus, no embolism problems are encountered. Humidity is also easily controlled by dehumidifying agents that keep the humidity at 36 to 70 percent. Contamination by trace constituents remains the increasing threat as more efficient methods of sealing space craft and removal of CO_2 and CO are found.

SUMMARY

1. A respiratory system exists to:
   a. Provide a surface for gas exchange.
   b. Ensure ventilation of that diffusing surface so as to maintain adequate diffusion gradients for gases.
   c. Protect the body from the entry of potentially dangerous materials in inhaled air.
   d. Provide control to assure ventilation compatible with body requirements for oxygen and carbon dioxide elimination.
2. The respiratory system consists of:
   a. A conducting division to transport gases; it constitutes a dead space filled with dead air.
   b. A respiratory division to permit gas exchange between lungs and bloodstream.
   c. The conducting division warms, moistens, and cleanses inhaled air; the respiratory division cleanses, using alveolar macrophages, and allows gas diffusion.
3. Breathing is divided into inspiratory and expiratory phases.
   a. Inspiration (air intake) is active, requiring muscular contraction that increases chest and lung volumes and decreasing pressure to allow inflow of air.
   b. Expiration is due primarily to elastic recoil of lung tissue stretched during inspiration that drives air from the lungs.
   c. Surface tension effects in the alveoli aids their becoming smaller during expiration. Surfactant (a phospholipid) lining the alveoli prevents their collapse during expiration.
4. The lungs contain certain volumes of air, which may be combined to give a series of capacities.
   a. Tidal volume is air exchanged during normal breathing. Average value in the adult is 500 ml.
   b. Inspiratory reserve volume is air inhaled above tidal volume. Average adult value is 3000 ml.
   c. Expiratory reserve volume is air exhaled beyond tidal volume. Average adult value is 1100 ml.
   d. Residual volume is air remaining in the lungs after forced expiration. Average adult value is 1200 ml.
   e. Minimal volume is air remaining in the lungs after their collapse. Average adult value is 600 ml.
   f. Vital capacity is the sum of tidal, inspiratory reserve, and expiratory reserve volumes, 4600 ml.
   g. Functional residual capacity is the sum of expiratory reserve and residual volumes, 2300 ml.

5. Respiratory minute volume (*RMV*) is the product of rate of breathing and tidal volume. A great reserve is found in the RMV to permit adjustment of ventilation to body demand for gas exchange.
6. The work of breathing is determined by resistance to air flow, compliance, and elasticity of the lungs.
   a. Air flow is determined by the diameter of the respiratory passageways. A smaller diameter requires more effort to move air through it.
   b. Compliance refers to the stiffness or ease with which the lungs are expanded.
   c. Elasticity refers to the elastic recoil of the lung that expels air.
   d. In some lung disorders, these factors are altered and the effort required to fill or empty the lungs is increased.
7. Pulmonary function may be assessed by tests that measure maximum ventilation, expiratory volumes, and expiratory flow rates.
8. Partial pressures of O_2 and CO_2 are such that diffusion of O_2 is from lungs to blood to tissues. Diffusion of CO_2 is in the opposite direction.
9. Rate of diffusion of gases is determined by volume, temperature, and partial pressure, as described by the gas laws.
10. Transport of gases is by the bloodstream.
    a. Oxygen is carried primarily on the hemoglobin in the erythrocytes. The oxygen dissociation curve shows hemoglobin saturation with oxygen as a function of P_{O_2}, blood acidity, temperature, and DPG concentrations.
    b. Carbon dioxide is carried as bicarbonate ion, on hemoglobin, and in solution in the plasma.
11. The lungs, by eliminating CO_2, are important organs aiding in regulation of acid-base balance.
12. Protection of the respiratory system and the body against particulate matter and foreign chemicals is provided by the mucociliary escalator, phagocytes, filtering devices, secretion of immune globulins, lymphatics, and reflex mechanisms.
13. Control of breathing (rate and depth) depends on the action of CO_2 and/or H^+ on nerve cells in the brainstem. Chemoreceptors and baroreceptors in the periphery modify basic breathing rhythms.
14. The fetus gains its oxygen and eliminates its carbon dioxide through the placenta.
15. The fetal lung is not capable of sustaining life until alveoli develop at about 25 weeks of intrauterine life.
16. The fetal circulation is provided with several shunts that bypass the nonfunctional lung.
17. At birth, various stimuli cause independent breathing on the part of the infant, expel fluid from the lungs, and seal the circulatory shunts to enable the lungs to assume their role as respiratory organs.
18. Abnormalities of respiration include respiratory insufficiency, hypoxia, and hypoxemia.
    a. Respiratory insufficiency usually is the result of poor ventilation and transport of gases.
    b. Hypoxia is insufficient O_2 at the tissue level and may result from inadequate ventilation, transport, or failure of cells to utilize O_2.
    c. Hypoxemia refers to low arterial P_{O_2} and may represent circulatory or gas supply and diffusion problems.
19. Excessive pressures of gases may prove to be toxic to the body.
    a. Oxygen toxicity causes irritation of the lungs, disturbance of central nervous system function, and eye problems.
    b. Excessive nitrogen in the body tissues results in euphoria, then depression and the "bends," if the escaping gas forms bubbles in the bloodstream.
    c. CO_2 excess causes depression of cellular activity.
    d. CO causes hypoxia and hypoxemia and may cause death by suffocation since the gas forms a stronger bond with hemoglobin than does oxygen.

QUESTIONS

1. Why do certain animal groups have a respiratory system? What does it provide for?
2. What are the "stages" or "phases" of activity a respiratory system supplies to ensure adequate oxygen supply and carbon dioxide removal?
3. Name the organs comprising the conducting and respiratory divisions of the respiratory system. What functions do each part have?
4. What is the difference between anatomical and physiological "dead spaces" and "dead air"?

5. How are inspiration and expiration achieved?
6. Name the volumes and capacities of the lung and give a volume for each.
7. What is the role of surfactant in pulmonary physiology?
8. What changes occur in pulmonary ventilation, and why, in asthma and emphysema?
9. What ensures an "outside to inside" movement of oxygen? An "inside to outside" movement of carbon dioxide?
10. How, and to what extent, are oxygen and carbon dioxide carried by the blood?
11. What factors determine the extent of oxygenation of the blood?
12. How is the lung involved in acid-base regulation?
13. What mechanisms exist to protect the body from inhaled toxins?
14. How is breathing controlled? What supplies the stimulus that leads to a rhythmical pattern of breathing?
15. How can minute volume of respiration be adjusted to increased demand for oxygen and/or carbon dioxide production?
16. How does the fetus acquire O_2 and eliminate CO_2? Are there "special" features of its blood that ensure adequate oxygenation; if so, what are they?
17. How is the fetal circulation adapted to bypass the nonfunctional lung?
18. How do respiratory and circulatory systems change at or after birth? What may cause "the first breath"?
19. What determines the effort required to breathe?
20. Compare the effects of excessive pressures of O_2, CO_2, and N_2 on a diver swimming at 200 feet without benefit of removal of CO_2 from his/her air.

READINGS

Avery, Mary E. "In Pursuit of Understanding the First Breath." *Ann. Rev. Resp. Dis. 100:*295. 1969.

Avery, Mary E., Nai-San Wang, and Wm. Taeusch, Jr. "The Lung of the Newborn Infant." *Sci. Amer. 228:*74. April 1973.

Berger, Albert J., Robert A. Mitchell, and John W. Severinghaus. "Regulation of Respiration." *New Eng. J. Med. 297:*92, July 14, 1977. *297:*138, July 21, 1977. *297:*194, July 28, 1977.

Cowlett, Richard M., and William Oh. "Respiratory Distress in Infants Delivered by Repeat Cesarean Section." *New Eng. J. Med. 295:*1222. Nov 25, 1976.

Cuestas, Raul A., Arnold Lindall, and Rolf R. Engel. "Thyroid Hormone and the Respiratory-Distress Syndrome of the Newborn." *New Eng. J. Med. 295:*297. Aug 5, 1976.

Darwin, Erasmus. *Zoonomia.* 3rd ed. London. J. Johnson, 1801, Vol III.

Fishman, Alfred P., and G. G. Pietra. "Handling of Bioactive Materials by the Lungs." *New Eng. J. Med. 291:*884. Oct 24, 1974. *291:*953. Oct 31, 1974.

Guilleminault, Christian, Rosa Peraita, Marianne Souquet, William C. Dement. "Apneas During Sleep in Infants: Possible Relationship to Sudden Infant Death Syndrome." *Science 190:*677. 14 Nov 1975.

Hachachka, P. W., and T. B. Storey. "Metabolic Consequences of Diving in Animals and Man." *Science 187:*613. 21 Feb 1975.

James, L. Stanley. "Perinatal Events and Respiratory Distress Syndrome." *New Eng. J. Med. 292:*1291. June 12, 1975.

Klotz, Irving M., Gerald L. Klippenstem, and Wayne A. Hendrickson. "Hemerythrin: Alternative Oxygen Carrier." *Science 192:*335. 23 April 1976.

Naeye, Richard L., Russell Fisher, Monique Ryser, and Philip Whalen. "Carotid Body in Sudden Infant Death Syndrome." *Science 191:*567. 13 Feb 1976.

Newhouse, M., J. Sanchis, and J. Bienenstock. "Lung Defense Mechanisms." *New Eng. J. Med. 295:*990. Oct 28, 1976. *295:*1045. Nov 4, 1976.

THE DIGESTIVE SYSTEM

OBJECTIVES

After studying this chapter, the reader should be able to:

☐ Give the general functions served by the digestive system.

☐ Explain how feeding is controlled, and what some of the possible stimuli triggering it are.

☐ List the major components of saliva, give their functions, and explain how its secretion is controlled.

☐ Show how digestion of carbohydrates is begun in the mouth.

☐ Explain how foods are swallowed and transported to the stomach.

☐ Diagram a gastric gland, list the cell types found therein, and give the secretions each produces.

☐ Describe digestion in the stomach.

☐ Explain why absorption from the stomach is limited in scope.

☐ Explain how gastric secretion is stimulated and inhibited.

☐ List the factors and their effects that govern emptying the stomach.

☐ List the major components of pancreatic juice, give their functions, and explain how their secretion is controlled.

☐ Explain the role of bile in the digestive process, and how its release from the gall bladder is controlled.

☐ Complete the digestion of carbohydrates and proteins by explaining the role of the intestinal epithelial cells in the process.

☐ Explain how secretion of the intestinal wall is controlled.

☐ Explain how the end products of digestion and other components of the intestinal fluid are absorbed.

☐ List the functions of the colon.

☐ Give the functions of the liver, other than bile secretion.

☐ Compare the infant digestive system to that of the adult.

☐ List the various types of motility occurring in the alimentary tract, and give the functions of each type.

☐ Explain how defecation occurs.

☐ List some common disorders of the digestive system and some causes for each.

A digestive system provides a means for food intake or INGESTION, devices to reduce foods to their constituent building blocks or DIGESTION, methods of ABSORBING those building blocks, and devices for eliminating or EGESTION of the unabsorbed residues of the digestive process.

These basic processes are provided by the organs of digestion (Fig. 23.1). The organs consist of a tubular portion, commencing with the mouth and terminating at the anus, known as the ALIMENTARY TRACT. In and from this tube, the processes of digestion and absorption occur. Lying outside of the alimentary tract are

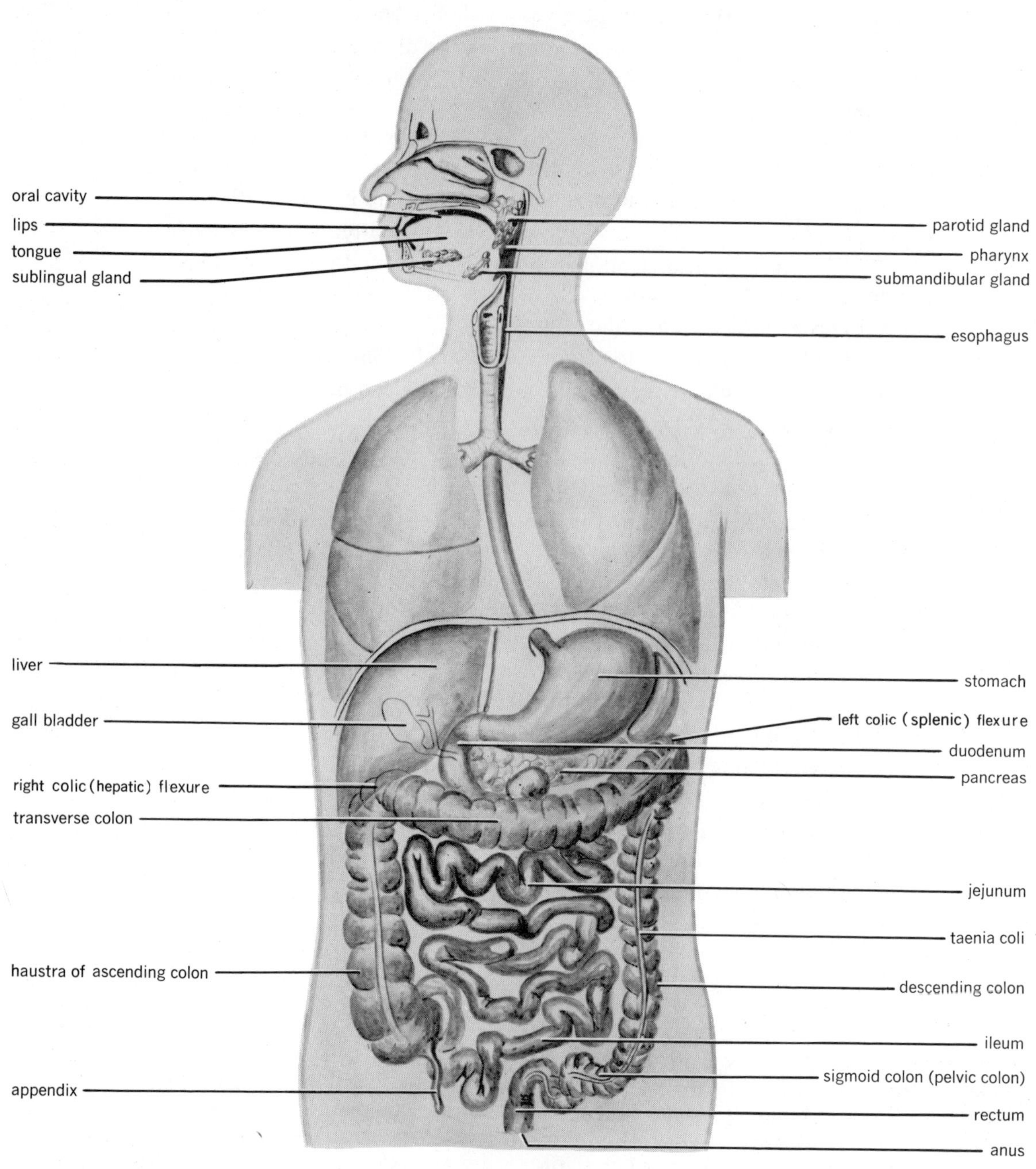

FIGURE 23.1
The organs of digestion.

the ACCESSORY ORGANS of digestion. These organs empty secretions into the tract by way of ducts; these secretions are necessary for digestion. Muscular tissue in the walls of the tract propels foodstuffs through the tube so that a sequential breakdown of large molecules to smaller ones occurs.

THE PROCESSING AND ABSORPTION OF FOODS

INGESTION

Feeding, or intake of food, appears to be controlled by neural mechanisms centered in the hypothalamus. The term HUNGER implies a need for food in general, and is satisfied by the act of FEEDING. APPETITE implies a desire for a specific food whose intake is generally associated with pleasant sensations. Hunger would appear to be instinctual, while appetite is learned.

As indicated in Chapter 7, the lateral hypothalamic area contains a region, which, if stimulated, causes food intake or feeding. This region has been designated as a FEEDING CENTER. The ventromedial hypothalamic area contains a region, which, if stimulated, causes cessation of feeding behavior. This region has been designated as a SATIETY CENTER that signals when food intake has satisfied the organism's hunger. In normal animals, the cooperation of these centers assures a food intake matched to activity levels and caloric need, and neither obesity nor starvation results.

Whether or not the stimuli for feeding and satiety *originate* within these hypothalamic areas continues to be a topic of research and debate. Whatever these stimuli are, and wherever they originate, they must respond to changes in the state of the body economy that are the result of presence or absence of foods or their products of digestion in the body. Several controlling mechanisms have been suggested.

Gastric sensation. Filling the stomach with food and liquid expands it and may trigger stretch receptors in the organ itself or in the walls of the abdominal cavity against which the filling stomach presses. These signals are then conveyed to the feeding center and it is inhibited. The feeding center, however, does not require an intact stomach, for removal of the stomach does not abolish desire for food.

Blood glucose. It is attractive to suggest that hunger is inversely related to blood glucose levels, and that satiety is directly related to blood glucose levels. Like a thermostat that triggers a furnace to go on when room temperature falls, a "glucostat" was hypothesized to trigger feeding when blood sugar levels fall. Rise of blood sugar as absorption occurred either inhibits the feeding center or stimulates the satiety center. The hypothesis breaks down when one considers that in diabetes mellitus, blood sugar levels are very high, but the victim eats ravenously. Thus, it may be the *availability of glucose to the cells* that determines feeding, rather than blood glucose level as such. In diabetes, lack of insulin results in failure of cellular uptake of glucose and the response is as though blood glucose was deficient.

Thermal effects. Intake of food is associated with a rise of metabolic activity and heat production as the body increases secretory, absorptive, and processing activities. This rise is designated the *specific dynamic effect (SDE)* of foods. Evidence exists to suggest that feeding is triggered when body temperature decreases with completion of these activities, and satiation when the SDE occurs. The magnitude of the SDE depends on magnitude of food intake, and the latter is in turn affected by environmental temperature (more food is eaten when it is cold). The type of food eaten also determines SDE (high when the diet contains either an excess or deficiency of protein). Thus, unlike blood glucose levels, food intake may be more closely matched to caloric need via body temperature.

General nutritional state. Loss of nutrients, such as in the milk of lactating animals, triggers a greater food intake with usually no change in body weight. In this case the centers may be responding to loss of a particular substance in the milk. What it is, is unknown.

Intake of specific foods. Hungry human subjects interviewed after intake of protein or intravenous injection of amino acids reported diminution of their sensations of hunger. It is known that certain amino acids may be converted to glucose, and it may be the elevation of blood sugar to which the body is responding.

Psychological status. It is well known that some individuals eat habitually or out of boredom or other factors.

All factors are probably interrelated in control of food intake. A reasonable position may be reached, but one not explainable by a single physiological mechanism.

Another interesting feature of feeding is that given a choice of a variety of foods in separate dishes, an animal tends to choose a diet that meets its needs at that time. If a specific nutrient is decreased, intake of an acceptable substitute will be increased. The sense of taste appears to be important in such adjustments. The human appears, over the long term, to eat a variety of items that meet the body needs. One should not be overly concerned about the consumption of specific foods at any one meal.

DIGESTION AND ABSORPTION IN THE MOUTH

Ingestion of food is followed by MASTICATION or chewing of the food. Subdivision of the food renders it more easily swallowed, and smaller particles are more easily digested by enzymatic action. The act of chewing is largely reflex in nature. A voluntary opening of the mouth for food reception is followed by contact of the food with the tissues lining the mouth cavity. Jaw closing muscles are stimulated and the food is compressed. Further contact inhibits the closing muscles, the jaw drops, and the

FIGURE 23.2
Names and locations of the major salivary glands in the human.

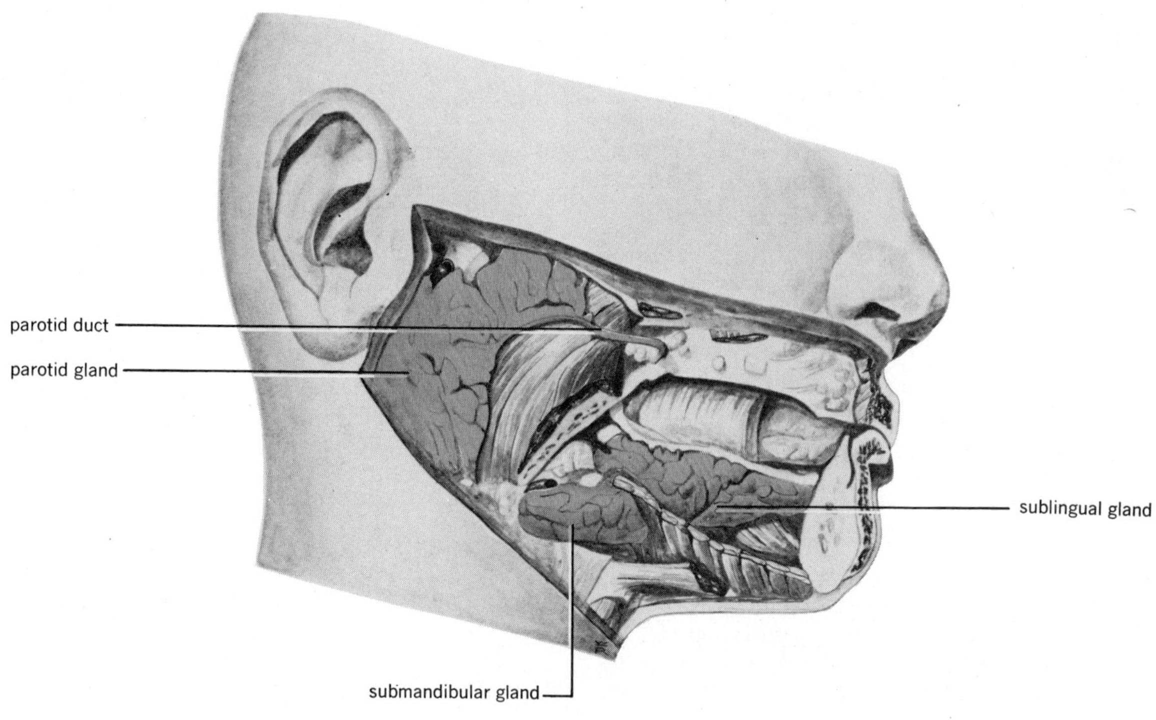

stretch applied to the muscles causes a reflex contraction. A rhythmical motion is created. Centers integrating chewing lie in the lower brainstem, and in the basal ganglia, and possibly in the hypothalamus and cerebral cortex. Maximum closing pressures exerted by the lower jaw may reach 90 kg. Converted to kg/cm^2, the force may reach several thousand kg. The chewing process mixes the foods with the saliva, secreted by several salivary glands emptying into the mouth (Fig. 23.2).

Saliva. The collective secretion of the salivary glands is the saliva. Human saliva contains about 99.5 percent water, and 0.5 percent solids. Characteristics and contained materials of human saliva are presented in Table 23.1.

The water content of saliva aids in DISSOLVING soluble food components, SOFTENS the food mass, MOISTENS the food and mouth linings, and DILUTES the food. The mucin (mucus) content LUBRICATES food for swallowing. CLEANSING of the mouth and teeth, PROTECTION against dental caries, and COMMENCING THE DIGESTION OF CARBOHYDRATES are other functions of saliva. In spite of these many activities, saliva is not essential for normal digestion and absorption of foods because there are enzymes and secretions in the small intestine that can do the same digestive job. Rate of salivary secretion increases when food is in the mouth due to the control mechanism to be described below. A resting rate averages about 0.05 ml/min, and may increase to 7.0 ml/min. A total production per day ranges from 500 to 1500 ml. Dry foods elicit the secretion of a more watery saliva, while coarse foods call forth a greater secretion of mucus for lubricating.

Control of salivary secretion. The control of salivary secretion is, unlike other glands of the system, exclusively neural. Nerve supply of the glands is primarily from parasympathetic fibers carried in facial and glossopharyngeal cranial nerves. These nerves (VII, IX) convey sensory information concerning taste to the brainstem and secretory (motor) impulses from the brainstem to the glands. Sympathetic nerves are less important and are derived from the superior cervical ganglion. The two nerve supplies *do not* control salivary secretion in a check-and-balance system; the parasympathetics are the primary controllers of secretion.

TABLE 23.1

SOME CONSTITUENTS AND CHARACTERISTICS OF ADULT HUMAN SALIVA

CONSTITUENT OR PROPERTY	CONCENTRATION OR VALUE
pH	5.8–7.1
Water	994 g/l
Sodium	6–23 meq/l
Potassium	14–41 meq/l
Bicarbonate	2–13 meq/l
Chloride	15–31.5 meq/l
Phosphorus (inorganic)	81–217 mg/l
(organic)	0–133 mg/l
Calcium	2.3–5.5 meq/l
Thiocyanate (—SCN)	24–380 mg/l
Magnesium	0.16–1.06 meq/l
Iodine	0.002–0.202 mg/l
Urea	140–750 mg/l
Ammonia	10–120 mg/l
Amino acids (products of bacterial action on proteins)	Variable—21 have been found
Uric acid	5–29 mg/l
Protein (albumin and globulin)	1.4–6.4 g/l
Lysozyme a (bacteriolytic enzyme)	to 0.15 g/l
Amylase (ptyalin)	0.38 mg/ml
Glucose	100–300 mg/l
Cholesterol	25–500 mg/l
Vitamin C	0.58–3.78 mg/l
Mucin	0.8–6.0 g/l

Control is exerted primarily by a reflex mechanism (Fig. 23.3) that utilizes the taste buds as the receptors. Dissolving of foods by the saliva results in material entering the taste buds of the tongue (primarily) and causing the generation of nerve impulses. These impulses are conveyed over the two cranial nerves to the salivatory centers in the brainstem, and impulses are then sent back over the same cranial nerves to the salivary glands. The latter act as effectors by producing saliva. "Cerebral influences" may cause the "mouth to water" as food is smelled or seen, or even thought of. Salivary

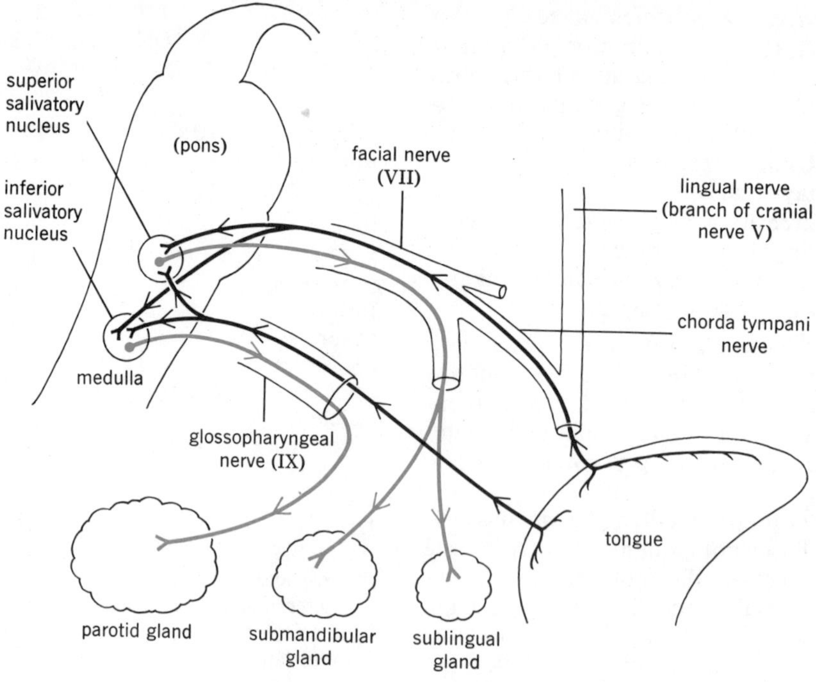

FIGURE 23.3
The control of salivary secretion. Sensory pathways (black), motor pathways (color).

secretion is an active process as evidenced by the dilute nature of the fluid as compared to plasma. A five-fold increase in oxygen consumption occurs when the glands become active. Some 5 to 8 kcal per day are estimated to be required to produce the average 1 to 1.5 L of saliva per day.

Digestion in the mouth. Digestion of carbohydrates is begun in the mouth by the action of an α-AMYLASE (*ptyalin*). This is a hydrolytic enzyme that cleaves the α (1–4) glucosidic linkages (Fig. 23.4) that link glucose molecules into starches. Optimum pH for the activity of the enzyme is in the range of 4 to 11, and the enzyme is not active unless Cl^- is present. The amylase hydrolyzes cooked starches in two stages: digestion to dextrins (units containing several to several hundred simple sugar molecules); digestion to disaccharides (units containing two simple sugar molecules). Because of the short time that foods remain in the mouth, only 3 to 5 percent of the starches will be converted to the disaccharide stage.

Absorption from the mouth. Absorption of materials from the mouth is limited. Certain medications, if given in the form of lozenges, may be held under the tongue, and the substances released as the lozenge dissolves pass through the mucous membranes lining the mouth to enter blood vessels. The process is slow and materials passing must be of low molecular weight. No absorption of digestion products occurs here; they are too large and do not remain in the mouth long enough to be absorbed.

TRANSPORT FROM MOUTH TO STOMACH

By the act of SWALLOWING (*deglutition*), food is transferred into the pharynx and esophagus and is transported to the stomach for further digestion. The tongue collects the food from between the teeth and in the mouth and forms it into a BOLUS, or segregated mass of food. The tip of the tongue is pressed against the hard palate and the sides elevated to form a groove into which the bolus collects. Progressively, the remaining portion of the tongue is pressed against the hard palate, pushing the bolus ahead of the organ into the pharynx. The soft palate is elevated to block the nasal cavities, and the tongue seals the oral cavity so that a positive pressure of 4 to 10 mm Hg may be created to force the bolus into the pharynx. Once into the pharynx, the pharyngeal constrictors move the bolus toward the esophagus as the larynx is elevated and the opening into the respiratory system is sealed. In the esophagus, contact of the bolus with the walls of the organ creates a PERISTALTIC WAVE that sweeps the bolus to the stomach in 5 to 6 sec. Peristalsis is a progressive, wavelike motion that occurs involuntarily in organs possessing circularly and longitudinally oriented smooth muscle coats. A wave of relaxation in the circular muscle layer precedes an area of constriction; this tends to collect materials ahead of the area that propels contents through the organs. PRIMARY PERISTALSIS is initiated by swallowing and is nervously (reflexly) controlled by medullary swallowing centers (Fig. 23.5). It depends on intact vagus nerves to serially innervate the musculature of the esophagus. SECONDARY PERISTALSIS originates within the esophagus itself by a sort of intrinsic nervous control. Liquid materials that are swallowed pass toward the stomach much more rapidly than solid boluses. No food will enter the stomach until a thickened muscular layer at the entrance of the stomach, known as the CARDIAC SPHINCTER, relaxes to allow passage. A liquid bolus thus must await entry into the stomach until the peristaltic wave accompanying swallowing relaxes the sphincter.

VOMITING involves passage of gastric contents upwards through the esophagus. Powerful contractions of the abdominal muscles compress the stomach that acts as a flaccid bag. Relaxation of the cardiac spincter and relaxation of the esophageal muscles create a "free passage" for ejection of material to the exterior. Coordination of these activities is by a medullary vomiting center, which appears to be triggered by impulses arriving from the pharynx, soft palate, and inner ear (semicircular canals). Chemicals (emetics), such as strong salt solutions, mustard, or apomorphine may also initiate vomiting by affecting the vomiting center.

DIGESTION AND ABSORPTION IN THE STOMACH

Digestion. The stomach contains many GASTRIC GLANDS (Fig. 23.6) that produce the GASTRIC JUICE. Food is stored in the stomach and the digestion of proteins is begun by PEPSIN, the major enzyme of the gastric juice. Pepsin requires an acid medium (pH 1.5 to 2.2) to exhibit its optimum activity, and this environment is provided by hydrochloric acid (*HCl*) also

FIGURE 23.4
Two glucose molecules linked by a 1–4 glucosidic bond. Carbons in glucose are numbered; site of amylase activity is shown by the large vertical arrow.

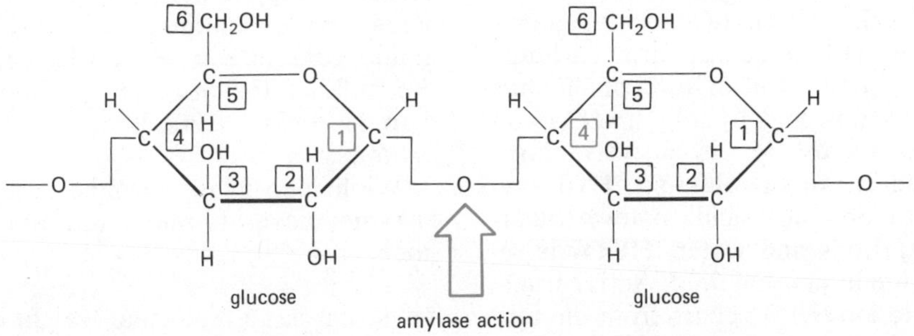

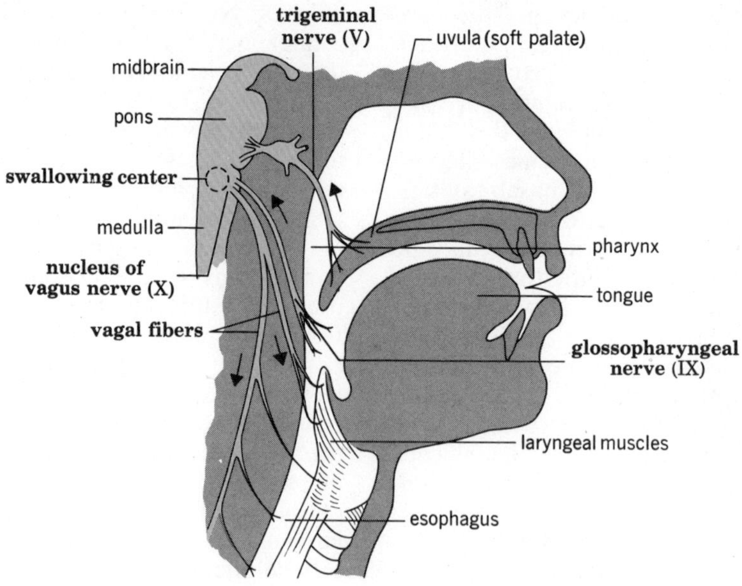

FIGURE 23.5
The basic nervous pathways responsible for swallowing and the development of peristalsis in the esophagus.

secreted by the gastric glands. The properties and constituents of human gastric juice are presented in Table 23.2.

Four types of cells are found in gastric glands. Fundic and body glands contain all four types; cardiac and pyloric glands usually contain only two. OXYNTIC (*parietal*) CELLS (Fig. 23.7), found in fundic and body glands, produce HCl from NaCl, H_2O, and CO_2, secreting the acid into the "intracellular canals" where it does not damage the cells themselves. Several theories as to how the acid is produced have been advanced. One mechanism is presented in Figure 23.8. Water and CO_2 enter the oxyntic cell. Aided by the action of carbonic anhydrase, CO_2 and H_2O react to form carbonic acid (H_2CO_3). In the wall of the canaliculus, water is broken down to hydrogen (H^+) and hydroxyl (OH^-) ions. H^+ is actively transported into the canalicular lumina. Hydroxyl ions react with carbonic acid to form bicarbonate (HCO_3^-) ions and water. HCO_3^- is reabsorbed into the interstitial fluid. Active transport of chloride ion (Cl^-) occurs from interstitial fluid to cell to canalicular lumen where HCl is formed.

ZYMOGENIC (*chief*) CELLS (Fig. 23.9) are also found in fundic and body glands, and produce a substance known as PEPSINOGEN. It is an inactive precursor of pepsin that is converted to pepsin in the presence of HCl. Pepsinogen has a molecular weight of about 42,500; HCl splits off a portion of the molecule, and pepsin, molecular weight 34,500, is produced. Pepsin is an *endopeptidase,* a term referring to the fact that it hydrolyzes the interior peptide bonds of a protein. It is most active on peptide bonds next to aromatic amino acids (phenylalanine, tyrosine). Some free amino acids are produced by this action, but the bulk of the products are units containing 4 to 12 amino acids, known as *proteoses* and *peptones*.

While pepsin is the most important gastric enzyme, several others are found in gastric juice.

PEPSIN B[5] has a molecular weight of 36,000 and

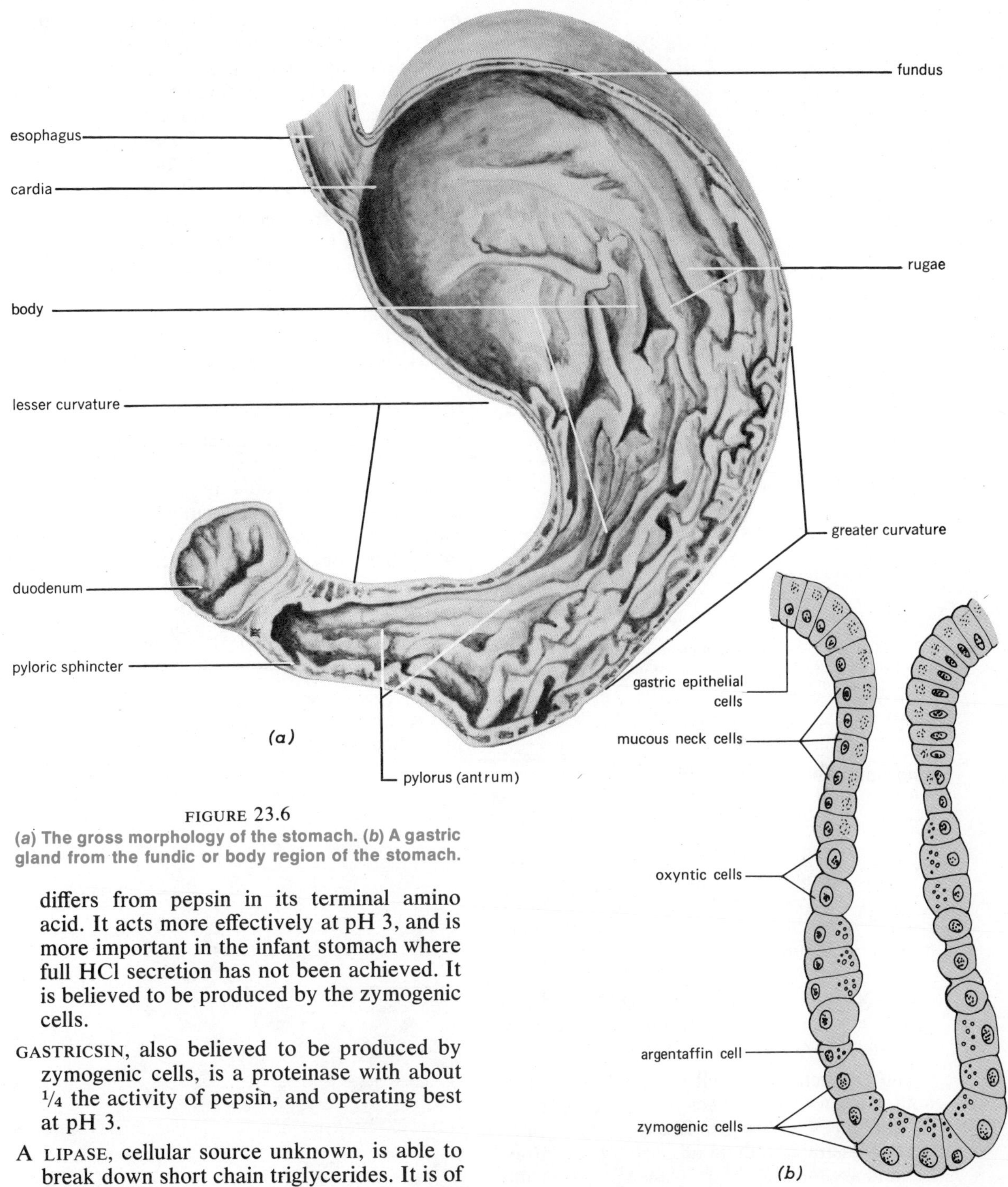

FIGURE 23.6
(*a*) The gross morphology of the stomach. (*b*) A gastric gland from the fundic or body region of the stomach.

differs from pepsin in its terminal amino acid. It acts more effectively at pH 3, and is more important in the infant stomach where full HCl secretion has not been achieved. It is believed to be produced by the zymogenic cells.

GASTRICSIN, also believed to be produced by zymogenic cells, is a proteinase with about 1/4 the activity of pepsin, and operating best at pH 3.

A LIPASE, cellular source unknown, is able to break down short chain triglycerides. It is of

TABLE 23.2

NORMAL CONSTITUENTS AND PROPERTIES OF ADULT GASTRIC JUICE

CONSTITUENT OR PROPERTY	CONCENTRATION OR VALUE
Fasting volume	50 ml
Secretion rate	74 ml/hr (fasting) 101 ml/hr (after a meal)
Water	994–995 g/l
pH	1.9–2.6
Bicarbonate	none
Chloride	75–150 meq/l
Phosphorus	Av. 70 mg/l
Potassium	6.4–16 meq/l
Sodium	19–69 meq/l
Calcium	2.0–4.8 meq/l
Magnesium	0.3–3.0 meq/l
Total nitrogen (amino acids, ammonia, urea, uric acid, creatinine)	752 mg/l
Proteins	2.8 g/l
Mucin	0.6–15.0 g/l
Carbohydrates (chiefly hexoses)	321 mg/l
Enzymes (pepsin)	28 KU/24 hr[a]

[a] KU—One unit (U) is the amount of any enzyme that will catalyze the transformation of 1 micromole (μM) of substrate per minute at standard conditions. A Kilounit (KU) is 1000 times as large, that is, it will catalyze 1 millimole (mM) under the same conditions.

no importance in the adult stomach because its activity is destroyed by the low gastric pH.

AMYLASE, from the salivary glands, is found in but not produced by, the stomach. It may continue its activity in the stomach for 15 to 30 minutes after food is swallowed, particularly if the stomach is full of food, and time is required for gastric juice to penetrate the mass. A digestion of 70 percent of the starches to disaccharides may occur in the stomach.

The two remaining cell types in the gastric glands produce no enzymes. NECK CHIEF CELLS (*mucous neck cells*) are found in the upper parts of gastric glands in all parts of the stomach, and are *the* cells of the glands in cardia and pylorus. They produce an alkaline mucus that protects the gland from the destructive action of pepsin and HCl. ARGENTAFFIN CELLS are most numerous in the fundus of the stomach and are scattered singly among the zymogenic cells in the lower portions of the gastric glands. They produce and store *serotonin,* a chemical having a vasoconstrictive effect.

Protection of the stomach lining. The lining of the stomach itself is also protected by mucus. The epithelial cells of the gastric mucosa have about the outer half of their height filled with alkaline mucus. The mucus aids in neutralization of HCl, and is only slowly digested by the gastric juice. The mucus is constantly renewed from below and normally forms a complete protective coating for the stomach wall.

Other functions of the stomach. The stomach produces INTRINSIC FACTOR, a substance that combines with and aids the absorption of vitamin B_{12}. The vitamin is essential for normal red blood cell development. Levels of intrinsic factor have been shown to decrease with age, and vitamin B_{12} injections are sometimes required to meet body needs for the vitamin.

Absorption. There is no evidence to suggest that great absorption of materials occurs

FIGURE 23.7

An oxyntic cell of the stomach as it would be seen through the electron microscope.

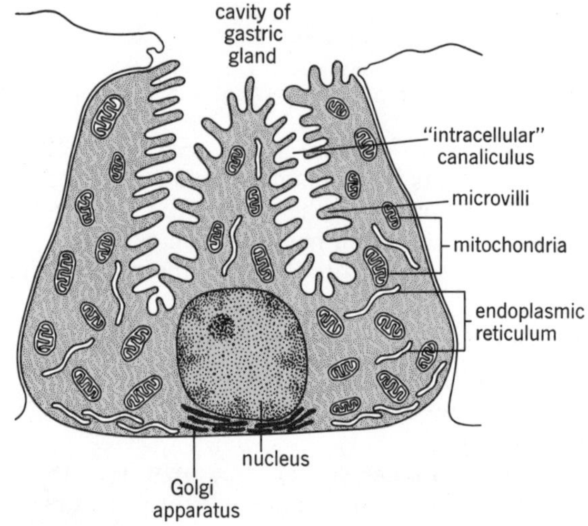

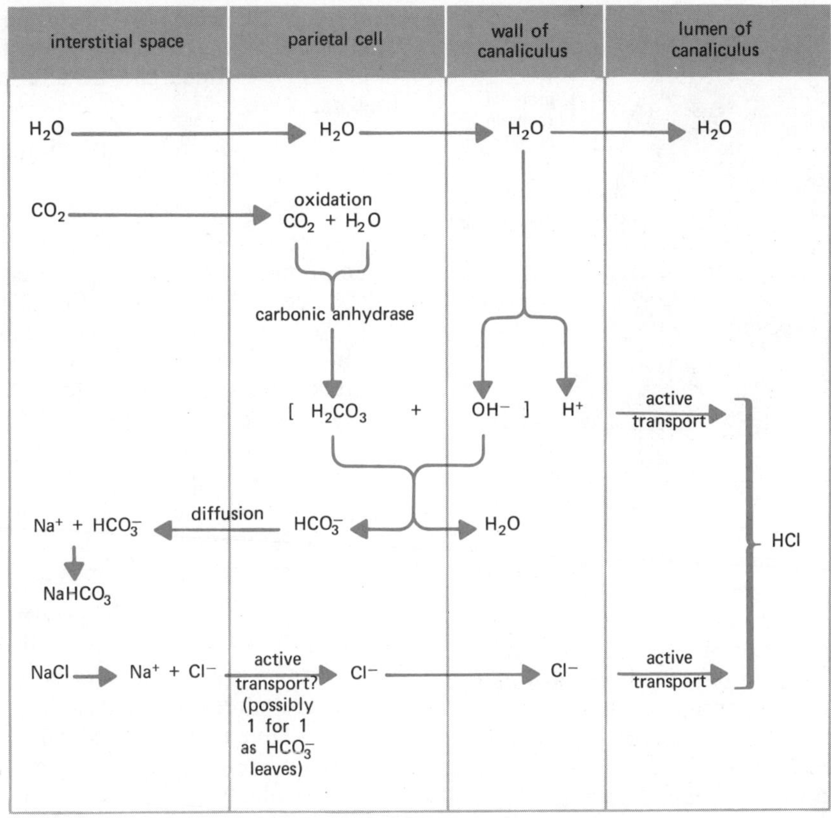

FIGURE 23.8
A theory of the mechanism by which hydrochloric acid is produced. (Modified from Hollander.)

through the stomach wall. Water and small molecules (e.g., alcohol) may penetrate the stomach at very slow rates, as may salts. The mucus coating of the stomach creates a nearly impenetrable barrier to the passage of materials.

Control of gastric secretion. Control of gastric secretion occurs in three phases.

The NEURAL OR CEPHALIC PHASE occurs when stimulation of the vagus nerve leads to HCl secretion. Nerve impulses from the taste buds pass to the brainstem and trigger vagal nuclei to send impulses to the stomach that cause secretion of gastric juice through acetylcholine release and release of a gastric hormone, *gastrin*. Hypoglycemia is also effective in causing vagal discharge. Some 50 to 150 ml of juice is produced as a result of this neural reflex, and it serves as an "ignition juice" to initiate digestion of proteins to proteoses and peptones.

The GASTRIC OR HUMORAL PHASE is occasioned by the gastric hormone GASTRIN. Gastrin is a polypeptide, containing 17 amino acids, that is produced when proteoses and peptones and distension trigger its release from its cell of production. The cell producing gastrin resembles the argentaffin cell described earlier. Gastrin passes into the venous drainage of the stomach and ultimately is distributed to all body organs including the stomach; the hormone also stimulates activity in nerve cells located in the stomach wall and which synapse with vagal efferents. The net result is the production by

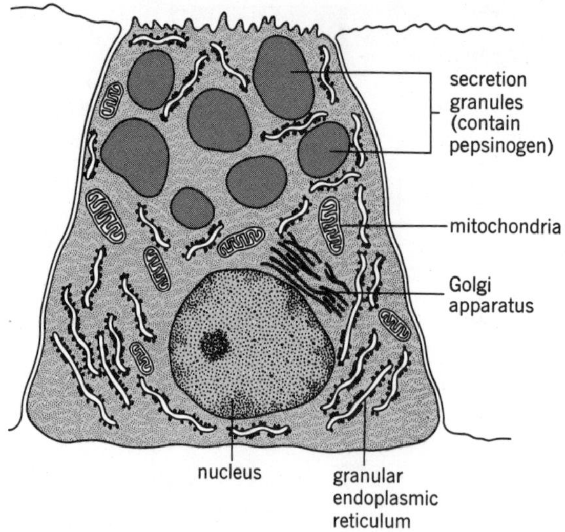

FIGURE 23.9
A zymogenic cell of the stomach as it would be seen through the electron microscope.

all gastric glands of 600 to 750 ml of gastric juice. This volume effectively occasions the digestion of the stomach contents.

The INTESTINAL PHASE occurs when a hormone is released as food enters the duodenum from the stomach. The nature of the hormone has not been specifically determined. Cholecystokinin has a gastric stimulating effect; it may be the effective agent, or something called "intestinal gastrin" has been described as the effective agent.

Stopping gastric secretion. Inhibition of gastric secretion as the stomach empties is provided by denervation (or decrease of vagal stimulation), and by the production of a humoral substance designated as ENTEROGASTERONE. Fats, acids, or hypertonic solutions entering the duodenum cause the production of the hormone, and also the release of *secretin.* Both materials suppress gastric secretion. Pancreatic *glucagon,* a true hormone, also inhibits gastric secretion, and has led to the postulation that an "intestinal glucagon" may be produced that leads to suppression of gastric secretion. Also, the removal of proteoses and peptones, as they pass from stomach to intestine, reduces the effect of the gastric phase.

Emptying the stomach. The time required to empty the stomach is normally three to four hours. These times can be altered by several factors.

Enterogasterone not only diminishes gastric secretion, but slows stomach motility and increases emptying time.

The TONICITY of the gastric contents influences emptying. If stomach contents are more, or less, concentrated than the plasma, motility is reduced until the contents have been rendered isotonic to plasma by further fluid intake or transfer of solutes. This action suggests the presence of osmoreceptors in the stomach wall.

FAT CONTENT of the meal, if elevated, increases emptying time. Presumably, this effect is mediated by the enterogasterone mechanism described above.

DECREASED VAGAL ACTIVITY also reduces motility, increasing emptying time.

SHORT-TERM STRESS increases emptying time; chronic stress decreases it. Acetylcholine production is associated with acute stress; adrenal steroids are elevated with chronic stress, and may be the cause of decreased emptying times.

DILATION of the stomach by foods produces an augmentation of motility. As foods are moved to the small intestine, this factor decreases in importance as a controller of gastric emptying.

DIGESTION AND ABSORPTION IN THE SMALL INTESTINE

Digestion. The PANCREAS is the primary source of digestive enzymes that act in the small intestine. It also provides the alkalinity necessary for creating the pH optimum required for enzyme activity in the small intestine.

The term PANCREATIC JUICE is applied to the exocrine secretion of the pancreas, and it contains enzymes active on carbohydrates, fats, proteins, elastin (a component of connective tissues) and nucleic acids. The general properties and constituents of pancreatic juice are presented in Table 23.3.

The bicarbonate content of the juice is what creates the alkaline environment for activity of the digestive enzymes. Secretion of the watery alkaline fluid appears to be relatively independent of enzyme secretion. There is evidence to suggest that the watery solution is produced by the small ducts of the gland, while the cells composing the *acini* (secretory units) provide the enzymes.

Proteolytic enzymes are secreted as *zymogens* or *proenzymes*; enzymes in inactive form. The enzymes, their activation, and actions are given as follows.

TRYPSINOGEN is converted to TRYPSIN by the action of ENTEROKINASE, an intestinal enzyme produced by epithelial cells lining the duodenum. Once formed, trypsin activates more trypsinogen in an autocatalytic process. Trypsin splits peptide bonds on the carboxyl side of lysine and arginine only.

TABLE 23.3

SOME PROPERTIES AND CONSTITUENTS OF PANCREATIC JUICE

CONSTITUENT OR PROPERTY	CONCENTRATION OR VALUE
pH	7.5–8.8
Secretion rate	6–36 ml/hr (700–2500 ml/day)
Water	987 g/l
Bicarbonate[a]	25–150 meq/l (amount proportional to secretory rate)
Chloride[a]	4–129 meq/l
Potassium	6–9 meq/l
Sodium	139–143 meq/l
Calcium	2.2–4.6 meq/l
Total nitrogen	0.76–0.98 g/l
Glucose	85–180 mg/l
Enzymes (amount/min after stimulation by cholecystokinin)	
Amylase	0.29–1.30 mg/min
Carboxypeptidase A	0.36–1.45 mg/min
Chymotrypsin	1.22–7.6 mg/min
Trypsin	0.38–1.42 mg/min
Lipase	0.78–3.50 KU/min

[a] Sum of HCO_3 and Cl^- concentrations is constant at about 154 meq/kg of water.

ELASTASE resembles trypsin in structure, and attacks the small uncharged side chains such as those found in the amino acids valine and leucine.

CHYMOTRYPSIN is secreted as CHYMOTRYPSINOGEN and is activated by trypsin. This enzyme hydrolyzes peptide bonds adjacent to aromatic (–⬡–) groups.

CARBOXYPEPTIDASE A is secreted as PROCARBOXYPEPTIDASE A and is also activated by trypsin or enterokinase. This enzyme hydrolyzes terminal amino acids with free carboxyl groups.

These enzymes provide a series that can hydrolyze nearly every peptide bond within a protein chain. The result of their activity is the production of many *free amino acids,* plus *dipeptides,* units containing two amino acids.

PANCREATIC α-AMYLASE, the same enzyme as found in saliva, hydrolyzes remaining dextrins to disaccharides. Several disaccharides are produced by this action, including *sucrose* (glucose and fructose), *lactose* (glucose and galactose), *maltose* (2 glucose), *trehalose* (2 glucose linked differently than in maltose), and *isomaltose* (another differently linked disaccharide composed of 2 glucose units).

PANCREATIC LIPASE hydrolyzes, in a single step, the bonds between glycerol and fatty acids in a triglyceride (Fig. 23.10). A molecule of glycerol and three fatty acid molecules result, and are ready for absorption.

NUCLEASES (ribonuclease, deoxyribonuclease) break nucleic acids into their constituent mononucleotides.

Control of pancreatic secretion. Control of pancreatic secretion is achieved by both nervous and humoral mechanisms. In the *gastric phase,* distention of the stomach fundus, by filling with food, causes augmentation of fluid and enzyme release by the pancreas. If the vagus nerve is severed, the response does not occur, indicating it depends on a nervous reflex. Distention of the gastric pyloric region, as mate-

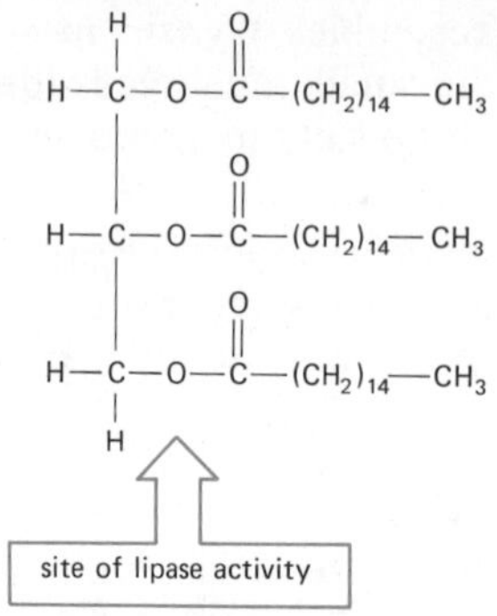

$$
\begin{array}{l}
\mathrm{H{-}\overset{H}{\underset{|}{C}}{-}O{-}\overset{O}{\overset{\|}{C}}{-}(CH_2)_{14}{-}CH_3} \\
\mathrm{H{-}\underset{|}{C}{-}O{-}\overset{O}{\overset{\|}{C}}{-}(CH_2)_{14}{-}CH_3} \\
\mathrm{H{-}\underset{H}{\underset{|}{C}}{-}O{-}\overset{O}{\overset{\|}{C}}{-}(CH_2)_{14}{-}CH_3}
\end{array}
$$

site of lipase activity

FIGURE 23.10
A triglyceride (tripalmatin) showing bonds attacked by lipase.

rial is moved toward the intestine, stimulates the enzyme output of the pancreas, and is probably due to gastrin release by the stomach. This response is *not* blocked by cutting the vagus nerves. The INTESTINAL PHASE is humoral, and depends on the release of 2 intestinal hormones, *secretin* and *cholecystokinin* (*CCK*).

SECRETIN is a hormone composed of 27 amino acids that is released from the mucosa of the upper intestine. It is absorbed into blood vessels, goes to the pancreas, and stimulates the release of fluid and bicarbonate from the pancreas. Acid entering the duodenum from the stomach is the most potent stimulant for its release, but partially digested proteins, fats, and amino acids also are effective.

CCK is a hormone composed of 33 amino acids, and stimulates the release of enzymes from the acini of the pancreas. Exact cell origin has not been determined, but a cell similar to those producing secretin is hypothesized. This hormone also passes to the pancreas via the bloodstream to exert its stimulating effect. The most effective stimuli for CCK release are protein hydrolysates, amino acids, and fatty acids. HCl from the stomach is a weak stimulator of hormone release.

The role of bile in digestion. BILE is produced by the liver, and is emptied into the duodenum. It undergoes a "circulation" (Fig. 23.11) that ensures availability of bile acids for the digestive process. Bile serves two important physiological functions: it provides the bile acids necessary for efficient absorption of lipids; it is an excretory pathway for products of red cell breakdown (e.g., bilirubin), and several organic anions. Its characteristics and constituents are presented in Table 23.4.

FIGURE 23.11
(*a*) The circulation of bile in the body. (*b*) Cholic acid, the substance to which materials are conjugated to form bile salts.

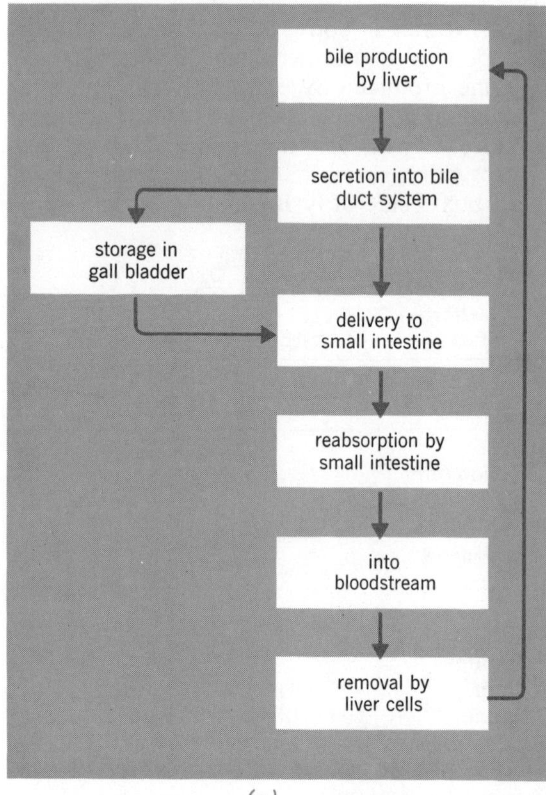

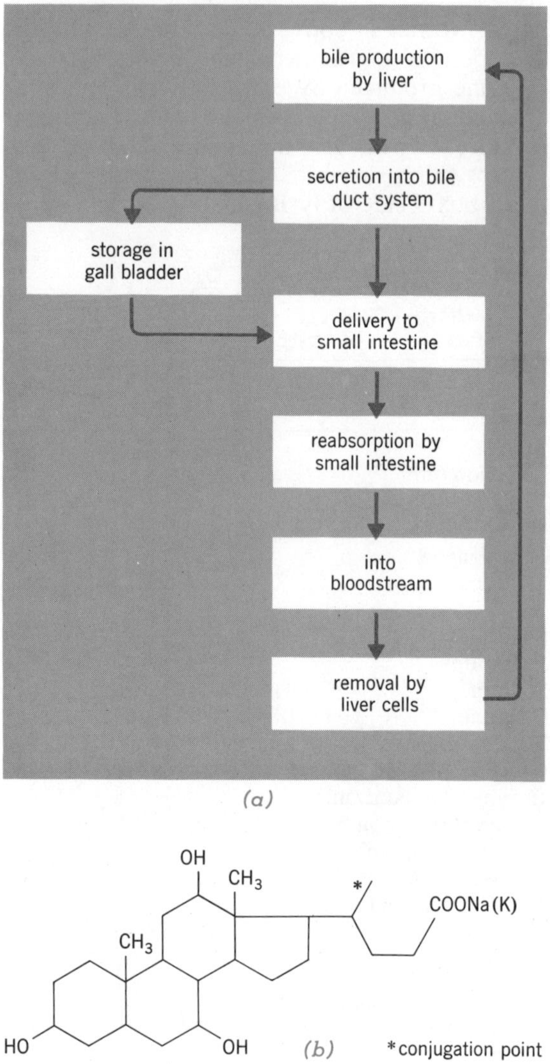

(*a*)

(*b*) *conjugation point

TABLE 23.4
SOME PROPERTIES AND CONSTITUENTS OF HEPATIC BILE[a]

CONSTITUENT OR PROPERTY	CONCENTRATION OR VALUE
pH	6.2–8.5
Secretion rate	250–1100 ml/day
Bicarbonate	30 meq/l
Chloride	89–118 meq/l
Potassium	2.6–12 meq/l
Sodium	131–164 meq/l
Calcium	3.3–4.1 meq/l
Iron	0.4–3.1 mg/l
Copper	0.35–2.05 mg/l
Total nitrogen	0.25–1.45 g/l
Urea	236 mg/l
Bilirubin (as glucuronide)	0.12–1.35 g/l
Protein	1.4–2.7 g/l
Cholesterol	0.8–1.8 g/l
Bile acids[b]	6.5–14.0 g/l

[a] In the gall bladder, bile is concentrated by removal of water; thus, constituents may increase in concentration by a factor of 2–10 times.
[b] See Figure 23.11.

Bile aids in buffering the intestinal fluid through its bicarbonate content, and lowers the surface tension of lipids allowing intestinal motility to break fat drops into minute droplets. This action is termed the EMULSIFYING ACTION of bile. Combination of bile acids (*see* Figure 23.11) with lipids forms bile salts that are water soluble and more easily absorbed. This action is termed the HYDROTROPIC ACTION of bile.

Although produced continually by the liver, bile is released intermittently into the intestine through the bile duct in those animals possessing a gall bladder. This organ *stores* bile (1 to 2 ml per kg of body weight) and *concentrates* the bile by removing water from it. Excessive removal of water may cause precipitation of bile solutes and form gallstones.

Bile is caused to leave the gall bladder when fats enter the small intestine. The CCK mechanism causes contraction of the bladder musculature, relaxes the sphincter (of Oddi) at the junction of the common bile duct with the small intestine. This allows bile to be delivered into the intestine.

Intestinal wall involvement in digestion. Completion of digestion of carbohydrates and proteins occurs by enzymes located in the membranes of the microvilli that cover the ends of the epithelial cells of the intestine. In human epithelial cells, it has been estimated that each cell contains about 1700 of these microvilli; they increase the surface area of the intestine 15 to 40 fold over what it would be without the projections. The foldings provide large surface area for contact of undigested substances with membrane-housed enzymes, and a large surface area for absorption of end products. Additionally, the cells are constantly renewed by migration toward the surface from the depths of the intestinal glands, a journey that takes two to five days. Additional surface area is provided by the VILLI, folds of connective tissue covered with epithelial cells, and by the PLICAE, folds of connective tissue. Blood vessels form networks within the villus, and a central lymphatic vessel called a *lacteal* is found in the villus. These features are shown in Figure 23.12.

Final hydrolysis of disaccharides occurs as they contact the microvilli of the epithelial cells. LACTASE splits lactose to its constituent monosaccharides; SUCRASE does the same for sucrose; MALTASE acts on maltose; ISOMALTASE on isomaltose; and TREHALASE on trehalose. The monosaccharides produced are then ready for absorption.

Dipeptides are digested by AMINOPEPTIDASE and DIPEPTIDASE found in the intestinal cell membranes and amino acids are produced.

In addition to the epithelial cells, the duodenal wall contains mucus secreting glands (of Brunner). These produce secretions that tend to aid in neutralization of gastric acid, and which accounts for a large part of the inorganic content of the intestinal juice (succus entericus).

Control of intestinal secretion. Control of intestinal secretion is believed to depend on humoral mechanisms. Mechanical stimulation of the duodenal wall, the presence of digesting foods or acid, is hypothesized to cause the release of ENTEROCRININ. This substance is circulated via blood vessels to the entire intestine and stimulates intestinal secretion of water and

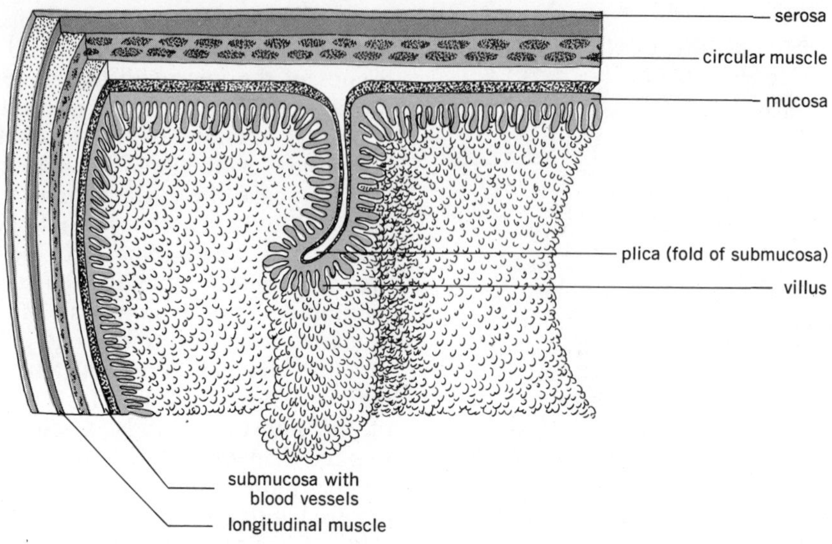

FIGURE 23.12
(*Above*) Villi and plicae of the small intestine. These devices increase surface area. (*Below*) The structure of a villus.

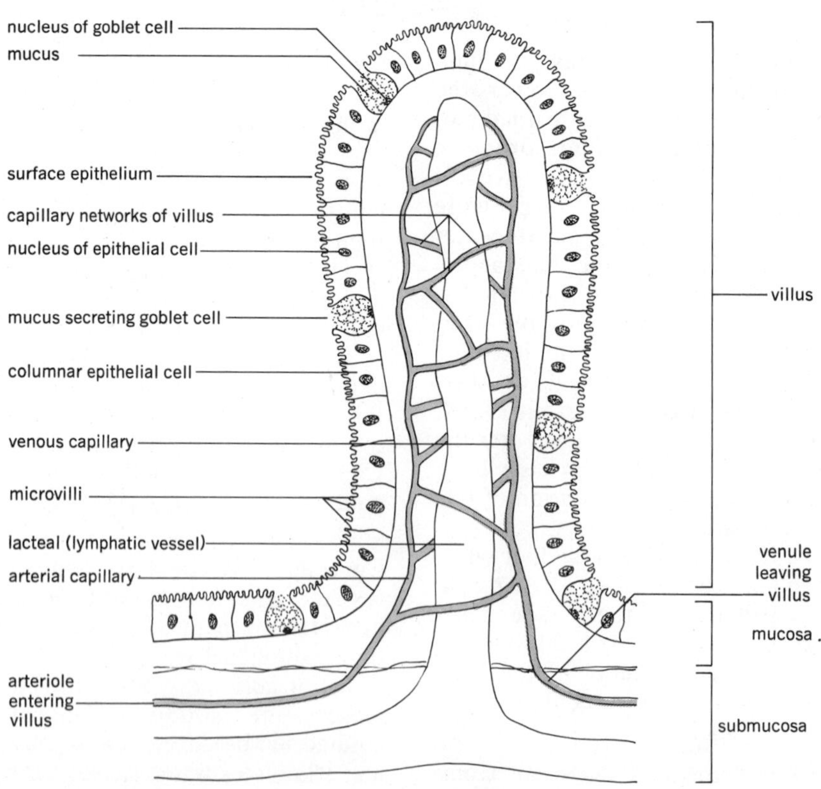

mucus. Additional work must be done to define the role of nervous and hormonal stimuli of intestinal secretion.

Absorption from the small intestine. Absorption of the digested and ingested materials occurs through the intestinal epithelial cells. Monosaccharides, water soluble (B, C) and fat soluble (A, D, E, K) vitamins, inorganic salts and water, and some protein are absorbed in the upper third of the small intestine. Fats are absorbed in the middle third, and bile salts and vitamin B_{12} in the lower third. The intestinal epithelial cells provide the route of absorption and, in general, products of carbohydrate and protein digestion, short chain fatty acids, vitamins, inorganic salts, and water pass to the villal blood vessels. Long chain fatty acids and sterols pass to the lacteals, that eventually empty into venous blood.

Carbohydrates. Absorption of monosaccharides is active, probably by active transport and facilitated diffusion. Galactose passes most rapidly, followed by glucose and fructose. Assigning glucose a value of 100, rates of monosaccharide absorption are given as

galactose	110–122
glucose	100 (70 gm/hr)
fructose	43–67
mannose	19
xylose	15
arabinose	9

Fructose cannot be absorbed against a concentration gradient and probably passes by facilitated diffusion, while glucose and galactose appear to be moved by active transport.

Amino acids and proteins. Amino acid absorption is by active transport with separate carrier systems for neutral amino acids, diamino acids, proline and hydroxyproline, and dicarboxylic acids being described. In the metabolic disorder CYSTINURIA, the carrier for cystine, lysine, arginine, and ornithine is absent, and none of the four acids is absorbed by intestine or kidney tubules. Some intact protein is absorbed by *pinocytosis,* especially in the infant.

Lipids. Lipid absorption occurs by dissolving of the lipids in the lipid component of the cell membrane, and they thus enter the epithelial cells by passive diffusion. Transport into the lacteal is believed to occur as *chylomicrons,* droplets of lipid coated with cell-produced protein envelopes. The process by which the chylomicrons enter the lacteal is unknown, but may be by exocytosis.

Ions are believed to be absorbed by active transport. Removal of the solutes described above causes water to follow by osmosis. Although not definitely known, vitamins appear to be absorbed by diffusion and facilitated diffusion.

THE FUNCTIONS OF THE COLON

On the average, 500 to 600 ml of fluid is delivered to the colon per day. The fluid is rich in Na^+, K^+, Cl^-, and HCO_3^-. About 300 to 400 ml of water is absorbed by the colon and 60 meq of Na^+ per day. Bicarbonate is removed to the extent of 30 meq per day, Cl^- 18 to 38 meq per day, and K^+ 36 to 66 meq per day. Thus, the colon is an important absorptive area for water and electrolytes. Colonic flora, consisting mainly of bacteria and yeasts, produce vitamins K, B_1, B_2, and B_{12}. These are also absorbed by the colon in nutritionally significant amounts. It may be recalled that vitamin K is a stimulant to prothrombin production, a clotting factor. Until an infant establishes its own colonic flora, umbilical stump bleeding is often controlled by giving the infant a "shot" of vitamin K at birth. This flora should be regarded as a balanced ecological system, disturbance of which as by laxatives, can upset the function of the entire organ.

THE LIVER; FUNCTIONS OTHER THAN BILE SECRETION

The structural unit of the liver is the LOBULE (Fig. 23.13). The functional unit is regarded as the PORTAL LOBULE. The lobules receive blood from two separate systems of vessels. Arterial blood is supplied through the hepatic artery at volumes of approximately 350 ml per minute. The arterial supply assures nutrition of the bile ducts and supporting elements of the organ. Flowing to liver through the portal vein is venous blood collected from the alimentary

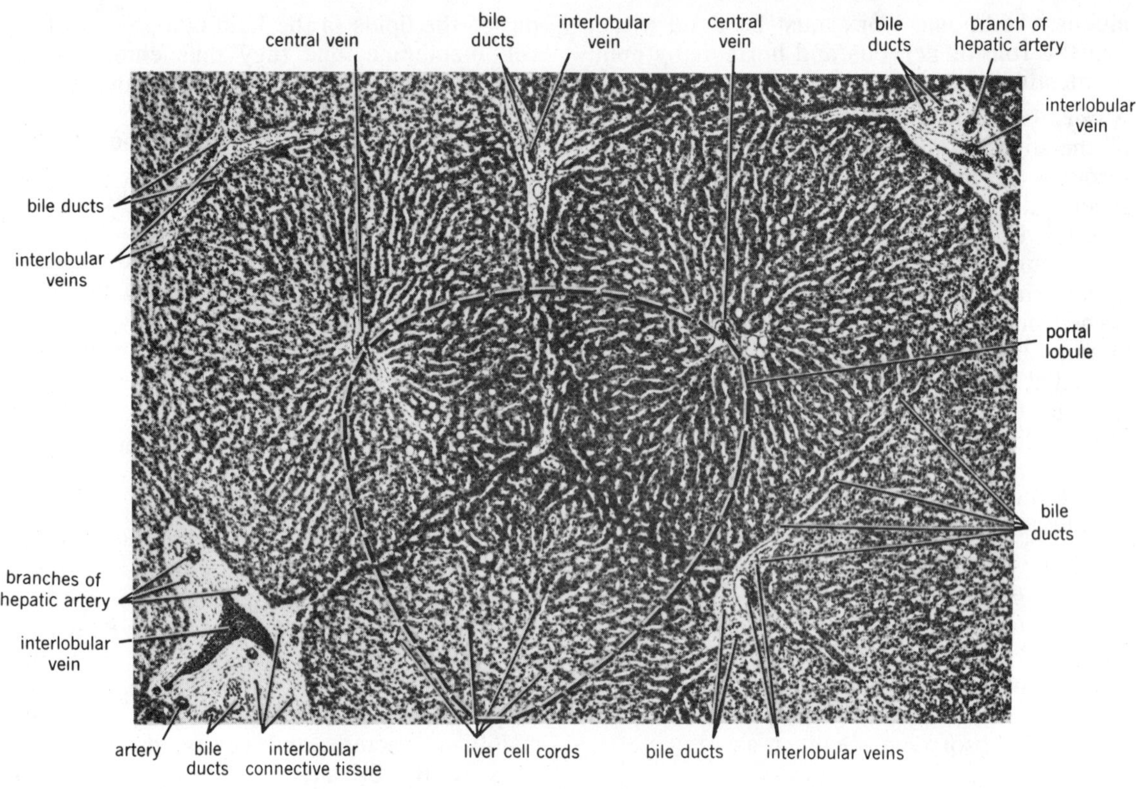

FIGURE 23.13
The lobules of the liver. (Courtesy W. B. Saunders Company. From Bloom and Fawcett. A Textbook of Histology, ninth edition. 1968.)

tract; this blood is rich in nutrients, digested foods, vitamins, salts, and water. Lacking is a high concentration of end products of fat digestion. Portal flow amounts to about 1100 ml per minute. Both supplies empty into the hepatic sinusoids. The functions of the liver may be grouped under eight headings.

IT ACTS AS A BLOOD RESERVOIR and as a MEANS OF TRANSFERRING BLOOD from portal to systemic circulations. A normal volume of about 350 ml of blood is found in the organ. An additional liter may be "stored" in the organ and transferred to the systemic circulation.

FILTERING OF THE PORTAL BLOOD occurs through the activity of phagocytic reticuloendothelial cells (*Kupffer cells*) found in the sinusoids. Bacteria, entering the blood through the intestine, are removed by these cells.

PRODUCTION OF BLOOD CELLS is a normal function of the fetal liver and occurs in the adult in certain abnormal states.

DESTRUCTION OF AGED ERYTHROCYTES occurs through the activity of the Kupffer cells.

EXCRETION OF BILE containing bile salts, cholesterol, acids, and various dyes occurs through liver activity.

The liver DETOXIFIES materials that might be

harmful to the body by combination with other materials (conjugation, methylation), by oxidation, and by reduction.

It is involved in METABOLIC REACTIONS including catabolism of glucose, glycogen, fatty acids, amino acids; synthesis of triglycerides, ATP, glycogen, urea.

It STORES iron, vitamins, and glycogen.

Because of the rather specific functions associated with the liver and the specific enzymes involved, there are a variety of tests available to assess normal or pathological conditions of the organ. The basis for the test and the tests themselves are described below.

Bile pigment metabolism produces bilirubin, which is then conjugated with glucuronic acid and excreted into the intestine. Reduction in the intestine produces urobilinogen. Reabsorption of bile by the intestine results in return of bilirubin and/or urobilinogen. Excessive accumulation of bile pigment results in jaundice, which results in yellowish coloration of the tissues. Several causes are described for jaundice.

OBSTRUCTIVE JAUNDICE is caused most commonly by blockage of the common bile duct (Fig. 23.14). It results in excessive passage of bile into the bloodstream.

FIGURE 23.14
The relationship of the bile ducts to the genesis of obstructive jaundice.

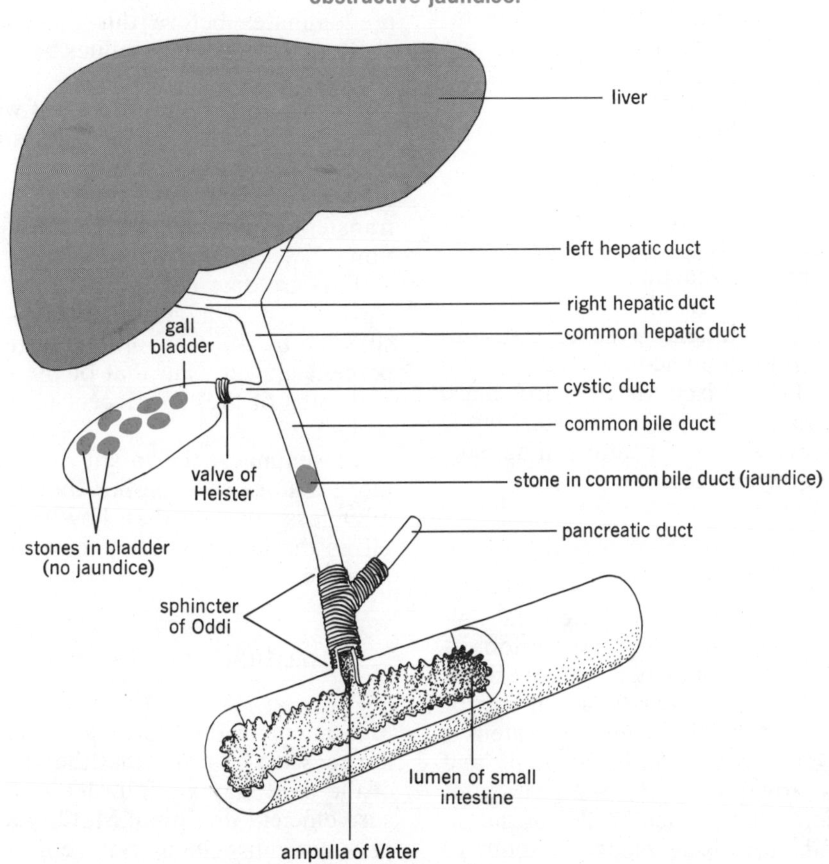

HEPATIC JAUNDICE is caused by hepatic cell damage due to toxins, viruses, and poisons with release of bile into the circulation.

HEMOLYTIC JAUNDICE results from excessive destruction of erythrocytes, with liberation of excessive amounts of pigment into the bloodstream.

The ICTERUS INDEX test determines the coloration of the plasma in comparison to the color of a solution (1:10,000 dilution) of potassium dichromate. The ICTOTEST produces a purple color on a test mat in the presence of excessive bilirubin in the urine.

Because certain dyes are excreted in the bile, the rate of their appearance in the bile is indicative of liver function. Administration of the chemical *bromsulphalein* results in a 10 to 15 percent removal from the blood per minute through the activity of liver cells. Injection of dye is followed by (1) removal of a blood sample at a later specified time (30 or 45 min) and (2) determination of plasma concentration of the dye.

The formation of hippuric acid depends on conjugation of benzoic acid and glycine. Administration of a derivative of benzoic acid (sodium benzoate) is followed by analysis of urine for benzoic acid. The test gives indication of overall hepatic function.

Certain enzymes, notably lactic dehydrogenase (LDH), glutamic-pyruvic transaminase (GPT), serum glutamic-oxaloacetic transaminase (SGOT), and isocitric dehydrogenase (ICD) are elevated in hepatic disease. This is interpreted as evidence of hepatic cell damage and release of enzymes into the plasma.

THE INFANT DIGESTIVE SYSTEM

The fetal digestive system is, like the fetal lung, presumed to be basically nonfunctional, although some processing and absorption of amniotic fluid materials may occur before birth. At birth, the system must handle a half liter or more of milk per day. The digestive system is relatively longer at birth than in the adult, and continues this growth for up to two years after birth. The microvillous brush border of the intestinal epithelial cells is developed at about 18 weeks of gestation, indicating at least the potential for terminal carbohydrate and protein digestion. Musculature is not as developed as the enzymatic capacity, and foods tend to move more slowly through the tract, possibly providing more time for digestion and absorption to occur.

Most of the carbohydrates ingested by the newborn are disaccharides. All disaccharidases are present by the third trimester of pregnancy, and all are presumably available to the newborn for disaccharide digestion. A major factor in developing carbohydrase activity appears to be the secretion of adrenal steroids. The ability to absorb monosaccharides is present from 16 weeks of gestation onwards. Ability to digest proteins occurs with the appearance of pepsin secretion at 4 to 5 months of gestation. HCl secretion does not reach its full potential until several months after birth. A pH of 4 to 5 pre-predominates before this time, and gradually decreases to the adult values by several months after birth.

Absorption of amino acids is well developed at birth. Some evidence suggests that the presence of amino acids in swallowed amniotic fluid may induce the development of amino acid transfer systems as early as 16 weeks of gestation.

Fats are digested by lipase that is present as early as 14 to 27 weeks of gestation. Fat absorption by the newborn's intestine is about 13 percent slower than that of the adult. By 3 to 6 months of life, 95 percent of the fats are absorbed.

In summary, the infant digestive system is more mature than earlier thought, and should offer less concern than how the infant metabolizes the nutrients after they are digested and absorbed.

MOTILITY IN THE ALIMENTARY TRACT

Movement of the food mass through the tract is necessary in order that the various enzymes of the tract may act in orderly sequence to ensure efficient digestion. Motility in the tract depends on muscular activity, controlled by direct

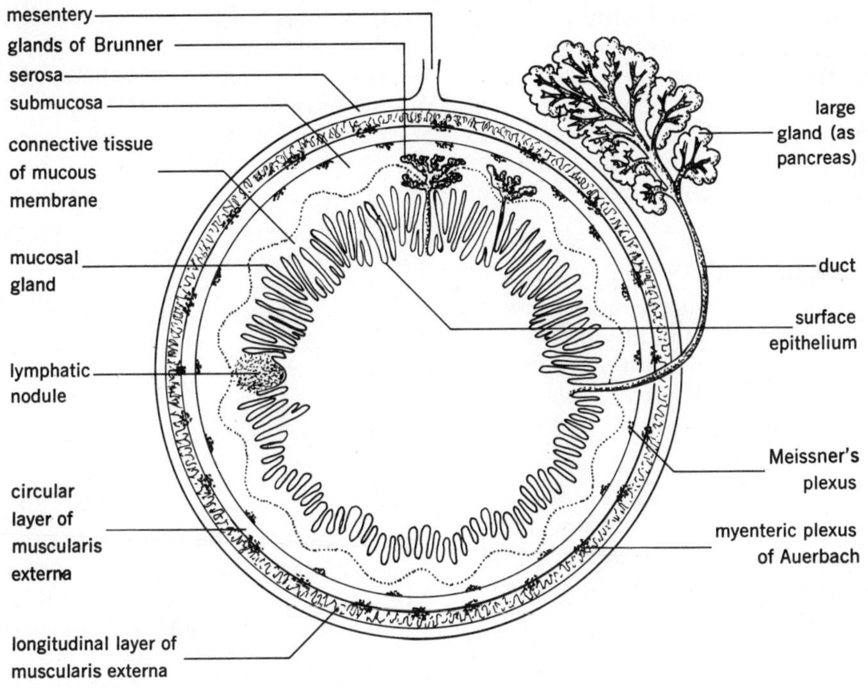

FIGURE 23.15
A diagram representing the tissue layers of the alimentary tract.

nervous stimulation, reflex mechanisms, and chemical factors. The tissue layers and intrinsic plexuses characteristic of the tract are shown in Figure 23.15. Movements in the oral cavity, pharynx, and esophagus were discussed earlier.

STOMACH

The stomach tends to act as a unit in terms of its motility since the muscle layers are essentially one continuous mass. When food enters the stomach under the impetus of the esophageal peristaltic wave, the stomach musculature undergoes a brief relaxation (*receptive relaxation*) followed by heightened activity. The volume of an empty stomach is about 50 ml, and the initial relaxation allows food to enter. Further movements of the stomach are of two types: TONIC CONTRACTIONS and PERISTALSIS. Tonic contractions occur at the frequency of 15 to 20 per minute and are found in all parts of the stomach. They serve to mix and churn the contents of the stomach. Peristaltic waves originate near the upper part of the organ and sweep toward the small intestine at the rate of 1 to 2 per minute. They extend only a few cm in distance. Strong contractions of the antral portion of the stomach move gastric contents against the partially open pyloric sphincter. Thus, mixing is promoted. Antral peristalsis also moves small amounts of the stomach contents into the small intestine for further digestion. The basic motility of the stomach depends upon inherent activity within the MYENTERIC (*Auerbach's*) PLEXUS and is only modified by the action of extrinsic nerves. Sympathetic stimulation inhibits, and parasympathetic stimulation increases, gastric motility. During fasting, the vagus nerve, in response to a lowered blood sugar level, stimulates the muscle of the stomach to very strong contractions. Such contractions are termed "hunger contractions." Feeding is the normal response to these contractions. With food intake the blood sugar level is elevated, and the contractions cease. The ENTEROGASTRIC REFLEX results when products of protein digestion, irritants, and hyper- or hypotonic fluids initiate a nervous reflex that diminishes gastric motility.

SMALL INTESTINE

Circular and longitudinal muscle layers operate much more independently in this organ than they do in the stomach. In addition, the muscularis mucosae plays an important role in one of the several types of movement that occurs in the intestine.

Peristalsis. Peristaltic waves occur at a frequency of 17 to 18 per minute. They depend on an intact myenteric plexus and occur primarily through activity of the circular muscle layer. The primary stimulus initiating peristalsis is a stretch of the intestinal wall. The reflex contraction of the muscularis that results is designated the MYENTERIC REFLEX. Examination of the fluids leaving the intestine after the myenteric reflex has been elicited shows a high content of a *chemical* substance, serotonin. This suggests that serotonin may play a role in the reflex, but the details are largely unknown.

Segmenting contractions. Segmenting contractions are ringlike local contractions of the circular muscular layer occurring at the rate of 12 to 16 per minute. The contractions appear to be due to inherent activity within the plexuses and muscle and serve to mix and churn the intestinal contents.

Pendular movements. Occurring primarily within the longitudinal muscle layer, pendular contractions cause a lateral to-and-fro movement of intestinal contents. They do not seem to have a particular frequency. They also mix the intestinal contents.

Villus contractions. Shortening of the villi, as well as waving motions, are observed in the intestine. A material designated VILLIKININ may be isolated from the blood leaving the intestine. Villikinin is produced when the upper part of the intestine is bathed by the digesting food mass, and it is thought to stimulate villus movements. Villus movements aid the absorption of materials by continually exposing the villus to "fresh" material.

Reflux of colon contents into the terminal ileum is prevented by a high pressure area in the ileum just proximal to the ileal-cecal junction. This area is commonly referred to as the ileocecal sphincter. Distension of the ileum above the sphincter causes it to relax; distension of the cecum increases contraction of the sphincter. Thus, a one-way movement of materials into the cecum is assured.

COLON AND RECTUM

Movements of the colon are similar to those seen in the small intestine. TONIC and SEGMENTING movements, similar to those of the stomach and small intestine, are present. Strong waves of PERISTALSIS move materials through the colon. Frequency of peristalsis is lower in the colon than elsewhere in the tract, varying between 3 and 12 per minute. A MASS MOVEMENT, or very strong peristaltic wave occurs three to four times a day and drives material into the pelvic colon.

Defecation. Material moving into the rectum from the pelvic colon distends the rectum and begins the act of defecation. Distention results in a reflex contraction (Fig. 23.16) of the rectal musculature that tends to expel the rectal contents. Expulsion cannot occur until the external sphincter, composed of voluntary striated muscle, is relaxed. Defecation is thus a reflex act that can be voluntarily inhibited. Another reflex can also stimulate defecation. The GASTROCOLIC REFLEX occurs when the stomach is distended with food and stimulates contraction of the rectal musculature. The uninhibited operation of the reflex is perhaps best seen in an infant, where feeding is almost invariably followed by defecation within 15 to 20 minutes. Local protection anywhere in the tract against hot or irritating foods is provided by the MUCOSAL REFLEX. Stimulation of the walls by irritants increases secretion and motility that tends

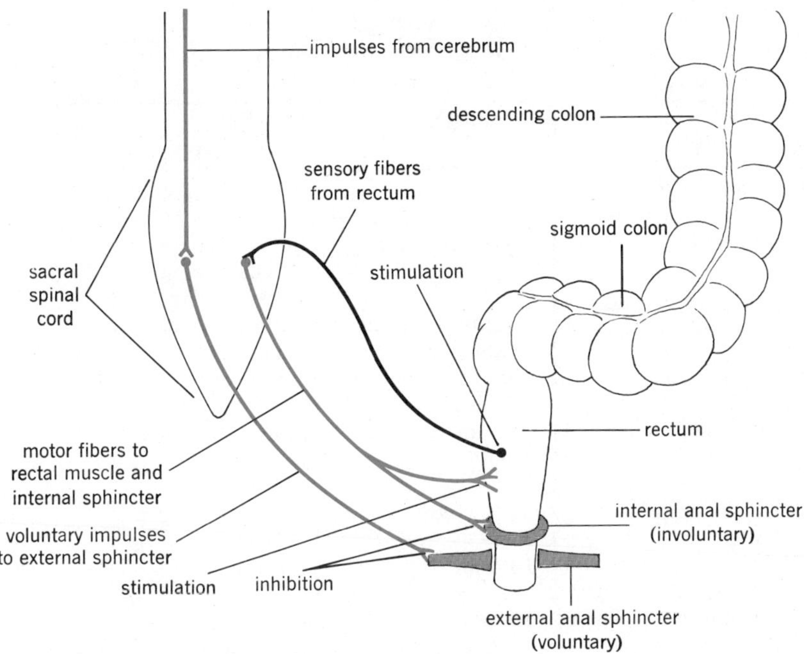

FIGURE 23.16
The reflex pathways involved in defecation. Stretching of the rectum causes a reflex contraction that empties the rectum if the external sphincter is relaxed.

to dilute and remove the irritating agent. The PERITONEAL REFLEX is often seen after peritoneal trauma, as following abdominal surgery or injury. This reflex inhibits motility in the entire gut. A summary of motility in the alimentary tract is given in Table 23.5.

DISORDERS OF THE DIGESTIVE SYSTEM

The digestive system appears to suffer from a wider variety of disorders than any other group of body organs. It is exposed to disruptions associated with excessive or unwise food intake and reflects the emotional state of the individual. Some of the more common disorders associated with the tract are presented in the following sections.

CLEFT PALATE is a common congenital (present at birth) defect that interferes with the swallowing mechanism. The disorder has a frequency of 1/1000 live births. Cleft palate is the result of the failure of the two palatine processes of the maxillary (upper jaw) bones to fuse in the midline and create a complete hard palate in the roof of the mouth. A communication between the oral and nasal cavities results and swallowing is impossible, since the oral cavity cannot be sealed to create the positive pressure necessary to move the foods to the throat. The infant cannot develop suction either because this requires sealing of the oral cavity. Speech problems will develop if the condition is not corrected before the child starts to speak.

ESOPHAGEAL ATRESIA, or failure of the normal esophageal-stomach communication to develop, is another disorder affecting the upper gastrointestinal tract. Surgery is necessary to allow nutrients to enter the stomach.

TABLE 23.5
A SUMMARY OF MOTILITY IN THE ALIMENTARY TRACT

AREA	TYPE OF MOTILITY	FREQUENCY	CONTROL MECHANISM	RESULT
Mouth	Chewing	Variable	Initiated voluntarily, proceeds reflexly	Subdivision, mixing with saliva
Pharynx	Swallowing	Maximum 20/min	Initiated voluntarily, reflexly controlled by swallowing center	Clears mouth of food
Esophagus	Peristalsis	Depends on frequency of swallowing	Initiated by swallowing	Transport through esophagus
Stomach	Receptive relaxation	Matches frequency of swallowing	Unknown	Allows filling of stomach
	Tonic contraction	15–20/min	Inherent by plexuses	Mix and churn
	Peristalsis	1–2/min	Inherent enterogastric reflex	Evacuation of stomach
	"Hunger contractions"	3/min	Low blood sugar level	"Feeding"
Small intestine	Peristalsis	17–18/min	Inherent	Transfer through intestine
	Segmenting	12–16/min	Inherent	Mixing
	Pendular	Variable	Inherent	Mixing
	Villus movements shortening and waving	Variable	Villikinin	Facilitates absorption
Colon	Peristalsis	3–12/min	Inherent	Transport
	Mass movement	3–4/day	Stretch	Fills pelvic colon
	Tonic	3–12/min	Inherent	Mixing
	Segmenting	3–12/min	Inherent	Evacuation of rectum
	Defecation	Variable 1/day–3/week	Reflex triggered by rectal distension	

ACHALASIA refers to a disorder wherein the cardiac sphincter fails to relax to permit entry of food into the stomach. The cause is attributed to lack of or damage to the myenteric plexus controlling coordination and activity of the external muscular coats of the esophagus. Consistent failure of the sphincter to relax may result in accumulation of food in the lower esophagus. The esophagus may enlarge and result in the condition known as MEGAESOPHAGUS.

GASTRITIS refers to inflammation of the stomach mucosa. Irritant foods, such as spices or alcohol, irritation by gastric juice with partial destruction of the mucosa, and bacterial inflammation, are possible causes of gastritis. If the stomach fails to secrete hydrochloric acid, ACHLORHYDRIA may result. Absence of hydrochloric acid means no pepsin action, and thus no gastric digestion.

PEPTIC ULCER is defined as erosion of the alimentary tract by the action of pepsin. Ulcerations may occur in the lower esophagus, stomach, or duodenum. The primary cause of the condition is the production of excess gastric juice when there is no food in the stomach for the juice to digest. Stress, tension, and chronic emotional upheavals seem to predispose to the condition by provoking sympathetic stimulation and decreased mucus secretion. Indeed, it has been characterized as a "disease of civilization" where the urge to "get ahead" results in continual stress. Abdominal pain, usually referred to the epigastric (above the umbilicus) region is characteristic, and vomiting may occur. The disorder requires management before perforation of the affected site occurs. A perforated ulcer, or one which penetrates the wall of the organ, may cause peritonitis, with massive accumulation of fluid in the abdominal cavity, and the development of shock.

PANCREATITIS, or inflammation of the pancreas usually occurs as a result of blockage of the pancreatic or bile duct with a resultant retention of trypsin. The pressure of accumulated secretions, including trypsin, prevented from passing into the small intestine, causes rupture of smaller ducts, and the secretions leak into the pancreatic tissue where the trypsin begins to digest the pancreas itself. Management of this disorder involves removal of the blocking agent.

PYLORIC STENOSIS results from hypertrophy of the pyloric sphincter and blocks exit of materials from the stomach. Vomiting is common.

FOOD POISONING results in inflammation of the stomach and/or small intestine (gastroenteritis). It is due to bacteria or their toxins. *Salmonella,* an organism most commonly found in foods as the result of careless habits of cleanliness, may cause inflammation. *Staphylococcal* gastroenteritis results from consuming food contaminated with the toxin of the staphylococcus organism. The organism grows well in mayonnaise, custards, milk, and similar foods, and is the most common type of food poisoning. *Botulism,* often fatal, is the result of eating foods contaminated with *Clostridium botulinum* toxin. The toxin is extremely potent and affects muscular activity, most noticeably in the diaphragm. Home canning methods may not destroy the organism or its spores, and this forms the major source of botulism.

All forms of food poisoning are associated with loss of appetite, chills, fever, headache, and vomiting and diarrhea. Untreated, the condition may lead to severe fluid and electrolyte loss (K^+, HCO_3^-), a fall in pH as the result of metabolic acidosis, and excessive Na^+ in the extracellular fluid. The latter may result in edema. Management of gastroenteritis involves careful monitoring of these materials and replacement where necessary.

APPENDICITIS results when the normally open lumen of the blind-ended appendix becomes obstructed and the organ is incapable of cleansing itself of accumulated fecal material. The organ may rupture creating PERITONITIS (inflammation of the membrane lining the abdominal cavity) and thus involve many other organs. Pain in the lower right portion of the abdomen, fever, nausea, and vomiting are usually present. Surgical removal of the inflamed organ is usually the treatment of choice.

The MALABSORPTION SYNDROME is a term applied to any disorder that results in malabsorption of fats, with passage of large, unformed, foul-smelling stools. Chief among the causes of the syndrome are genetically dependent malformations resulting in decreased number, or small villi, thereby reducing the absorptive surface, and causing the symptoms listed above. Other causes include the surgical removal of excessive amounts of intestine and radiation injury. If of genetic origin, little can be done but to feed small, low fat meals often to the affected person.

PARASITIC INFECTIONS are too numerous to consider other than to say they probably form the most common disorder in "underdeveloped countries." Primitive sanitary practices probably contribute greatly to the spread of parasites from animals to humans.

TUMORS of the tract may occur in any part. Malignancies are most common in persons over 40 and typically are more common in the lower portions of the tract. Cancer of the rectum is one of the more common malignancies but, if discovered early, is one that has a high recovery rate. Primary signs of malignancy in the tract include abdominal pain, nausea, vomiting, bloody or black (indicating blood from higher in the tract) stools, and any change in bowel habits, particularly persistent diarrhea.

DIARRHEA is the passage of a watery unformed stool. Inflammation, stress, and use of laxatives may result in excessively rapid passage of stools without time for water reabsorption to occur or in osmotic flow of water into the colon with dilution of fecal matter. Diarrhea may affect pH and fluid balance (*see* Chapter 16).

CONSTIPATION refers to the difficult passage of hard, dry feces, regardless of frequency. Slow passage of material through the colon appears to result in increased water reabsorption, with hardening of the fecal matter.

MEGACOLON is enlargement of the colon due to retention of feces because of congenital absence of the myenteric plexus (*aganglionic colon*) in all or part of the organ or from excessive use of laxatives (tone disappears) or enemas. If the entire colon is not involved, removal of the aganglionic portion and anastamosis (attachment of the cut ends) may restore normal function. If the entire organ is involved, a colostomy (creating an opening of the bowel to the outside surface of the abdomen) may be necessary.

VIRAL HEPATITIS is a disease that seems to be reaching near-epidemic proportions. Infectious hepatitis is transmitted mainly by the use of virus-contaminated tools or consumption of tainted food and water. Serum hepatitis is the result of transmission by transfusion of infected blood and by the sharing of contaminated needles by drug addicts. The liver enlarges, and signs of hepatic malfunction (jaundice, abnormal liver function tests) appear.

CIRRHOSIS of the liver may follow hepatic cell destruction, regardless of cause. Fibrous tissue replaces the destroyed cells, and, if many cells have been lost, remaining liver activity cannot meet body demands for detoxification, removal of erythrocytes, and metabolic activity.

FLATUS, or excessive gas in the tract, may result from excessive swallowing of air (this results in eructation or belching), gas formation from bacterial action (chiefly hydrogen sulfide, methane, and carbon dioxide), or by passage of gases from blood to intestine. Although gas is a normal constituent of the gut, its expulsion may be stimulated by certain foods such as beans, cabbage, vinegar, and others that stimulate peristaltic action.

INTESTINAL OBSTRUCTION results in failure to move intestinal contents onwards. Hernias (femoral and inguinal), peritoneal adhesions, handling of the gut during abdominal operations, tumors that obstruct the gut, vascular insufficiency, and deficiency of nervous stimulation to the musculature (*paralytic ilius*) are some of the causes of loss of peristalsis and obstruction. Stasis of the intestinal contents results in inflation of the gut by fluid accumulation, and intestinal bacteria multiply, producing large amounts of gas. Treatment is aimed at decompression and removal of the cause of the obstruction.

SUMMARY

1. A digestive system provides:
   a. A route for ingestion of foods.
   b. Devices for digestion of foods.
   c. Surface for absorption of foods.
   d. A means of egestion of residues of the digestive process.
2. Feeding is hypothalamically controlled. Some theories as to control mechanisms include:
   a. Filling of the stomach or stretching of the abdominal wall.
   b. Blood glucose levels (glucostat).
   c. Specific dynamic effect of foods (thermostat).
   d. General nutritional state.
   e. Intake of specific foods.
   f. Psychological status.
3. Digestion in the mouth follows chewing of the food. It is due to saliva that has the following functions.
   a. Dissolves, softens, moistens the foods; cleanses and moistens the mouth linings; protects against tooth decay.
   b. Salivary amylase breaks starches to dextrins and some disaccharides.
4. Control of salivary secretion is neural, involving taste buds as receptors, VII and IX cranial nerves as sensory and motor pathways, salivatory centers in the brainstem, and the salivary glands as the effectors.
5. No absorption of foods occurs in the mouth.

6. Foods are transported from the mouth to the stomach by swallowing, which requires a positive pressure, and by esophageal peristalsis.
7. Digestion in the stomach is carried out by gastric juice, a watery solution of hydrochloric acid and pepsin.
   a. HCl is produced by oxyntic cells, by active processes. It creates the environment necessary for pepsin action.
   b. Zymogenic cells produce pepsinogen that is activated to pepsin by HCl; pepsin breaks proteins to proteoses and peptones and a few amino acids.
   c. Other enzymes may be found in the stomach; their importance to digestion is minimal.
8. Little absorption occurs from the stomach. Water, salts, and alcohol pass slowly.
9. Control of gastric secretion occurs in three phases.
   a. The neural phase is triggered by food in the mouth and produces about 100 ml of juice that starts the second phase.
   b. The gastric phase is caused by gastrin, a hormone produced by the stomach in response to partially digested proteins.
   c. The intestinal phase is controlled by an uncharacterized gastrin-like hormone.
10. Cessation of gastric secretion occurs by enterogasterone, and movement of materials from the stomach.
11. Emptying of the stomach is controlled by several factors.
    a. Enterogasterone diminishes motility and lengthens emptying time.
    b. Stomach contents must be rendered isotonic; motility is reduced until this happens.
    c. High fat content in the meal lengthens emptying time.
    d. Vagal stimulation shortens emptying time.
    e. Stress may shorten or lengthen emptying time.
    f. Dilation of the stomach increases motility and shortens emptying time.
12. The pancreas secretes pancreatic juice containing enzymes that act on many materials. Protein digestion occurs by the following.
    a. Trypsinogen is converted to trypsin by enterokinase and trypsin itself. It splits specific peptide bends.
    b. Elastase acts on certain side chains in amino acids.
    c. Chymotrypsinogen is converted to chymotrypsin and attacks peptide bonds different than those attacked by trypsin.
    d. Procarboxypeptidase A is converted to carboxypeptidase A by trypsin and enterokinase. It splits terminal amino acids from proteins.
    e. Free amino acids and dipeptides result from the action of these enzymes.
13. Pancreatic amylase hydrolyzes dextrins to disaccharides such as sucrose, maltose, lactose, and others.
14. Pancreatic lipase hydrolyzes triglycerides to glycerol and fatty acids.
15. Pancreatic secretion involves two mechanisms.
    a. Neural (vagus) stimulation increases secretion of a watery buffered solution deficient in enzymes, as does release of secretin when acid enters the duodenum.
    b. Enzyme secretion is stimulated by CCK, released when fats and other substances (not acid) enter the duodenum.
16. Bile, produced by the liver aids fat digestion and absorption. It is not an enzyme.
    a. Bile constituents are secreted and absorbed in a "circulation."
    b. Bile exerts an emulsifying action on fats.
    c. Its acids react with fats to make them more easily absorbed (hydrotropic action).
    d. Bile is stored in the gall bladder, and is emptied from that organ by CCK.
17. Completion of carbohydrate and protein digestion occurs at the surface of the intestinal epithelial cells by enzymes located in the cell membranes.
    a. Maltase, lactase, sucrase and other disaccharidases act on their respective disaccharides and split them to monosaccharides.
    b. Aminopeptidase and dipeptidase split dipeptides to amino acids.
18. Little is known about control of intestinal secretion. A hormone, enterocrinin, produced by the intestinal wall, is believed to stimulate intestinal secretion.
19. Absorption of end products of digestion and ingested substances occurs through the epithelial cells of the small intestine.
    a. Monosaccharides are actively absorbed and pass to the blood vessels of the villi.
    b. Amino acid and some whole protein absorption occurs actively and these pass to the blood vessels.

c. Lipids are absorbed passively and pass to the lacteals.
d. Ions and vitamins are actively absorbed; water passes by osmosis as all solutes are removed.

20. The colon is a site of water and electrolyte absorption, absorption of vitamins B and K that are produced by intestinal flora, and forms the feces.

21. The liver serves many functions besides secretion of bile.
   a. It acts to "store" and transfer blood from portal to systemic circulations.
   b. It cleanses the blood from the gut, of bacteria.
   c. It produces blood cells in the fetus.
   d. It destroys aged erythrocytes.
   e. It excretes products of metabolism through the bile.
   f. It detoxifies substances.
   g. It metabolizes and synthesizes many substances.
   h. It stores substances.

22. Tests of liver function involve determinations of dye excretion, liver enzymes in the blood, and bilirubin levels. Jaundice may develop from bile duct blockage, liver damage, and excess red cell destruction.

23. The infant digestive system has developed most of the adult capabilities by the time of birth. Full HCl production requires several months.

24. Motility (movement) in the tract involves peristalsis, tonic contractions, segmenting, and pendular movements for moving materials through the tube and mixing foods with secretions.

25. Defecation is basically a reflex emptying of the rectum.

26. A wide variety of disorders affect the digestive system. The most common are described.

QUESTIONS

1. What are the basic functions of a digestive system?
2. What is indicated by the phrase, that, in its lumen each digestive organ has its "own peculiar ecology"? Illustrate, using the human digestive system.
3. Suppose that a lesion was placed in the ventromedial hypothalamus. What would happen, and why, to the weight of an animal?
4. What are some of the possible causes for cessation of feeding?
5. Starting with the form in which carbohydrates, proteins, and fats are usually ingested, trace the digestion of each. Include enzymes involved, where they are produced, pH required for optimum action, and end products of each enzyme's action.
6. What function other than digestion does saliva serve?
7. Compare control of secretion of the salivary glands, stomach, and pancreas.
8. How would a cleft palate interfere with swallowing?
9. Looking at the enzymes involved, in what way or ways does the final hydrolysis of carbohydrates and proteins differ from the earlier stages of digestion?
10. What does bile contribute to the digestive process? What other functions does the liver have?
11. How, and by what routes, are the following absorbed and leave the intestine?
    a. Water
    b. Salts
    c. Amino acids
    d. Triglycerides
    e. Glucose
12. Does digestion occur in the colon? What functions does the organ have?
13. In what ways is the infant digestive system functionally different from that of the adult?
14. What types of motility are common to most organs of the alimentary tract? Which are unique to a particular organ?
15. What do you consider to be the most common disorders of the digestive system? Describe each regarding the physiological mechanism that has been interfered with.

READINGS

Deutsch, J. A., and M.-L. Wang. "The Stomach as a Site for Rapid Nutrient Reinforcement Sensors." *Science 195:*89. 7 Jan 1977.

Gray, G. M. "Carbohydrate Digestion and Absorption: Role of the Small Intestine." *New Eng. J. Med. 292:*1225. June 5, 1975.

Javitt, Norman B. "Hepatic Bile Formation." *New Eng. J. Med. 295:*1464. Dec 23, 1976. *295:*1511. Dec 30, 1976.

Kappas, Attallah, and Alvito P. Alvares. "How the Liver Metabolizes Foreign Substances." *Sci. Amer. 232:*22. June 1975.

Rayford, Phillip L., Thomas A. Miller, and James C. Thompson. "Secretin, Cholecystokinin and Newer G-I Hormones." *New Eng. J. Med. 294:*1093. May 13, 1976. *294:*1157. May 20, 1976.

Rothman, S. S. "Protein Transport by the Pancreas." *Science 190:*747. 21 Nov 1975.

Satir, Birgit. "The Final Steps in Secretion." *Sci. Amer. 233:*28. (Oct) 1975.

Stroud, Robert M. "A Family of Protein-cutting Proteins." *Sci. Amer. 231:*74. (July) 1974.

Walsh, John H., and Morton I. Grossman. "Gastrin." *New Eng. J. Med. 292:*1324. June 19, 1975. *292:*1377. June 26, 1975.

24

METABOLISM AND NUTRITION

OBJECTIVES

After studying this chapter, the reader should be able to:

☐ Define a calorie (large and small), and give the heat values (kcal/gm) for the three basic foodstuff groups.

☐ Describe the role of ATP and other "high energy phosphate compounds" in the body economy.

☐ Discuss the roles of cyclic nucleotides, dinucleotides, and coenzyme A in intermediary metabolism.

☐ Trace a glucose molecule from uptake by a cell to complete degradation to CO_2 and H_2O, naming the metabolic cycles involved and their contributions to the degrading process.

☐ Show how it is possible to interconvert carbohydrates to fats and amino acids and vice versa. List cycles involved and what they do.

☐ Describe the role of lipoproteins in the genesis of atherosclerosis, and explain atherosclerosis.

☐ Define what an essential fatty acid is, and give its functions.

☐ Define what the ketone bodies are.

☐ Show how the body disposes of the ammonia created by amino acid metabolism.

☐ Define respiratory quotient, and show how it aids in determination of caloric balances in the body.

☐ Define basal metabolic rate and the factors controlling it.

☐ Recommend a diet that is adequate in calories, vitamins, and minerals.

☐ Comment on the dangers of hyperalimentation.

GENERAL CHARACTERISTICS OF INTERMEDIARY METABOLISM

Simple sugars, amino acids, glycerol, and fatty acids form the major sources of energy for running the body's physiological mechanisms. A variety of METABOLIC CYCLES degrade such substances in a stepwise fashion because cells have no mechanisms by which large single outputs of energy can be captured and directed to particular processes. Energy released may also be used in the synthesis of new substances, and the reactions may also require accessory factors such as hormones, vitamins, and metals.

The various cycles that enable degradation and synthesis of substances within the body cells constitute that area of physiology known as intermediary metabolism.

EXPRESSION OF ENERGY CONTENT

The conventional unit used to express energy content of a foodstuff is the CALORIE. There are two designations of calories:

A CALORIE (*small c*) is defined as the amount of heat required to raise the temperature of 1 gram of water 1°C (as from 15° to 16°C).

A CALORIE (*capital C*) or *kilocalorie* (kcal) is 1000 times larger than the small calorie and is the unit generally used in describing the energy changes of intermediary metabolism. It is this Calorie that is commonly used when talking about diets.

The symbol ΔG is employed to express the energy required for or released from a given reaction. If $A \rightarrow B$ represents a chemical reaction, and B has a lower energy content than

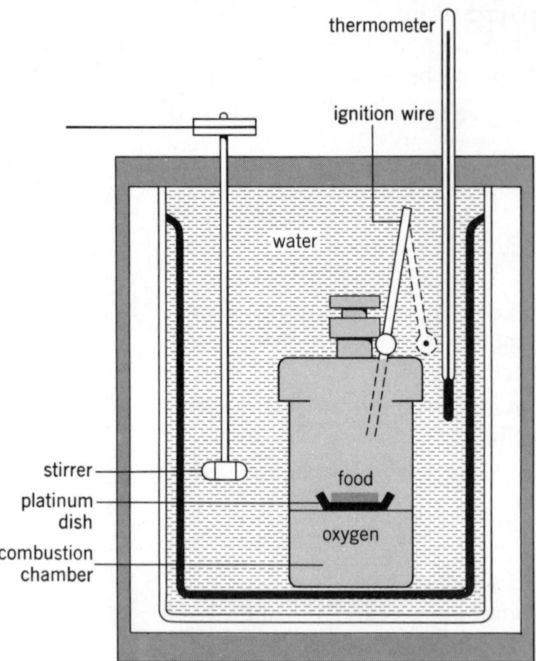

FIGURE 24.1
A diagram of a bomb calorimeter.

A, ΔG is negative. The reverse situation results in a positive ΔG.

Determination of the HEAT VALUES of the foodstuffs may be made by igniting specified quantities of the foods in a bomb calorimeter (Fig. 24.1) and measuring the elevation of temperature of the water around the ignition chamber. For carbohydrate, fat, and protein, the heat values are 4.1, 9.4, and 5.6 kcal per gram. In the body, protein is not completely combusted, and the energy content of the urea formed from its metabolism must be subtracted from the figure given above. This reduces the heat value of protein to 4.1 kcal/g. For ease of calculation, the heat values given above are commonly "rounded off" to 4 KCAL/G OF CARBOHYDRATE, 9 KCAL/G OF FAT, and 4 KCAL/G OF PROTEIN.

COUPLING ENERGY TO PHYSIOLOGICAL MECHANISMS

Although foods are consumed for their energy content, the energy in their chemical bonds is not directly available to such physiological mechanisms as muscle contraction, membrane transport systems, or glandular secretion. The foodstuff energy, released by metabolic cycles, must be transferred to a compound that can pass the energy on to the reaction. A group of large organic molecules known as HIGH-ENERGY PHOSPHATE COMPOUNDS are formed during intermediary metabolism and serve as the immediate sources of energy for running the body machinery. Among these compounds are the nucleotides adenosine triphosphate (ATP), guanine triphosphate (GTP), uridine diphosphate (UDP), and the molecule creatine phosphate (CP) (Fig. 24.2). These compounds have in common one or more phosphate groups that are held to the rest of the molecule by chemical bonds (high-energy phosphate bonds or ~) that release more than the usual amount of energy when they are broken. By a process known as PHOSPHORYLATION, these phosphate groups may be transferred to other compounds, and the transfer may result in some visible activity or it may raise the energy level of the phosphorylated molecule, enabling it to begin or continue its progress in a metabolic scheme.

In the last analysis, the function of foods we eat is to be metabolized to supply energy for the synthesis of these energy-rich intermediaries in the hierarchy of body function. Some 90 percent of foodstuff energy follows this fate.

COMPOUNDS ESSENTIAL TO INTERMEDIARY METABOLISM

In addition to the nucleotides and phosphate compounds mentioned above, several other compounds appear necessary in order that intermediary metabolism may proceed.

CYCLIC MONONUCLEOTIDES (Fig. 24.3) are produced by the action of cyclase enzymes on a basic "straight" nucleotide such as ATP or GTP. In the process, a double-phosphate group known as a pyrophosphate is split out, and the original three-phosphate compound (e.g., ATP–T for tri-) is reduced to a one-phosphate compound (e.g., AMP–M for mono-). Cyclic mononucleotides have been implicated as activators of enzyme systems that bring about chemical reactions. In turn, it has been shown that the production of cyclic mononucleotides is influenced by hormones, so that a basis for the ability of hormones to alter chemical reactions may be established. Table 24.1 presents some of the processes affected by cyclic nucleotides, with cyclic AMP as the example.

DINUCLEOTIDES (Fig. 24.4) are formed by the linking of two nucleotides through their phosphate groups; they are important in oxidation of foodstuff molecules. One form that oxidation takes is through the removal of hydrogen from the molecules of foodstuffs; the hydrogens are transferred to the dinucleotides, which act as HYDROGEN ACCEPTORS, and are ultimately carried to a scheme that forms water and ATP. Three dinucleotides are used as hydrogen acceptors. They are nicotinamide adenine dinucleotide (*NAD*), nicotinamide adenine dinucleotide phosphate (*NADP*), and

FIGURE 24.2

The formulas of several high-energy phosphate compounds.

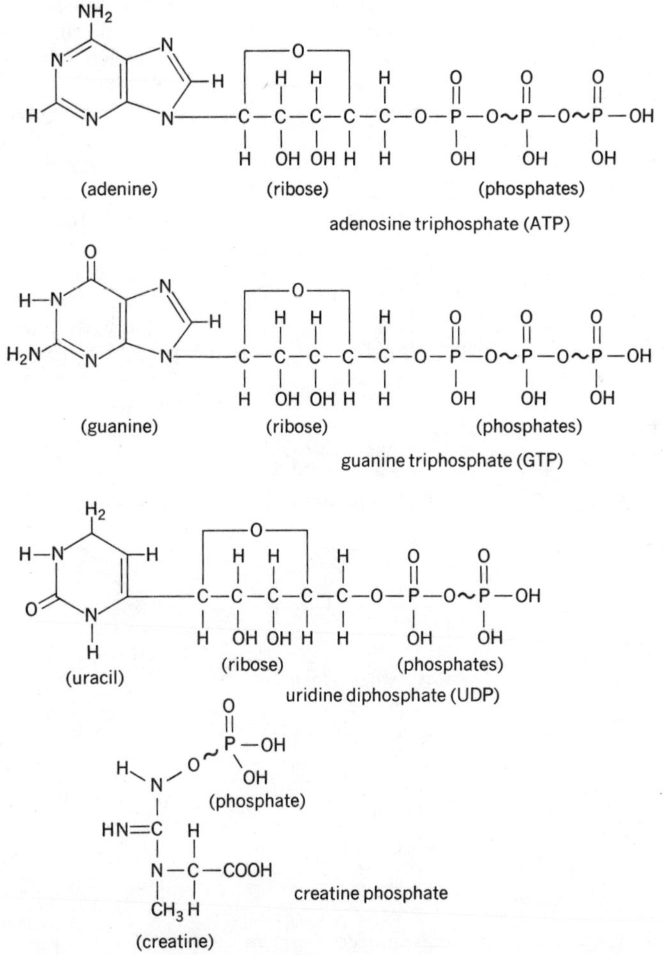

TABLE 24.1

SOME EFFECTS OF CYCLIC AMP INCREASE

REACTION, TISSUE, OR PROCESS AFFECTED	RESULT OR EFFECT	COMMENTS
Glycogenolysis (liver)	Stimulated	Increase of cAMP caused by glucagon and epinephrine. Insulin decreases cAMP and decreases all these processes
Gluconeogenesis (liver)	Stimulated	
Ureogenesis (liver) (Urea formation)	Stimulated	
Ketogenesis (liver) (Ketone formation)	Stimulated	
Hepatic enzyme synthesis	Increased synthesis	Transaminase and carboxykinase are increased most
Histone (a type of protein) phosphorlylation	Stimulates	Activates histone kinase; insulin stimulates reaction
Enzymes of prostate gland	Activates and elevates levels	Control some phases of gland's CH_2O metabolism
Insulin release	Stimulated	Decreased blood sugar increases cAMP, which stimulates insulin release
Release of anterior lobe hormones	Increased secretion	Thyrotropin releasing factor and T_3 increase cAMP and increase TSH secretion
Thyroxin secretion	Increased secretion	TSH activates cAMP in thyroid; more thyroxin released
Calcitonin secretion	Increased secretion	Calcitonin increases Ca++ deposition in bone
Prostaglandin secretion	Increased secretion	Prostaglandins activate many enzymes in body cells
Water movement across membranes	Increased movement	cAMP probably stimulates active solute movement and therefore water

FIGURE 24.3

Cyclic nucleotides.

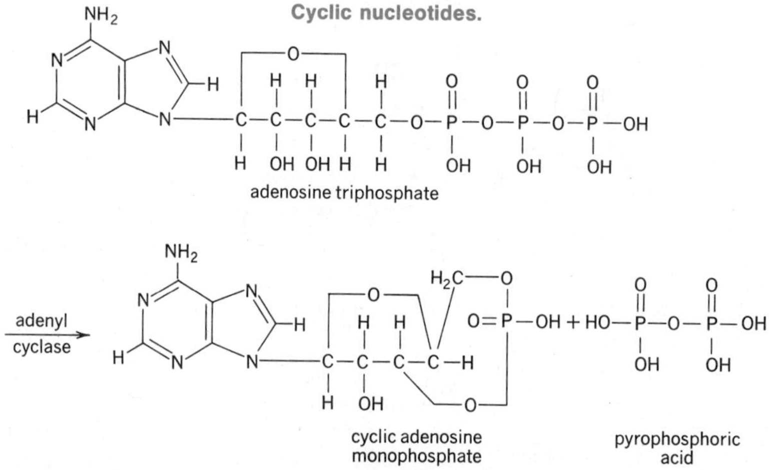

adenosine triphosphate

adenyl cyclase →

cyclic adenosine monophosphate + pyrophosphoric acid

TABLE 24.1 (continued)

SOME EFFECTS OF CYCLIC AMP INCREASE

REACTION, TISSUE, OR PROCESS AFFECTED	RESULT OR EFFECT	COMMENTS
Sodium transport across membranes	Increased transport	Probably stimulates active transport
Gastric secretion	Inhibits secretion	cAMP reduces blood flow; may also increase release of insulin, glucagon and prostaglandins, all of which decrease gastric secretion
Immune response	Activation of of immunocompetent cells	
Skeletal muscle	Initiates contraction	cAMP increases permeability of sarcoplasmic reticulum to Ca^{++}; therefore, it initiates contraction
Smooth muscle	Inhibition of contraction	
Platelets	Increased activity	Increase of cAMP stimulates cyclase and increases activity in platelets
Pineal gland	Increased secretion of melatonin	Effect mediated by increased light
Nerve tissue	Increases protein kinase activity	
Parathyroid effects on bone and kidney	Increases kidney reabsorption of Ca^{++}; increases resorption of bone	Stimulates active transport by kidney; activate enzymes of osteoclasts

flavine adenine dinucleotide (*FAD*). After receiving hydrogen, the acceptors are written as $NADH + H^+$, $NADPH + H^+$, and $FADH_2$.

COENZYME A (Fig. 24.5) is an energy-rich compound that combines with and activates compounds so that they may enter a metabolic scheme or continue their process through a cycle. It activates acetic acid for entry into the Krebs cycle, and fatty acids for entry into beta oxidation. The coenzyme forms a pivotal compound in intermediary metabolism.

CARBOHYDRATE METABOLISM

TRANSPORT INTO CELLS AND PHOSPHORYLATION

It may be recalled that the processes of digestion produce a variety of monosaccharides, including glucose, fructose, and galactose. These are absorbed into the bloodstream and distributed to body cells. To be utilized by cells, these sugars must pass through the cell membrane. Passage is by means of active transport

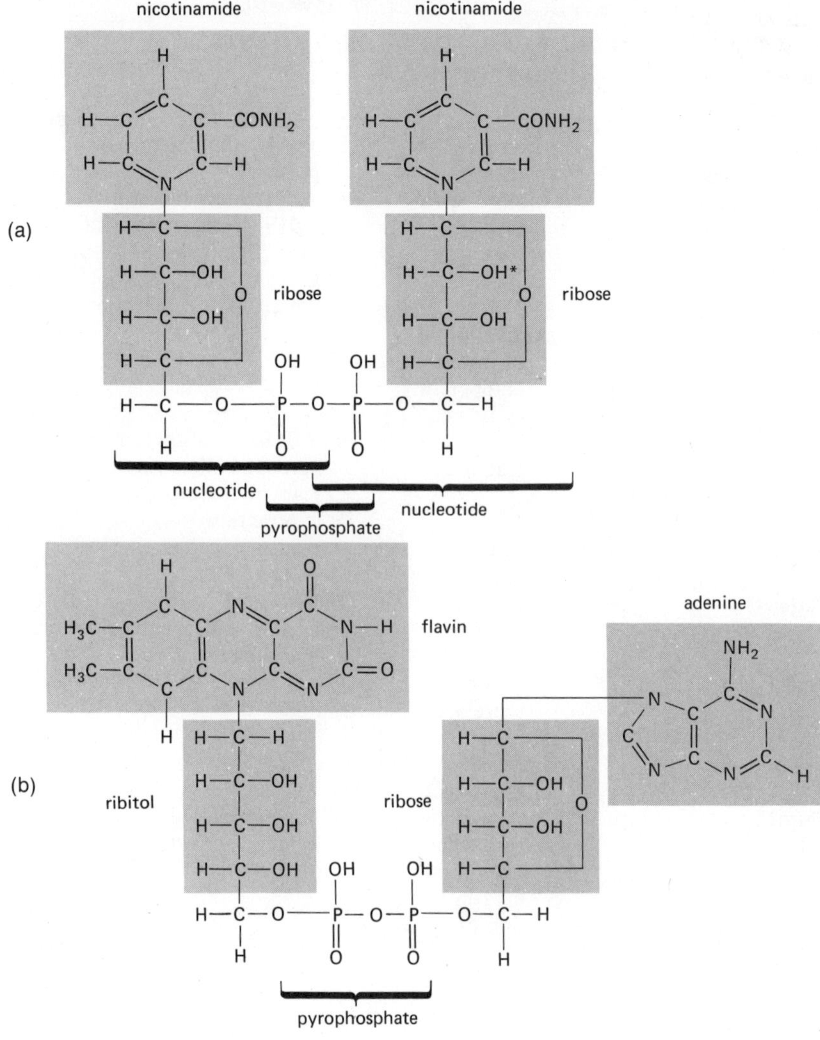

FIGURE 24.4
The dinucleotides (*a*) NAD (NADP has an extra phosphate at the *). (*b*) FAD)

and facilitated diffusion, and, once into the cell, the sugar is PHOSPHORYLATED in the presence of ATP and specific enzymes, as shown in Fig. 24.6.

The phosphorylation of the sugar is irreversible except in liver and kidney tubule cells. In these areas, a phosphatase enzyme is present that splits the phosphate group from the sugar. This particular reaction frees the sugar to leave the cell and go elsewhere in the body, as shown in Fig. 24.7.

INTERCONVERSION OF SIMPLE SUGARS

Glucose and fructose form the major sources of energy for cellular activity. There are no enzyme systems capable of metabolizing galac-

```
    H  CH3  OH  O  H  H  H  O  H  H  H
    |   |   |   ‖  |  |  |  ‖  |  |  |
H — C — C — C — C — N — C — C — C — N — C — C — SH
        |   |           |  |           |  |
       CH3  H           H  H           H  H
    |
    O                                       NH2
    |                                        |
O = P — OH                                   C
    |                                  N — C    N
    O        H   H   H     H   H     C     ‖    |
    |        |   |   |     |   |           C    C — H
O = P — O — C — C — C  —  C — C — N — C — N
    |        |   |   |     |   |
   OH        H   |  OPO3  OH   |
                 |_____O_______|
```

FIGURE 24.5

Coenzyme A. The molecule attaches to other compounds at the -SH group.

tose efficiently. Galactose forms about one-third of the monosaccharides liberated by carbohydrate digestion in the alimentary tract. In order to use galactose in the body economy, it must be converted to a form that is metabolizable—that is, glucose. Liver cells possess a system of enzymes capable of achieving this conversion. An energy-rich phosphate compound, uridine diphosphoglucose (UPDG) is required in addition to the enzymes. The reactions occurring are shown in Fig. 24.8.

Reaction 1 activates the galactose, using ATP. In reaction 2, galactose replaces glucose on the uridine compound. In reaction 3, conversion of galactose to glucose occurs by rearrangement of atoms in the sugar molecule. In reaction 4, glucose is split from the uridine molecule with formation of another high-energy compound, UTP.

In the inherited metabolic disease GALACTOSEMIA, there is accumulation of galactose in the blood due to an apparent inability to convert it to glucose. Deficiency of the enzyme catalyzing step 2 may be demonstrated.

FIGURE 24.6

The phosphorylation of several simple sugars.

glucose —(ATP → ADP; glucokinase)→ glucose-6-phosphate

fructose —(ATP → ADP; fructokinase)→ fructose-6-phosphate

galactose —(ATP → ADP; galactokinase)→ galactose-1-phosphate

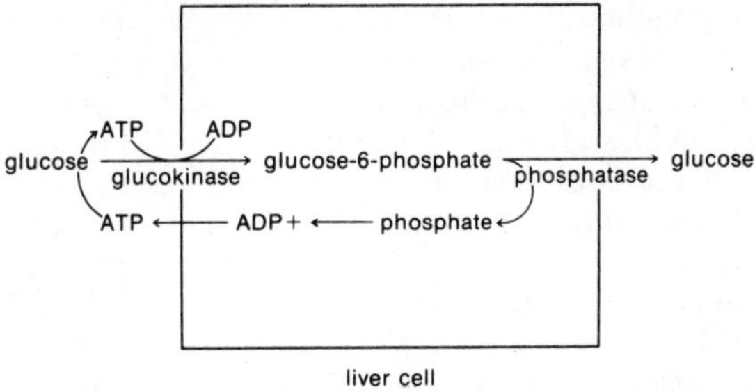

FIGURE 24.7

The dephosphorylation of glucose in the liver cell.

In infants, galactosemia intake is reduced by eliminating milk and milk products from the diet. A milk substitute is provided to supply essential proteins, lipids, carbohydrates, and minerals and vitamins. On reaching adolescence, or later in life, patients with galactosemia may ingest varying amounts of galactose without apparent side effects. This apparent reversal of galactose tolerance may be related to these factors: improvement in galactose utilization; ingested galactose in later years forms a smaller portion or fraction of body weight and can be tolerated better; an increasing efficiency and utilization of alternative pathways for galactose metabolism.

Fructose may be metabolized by the same series of enzymes metabolizing glucose, and thus does not require conversion.

FORMATION OF GLYCOGEN

If the body is provided with more glucose than it can metabolize to meet its needs at a given moment, the excess is formed into glycogen

FIGURE 24.8

The conversion of galactose into glucose.

galactose —(① galactokinase; ATP → ADP)→ galactose-1-phosphate

galactose-1-phosphate + UDP-glucose —(② galactose-1-phosphate uridyl transferase)→ glucose-1-phosphate + UDP-galactose

UDP-galactose —(③ epimerase)→ UDP-glucose

UDP-glucose —(④ pyrophosphorylase; P.P. → UTP)→ glucose-1-phosphate

P.P.=pyrophosphate (2 phosphate groups)
UTP=uridine triphosphate

(animal starch) by the process known as GLYCOGENESIS. The process involves polymerization (putting together) of glucose molecules into a treelike glycogen molecule. The "stem" of the tree is elongated by an enzyme called *glycogen synthetase* until 7 to 21 glucose units have been polymerized, then another enzyme, called *"brancher enzyme"* creates a side chain on the stem. The elongation of the stem of the glycogen molecule is aided by the presence of the hormone *insulin*. The enzymes necessary for glycogenesis are found in quantity only in liver and muscle cells thus these are the only cells capable of storing significant quantities of glucose as glycogen.

RECOVERY OF GLUCOSE FROM GLYCOGEN

Glycogen may be split into its constituent glucose molecules in the process known as GLYCOGENOLYSIS. Linkages between glucose units in side branches of the glycogen molecule are broken by an enzyme known as *phosphorylase*. The glucose thus released is utilized to maintain blood glucose levels and as a source of energy for cellular activity. When a side chain connection to the main stem of the glycogen molecule is reached, another enzyme, called *"debrancher enzyme"* is required. It splits the bond of the side chain and thus exposes another series of "stem" molecules to the action of phosphorylase. The degradation of glycogen to glucose is aided by the presence of the hormones epinephrine and glucagon. These hormones appear to increase cyclic AMP, which then increases the activity of phosphorylase. These reactions are summarized below.

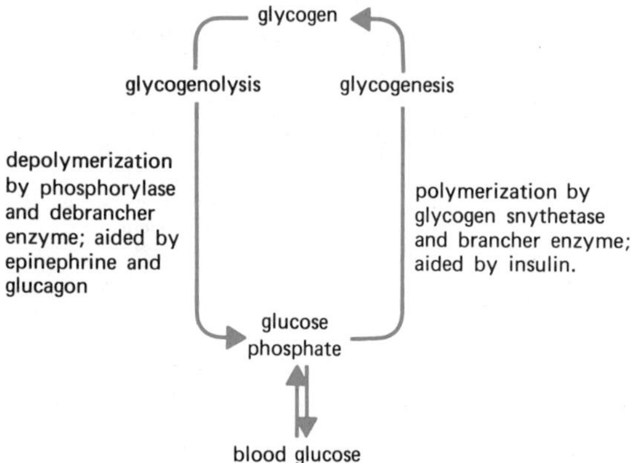

By these reactions, the amounts of blood glucose made available to cells may be kept within the normal range of 80 to 120 mg per 100 ml of blood, in the face of demands for exercise, and food intake. As one of, if not the main consumer of blood glucose, the skeletal muscles become important as regulators of blood sugar levels. Their activity may thus determine how much glucose is stored in or released from the liver. The interrelationships between liver and muscle glycogen are shown by the Cori cycle, diagrammed opposite.

Blood glucose enters muscle cells, and is stored as glycogen, or degraded to pyruvic acid. If insufficient oxygen is available to completely metabolize all of the pyruvic acid, the excess may be converted to lactic acid. Both pyruvic and lactic acids may be put into the bloodstream to serve as sources of glucose and glycogen synthesis by the liver. From the liver, glucose may enter the blood. One may thus appreciate that availability of glucose for cellular activity is governed by a variety of interlocking factors.

Figure 24.9 shows the reactions of glycogenesis and glycogenolysis. So far, the following facts have been established:

The main use to which carbohydrates and fats are put in the body is to release energy to synthesize ATP, the directly utilizable source of energy for cellular reactions.

Transport of sugars into cells requires ATP in order to activate the sugar.

Sugars other than glucose, specifically galactose, may be converted into glucose.

Glycogen may either be formed from glucose or may be broken down into glucose. Both conversions are hormone-dependent.

Specific enzymes are required for each step in these reactions.

The interrelationships of the reactions presented so far are shown in Fig. 24.10. Enzymes are not included.

METABOLIC DISORDERS ASSOCIATED WITH GLYCOGEN METABOLISM

The genes in the nucleic acids of the nuclear DNA determine the synthesis of specific en-

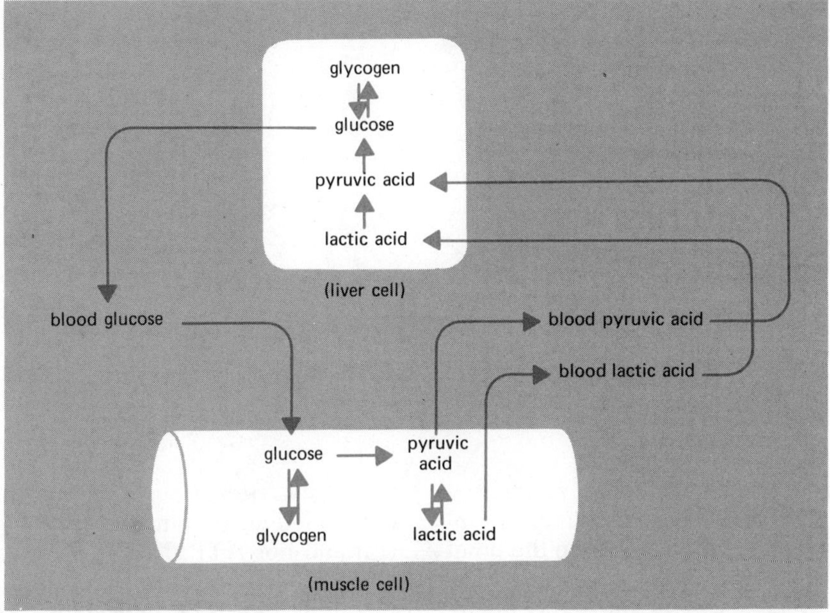

zymes involved in control of metabolic cycles. Thus, mutation or alteration in the genes concerned with production of enzymes necessary for glycogenesis and glycogenolysis may result in missing or defective enzymes. Seven types of GLYCOGENOSES (glycogen storage diseases) are described in Table 24.2. All involve the formation of an abnormal glycogen, or an abnormal amount of glycogen is deposited in muscle and liver cells. Of particular interest are the disorders associated with muscle phosphorylase deficiency, brancher enzyme deficiency, de-

FIGURE 24.9
Glycogenesis and glycogenolysis.

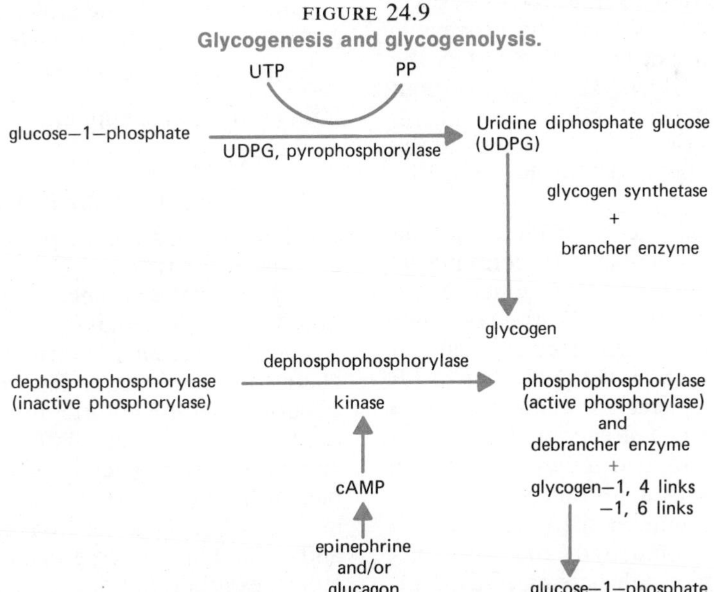

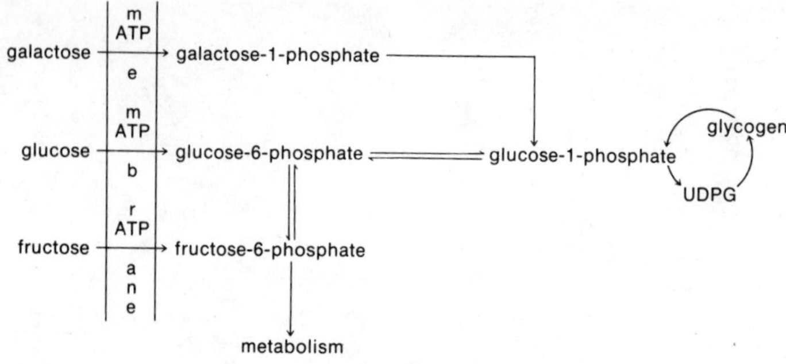

FIGURE 24.10
The interrelationships of simple sugars and glycogen.

brancher enzyme deficiency, and synthetase deficiency. These cause, respectively, an inability to raise blood glucose levels to the point where useful muscular work may be done, little storage of glycogen in the liver, an inability to degrade glycogen, and little or no formation of glycogen in the liver and muscles. While occurrence of these disorders is described as "rare," when they occur, there is a great effect on carbohydrate metabolism.

GLUCOSE AND FRUCTOSE METABOLISM

Conversion of glucose and fructose to pyruvic acid is known as GLYCOLYSIS. The process is anaerobic; that is, it can be carried out in the absence of oxygen. Glycolysis releases about 56 kcal of energy per mol of glucose, and results in a net gain of two ATP molecules to the body. The reactions involved are shown in Fig. 24.11.

Reaction 1 utilizes one ATP molecule and activates the sugar to enable it to continue its metabolism. Reaction 2 rearranges glucose into fructose. It is at this step that absorbed fructose may enter the scheme without being changed into glucose. Reaction 3 uses another ATP molecule, forming a 6-carbon compound with phosphate on either end. Reaction 4 splits the 6-carbon fructose into two different 3-carbon compounds, each with a phosphate attached. The dihydroxy compound is then changed into the glyceraldehyde compound, and from this point onward, two molecules are carried simultaneously. Reaction 5 phosphorylates each compound again, but uses inorganic phosphate (Pi) and not ATP. Reaction 6 releases enough energy to resynthesize two ATP molecules. To this point two ATP molecules have been used, so the amount is balanced. Reaction 9 again results in enough energy release to resynthesize two more ATP molecules per molecule of glucose. This represents a net gain. Reaction 9 occurs without enzymes and forms pyruvic acid. These reactions of glycolysis are reversible.

Since the scheme releases 56 kcal and 2×7.7 kcal is trapped in the two ATP molecules formed, the efficiency of the operation in terms of energy trapping is $\frac{15.4}{56.}$ kcal $\times$ 100, or 27.5 percent.

LACTIC ACID FORMATION

Further combustion of pyruvic acid to form, eventually, CO_2 and H_2O requires the presence of oxygen. If the O_2 necessary for this combustion is not immediately available, as during strenuous muscular exertion, pyruvic acid is changed to lactic acid. The lactic acid acts as a temporary storage form for the pyruvic acid, until O_2 becomes available, whereupon lactic acid is converted back to pyruvic acid and combusted. Lactic acid formation occurs as shown in Fig. 24.12. The hydrogens are contributed and picked up by hydrogen acceptors, molecules described previously.

PYRUVIC ACID DEGRADATION; THE KREBS CYCLE (FIG. 24.13)

In order to enter the cycle that further combusts it, pyruvic acid must be changed to a 2 carbon molecule. This is accomplished by removal of CO_2 (*decarboxylation*) from the pyruvic acid molecule. In the transformation, a pair of hydrogens is released.

The 2 carbon molecule then reacts with another molecule, coenzyme A (CoA), to form acetyl CoA.

$$\text{acetic acid} + \text{CoA} \rightarrow \text{acetyl CoA}$$

Acetyl CoA then enters the Krebs cycle (citric acid cycle, tricarboxylic acid or TCA cycle). The main job that the Krebs cycle carries out is to produce 1 ATP molecule, 2 CO_2, and 8 H^+ per revolution.

Reaction 1 couples the 2-carbon acetyl CoA to the 4-carbon oxaloacetic molecule to form a 6-carbon unit, citric acid. Reaction 2 rearranges the molecule. Reaction 3 removes 2 H. Reaction 4, removing 1 CO_2, reduces the 6-carbon oxalosuccinic acid to the 5-carbon alpha-ketoglutaric acid. Reaction 5 removes an additional CO_2, as well as 2 H, and creates the 4-carbon compound, succinyl CoA. Reaction 5 has a very high ΔG, and makes this step hard to reverse. Studies involving isotope-labeled acetyl CoA have proved that the 2 CO_2 molecules released originate from the acetyl CoA. We can therefore state that the acetyl CoA (derived from pyruvic acid) is undergoing degradation. Reaction 6 produces enough energy to synthesize a molecule of ATP from ADP. Reaction 7 produces 2 H. Reaction 8 is a rearrangement of the molecule. Reaction 9, in addition to releasing 2 H, regenerates oxaloacetic acid, which is then ready to accept another acetyl CoA molecule and repeat the cycle. The cycle would be traversed twice in combusting the 2

(*Text continued on page 451.*)

TABLE 24.2
GLYCOGENOSES (GLYCOGEN STORAGE DISEASES)

TYPE AND/OR NAME OF DISORDER	CHARACTERISTICS	ENZYME ABNORMALITY	HEREDITY
Type I, glucose-6-phosphatase deficiency	Enlargement of liver due to glycogen accumulation–glycogen is normal; cannot convert G-6-P to glucose ∴ accumulates G-1-P	G-6- phosphatase deficiency	Autosomal recessive
Type II, generalized glycogenosis	Cardiac enlargement due to myocardial glycogen deposits	Lack of glucosidase to degrade glycogen	Autosomal recessive
Type III, limit dextrinosis	Cells packed with glycogen–can degrade glycogen at –1,4– links; not at –1,6– links	Deficiency of debrancher enzyme	Autosomal recessive
Type IV, glycogenosis	Cirrhosis of liver; large amounts of abnormal glycogen (fewer branches and longer chains)	Deficiency of brancher enzyme	Unknown
Type V, myophosphorylase deficiency glycogenosis	Increased amounts of normal glycogen in skeletal muscle; cannot perform work, no glucose available in the muscle	Muscle phosphorylase deficiency	Autosomal recessive
Type VI, hepatophosphorylase deficiency glycogenosis	No description	Liver phosphorylase deficiency	Autosomal recessive
Type VII, UDP–glucose-glycogen synthetase deficiency	Hypoglycemia and convulsions; little or no glycogen in liver	Synthetase deficiency	Unknown

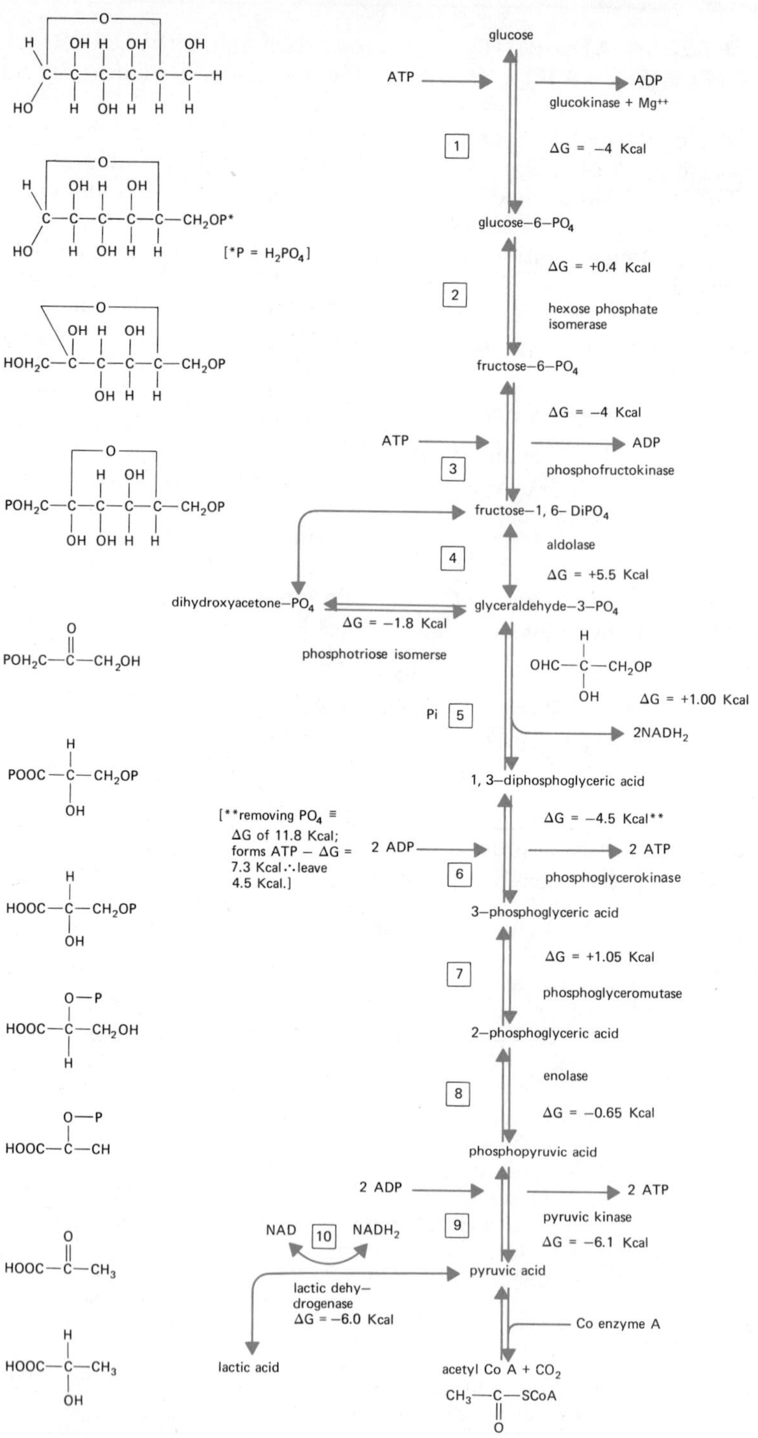

FIGURE 24.11
Glycolysis, the anaerobic breakdown of glucose (Embden-Meyerhof pathway).

$$\underset{\text{pyruvic acid}}{CH_3-\overset{O}{\overset{\|}{C}}-COOH} \xrightarrow[NADH_2 \;\rightarrow\; NAD]{+2H} \underset{\text{lactic acid}}{CH_3-\overset{OH}{\overset{|}{\underset{H}{\underset{|}{C}}}}-COOH}$$

FIGURE 24.12

The reaction forming lactic acid from pyruvic acid. (The reaction is reversible, by removal of 2H from lactic acid.)

FIGURE 24.13

The Krebs cycle, the aerobic breakdown of pyruvic acid.

Krebs Cycle

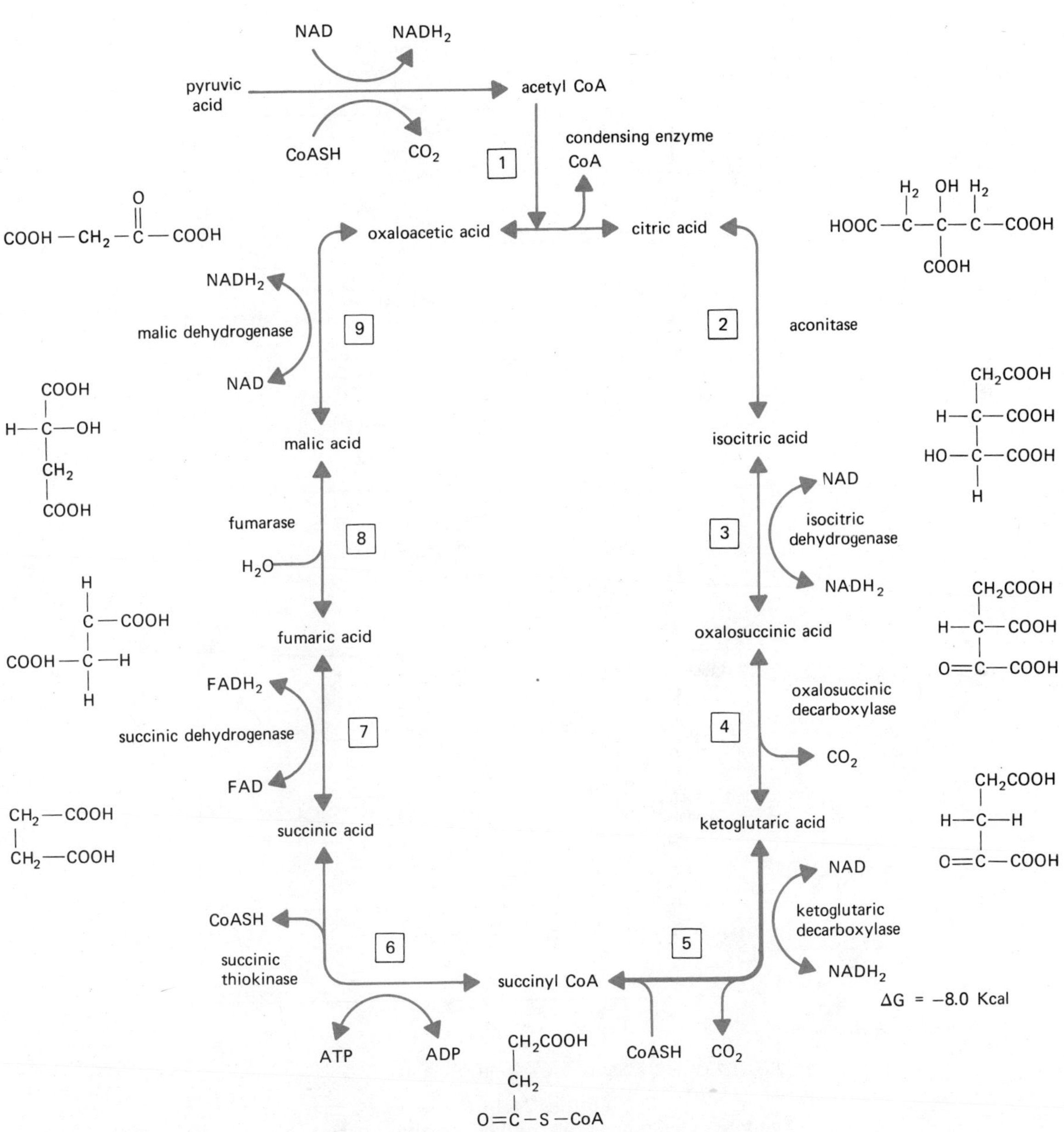

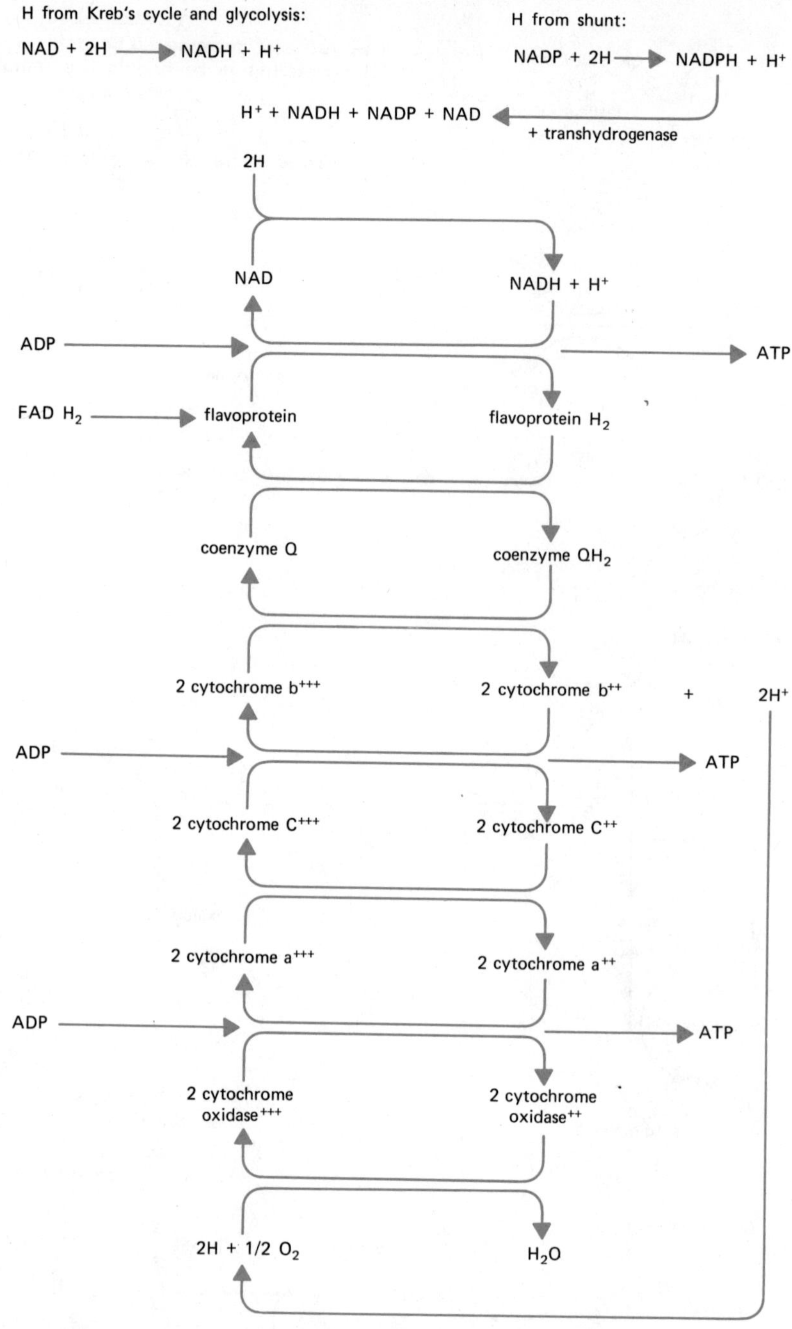

FIGURE 24.14
The fate of hydrogens released by metabolic cycles.

pyruvics resulting from glycolysis, with a total production of 16 H, 2 ATP, and 4 CO_2 molecules.

FATE OF HYDROGENS RELEASED IN THE VARIOUS CYCLES

Examination of the schemes of glycolysis, pyruvic → acetyl CoA transformations, and Krebs cycle discloses that a total of 24 hydrogens have been liberated from one molecule of glucose (4 from glycolysis, 4 from pyruvic acid → acetyl CoA transformations, and 16 by two revolutions of Krebs). The hydrogens are picked up by molecules known as hydrogen acceptors, described earlier.

Hydrogens released by glycolysis and the pyruvic → acetyl CoA transformation are picked up by NAD, as are 6 of the 8 released by one revolution of Krebs cycle. FAD accepts hydrogens only from step 7 of the Krebs cycle. Remembering that to combust 2 pyruvic acid molecules requires two revolutions of the Krebs cycle, we can state that 20 of the 24 hydrogens produced are accepted by NAD, and 4 by FAD. The hydrogens are then carried to still another cycle known as OXIDATIVE PHOSPHORYLATION (*see* Fig. 24.14), wherein hydrogen reacts with oxygen to form water, and releases energy for ATP formation.

Protons (H^+ gathered from glycolysis, Krebs cycle, beta oxidation, and other sources) are picked up by large molecules known as hydrogen acceptors. The acceptor in turn has the protons and their accompanying electrons catalytically removed. The electrons are then transported over a series of coupled oxidation-reduction reactions utilizing compounds known as cytochromes. The electrons are ultimately transferred to molecular oxygen, and the hydrogens combine with oxygen to form water. Energy released during the electron transfer is used to phosphorylate (add phosphate to) ADP, creating ATP. Since the energy is stored by synthesis of ATP, the overall coupled process is termed oxidative phosphorylation (Fig. 24.14). If oxygen is not present as the final acceptor in this system, the production of energy (ATP) will be insufficient to sustain life.

We may note that each compound on the left side of the scheme accepts hydrogens (or electrons), and each one on the right has hydrogen (or electrons) on it. Also, if hydrogens enter this scheme on NAD, they begin at the top and thus give rise to 3 ATP molecules for each pair of hydrogens. Remembering that 20 hydrogens arrive at the scheme on NAD, we can calculate that 30 ATP molecules (20 H = 10 pairs; 3 ATP per pair or 3 × 10) will be produced. Four hydrogens come to this cycle on FAD. These give rise to 2 ATP per pair, a total of 4 ATP. A total of 34 ATP are thus produced by oxidative phosphorylation from the hydrogens released in other cycles.

Recalling that the starting material was glucose, and remembering that 2 ATP were produced by glycolysis and 2 by the Krebs cycle in combusting 2 acetyl CoA molecules, we can conclude that the complete combustion of a molecule of glucose will create a net total of 38 ATP molecules to the body. These 38 molecules represent a storage of 292.6 kcal of energy (38 × 7.7 kcal per phosphate bond) in the body cells. Because there is inherent in a glucose molecule 686 kcal of energy, the trapping represents an efficiency of 42.6 percent $\left(\frac{292.6}{686}\text{ kcal} \times 100\right)$ – a good figure. Recall again the statement that a main aim of combusting foodstuffs is to release energy for ATP synthesis. The combustion of a single glucose molecule creates a significant increase in the body's store of ATP, if oxygen is present.

ALTERNATIVE PATHWAYS OF GLUCOSE METABOLISM

Two additional metabolic pathways exist for the degradation of glucose. The DIRECT OXIDATIVE PATHWAY (hexose monophosphate shunt or HMS) degrades glucose into 5-carbon sugars, glyceraldehyde, a 2-carbon fragment known as active glycolaldehyde, and utilizes NADP as the hydrogen acceptor. The 5-carbon sugars are utilized in nucleic acid synthesis. $NADPH_2$ serves as the hydrogen donor for fatty acid synthesis. The reactions of the direct pathway are shown in Fig. 24.15.

The GLUCURONIC ACID PATHWAY forms UDPG for utilization in galactose → glucose conversion and glycogenesis, and creates glu-

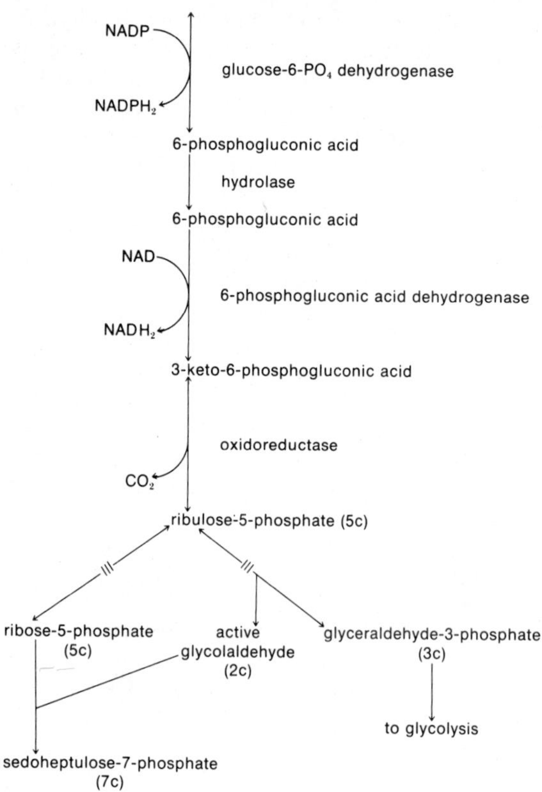

FIGURE 24.15
The direct pathway of glucose oxidation; the hexose monophosphate shunt.

curonic acid for utilization into connective tissue ground substance. This is shown in Fig. 24.16.

Of the two pathways, the shunt must be regarded as more important as a metabolic pathway for glucose degradation. It is an important route of glucose metabolism in the lactating mammary gland. It can handle up to 30 percent of the glucose metabolism in other cells.

LIPID METABOLISM

Digestion of fats in the intestine results in the production of fatty acids and glycerol as the major end products. Absorption of these end products occurs primarily by passive processes, and the products are carried from the gut by way of the lymphatics (lacteals) to venous blood. As the fatty acids and glycerol pass through the intestinal mucosa, they are resynthesized into triglycerides (3 fatty acid molecules plus 1 glycerol molecule) and enter the lacteal as tiny droplets of fat known as CHYLOMICRONS. Coalescence of the chylomicrons is prevented by adsorption of lacteal protein upon the fat, creating a lipoprotein. The chylomicrons are then transferred to the venous circulation via the thoracic duct. After consumption of a meal rich in fat, the fat content of the blood may rise to where the blood becomes "milky." A LIPEMIA has thus occurred. Over a period of 4 to 6 hours, the lipemia will disappear. Disappearance is due to absorption of chylomicrons by liver cells and subsequent metabolism, and to the action of a plasma enzyme, LIPOPROTEIN LIPASE (*clearing factor*), that hydrolyzes the fats again into free fatty acids and glycerol. The acids and glycerol may then enter storage depots (adipose tissue) in the body. The chief storage areas are in the subcutaneous tissue (*panniculus adiposus*), in the omentum, and in muscles.

THE FATE OF GLYCEROL

Glycerol, a 3-carbon compound, is easily converted to glyceraldehyde-3-phosphate, a compound found in the scheme of glycolysis, as shown in Fig. 24.17.

The reactions are reversible, and because the compound glyceraldehyde-3-phosphate is common to the metabolism of two different foodstuff groups, it forms a method of linking the two schemes. In short, it is possible to convert lipid sources to carbohydrate and vice versa.

The ultimate fate of the glyceraldehyde will be either conversion to glucose or metabolism via glycolysis. Formation of glucose from *any* noncarbohydrate precursor is called *gluconeogenesis*.

THE FATE OF FATTY ACIDS

Fatty acids undergo synthesis into triglycerides in depot areas, or they undergo degradation via

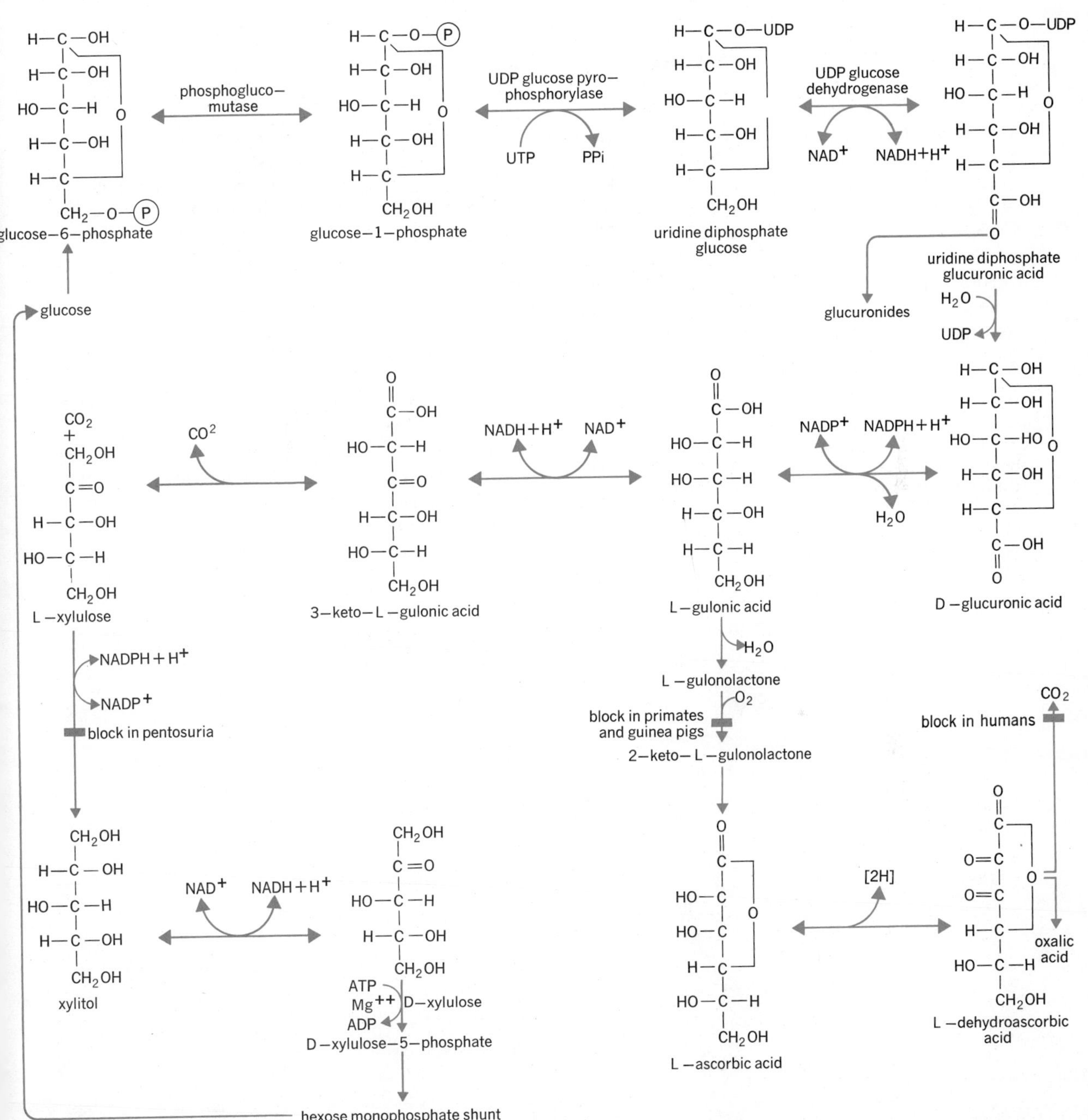

FIGURE 24.16
The glucuronic acid pathway.

$$\text{glycerol} \xrightleftharpoons[\text{phosphatase}]{} \text{dihydroxyacetone phosphate} \xrightleftharpoons[\text{triose isomerase}]{} \text{glyceraldehyde-3-phosphate}$$

FIGURE 24.17
The conversion of glycerol to glyceraldehyde.

a process known as BETA OXIDATION.* In general terms, the process involves the degradation of a fatty acid by the progressive removal of 2-carbon segments in the form of acetyl CoA. Five steps occur, as shown in Fig. 24.18.

Acetyl CoA is a compound found in the carbohydrate metabolism scheme at the beginning of the Krebs cycle. Again, a compound linking two foodstuff metabolism schemes is present. Accordingly, AcCoA formed from fats may enter the Krebs cycle for combustion or be synthesized into glucose.

Glucose that is not stored as glycogen is converted to lipids. The liver and muscles contain only a day or two energy supply of glycogen. About two weeks energy supply is stored in adipose tissue.

SYNTHESIS OF FATS (TRIGLYCERIDES)

To synthesize a triglyceride, glycerol and fatty acids are required. Glycerol may be made from glyceraldehyde-3-phosphate. Fatty acids are synthesized by two routes.

BETA REDUCTION is a reversal of the steps in beta oxidation. $NADPH_2$ from the shunt provides the source of hydrogen needed to create the fatty acid (the step when FAD picked up H_2). This pathway is found in the mitochondria.

* A fatty acid is an even-numbered (16 or 18) chain of carbon atoms. At one end is a carboxyl (–COOH) group. Carbons in the chain are named alpha, beta, gamma, delta, etc., carbons, commencing with the first carbon beyond the carboxyl carbon. The changes occurring in beta oxidation take place on the beta carbon.

FIGURE 24.18
Beta oxidation, the degradation of a fatty acid.

1. $R{-}CH_2{-}CH_2{-}\overset{O}{\overset{\|}{C}}{-}OH + CoA{-}SH \xrightarrow[Mg^{++}]{ATP\ +\ thiokinase} R{-}(CH_2)_2{-}\overset{O}{\overset{\|}{C}}{\sim}SCoA + AMP + PP^{*}$

*pyrophosphate

2. $R{-}(CH_2)_2{-}\overset{O}{\overset{\|}{C}}{\sim}SCoA \xrightleftharpoons[FAD \longrightarrow FADH_2]{\text{acyl dehydrogenase}} R{-}CH{=}CH{-}\overset{O}{\overset{\|}{C}}{\sim}SCoA$

3. $R{-}CH{=}CH{-}\overset{O}{\overset{\|}{C}}{\sim}SCoA \xrightleftharpoons{\text{crotonase}} R{-}\overset{OH}{\overset{|}{C}H}{-}CH_2{-}\overset{O}{\overset{\|}{C}}{\sim}SCoA$

4. $R{-}\overset{OH}{\overset{|}{C}H}{-}CH_2{-}\overset{O}{\overset{\|}{C}}{\sim}SCoA \xrightleftharpoons[NAD \longrightarrow NADH_2]{\text{dehydrogenase}} R{-}\overset{O}{\overset{\|}{C}}{-}CH_2{-}\overset{O}{\overset{\|}{C}}{\sim}SCoA$

beta-keto fatty acid

5. $R{-}\overset{O}{\overset{\|}{C}}{-}CH_2{-}\overset{O}{\overset{\|}{C}}{\sim}SCoA + CoA{-}SH \xrightarrow{\text{thiolase}} \boxed{R{-}\overset{O}{\overset{\|}{C}}{-}SCoA} + CH_3{-}\overset{O}{\overset{\|}{C}}{\sim}SCoA$

acetyl CoA

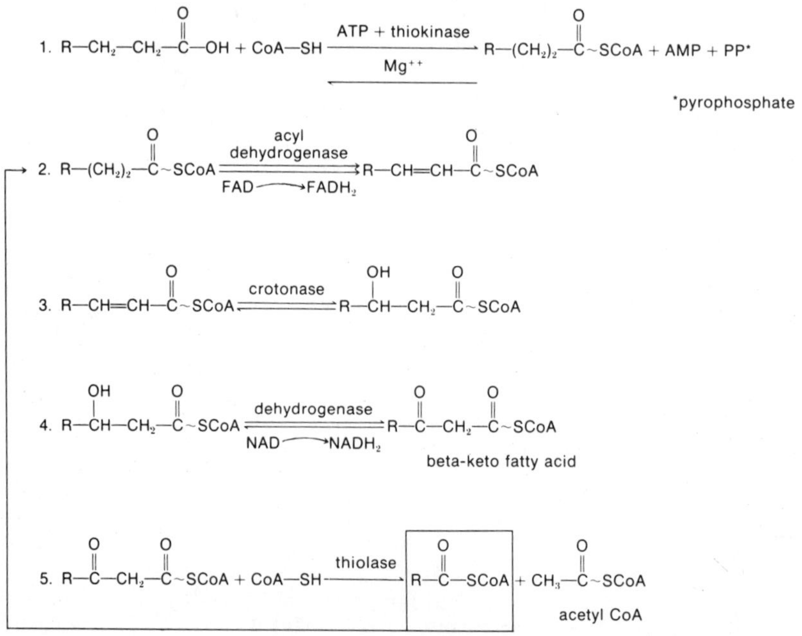

A CYTOPLASMIC SYSTEM, independent of the mitochondria, carboxylates (adds CO_2) to acetyl CoA to form a 3-carbon unit, malonyl CoA. Malonyl CoA is then added to a fatty acid chain (even number of carbons) and, during the addition, CO_2 is removed. A beta-keto fatty acid results. From here, reversal of steps 4 to 1 in beta oxidation occurs.

PHOSPHOLIPID AND OTHER COMPLEX LIPID METABOLISM

The major phospholipids consist of glycerol, two fatty acids, and a phosphate that may be present by itself or in combination with choline or ethanolamine.

If phosphate alone is present, phosphatidic acid is formed.

If phosphate and choline are present, lecithin is formed.

If phosphate and ethanolamine are present, cephalin is formed.

Sphingomyelins are compounds of fatty acids, phosphate choline, and an alcohol sphingosine. Cerebrosides are composed of sphingosine, fatty acid, and galactose. The last two major categories of materials are important constituents of the nervous system. Degradation of these substances is accomplished by appropriate enzyme systems that reduce the materials to their constituent building blocks.

In certain disease states, there may be defective formation or degradation of these materials.

MULTIPLE SCLEROSIS is associated with excessive loss of phospholipids and sphingolipids from myelinated portions of the central nervous system.

In GAUCHER'S DISEASE, NIEMANN-PICK DISEASE, and TAY-SACHS DISEASE, deposition of abnormal phospholipids occurs in the cells of the cerebral cortex, spleen, liver, and lymph nodes. A disorder of synthesis is suspected. All of these conditions are characterized by developmental and mental retardation, changes in neurons, and enlargement of liver and spleen.

Lipoproteins have attracted considerable attention recently because of their involvement in atherosclerosis and coronary heart disease. Four major groups of lipoproteins have been separated and identified by electrophoresis and ultracentrifugation. These are commonly named according to their densities, as measured in the ultracentrifuge, and are referred to as chylomicrons (lowest density), very low density lipoproteins (VLDL), low density lipoproteins (LDL), and high density lipoproteins (HDL). Increase in density is associated with an increase in protein content of the lipoprotein.

VLDL are formed primarily in the liver, and chylomicrons and some VLDL are formed in intestinal mucosa cells, perhaps in response to the fat load being absorbed. LDL and HDL synthesis areas are not known, but they may be metabolic products of chylomicrons and VLDL. The term *apolipoprotein* (apoproteins) is applied to the proteins that combine with the lipids. They are relatively few in number, and differ from one another in their terminal amino acids.

If lipoprotein levels are low in the bloodstream, a HYPOLIPOPROTEINEMIA develops. In general, this is associated with elevated levels of free fatty acids (FFA) or triglycerides in the blood, and with depressed cholesterol levels in the blood. High FFA levels may contribute to the development of atherosclerosis, a disease associated with lipid deposition in arterial walls and development of an abnormal smooth muscle cell.

Three inherited *hypo*lipoproteinemias are described.

TANGIER DISEASE (HDL deficiency) is inherited as an autosomal recessive condition, and is associated with high blood triglyceride levels and low cholesterol levels.

ABETALIPOPROTEINEMIA (LDL, VLDL, and chylomicron absence) is associated with malabsorption of fats, low serum triglyceride levels, and "thorny" appearing erythrocytes (*acanthocytosis*). It is inherited as an autosomal recessive characteristic.

HYPOBETALIPOPROTEINEMIA (LDL deficiency, normal HDL, VLDL slightly reduced), is inherited as an autosomal dominant characteristic, and causes only mild disturbances in body function.

The *hyper*lipidemias and *hyper*lipoproteinemias are a greater threat because they indicate a greater fat load in the bloodstream. In these conditions, blood cholesterol and triglyceride levels are elevated with a potentially greater possibility for deposition in blood vessel walls. Cause may be genetic, or secondary to hypothyroidism, excessive intake of lipids in the diet, alcohol consumption, and use of contraceptive "pills." Normal values for triglycerides, cholesterol, and lipoproteins in the blood may be given as:

in the vessel wall. Smooth muscle cells accumulate LDL, forming what are called "foam cells." In tissue culture, such cells are stimulated to proliferate by LDL, and to secrete elastic fiber protein, collagen and protein-polysaccharides that further contribute to plaque formation.

CHOLESTEROL METABOLISM

Some 1.3 grams of cholesterol is added to the body economy each day, with approximately 1 gram of that being synthesized and the remainder entering via the diet. It is eliminated from the body by way of the bile acids and in the feces.

Synthesis of cholesterol occurs from acetyl

	TRIGLYCERIDES (mg/dL)	CHOLESTEROL (mg/dL)	LIPOPROTEINS (ACTUAL AMOUNTS VARY, BUT RELATIVE % OF TOTAL IS FAIRLY CONSTANT)—RELATIVE %			
			CHYLOMICRONS	HDL	LDL	VLDL
Adult	10–150	130–250	0–2	10–38	38–74	1–37
Child	10–140	120–230	Levels in children are about 1/3 that of the adult, and reach adult values about 13 yrs. of age.			
Infant	10–140	45–170				

CoA, followed by ring formation to give the characteristic four-ring formation of the compound.

The suggestion has been made that hyperlipidemia results where there is obesity, a high dietary intake of lipids, and a genetic predisposition for elevated blood lipid levels. Dietary restriction of both lipid and carbohydrate reduces the effect of the first two factors, and minimizes the operation of the third.

Atherosclerosis cannot be guaranteed not to occur if blood lipid levels are kept low; neither is there an upper limit beyond which it is inevitable. The changes characteristic of atherosclerosis begin early in life with "fatty streaks" accumulating within the inner layer of a blood vessel wall. This accumulation is universal, but is reversible. Continued accumulation of fats in the vessel wall leads to the formation of a fatty plaque that encroaches on the vessel lumen, reducing blood flow. Progression to plaque formation can be slowed or stopped by dietary measures. An additional factor in plaque development is proliferation of smooth muscle cells

Dietary cholesterol is absorbed from the intestine into the lacteals and forms chylomicrons. From the blood it is removed chiefly by the liver and is incorporated into bile acids. Thyroid hormone speeds the removal of cholesterol from the blood and its incorporation into bile acids. In hypothyroidism, blood cholesterol levels are higher than normal.

Cholesterol has been implicated in the genesis of coronary heart disease and atherosclerosis. Elevated blood cholesterol levels contribute to deposition of the lipid in the walls of blood vessels (atherosclerosis), narrowing their lumina and decreasing blood flow beyond the area of deposition. If narrowing of the vessels of the heart occurs, diminished coronary flow and heart attacks may result. The substitution of unsaturated fats for cholesterol in the diet has been advanced as a possible remedy for high blood cholesterol and as a means to slow deposition of fat in blood vessel walls. These fats are supposed to hasten the metabolism of cholesterol and to slow its synthesis. Such measures may temporarily reduce blood cholesterol levels, but the body quickly resumes its production of cholesterol, essential for synthesis of hormones and bile acids.

ESSENTIAL FATTY ACIDS

Certain fatty acids present in the body contain —C=C— bonds (double bonds) and are known as UNSATURATED FATTY ACIDS. They differ from the saturated fatty acids thus far described in that they contain less hydrogen than the saturated acids and are found in animal and vegetable oils, rather than adipose tissue. Some examples are as follows:

Palmitoleic acid
$CH_3(CH_2)_5CH{=}CH(CH_2)_7COOH$

Oleic acid
$CH_3(CH_2)_7CH{=}CH(CH_2)_7COOH$

Linolenic acid
$CH_3CH_2CH{=}CHCH_2CH{=}CHCH_2CH{=}CH(CH_2)_7COOH$

Linoleic acid
$CH_3(CH_2)_4CH{=}CHCH_2CH{=}CH(CH_2)_7COOH$

Arachidonic acid
$CH_3(CH_2)_4(CH{=}CHCH_2)_4(CH_2)_2COOH$

Of these, palmitoleic and oleic acids may be synthesized in the body and are not essential acids. The others cannot be synthesized, must be taken as such in the diet, and are thus known as essential fatty acids. These acids function in the prevention of fatty liver and aid in the degradation of cholesterol. They also may be used in the treatment of skin lesions in infants on low-fat diets.

ATP PRODUCTION DURING DEGRADATION OF FATTY ACIDS

Degradation of fatty acids produces a net gain of far more ATP to the organism than does combustion of carbohydrate. The main reason for this is the production of more acetyl CoA molecules and hydrogens than are produced during carbohydrate breakdown. Taking as an example the breakdown of stearic acid (18 carbons), we can calculate ATP production.

Four hydrogens are produced for each CoA formed in beta oxidation (see steps 2 and 4). Two of these are picked up by FAD, two by NAD.

To degrade an 18-carbon acid into 2-carbon units, 8 splits are required. Therefore, 8×4 or 32 hydrogens will be produced, 16 going to FAD, 16 to NAD.

Nine acetyl CoA molecules will be produced, which, going through the Krebs cycle, will produce 8 H and 1 ATP per molecule. Nine revolutions will take place, producing 72 H (9×8) and 9 ATP (9×1). Of the hydrogens, 18 will be picked up by FAD and 54 by NAD.

A total of 34 H (16 + 18) will be carried on FAD to oxidative phosphorylation. The 34 H on FAD will give 2 ATP per pair, or 34 ATP.

The 70 H (16 + 54) on NAD will give 3 ATP per pair, or 105 ATP. A total of 148 ATP will be produced (9 from Krebs, 34 on FAD, 105 on NAD).

FORMATION OF KETONE BODIES

The combustion of excessive amounts of fat may cause "oversaturation" of the Krebs cycle. If the Krebs cycle cannot combust the acetyl CoA as fast as it is produced, a compound known as acetoacetic acid is formed, and, from this, acetone, and beta-hydroxybutyric acid will be produced, as shown in Fig. 24.19.

In uncontrolled diabetes mellitus (sugar diabetes), KETOSIS is common due to the reliance of the body on the combustion of fats, since glucose cannot enter the cells for energy production. These acids (acetoacetic and beta-hydroxybutyric) are also threats to the body's buffering capacity. Buffering the organic acids may not leave enough buffer to handle the normal H_2CO_3 production. Hence, ketosis and ACIDOSIS are commonly seen to occur together.

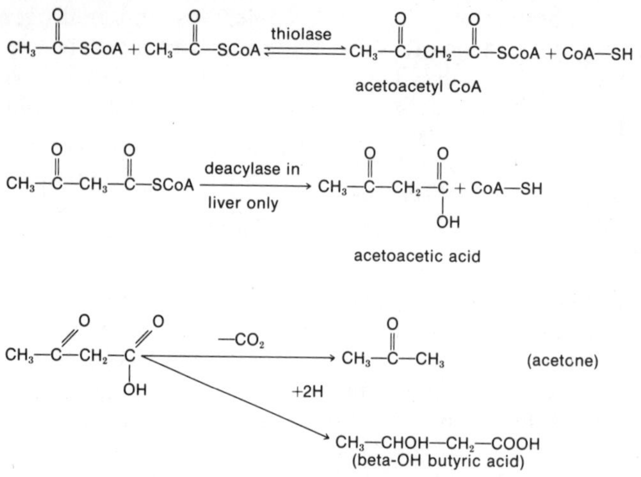

$$CH_3-\overset{O}{\overset{\|}{C}}-SCoA + CH_3-\overset{O}{\overset{\|}{C}}-SCoA \underset{}{\overset{\text{thiolase}}{\rightleftharpoons}} CH_3-\overset{O}{\overset{\|}{C}}-CH_2-\overset{O}{\overset{\|}{C}}-SCoA + CoA-SH$$

acetoacetyl CoA

$$CH_3-\overset{O}{\overset{\|}{C}}-CH_2-\overset{O}{\overset{\|}{C}}-SCoA \xrightarrow[\text{liver only}]{\text{deacylase in}} CH_3-\overset{O}{\overset{\|}{C}}-CH_2-\overset{O}{\overset{\|}{C}}(OH) + CoA-SH$$

acetoacetic acid

$$CH_3-\overset{O}{\overset{\|}{C}}-CH_2-\overset{O}{\overset{\|}{C}}-OH \xrightarrow{-CO_2} CH_3-\overset{O}{\overset{\|}{C}}-CH_3 \quad \text{(acetone)}$$

$$CH_3-\overset{O}{\overset{\|}{C}}-CH_2-\overset{O}{\overset{\|}{C}}-OH \xrightarrow{+2H} CH_3-CHOH-CH_2-COOH$$

(beta-OH butyric acid)

FIGURE 24.19
The formation of the ketone bodies.

AMINO ACID AND PROTEIN METABOLISM

REACTIONS OF AMINO ACIDS

Amino acids that are taken into cells may suffer one of two fates: they may be synthesized into proteins, or they may undergo reactions that lead ultimately to their degradation or conversion into other compounds.

PROTEIN SYNTHESIS

Synthesis of proteins was discussed in Chapter 4. The five basic steps are summarized below.

By the process of TRANSCRIPTION, nuclear DNA is used to synthesize m-RNA and t-RNA.

m-RNA moves to the ribosomes, and t-RNA moves to the cytoplasm.

t-RNA attaches to activated amino acids and carries them to the m-RNA "on" a ribosome.

The ribosome moves along the m-RNA, attracting t-RNA molecules with attached amino acids.

The amino acids are joined by peptide bonds to form proteins that are released into the cytoplasm, along with free t-RNA molecules.

While many cells synthesize proteins for their own specific use, the liver is the major organ that supplies the proteins circulating in the bloodstream.

AMMONIA AND THE FORMATION OF UREA

If not synthesized into proteins, amino acids undergo deamination in the liver. Deamination is a process whereby a hydrogen and the amine group of the amino acid are removed and formed into ammonia (NH_3) and a residue called an alpha-keto acid is formed (Fig. 24.20). The alpha-keto acid may be one that is a component of glycolysis or the Krebs cycle, and it may follow, either in a degradative or synthetic course, the fate of the molecules of those cycles. For example, the relationships of several amino acids to components of glycolysis and the Krebs cycle are

$$\left.\begin{array}{l}\text{glycine} \xrightarrow[+\frac{1}{2}O_2]{-NH_3} \text{acetic acid} \\ \text{alanine} \xrightarrow[+\frac{1}{2}O_2]{-NH_3} \text{pyruvic acid}\end{array}\right\} \text{glycolysis and conversion to Krebs cycle compound}$$

$$\left.\begin{array}{l}\text{glutamic acid} \xrightarrow[+\frac{1}{2}O_2]{-NH_3} \alpha-\text{ketoglutaric acid} \\ \text{aspartic acid} \xrightarrow[+\frac{1}{2}O_2]{-NH_3} \text{oxaloacetic acid}\end{array}\right\} \text{Krebs cycle}$$

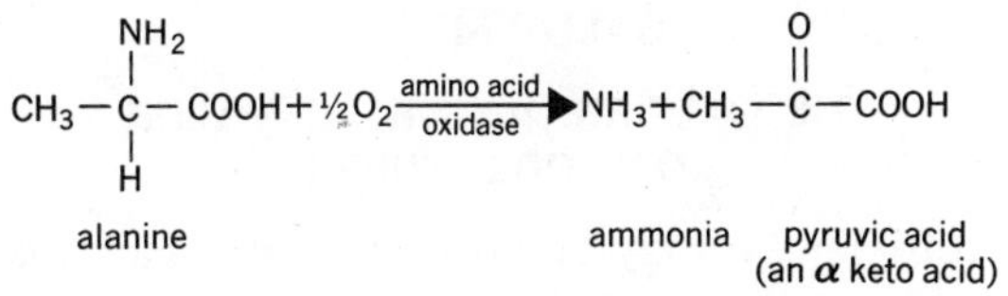

$$CH_3-\underset{H}{\overset{NH_2}{C}}-COOH + \tfrac{1}{2}O_2 \xrightarrow[\text{oxidase}]{\text{amino acid}} NH_3 + CH_3-\overset{O}{\overset{\|}{C}}-COOH$$

alanine — ammonia — pyruvic acid (an α keto acid)

FIGURE 24.20
The deamination of an amino acid.

The ammonia produced by deamination is an extremely toxic substance that would require a large volume of fluid to dilute it effectively to nontoxic levels for excretion. The problems of toxicity and water conservation are solved by way of the ORNITHINE CYCLE (Fig. 24.21). This cycle, which occurs in the liver, involves the formation of urea, which is less toxic than ammonia. Two hereditary disorders of the cycle are recognized.

CITRULLINURIA is characterized by elevated blood, urine, and cerebrospinal fluid levels of citrulline with occurrence of mental retardation, alkalosis, and disturbance of blood ion levels. Although the exact site of the defect is not known, a deficiency of condensing enzyme is the result.

ARGINOSUCCINIC ACIDURIA is characterized by accumulation of arginosuccinic acid in the blood and urine. A deficiency of arginosuccinase is the cause.

If certain alpha-keto acids are available, synthesis of the corresponding amino acids can take place by the process of TRANSAMINATION. This process involves the addition of an amine group, derived from an amine donor, to the keto acid. The nonessential amino acids (*see* Fig. 3.1) are synthesized in the body in this manner.

Because specific enzymes are required for the metabolism of amino acids, the possibility of genetically determined disorders of metabolism exists.

MAPLE SYRUP URINE DISEASE. Leucine, valine, and isoleucine, all branched-chain amino acids, fail to be decarboxylated during their metabolism and appear in the urine, giving it a characteristic maple syrup odor. A decarboxylase enzyme is deficient.

In metabolism of histidine, HISTIDINEMIA results when deficiency of histidinase is present.

In the metabolism of phenylalanine and tyrosine, several enzyme defects are possible (Fig. 24.22).

ALKAPTONURIA results when homogentisic acid oxidase is deficient and homogentisic acid is excreted in the urine. The urine oxidizes and darkens on exposure to air.

FIGURE 24.21
The formation of urea; the ornithine cycle.

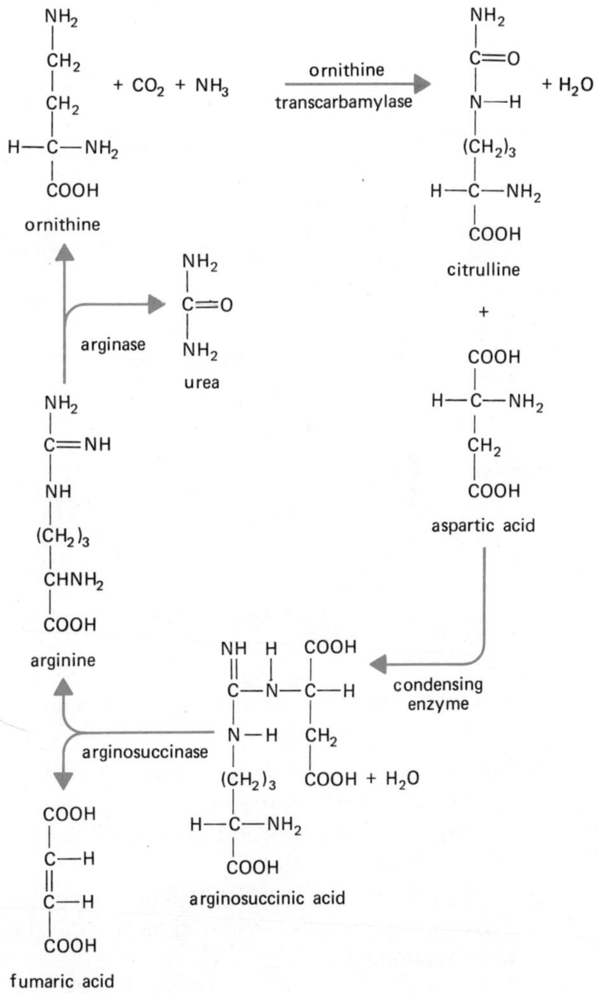

the metabolism of phenylalanine and tyrosine

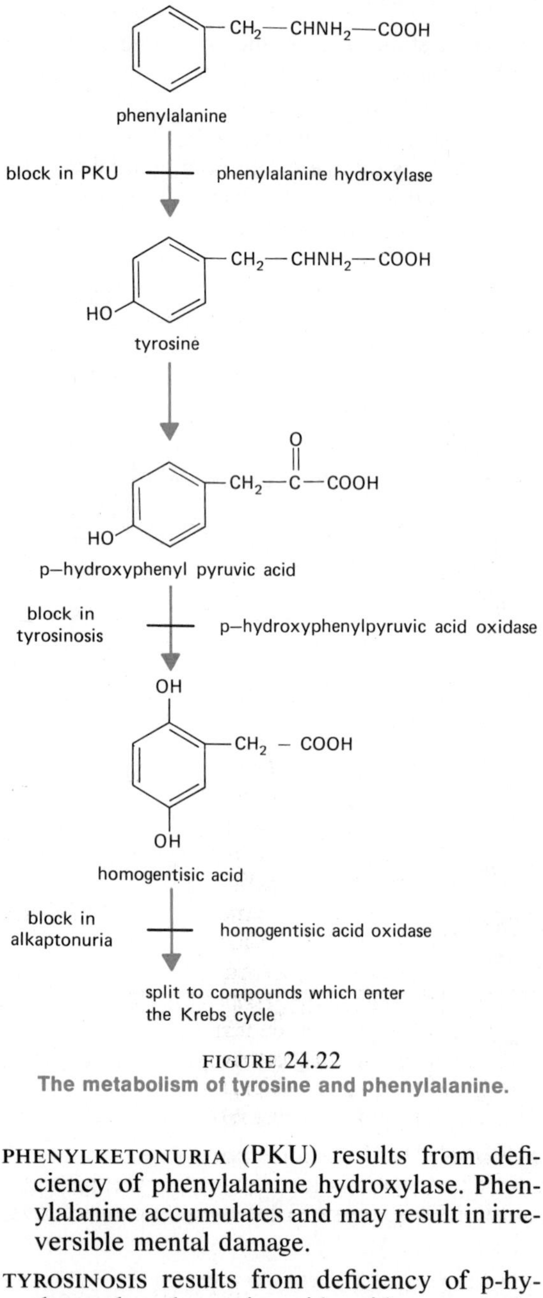

FIGURE 24.22
The metabolism of tyrosine and phenylalanine.

PHENYLKETONURIA (PKU) results from deficiency of phenylalanine hydroxylase. Phenylalanine accumulates and may result in irreversible mental damage.

TYROSINOSIS results from deficiency of p-hydroxyphenylpuruvic acid oxidase, and the acid accumulates.

CALORIMETRY

THE CALORIE AND HEAT VALUE OF FOODSTUFFS

To review, it may be useful to state again that energy is most easily measured by the large Calorie (kilocalorie), which is that amount of heat necessary to raise the temperature of 1 kg of water 1°C. To determine the heat content of any given amount of a foodstuff, it may be ignited or combusted in a bomb calorimeter, and the heat liberation is measured directly. Such determinations made for carbohydrate, fat, and protein show heat values of 4.1, 9.4, and 5.6 kcal/g of substance. Within the body, protein is not completely combusted, the urea being excreted. Because urea has a small energy content that is not available for use, the figure of 5.6 kcal/g must be reduced to 4.1 kcal/g of protein. For ease of calculation, the heat values are commonly rounded off to 4 kcal/g carbohydrates, 9 kcal/g fat, and 4 kcal/g protein.

DIRECT AND INDIRECT CALORIMETRY

Because all of the energy produced by body activity is ultimately dissipated as heat, measurement of heat production by a living organism can give clues as to the amount of activity occurring and the caloric requirements that must be met to keep the animal in a state of caloric equilibrium. In short, caloric intake must balance output. Direct calorimetry involves placing an animal in a chamber similar to that shown in Fig. 24.23, measuring its heat production, O_2 consumption, and CO_2 production, and determining the energy content of feces and urine. The apparatus is expensive and cumbersome. Indirect calorimetry is based on the fact that the combustion of foodstuffs is attended by a more or less fixed requirement for oxygen to be used in the combustion, and by the production of a fixed amount of CO_2 by the combustion. Knowing the volume of O_2 required and the volume of CO_2 produced, one may calculate the RESPIRATORY QUOTIENT (RQ). $RQ = CO_2/O_2$.

Carbohydrate (glucose) combustion may be represented as

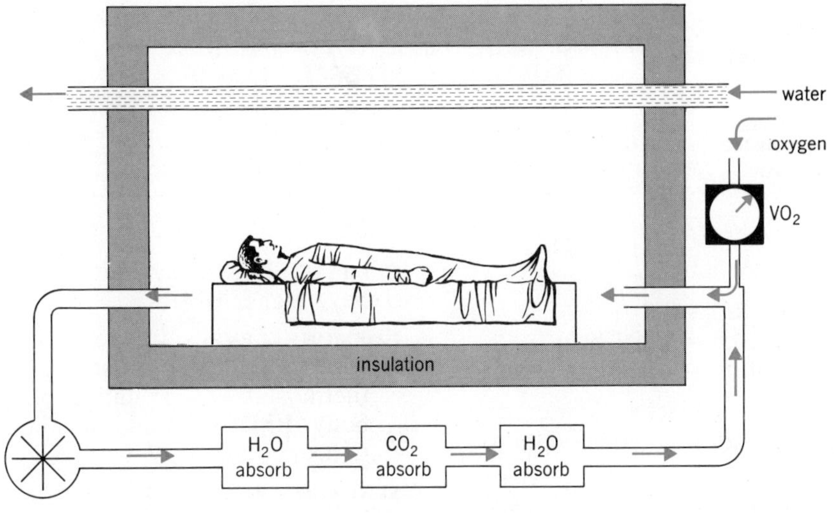

FIGURE 24.23
A diagram illustrating the basics of direct calorimetry.

$$C_6H_{12}O_6 + 6\ O_2 \rightarrow 6\ CO_2 + 6\ H_2O$$
glucose

$$RQ = \frac{6}{6} = 1$$

Fats, having less oxygen in the molecule, require more from the outside in order to liberate CO_2.

$$2\ C_{57}H_{110}O_6 + 163\ O_2 \rightarrow 114\ CO_2 + 110\ H_2O$$
tristearin

$$RQ = 114/163 = 0.70$$

Proteins have a calculated average RQ of 0.8. Because of unknown structure, most proteins cannot be specifically represented as to O_2 and CO_2 required and produced.

The particular RQ exhibited by the body as a whole is a reflection of the relative amounts of basic foodstuffs that are being combusted. If an RQ at rest was measured as 1.0, one could assume that the foodstuff being combusted was entirely carbohydrate. A similar conclusion could be reached for fats if the resting RQ was 0.7. A normal RQ for the body is about 0.82, indicating that a mixture of carbohydrate and fat is being combusted. Protein is not normally utilized for energy to a great extent if carbohydrate and fat are present, so that one may assume that a given RQ is the result of CH_2O and fat combustion. Tables have been worked out relating the percentage of CH_2O and fat combusted at different RQs. Table 24.3 presents such data.

Knowing what the RQ is, we can next estimate a quantity known as the CALORIC EQUIVALENT OF OXYGEN. It represents the number of calories produced by fuel combustion per liter of O_2 at a given RQ. This quantity, once determined, will enable estimation of caloric production if only O_2 consumption is known, and thus we can express energy production and requirements in terms of calories. Again, tables have been produced relating oxygen equivalents to calories. Table 24.4 presents such data.

A determination of heat production by the indirect method would therefore require the following information:

The RQ under the particular conditions specified.

The caloric equivalent of O_2 at that RQ.

The liters of O_2 consumed during the time the experiment was proceeding.

TABLE 24.3
THE PERCENTS OF CARBOHYDRATE AND FAT COMBUSTED AT DIFFERENT RQs

Nonprotein RQ	1.0	0.95	0.9	0.85	0.8	0.75	0.71
Percent CH_2O combusted	100	82	65	47	29	11	0
Percent fat combusted	0	18	35	53	71	89	100

As a sample calculation:

RQ = 0.8
caloric equivalent at 0.8 = 4.801 kcal/L
O_2 consumption = 20L/hr

Question: What is subject's caloric output per day, assuming an O_2 consumption of 20L/hr?

O_2 consumption of 20L/hr = 20 × 24 hr or 480L/day
at RQ of 0.8, 1L of O_2 = 4.801 kcal/L
480L/day × 4.801 kcal/L = 2304.48 kcal/day

THE BASAL METABOLIC RATE

If determinations, as have just been illustrated, were to be taken with the subject at complete mental and physical rest, at least 12 hours after the last meal, the resulting number of calories could be said to represent the subject's basal metabolic rate (BMR), or the rate of calories produced to maintain all basic body processes, extra activity eliminated. Actual BMR is determined by several factors.

AGE. There is a progressive decrease in rate associated with aging.

SEX. Females have about 10 percent lower BMR than equivalent-sized males. This represents a smaller percentage of muscle in the female body.

RACE. Orientals have been shown to have a slightly lower BMR than Caucasians.

EMOTIONAL STATE. Anxiety or stress elevates BMR.

CLIMATE. Individuals adapted to cold climates have higher BMRs.

HORMONE LEVELS in bloodstream. Thyroxin appears to be a substance that "sets the thermostat" of oxidative processes in the body. BMR rises and falls in direct proportion to circulating levels of this substance.

SURFACE AREA. Caloric expenditure appears best related to body surface, inasmuch as the skin represents a radiation surface for heat loss. Caloric release is usually expressed in kilocalories per square meter of surface per hour (kcal/m²/hr).

NUTRITIONAL CONSIDERATIONS

Maintenance of adequate daily caloric intake as well as providing proper balance of nutrients may be assured by choosing a diet that contains a daily intake of 33 to 42 g of protein, 250 to 500 g of carbohydrate, and 66 to 83 g of fat. Nutritional authorities recommend that selection be made from the categories of milk and milk products, meats, poultry and fish, vegetables and fruits, and breads and cereals. Table 24.5 presents recommendations for daily intake of calories and protein, according to several factors.

INTAKE OF SPECIFIC FOODS

Carbohydrates. After weaning, cereals (in bread and breakfast foods) and fruits and vegetables provide the major sources of carbohydrate. Sufficient carbohydrate should be

TABLE 24.4
CALORIC VALUE PER LITER OF O_2

RQ	0.707	0.75	0.80	0.85	0.90	1.0
kcal	4.686	4.739	4.801	4.862	4.924	5.047

TABLE 24.5

DAILY RECOMMENDATIONS FOR CALORIES AND PROTEIN INTAKE ACCORDING TO SEX, AGE, WEIGHT, AND HEIGHT[a]

	AGE (YEARS)	WEIGHT KG	(LB)	HEIGHT CM	(IN.)	TOTAL CALORIES	PROTEIN (G)
No difference by sex	0–1/6	4	(9)	55	(22)	kg × 120	kg × 2.2
	1/6–1/2	7	(15)	63	(25)	kg × 110	kg × 2.0
	1/2–1	9	(20)	72	(28)	kg × 100	kg × 1.8
	1–2	12	(26)	81	(32)	1100	25
	2–3	14	(31)	91	(36)	1250	25
	3–4	16	(35)	100	(39)	1400	30
	4–6	19	(42)	110	(43)	1600	30
	6–8	23	(51)	121	(48)	2000	35
	8–10	28	(62)	131	(52)	2200	40
Males	10–12	35	(77)	140	(55)	2500	45
	12–14	43	(95)	151	(59)	2700	50
	14–18	59	(130)	170	(67)	3000	60
	18–22	67	(147)	175	(69)	2800	60
	22–35	70	(154)	175	(69)	2800	65
	35–55	70	(154)	173	(68)	2600	65
	55–75+	70	(154)	171	(67)	2400	65
Females	10–12	35	(77)	142	(56)	2250	50
	12–14	44	(97)	154	(61)	2300	50
	14–16	52	(114)	157	(62)	2400	55
	16–18	54	(119)	160	(63)	2300	55
	18–22	58	(128)	163	(64)	2000	55
	22–35	58	(128)	163	(64)	2000	55
	35–55	58	(128)	160	(63)	1850	55
	55–75+	58	(128)	157	(62)	1700	55
	Pregnant					+200	65
	Lactating					+1000	75

[a] The table assumes sufficient calories to permit normal growth in subadults and to maintain weight in an adult. Normal activity is assumed.

provided to assure normal weight gain or maintenance according to the individual's activity levels. Table 24.6 is presented as a guide to desirable adult weight.

Protein. As structural materials, components of enzymes, and contractile tissues, proteins assume primary importance in the body economy. Egg and milk sources provide the most complete proteins in terms of essential and nonessential amino acids, closely followed by meats, poultry, and fish. Plant proteins are deficient in some amino acids or contain too low an amount to sustain optimum nutrition. Determination of the nitrogen balance of the individual (amount of nitrogen intake as compared to excretion) may indicate if protein intake is adequate. (Starvation is associated with greater output than intake; growth, just the reverse.)

Fats. Fats form important energy sources and aid the absorption of fat-soluble vitamins (A, D, E, K). Intake must include the essential fatty acids to ensure proper metabolism of other lipids. Evidence is gathering to suggest that heavy fat intake early in life results in the formation of larger-than-normal numbers of adipose cells that accumulate fat and may lead to obesity. Insufficient intake of fats may retard

TABLE 24.6
DESIRABLE WEIGHTS IN POUNDS

	HEIGHT FT.	HEIGHT IN.	SMALL BONES	MEDIUM BONES	HEAVY BONES
Male (in typical indoor clothing; shoes with 1-in. heels)	5	2	112–120	118–129	126–141
	5	3	115–123	121–133	129–144
	5	4	118–126	124–136	132–148
	5	5	121–129	127–139	135–152
	5	6	124–133	130–143	138–156
	5	7	128–137	134–147	142–161
	5	8	132–141	138–152	147–166
	5	9	136–145	142–156	151–170
	5	10	140–150	146–160	155–174
	5	11	144–154	150–165	159–179
	6	0	148–158	154–170	164–184
	6	1	152–162	158–175	168–189
	6	2	156–167	162–180	173–194
	6	3	160–171	167–185	178–199
	6	4	164–175	172–190	182–204
Female (in typical indoor clothing in shoes with 2-in. heels)	4	10	92– 98	96–107	104–119
	4	11	94–101	98–110	106–122
	5	0	96–104	101–113	109–125
	5	1	99–107	104–116	112–128
	5	2	102–110	107–119	115–131
	5	3	105–113	110–122	118–134
	5	4	108–116	113–126	121–138
	5	5	111–119	116–130	125–142
	5	6	114–123	120–135	129–146
	5	7	118–127	124–139	133–150
	5	8	122–131	128–143	137–154
	5	9	126–135	132–147	141–158
	5	10	130–140	136–151	145–163
	5	11	134–144	140–155	149–168
	6	0	138–148	144–159	153–173

myelination of nerves and cause retarded neural development.

Vitamins. Vitamins are substances essential to normal cellular metabolism. None are synthesized by human cells in the body, which thus depend on dietary intake. Two groups of vitamins are recognized.

FAT-SOLUBLE VITAMINS (A,D,E,K) are ingested and absorbed with fats in the diet. WATER-SOLUBLE VITAMINS (B complex, C, folic acid, pantothenic acid, biotin) are widely distributed in foods, but are easily destroyed by heat and oxidation. Table 24.7 presents facts related to the vitamins and certain other food factors considered vital to normal cellular function.

Inorganic substances. Fourteen elements of the periodic table have been shown to be essential for health. Table 24.8 presents information concerning these elements.

(*Text continued on page 469.*)

TABLE 24.7

VITAMINS

DESIGNATION LETTER AND NAME	MAJOR PROPERTIES	REQUIREMENT PER DAY	MAJOR SOURCES	METABOLISM	FUNCTION	DEFICIENCY SYMPTOMS
A–Carotene	Fat soluble yellow crystals, easily oxidized	5000 I.U.	Egg yolk, green or yellow vegetables and fruits	Absorbed from gut; bile aids, in liver	Formation of visual pigments; maintenance of normal epithelial structure	Night blindness, skin lesions
D_3–Calciferol	Fat soluble needlelike crystals, very stable	400 I.U. much made through irradiation of precursors in skin	Fish oils, liver	Absorbed from gut; little storage	Increase Ca absorption from gut; important in bone and tooth formation	Rickets (defective bone formation)
E–Tocopherol	Fat soluble yellow oil, easily oxidized	Not known for humans	Green leafy vegetables	Absorbed from gut; stored in adipose and muscle tissue	Humans–maintain resistance of red cells to hemolysis Animals–maintain normal course of pregnancy	Increased RBC fragility Abortion, muscular wastage
K–Naphthoquinone	Fat soluble yellow oil, stable	Unknown	Synthesis by intestinal flora; liver	Absorbed from gut; little storage; excreted in feces	Enables prothrombin synthesis by liver	Failure of coagulation
B_1–Thiamine	Water soluble white powder, not oxidized	1.5 mg	Brain, liver, kidney, heart; whole grains	Absorbed from gut; stored in liver, brain, kidney, heart; excreted in urine	Formation of cocarboxylase enzyme involved in decarboxylation (Krebs cycle)	Stoppage of CH_2O metabolism at pyruvate, beriberi, neuritis, heart failure, mental disturbance
B_2–Riboflavin	Water soluble; orange-yellow powder; stable except to light and alkalies	1.5–2.0 mg	Milk, eggs, liver, whole cereals	Absorbed from gut; stored in kidney, liver, heart; excreted in urine	Flavoproteins in oxidative phosphorylation (hydrogen transport)	Photophobia; fissuring of skin
Niacin	Water soluble; colorless needles; very stable	17–20 mg	Whole grains	Absorbed from gut; distributed to all tissues; 40% excreted in urine	Coenzyme in H transport, (NAD, NADP)	Pellagra; skin lesions; digestive disturbances, dementia

TABLE 24.7 (continued)

VITAMINS

DESIGNATION LETTER AND NAME	MAJOR PROPERTIES	REQUIREMENT PER DAY	MAJOR SOURCES	METABOLISM	FUNCTION	DEFICIENCY SYMPTOMS
B_{12}–Cyanocobalamin	Water soluble; red crystals; stable except in acids and alkalies	2–5 mg	Liver, kidney, brain. Bacterial synthesis in gut	Absorbed from gut; stored in liver, kidney, brain; excreted in feces and urine	Nucleoprotein synthesis (RNA), prevents pernicious anemia	Pernicious anemia; malformed erythrocytes
Folic acid (Vitamin B_c, M, pteroyl glutamate)	Slightly soluble in water; yellow crystals; deteriorates easily	500 micrograms or less	Meats	Absorbed from gut; utilized as taken in	Nucleoprotein synthesis; formation of erythrocytes	Failure of erythrocytes to mature; anemia
Pyridoxine (B_6)	Soluble in water; white crystals; stable except to light	1–2 mg	Whole grains	Absorbed from gut; one-half appears in urine	Coenzyme for amino acid metabolism and fatty acid metabolism	Dermatitis; nervous disorders
Pantothenic acid	Water soluble; yellow oil; stable in neutral solutions	8.5–10 mg	?	Absorbed from gut; stored in all tissues; urine	Forms part of coenzyme A (CoA)	Neuromotor disorders, cardiovascular disorders GI distress
Biotin	Water soluble; colorless needles; stable except to oxidation	150–300 mg	Egg white; synthesis by flora of GI tract	Absorbed from gut; excreted in urine and feces	Concerned with protein synthesis, CO_2 fixation and transamination	Scaly dermatitis; muscle pains, weakness
Choline (maybe not a vitamin)	Soluble in water; colorless liquid; unstable to alkalies	500 mg	?	Absorbed from gut; not stored	Concerned with fat transport; aids in fat oxidation	Fatty liver; inadequate fat absorption
Inositol	Water soluble; white crystals	No recommended allowance	?	Absorbed from gut; metabolized	Aids in fat metabolism; prevents fatty liver	Fatty liver
Para-amino benzoic acid (PABA)	Slightly water soluble; white crystals	No evidence for requirement	?	Absorbed from gut; little storage; excreted in urine	Essential nutrient for bacteria; aids in folic acid synthesis	No symptoms established for humans
Ascorbic acid (Vitamin C)	Water soluble; white crystals; oxidizable	75 mg/day	Citrus	Absorbed from gut; stored; excreted in urine	Vital to collagen and ground substance	Scurvy–failure to form c.t. fibers

TABLE 24.8

INORGANIC SUBSTANCES

SUBSTANCE	REQUIREMENTS PER DAY	HIGH LEVEL SOURCES	WHERE ABSORBED	WHERE FOUND IN BODY	FUNCTIONS	EFFECTS OF EXCESS	DEFICIENCY
Calcium	About 1 gm	Dairy products, eggs, fish, soybeans	Small intestine	Bones, teeth, nerve, bloodstream, muscle	Bone structure, blood clotting, muscle contraction, excitability, synapses	None	Tetany of muscles, loss of bone minerals
Phosphorus	About 55 mg/kg	Dairy products, meat, beans, grains	Small intestine	Bones, teeth, nerve, bloodstream, muscle, ATP	Bone structure, intermediary metabolism, buffers, membranes	None	Unknown; related to rickets, loss of bone mineral
Magnesium	Estimated 13 mg	Green vegetables, milk, meat	Small intestine	Bone, enzymes, nerve, muscle	Bone structure, factors with enzymes, regulation of nerve and muscle action	None	Tetany
Sodium	Newborn 0.25 gm Infant 1 gm Child 3 gm Others 6 gm	All foods, table salt	Stomach, small and large intestine	Extracellular fluids	Ionic equilibrium, osmotic gradients, excitability in all cells	Edema, hypertension	Dehydration, muscle cramps, renal shutdown
Potassium	1–2 gm	All foods, meats, vegetables, milk	Stomach, small and large intestine	Intracellular fluids	Buffering, muscle and nerve function	Heart block (> 10 meq/L)	Changes in ECG, alteration in muscle contraction
Sulfur	0.5–1 gm ?	All protein containing foods	Small intestine, as amino acids primarily	Amino acids, bile acids, hormones, nerve	Structural as amino acids are made into proteins	Unknown	Unknown
Chlorine	2–3 gm	All foods, table salt	Stomach, small and large intestine	Extracellular fluids	Acid-base balance, osmotic equilibria	Edema	Alkalosis, muscle cramps

TABLE 24.8 (continued)

INORGANIC SUBSTANCES

SUBSTANCE	REQUIREMENTS PER DAY	HIGH LEVEL SOURCES	WHERE ABSORBED	WHERE FOUND IN BODY	FUNCTIONS	EFFECTS OF EXCESS	DEFICIENCY
Iron	Infant 0.4–1 mg/kg Child 0.4 mg/kg Adult 16 mg	Liver, eggs, red meat, beans, nuts, raisins	Small intestine	Respiratory proteins (hemoglobin, myoglobin, cytochromes)	O_2 and electron transport	May be toxic	Anemia (insufficient hemoglobin in red cells)
Copper	Infant and Child 0.1 mg/kg Adult 2 mg	Liver, meats	Small intestine	Bone marrow	Necessary for hemoglobin formation	None	Anemia
Cobalt	Unknown	Meats	Small intestine	Liver (Vitamin B_{12})	Essential to hemoglobin formation	None	Pernicious anemia
Iodine	Children 40–100 micrograms Adult 100–200 micrograms	Iodized table salt, fish	Small intestine	Thyroid hormone	Synthesis of thyroid hormone	None	Goiter, cretinism
Manganese	Unknown	Bananas, bran, beans, leafy vegetables, whole grains	Small intestine	Bone marrow, enzymes	Formation of hemoglobin, activation of enzymes	Muscular weakness, nervous system disturbance	Subnormal tissue respiration
Zinc	Unknown	Meat, eggs, legumes, milk, green vegetables	Small intestine	Enzymes, insulin	Part of carbonic anhydrase, insulin, enzymes	Unknown	Unknown
Fluorine	0.7 part/million in water is optimum	Fluoridated water, dentifrices, milk	Small intestine	Bones, teeth	Hardens bones and teeth, suppresses bacterial action in mouth	Mottling of teeth	Tendency to dental caries

HYPERALIMENTATION AND OBESITY

Malnutrition is usually associated with undernutrition. A greater problem in many parts of the world is overnutrition, or the intake of calories in excess of body needs, leading to the development of obesity. Obesity is a more threatening form of mal- (poor) nutrition than deficiency of caloric intake. A definition of obesity is that a person is 20 percent over "ideal weight" (see Table 24.6). In the United States it has been estimated that 15 million people are 20 percent or more over their ideal weight and are thus obese. What causes obesity? Overeating because of boredom, unhappiness, or other emotional problems accounts for 95 percent of obesity. Genetic causes account for less than 5 percent. The penalties paid for obesity include a higher mortality (death rate) from heart disease, the development of diabetes mellitus, digestive disorders, cerebral hemorrhage, and a higher morbidity (sickness) rate.

Basically, the control of obesity involves motivation to reduce weight and a caloric intake less than that required for activity and general living. A person serious about losing weight must have an appreciation of the caloric costs of activity and adjust calorie intake accordingly. A loss of one pound per week requires about 500 kcal/day reduction in intake, or an increase in activity sufficient to increase caloric consumption by 500 kcal per day. All in all, it seems easier to reduce the caloric intake than to increase the exercise level. For example, dietary omission of an ice cream sundae, or a piece of pie à la mode, eliminates the necessary calories; to burn up an equivalent amount of calories would involve running for an hour, walking (at four miles per hour) for nearly two hours, or washing dishes for over seven hours.

The use of diet pills, usually amphetamines, is effective for about two weeks. Such drugs diminish appetite and elevate the spirits. After two weeks, the body adjusts to the drug, and appetite tends to return to previous levels. To get the effect again, a user may take more drug or take it more often, leading to a dependency on it (see Appendix).

SUMMARY

1. A variety of metabolic cycles are available within body cells to combust foods and synthesize new substances from basic building blocks. These reactions constitute intermediary metabolism.
   a. Such cycles synthesize or degrade products in a steplike fashion to avoid large energy releases or demands.
   b. Energy content of a food is conventionally expressed by the Calorie (kilocalorie) or the amount of heat required to raise the temperature of 1 kg of water 1°C.
2. A variety of large molecules are essential to intermediary metabolism.
   a. Nucleotides, combinations of nitrogenous bases, pentose, and phosphate, serve as energy sources for physiological activities and as sources of cyclic nucleotides to control chemical reactions. ATP is the best known nucleotide; cyclic AMP (cAMP) is the best known cyclic nucleotide.
   b. Dinucleotides act as hydrogen acceptors in various metabolic schemes. There are three: NAD, NADP, and FAD.
   c. Coenzyme A (CoA) is an activator of chemical compounds and ensures intermediary metabolism of most carbohydrates and lipids.
3. Carbohydrate metabolism involves the utilization of at least nine different metabolic pathways or reactions.
   a. Phosphorylation is a necessary step for any monosaccharide as it enters a cell. This assures further metabolism.
   b. Galactose and fructose may be converted to glucose for cell use. This assures that the energy in these sugars is not lost.
   c. Glycogen may be formed (glycogenesis) from glucose by liver and muscle cells.
   d. Glucose may be released from glycogen (glycogenolysis) by enzymatic action. (3c and 3d aid in regulating blood sugar levels.)

e. Glycolysis is a cycle not requiring oxygen, and it reduces glucose to two pyruvic acid molecules, four hydrogens, two ATP, and about 56 kcal of energy. The two new ATP molecules trap about 15 kcal of energy; about 41 kcal is released as heat.
f. The hexose monophosphate shunt is an alternate path for glucose combustion; it also produces $NADPH_2$ for triglyceride synthesis; it produces five carbon sugars (e.g., ribose) for RNA synthesis. The glucuronic acid pathway is another alternate pathway for glucose metabolism. It forms UDPG.
g. The Krebs cycle degrades pyruvic acid to CO_2 and produces ATP, hydrogen, and over 600 kcal of energy. It requires oxygen.
h. Oxidative phosphorylation is a cycle that utilizes hydrogens released from other cycles, and converts these to water and ATP.

4. Gluconeogensis is a scheme involving the synthesis of glucose from amino acids, or fats.
   a. An amino acid has its amine group removed and then enters glycolysis and is converted to glucose by a reversal of the glycolysis steps.
   b. Amino acids are not normally combusted for energy in any quantity. They are "spared" by the presence of glucose and fatty acids.
   c. Glycerol or acetylCoA may be converted to glucose by a reversal of glycolysis.
5. Triglyceride metabolism involves three basic schemes.
   a. Glycerol may be converted into compounds in glycolysis and made into glucose. Fats and carbohydrates are interconvertible.
   b. Beta oxidation degrades a fatty acid by the progressive removal of two carbon units (acetyl CoA) from the fatty acid.
   c. Beta reduction results in fatty acid production by putting acetyl CoA molecules together.
6. Phospholipids are found in cell membranes and in nerve cells, and include lecithins, cephalins, myelin, and cerebrosides. Lipoproteins, the form in which lipids are transported in the blood, are involved in the generation of atherosclerosis and inherited lipoprotein dysfunctions. Four categories are recognized, and hypo- and hyperlipoproteinemias may result. High blood lipid levels are associated with higher incidence of atherosclerosis, a disorder involving fat deposition in blood vessel walls.
7. Cholesterol is a lipid that forms the basis for bile acids and steroid hormones.
   a. It is synthesized from acetyl CoA.
   b. It has been implicated in heart disease and in vascular occlusion.
8. Essential fatty acids are unsaturated fatty acids; they aid in the metabolism of other lipids.
9. Ketone bodies are formed from acetyl CoA when the Krebs cycle is not able to handle all the acetyl CoA produced by metabolism. Their production may result in ketosis, ketonemia, ketonuria, and acidosis.
10. Amino acids, if metabolized and not synthesized into proteins are usually deaminated and enter glycolysis or the Krebs cycle.
    a. Amine groups may form ammonia, which is converted to urea by the ornithine cycle.
    b. Synthesis of amino acids occurs by transamination of keto-acids.
11. A variety of genetically determined metabolic disorders are associated with the metabolism of carbohydrates, lipids, and amino acids.
12. Calorimetry is the study of energy requirements and energy production.
    a. Each basic foodstuff has a per gram calorie release when combusted. This is known as its heat value. Carbohydrates release 4 kcal/g, fats 9 kcal/g, and protein 4 kcal/g.
    b. The respiratory quotient (RQ) relates the O_2 required to combust a food, to the CO_2 produced in the combustion and is related to the type of foodstuff being combusted. Carbohydrates have an RQ of 1.0, fats 0.7, and proteins 0.8.
    c. The caloric equivalent of oxygen relates the calories produced when 1 liter of O_2 is used to combust a foodstuff. It varies with RQ.
    d. The basal metabolic rate (BMR) reflects energy production associated with maintenance of basic body activity, with no exercise involved. It varies according to many factors (age, sex, climate, race, emotional state, hormones, skin surface area, chemicals).
13. Nutrition is concerned with quality and quantity of food intake as related to normal and optimal growth, development, metabolism, and weight.
    a. Specific recommendations are presented as to intake of carbohydrates, fats, proteins, vitamins, and minerals.
    b. Overnutrition and obesity and their consequences of greater morbidity and mortality are considered.

QUESTIONS

1. What purposes do the cellular metabolic cycles serve in terms of assuring continuance of cellular activity?
2. What compounds must be present in order for intermediary metabolism to proceed? What function(s) does each serve?
3. In outline form, discuss the metabolic cycles involved in the complete degradation of a glucose molecule. Specify the contribution of each.
4. How is it possible to interconvert carbohydrates, amino acids, and fats?
5. What is the importance of cholesterol to the metabolism and pathology of the body?
6. What happens to the ammonia produced from amino acid metabolism? Why must it be formed into a new compound?
7. Describe the genesis of at least one metabolic disorder of carbohydrate, fat, and amino acids.
8. Define and give the importance of the following:
   a. Heat value of each foodstuff.
   b. RQ of each foodstuff.
   c. BMR.
9. From the facts presented in this chapter, construct one day's menu designed to meet requirements for all essential nutrients.
10. What are the roles of lipoproteins in the body economy? How are they involved in the generation of atherosclerosis?

Brady, Roscoe O. "Hereditary Fat-metabolism Diseases." *Sci. Amer. 229:*88. (Aug) 1973.

Brown, Michael S., and Joseph L. Goldstein. "Receptor-Mediated Control of Cholesterol Metabolism." *Science 191:*150. 16 Jan 1976.

Charney, Evan, Helen Chamblee Goodman, Margaret McBride, Barbro Lyon, and Rosalie Pratt. "Childhood Antecedents of Adult Obesity." *New Eng. J. Med. 295:*6. July 1, 1976.

Fishman, Peter H., and Roscoe O. Brady. "Biosynthesis and Function of Gangliosides." *Science 194:*906. 26 Nov 1976.

Kolati, Gina Bari. "Atherosclerotic Plaques: Competing Theories Guide Research." *Science 194:*592. 5 Nov 1976.

Lieber, Charles S. "The Metabolism of Alcohol." *Sci. Amer. 234:*25. (Mar) 1976.

Marx, Jean L. "Atherosclerosis: The Cholesterol Connection." *Science 194:*711. 12 Nov 1976.

Miller, Z., E. Lovelace, M. Gallo, and I. Pastan. "Cyclic Guanosine Monophosphate and Cellular Growth." *Science 190:*1213. 19 Dec 1975.

Motulsky, Arno G. "Current Concepts in Genetics: The Genetic Hyperlipidemias." *New Eng. J. Med. 294:*823. April 8, 1976.

Nicolaidis, Stylianos, and Neil Rowland. "Intravenous Self-Feeding: Long-Term Regulation of Energy Balance in Rats." *Science 195:*589. 11 Feb 1977.

Ravelli, Gian-Paolo, Zena A. Stein, and Mervyn W. Susser. "Obesity in Young Men after Famine Exposure in Utero and Early Infancy." *New Eng. J. Med. 295:*349. Aug 12, 1976.

Ross, Russell, and John A. Glomset. "The Pathogenesis of Atherosclerosis." *New Eng. J. Med. 295:*369. Aug 12, 1976. *295:*420. Aug 19, 1976.

Scientific American. (Entire issue on Nutrition.) (Sep) 1976.

West, E. E., and T. G. Redgrove. "Reservations on the Use of Polyunsaturated Fats in Human Nutrition." *American Laboratory.* (Jan) 1975.

READINGS

THE URINARY SYSTEM

OBJECTIVES

After studying this chapter, the reader should be able to:

- ☐ Explain what the various pathways are that the body uses to eliminate wastes of metabolism.
- ☐ Give the functional anatomy of the units of the kidney that form urine and regulate blood composition.
- ☐ List the processes that are employed by the kidney in carrying out its functions.
- ☐ Define glomerular filtration, explain the process causing it, and what sets the limits of what will be filtered.
- ☐ Explain what is meant by effective filtration pressure, and what may happen if it is not maintained.
- ☐ Define what tubular reabsorption and secretion are, where they occur in the nephron, and the result of their operation.
- ☐ Explain what clearance is and how it may be used to predict kidney function.
- ☐ Explain how the countercurrent multiplier operates, and what the result of its operation is.
- ☐ Explain the role of the countercurrent exchanger in maintaining the osmotic gradient in the renal interstitial fluid.
- ☐ Show how the nephron acidifies the fluid that passes through it.
- ☐ Explain how a concentrated urine is produced using the countercurrent multiplier and the antidiuretic hormone mechanism.
- ☐ List the factors involved in determining urine volume and explain how they operate.
- ☐ Relate hydrogen ion secretion to the role of the kidney in maintaining acid-base balance of the body.
- ☐ Characterize the urine and compare it to the plasma from which it is produced.
- ☐ Discuss what signs will develop in renal malfunction and how those signs occur.
- ☐ Present some causes for renal dysfunction.
- ☐ Describe some functions carried out by the kidney that do not bear directly on urine formation or regulation of blood composition.
- ☐ Explain how micturition occurs from the neural point of view.
- ☐ Explain how the common types of micturitional dysfunction are brought about.

PATHWAYS OF EXCRETION

Metabolism of nutrients by the body cells results in the production of a great variety of substances. Carbon dioxide, nitrogenous byproducts (such as ammonia, urea, and uric acid), heat, and excess water must be eliminated lest they disrupt the homeostasis of the body. The

TABLE 25.1
EXCRETORY ORGANS OF THE BODY AND SUBSTANCES WITH WHICH THEY DEAL

ORGAN	SUBSTANCE(S)	COMMENTS
Kidney	Water	Regulates body hydration
	Nitrogen containing wastes of metabolism (NH_3, urea, uric acid)	Originate from protein metabolism. Toxic if retained.
	Inorganic salts (NA^+, Cl^-, K^+, PO_4, SO_4), H^+, HCO_3	Regulates osmotic pressure, pH of ECF; maintains excitability of cells.
	Drugs	
	Detoxified substances	Detoxification occurs in the liver; kidney rids body of end product.
Lungs	Carbon dioxide	Regulation of acid-base balance
	Water	Fixed loss
	Heat	Fixed loss
Skin	Heat	Regulates temperature
	Water	Cooling
Alimentary tract	Digestive wastes	—
	Salts (Ca^{++})	—

lungs account for elimination of the greater part of the carbon dioxide, along with small quantities of water and heat. The alimentary tract accounts for the loss of some carbon dioxide, water, heat, some salts, and the unabsorbed secretions of the digestive glands. The skin plays the major role in the elimination of excess body heat, but a minor role in elimination of solid wastes. Also treated as wastes, in that they will ultimately be eliminated from the body, are the substances present in greater amounts than are required for normal cellular function; they have, for the most part, entered the body in the diet. Water, salts, hydrogen ion, sulfates, and phosphates are examples of physiologically important substances often present in excess amounts. The kidney is the most important organ in regulating excretion of solutes resulting from metabolism and in regulating composition of the blood and hence ECF. Table 25.1 summarizes the excretory organs of the body and the substances with which they deal.

THE PHYSIOLOGICAL ANATOMY OF THE URINARY SYSTEM

THE ORGANS OF THE SYSTEM

The urinary system (Fig. 25.1) includes two KIDNEYS, two URETERS, a single URINARY BLADDER, and a single URETHRA to the exterior.

THE KIDNEYS

Gross anatomy. A medially directed concavity is termed the HILUS; it leads to the RENAL SINUS. The blood vessels, nerves, and ureters enter and exit the organ at this region. Affixing the kidneys in position behind the parietal peritoneum are the ADIPOSE CAPSULE (*perirenal fat*) and the double layers of the SUBSEROUS (*renal*) FASCIA. A FIBROUS CAPSULE forms the external covering of the kidney itself. A frontal (longitudinal) section of the organ (Fig. 25.2) reveals the upper expanded end of the ureter (the RENAL PELVIS) and the calyces lying within the renal sinus. The MAJOR and MINOR CALYCES are the primary and secondary subdivisions of the pelvis. An inner MEDULLA is composed of 8 to 18 RENAL (*medullary*) PYRAMIDS, with bases directed toward the periphery of the kidney, and apices or PAPILLAE projecting into the minor calyces.

The CORTEX arches around the bases of the pyramids and projects between the pyramids, where it is designated as the RENAL COLUMNS. Cortical and medullary portions together constitute the PARENCHYMA or cellular portion of the kidney. The parenchyma is formed, in each kidney, of *at least* one million microscopic nephrons (renal tubules) with their associated blood supply. The nephron carries out the functions of the kidney—that is, formation of urine and regulation of the blood composition.

Blood supply (Fig. 25.3). Originating from the aorta and passing to each kidney is a large RENAL ARTERY. These arteries carry to the kidneys approximately one-fourth of the total cardiac output, a quantity (renal blood flow or RBF) averaging 1200 ml of blood per minute. Of this volume, 650 ml is plasma (renal plasma flow or RPF). Upon entering the kidney at the hilus, the renal artery branches to form several INTERLOBAR ARTERIES that enter the parenchyma. The vessels pass between the pyramids and, near their bases, form a series of curved vessels, the ARCUATE ARTERIES. The arcuate arteries in turn give rise to a series of radially directed vessels running through the cortical region of the kidney. These vessels are known as INTERLOBULAR ARTERIES. The complex of arcuate arteries rising from any one given interlobar artery does not interconnect with similar vessels rising from other interlobar arteries. Because of this arrangement, the arcuate arteries are known as *end arteries*. This arrangement is

FIGURE 25.1
The organs of the urinary system and associated structures.

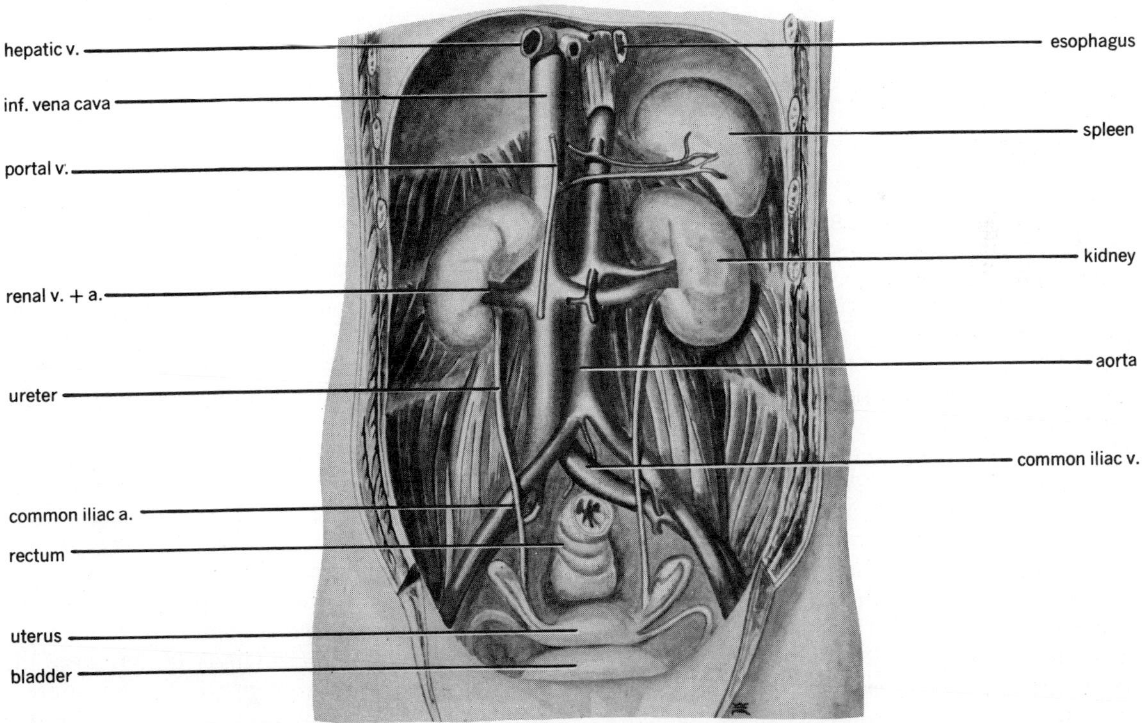

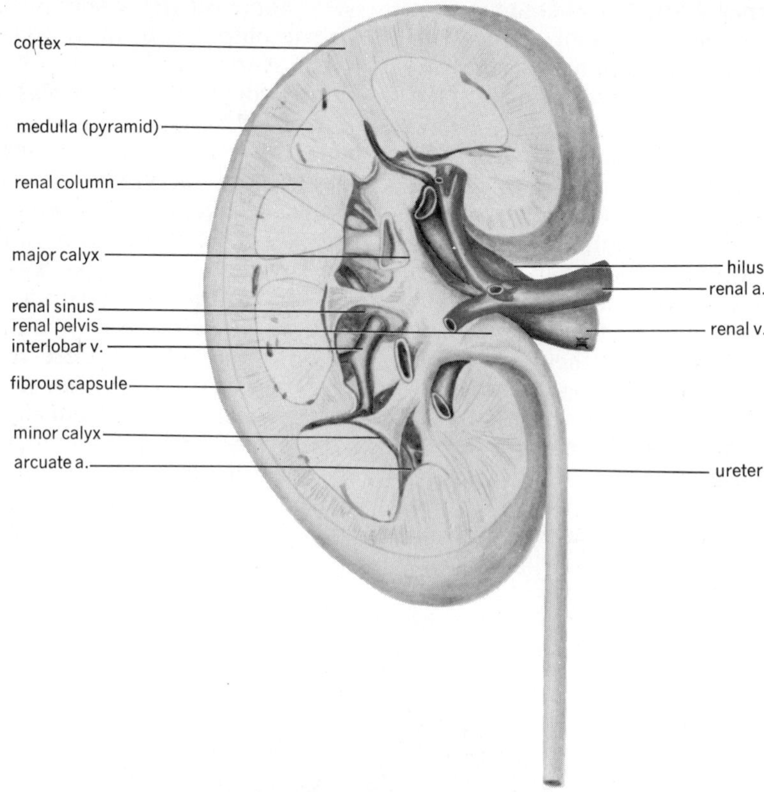

FIGURE 25.2
A frontal section of the kidney.

one factor ensuring the maintenance of a high level of pressure on the blood as it reaches the outer portions of the cortex. The interlobular arteries give off nutrient branches to the cortical substance and to the capsule of the kidney. Also, and most important, branches called AFFERENT ARTERIOLES go to the ball-like networks of capillaries called the GLOMERULI. From the glomeruli, EFFERENT ARTERIOLES lead out to form either a network of capillaries (peritubular capillary network) around the renal tubules or a looped blood vessel (vasa recti) that dips deeply into the medullary portion of the kidney and then returns toward the cortex. The efferent arteriole carries a smaller volume of blood than the afferent and is smaller in diameter. From here the blood vessels form a series of veins named in the same manner as the incoming arteries. INTERLOBULAR VEINS drain into ARCUATE VEINS lying along the bases of the medullary pyramids. Unlike the corresponding arteries, the arcuate veins do connect with one another. These in turn form INTERLOBAR VEINS that form a single RENAL VEIN from each kidney, the latter emptying into the vena cava.

The nephrons. The nephrons (Fig. 25.4) are the microscopic, and mainly tubular, structural and functional units of the kidneys. It is estimated that there are 1 to 1.5 million nephrons in each kidney.

Two types of nephrons are generally recognized. Both have similar parts and differ from one another only in the length of certain of the tubules and in what happens to the efferent arteriole as it leaves the glomerulus. The two types are known as CORTICAL and JUXTAMEDULLARY NEPHRONS. The cortical nephron usually has its glomerulus in the outer portion of the cortex, and the rest of its parts rarely reach deeply into the medulla. The juxtamedullary nephron, on the other hand, has its glomerulus close to the corticomedullary junction, and has part of its structure (the loop of Henle) located deep within the medulla.

A nephron begins with a double-walled cup termed the GLOMERULAR (*Bowman's*) CAPSULE. The inner wall of this capsule is known as the VISCERAL LAYER, and it follows intimately the twists and turns of the glomerular capillary network. It is formed of highly modified cells known as PODOCYTES. The outer or PARIETAL LAYER of the capsule is a simple squamous epithelium and lies a short distance from the visceral layer so that an actual space between the two layers is created. The capsule and the contained glomerulus form a unit designated as the RENAL CORPUSCLE. Leading from the capsule is a PROXIMAL CONVOLUTED TUBULE in which the cells are cuboidal, have central nuclei, and have a "brush border" on the lumen side. The brush border consists of minute cytoplasmic extensions of the cell called microvilli. These serve to increase the surface area for reabsorption and/or secretion of materials by this part of the nephron. The proximal tubule then becomes straight as it nears the medullary region. The cells become flat, and the tube narrows and dips toward or into a pyramid as the DESCENDING LOOP OF HENLE. Then the tube bends back upon itself, enlarges, its cells become rectangular, and, as the ASCENDING LOOP OF HENLE, it returns to or toward the cortical region. In the cortex, the tube again becomes straight and then convoluted and is known as the DISTAL CONVOLUTED TUBULE. Its cells are cuboidal, have central nuclei, have lighter-staining cytoplasm than the proximal tubule, and carry no brush border. This portion joins a COLLECTING TUBULE. The collecting tubules receive the distal terminations of many nephrons. They open into the

FIGURE 25.3
The blood vessels of the kidney.

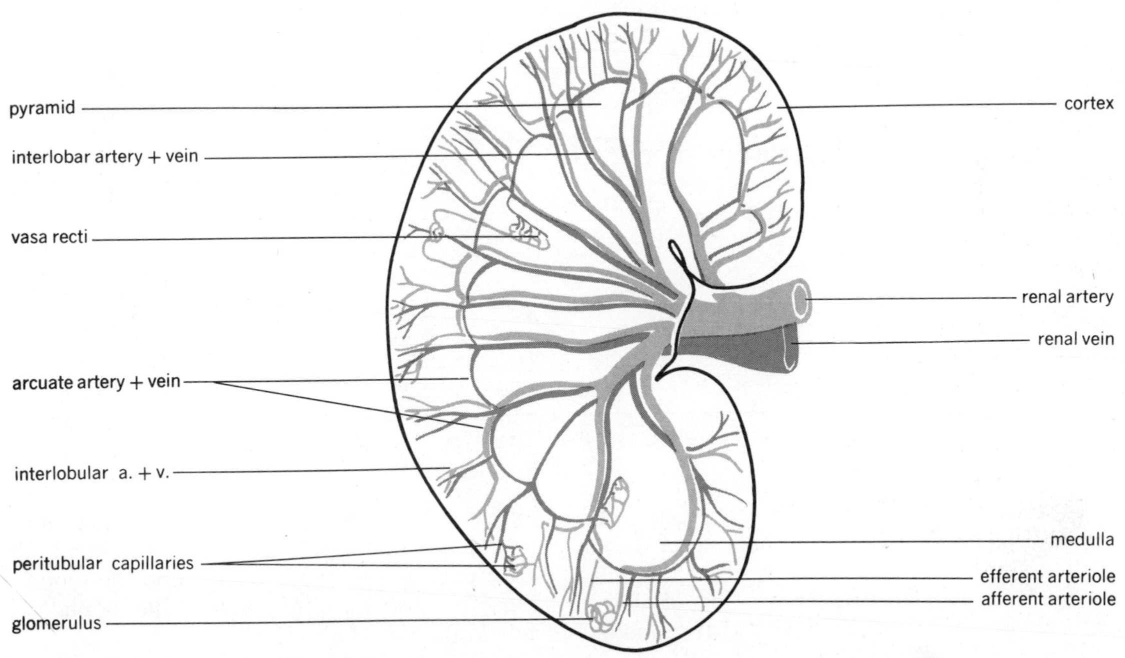

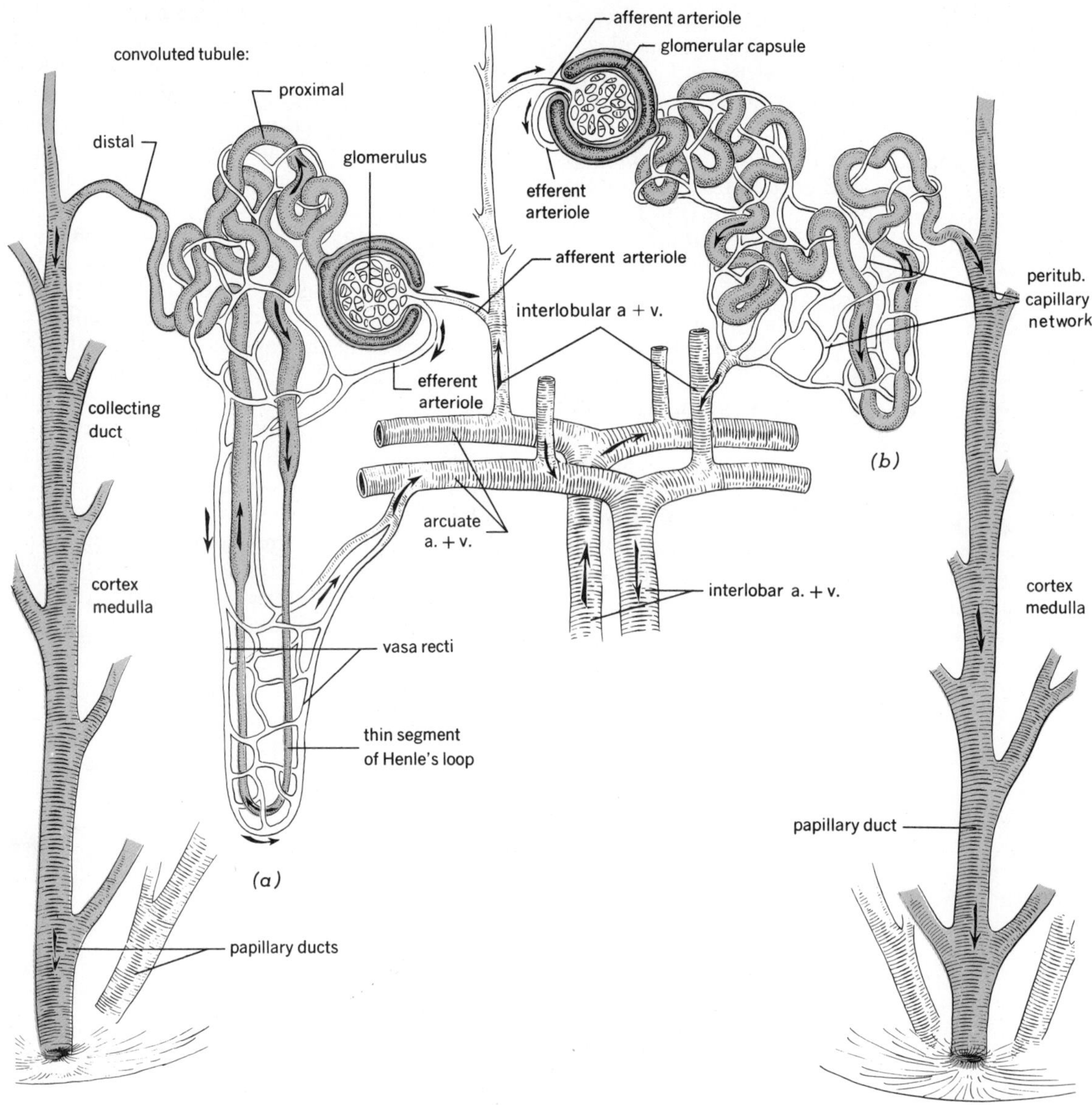

FIGURE 25.4
The nephrons of the kidney and their associated blood vessels. (*a*) **Juxtamedullary nephron** (*b*) **Cortical nephron.**

calyces of the pelvis through the PAPILLARY DUCTS. In a cortical nephron, the loop of Henle is quite short and typically does not reach into the medulla. It is around the tubules of such a nephron that we find the peritubular capillary network. The loops of Henle of the juxtamedullary nephrons, on the other hand, dip very deeply into the medulla, and the loops of these nephrons are followed by the looped vessels known as the vasa recti.

An anatomical difference such as described obviously suggests a functional difference between the two types of nephrons and, indeed, such a difference is found to exist. In general, it may be said that cortical nephrons are reabsorptive and secretory structures. Juxtamedullary nephrons also carry out these functions; in addition, they are responsible for creating an osmotic gradient within the parenchyma (cellular part of the kidney), which makes it possible to elaborate a hypertonic urine.

THE FORMATION OF URINE AND REGULATION OF BLOOD COMPOSITION

The nephron forms urine, which is HYPERTONIC and usually more ACID than the plasma. The nephron also governs, within narrow limits, the composition of the blood leaving the kidney. In accomplishing these tasks, the kidney is aided by hormones secreted by certain endocrine glands and by chemical substances originating within the organ itself.

The processes employed by the nephron in forming urine are FILTRATION, TUBULAR TRANSPORT, the COUNTERCURRENT MULTIPLIER and EXCHANGER, ACIDIFICATION (a result of tubular transport), and CONCENTRATION of the filtrate.

GLOMERULAR FILTRATION

The blood, with its contained load of wastes, excess materials, and substances the body wishes to conserve, is delivered to the glomerulus under a hydrostatic blood pressure (P_b) of 60 to 75 mm Hg. The end artery structure of the arcuate arteries is instrumental in assuring maintenance of high pressure as the vessels branch and penetrate into the kidney. The glomerular capillaries and the closely applied visceral layer of the glomerular capsule act as a coarse sieve. Providing the blood pressure is higher than the sum of all opposing pressures (see below), all materials small enough to pass through these membranes will be forced through by the pressure differential in a process of filtration, and the fluid formed is called the FILTRATE. The capillaries behave as though they possess pores of small diameter (100Å), and the cells of the visceral layer (podocytes) do not form a continuous stratum (Fig. 25.5). The foot processes of the podocytes as they are applied to the capillaries form a series of "slit pores" that govern the passage of materials into the capsule. In general, it may be stated that substances having a molecular weight up to 10,000 pass easily through these layers of tissue. Above this figure, materials pass with increasing difficulty, and a probable upper limit is reached at a molecular weight of about 200,000. Thus, nearly all materials present in the plasma, except the formed elements, are capable of being filtered into the cavity of the glomerular capsule. The filtrate contains protein, chiefly albumin, to the extent of 10 to 20 mg/100 ml. A mathematical calculation would indicate that some 30 g of protein is thus filtered per day. The material (filtrate) formed by this process resembles plasma (see Table 25.5), except that it lacks formed elements and has a lower protein concentration. Because the greater concentration of plasma proteins does not filter through, it exerts an osmotic pull, or back pressure, that tends to cause water movement back into the glomerulus. This osmotic back pressure (P_o) normally amounts to about 30 mm Hg. Also opposing the filtration of material through the glomerulus is the renal interstitial pressure (P_{rip}) or the pressure within the interstitial space around the kidney tubules, and the intratubular pressure (P_{it}). This latter quantity is the resistance the filtrate encounters as it attempts to push into the kidney tubules from the glomerulus. Because the tubules are already full of fluid, a resistance to the further movement of fluid is encountered. P_{rip} and P_{it} amount to about 10 mm Hg each. By adding together the opposing forces and subtracting them from the original blood pressure, a quantity called the effective filtration pressure (P_{eff}) may be calculated. This pressure represents the net pressure, or the force actually causing filtration through the glomerular capsule. Filtration will not proceed normally, nor will the wastes be removed effectively from the blood, unless a normal effective filtration pressure is maintained. The relationships de-

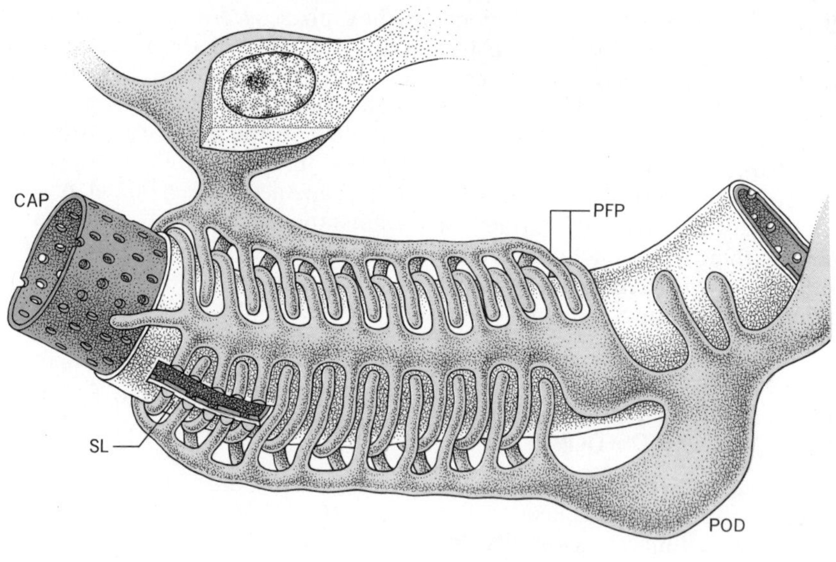
CAP
PFP
SL
POD

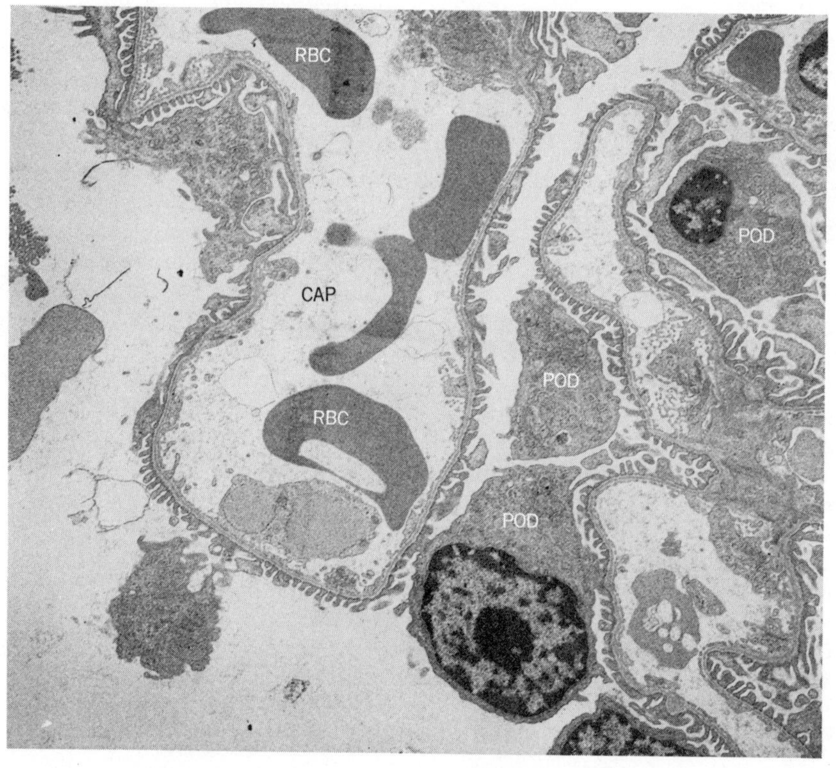
RBC
CAP
RBC
POD
POD
POD

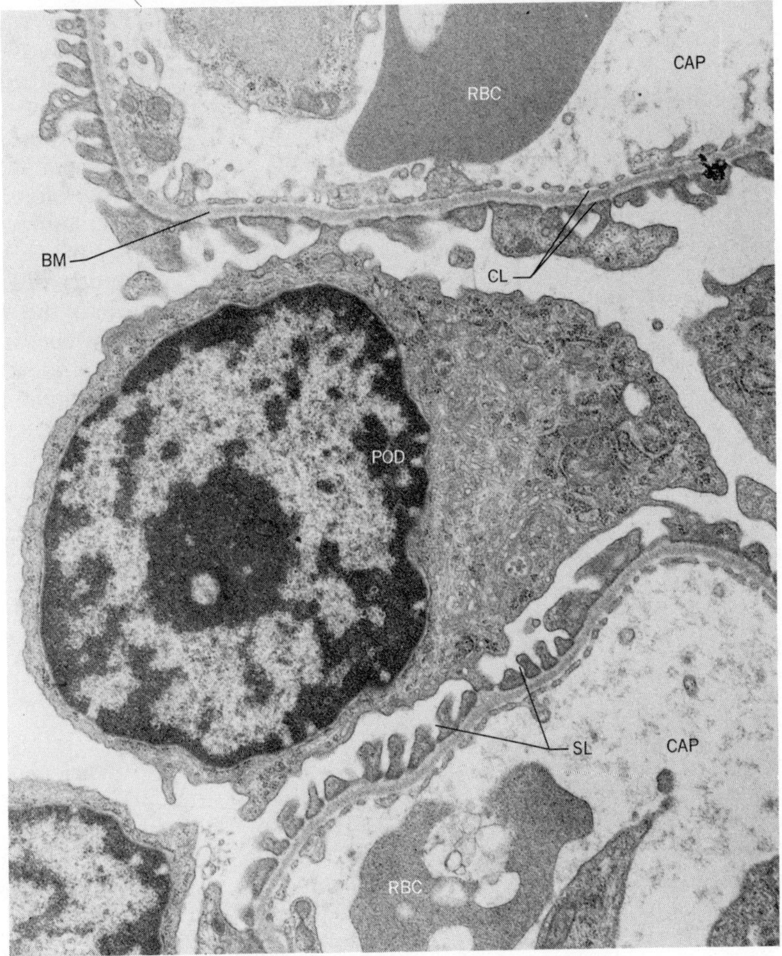

FIGURE 25.5
The ultrastructure of the renal corpuscle. CAP = capillary, POD = podocyte of visceral layer of capsule, RBC = red blood cell, BM = basement membrane, CL = cement layers, PFP = podocyte foot processes, SL = slit pores (s). (Electron micrographs courtesy Norton B. Gilula, The Rockefeller University.)

scribed above may be expressed mathematically by the following formula:

$$P_{eff} = P_b - (P_o + P_{rip} + P_{it})$$

$$P_{eff} = (60\text{–}75) - (30 + 10 + 10)$$

$$P_{eff} = 10\text{–}25 \text{ mm Hg}$$

The rate of glomerular filtration (GFR) can be measured experimentally and averages 120 ml/min for both kidneys. The total plasma flow was 650 ml/min. If only 120 ml/min is filtered, a filtration fraction (FF) of 18.5 percent may be calculated ($120/650 \times 100$). That is, only about one-fifth of the fluid arriving in the glomerulus is actually filtered.

TUBULAR TRANSPORT

The filtrate contains not only the waste materials that must ultimately be eliminated from the body, but also quantities of substances the body finds useful, such as water, hormones, vitamins, enzymes, glucose, inorganic salts, proteins, and amino acids. In order that these materials not be lost from the body, a second process must occur at this time—tubular transport, which involves the use of active processes to transport through the cells of the kidney tubule materials that are of value to the body. If the direction of tubular transport is from the lumen or cavity of the tubule through the cell and ultimately into the surrounding blood vessels, it is called REABSORPTION. The materials the body finds useful are reclaimed in this fashion. Some materials considered as wastes are not removed completely enough from the blood by filtration and are actively transported from the blood through tubule cells and into the tubule lumen. Movement in this direction is termed SECRETION.

In the proximal convoluted tubule an average reabsorption of 80 to 90 percent of the physiologically useful solutes is achieved. Among the materials actively transported through this region are glucose, amino acids and protein, phosphate, sulfate, uric acid, vitamin C, beta-hydroxybutyric acid, and calcium, sodium, and potassium ions. As these solutes are reabsorbed by active processes, water follows osmotically, so that it too achieves an 80 to 90 percent reabsorption in this area. Chloride, a negatively charged ion, follows positive ions (Na^+ and K^+) out of the tubule by electrostatic attraction. Because of the water movement accompanying the solute transport, the NET TONICITY OF THE FLUID AT THIS POINT DOES NOT CHANGE. Its volume has been reduced, but its osmolarity and pH have not been affected.

The reabsorption of sodium ion by the proximal tubule is directly proportional to the concentration of aldosterone secreted by the adrenal cortex. Reabsorption of calcium ion is increased by parathyroid hormone. Glucose usually achieves a 100 percent removal in the proximal tubule because the carrier system transporting glucose has a transport maximum (T_{M_G}) of 300 to 375 mg glucose per minute. The amount of glucose in the filtrate is normally 80 to 120 mg/100 ml. Therefore, the carrier system has no difficulty in removing all the filtered glucose. In certain diseases where the concentration of glucose in the blood is high, such as diabetes mellitus, the amount of filtered glucose may exceed the ability of the transport system to reabsorb it, and under these conditions glucose will spill over into the urine. Protein is also absorbed completely in the proximal tubule, probably by pinocytosis. As water passes out of the tubule, an increase in the concentration of urea in the tubule occurs. Urea (a nitrogenous waste) does not possess a carrier system and would not normally be reabsorbed. However, its concentration rises to a level at which some will pass out of the tubule by simple diffusion. The reabsorption of urea is an example of *obligatory* ("have to") *reabsorption* over which the nephron has no control. The movement by carriers can be controlled and is referred to as *facultative* (optional) *reabsorption.*

The proximal tubule is also an important area for secretion. A variety of organic acids (phenol red, hippuric acid, creatinine, para-aminohippuric acid, penicillin, Diodrast) and strong organic bases (choline, guanidine, histamine) are secreted into the filtrate by the proximal tubule cells.

The concept of clearance. The term *clearance* implies the ability of the kidney to eliminate a given substance from the plasma in a specific length of time. It is expressed as ml/min and is a hypothetical volume equivalent to the amount of a substance that would be contained in that volume of blood. For example, if a clearance is 100 ml/min, this implies that the amount of that substance removed per minute would be equivalent to that contained within 100 ml of blood. Additionally, clearance of certain substances may give an indication of how the kidney is treating a substance. For example, the glomerular filtration rate (GFR) may be measured by a substance that meets the following criteria: it is cleared or removed from the blood only by filtration by the kidney; it is not stored or metabolized by any body cell; it is not bound to proteins in the plasma; it is not toxic to the body. Such a substance is the polysaccharide *inulin*. The material is infused intravenously in a large initial dose, followed by a continuous infusion to keep the plasma level constant. A timed urine specimen is then collected and analyzed for inulin. A plasma sample is also collected and its inulin level determined. The GFR is calculated by the equation:

$$C_{in} = \frac{U_{in}V}{P_{in}}$$

where C_{in} = clearance of inulin (GFR), or rate of removal from the blood (ml/min)
U_{in} = concentration of inulin in urine (mg/ml)
V = rate of urine formation (ml/min)
P_{in} = plasma concentration of inulin (mg/ml)

With typical values substituted in the equation, inulin clearances (and GFR) are about 120 ml/min, or about 173 liters per day. If a material has a clearance equal to that of inulin, it may be concluded that it is a small water-soluble molecule and that the kidney is treating it in the same manner as inulin—that is, by filtration only. If a substance has a clearance less than that of inulin, it must have been removed (i.e., reabsorbed) from the filtrate because it appears in the urine at a slower rate than did inulin. If a substance has a clearance greater than that of inulin, it is being added to the filtrate (i.e., secreted) to give a greater concentration than could have appeared due to filtration alone. Clearance tests can thus be performed to assess the level of kidney function.

The filtrate next enters the loop of Henle, and here, in the juxtamedullary nephron, is exposed to the countercurrent multiplier and the countercurrent exchanger. These mechanisms create the conditions necessary for achieving a concentrated or hypertonic urine.

THE COUNTERCURRENT MULTIPLIER

This physiological mechanism is so named because of the hairpinlike arrangement of Henle's loop and the fact that fluid flows in opposite directions in the two limbs (countercurrent). *It has as its purpose the creation of an increasing concentration of solute as one progresses from the periphery to the tips of the medullary pyramids.* As the mechanism operates, the concentration of solute is increased (multiplied) in the descending limb and renal interstitial fluid (interstitium) by the activity of the ascending limb. In order for the multiplying effect to be achieved, several conditions must be met. First, the ascending portion of the loop of Henle must be capable of actively transporting ions from the filtrate in its lumen into the interstitium and then diffusion will carry them into the descending portion of the loop. A carrier system for chloride ion exists in these cells, and the predominant extracellular anion is chloride. Therefore, one may say that the creation of the osmotic gradients depends on the transport of chloride ion. Second, the ascending portion of the loop must be impermeable to water, so that as chloride, with sodium following passively by electrostatic attraction, is transported out, water does not follow it osmotically. Third, the ascending portion of the loop must be capable of transporting chloride to the extent that it can achieve a 200 milliosmolar (mOs) difference in the concentration of chloride between it and the descending portion of the loop. Assuming an initial input of fluid to the top of the descending portion of the loop of 300 mOs concentration, and viewing what happens next much in the same fashion as a movie film run slowly enough that one can see the individual frames, what happens is described below (Fig. 25.6).

Assuming the entire loop to first be filled with 300 mOs fluid,

a. The ascending loop will transport sodium chloride into the descending loop until a 200 mOs difference has been attained.
b. This activity will place a 400 mOs solution in the descending loop opposite a 200 mOs solution in the ascending loop.

If we now advance the whole mechanism one step,

c. A new mass of fluid of 300 mOs concentration will enter the top of the descending loop,

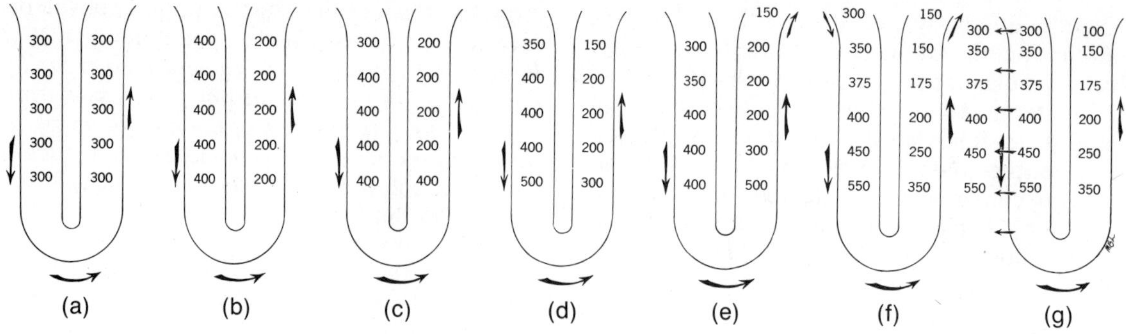

FIGURE 25.6
The operation of the countercurrent multiplier. (see text for explanation).

and a 200 mOs solution will exit from the top of the ascending loop. In the lower part of the loop, moving a 400 mOs area around the tip and into the ascending loop will place it opposite another 400 mOs region, which descends into the space vacated by the one that moves around the tip of the loop. The 200 mOs difference has been lost at the top and tip of the loop. Ions will again be transported,

d, until a 200 mOs difference has been attained. In the lower part of the descending loop, a concentration of 500 mOs will be attained, with a 300 mOs area opposite it. Advancing mechanism still another step,

e, will place a 300 mOs solution opposite a 200 mOs solution in the ascending portion of the loop. Again, ions will be transported until the 200 mOs difference is attained.

f,g. As this process progresses, it can be seen that a gradually increasing milliosmolar concentration will be created as the descending loop is traversed, and a gradually decreasing milliosmolar concentration will result in the ascending portion.

Remembering that the descending portion of the loop permits free diffusion of materials, the changes created in the descending loop will result in similar changes in the interstitium. Therefore, an increasing concentration of solutes will be found the deeper one penetrates into the medulla. The maximum figure to which the concentration can rise, assuming that the previously mentioned conditions are met, is 1200 to 1400 mOs (Fig. 25.7). The material thus delivered to the distal convoluted tubule is hypotonic, containing about one-third the solute concentration originally fed into the loop.

This generally accepted description of the operation of the countercurrent multiplier has been modified to include a role for urea. According to this idea, as sodium chloride removal occurs in the loop of Henle, urea becomes concentrated in the tubular fluid. It will subsequently pass from the collecting tubules along with water, and will reenter the loop of Henle, to pass out again through the collecting tubules. This medullary recycling of urea causes it to accumulate in large quantities in the medullary interstitial fluid where it aids in extracting water from the descending limb of the loop and thereby concentrates NaCl in the ascending limb. Thus, passage of NaCl from the ascending limb is aided by diffusion as well as active Cl^- movement, with sodium following. These steps are presented in Figure 25.8.

THE COUNTERCURRENT EXCHANGER

If the sodium chloride and urea that diffused or was actively transported into the interstitium is to exert any osmotic activity, it must remain within this area of the kidney. The countercurrent exchanger operates between the interstitial fluid of the kidney and the vasa recti. Blood in the vasa recta is forced to pass through the medullary region, which has the increasing concentration of solutes created by the multiplier. Into this vascular loop will diffuse sodium and chloride ions and urea and water will diffuse out, so that the changes occurring in the osmolarity of this blood follow the same pattern as the fluid in the loop of Henle. As the vascu-

FIGURE 25.7
The end result of the countercurrent multiplier's operation.

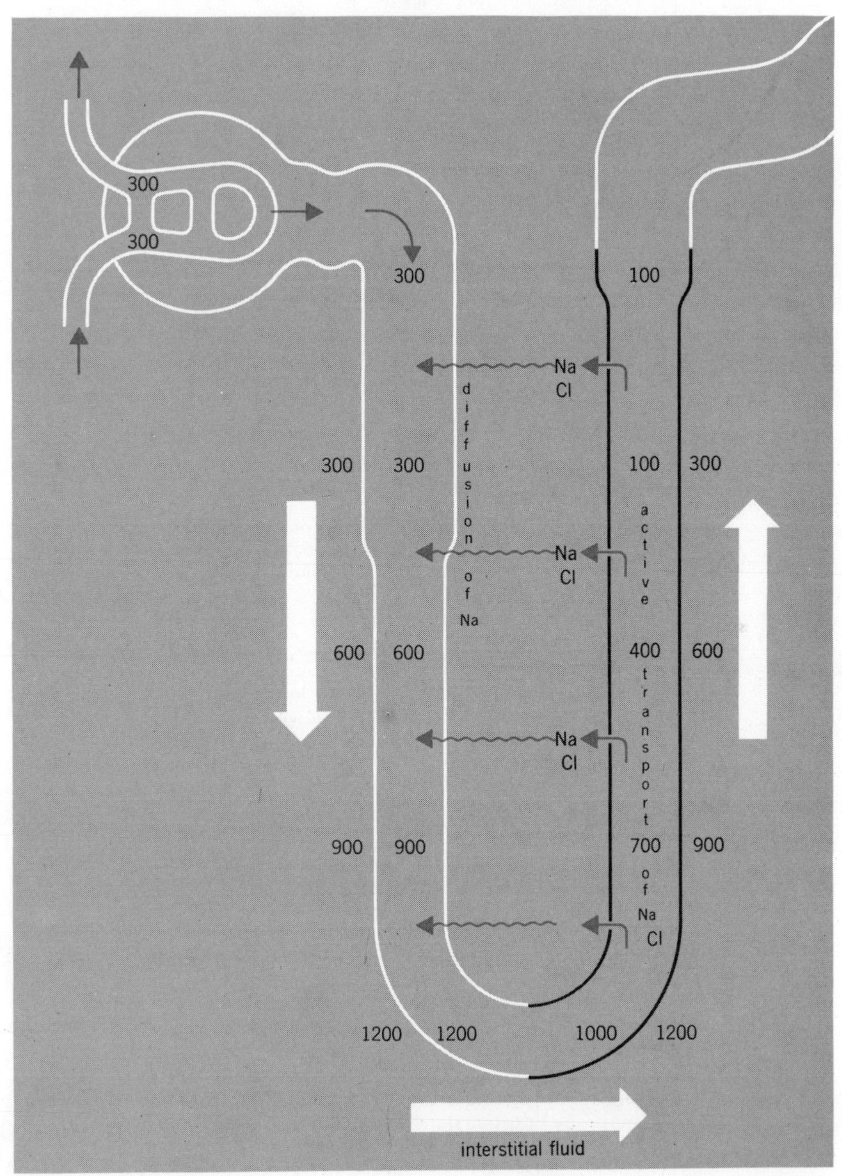

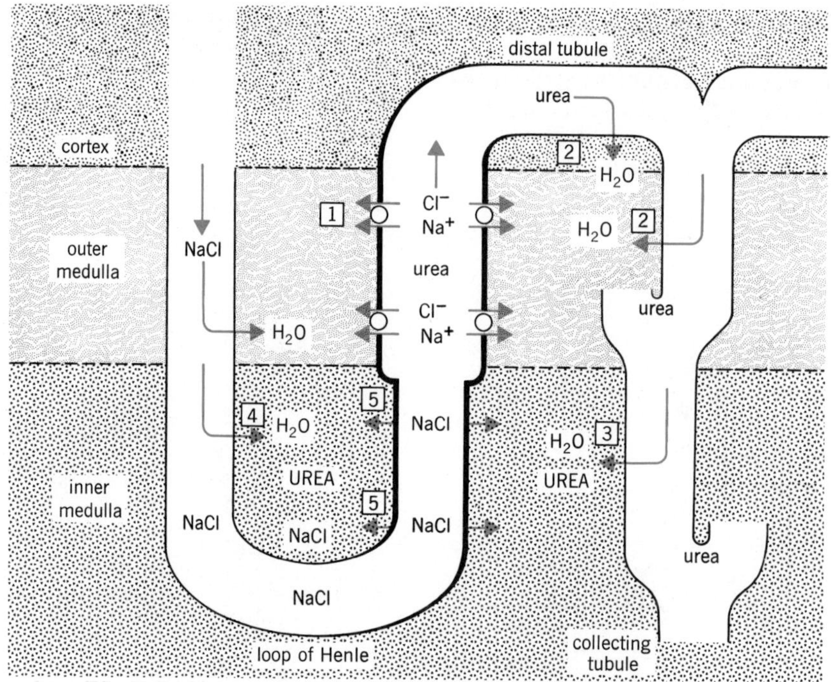

FIGURE 25.8

Recent Modifications of the Countercurrent Hypothesis by Stephenson and Koko and Rector.

Both the thin ascending limb in the inner medulla, as well as the first part of the distal tubule, are impermeable to water, as indicated by the thickened lining. In the thick ascending limb, active chloride reabsorption, accompanied by passive sodium movement (1), renders the tubule fluid dilute and the outer medullary interstitium hyperosmotic. In the last part of the distal tubule and in the collecting tubule in the cortex and outer medulla, water is reabsorbed down its osmotic gradient (2), increasing the concentration of urea that remains behind. In the inner medulla both water and urea are reabsorbed from the collecting duct (3). Some urea re-enters the loop of Henle (not shown). This medullary recycling of urea, in addition to trapping of urea by countercurrent exchange in the vasa recta (not shown), causes urea to accumulate in large quantities in the medullary interstitium (indicated by the large type), where it osmotically extracts water from the descending limb (4) and thereby concentrates sodium chloride in descending-limb fluid. When the fluid rich in sodium chloride enters the sodium-chloride-permeable (but water-impermeable) thin ascending limb, sodium chloride moves passively down its concentration gradient (5), rendering the tubule fluid relatively hypo-osmotic to the surrounding interstitium. (Reprinted by permission from R. Jamison, and R. Maffly, *The New England Journal of Medicine, 295:*1059, November 4, 1976.)

lar loop climbs back out of the medulla, sodium and chloride ions and urea diffuse out and water in. *This tends to keep the sodium chloride and urea within the medullary region and maintains the concentration gradient.*

ACIDIFICATION

The distal convoluted tubule receives a hypotonic fluid from which it may reabsorb any remaining physiologically useful solutes. Recall that the proximal tubule achieved only 80 to 90 percent reabsorption of materials. The distal tubule is also an important area of secretion, primarily of hydrogen ion, ammonia, and potassium ion. The secretion of these three materials is important in the ability of the kidney to participate in the regulation of acid-base balance.

Acid-producing foods predominate over alkali-producing foods under normal dietary conditions. Much free hydrogen ion also results from the production of HCl by the stomach cells and subsequent reabsorption lower in the tract. The body therefore faces the problem of maintaining average pH (7.4) in the face of forces tending primarily to lower the pH of ECF. Acids entering the extracellular fluids first react with the chief buffer in the extracellular fluid, sodium bicarbonate. Carbonic acid is formed, which is subsequently carried to the lungs and disposed of as carbon dioxide, as shown in the following:

$$HA + NaHCO_3 \rightarrow NaA + H_2CO_3 \xrightarrow[\text{in lungs}]{\text{carbonic anhydrase}} H_2O + CO_2 \uparrow$$

A = unspecified anion ($SO_4^=$, $PO_4^{\equiv}$, Cl^-)

The anion combined with sodium will eventually be excreted in the urine. It would be disadvantageous to the body to lose the anion as the sodium salt since sodium, in the form of sodium bicarbonate, is the primary buffering agent of the extracellular fluid. The kidney rids the body of the anion and reabsorbs the sodium, exchanging it for a hydrogen ion.

Ammonia, derived from the deamination of amino acids and glutamine, diffuses into the tubule lumen and there reacts with a hydrogen ion to form an ammonium ion. The ammonium ion may then react with a variety of anions and thus carry out not only an excess hydrogen ion, but an anion as well. Ammonia production normally occurs at a relatively low level unless the body is presented with large amounts of hydrogen ion. Ammonia production therefore acts as an additional means of disposing of hydrogen ion, in the form of ammonium ion. Figure 25.9 illustrates the operation of some of these processes.

The urine leaving the distal tubule has thus been acidified, but is still hypotonic, since any exchange of materials has been made primarily on a one-for-one basis.

PRODUCTION OF A CONCENTRATED URINE

The collecting tubules receive the hypotonic solution from the distal tubule. These tubules run through the medulla toward the tips of the pyramids. Remember that in the medulla the countercurrent multiplier has operated to create an increasing degree of solute concentration

FIGURE 25.9
Methods of increasing H^+ secretion by the kidney. (*a*) Secretion of ammonia, (*b*) Secretion of potassium.

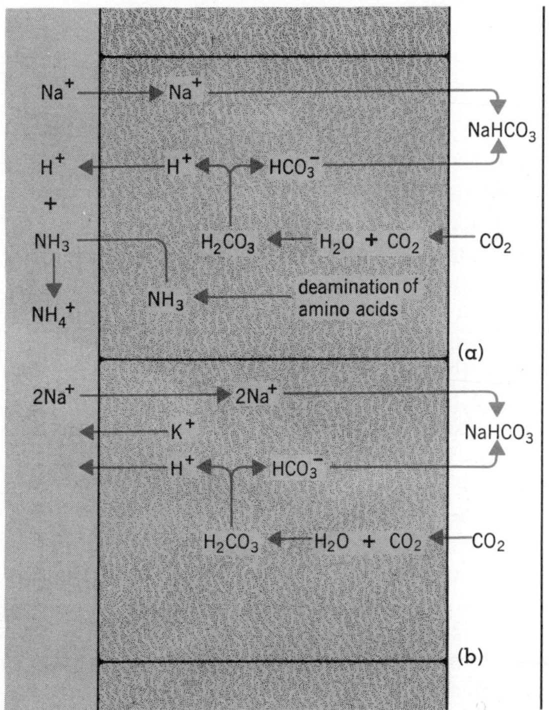

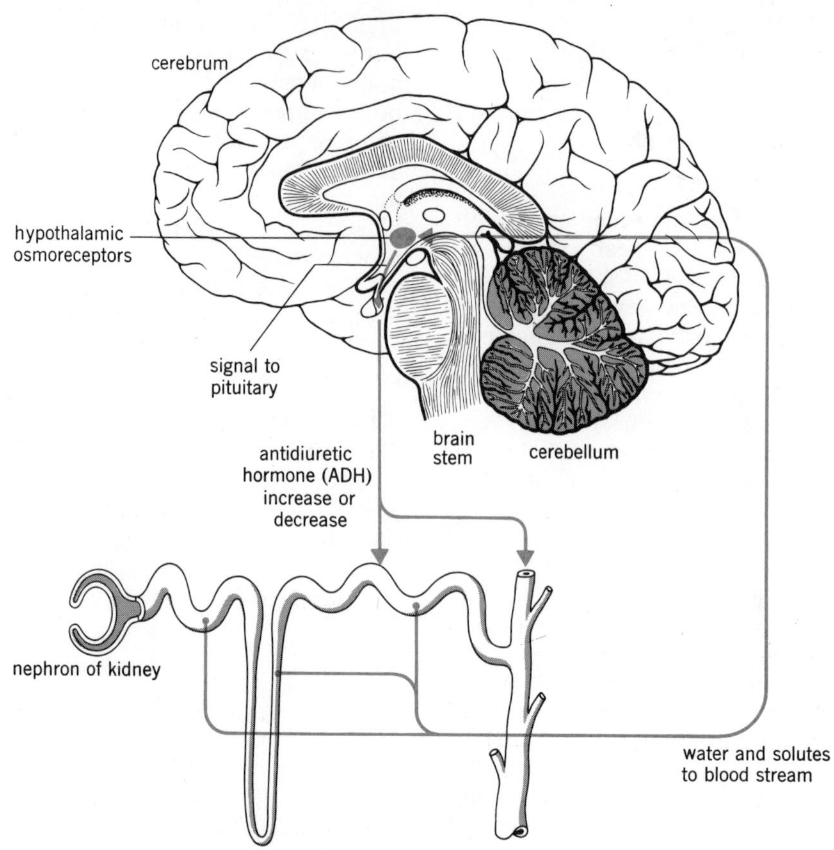

FIGURE 25.10
The ADH mechanism for control of extracellular fluid volume and osmotic pressure.

toward the pyramid tip. The tubule therefore has around it a solution of much greater solute concentration than the fluid in the tubule. There is thus a tendency for water to leave the collecting tubule by osmosis, but this is not permitted unless the tubule becomes permeable to water.

The tubule becomes permeable to water in the presence of ADH or ANTIDIURETIC HORMONE. This substance is produced in the hypothalamus and is stored in and released from the posterior lobe of the pituitary gland (Fig. 25.10). It exerts a permissive effect on the cells, causing them to allow passage of water into the interstitium. The amount of ADH secreted is determined by the osmotic pressure of the blood reaching the hypothalamus. As the fluid passes through the collecting tubules and into the papillary ducts, it loses more and more water and will become hypertonic. The highest concentration to which the fluid can rise is 1200 to 1400 mOs, which is the maximum concentration that the multiplier can create. Normally, a urine of about 800 mOs is produced, indicating that the cells of the collecting tubules never become freely permeable to water. Table 25.2 summarizes the processes involved in urine formation.

FACTORS CONTROLLING URINE VOLUME

The volume of urine produced by the kidneys depends on the balance between several of the basic physiological processes discussed in the previous sections.

The amount of filtrate formed varies directly with the HYDROSTATIC PRESSURE of the blood. Assuming that osmotic back pressure remains constant, then the higher the blood pressure, the greater the rate of filtration. In shock or cardiac failure the blood pressure may fall below that required for adequate filtration. Uremia may result and require artificial removal of wastes. The principle of the artificial kidney is shown in Fig. 25.11.

The volume of the filtrate may be increased by an INCREASE IN CONCENTRATION OF SOLUTES in the filtrate, or by a decrease in plasma protein content. In diabetes mellitus, for example, the large amount of glucose present in the filtrate draws water osmotically into the tubule, or prevents its loss as sodium is transported out.

The volume of urine formed varies inversely with the amount of ANTIDIURETIC HORMONE (ADH) secreted, and in turn, ADH secretion varies directly with the solute concentration of the blood reaching the hypothalamus. Excess fluid intake poses a threat of dilution of the internal environment and is relieved by decrease in water reabsorption. Diabetes insipidus results from loss or greatly diminished production of ADH.

The volume of fluid excreted by the urinary system varies inversely with WATER LOSS BY OTHER SYSTEMS, especially the digestive system, and water loss through perspiration. The body must maintain a reasonable balance between fluid intake and loss, and if vomiting, diarrhea, or excessive sweating increase fluid loss, urine volume will decrease. This again is effected mainly by variation in ADH secretion.

Alteration in the efficiency of operation of the countercurrent multiplier, particularly in the transport of chloride, will change the urine volume. Certain drugs, such as diuretics, increase urine volume by decreasing the active removal of chloride ion from the filtrate. This allows these ions to remain as osmotically active particles in the filtrate. Diuretics (Table 25.3) also cause the countercurrent multiplier to equilibrate at a lower maximum concentration, so that the amount of water passing through the walls of the collecting tubules will be less. Diuretics may cause excessive loss of K^+, and blood K^+ levels should be monitored as diuretics are used.

TABLE 25.2
SUMMARY OF THE PROCESSES OCCURRING IN URINE FORMATION

PROCESS	WHERE OCCURRING	FORCE RESPONSIBLE	RESULT
Filtration	Renal corpuscle	Blood pressure, opposed by osmotic, interstitial, and intratubular pressures	Formation of fluid having no formed elements and low protein concentration
Tubular transport Reabsorption	Proximal tubule Distal tubule Loop of Henle	Active transport	Return to bloodstream of physiologically important solutes
Secretion	Proximal tubule Distal tubule	Active transport	Excretion of materials Acidification of urine
Acidification (acid and base regulation)	Distal tubule	Active transport and exchange of alkali for acid	Excretion of excess H^+ Conservation of base (Na^+ and HCO_3)
Countercurrent multiplier and exchanger	Loop of Henle and vasa recta	(Multiplier) active transport (Exchanger) diffusion	Creates conditions for hypertonic urine formation
ADH mechanism	Collecting tubule and papillary ducts	Osmosis of water under permissive action of ADH	Formation of hypertonic urine

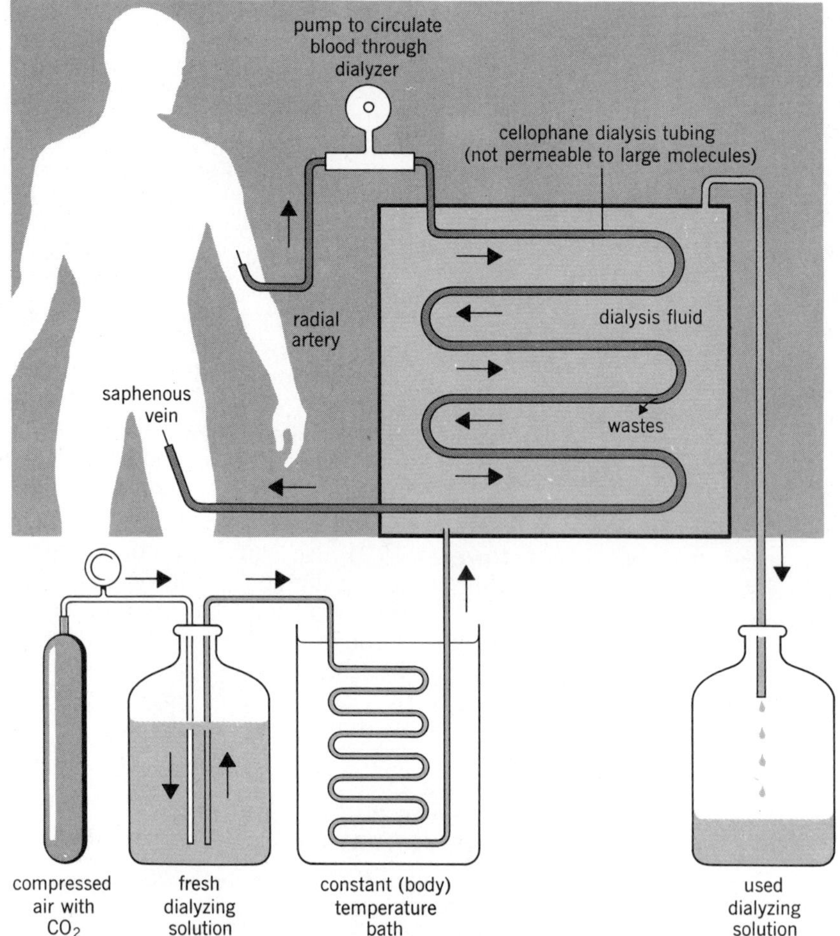

FIGURE 25.11
The principle of operation of the artificial kidney (hemodialysis unit). The dialysis fluid contains electrolytes and essential substances in the same proportions as in the blood.

THE KIDNEY AND ACID-BASE BALANCE

The kidneys play a vital role in the regulation of acid-base homeostasis by their ability to secrete H^+ in the proximal and distal convoluted tubules. Hydrogen ion accumulates in the body as a result of the production of CO_2 from metabolic processes, and the reaction of CO_2 with H_2O is according to the following equation:

$$CO_2 + H_2O \rightleftharpoons H_2CO_3 \rightleftharpoons H^+ + HCO_3^-$$

The body normally is capable of preventing the accumulation of excessive H^+ in the form of H_2CO_3 by elimination of CO_2 through the lungs. Additional sources of H^+ include such metabolic sources as the production of organic acids during incomplete combustion of foodstuffs, the production of acids from dietary intake of phosphorous (as phosphate), sulfur (as sulfate), and the metabolism of compounds containing these elements. The lungs play only a minor role in the elimination of H^+ from these sources; elimination depends on the kidney.

Hydrogen ion is secreted by the proximal and distal tubule cells in exchange for sodium ion, the latter being returned to the bloodstream. The exchange is made on a one-for-one basis. Secreted H^+ reacts with bicarbonate in the tubular lumen to form carbonic acid; carbonic acid dissociates to form CO_2 and H_2O, with the CO_2 diffusing into the tubule cells according to its concentration gradient where carbonic anhydrase catalyzes the formation of carbonic acid. The dissociation of this carbonic acid provides more H^+ for secretion, and the bicarbonate that results is transported into the bloodstream. The process was illustrated in Fig. 25.9 and results in maintenance of body bicarbonate and sodium reserves at the same time it rids the body of H^+. When H^+ excretion and $NaHCO_3$ reabsorption alone are insufficient to maintain acid-base balance, the kidney may secrete ammonium ion (NH_4^+) in addition to H^+, and it may secrete potassium ion (K^+) in addition to H^+. Deamination of amino acids in the tubule cells produces ammonia, which combines with an H^+ to form the ammonium ion, according to the equation

$$NH_3 + H^+ \longrightarrow NH_4^+$$

The ammonium ion then diffuses into the tubule lumen. Active potassium secretion results in the passage of two positive ions (H^+ and K^+) into the tubule lumen and results in the reabsorption of 2 Na^+ in exchange. Both mechanisms essentially double the number of $NaHCO_3$ molecules reabsorbed from the tubule lumen, restoring the alkali of the bloodstream to a normal ratio of H^+. These processes were also depicted in Fig. 25.9.

The reactions described above do not occur instantaneously; a day or two is usually required to restore normal pH in the face of acid-base disturbances. As presented in Chapter 17, there are four major disturbances of acid-base balance. RESPIRATORY ACIDOSIS occurs from diminished elimination of CO_2 through the lungs. Accumulation of H^+ stimulates breathing, which may suffice to eliminate the excess CO_2 producing the disorder. Additionally, increased kidney secretion of H^+ aids in returning the pH to normal. On its own, over a period of about two days, the kidney can fully compensate the condition. RESPIRATORY ALKALOSIS results from an excessive loss of CO_2 that may occur during hyperventilation. Loss of CO_2 lowers the H^+ available from the reaction of CO_2 and H_2O. Loss of CO_2 slows breathing, and kidney secretion of H^+ is lowered. Diminished secretion of H^+ results in less HCO_3^- available for reabsorption, and the pH falls. METABOLIC ACIDOSIS develops from loss of bicarbonate from the body, as in diarrhea, or from excessive H^+ produced by exaggerated or abnormal metabolic processes, as in diabetes mellitus. The kidney increases its secretion of H^+ and elimination of organic acids and reabsorbs increased amounts of bicarbonate ion.

TABLE 25.3

SOME DIURETICS AND THEIR MECHANISMS OF ACTION

MECHANISM OF ACTION	EXAMPLES	SITE OF ACTION AND RESULT
Inhibition of ADH secretion	Water	Hypothalamus; urine volume ↑
	Ethyl alcohol	Hypothalamus; urine volume ↑
Inhibition of Na^+	Caffeine	Glomerulus and tubules; GFR; ↓ tubular Na^+ reabsorption ∴ volume
reabsorption	Mercurial salts	Beyond proximal tubules; volume ↑
	Thiazides (Diuril)	Loop of Henle and distal tubule. Volume ↑
	Furosemides (Lasix)	Loop of Henle. Volume ↑
Decrease H^+ secretion and thus Na^+ reabsorption	Carbonic anhydrase inhibitors (Diamox)	Proximal and distal tubules; since H^+ and Na^+ are exchanged one for one, results in Na^+ remaining in tubules. Volume ↑

METABOLIC ALKALOSIS may result from loss of H^+, as in vomiting, or as a result of excessive intake of alkalis, as in antacid medications. Here again, decreased secretion of H^+ by the kidney tubules and lowered reabsorption of bicarbonate ion aids in returning pH to normal. Therefore, the kidney is the last—but most powerful—line of defense in control of acid-base homeostasis.

THE URINE

COMPOSITION AND CHARACTERISTICS

Normal urine is an amber or yellow transparent fluid having a pH between 5 and 7, a characteristic odor, and a specific gravity varying between 1.015 and 1.025.

The COLOR is due to the presence of urobilinogen, a pigment derived from the destruction of hemoglobin by cells of the reticuloendothelial system, as follows:

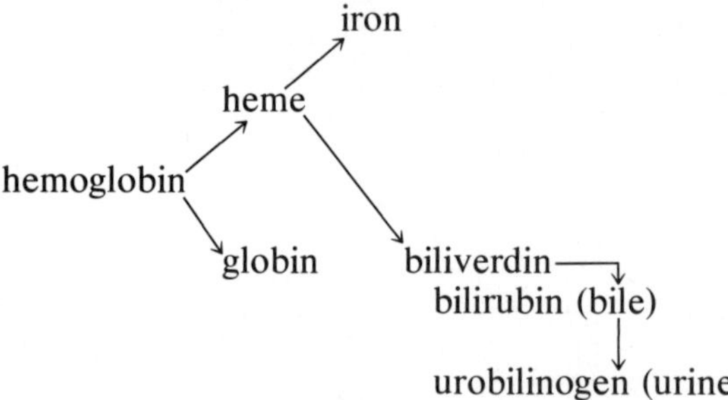

The ACIDIC REACTION is due primarily to secretion of H^+ into the filtrate by the proximal and distal tubules. The ODOR of freshly voided urine is due chiefly to the presence of organic acids, while the odor of stale urine is ammoniacal due to the decomposition of urea with release of ammonia. The specific gravity depends on the total solute concentration per unit volume of fluid and accounts for the variability of this characteristic.

The composition of normal urine varies little in terms of materials present. Tables 25.4 and 25.5 indicate some comparisons between plasma, filtrate, and urine and some normal constituents of urine, their origins, and amounts excreted per day.

Glucose normally does not appear in the urine except after a meal rich in carbohydrates because it is usually completely reabsorbed.

Protein is filtered only to a limited extent and is also normally completely absorbed.

The ketone bodies, acetone and beta-hydroxy-butyric acid, are present in the urine only when large amounts of fatty acids are being catabolized, as in diabetes mellitus. Only when these materials are present in the urine above trace levels can their presence be considered abnormal.

Casts may be found in the urine and are evidence of renal lesions. They are named according to composition, and include epithelial, fatty, pus, blood, or hyaline (clear) casts. The casts maintain the shape of the kidney tubules where they are formed.

CALCULI (*stones*) are precipitated masses of inorganic (e.g. calcium) or organic (e.g. urea) material. They may grow to large size and bizarre shapes, and may require surgery for their removal.

DISORDERS OF THE KIDNEY

SIGNS OF DISORDERED RENAL FUNCTION

Prolongation of POLYURIA (excessive urine volume), OLIGURIA (scanty urine volume), or ANURIA (cessation of urine production) beyond a few hours is usually indicative of abnormal renal function. DECREASED CLEARANCES of inulin or other test materials and increased blood urea levels (uremia) also indicate malfunction. In damage to the nephrons, PROTEINURIA typically occurs due to increased permeability of glomerular membranes or trauma and destruction of tubule cells. The proteins appearing in the urine are most commonly albumins from the blood. They are the smallest and pass through membranes most easily. Accumulation of urea as uremia results in lethargy, pruritus (itching), muscle twitching, mental confusion, and coma. If not lowered, urea may cause death through toxic effects on the brain and liver. The artificial kidney may be employed in the process of HEMODIALYSIS. EDEMA or excessive fluid retention is also characteristic of certain renal disorders, particularly those secondary to heart failure and

TABLE 25.4

COMPARISONS OF THE PROPERTIES OF PLASMA, FILTRATE, AND URINE[a]

PROPERTY OR SUBSTANCE	PLASMA	FILTRATE	URINE	DEGREE OF CHANGE BETWEEN FILTRATE AND URINE
Osmolarity (mOs)	300	300	600–1200	2–4 fold
pH	7.4 ± 0.02	7.4 ± 0.02	4.6–8.2	300–30,000 fold
Specific gravity	1.008–1.010	1.008–1.010	1.015–1.025	—
Sodium	0.3	0.3	0.35	—
Potassium	0.02	0.02	0.15	7 fold
Chloride	0.37	0.37	0.60	2 fold
Ammonia	0.001	0.001	0.04	40 fold
Urea	0.03	0.03	2.0	60 fold
Sulfate	0.002	0.002	0.18	90 fold
Creatinine	0.001	0.001	0.075	75 fold
Glucose	0.1	0.1	0	—
Protein	7–9	Trace	0	—

[a] Values, unless otherwise indicated are grams %.

decreased filtration of solutes. ACIDOSIS, secondary to decreased H^+ secretion, may also occur in kidney malfunction.

The artificial kidney (Fig. 25.11) relies on the principle that individual chemical substances in the bloodstream will diffuse through selective membranes if a gradient is created to cause that diffusion. In the artificial kidney are many feet of cellophane dialyzing tubing, surrounded by a fluid having a composition similar to blood except for zero concentrations of wastes (e.g., urea, creatinine, uric acid). The machine re-

TABLE 25.5

SOME NORMAL CONSTITUENTS, ORIGINS, AND AMOUNTS EXCRETED PER DAY IN NORMAL URINE

CONSTITUENT	ORIGIN	AMOUNT PER DAY
Water	Diet and metabolism	1200–1500 ml
Urea	Ornithine cycle	30 g
Uric acid (purine)	Catabolism of nucleic acids	0.7 g
Hippuric acid	Liver detoxification of benzoic acid	Trace
Creatinine	Destruction of intracellular creatine phosphate of muscle	1–2 g
Ammonia	Deamination of amino acids	0.45 g
Chloride (as NaCl)	Diet	12.5 g
Phosphate	Diet and metabolism of phosphate containing compounds	3 g
Sulfate	Diet, metabolism of sulfate containing compounds, formation of H_2SO_4 in kidney tubules	2.5 g
Calcium	Diet	200 mg

TABLE 25.6
SOME DISORDERS OF RENAL TRANSPORT MECHANISMS

CONDITION	HEREDITY	DISTURBANCE OF	CONTROLLED BY	COMMENTS
Nephrogenic diabetes insipidus	Autosomal dominant	Water excretion; tubule cells fail to respond to ADH	Reduction of dietary solutes and protein intake	Control measures extend 1–2 years; then child can usually control own diet
Renal glucosuria	Autosomal dominant	Glucose reabsorption; failure to reabsorb normal amounts of glucose in the kidney	None required	Glucose appears in urine when blood sugar level is normal
Hartnup disease	Autosomal recessive	Tryptophane (an essential amino acid) reabsorption and metabolism; gross amino aciduria	Administration of nicotinamide	May lead to shortage of nicotinic acid for NAD and NADP synthesis
Cystinuria	Autosomal recessive	Transport of the four dibasic amino acids (cystine, lysine, arginine, ornithine)	Administration of bicarbonate to prevent stone formation in kidney; diuresis	Cystine stones are most common complication; treatment minimizes formation

ceives its blood from the radial artery of the patient. Blood entering the "kidney" is infused with heparin to keep it from coagulating in the machine. Diffusion of wastes into the dialyzing fluid occurs in the machine. Outflowing blood is infused with an antiheparin substance (e.g., protamine) to minimize bleeding in the patient as blood is returned to the saphenous vein of the lower limb. About 500 ml (1 pint) of the patient's blood is in the machine at any given time, a volume loss to which the body adjusts easily. Usually, a patient is "on the machine" for 8 to 12 hours every three to four days. Actual time spent depends on the severity of damage to the patient's kidneys or if the organs are present. The machine is thus used to "rest" damaged kidneys or to substitute entirely for missing organs.

The most common functional disorders affecting the kidney are associated with genetic and infectious causes.

GENETIC CAUSES

At least seven errors of metabolism associated with defects in renal transport systems are recognized. Gene defects, which result in an abnormal enzyme or carrier involved in an active transport system, are postulated as the causes of the defects. Several of the more important disorders are presented in Table 25.6.

WILMS' TUMOR is a malignant embryonic renal tumor of genetic origin that forms the most common malignant abdominal tumor of infancy and childhood. It usually occurs in only one kidney, and removal of the affected organ, followed by radiation and/or chemotherapy, results in 70 to 90 percent survival rates.

INFECTIOUS AND INFLAMMATORY CAUSES

PYELONEPHRITIS is a bacterial infection of the kidney in which the causative organism is delivered to the kidney via the bloodstream, or there is a retrograde infection from the bladder up the ureters to the kidney. Antibiotic therapy depends on the sensitivity of the causative organism to the drug. GLOMERULONEPHRITIS results from an infectious agent, or its product(s), that damages the glomerular basement membrane. Chemical alterations in its structure render it antigenic, antibodies are produced, and the resultant antigen-antibody reaction results in cracking and splitting of the basement membrane. Loss of plasma albumin is common in this disorder. The causative organism involved is a streptococcus. Infections of the kidney are the most common causes leading to renal failure, which implies an inability of the kidney to perform its normal tasks adequately. Filtration rate falls, urea elimination is decreased, and signs of uremic poisoning ("odor of urine" to the perspiration, generalized itching, low urine output) appear. The NEPHROTIC SYNDROME is the result of glomerular damage and tubular necrosis (death of the tissues). The primary symptom is a massive edema that develops secondarily to plasma protein loss through the damaged nephrons. Treatment is directed toward prevention of secondary complications (e.g., infections) while the damage is repaired by the kidney itself.

OTHER FUNCTIONS OF THE KIDNEY

At the point shortly before the afferent arteriole enters the glomerulus, there is an interesting group of modified smooth muscle cells called *the juxtaglomerular cells.* An adjacent region on the distal tubule, called the *macula densa,* consists of modified tubule cells. The two areas together constitute the JUXTAGLOMERULAR APPARATUS (see Fig. 16.7). This particular region of the arteriole has been shown to be sensitive to changes in blood flow. It has been hypothesized that if the kidney becomes ischemic, the apparatus secretes an enzymelike material called *renin.* This particular substance works on a substrate in the plasma called hypertensinogen (angiotensinogen) and converts it into a weakly active material, hypertensin I. Hypertensin I is in turn converted to hypertensin II by a plasma enzyme. Hypertensin II is an active vasoconstrictor, bringing about a narrowing of arterioles over the entire body. This particular effect raises the blood pressure and ensures a continuance of high blood pressure to the kidney, which is a necessary requisite for filtration to occur. Renal hypertension, a chronic elevation of the blood pressure, may occur if the kidney suffers continual ischemia and renin production is not diminished. Hypertensin II also affects the outer zone of the adrenal cortex that secretes aldosterone. The hormone is instrumental in governing reabsorption of sodium by the kidney tubules. Therefore, the kidney governs its own activity to a small degree.

A great degree of AUTOREGULATION is demonstrated by the kidney in controlling blood flow to the glomerulus. The afferent arteriole has been shown to undergo constriction or dilation in response to a wide range (90 to 220 mm Hg) of renal artery pressure so as to maintain renal blood flow constant. The response occurs in the absence of all nerves to the kidney and in a kidney removed from the body and artificially perfused. It is therefore an inherent mechanism within the kidney itself.

Three theories have been advanced to explain autoregulation.

The THEORY OF CELL SEPARATION suggests that as blood passes through the interlobular arteries into the afferent arterioles, a separation of red cells and plasma occurs that allows greater perfusion of some glomeruli, while others suffer a reduction. Perfusion balances out to maintain an overall nearly constant rate of filtration.

The THEORY OF INTRARENAL PRESSURE suggests that increases in renal arterial pressure to the kidney are matched by increases in the pressure of the interstitial fluid of the kidney. This compresses the capillary networks and veins of the kidney to impede flow and maintain pressure.

The MYOGENIC THEORY suggests that all autoregulation is the result of changes in tone of the smooth muscle of the afferent and efferent arterioles. It is known that smooth muscle cells respond to stretching (as with an increased perfusion pressure) by a stronger contraction. This maintains vessel diameter, flow, and pressure. Decrease of pressure causes relaxation that increases diameter and maintains flow and pressure. If the afferent arteriole dilates and the efferent arteriole does not change, flow and pressure to the glomeruli will increase. Dilation of the efferent arteriole or constriction of the afferent arteriole reduces flow and pressure in the glomeruli. Also, pressure changes in the afferent arteriole may influence the juxta-

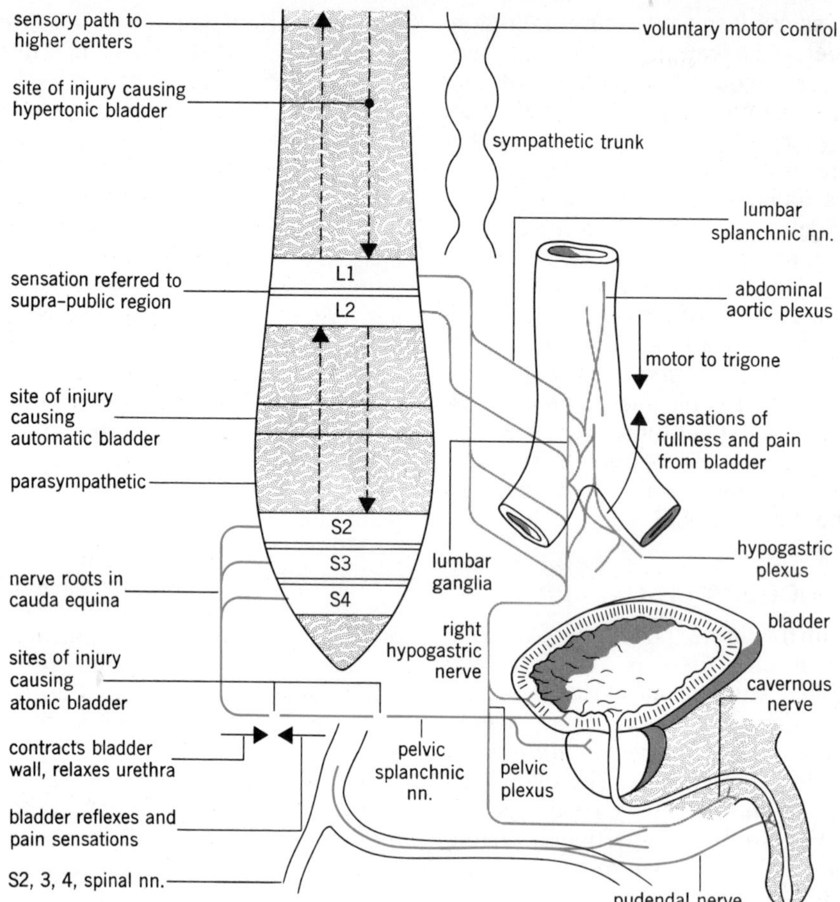

FIGURE 25.12
The innervation of the bladder illustrating the genesis of bladder dysfunction.

glomerular apparatus to secrete chemicals (e.g., renin) that result in vascular diameter changes. The myogenic theory would appear to have more direct proof to sustain it because autoregulation is eliminated when the kidney is perfused by substances (cyanide, procaine) that paralyze smooth muscle.

MICTURITION

The efferent nerves to the bladder and urethra (Fig. 25.12) are from both sympathetic and parasympathetic divisions. The sympathetic fibers furnish inhibitory fibers to the muscle of the bladder; they furnish motor fibers to the trigone and internal sphincter, and to the muscle of the upper part of the urethra. These fibers rise in the lumbar spinal segments and pass to the bladder via the inferior hypogastric plexus. The parasympathetic nerves supply motor fibers to the detrusor muscle (the muscle of the bladder) and inhibitory fibers to the internal sphincter. The desire to urinate occurs when a volume of 200 to 300 ml of urine has accumulated in the bladder. Stretch on the muscle of the bladder, brought about by filling, evokes an afferent impulse in the pelvic nerves. This impulse ascends through the spinal cord to a center in the hindbrain. An efferent discharge of motor impulses to the muscle of the bladder is accomplished through descending pathways in the cord and through the pelvic motor nerves. The same efferent or motor fibers bring about a simultaneous relaxation of the internal sphincter so that urine may be emptied from the bladder into the urethra (micturition). Urination may occur by reflex action not involving the hindbrain center. Filling evokes a reflex contraction of the detrusor and relaxation of the sphincter, which is served by lower cord segments only. The reflex is seen in infants, where the voluntary control over sphincters has not yet been achieved. Nerve injuries produce three types of bladder dysfunction.

ATONIC BLADDER. Interruption of sensory supply results in loss of bladder tone, and the organ may become extremely distended with no development of an urge to urinate.

HYPERTONIC BLADDER. Interruption of the voluntary pathways results in excessive tone, and very small distentions create an uncontrolled desire to urinate.

AUTOMATIC BLADDER. Complete section of the cord above the first sacral nerve exit (S1) produces automatic emptying in response to filling, by the above-described cord reflex.

SUMMARY

1. The body utilizes several pathways to rid itself of products of metabolism.
   a. Heat is eliminated mainly by the skin.
   b. Carbon dioxide is eliminated mainly by the lungs.
   c. The alimentary tract eliminates wastes of digestion and some minerals.
   d. The kidney eliminates excess water, nitrogenous wastes, excess inorganic substances, and a variety of detoxified materials.
2. The kidneys, ureters, urinary bladder, and urethra comprise the urinary system.
3. The kidneys are bean-shaped, and show a variety of gross features that are correlated with their major functions (*see* Fig. 25.2).
4. The nephrons of the kidney are the functional units that carry out the functions given in 1.d. above. They consist of a capsule and tubules.
5. There are two types of nephrons.
   a. Cortical nephrons are primarily reabsorptive in function.
   b. Juxtamedullary nephrons do these jobs *and* operate the countercurrent systems.
6. Several processes are utilized as the nephrons function.
   a. Glomerular filtration moves materials from plasma to nephron. It requires a minimal pressure (effective filtration pressure).
   b. Tubular reabsorption reclaims physiologically important solutes.
   c. Tubular secretion moves materials from plasma to tubule.
   d. The proximal convoluted tubule carries out 80 to 90 percent of reabsorption and also secretes.
   e. The countercurrent multiplier moves solutes, chiefly sodium chloride, into renal ECF, to create an osmotic gradient for water to leave the collecting tubules. Urea also moves from the collecting tubules to aid in the creation of the gradient.
   f. The countercurrent exchanger maintains the solutes in the renal ECF to keep the osmotic gradient high.
   g. Acidification occurs in the distal convoluted tubule by H^+ secretion.
   h. A concentrated urine is produced by osmotic flow of water from collecting tubules in the presence of antidiuretic hormone.
7. Urine volume is controlled by several factors.
   a. Increased blood pressure increases volume.
   b. Increase of filtered solutes increases volume by osmotic "holding" of water in the tubules.
   c. ADH increase decreases volume by causing more water reabsorption.
   d. Loss by any other route decreases volume.
   e. Diuretics increase volume by interfering with solute reabsorption, enzyme inhibition, and other mechanisms (*see* Table 25.3).
8. The kidney is a major factor in maintenance of acid-base balance.
   a. The nephron secretes hydrogen ion.
   b. The nephron reabsorbs bicarbonate ion.
   c. The kidney can compensate acidosis and alkalosis over a period of a day or two.

9. The urine is the end product of nephron function.
   a. It is an acidic, amber colored fluid, with an osmolarity averaging about 900 mOs.
   b. Its major constituents are water, sodium chloride, and urea.
10. Renal disorder is commonly associated with prolonged alteration in urine volume, decreased clearances, proteinuria, and plasma urea accumulation.
11. Genetic and inflammatory or infectious causes account for many kidney disorders.
    a. Table 25.6 presents several genetic disorders affecting kidney function.
    b. Glomerulonephritis results from a streptococcal infection.
    c. The nephrotic syndrome results from glomerular and tubular necrosis.
12. The kidney produces renin when it is ischemic. Renin triggers the angiotensin mechanism. Autoregulation of blood flow is a property of the kidney. Several theories to account for autoregulation are presented.
13. Micturition is basically a reflex process that can be voluntarily modified.
    a. Atonic bladder results from interruption of sensory nerves.
    b. Hypertonic bladder results from interruption of cerebralspinal pathways.
    c. Automatic bladder results from complete section of the cord above spinal level S-1.

QUESTIONS

1. What pathways are employed by the body to eliminate the wastes of metabolic activity, and what products are eliminated by each route?
2. How does the blood supply of the nephrons ensure the high blood pressure necessary for filtrate formation?
3. What anatomical and functional differences exist between the two types of nephrons in the kidney?
4. What factors determine what the effective filtration pressure of the glomerulus will be?
5. What are tubular reabsorption and secretion, and what substances are transported by each process?
6. What is the significance of reabsorption and secretion to regulation of ECF composition?
7. What functions do the countercurrent multiplier and exchanger serve?
8. How is the osmotic gradient that draws water from the tubules created by the multiplier and maintained by the exchanger?
9. What is the role of antidiuretic hormone in creation of a concentrated urine?
10. How do the kidneys contribute to the body's acid-base balance?
11. What factors determiné the volume of urine produced? How is the effect of each brought about?
12. How does the urine compare to plasma in terms of characteristics and components? What accounts for the differences exhibited?
13. What are the common symptoms of renal disorder regardless of cause?
14. Explain the mechanism that brings about micturition.

READINGS

Burg, M. B. "Tubular Chloride Transport and the Mode of Action of Some Diuretics." *Kidney Int. 9*:189, 1976.

Jamison, Rex L., and Roy H. Maffly. "The Urinary Concentrating Mechanism." *New Eng. J. Med. 295*:1059. Nov 4, 1976.

Kokko. J. P., and F. C. Rector, Jr. "Countercurrent Multiplication System without Active Transport in Inner Medulla." *Kidney Int. 2*:214. 1972.

Lassiter, William E. "Kidney." *Ann. Rev. Physiol.* Vol *37*. Palo Alto, Calif. 1975.

Nicoll, P. A., and T. A. Cortese, Jr. "The Physiology of Skin." *Ann. Rev. Physiol.* Vol *34*. Palo Alto, Calif. 1972.

Schrier, R. S., and Thomas Berl. "Nonosmolar Factors Affecting Renal Water Excretion." *New Eng. J. Med. 292*:81. Jan 9, 1975. *292*. Jan 16, 1975.

Stephenson, J. L. "Concentration of Urine in a Central Core Model of the Renal Counterflow System." *Kidney Int. 2*:85. 1972.

THE ENDOCRINE SYSTEM

(continued on next page)

OBJECTIVES

After studying this chapter, the reader should be able to:

☐ Give a general review of from where the various endocrine organs take their origins.

☐ List the criteria used to determine endocrine status.

☐ Comment generally on the "life history" of hormones (production, utilization, fate).

☐ List the methods by which endocrine secretion is controlled.

☐ Describe, for each endocrine, the following:
The hormone(s) produced
The effect of those hormone(s) on body physiology
The method of control of hormone secretion
A major disorder of over and under production, with obvious symptomology

☐ Describe the involvement of adrenal cortical hormones in inflammation and stress.

☐ Describe prostaglandins as to sites of production, effects on the body, and possible therapeutic uses.

☐ Discuss the status of the pineal gland as an endocrine organ, with evidence to support or refute that status.

DEVELOPMENT OF THE MAJOR ENDOCRINES

During the fourth to fifth week of development, the embryo shows a series of cartilagenous arches, the BRANCHIAL ARCHES, in the neck region. Outpocketings of the lining of the foregut, the PHARYNGEAL POUCHES, will develop between arches. In the roof of the foregut, a single median outpocketing, RATHKE'S POUCH, appears. The HYPOPHYSIS (pituitary) is a gland whose anterior portion (anterior lobe) is derived from Rathke's pouch. This portion migrates up and backward to fuse with a downgrowth of the hypothalamus, the latter representing primarily the posterior lobe of the gland.

The THYROID GLAND is derived from a single medial outgrowth of the floor of the foregut at the level of the second pharyngeal pouch. It migrates inferiorly to assume a final position around the larynx and upper trachea.

The PARATHYROIDS develop from the anterior portions of the third and fourth pharyngeal pouches. These migrate inferiorly to assume a final position on the dorsal side of the thyroid.

The THYMUS originates from the posterior portions of the third and fourth pharyngeal pouches and migrates to lie behind the sternum.

The ADRENAL MEDULLA is a derivative of neural crest material. A covering of mesoderm forming the CORTEX of the gland is applied to the adrenal medulla.

The endocrine tissue of the PANCREAS (islets) arises from the ends of the endodermal outpocketings of the gut. The outpocketings form the ducts, and the islets lose their connection with the ducts.

The endocrine cells of TESTIS and OVARY differentiate from the mesodermal cells as the gonads develop.

The diversity of origins of the endocrines, in terms of both site and germ layer is impressive. The origins are shown in Fig. 26.1.

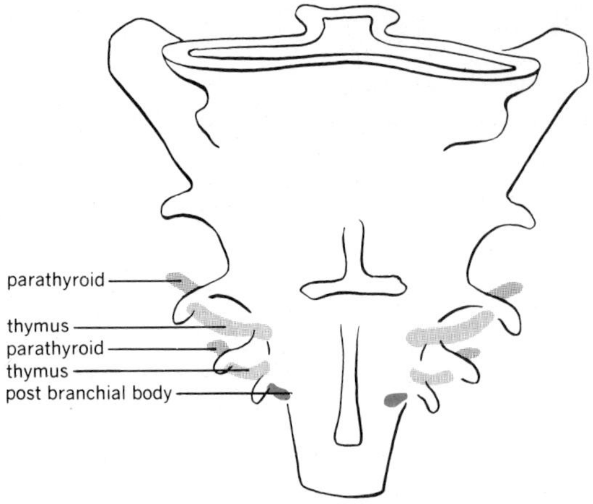

FIGURE 26.1
Origins of some of the endocrine organs from branchial pouches.

GENERAL PRINCIPLES OF ENDOCRINOLOGY

We have examined the role of the nervous system in stimulus-response and homeostatic control. The endocrine system is primarily concerned with GOVERNING BODY PROCESSES such as growth, differentiation, and metabolism. Endocrine glands utilize the bloodstream to distribute their chemical messengers, the HORMONES, and in this way achieve contact with all body cells. It is becoming increasingly evident that the two "control systems" operate together to control body activity. The nervous system may, for example, respond to environmental changes and influence an endocrine to secrete a product that has the effect of restoring homeostasis.

CRITERIA FOR DETERMINING ENDOCRINE STATUS

Endocrine glands are designated as being CIRCUMSCRIBED, or specific groups of cells, that PRODUCE SPECIFIC CHEMICAL SUBSTANCES having well-defined effects on body function. Removal of a group of cells suspected of being an endocrine causes clearly defined alterations of body function, and administration of the active chemical of the gland likewise alters body function. Utilizing these criteria, we recognize the endocrine structures shown in Fig. 26.2.

HORMONES AND PARAHORMONES

The product of an endocrine gland (as defined above) may properly be called a HORMONE. There are many chemical substances that have specific effects on the body, but are not traceable to specific cells. Examples include carbon dioxide, gastrin, secretin, enterocrinin, and the prostaglandins. Such substances are commonly called "hormones"; the term PARAHORMONE, however, serves to distinguish such substances from true hormones.

CHEMICAL NATURE OF HORMONES

Mammalian hormones fall into three classes: PROTEINS, POLYPEPTIDES, or compounds such as GLYCOPROTEINS; AMINES; and STEROIDS. A general correlation exists between the germ layer of origin of the endocrine and the class of hormone produced. Most amine hormones are produced by ectodermally derived endocrines. Steroids are produced by mesodermal derivatives. Polypeptide or protein hormones are produced by endodermal derivatives.

SYNTHESIS, STORAGE, TRANSPORT, AND USE OF HORMONES

A given endocrine cell synthesizes its hormones from raw materials present in the blood reaching the gland. Some glands (e.g., thyroid) produce more hormone than the body requires at a time and may store excess hormone within the gland. Secretion into the bloodstream occurs as the result of a chemical or nervous stimulus. In the bloodstream, most steroid and amine hormones attach to a specific plasma protein "carrier" and are transported to the body cells in this combined fashion. At the cell, the hormone may attach to the cell surface and enter by pinocytosis, or pass directly through the membrane to influence enzymes, the nucleus, or cellular organelles. As the cell utilizes the hormone, it generally alters the hormone's structure, partially or completely destroying the physiological activity of the chemical. This process is termed "inactivation." The inactivated product passes from the cell to the bloodstream and is usually excreted. Inactivation of hormones by the organ it affects (*target organ*) results in the requirement that hormones be continually secreted by the endocrine gland. In short, hormones are not reused, but are used up by the target organ.

FUNCTION, EFFECT, AND ACTION OF HORMONES

The FUNCTION of a hormone is what we conclude the purpose of a hormone to be, in terms of whole body function. EFFECT refers to measurable alterations in function occasioned by deprivation or administration of a hormone. ACTION refers to the cellular mechanism influenced by the hormone. For example, thyroxin, the product of the thyroid gland, is known to increase glycolysis within the mitochondria (action) and to bring about an increase in BMR and heat production (effect). It is concluded that the hormone controls BMR (function).

Hormone action appears to depend on the presence within the cell of a chemical reaction whose rate can be influenced by the hormone. Some hormones are without action on certain cells, suggesting the absence of an influencable reaction.

CONTROL OF SECRETION OF HORMONES

In general terms, SIX DIFFERENT METHODS exist by which both rate and quantity of hormone secretion may be determined.

NEUROHUMORS. A nerve cell or group of nerve cells produces a neurohumor or neurosecretion that reaches the endocrine gland via blood vessels or nerve fibers. Hypothalamic production of neurohumors is an example of this type of control.

FIGURE 26.2
The locations and names of the major endocrine organs.

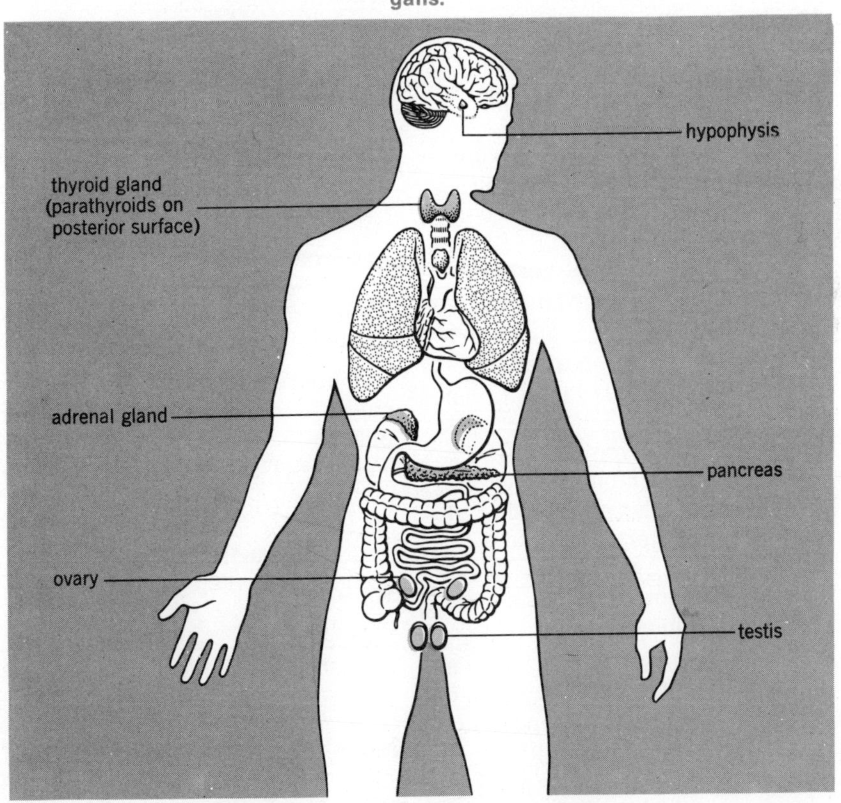

TABLE 26.1

GENERAL SUMMARY OF ENDOCRINE GLANDS

ITEM	COMMENTS
Functions of endocrines	To integrate, correlate, and control body processes by chemical means.
Criteria for establishing function as endocrine	Cells are morphologically distinct. Cells produce specific chemicals not produced elsewhere. Chemicals exert specific effects; effect is lost if gland removed, restored if chemical administered.
Hormones	Hormones are produced by cells qualifying as endocrines.
Chemical nature of hormones	Steroid—mesodermal origin (e.g.: adrenal cortex, gonads). Polypeptides—endodermal origin (e.g.: pancreas). Small MW amines—ectodermal origin (e.g.: posterior pituitary, adrenal medulla).
Methods of control available	
1. Neurohumor	Nerve cells produce a chemical, it goes to endocrine and controls secretion. Stimulus is chemical.
2. Nerves	Nerve fibers pass to endocrine. Stimulus is electrical (nerve impulse).
3. Feedback	Target organ hormone influences secretion of another endocrine which stimulated target organ.
4. Nonhormonal organic substances in blood	Glucose acting on pancreatic islets. Rise causes increased secretion (usually).
5. Total osmolarity of blood	Requires nervous system to detect it. Nervous system then signals endocrine.
6. Inorganic substances in blood	Ca^{++} on parathyroid. Effect usually direct, not inverse.

DIRECT INNERVATION. Usually done by autonomic fibers, this results in rapid secretion as a result of nerve impulses. Sympathetic control of the adrenal medulla, and hypothalamic control of the release of hormones from the posterior pituitary are examples.

FEEDBACK CONTROL. One endocrine gland produces a hormone that affects another endocrine gland (*target organ*). The target organ produces its hormone that influences the secretion by the first endocrine gland. If the first endocrine gland is inhibited, negative feedback has occurred. The relationship of trophic anterior pituitary hormones and their target organ hormones are examples. These relationships are excellent examples of the cybernetic control of homeostasis.

BLOOD LEVELS OF ORGANIC SUBSTANCES OTHER THAN HORMONES. Blood levels of glucose and the secretion of pancreatic hormones is the best example of this control method.

BLOOD LEVELS OF SPECIFIC INORGANIC SUBSTANCES. The relation of blood calcium level to parathyroid and thyroid secretion of hormones is an example.

BLOOD LEVELS OF SOLUTES IN GENERAL. The hypothalamus monitoring the osmotic pressure of the blood for ADH secretion is the obvious example of this means of control.

Understanding of the discussions of individual endocrine glands will be aided if the remarks on the previous pages are well understood. Table 26.1 summarizes the preceding sections.

TABLE 26.2

DIVISIONS OF THE HYPOPHYSIS

MAJOR DIVISIONS	SUBDIVISIONS
Adenohypophysis	Pars distalis (anterior lobe) Pars tuberalis Pars intermedia
Neurohypophysis	Infundibulum (stalk) Pars nervosa (posterior lobe)

THE HYPOPHYSIS (PITUITARY)

The hypophysis is a gland of double origin lying in the sella turcica of the sphenoid bone. It measures about 10 × 13 mm and weighs approximately 1/2 g. The gland is composed of a portion derived from the roof of the oral cavity, the adenohypophysis, and a portion derived from the hypothalamus, the neurohypophysis. Each region has several subdivisions, as shown in Fig. 26.3 and Table 26.2.

Both major divisions of the gland maintain connections with the hypothalamus (*see* Fig. 26.3). The ADENOHYPOPHYSIS is connected to the hypothalamus by a system of blood vessels originating in the hypothalamus and terminating as sinusoids in the pars distalis. This system is designated as the HYPOTHALAMICO-HYPOPHYSEAL PORTAL SYSTEM (HHPS) or PITUITARY PORTAL SYSTEM. The NEUROHYPOPHYSIS is connected to the hypothalamus by nerve fibers originating in the hypothalamus and terminating in the pars nervosa. These fibers form the HYPOTHALAMICO-HYPOPHYSEAL TRACT (HHT).

FIGURE 26.3
The hypothalamic-hypophyseal connections, and the parts of the gland.

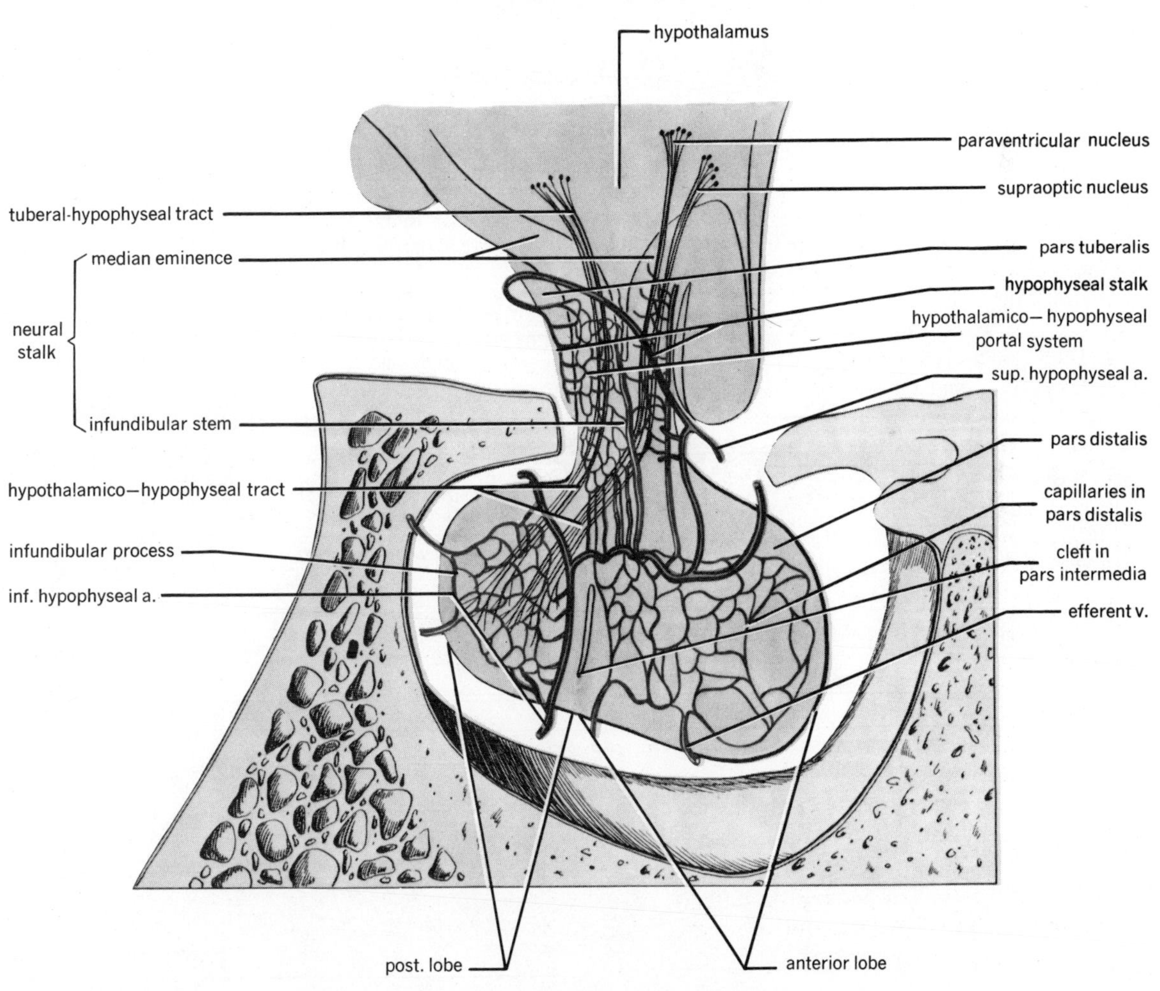

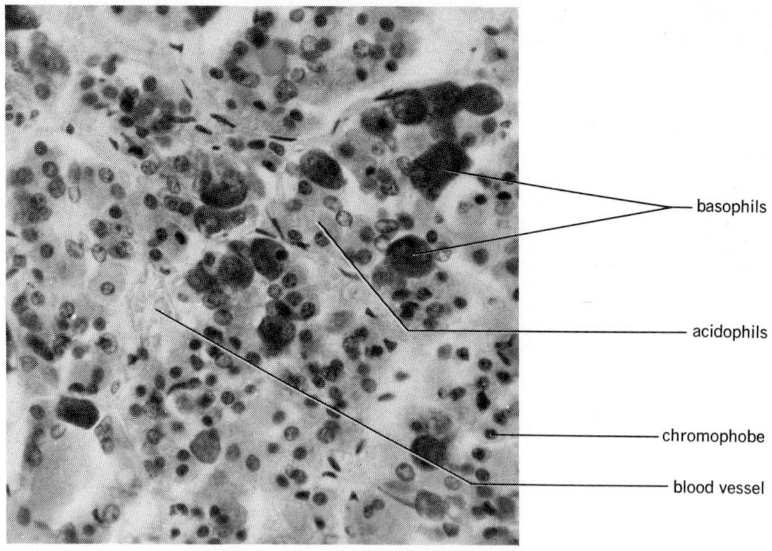

FIGURE 26.4
The cells of the pars distalis. With special stains: the STH-o-cyte stains orange-yellow and is irregular in shape; the Prolactin-o-cyte stains bright red and is irregular in shape; the ACTH-o-cyte stains reddish and is irregular in shape; the TSH-o-cyte stains bluish and is irregular in shape; the LH-o-cyte stains bluish and is rounded in shape; the FSH-o-cyte stains reddish and is rounded in shape. Chromophobes stain lightly or not at all.

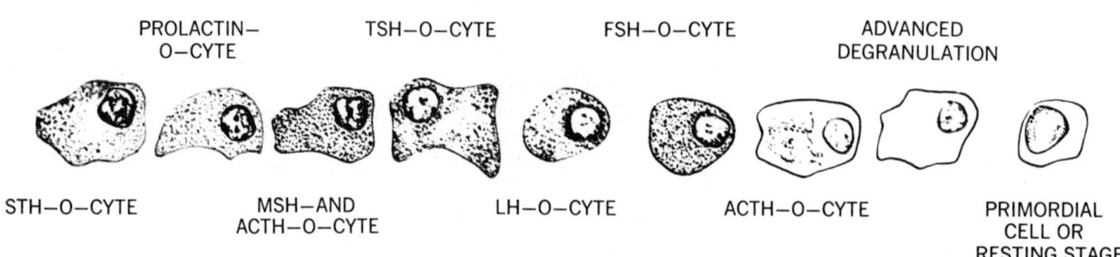

azocarmine–orange–G stain

These blood and nervous connections emphasize at least two important facts relative to the activity of the hypophysis: first, these connections enable nervous and endocrine systems to work in an integrated fashion to control body activity; second, endocrine response to nerve input may be achieved rapidly.

THE PARS DISTALIS

With ordinary stains, three cell types, designated ACIDOPHILS, BASOPHILS, and CHROMOPHOBES, may be recognized in the distalis. These cells are named according to their preference for red, blue, or no stain. With special stains, particularly PAS (periodic acid Schiff, a glycoprotein stain) and aldehyde thionine (protein polysaccharide stain), seven cell types may be differentiated. The production of at least six hormones has been attributed to the distalis, and a tentative assignment of hormones to a particular cell type has been made. The morphology and function of these cells is shown in Fig. 26.4 and described in Table 26.3.

HORMONES OF THE DISTALIS

Growth hormone [*somatotropin, somatotrophic hormone (STH), human growth hormone (HGH)*]. Growth hormone from human hypophysis is a single chain of 188 amino acids linked in two places by disulfide (—S—S—) bonds (Fig. 26.5). It has been synthesized in the laboratory. The normal pituitary contains 4 to 10 mg of hormone, and there is little difference in hormone content with age. Secretion of growth hormone is such as to maintain a blood level of about 3 μg (3/1,000,000 g) per ml.

The hormone exerts its effect primarily on the hard tissues of the body, secondarily on the soft tissues. In these two tissues, growth hormone increases rate of growth and maintains the size of body parts once maturity has been reached. Metabolic effects include: increasing the rate of conversion of carbohydrates to amino acids (transamination); increasing the cellular uptake of amino acids; mobilization of

TABLE 26.3
DISTALIS CELL TYPES

NAMES OF CELLS UNDER: ORDINARY STAIN	SPECIAL STAIN	DESCRIPTION	STAINING PREFERENCE	HORMONE PRODUCED
Acidophils	Somatotropic cell (α_1-orangophil)	Round or ovoid shape with small round acidophilic cystoplasmic granules	Orange G	Human growth hormone (HGH)
	Lactotropic cell (α_2-carminophil)	Round or ovoid, contains large round or oval acidophilic cytoplasmic granules	Azocarmine	Prolactin
Basophils	Corticotropic cell (β1)	Irregular shape, eccentric nucleus	PAS	Adrenocorticotropic hormone (ACTH)
	Thyrotropic cell (β2)	Irregular shape, eccentric nucleus	PAS and aldehyde thionine	Thyroid-stimulating hormone (TSH)
	Luteotropic cell (Δ_1)	Round outline, nearly central nucleus	PAS and aldehyde thionine	Luteinizing hormone (LH)
	Follicle stimulating hormone cell (Δ_2)	Round outline, nearly central nucleus	PAS	Follicle-stimulating hormone (FSH)
Chromophobe	Chromophobe	Irregular outline, eccentric nucleus, very pale staining cytoplasm	None, stains little if at all	ACTH?

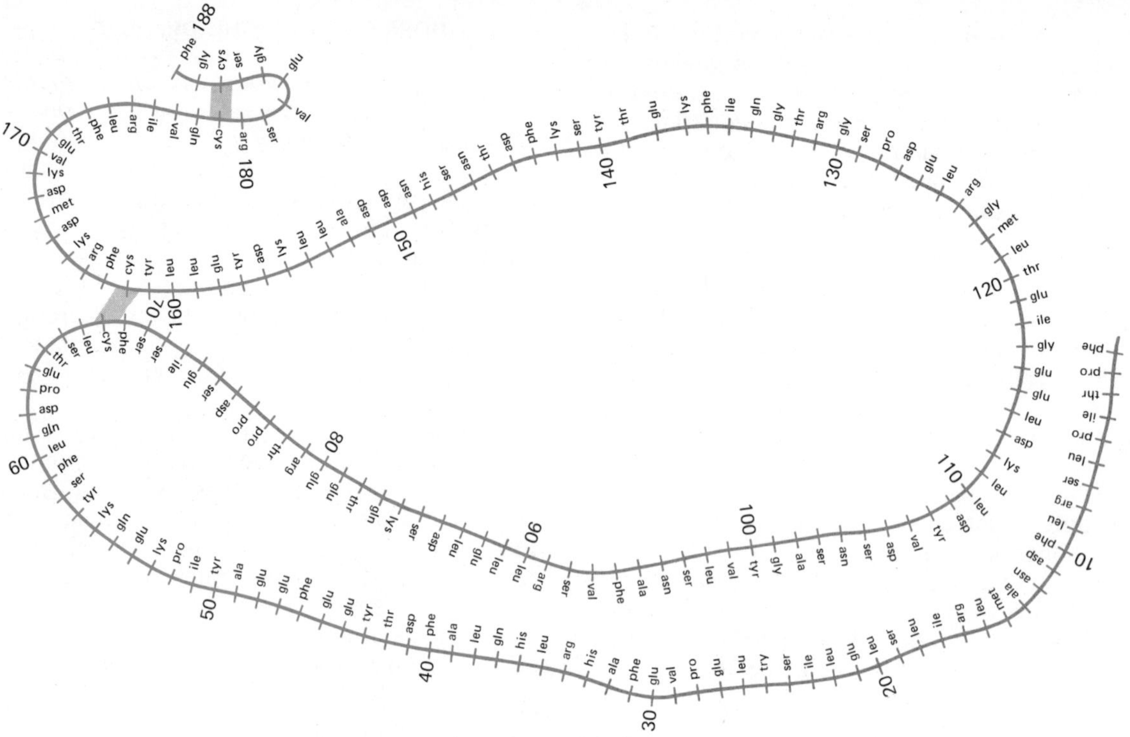

FIGURE 26.5
The amino acid sequence in human growth hormone.

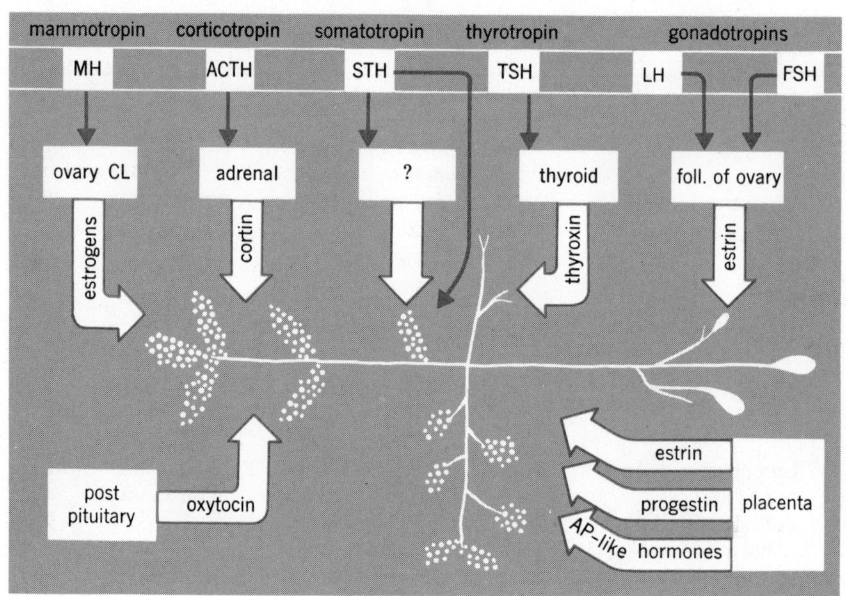

FIGURE 26.6
The hormones involved in lactation.

fats from storage areas; increasing fat metabolism, as evidenced by a fall in respiratory quotient (RQ); increasing blood sugar level (diabetogenic effect). There is some evidence to suggest that the fat-mobilizing effect may be separated from the other effects; some investigators claim a separate lipid mobilizing hormone exists.

Prolactin (*lactogenic hormone, luteotropic hormone LTH*). Prolactin is a polypeptide hormone of 205 amino acids. It is one of several hormones involved in milk production by the mammary glands. If the gland has been "primed" by sex hormones, thyroid hormone, and adrenal hormones, prolactin will cause milk secretion (Fig. 26.6). In birds, other effects of the hormone include aiding the expression of maternal behavior (nesting, egg hatching).

Adrenocorticotropic hormone (*ACTH, corticotropin*). ACTH is a single unbranched chain of 39 amino acids. It is synthesized as the body requires it, since the gland has less than 1/4 mg in it. ACTH controls the synthetic and secretory activity of the two inner zones of the adrenal cortex. Metabolic effects on cells in general include mobilization of fats, production of hypoglycemia, and increase in muscle glycogen. Acting through the adrenal cortex, ACTH is also involved in body resistance to stress.

Thyroid-stimulating hormone (*thyrotropin, TSH*). TSH is a glycoprotein having a molecular weight of about 25,000. TSH influences all phases of thyroid gland activity: accumulation of material, synthesis, and secretion. Acting through the thyroid, TSH becomes involved in regulation of basal metabolic rate. Effects on other tissues include increasing the breakdown of fat, working with growth hormone, and increasing water content of loose connective tissue.

TABLE 26.4
HORMONES OF THE PARS DISTALIS

HORMONE	CHARACTERISTICS	EFFECTS ON BODY
Growth Hormone (GH)	Molecular weight 21,500; unbranched peptide chain	Controls body hard and soft tissue growth
Prolactin	Molecular weight about 23,500	Causes milk secretion in "primed" gland
Adrenocorticotropic hormone (ACTH)	Molecular weight 4567; contains 39 amino acids in single peptide chain	Controls activities of adrenal cortex
Thyroid stimulating Hormone (TSH)	Molecular weight 25,000, composed of 200 to 300 amino acids	Controls all aspects of thyroid activity
Luteinizing hormone (LH) or Interstitial cell stimulating hormone (ICSH)	Molecular weight about 30,000; a glycoprotein	Female: Corpus luteum formation, ovulation, implantation Male: Controls interstitial cell activity
Follicle stimulating hormone (FSH)	Molecular weight about 20,000; a glycoprotein	Female: Controls follicular maturation Male: Controls spermatogenesis

Luteinizing hormone (*LH, luteotropin; in the male, it is sometimes called ICSH for interstitial-cell-stimulating hormone*). LH is a glycoprotein with a molecular weight of about 30,000. In the female LH occasions changes leading to the formation of the corpus luteum in the ovary and is involved in readying the mammary gland for secretion. It is necessary for ovulation and implantation of the zygote. In the male, LH controls secretion of the interstitial cells of the testis.

Follicle-stimulating hormone (*FSH*). FSH is also a glycoprotein with molecular weight about 30,000. In the female, FSH controls the maturation of primordial follicles to vesicular follicles; in the male, it controls spermatogenesis. Table 26.4 presents a summary of the characteristics and effects of pars distalis hormones.

CONTROL OF SECRETION OF PARS DISTALIS HORMONES

The role of the hypothalamus in maintenance of homeostasis was discussed in Chapter 7. It was emphasized that many other areas of the brain have nervous connections with the hypothalamus, and that the bloodstream is also monitored and can influence hypothalamic activity. One of the efferent connections of the hypothalamus is to the pituitary gland, and it has been shown that a system of blood vessels and nerve tracts connects the hypothalamus and the pituitary. Such systems of connections enable emotional disturbances, sensory input and environmental changes to be translated into endocrine response.

Control of secretion and synthesis of hormones by the adenohypophysis of the pituitary is believed to be mediated by chemical factors produced by the hypothalamus in neurosecretory cells and placed in the pituitary portal system for delivery to the adenohypophysis. Such chemicals are most properly termed REGULATING FACTORS, for some may stimulate release of adenohypophyseal hormones (releasing factors), and others may inhibit secretion (inhibiting factors). Some investigators believe these factors qualify as true hormones.

Today, many hypothalamic factors have been isolated or postulated to exist, and in several cases their chemical structure has been determined and the compound synthesized (Table 26.5).

Other factors than the hypothalamus appear to be involved in distalis secretion. For example, growth hormone secretion is increased by a fall in blood sugar level. This effect may be direct on the gland or through the hypothalamus. Negative feedback exists between thyroxin and thyroid-stimulating hormone, cortical steroids and adrenocorticotropic hormone, progestin and luteinizing hormone, and estrogen and follicle-stimulating hormone. Again, the problem is whether the effect is direct on the distalis or through the hypothalamus.

THE PARS INTERMEDIA

In the human hypophysis, the pars intermedia is virtually nonexistent, being reduced to a few cells and spaces (cysts) between the distalis and neural lobe. A hormone consisting of 22 amino acids and differing from those of lower animals is secreted by the human intermedia. It is called beta-MSH, and its physiological significance is uncertain. In lower vertebrates (fishes and amphibians), the intermedia is the source of MELANOCYTE-STIMULATING HORMONE (*MSH, intermedin*), which causes expansion of pigment cells (melanophores) in the skin of such animals. A protective effect of "matching" the color of the animal to the environment is the result. In man, adrenocorticotropic hormone has a function similar to MSH. Release of MSH is controlled by hypothalamic regulating factors (see Table 26.5).

THE PARS TUBERALIS

No known hormone is produced by the cells of the tuberalis.

NEUROHYPOPHYSIS

The neurohypophysis cannot be considered a separate endocrine gland, for it produces no hormones. The hypothalamus produces the hormones of the neural lobe and passes them over the hypothalamico-hypophyseal nerve tract merely to be stored in and released from the nerve terminals in the nervosa.

The characteristic cell of the neural lobe appears to be the pituicyte (Fig. 26.7).

TABLE 26.5
HYPOTHALAMIC FACTORS CONTROLLING SECRETION OF ADENOHYPOPHYSEAL HORMONES

HYPOTHALAMIC FACTOR	DESIGNATION/ ABBREVIATION	AMINO ACID SEQUENCE (IF KNOWN)	COMMENTS
Corticotropin (ACTH) releasing factor (or hormone)	CRF (CRH)	Ac-Ser-Tyr-Cys-Phe-His ($AspNH_2$-$GluNH_2$) Cys (Pro-Val)-Lys-$GlyNH_2$	Chemical instability and difficulty of assay make chemical formula uncertain
Thyrotropic (TSH) releasing factor (or hormone)	TRF (TRH)	Glu-His-$ProNH_2$	It is believed that the factor causes increase in cAMP in pituitary cells, and this increases release of TSH. TSH in turn increases thyroid activity by increasing thyroid cAMP
Luteinizing hormone (LH) releasing factor (or hormone) and Follicle-stimulating hormone (FSH) releasing factor (or hormone)	LH-RF (LH-RH) and FSH-RF (FSH-RH)	Glu-His-Trp-Ser-Tyr-Gly-Leu-Arg-Pro-$GlyNH_2$	One decapeptide appears to have both LH and FSH releasing ability. Effects not yet separated
Growth hormone (GH) releasing factor (or hormone) or Somatotropic hormone (STH) releasing factor (or hormone)	GH-RF (GH-RH) or STH-RF (STH-RH)	Val-His-Leu-Ser-Ala-Glu-Glu-Lys-Glu-Ala	May be species specific
Growth hormone (GH) inhibiting factor or (hormone) or Somatotrophic hormone (STH) release inhibiting factor	GIF (GH-RIF) or STH-RIF	HAla-Gly-Cys-Lys-Asn-Phe-Phe-Trp-Lys-Thr-Phe-Thr-Ser-CysOH	—

TABLE 26.5 continued
HYPOTHALAMIC FACTORS CONTROLLING SECRETION OF ADENOHYPOPHYSEAL HORMONES

HYPOTHALAMIC FACTOR	DESIGNATION/ ABBREVIATION	AMINO ACID SEQUENCE (IF KNOWN)	COMMENTS
or Somatostatin			
Prolactin-inhibiting factor (or hormone) or Prolactin-releasing inhibiting factor (or hormone)	PIF (PIH) or PRIF (PRIH)	Unknown	Believed to be a small MW substance, similar to oxytocin. Concentration reduced by suckling
Prolactin-releasing factor (or hormone)	PRF (PRH)	Unknown	Produced after birth of offspring. TRH also has a prolactin stimulating effect
Melanocyte-stimulating hormone (MSH) inhibiting factor (or hormone)	MIF (MIH)	Pro-Leu-GlyNH_2 and/or Pro-His-Phe-Arg-GlyNH_2	—
Melanocyte-stimulating hormone (MSH) releasing factor (or hormone)	MRF (MRH)	Cys-Tyr-Ile-Glu-Asn-CysOH	—

Hormones. Two hormones may be isolated from the neural lobe, VASOPRESSIN-ADH (*VADH*) and OXYTOCIN. Both are small molecules containing nine amino acids (Fig. 26.8).

Vasopressin-ADH exerts a mild stimulating effect on smooth muscle of arteries, causing vasoconstriction and an elevation of blood pressure. The hormone also exerts an antidiuretic effect (hence ADH for antidiuretic hormone) on the kidney, resulting in increased reabsorption of water from the kidney tubules. Control of VADH secretion is determined by osmosensitive cells within the hypothalamus that continually monitor blood osmotic pressure, and which then send nerve impulse to the posterior lobe to cause hormone release.

Oxytocin exerts its effect primarily on the smooth muscle of the pregnant uterus and the contractile cells (*myoepithelial cells*) around the ducts of the mammary glands. It stimulates the contraction of both types of cells, and is the active principle given in injections to induce or speed labor. Oxytocin is not essential to either function, but apparently aids in childbirth and milk ejection. Control of secretion has not been definitely established. Suckling by the infant increases secretion, as does the sex act, and dilation of the cervix prior to childbirth.

DISORDERS OF THE HYPOPHYSIS

Excess secretion of hormones from the adeno-

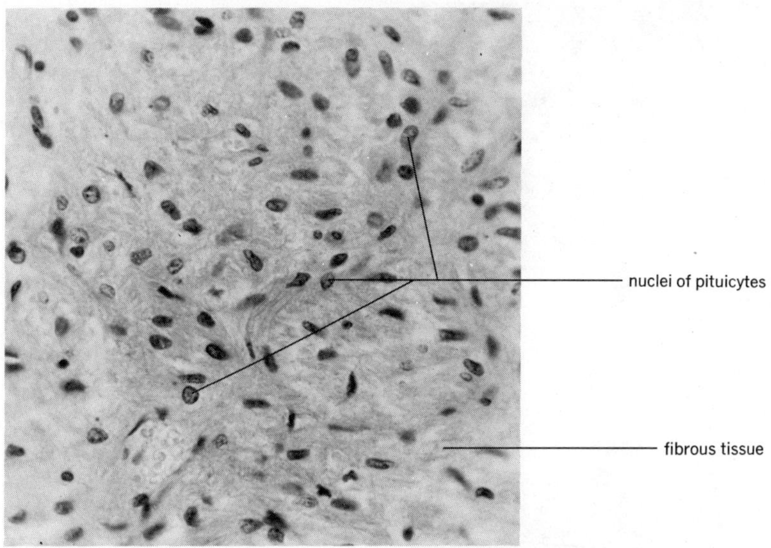

FIGURE 26.7
The appearance of the pars nervosa.

hypophysis may, in theory, result in overproduction of all hormones, or of only one. The usual cause of oversecretion of any hormone is a tumor involving the cells in which a particular hormone is produced. Chromophobe tumors account for 85 percent of all hypophyseal tumors; alpha cell tumors account for 10 to 14 percent; beta and delta cell tumors are rare. Visual disturbances may be the first sign of a tumor, as the enlarging gland presses on the neighboring optic tract.

GIANTISM (Fig. 26.9) is the result of excess growth hormone secretion occurring before skeletal maturity has been achieved. The body continues to grow and may reach heights in excess of 8 feet and weights in excess of 400

FIGURE 26.8
The amino acid sequences in ADH and oxytocin.

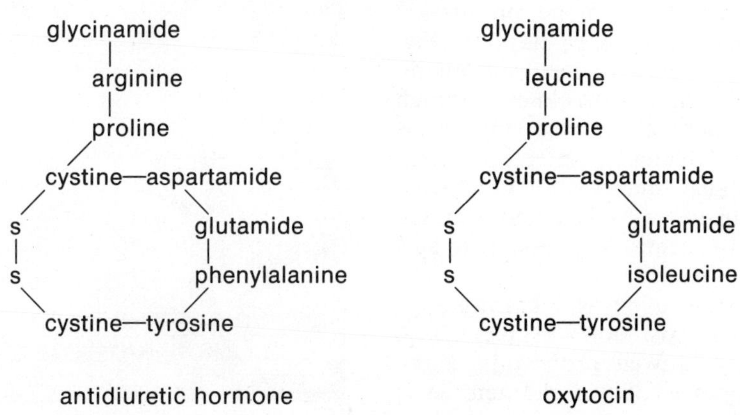

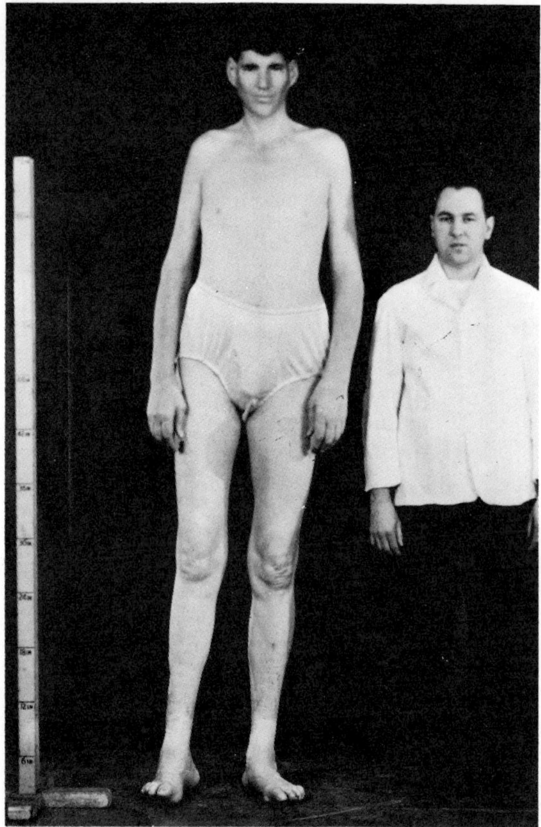

FIGURE 26.9
Giantism in a 42-year-old man. He is 7 feet, 6 inches tall. His companion is normal height. The stick is 6 feet tall. (From the teaching collection of the late Dr. Fuller Albright. Courtesy Endocrine Unit and Department of Medicine, Massachusetts General Hospital.)

pounds. Excess growth hormone production after maturity results in ACROMEGALY (Fig. 26.10). The bones can no longer grow in length, but can increase in width or thickness through the activity of the periosteum. The jaw, hands, and feet are most commonly affected. Treatment of hypophyseal tumors may be carried out by radiation delivered to the gland from the outside; the gland is nearly inaccessible to surgery.

Deficient secretion of growth hormone results in HYPOPHYSEAL INFANTILISM (Fig. 26.11). The body is well proportioned but juvenile in appearance. Sexual characteristics are minimally developed and stature is small. There is apparently no effect on mental capacity. Treatment of this condition requires administration of human growth hormone. Hormones from other species are ineffective.

Excess secretion of vasopressin, secondary to hypothalamic tumor, results in EXCESSIVE WATER RETENTION, dilution of body fluids, and weight gain. Limitation of water intake and administration of diuretics aid treatment. Failure of vasopressin secretion as a result of hypothalamic damage results in DIABETES INSIPIDUS. Excessive dilute urine production (to 20 to 30 liters per day in severe cases) is characteristic. Apparently there are no disorders associated with oxytocin. Table 26.6 presents a summary of hypophyseal disorders. The table includes disorders not described in the text.

THE THYROID GLAND

The thyroid gland (Fig. 26.12) is a TWO-LOBED

FIGURE 26.10
Acromegaly in an adult. Note the coarseness of the facial features, the enlarged mandible, and the large hands with thick, blunt fingers. (Armed Forces Institute of Pathology.)

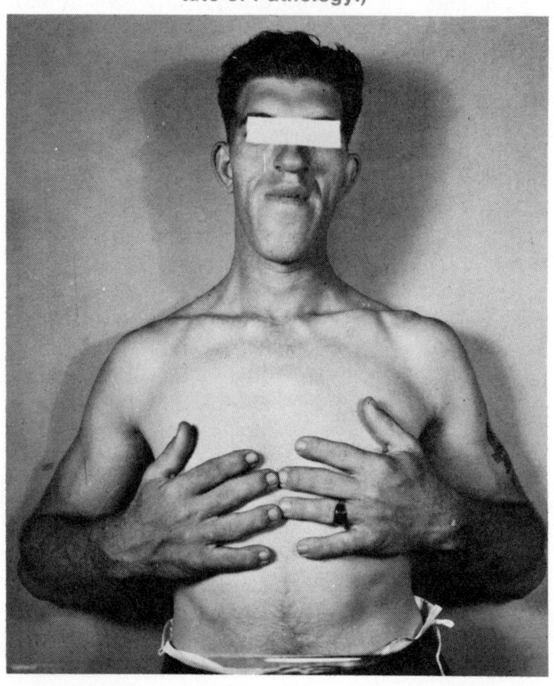

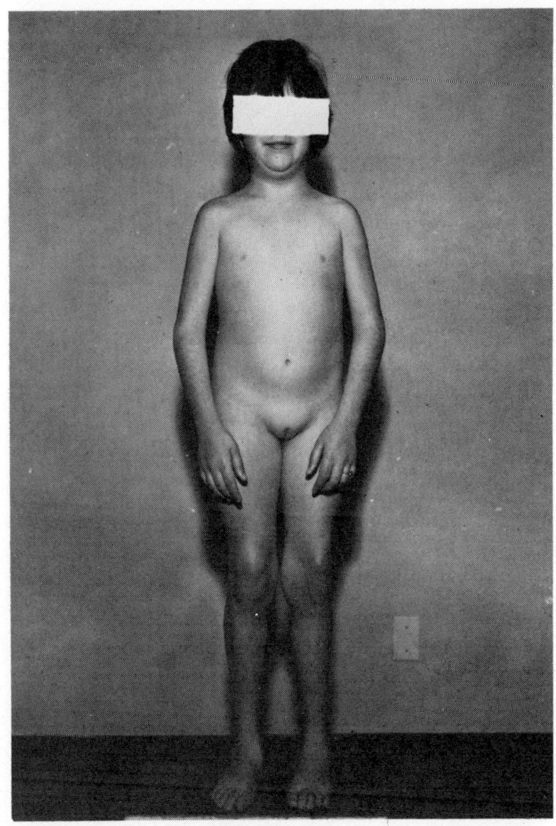

FIGURE 26.11
Hypophyseal infantilism in a 15-year-old female. Note shortness of stature (4 feet, 3¾ inches tall) and failure to develop sexually. (Armed Forces Institute of Pathology.)

organ lying on the lateral aspect of the lower larynx and upper trachea. A small ISTHMUS of thyroid tissue connects the two lobes across the midline. The gland weighs about 20 g in the adult and is one of the most vascular of all endocrines. The gland receives 80 to 120 ml of blood per minute, a fact important in maintaining adequate supplies of building blocks to the organ. The gland is composed of many small spherical units known as THYROID FOLLICLES. Each follicle is lined with a simple cuboidal epithelium carrying microvilli, and is filled with THYROID COLLOID. The colloid represents a storage form of hormone. Parafollicular cells (C cells), distinguished by a lighter-staining cytoplasm, are interspersed with the follicular epithelial cells and may be found between follicles. These features are shown in Fig. 26.13.

HORMONE SYNTHESIS

The substance generally regarded as the true hormone of the thyroid is THYROXIN. Thyroxin (tetraiodothyronine, T_4) is an iodinated amino acid with the formula shown in Fig. 26.14. Removal of one iodine results in triiodothyronine or T_3, even more active physiologically than T_4. Thyroxin and T_3 are synthesized (see Fig. 26.14) only after the gland has accumulated iodide and the amino acid tyrosine. By iodinating the molecule of tyrosine, then coupling two acids together, thyroxin is formed. Depending on body demands for hormone, the hormone may be released into the bloodstream, or linked to the protein thyroglobulin in the colloid. If linked, the thyroxin must be hydrolyzed from the thyroglobulin before entering the blood. Once into the blood, thyroxin combines with a specific plasma protein designated thyroid-binding globulin (TBG) for transport.

FIGURE 26.12
Gross anatomy and location of the thyroid gland.

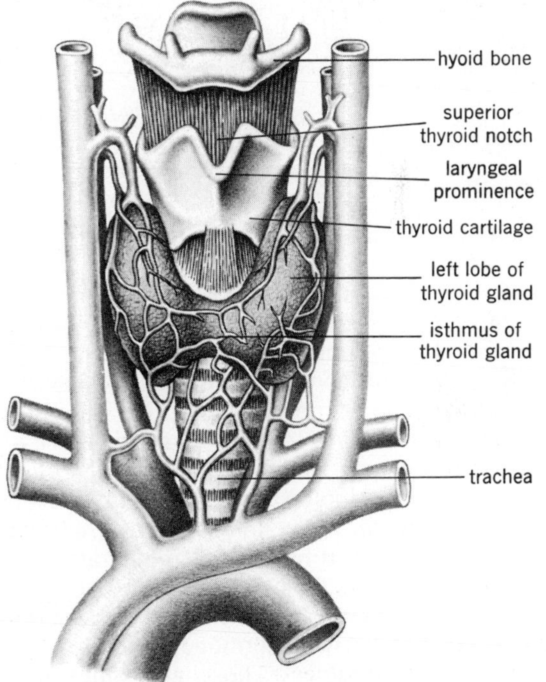

TABLE 26.6
SOME DISORDERS OF THE HYPOPHYSIS

CONDITION	CAUSE	HORMONE(S) INVOLVED	SECRETION IS: EXCESS	DEFICIENT	CHARACTERISTICS	COMMENTS
Giantism	Tumor before maturity	Growth hormone (GH)	X		Large body size	Proportioned but large
Acromegaly	Tumor after maturity	Growth hormone (GH)	X		Misshapen bones, hands, face, feet	Disproportional growth
Hypophyseal infantilism	Destruction of acidophils by disease, accident, etc.	Growth hormore (GH)		X	Juvenile appearance, sexually and in height	Body properly proportioned
Water retention + *demonstration of high VADH blood levels*	Hypothalamic tumor	Vasopressin-antidiuretic hormone	X		Dilution of body fluids, weight gain	*Must* differentiate from H_2O retention due to other causes
Diabetes insipidus	Hypothalamic damage	Vasopressin-antidiuretic hormone		X	Excessive urine excretion	No threat to life
Sheehan's syndrome	Atrophy of distalis	All		X	Sexual characteristics involute, no menses, anemic	Occurs mostly in the female. Give target organ hormones
Hypophyseal myxedema	Basophil degeneration	Thyroid-stimulating hormone		X	Dry flaky skin, lethargy, anemia	May be controlled by thyroxin

A test for thyroxin levels in the bloodstream is determination of the protein-bound iodine (PBI) levels of the blood. Since thyroxin and T_3 are carried bound on protein, the test is interpreted as reflecting T_4 and T_3 levels. Actually, *all* protein-bound iodine is determined by the test, a small fraction reflecting inorganic iodide on protein. Normal values for the PBI are 4 to 8 μg per 100 ml.

A more specific test for T_3 and T_4 levels in the bloodstream is the T_3 or T_4 RIA (*r*adio*im*muno*a*ssay) test. It has been demonstrated that protein hormones (or other hormones attached to proteins) of one species can cause formation of antibodies to those hormones when injected into another species. Those antibodies then bind only to the hormone(s) that caused their formation. To perform the assay, radioactive iodine (an easily measured isotope) is chemically attached to a sample of pure hormone. This labels the hormone. A constant amount of labeled hormone is added to a constant amount of antibody and unknown plasma in a test tube. Labeled and unlabeled hormone compete for binding sites on the antibody molecules. The more unlabeled hormone in the unknown plasma, the less labeled hormone will be bound to the antibody molecules. Bound and unbound radioactivity is then separated (by chromatography, electrophoresis, or other means) and the ratio of bound to unbound hormone is related to carefully worked out tables. By such a pro-

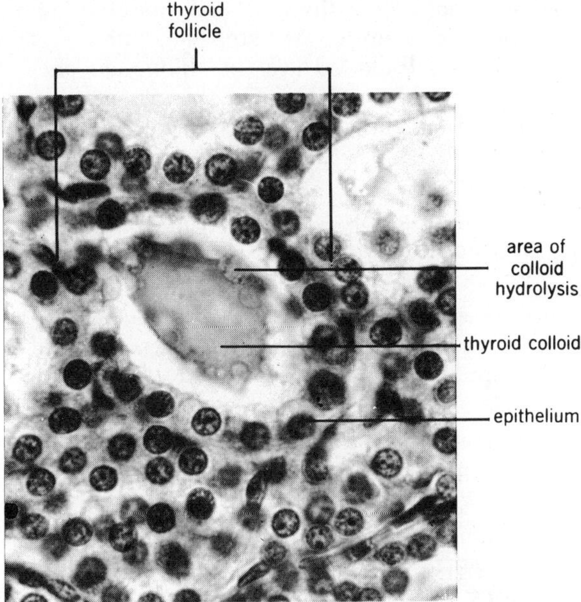

FIGURE 26.13
Microscopic anatomy of the thyroid gland.

cedure, T_4 RIA values are of the order of 4 to 10 μg/100 ml, and T_3 RIA values are 100 to 200 ng (av. 140 ng)/100 ml.

CONTROL OF THE THYROID

Accumulation of the necessary building blocks, the synthesis, and the release of thyroxin are all controlled by pituitary THYROID-STIMULATING HORMONE (TSH). Thyroxin in turn exerts a negative feedback on TSH production.

EFFECTS OF THYROXIN

Thyroxin has no particular target organ; those cells possessing catabolic systems of enzymes respond to the hormone. Its effects may be divided into four general categories.

Calorigenic effect. Thyroxin accelerates the catabolic reactions of glycolysis, Krebs cycle, beta oxidation and oxidative phosphorylation. One mg of thyroxin can raise heat production by 1000 Calories and CO_2 production by 400 g.

Growth and differentiation. Working with growth hormone, thyroxin ensures proper development of the brain. Deficiency of thyroxin

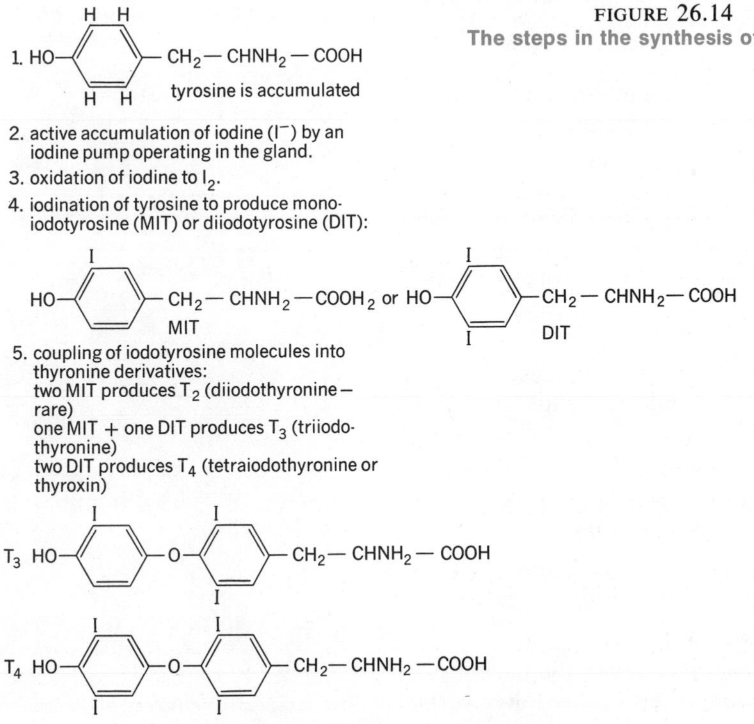

1. HO—(benzene ring, H H / H H)—CH_2—$CHNH_2$—COOH
   tyrosine is accumulated
2. active accumulation of iodine (I^-) by an iodine pump operating in the gland.
3. oxidation of iodine to I_2.
4. iodination of tyrosine to produce monoiodotyrosine (MIT) or diiodotyrosine (DIT):

   HO—(ring, I)—CH_2—$CHNH_2$—$COOH_2$ (MIT) or HO—(ring, I, I)—CH_2—$CHNH_2$—COOH (DIT)

5. coupling of iodotyrosine molecules into thyronine derivatives:
   two MIT produces T_2 (diiodothyronine—rare)
   one MIT + one DIT produces T_3 (triiodothyronine)
   two DIT produces T_4 (tetraiodothyronine or thyroxin)

T_3 HO—(ring, I)—O—(ring, I, I)—CH_2—$CHNH_2$—COOH

T_4 HO—(ring, I, I)—O—(ring, I, I)—CH_2—$CHNH_2$—COOH

FIGURE 26.14
The steps in the synthesis of T_3 and T_4.

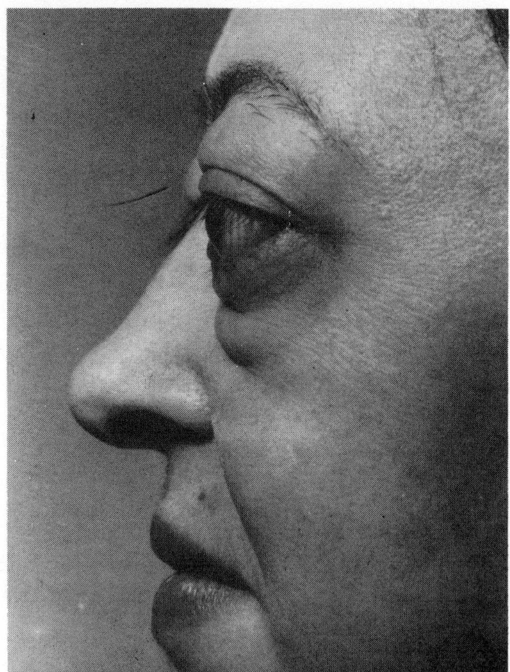

FIGURE 26.15
Hyperthyroidism. Note the marked protrusion of the eyeball (exophthalmos) that may accompany the disease. (R. W. Carlin, Medical Photographer.)

during development leads to smaller and fewer neurons and defective myelination.

Metabolic effects. Thyroxin acts as a diuretic, increasing urine production. It increases protein breakdown when present in greater than normal quantities, increases uptake of glucose by cells (including intestinal absorptive cells), enhances glycogenolysis, and depresses blood cholesterol levels.

Muscular effects. Thyroxin in excess interferes with ATP synthesis and may thus speed exhaustion of energy in muscle. Both skeletal and cardiac muscle are affected in this manner.

OTHER FACTORS INFLUENCING THYROID ACTIVITY

A wide variety of influences may alter the basic activity of the gland. Among these are the following:

ANTITHYROID AGENTS. Generally called GOITROGENS, because they cause goiter or enlargement of the thyroid, such materials inhibit one or more of the steps in synthesis of hormone. Included are the thioureas, phenols, and thiocyanates. The thiocyanates are present in cabbage, turnips, rutabaga, and mustard; hence, these foods are said to be goitrogenic.

GONADAL HORMONE LEVELS. Excessive gonadal hormone appears to decrease thyroxin transport.

PREGNANCY. Pregnancy increases all aspects of thyroid activity due, apparently, to fetal competition for maternal iodide and amino acids.

AGE. Activity of the thyroid decreases with age, although the decrease is small.

STRESS. Stress, particularly that of a cold environment, increases thyroid activity.

LONG-ACTING THYROID STIMULATOR (*LATS*). LATS is a gamma globulin antibody produced during certain disease states of the

FIGURE 26.16
Childhood myxedema; note protruding umbilicus and puffy appearance of face.

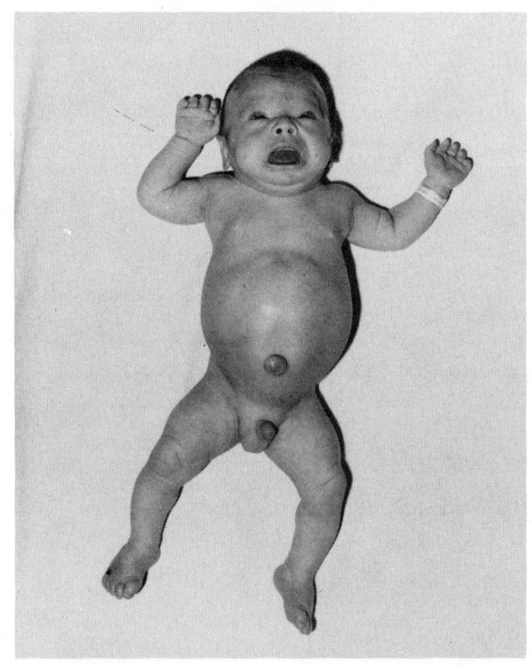

thyroid. It originates from plasma cells in autoimmune disorders. It acts on the thyroid in a manner similar to TSH, resulting in stimulation of catabolic reactions.

THYROCALCITONIN

In 1962 a hormone from the thyroid gland was shown to lower plasma concentrations of calcium and phosphate. It is secreted in response to a high blood calcium level. The name calcitonin was applied to the substance, and it was originally thought to be secreted by the parathyroids. Subsequently, the name thyrocalcitonin was applied, reflecting its thyroidal origin. Thyrocalcitonin apparently works with parathyroid hormone to control body calcium and phosphate balance. The effect is mediated primarily through the kidneys (increased Ca^{++} absorption) and possibly on bones (increased mineralization). It is believed to be produced by the parafollicular cells. Its secretion is stimulated by a rise of blood calcium levels.

DISORDERS OF THE THYROID

Overproduction of thyroxin due to thyroid tumor, or overstimulation by TSH, brings about the condition known as HYPERTHYROIDISM (*Grave's disease*) (Fig. 26.15). There are different degrees of a basically similar set of symptoms in the disease, so the following list will not attempt to separate the various specific diseases.

Excessive nervousness and excitability.

Elevated heart rate and blood pressure.

Elevated BMR (40 to 100 percent).

Weakness.

Weight loss.

Bulging of the eyes; exophthalmos.

The first five symptoms may be accounted for by considering the catabolic stimulating effect of the hormone; in short, everything is accelerated. The EXOPHTHALAMOS (see Fig. 26.15) is due to increases in mass of tissue behind the eye, pushing it forward. Treatment of hyperthyroidism involves surgical removal of excess tissue, or the use of radioactive iodine (^{131}I). Because of the uptake of iodine by the thyroid, the ^{131}I will be concentrated in the organ, and will irradiate the gland from inside, destroying cells.

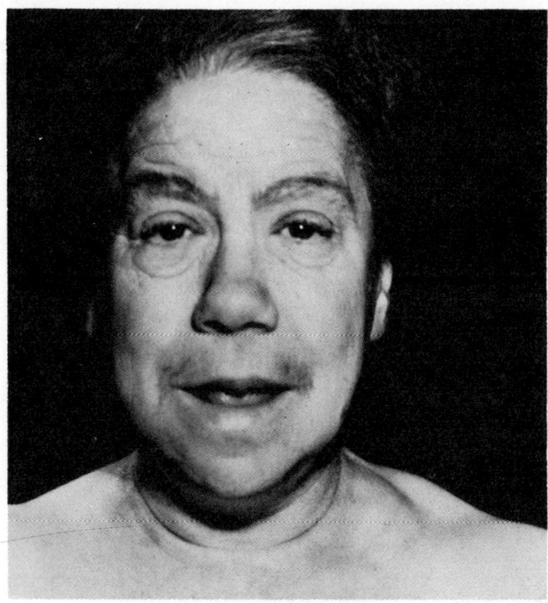

FIGURE 26.17
Myxedema in a 55-year-old female. Her PBI was less than 1 μg/100 ml (normal 4–8), and metabolic rate was 51 percent below normal. Note puffy appearance of face and generalized fatigued appearance. (From the teaching collection of the late Dr. Fuller Albright. Courtesy Endocrine Unit and Department of Medicine, Massachusetts General Hospital.)

Deficient thyroid activity results in a wider variety of disorders, depending on the cause of the deficiency. SIMPLE GOITER is generally the result of insufficient dietary iodide. The gland enlarges, "hoping" to trap more of the available iodide, which, of course, creates more demand, a greater enlargement, and so on. Untreated goiters can reach extreme sizes. Dietary deficiency of iodide in a pregnant woman may result in insufficient thyroxin production by the embryo, and cretinism may occur.

CRETINISM results in idiocy (remember the role of thyroxin in nervous system development) and retarded growth. The mental damage cannot be overcome. MYXEDEMA is seen after birth, in both children and adults (Figs. 26.16 and 26.17). Myxedema is associated with dry skin, low BMR, and intolerance to cold.

TABLE 26.7
SUMMARY OF THE THYROID

ITEM	COMMENTS
Location and parts; weight	Two-lobed, plus connecting isthmus; located on lower larynx and upper trachea; 20 grams in weight
Requirements for hormone synthesis	The amino acid tyrosine; iodide in diet; TSH to control all steps in synthesis
Hormones produced and effects:	
Thyroxin	Main controller of catabolic metabolism
Thyrocalcitonin	Lowers blood Ca++ and PO_4 levels
Control of hormone secretion:	
Thyroxin	By thyroid stimulating hormone
Thyrocalcitonin	By blood Ca++ level
Disorders	
Hypersecretion	Creates hyperthyroidism: Elevated BMR Elevated heart action Exophthalmic goiter
Hyposecretion	Creates: Goiter—enlarged gland Hypothyroidism (cretinism, myxedema) Low BMR Low heart rate, blood pressure, body temperature

Treatment of these hypothyroid conditions usually involves the administration of dried thyroid gland, by mouth, in the form of a tablet. Only cretinism will fail to respond to such treatment.

Table 26.7 summarizes facts about the thyroid gland.

THE PARATHYROID GLANDS

There are usually four parathyroid glands in the human located on the posterior aspect of the thyroid (Fig. 26.18). They measure about 5 mm in diameter, and have a combined weight of about 120 mg. Histologically, the glands consist of densely packed masses of principal (chief) cells and oxyphil cells (see Fig. 26.18). The PRINCIPAL CELLS are regarded as the source of parathyroid hormone (PTH), while the OXYPHIL CELLS are thought to be reserve cells, capable of assuming hormone production if need arises. PTH is a polypeptide chain containing 74 to 80 amino acids. Regulation of PTH secretion is determined by the blood calcium level, a fall of blood calcium increasing PTH secretion, and vice versa.

THE PARATHYROIDS AND ION METABOLISM

Calcium, phosphate, magnesium, and citrate are ions that assume an extremely important role in the body economy. As constituents of the bones and teeth, these ions play structural roles in the body. The phenomenon of muscular contraction apparently depends on calcium ion, as does normal membrane permeability. Calcium is also involved in nervous irritability and in the clotting of the blood. Regulation of the body content of these ions therefore becomes very important in terms of whole body function.

Calcium. The average adult body contains about 1100 g of calcium, 99 percent of which is found in bone. Intake occurs by absorption of the ion from the gut, and it is then placed in the extracellular fluids (plasma, interstitial fluid). Removal of calcium from extracellular fluid occurs by filtration in the kidney (99 percent

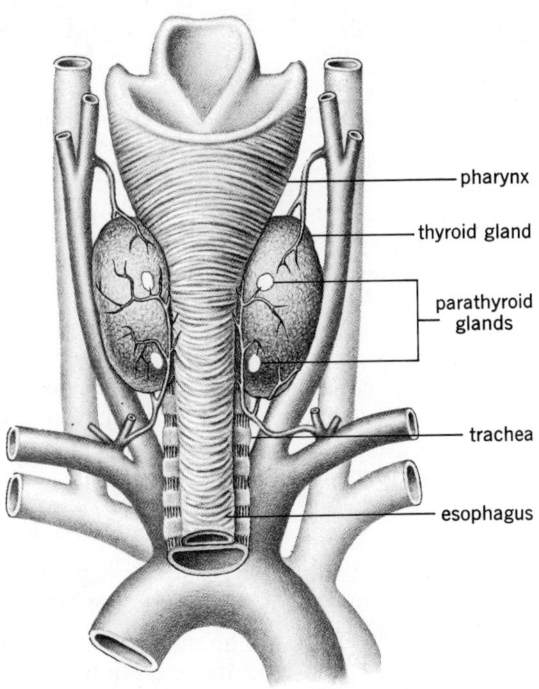

FIGURE 26.18
The location and microscopic appearance of the parathyroid glands.

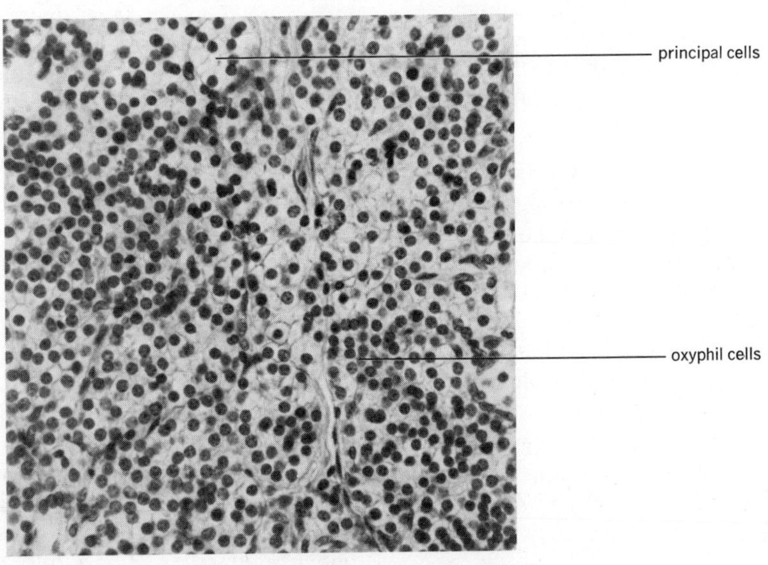

reabsorbed), exchange with tissues (chiefly muscle and bone), and secretion of digestive fluids into the gut and out via the feces. In the adult, about 1 g of calcium per day will keep all forces in balance.

Phosphate. The body contains about 500 g of phosphate, 85 percent of which is in the skeleton. Other compounds containing phosphate include ATP, ADP, and creatine phosphate, all important energy sources. DNA and RNA also contain phosphate. Phosphate is absorbed through the gut, apparently in inverse proportion to calcium. The absorbed phosphate is also placed into the extracellular fluids, from which it is withdrawn by cells, bony tissue, and the kidney. The kidney route constitutes the major path for excretion of phosphate.

Magnesium. Total body content of magnesium is about 20 g with 50 percent of this in the skeleton. Absorption occurs in the gut, apparently in inverse proportion to calcium absorption. Magnesium is essential as a cofactor in many enzymatic reactions, particularly those involving ATP synthesis and degradation and ribosomal protein synthesis. Renal excretion is the major pathway for magnesium loss.

Citrate. Citrate ion is an important intermediate in glycolysis and the Krebs cycle. It forms a major source of energy for ATP synthesis.

This brief discussion of these ions indicates their importance in body function. The regulation of cellular fluid levels of these substances is under the primary control of the parathyroid glands.

EFFECTS OF THE HORMONE

If the hormone controls the metabolism of the previously mentioned ions, it must be involved in at least three phases of this metabolism: absorption, exchange between cells and body fluids, and excretion. Accordingly, it is suggested that the gut, bones, and kidney are the three tissues or organs primarily affected by PTH.

Effects on the gut. PTH INCREASES ABSORPTION of calcium and magnesium from the intestine, provided adequate amounts of vitamin D are present. Phosphate absorption is also increased by the hormone.

Effects on bone. PTH acts directly on bone. The effect is thought to be mediated via osteoclasts (bone-destroying cells) and the connective tissue fibers of bone. PTH causes increase in numbers of osteoclasts and occasions proliferation of fibers. The result will be removal of the inorganic phase of bone (with consequent RISE OF BLOOD Ca^{++} and $PO_4^{\equiv}$) and increase in organic content. Action of the hormone on bone may revolve around lysosome dissolution in bone cells and consequent enzymatic destruction of bony tissue.

Effects on the kidney. PTH controls REABSORPTION of calcium and magnesium in direct proportion to PTH levels, while increasing phosphate excretion.

The combination of effects on intake, use, and excretion tends to keep plasma levels of these materials within normal limits.

EFFECTS ON OTHER ORGANS AND CELLS

The mammary gland responds to PTH by lowering secretion of calcium in milk. In cells such as those of kidney, liver, or intestine grown in tissue culture, PTH increases decarboxylation, releases H^+, causes swelling of mitochondria, and stimulates the Krebs cycle. A caution: PTH is not the only hormone exerting an effect on Ca^{++} metabolism. It might be predicted that any hormone concerned with growth would lead to, at the very least, increased absorption of Ca^{++} and $PO_4^{\equiv}$. Thus, growth hormone, thyroxin, and certain sex hormones work with PTH on the absorptive aspect. If one is concerned with maintenance, rather than growth, PTH and thyrocalcitonin are the two hormones most concerned with controlling metabolism of these ions.

DISORDERS OF THE PARATHYROIDS

HYPERPARATHYROIDISM (*von Recklinghausen's disease*) usually results from tumor formation in the glands. Excess PTH results in destruction of bone tissue and subsequent formation of excess fibrous tissue and cysts in the bones (*osteitis fibrosa cystica*). The bones may become softened to the point that they collapse. Blood Ca^{++} levels are very high, resulting in muscular weakness, mental disorder, and cardiac irregularities. The kidneys may form cal-

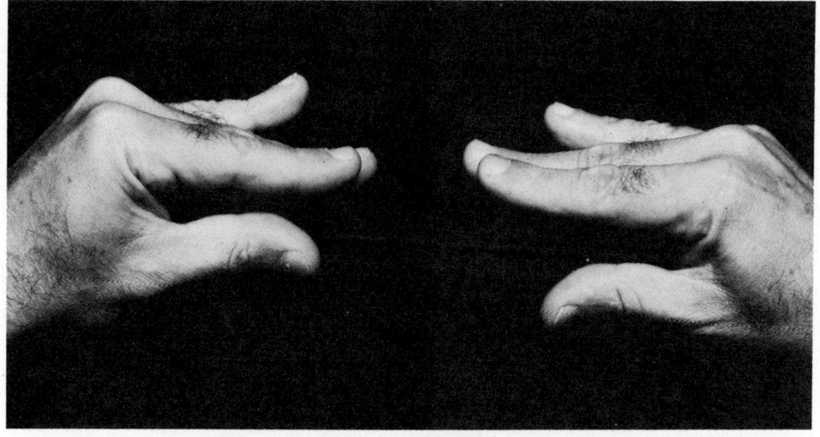

FIGURE 26.19
Carpospasm (Trousseau's sign) in hypoparathyroidism. Occluding the blood flow to the hands, with a blood pressure cuff pumped to about 100 mm Hg causes muscle spasm of the hands. Position is charactertistic for the disease. (Camera, M.D. Studios.)

culi, and vomiting, constipation, and other intestinal disturbances are usually present.

HYPOPARATHYROIDISM (*tetany*) usually results from damage to or removal of parathyroids during thyroid surgery or destruction of parathyroids by disease, infection, or hemorrhage. The term *tetany* reflects the primary symptom of the disease; muscular rigidity and paralysis due to low blood calcium levels. Characteristic positions of the hand (*carpospasm*) are seen (Fig. 26.19). The primary consideration in treatment is to achieve an elevation of the blood calcium level. This may be achieved immediately by intravenous injection of a solution such as 5 percent calcium gluconate. This relieves the symptoms, and further treatment may then be instituted for control of the disease.

THE PANCREAS

The endocrine portion of the pancreas, the ISLETS (*of Langerhans*), develops from the terminal portions of the ducts of the exocrine (digestive portion) and subsequently separates from the ducts. Estimates of the number of islets vary from 500 thousand to 2 million. The tissue becomes functional at about three months *in utero*. Microscopically, the islets show two major cell types, designated alpha and beta. ALPHA CELLS comprise about 25 percent of the islet population, and to them is attributed the production of the hormone GLUCAGON (hyperglycemic glycogenolytic factor or HGF). Glucagon is a single, long-chain polypeptide with a molecular weight of 3485. Its secretion from the alpha cells is caused by a fall in blood sugar and by a rise of growth hormone secretion. The hormone exerts its action on the enzymes stimulating conversion of glycogen to glucose (glycogenolysis) by stimulating formation of cAMP that increases the activity of phosphorylase. Such conversion raises the concentration of blood glucose and may, if secretion continues, create a hyperglycemia (excess blood sugar level).

The BETA CELLS, constituting about 75 percent of the islet cells, produce the more familiar hormone INSULIN. Insulin is a complex protein hormone consisting of two protein chains of two types. It has a molecular weight of about 6000. Insulin is produced as proinsulin and is hydrolyzed to insulin prior to secretion (Fig. 26.20). Insulin is secreted in response to a rise in blood sugar level (as after a meal); it lowers the blood sugar by stimulating

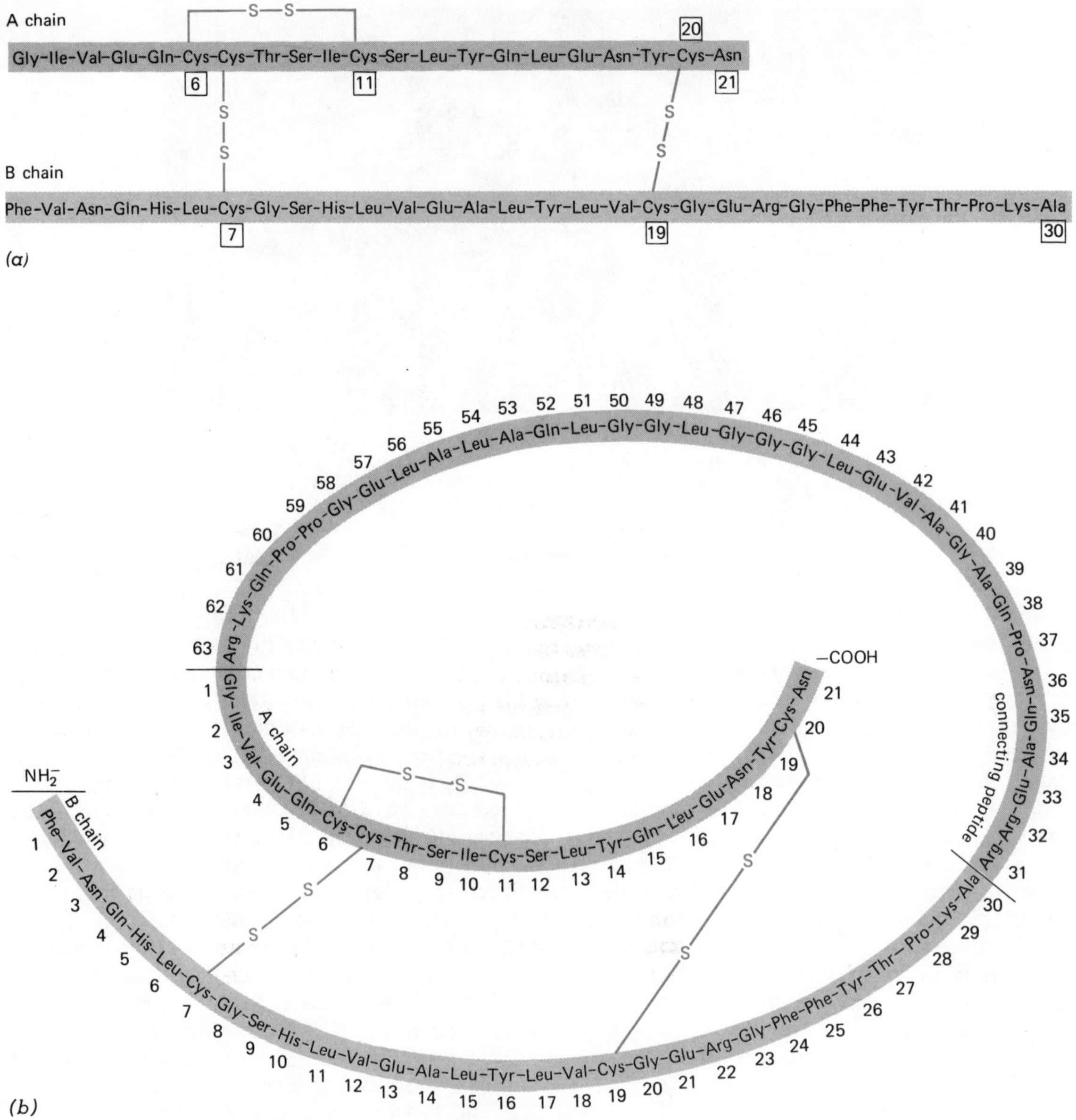

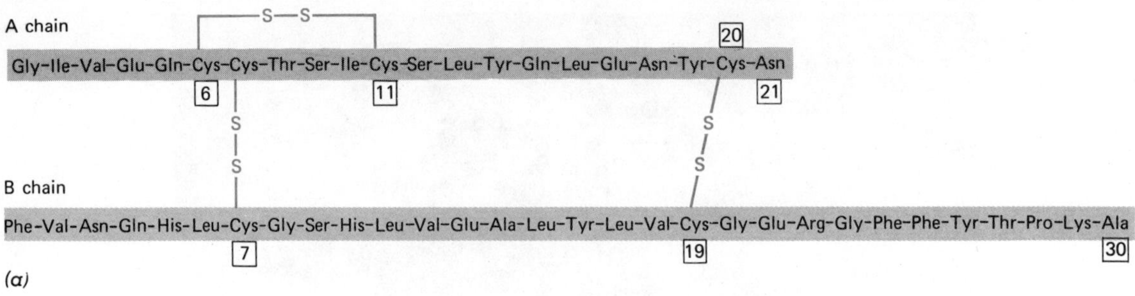

FIGURE 26.20
(*a*) The amino acid sequence in insulin. (*b*) The structure of proinsulin.

the conversion of glucose to glycogen (glycogenesis) and by stimulating uptake of glucose by certain cells (Tables 26.8 and 26.9). It also promotes protein and lipid synthesis.

Two conclusions may be drawn at this point: the blood sugar level reflects a balance of the effect of the two hormones, and the effects of both hormones are exerted primarily on the metabolism of carbohydrate. Recalling that the metabolism of carbohydrates, fats, and proteins is interrelated, we may further conclude that a disturbance of carbohydrate metabolism

TABLE 26.8
TISSUE IN WHICH INSULIN AFFECTS GLUCOSE UPTAKE

TISSUE AFFECTED	TISSUE NOT AFFECTED
Skeletal, cardiac, and smooth muscle	Brain
Leucocytes	Intestinal epithelium
Lens	Erythrocytes
Hypophysis	Testis
Mammary	Islet cells
Liver	Kidney tubules
Adipose tissue	

must cause disruption in the metabolism of the other substances as well.

INSULIN DEFICIENCY; DIABETES MELLITUS

Insulin deficiency results from the inability of the beta cells to produce enough insulin for body requirements. A series of reactions is set in motion by the deficiency.

BLOOD SUGAR LEVEL RISES (*hyperglycemia*). Glucose polymerization into glycogen is re-reduced, as is cellular uptake.

More sugar is filtered in the kidney than the tubule cells can reabsorb. GLUCOSE APPEARS IN THE URINE (*glucosuria*).

The cells of the body, unable to acquire glucose, shift their energy source to FATS. The RQ decreases; excess acetyl-CoA (from beta oxidation) is produced; the Krebs cycle cannot handle it all, and acetoacetic acid production is increased. The ketone bodies are increased in concentration, and KETOSIS may result. Also, the buffering systems of the body may be overtaxed and ACIDOSIS may occur.

Glucose eliminated in the urine osmotically carries excess quantities of water with it. URINE VOLUME IS INCREASED (*polyuria*).

Excess urinary loss of water dehydrates the body, and a great thirst develops, with RISE OF WATER INTAKE (*polydipsia*).

Glucose loss or its inability to enter cells, triggers the desire to eat, and EXCESS QUANTITIES OF FOOD may be consumed (*polyphagia*).

High levels of fats in the blood may lead to narrowing of blood vessel caliber through ATHEROSCLEROSIS. Occlusions of vessels in the legs may give rise to diabetic gangrene as tissues are deprived of blood, die, and become infected. The events described, along with their causes, constitute the symptoms of diabetes mellitus.

Treatment of diabetes may depend on the severity of the disease. Mild disease may often be controlled by dietary measures, while severe disease may require insulin administration.

Diabetes is one of the oldest-known endocrine disorders, the sweetness of the urine having been described in Egypt in 1500 B.C. Its major symptoms were described at the beginning of this section. Two types are recognized today. Juvenile-onset type occurs usually before 15 years of age, is of rapid onset, is the more severe type, and results in dependence on insulin for control. The beta cells are

TABLE 26.9
FACTORS AFFECTING INSULIN SECRETION

FACTORS INCREASING INSULIN SECRETION	FACTORS DECREASING INSULIN SECRETION
Glucose intake (hyperglycemia)	Epinephrine
Increase in intake of certain amino acids	Certain diuretics (thiazides)
cAMP	Glucose metabolism blocking agents (e.g., 2-deoxyglucose, mannoheptulose)
Vagal stimulation	
Intestinal parahormones	
Sulfonylureas	
Glucagon	

TABLE 26.10

TYPES OF INSULIN AVAILABLE FOR USE IN DIABETES

TYPE	pH[d]	TIME OF MAXIMUM EFFECTIVENESS (HOURS)	DURATION OF EFFECTS (HOURS)
Semilente	7.7–7.5	4–6	12–16
Lente	7.1–7.5	8–12	18–24
Ultralente	7.1–7.5	16–18	30–36
Crystalline Zn[a]	3.0	4–6	6–8
NPH[b]	7.1–7.4	8–12	18–24
Protamine Zn[c]	7.1–7.4	14–20	24–36
Globin insulin	3.6	6–10	12–18

[a] Almost pure insulin—quick results, often used in combination with longer acting form.
[b] "*N*eutral *P*rotamine *H*agedorn"—insulin and equivalent amount of protein.
[c] Excess protein; stretches effect.
[d] pH determines the state of the insulin. At low pH, the insulin goes into solution, is absorbed from the injection site rapidly, and is shorter in duration of effect. At higher pH, the insulin tends to remain crystalline and is absorbed more slowly with longer duration of effect.

atrophic (wasted or degenerated). Maturity-onset type occurs usually after age 35 to 40. It develops slowly, is usually not severe, may be controlled by diet alone in some cases, and usually is associated with vascular complications (atherosclerosis, arteriosclerosis, hemorrhage). Several factors appear to predispose to the development of diabetes.

Heredity. Twenty to fifty percent of individuals developing diabetes show a history of the disorder in their family. Some geneticists believe the disorder is inherited as an autosomal recessive trait. Others believe it to be multifactored.

Obesity. If an individual is predisposed to the development of diabetes, any condition that stresses the beta cells may trigger the appearance of symptoms. Such an individual is often on the border line of deficiency and is "tipped over" by a stressor. Obesity may be one such factor.

Aging. Tolerance to intake of glucose diminishes with age as if the beta cells' capacity to produce insulin diminishes with age. Dietary restriction of carbohydrate with advancing age would thus seem to be a reasonable suggestion. It might also reduce the tendency of older individuals to put on weight.

Influence of other hormones. Growth hormone is known to increase an existing diabetes, as do adrenal cortical hormone (cortisol) and thyroxin. They cannot, however, produce diabetes in the absence of beta-cell weakness.

Infection. Infections in general appear to act as stressors and may cause appearance of diabetic symptoms in a latent diabetic. The presence of diabetes seems to reduce resistance to infection.

The development of diabetes appears to proceed in STAGES. Three to seven stages are described according to the authority involved. The sequence presented below is that according to Danowski.

Stage I. Prediabetes. No symptoms are shown; a predisposition to develop diabetes is present.

Stage II. Stress diabetes. Appearance of diabetic symptoms with stress; disappearance with removal of stressor.

Stage III. Appearance of abnormal glucose tolerance tests (prolonged elevation of blood sugar after ingestion of glucose solution). Fasting blood glucose levels are normal. Normally, blood sugar rises after glucose ingestion and returns to normal (80 to 120 mg percent) within three to four hours.

Stage IV. Fasting blood glucose is elevated and there is prolongation of the glucose tolerance test.

Stage V. Ketosis appears. Ketone bodies appear in the urine (ketonuria).

Stage VI. Ketonuria is marked, with ketonemia developing.

Stage VII. Acidosis appears, coma may develop, and ketone levels are high in blood and urine.

Children are usually in stages V, VI, or VII when diagnosed. Adults are usually in stages II, III, or IV when diagnosed.

VASCULAR COMPLICATIONS in diabetes occur mainly as a result of deposition of lipids in the intima of the blood vessels. This narrows luminal diameter, reduces blood flow, and starves tissues. Often, the tissues become gangrenous. Examination of the retinal vessels by the ophthalmoscope may discover the presence of diabetic changes in blood vessels (diabetic retinopathy) that can lead to blindness.

INSULIN AS THE TREATMENT FOR DIABETES

Table 26.10 summarizes some types of insulin available for treatment of diabetes.

Insulin must be given by injection inasmuch as it is a protein and would be destroyed by the digestive enzymes if given by mouth. One must also recognize that it is virtually impossible to obtain normal insulin levels by injection, and that only control, rather than cure, can be achieved in the disease.

At least one investigator has developed an "artificial pancreas" that has been implanted in animals. The device consists of a detector that monitors blood glucose levels, and releases stored insulin in proportion to blood glucose elevation. An obvious limitation of the device is in its store of insulin; after it is gone, the device must be recharged. A connection to the external body surface is thus required, with attendant chances of septicemia.

ORAL TREATMENT OF DIABETES

If any beta cells survive in the islets, the administration by mouth of certain substances may control diabetes. Most belong to the group of chemicals known as SULFONYLUREAS (Fig. 26.21). They act in several ways:

Tend to INHIBIT GLYCOGEN BREAKDOWN; an effect exerted directly on the liver.

INCREASE SECRETION OF INSULIN by any surviving beta cells.

INCREASE CELLULAR UPTAKE of glucose, possibly by increasing the production of insulin receptors on cell membranes.

Side effects of the sulfonylureas include gastrointestinal disturbances, decrease of white

FIGURE 26.21

Formulas of compounds orally effective in control of some types of diabetes. Active portion of the molecule is enclosed in the rectangle.

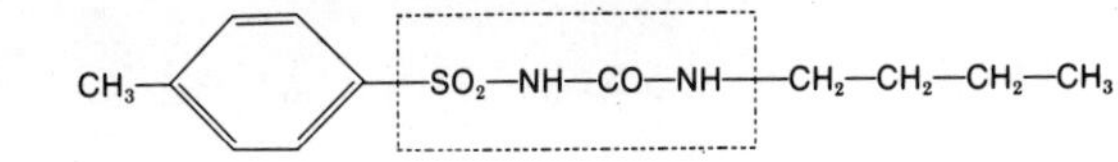

CH_3—(benzene ring)—[SO_2—NH—CO—NH]—CH_2—CH_2—CH_2—CH_3

tolbutamide
(1-butyl-3-p-tolylsulfonylurea)

CH_3—(benzene ring)—[SO_2—NH—CO—NH]—N<(CH_2—CH_2—CH_2, CH_2—CH_2—CH_2)

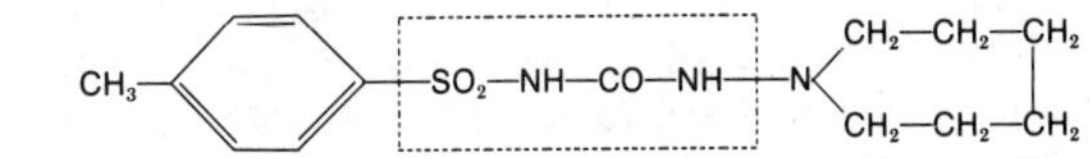

tolazamide
(1-(hexahydro-1-azepinyl)-3-p-tolylsulfonylurea)

CH_3—CO—(benzene ring)—[SO_2—NH—CO—NH]—(ring)

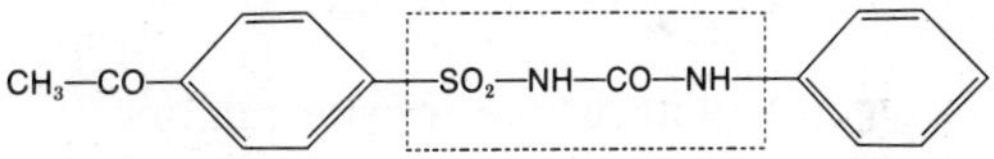

acetohexamide
(N-(p-acetylbenzenesulfonyl)-N^1-cyclohexylurea)

Cl—(benzene ring)—[SO_2—NH—CO—NH]—CH_2—CH_2—CH_3

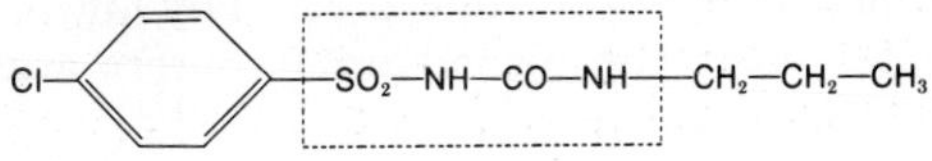

chlorpropamide
(1-propyl-3-p-chlorobenzenesulfonylurea)

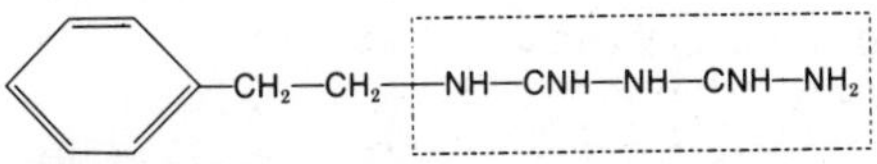

(benzene ring)—CH_2—CH_2—[NH—CNH—NH—CNH—NH_2]

phenformin
CN^1-B-phenethylformamidinyliminourea
or phenethylbiguanide)

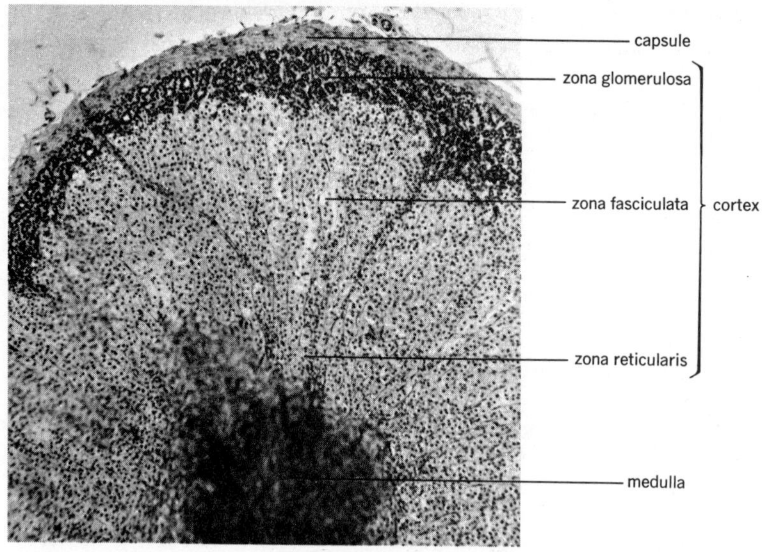

FIGURE 26.22
The structure of the adrenal gland.

cells and platelets, jaundice, hypoglycemia, and a cumulative effect on lowering blood sugar in the presence of alcohol. A prospective patient for oral medication should be carefully evaluated as to which preparation and what dosage is used.

HYPOGLYCEMIA

Low blood sugar levels may result from starvation, failure of sugar absorption, high activity levels, hyperinsulinism due to islet tumors (insulin shock), or overdosing with insulin or oral medication. Regardless of cause, hypoglycemia affects one body area primarily—the brain. Brain cells normally utilize only glucose and are therefore immediately affected by low blood sugar levels. The higher brain centers are affected first, causing:

Disturbances in locomotion.

Mental confusion.

"Dull-wittedness."

Decreased muscle tone.

Effects on lower centers include:

Disturbances in respiration.

Loss of consciousness.

Depression of reflexes.

Depression of body temperature.

If the blood sugar level continues to decline, death will result. Treatment involves raising the blood sugar level. If the individual can recognize the onset of symptoms, consumption of sugar or a candy bar may arrest their development. If unconscious, intravenous injection of glucose solution is necessary.

THE ADRENAL GLANDS

The adrenal glands are paired, hat-shaped organs located superior to the kidneys. Each gland consists of an inner MEDULLA of ectodermal origin and an outer CORTEX of mesodermal origin (Fig. 26.22). The medulla is not essential for life; at least one-sixth the total cortical substance is necessary to maintain life.

THE ADRENAL MEDULLA

The hormone-producing cells of the medulla are known as the chromaffin cells and are arranged in the form of cords. The cords of cells are separated by large venous sinuses. Upon stimulation, the chromaffin cells secrete into the sinuses, which empty very rapidly into the

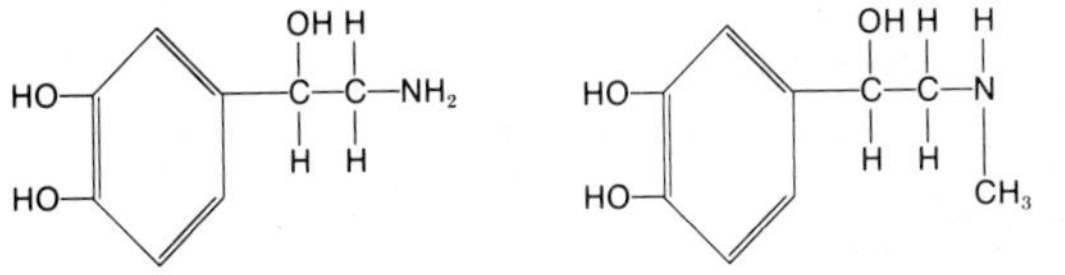

norepinephrine　　epinephrine

FIGURE 26.23
The chemical formulas of epinephrine and norepinephrine.

circulation. Control of medullary secretion is by way of preganglionic neurons of the sympathetic nervous system that terminate without synapse on the medullary cells. Response is therefore very rapid.

HORMONES OF THE MEDULLA

Two active substances, NOREPINEPHRINE and EPINEPHRINE (adrenalin) (Fig. 26.23), may be isolated from the medulla. Norepinephrine composes about 20 percent of the medullary secretion, epinephrine the remaining 80 percent. Both hormones are "sympathomimetic"; that is, their effects are similar to those obtained by stimulation of the sympathetic nervous system. The effects of the two hormones are presented in Table 26.11.

The list of effects of epinephrine appears to create a set of conditions to enable the body to "run or fight" (the "fight or flight" reaction).

Fortunately, the effects, although great, are short-lived. The hormone is inactivated by the liver in about three minutes.

DISORDERS OF THE MEDULLA

No disease of hypofunction exists. As stated before, the medulla is not essential for life. Its functions can be taken over by the sympathetic nervous system. Occasionally, a medullary tumor (phaeochromocytoma) may occur, resulting in overproduction of hormone. This is a very dangerous condition, because of the extremely high blood pressures that may be achieved (to 300 mm Hg systolic pressure).

TABLE 26.11
EFFECTS OF ADRENAL MEDULLARY HORMONES

HORMONE	GENERAL EFFECT	MECHANISM OF PRODUCTION OF EFFECT
Norepinephrine	Raises blood pressure	Causes constriction of muscular arteries
Epinephrine	Raises blood pressure	Increases rate and strength of heartbeat Causes constriction of all muscular arteries except those to heart and skeletal muscles, the latter dilating Causes contraction of spleen
	Stimulates respiration	Increases rate and depth of breathing by direct effect on respiratory centers
	Dilates bronchi and bronchioles	Causes relaxation of smooth muscle of respiratory system
	Slows digestive process	Inhibits muscular contraction of stomach, intestines
	Postpones fatigue in skeletal muscle, and increases efficiency of contraction	Increases muscle glycogenolysis Creates more ATP
	Causes hyperglycemia and glucosuria	Increases liver glycogenolysis; kidney tubule Tm_g is exceeded
	Increases O_2 consumption, CO_2 production	Stimulates general metabolic activity of cells
	Miscellaneous effects:	
	Stimulates sweating	Stimulates eccrine secretion
	Increases lacrymation	Stimulates secretion of lacrimal glands
	Dilates pupil	Causes contraction of dilator pupillae
	Increases salivary secretion	Stimulates secretion (nervous)
	Increases coagulability of blood	Accelerates clotting reaction

TABLE 26.12
ADRENAL CORTICAL HORMONES

ZONE OF PRODUCTION	GENERAL CATEGORY OF HORMONES	SPECIFIC HORMONAL EXAMPLES	EFFECTS OF HORMONE—HOW PRODUCED
Zona glomerulosa	Mineralocorticoids	Aldosterone	Increases tubular transport of sodium, decreases transport of potassium; sodium reabsorption causes chloride, bicarbonate, and H_2O reabsorption by passive attraction
Zona fasciculata	Glucocorticoids	Cortisone Cortisol Corticosterone	Increases gluconeogenesis, stimulates deamination Maintains muscular strength; keeps adequate glucose and ATP available Maintains proper nerve excitability Exerts anti-inflammatory effects, possibly by rendering lysosome more resistant to disruption by disease Increases resistance to stress
Zona reticularis	Cortical sex	Androgens Estrogens	Androgens exert anti-feminine effects; accelerate maleness; estrogen, just the reverse

Surgical removal of the tumor is mandatory before circulatory damage occurs.

THE ADRENAL CORTEX

The adrenal cortex is subdivided into three zones, based primarily on cellular arrangement (*see* Fig. 26.22). The outer ZONA GLOMERULOSA contains cortical cells in ball- or knotlike masses (glomeruli). The central ZONA FASCICULATA contains radially arranged cords (fascicles) of cells. The inner ZONA RETICULARIS has branching cords of cells. Each zone produces a different type of steroid hormone having widely different effects on the body. The glomerulosa produces MINERALOCORTICOIDS, affecting sodium metabolism; the fasciculata produces primarily GLUCOCORTICOIDS, affecting foodstuff metabolism (chiefly carbohydrates); the reticularis produces primarily SEX HORMONES, chiefly androgens (male).

Table 26.12 presents some of the specific effects of these hormones. Fig. 26.24 shows the formulas of the more important hormones in each group.

GLUCOCORTICOIDS AND STRESS

Chronic stress of any sort causes increased production of glucocorticoids. This increase in hormone levels may serve as a call to arms to equip the body to resist the stress. As a result of the increased production, gluconeogenesis is increased, providing more energy to resist the stress, circulatory and cell membrane integrity is promoted, amino acids are redistributed in the body, antiinflammatory effects are produced, and all of the repair mechanisms of the body are stimulated. Continued high levels of glucocorticoids are damaging to the body. For example, ulcer formation, high blood pressure, and atrophy of lymphatic tissue will occur. The individual may, in the long run, be rendered more susceptible to disease. In conclusion, increased hormone production is observed during stress, but the exact role this plays in the body's adaptive mechanisms is in doubt.

In 1953, Hans Selye proposed that such stimuli be called *stressors* and advanced the theory that the body responded to any or all stressors by a more-or-less consistent series of reactions he termed the GENERAL ADAPTATION SYNDROME (*G.A.S.*). Selye divided the G.A.S. into three phases.

The ALARM REACTION, in which the stressor causes initial nervous and circulatory depression, followed by ACTH secretion and development of resistance to the stressor.

The STAGE OF RESISTANCE, in which full resistance to the stressor has been developed as cortisol secretion is elevated, and the body is functioning at a higher level than before.

The STAGE OF EXHAUSTION, in which high levels of cortisol in the body begin to wreak havoc on digestive, circulatory, and immune systems. ULCERS, DECREASED RESISTANCE TO INFECTION, HEMOCONCENTRATION, and SHOCK appear as the adaptation can no longer be maintained. Although there is no strict proof of Selye's hypothesis, it is an attractive one to explain the "diseases of civilization" present in today's world.

FIGURE 26.24
The formulas of some important adrenal steroids.

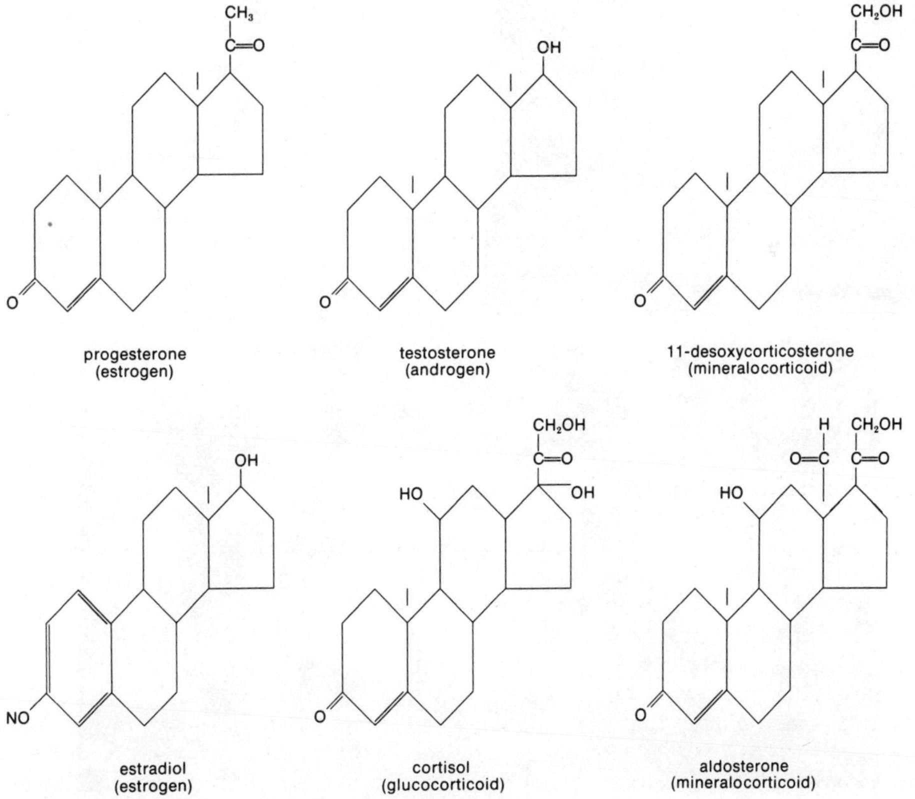

CH_3 C=O O

progesterone (estrogen)

OH O

testosterone (androgen)

CH_2OH C=O O

11-desoxycorticosterone (mineralocorticoid)

OH NO

estradiol (estrogen)

CH_2OH C=O HO OH O

cortisol (glucocorticoid)

H CH_2OH O=C C=O HO O

aldosterone (mineralocorticoid)

Glucocorticoids inhibit inflammatory responses to injury when given in large doses. The mechanism appears to be one of stabilization of lysosome breakdown, reduction of fibroblast activity, and reduction of swelling. However, the amounts required to produce the antiinflammatory effect are so large that they produce the typical effects of glucocorticoid excess. Additionally, corticoid use may mask the appearance of symptoms that would call for further treatment. The use of glucocorticoids should thus be tempered by the knowledge that they are dangerous.

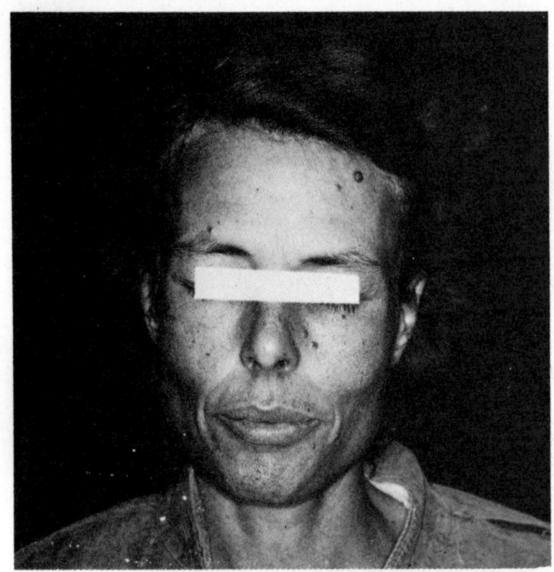

(a)

CONTROL OF CORTICAL SECRETION

Activity of the glomerulosa appears to be controlled by the BLOOD SODIUM LEVEL and by ANGIOTENSIN. Increase in either factor stimulates aldosterone secretion, leading to increased sodium reabsorption, water retention, and chloride and bicarbonate reabsorption. ACTH plays little or no role in glomerulosa activity. The fasciculata and reticularis are under ACTH control. The hormones produced from the latter two zones also exert a negative feedback upon ACTH production.

DISORDERS OF THE CORTEX

HYPOFUNCTION, particularly of the fasciculata, results in ADDISON'S DISEASE (Fig. 26.25). Anemia, muscular weakness, fatigue, elevated blood potassium, and lower blood sodium evidence deficiency of glucocorticoids and mineralocorticoids. A peculiar bronzing of the skin is apparently due to the excess ACTH production occasioned by the removal of the negative feedback system as cortical hormone production decreases. (ACTH has melanocyte-expanding properties in the human.) Administration of cortisol controls the condition.

FIGURE 26.25
Addison's disease. (*a*) Shows generalized fatigued and dehydrated appearance of a patient having the disease. (Armed Forces Institute of Pathology.) (*b*) The lips and gums of an Addison's patient showing characteristic dark (actually brown) pigment deposits. (The Center for Disease Control, Atlanta, Georgia.)

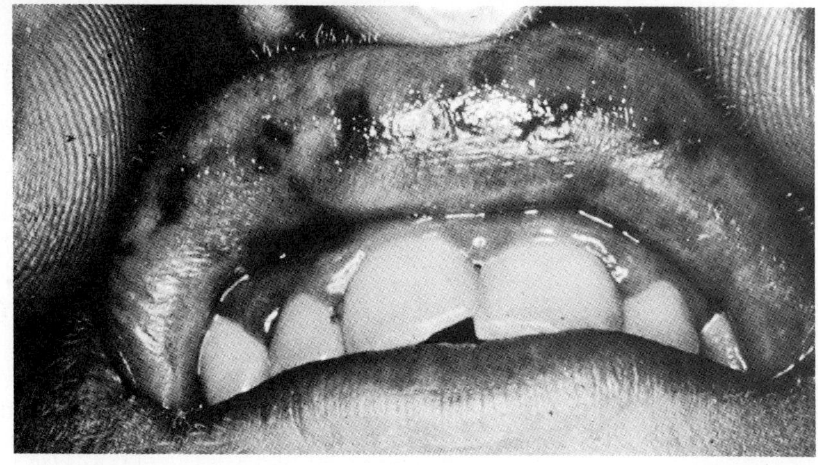

(b)

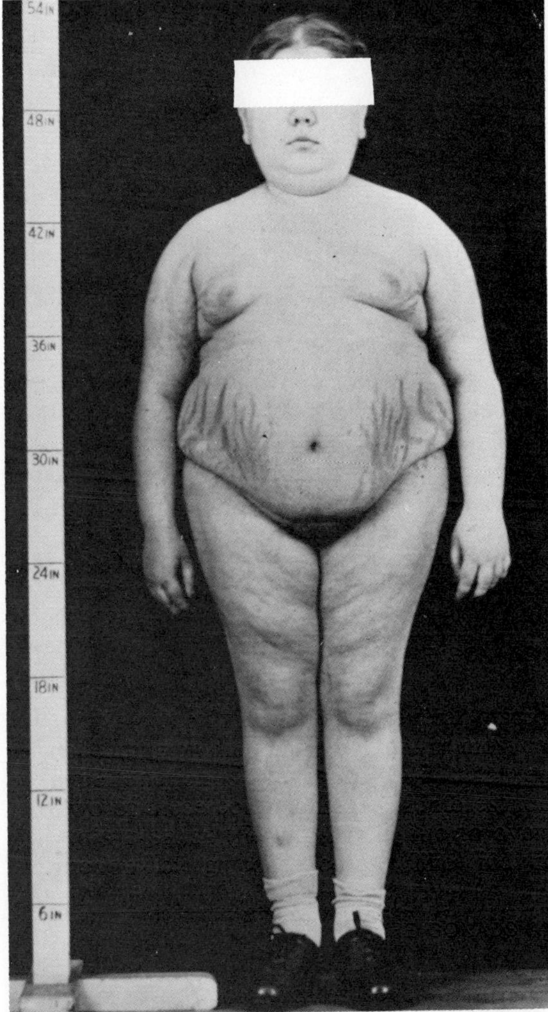

FIGURE 26.26
Cushing's syndrome in a 12-year-old female. Note the moon face, pendulous abdomen, abdominal striae, and concentration of adiposity on the trunk. (From the teaching collection of the late Dr. Fuller Albright. Courtesy Endocrine Unit and Department of Medicine, Massachusetts General Hospital.)

HYPERFUNCTION creates several disorders, depending on the zone involved. In most cases, hyperfunction is the result of a cortical tumor. Two of these conditions are of interest. CUSHING'S DISEASE (Fig. 26.26) is due to overproduction of glucocorticoids. The individual shows regional adiposity sparing the extremities, stripes on the abdomen, and diabetic tendencies. Surgical removal of the tumor followed by cortisol administration controls the disease. The ADRENOGENITAL SYNDROME results from a reticularis tumor with resultant overproduction of sex hormones. In general, overproduction of this category of hormones will masculinize a female, or accelerate the sexual development of the male. If the excess production occurs *in utero,* when the child's reproductive organs are developing, bizarre alterations in appearance of the organs may result. For example, if a genetic female is subjected to excess male hormone, the result will be conflicting orders to the developmental mechanism. Figure 26.27 shows a sufferer from adrenogenital syndrome.

THE GONADS AND SEXUAL DIFFERENTIATION

THE BASIS OF SEX

The ultimate maleness or femaleness of an individual depends on the simultaneous action of several factors.

The GENETIC CONTRIBUTION, determined by the sex chromosomes received from the parents (XX or XY).

The GONADAL CONTRIBUTION, expressed by the appearance of the gonads and by the development of the external genitalia and secondary sex characteristics.

The HORMONAL CONTRIBUTION, determined by the secretion of estrogens or androgens by the developing mature gonad and, to a degree, by the influence of hormones on the sexual organs.

The SOCIAL CONTRIBUTION, determined by how the parents view and treat the offspring and by the offspring's view of itself. Often termed the gender role, this component is neither male nor female at birth and is a result of environment alone.

The genetic contribution. Human chromosomes normally consist of two sex chromosomes and 22 pairs of autosomes. Two sex chromosomes

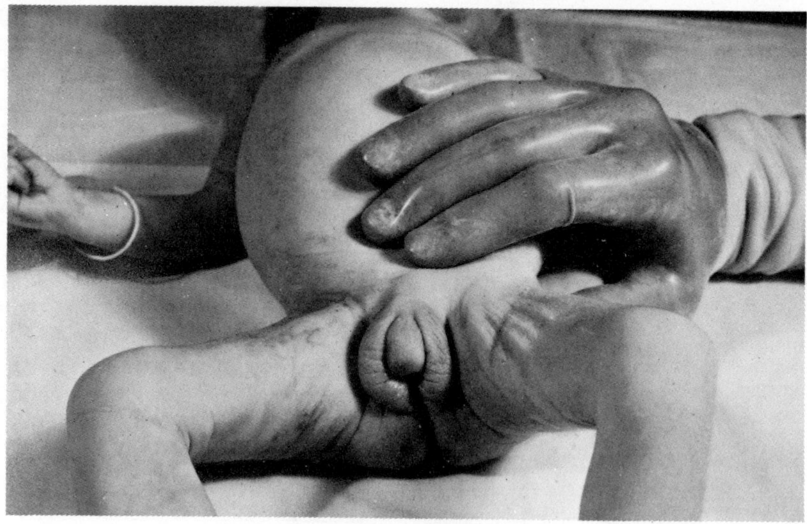

FIGURE 26.27
Adrenogenital syndrome in a 24-day-old female. Note masculinization, with enlarged clitoris (center of genitalia) and scrotumlike development of the labia. (Armed Forces Institute of Pathology.)

of the same type (XX) are characteristic of the female, while two different sex chromosomes (XY) are characteristic of the male. During meiosis, production of ova and sperm normally results in a halving of the diploid chromosome number and the production of cells having 22 plus X in ova and one-half of the sperm, or 22 plus Y chromosomes in the other one-half of the sperm. In some individuals, failure of chromosomes to separate (nondisjunction) during meiosis may result in a cell receiving both members of a chromosome pair, while another cell receives none. Deletion may result from a chromosome break followed by a disjunction that carries most of the chromosome to one cell, with only a fragment to another. Nondisjunction may occur in either autosomes or sex chromosomes. Congenital disorders may result that are reflected in the body generally or in sexual development. Some of the conditions that may result from such disorders are presented in Table 26.13. A normal human karyotype is shown in Fig. 26.28. Such a karyotype arranges homologous chromosomes in decreasing size or length of arms and numbers the resulting groups.

The hormonal contribution. Development of the accessory organs of the female reproductive system depends primarily on genetic influences. Development of the male organs requires an *in utero* production of testosterone by the fetal gonad at about one month of development. Subsequent secretion does not occur until puberty at approximately 12 to 17 years of age. Thus, the appearance of the male external genitalia depends on embryonic secretion of sex hormone.

The social contribution. Included here are the psychosocial factors associated with the relationships between the individual and members of the same or opposite sex, mannerisms of dress, and orientation of sexual impulses. At present, all attempts have failed to show any correlation between chromosome abnormalities and gender role; thus, as stated earlier, gender role seems to depend solely on environment. The gender role is usually firmly established between 18 to 30 months of age.

Puberty. The assumption by the female of secondary sex characteristics and menses (menstruation) usually marks her first "change of life." It is associated with hypophyseal output of FSH and LH, with the maturation of ovarian follicles and production of ovarian hormones. In the male, sperm production and assumption of male secondary sex characteristics mark onset of puberty. Age of onset is 12 to 17 years in females and 13 to 16 years in males. Resultant production of gonadal hormones causes maturation of the sex organs.

THE OVARIAN FOLLICLES

The follicles are estimated to number about 400,000 in the ovaries prior to birth. They remain in a quiescent condition until puberty, when hormonal changes initiate follicle development and menstruation. A total of about 400 ova will develop and be discharged over the reproductive life of the female. The remainder will undergo a degenerative process known as atresia.

PRIMORDIAL FOLLICLES are the most numerous of the follicles and are also the smallest. These follicles are 40 to 50 μm in diameter and

TABLE 26.13

AN ANNOTATED LIST OF CHROMOSOMAL DISORDERS

DISORDER AND FREQUENCY / # LIVE BIRTHS	CHROMOSOME (€)		MECHANISM OF PRODUCTION	CHARACTERISTICS
	NO.	ABNORMALITY		
Autosomal disorders				
Down's syndrome (Mongolism); 1:600	47	3 € in group 21 (trisomy 21)	Nondisjunction	Simian (monkey) type finger and hand prints; low set ears
Trisomy of group 18; 1:4,000	47	3 € in group 16–18	Nondisjunction	Index finger crossed over middle finger, mental retardation, do not thrive, renal and cardiac abnormalities
Deletion of group 17–18	46	Deletion of long or short arms € 17–18	Chromosome break	Small head, short stature, receding chin, gamma globulin deficiency
Trisomy 13–15 syndrome	47	Extra € in group 13–15	Nondisjunction	Cleft palate and lip, deafness, mental retardation, extra digits, simian lines
Cri-du-chat (cat cry) syndrome	46	Deletion of part of € 5	Deletion	Catlike cry in infancy, do not thrive, mental retardation
Sex chromosome disorders				
Turner's syndrome 1:10,000	45/46	45 O or 45 X	Nondisjunction	Cardiac abnormalities, webbed neck, sexual infantilism
Kleinfelter's syndrome; 1:500	47	XXY (most common)	Nondisjunction	Small nonfunctional gonads, sterility
Pseudohermaphroditism	46XY or 45XO	Missing X €	Nondisjunction	Masculinization of genitalia
True hermaphrodite	46	46XX/46XY	Ovum fertilized by two sperm	Presence of both functioning ovarian and testicular tissue

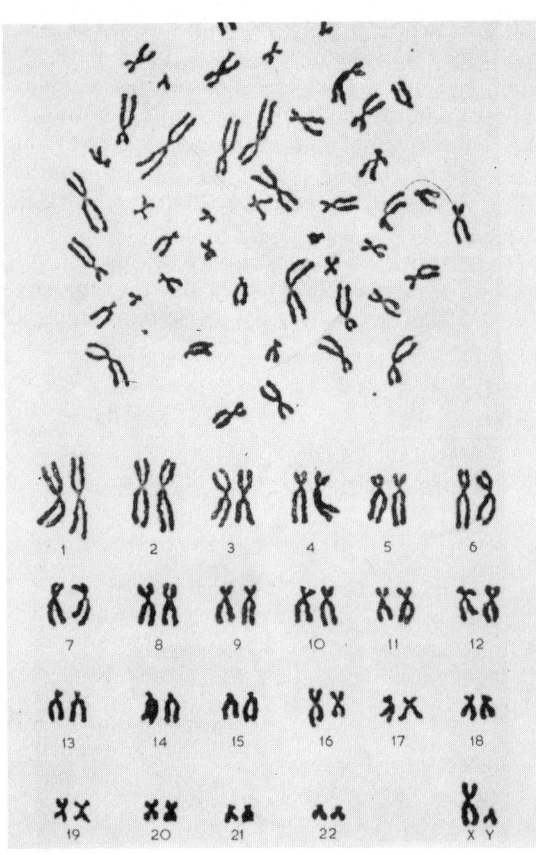

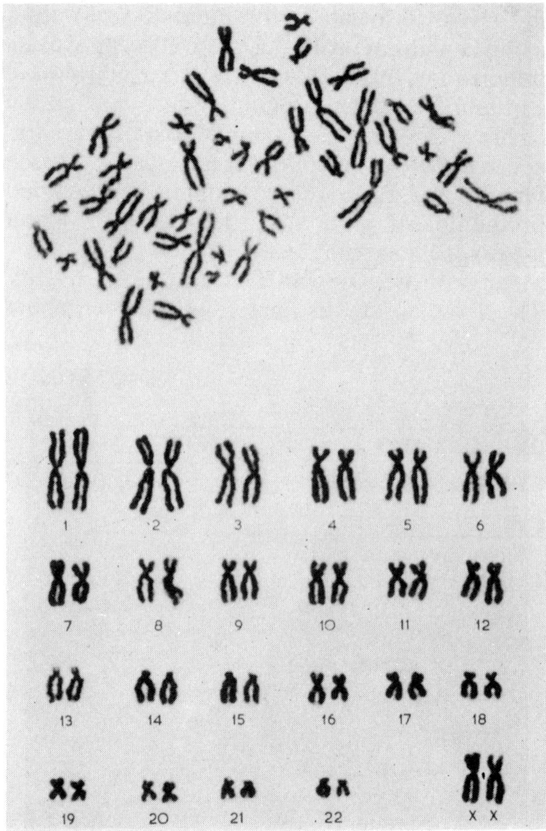

FIGURE 26.28
Normal human chromosomes (*top*), and karyotype (*bottom*). Male to the left; female to the right. (Courtesy Little Brown & Company, from L. S. Penrose, *Recent Advances in Human Genetics,* Little Brown, Boston, 1961.)

may be found in the external portions of the cortex. They consist of an ovum surrounded by a single layer of stromal cells. Further development of the follicle results in the formation of a PRIMARY FOLLICLE. Primary follicles are larger, about 80 μm in diameter, and have several layers of what may now be designated as follicular cells around the ovum. Next, several small cavities develop within the layers of follicular cells, and the unit is called a SECONDARY FOLLICLE. These cavities enlarge, fuse, and eventually form a single follicular cavity. The ovum is forced to the periphery of the follicular cavity. The entire follicle increases to a maximum size of 10 to 13 mm and is known as a VESICULAR or GRAAFIAN FOLLICLE.

The developing follicle is the source of ESTRADIOL, a female sex hormone responsible for the development of female secondary sex characteristics and for the development of several of the organs of the female system. OVULATION, release of the ovum, occurs from the vesicular follicle. The follicle collapses and a small amount of bleeding may occur into the follicle, forming a CORPUS HEMORRHAGICUM. The tissues remaining in the ovary are next caused to form a CORPUS LUTEUM, the source of the second hormone, PROGESTIN (*pro-*

gesterone). The latter hormone maintains and advances the changes initiated by estrogen. The fate of the corpus luteum will be the same regardless of whether or not the ovum is fertilized. With no fertilization, the luteum undergoes degeneration in about 2 weeks; with fertilization and successful uterine implantation, the luteum remains functional for about 3 months of the pregnancy before degenerating. In either case, the luteum is replaced by collagenous connective tissue (scar tissue) and a CORPUS ALBICANS is formed.

OVARIAN HORMONES

The ovary produces three hormones.

ESTROGENS (*estrin, estrone, estradiol*) are steroid hormones produced by the vesicular follicle. They cause the proliferative phase of the menstrual cycle, increase mammary growth, stimulate contraction of the uterus, prevent atrophy of the accessory organs, and develop the female secondary characteristics. They also prime the body for progestin.

PROGESTINS (e.g., *progesterone*) are produced by the corpus luteum. Progesterone is responsible for the secretory phase of the menstrual cycle, further develops the mammary glands, and is necessary for placenta formation.

RELAXIN, produced in very small quantities before birth, softens the pelvic ligaments. This softening theoretically allows enlargement of the birth canal. In the human, the effect is negligible.

FIGURE 26.29
The relationships of hormones to ovarian and uterine activity during the menstrual cycle.

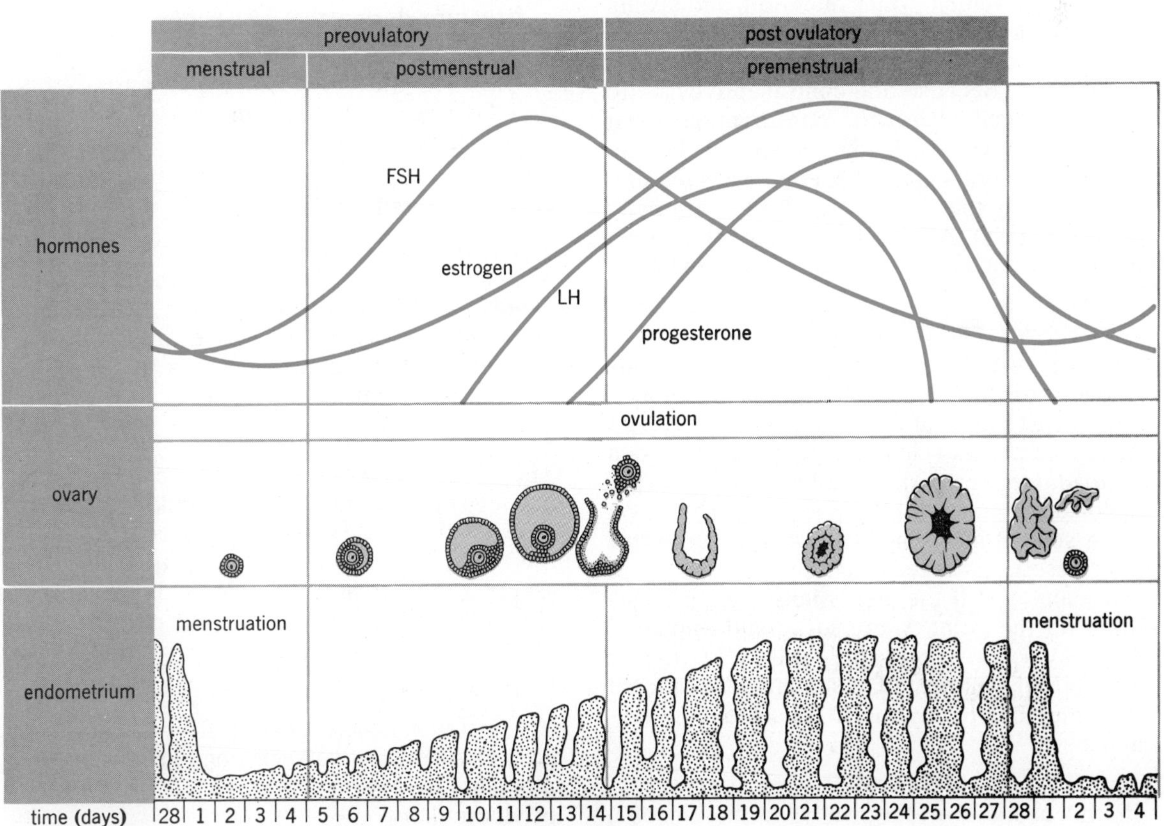

CONTROL OF OVARIAN SECRETION

FOLLICLE-STIMULATING HORMONE (FSH) from the anterior pituitary initiates development of follicles, and, as a consequence, estrogen levels rise. Estradiol inhibits FSH production.

LUTEINIZING HORMONE (LH) from the pituitary causes corpus luteum formation with progestin production. Progestin also inhibits FSH production. The four hormones work together, as shown in Fig. 26.29.

ORAL CONTRACEPTION AND BIRTH CONTROL

The use of "the Pill" to control conception is based on the above effects of estrogen and progestin in inhibiting FSH production. The first substances utilized were similar to progestin. Although they inhibited FSH, and therefore follicle development, they also created a number of uncomfortable side effects, including mammary swelling, water retention, and a state of "pseudo-pregnancy." Newer compounds have the effects of estrin, creating far fewer side effects. Because oral contraceptives act to inhibit follicular growth, consumption must begin after menstrual flow, when follicles are prone to begin their growth, and continue uninterrupted for the full course of pills (normally about 21 days). A pill taken before intercourse inhibits nothing and is ineffective in preventing conception.

FERTILITY DRUGS

"Fertility drugs" are estrogens, progesteroids (progesteronelike substances), or gonadotropins. Most cases of failure to conceive result from menstrual cycles that do not release ova (*anovulatory cycles*). Administration of gonadotropins causes ripening and release of ova, often several at a time, so that multiple births may result. Estrogens and progesteroids are also employed if the uterus fails to develop its lining to the point where successful implantation may occur. Tables 26.14 and 26.15 are presented in the interest of acquainting the student with methods of contraception currently available.

THE PLACENTA

In the event of a pregnancy, formation of a placenta and hormone secretion by the placenta ensure the continued secretion of progesterone by the corpus luteum and produce female sex hormones that can, if necessary, sustain the pregnancy alone after three months. At least five hormones are known to be secreted by the placenta.

ESTRIN ensures the growth of the uterine lining, the development of the mammary duct tissue, and the sustenance of secondary characteristics.

PROGESTERONE ensures the vascularity and glandularity of the uterine wall, ensures the development of the secretory tissue of the mammary glands, and quiets the muscular activity of the uterus.

HUMAN CHORIONIC GONADOTROPIN (HCG) is similar to LH and ensures stimulation of the corpus luteum, therefore resulting in continued secretion of progesterone by that structure during the first trimester of pregnancy.

CHORIONIC GROWTH-HORMONE-PROLACTIN (CGP) has lactogenic and growth-stimulating activity.

An unnamed hormone with TSH ACTIVITY is also secreted by the placenta.

The placenta seems to be insurance that the pregnancy will not suffer deficiencies of hormones necessary to continuation of development. The estrogens and progestins necessary to maintain the last 2 trimesters of the pregnancy come from the placenta.

THE TESTIS

The INTERSTITIAL CELLS of the testis, located between the seminiferous tubules, produce TESTOSTERONE, a steroid. This hormone serves the same function in the male that estrogen serves in the female; that is, it develops and maintains the accessory organs and the secondary sexual characteristics.

CONTROL OF TESTICULAR SECRETION

LUTEINIZING HORMONE (LH) from the pituitary governs testosterone secretion. The latter hormone in turn exerts a negative feedback on LH production.

TABLE 26.14

METHODS OF CONTRACEPTION CURRENTLY AVAILABLE

METHOD	MODE OF ACTION	EFFECTIVENESS IF USED CORRECTLY	ACTION NEEDED AT TIME OF COITUS	REQUIRES RESUPPLY OF MATERIALS USED	REQUIRES INSTRUCTION IN USE	REQUIRES SERVICES OF PHYSICIAN	SUITABLE FOR MENSTRUALLY IRREGULAR WOMEN	SIDE EFFECTS
Oral pill 21 day administration	Prevents follicle maturation and ovulation	Highest	None	Yes	Yes—timing	Yes—prescription	Yes	Early—some water retention, breast tenderness. Late—possible embolism, hypertension
Intrauterine device (coil, loop)	Prevents implantation	High	None	No	No	Yes—to insert	Yes	Some do not retain device. Some have menstrual discomfort
Diaphragm with jelly	Prevents sperm from entering uterus, plus jelly spermicidal	High	Previous to coitus	Yes	Yes—must be inserted correctly each time	Yes—for sizing and instruction on use	Yes	None
Condom (worn by male)	Prevents sperm entry into vagina	High	Yes	Yes	Not usually	No	—	Some deadening of sensation in male
Temperature rhythm	Determines ovulation time by noting body temperature at ovulation. T ↑	Medium	No	No	Definitely. Must learn to interpret chart correctly	No. Physician should advise	Yes, if skilled in reading graph	None. Requires abstinence during part of cycle
Calendar rhythm	Abstinence during part of cycle	Medium to low	No—no coitus	No	Definitely. Must know when to abstain	No. Physician should advise	No!	None (pregnancy?)

TABLE 26.14 (continued)

METHODS OF CONTRACEPTION CURRENTLY AVAILABLE

METHOD	MODE OF ACTION	EFFECTIVENESS IF USED CORRECTLY	ACTION NEEDED AT TIME OF COITUS	REQUIRES RESUPPLY OF MATERIALS USED	REQUIRES INSTRUCTION IN USE	REQUIRES SERVICES OF PHYSICIAN	SUITABLE FOR MENSTRUALLY IRREGULAR WOMEN	SIDE EFFECTS
Vaginal foams	Spermicidal	Medium to low	Yes. Requires application before coitus	Yes	No	No	Yes	None usually. May irritate
Withdrawal	Remove penis before ejaculation	Low	Yes. Withdrawal	No	No	No	Yes	Frustration in some
Douche	Wash out sperm	Lowest	Yes. Immediately after	No	No	No	Yes	None

Vasectomy (Tying or cutting the vas deferens) or tubal ligation (tying or cutting the uterine tubes) affords additional means of preventing fertilization. These procedures are generally not reversible, and sterility will result.

TABLE 26.15

OTHER METHODS OF CONTRACEPTION UNDER INVESTIGATION

METHOD	MODE OF ACTION	EFFECTIVENESS IF USED CORRECTLY	ACTION NEEDED AT TIME OF COITUS	REQUIRES RESUPPLY OF MATERIALS USED	REQUIRES INSTRUCTION IN USE	REQUIRES SERVICES OF PHYSICIAN	SIDE EFFECTS
"Mini-pill"—very low content of progesterone (¼ mg.)	Inhibits follicle development	High	No	Yes	Yes	Yes—prescription	Irregular cycles and bleeding (25%)
"Morning-after pill" (DES, diethylstilbesterol)	Arrests pregnancy probably by preventing implantation. 50 × normal dose of estrogenic material	By currently available data, high	No. For 1–5 days after coitus	Yes	No	Yes	Breast swelling, nausea, water retention. Use by mother can cause rare type of vaginal cancer in female offspring
Vaginal ring—inserted in vagina; contains progesteroid in it	"Leaks" progesteroid into bloodstream through vagina at constant rate. Thereby inhibits follicle maturation	Studies are "promising"	No	Yes. Perhaps at yearly intervals	Yes	Yes	Spotting, some discomfort
Once-a-month pill	Injected in oil base into muscle. Slow passage of birth control drug into circulation inhibits follicle maturation	Said to be 100 percent	No	Yes. On monthly basis	No	Yes	Similar to oral pill
Depo-Provera (DMPA) 3 month injection	Injected. Inhibits follicle development	By currently available data, high	No	Yes. On 3 month basis	Yes	Yes	Similar to oral pill
Antipregnancy vaccine	Antibodies to HCG cause pregnant female to menstruate	Said to be high	No	No	No	Yes	Unknown
Male "pill"	Inhibition of LH secretion by drug (denezol) given orally	85%	No	Yes	Yes	Yes	Unknown

DISORDERS OF THE OVARIES AND TESTES

Both sexes may suffer from deficient production of hormones in the condition known as EUNUCHOIDISM (Fig. 26.30). As expected, the most obvious symptoms will be atrophy of accessory organs and disappearance of secondary characteristics. Complete loss of hormones produces eunuchism, similar to eunuchoidism, but presenting more severe symptoms. Treatment with appropriate hormone will control symptoms.

HYPERFUNCTION, due to gonadal tumors, creates PRECOCIOUS PUBERTY. In this condition, accessory organ development and characteristics appear at earlier-than-normal ages. Precocious females may actually become fer-

FIGURE 26.30
Eunuchoidism in a 40-year-old male. Note disproportionately long arms and legs for height (6 feet, 1 inch) and failure of the genitalia and secondary characteristics to develop normally. (Lester V. Bergmann & Associates.)

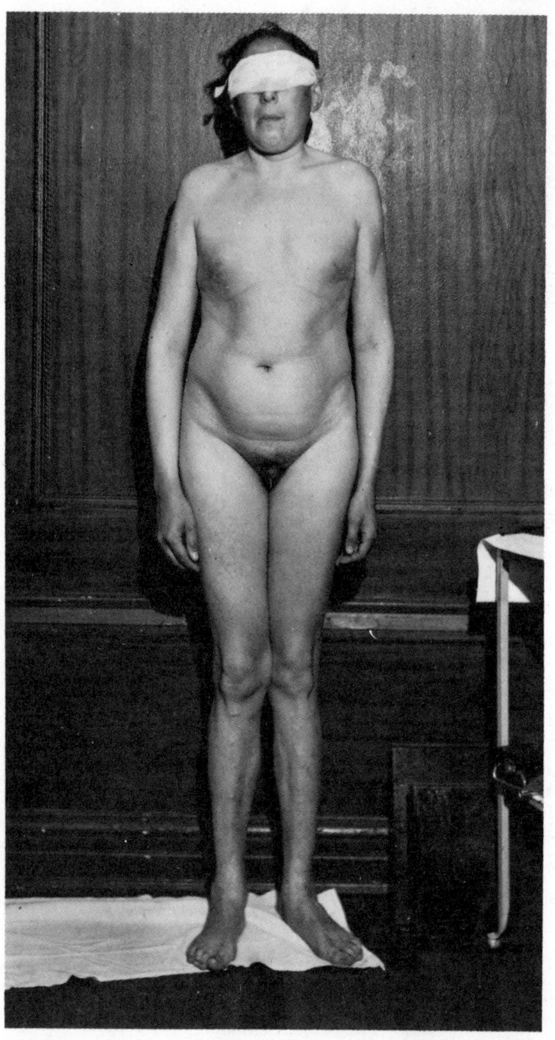

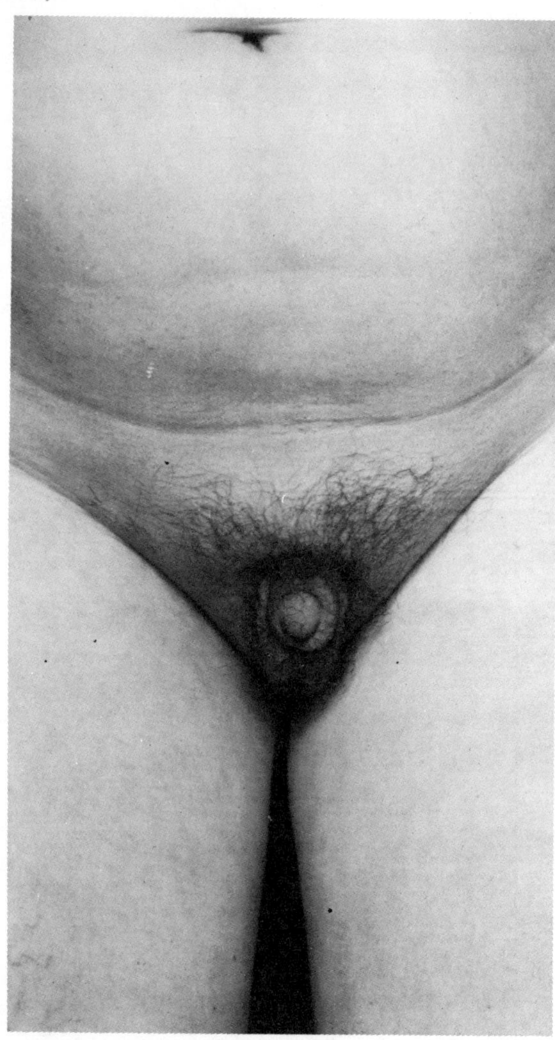

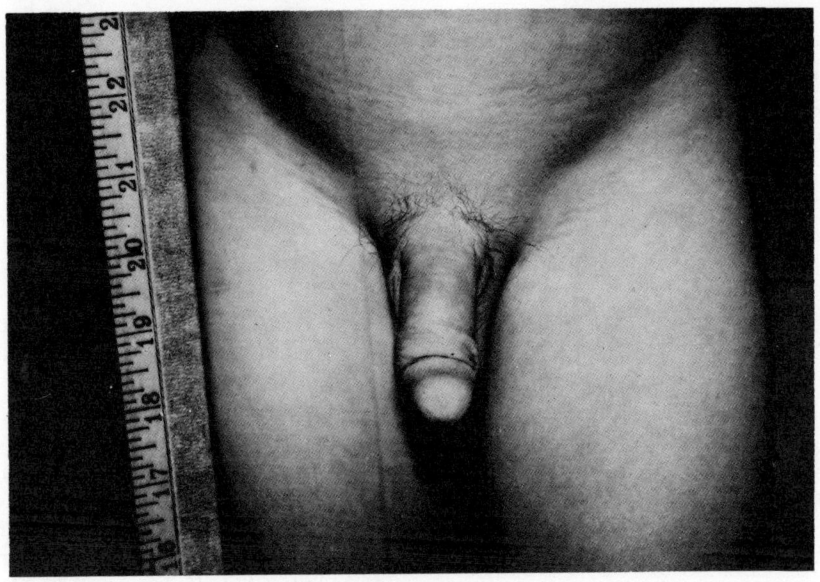

FIGURE 26.31
Precocious puberty in a 3-year-old male. Note development of genitalia equivalent to that of an adolescent, and appearance of pubic hair. (Armed Forces Institute of Pathology.)

tile at very tender ages (the earliest on record became pregnant at five years old). Figures 26.31 and 26.32 show male and female precocious puberty patients. Surgical removal of the tumor is required, and the development may be arrested and partially reversed.

CLIMACTERICS

As they age, both males and females undergo a cessation or diminution of gonadal hormone production. In the female, this period is often called the MENOPAUSE, and it usually occurs between 45 and 50 years of age. In males, similar slowdown is termed the male CLIMACTERIC, and it occurs gradually, beginning at about 60 years of age.

In both sexes the following symptoms may develop:

PSYCHIC SYMPTOMS (more common in female). Nervousness, irritability, crying spells, some loss of mental acuity.

VASOMOTOR SYMPTOMS. Sweating, headache, hot and cold flashes.

CONSTITUTIONAL SYMPTOMS. Fatigue, muscular weakness, "lack of ambition."

SEXUAL SYMPTOMS. Loss of sex drive; in female, ultimate sterility.

It should be emphasized that the climacterics will occur as aging proceeds. Understanding of the condition and acting accordingly is often the most important factor in getting through the period. Appropriate sex hormones are often used to treat those individuals in whom these symptoms are causing problems in their activities, the general idea being to make the disappearance of the hormones more gradual and less disturbing.

THE PROSTAGLANDINS

The prostaglandins (PGs) are a series of cyclic, oxygenated derivatives of prostanoic acid (Fig. 26.33). They are formed by the action of a prostaglandin synthetase enzyme system located on the microsomal (ER fragments) fraction of the cellular organelles of a wide variety

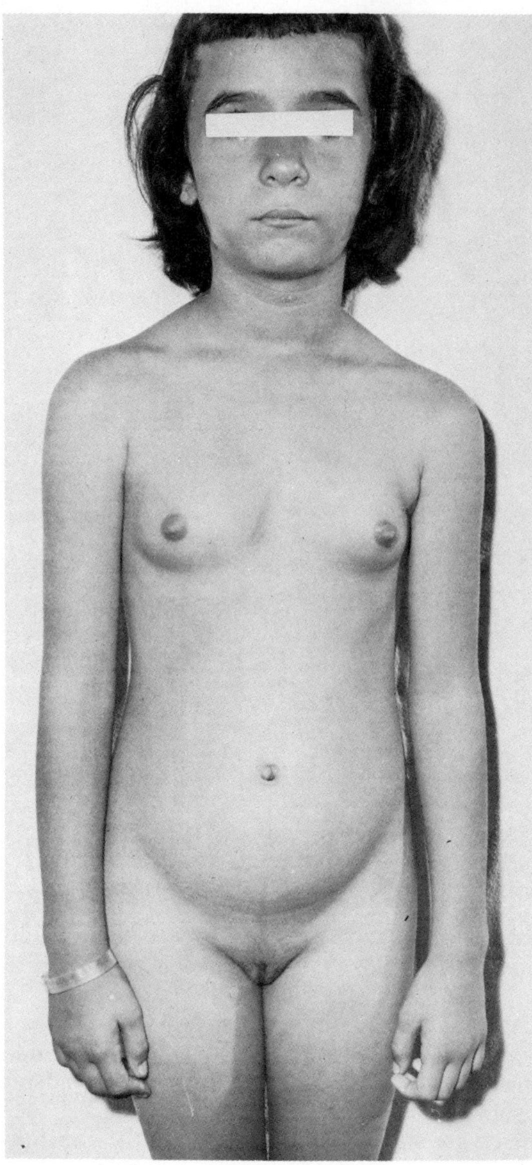

FIGURE 26.32
Precocious puberty in a 7-year-old female. Height was 4 feet, $3^1/_2$ inches. Note development of pubic hair and mammary glands. (Lester V. Bergmann & Associates.)

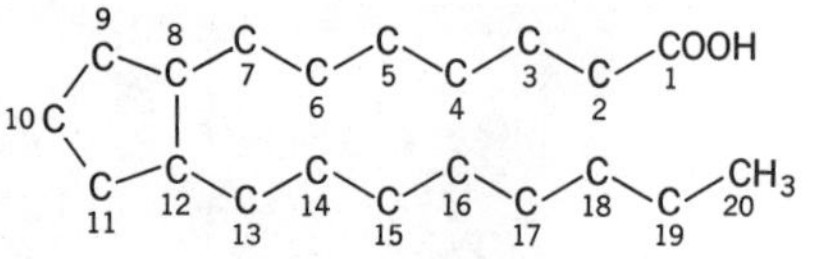

FIGURE 26.33
The chemical structure of prostanoic acid.

of body cells (e.g., lungs, liver, muscle, reproductive organs). A long-chain polyunsaturated fatty acid is caused to assume a cyclic form under the influence of the synthetase.

The PGs are categorized by the chemical groups found attached to the cyclic portion of the molecule, and by their degree of unsaturation. Utilizing the chemical groups on the ring, four basic types may be distinguished, as shown in Fig. 26.34.

Further description of the molecule is provided by a subscript number that indicates the number of unsaturated carbon-carbon bonds in the molecule. Additionally, the designation α (alpha) or β (beta) is used to describe the direction of projection of the chemical groupings at carbons 8 or 9, 11, and 15. Alpha designates a bond or bonds projecting "below" the average plane of the 5-carbon ring; beta implies bond projection "above" the average plane of the 5-carbon ring. The primary PGs and their precursors are presented in Fig. 26.35.

Prostaglandins are widely distributed in the animal kingdom. They have been found in corals and within the "higher" animal organisms, and have been demonstrated in most animal cells and tissues. They were first demonstrated in human seminal fluid in 1930. Since that time, their distribution has been shown to be nearly ubiquitous.

With the discovery of the presence of cyclic nucleotides and their role in control of enzymatic reactions, PGs have been investigated as controllers of the adenylcyclase or guanylcyclase enzyme systems, and therefore the amount of cyclic nucleotides in cells. Evidence to date indicates that PG effect is indeed exerted via the cyclase enzymes, and may be either stimulatory or inhibitory in effect. The inhibitory effect may additionally be exerted through the mechanism of blocking the receptor site for a hormone that normally increases cyclic nucleotide formation. All this suggests that PGs may act as "intracellular switches" to turn chemical reactions off or on.

Tables 26.16 and 26.17 show some tissues in

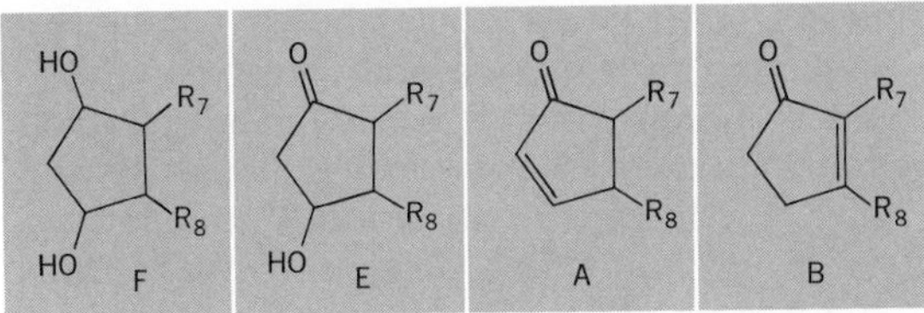

FIGURE 26.34
The four basic ring structures of the prostaglandins. (Courtesy N. Anderson and H. Benson, *Annals, N. Y. Acad. Sci.,* Vol. 180, p. 15, Fig. 1.)

which PGs increase cAMP concentrations, or where hormone effect is inhibited.

Release of PGs from tissues or cells is caused by a wide variety of stimuli, as shown in Table 26.18, and may be associated with disturbances of any cell membrane by normal or pathological processes. Cells may release PGs to resist change, as when tension is applied to smooth muscle or when inflammation occurs. A primary result appears to be to resist tearing or rupture of the cell membrane.

Effects of PGs are nearly as widespread as the tissues in which they are found. The F and E groups have been most extensively investigated as to effects. The list presented below indicates some, but certainly not all, of the effects of PGs on the body.

Smooth muscle. Uterine smooth muscle is stimulated to increase its contractions by PGE. Vascular smooth muscle is relaxed by PGE and PGA_2. PGF causes contraction of vascular smooth muscle.

Sodium excretion. PGA, PGA_2, and PGE_2 are all natriuretic and can lower blood pressure by decreasing blood volume via Na and water excretion. The effect appears to be mediated via the kidney.

Ion and water exchange. PGE increases calcium release from binding sites and inhibits the ADH effect on toad bladder and perhaps also on kidney tubules.

Skin. The highest concentrations of PGs are found in the epidermis, where they are postulated to be involved in the synthesis of the waterproofing substances of the upper epidermal cell layers. If deficient in PG, these cells have been shown to become more permeable to water.

FIGURE 26.35
The primary prostaglandins and their precursors. α bond configuration is indicated by dashed lines, β by solid lines.

prostanoic acid

8, 11, 14, — eicosatrieonic acid → PGE_1 + $PGF_1\alpha$

5, 8, 11, 14, — eicosatetraeonic acid → PGE_2 + $PGF_2\alpha$

5, 8, 11, 14, 17, eicosapentaenoic acid → PGE_3 + $PGF_3\alpha$

TABLE 26.16

TISSUES WHERE PROSTAGLANDINS INCREASE ACCUMULATION OF CYCLIC 3′5′-AMP

TISSUE	SPECIES	PROSTAGLANDIN (μM) MOST ACTIVE	
Lung	Rat	PGE_1	2.8
Spleen	Rat	PGE_1	2.8
Diaphragm	Rat	PGE_1	2.8
Adipose	Rat	PGE_1	2.8
Leucocytes	Human	PGE_1	—
Platelets[a]	Human	PGE_1	0.28
	Human	PGE_1	0.01
	Human	PGE_1	0.1
	Human	PGE_1	0.1
	Rabbit	PGE_1	0.1
	Rabbit	PGE_1	0.15
Liver	Rat	PGE_1	—
Anterior pituiary[a]	Rat	PGE_1	2.8
Aorta	Rat	PGE_1	2.8
Bone	Rat	PGE_1	—
Gastric mucosa[a]	Guinea pig	PGE_1	—
Kidney	Dog, rat	PGE_1	—
Heart	Guinea pig	PGE_1	0.01
		$PGF_{1\alpha}$	0.01
Corpus luteum[a]	Bovine	PGE_2	28
Thyroid	Dog	PGE_2	—
Erythro-cytes[a]	Rat	PGE_2	0.03

[a] Indicates those tissues where prostaglandins increase adenylcyclase activity.
Source: J. E. Shaw, Annals, N.Y. Acad. Sci., Vol. 180, p. 242, Table 1.

The ovary. $PGE_{2\alpha}$ has been shown to be luteolytic—that is, capable of inducing the destruction of the corpus luteum.

Gastric secretion. PGA_1 and PGE_1 reduce gastric secretion.

Inflammatory response. PGE appears to increase the response of leucocytes in reaching an area of inflammation or injury. This substance also appears to increase blood flow and the edematous response that usually follows trauma to a tissue. It also causes increased breakdown of tissue basophils (mast cells).

Neurons. PGF_2 has been shown to depress synaptic transmission and elevate discharge rates of neurons. It is not a transmitter itself, but influences those processes that synthesize or degrade the normal transmitters. PGE also elevates discharge by neurons, but has no effect on synaptic transmission.

Therapeutic uses of PGs will probably revolve around their effects on smooth muscle and the ovary. The use of PGs to CONTROL HYPERTENSION (through vasodilating effects), nasal congestion and asthma (through vasoconstrictive and smooth muscle relaxing effects respectively), ULCERS (through decreased gastric secretion), and as ABORTIFACIENTS (through luteolysis) has reached the experimental stage in humans. Minimal side effects appear to be produced, and the PGs may, in the future, achieve the fame accorded the antibiotics at their appearance.

THE PINEAL GLAND

The pineal gland, located on the posterior aspect of the midbrain (Fig. 26.36), has, in recent years, acquired new status as an endocrine structure. Several chemically active substances

TABLE 26.17

TISSUES WHERE PROSTAGLANDINS[a] INHIBIT HORMONALLY INDUCED RESPONSES

TISSUE	HORMONE	RESPONSE
Toad bladder	Vasopressin	Water transport
Rabbit kidney tubules	Vasopressin	Water transport
Rat adipocytes[b]	Epinephrine ACTH TSH Glucagon Growth hormone	Lipolysis
Cerebellar Purkinje cells	Norepinephrine	Inhibition of discharge frequency

[a] PGE_1 most effective at <0.28 μM.
[b] Inhibition associated with decreased cyclic AMP accumulation.
Source: J. E. Shaw, Annals, N.Y. Acad. Sci., Vol. 180, p. 243, Table 2.

TABLE 26.18
THE RELEASE OF PROSTAGLANDINS BY VARIOUS STIMULI

SPECIES	TISSUE	STIMULUS
Rabbit	Eye (ant. chamber)	Mechanical
	Spleen	Catecholamines
		Serotonin
	Epigastric fat pad	Hormones
	Somatosensory cortex	Neural
		Analeptics, etc.
		Reticular formation
	Spleen	Catecholamines
	Adrenal	Acetylcholine
Dog	Spleen	Neural
		Catecholamines
		Colloids
	Bladder	Distension
	Cerebral ventricles	Serotonin
Rat	Phrenic diaphragm	Neural
		Biogenic amines
	Epididymal fat pad	Neural
		Biogenic amines
	Stomach	Neural
		Stretch
		Secretagogues
	Skin	Inflammation
	Liver	Glucagon
	Lung	Air embolus
		Infusion of particles
Guinea pig	Lung (whole, perfused)	Anaphylaxis
		Particles
		Histamine
		Tryptamine
		Serotonin
		Massage
		Air embolus
		Distension
	Lung (chopped)	Stirring
Human	Thyroid	Medullary carcinoma
	Uterus	Parturition
		Distension
	Platelets	Thrombin
Frog	Intestine	Distilled water
	Skin	Isoproterenol
	Spinal cord	Neural
		Analeptics

(Courtesy P. J. Piper and J. Vane, Annals, N.Y. Acad. Sci., Vol. 180, pp. 376–377, Table 2.)

FIGURE 26.36
The location of the pineal gland as shown on a mid-sagittal section of the brain.

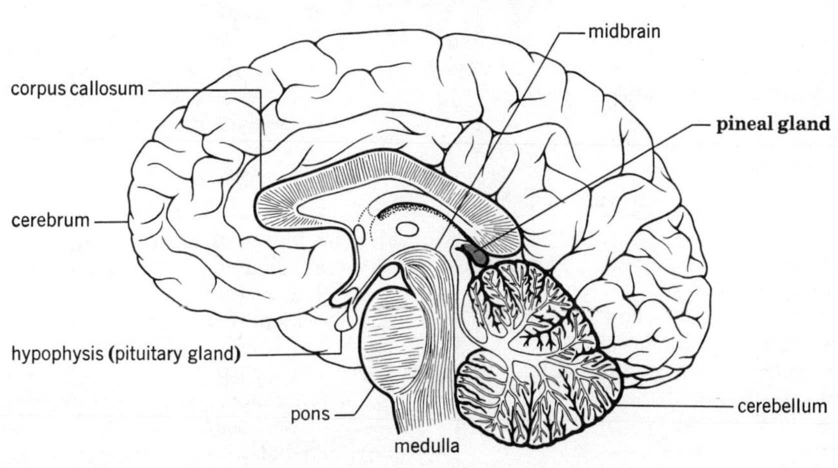

have been isolated from the pineal, including MELATONIN, norepinephrine, serotonin, and histamine. Of these substances, melatonin is synthesized only in the pineal. Melatonin has the property of causing agglomeration of pigment granules in pigment cells (melanophores) in amphibian skin. In such animals, the secretion of melatonin may be greatly diminished or abolished by enucleation of the eye. Secretion of the substance thus appears to respond to changes in light as detected by the eye and may have protective value to the organism in terms of camouflaging it from predators. Effects of the hormone in humans are uncertain. Pineal secretion appears to affect gonadotropin secretion, and melatonin may inhibit the ovary in its hormone secretion. The role of the pineal in the human body remains to be determined.

SUMMARY

1. The endocrine glands develop as outgrowths of the pharynx, gut, or from mesoderm or neural crest material.
2. Endocrines provide a means of controlling homeostasis, growth, development, and metabolism by blood-borne hormones.
3. Endocrines are specific groups of cells that secrete specific chemicals.
4. There are three classes of hormones: proteins, amines, and steroids.
5. Hormones are synthesized by endocrines, are produced in amounts as required, are inactivated by their use, and influence chemical reactions in the body.
6. There are six basic control devices for governing secretion of hormones.
7. The pituitary has two main parts: anterior and posterior lobes.
   a. The anterior lobe produces six hormones (GH, prolactin, ACTH, TSH, LH, FSH).
   b. The anterior lobe is controlled by hypothalamic regulating factors and feedback mechanisms.
   c. The posterior lobe makes no hormones, but stores ADH (antidiuretic hormone) and oxytocin (smooth muscle stimulant) produced by the hypothalamus.
   d. The release of posterior lobe hormones is controlled by nerves.
   e. Disorders include giantism, acromegaly, and diabetes insipidus.
8. The thyroid gland lies in the neck and produces thyroxin and calcitonin.
   a. Thyroxin controls BMR and is involved in growth, especially of the brain cells.
   b. Synthesis of thyroxin is controlled by TSH.
   c. Calcitonin influences metabolism of calcium and phosphate, and aids their deposition in bony tissue.
   d. Calcitonin production is directly related to blood calcium level.
   e. Thyroid disorders include hyperthyroidism, goiter, and cretinism.
9. The four parathyroid glands lie on the posterior aspect of the thyroid.
   a. The hormone (PTH) influences metabolism of Ca, PO_4, Mg, and citrate by increasing absorption and removal from bony tissue.
   b. Secretion is inversely related to blood calcium level.
   c. Disorders of the parathyroids include hyperparathyroidism and tetany.
10. The pancreas islets produce two hormones: glucagon and insulin.
    a. Glucagon increases conversion of glycogen to glucose and raises blood sugar levels.
    b. Insulin increases conversion of glucose to glycogen, increases glucose uptake by certain cells, and thus lowers blood sugar levels.
    c. Insufficient insulin production results in diabetes mellitus.
    d. Diabetes results in hyperglycemia, glucosuria, and other characteristic symptoms; it develops in stages.
    e. Diabetes may have a hereditary basis that results in disease production when the body is stressed by a variety of factors.
    f. Hypoglycemia results in abnormal brain function.
    g. Secretion of both hormones is controlled by blood sugar levels.
11. The adrenal glands contain two functional areas: medulla and cortex.
    a. The medulla produces epinephrine and norepinephrine that accelerate body activity and energy expenditure.
    b. The cortex produces: mineralocorticoids (e.g., aldosterone) that deals with Na metabolism, glucocorticoids (e.g., cortisol) that are anabolic hormones, and sex hormones.

c. Glucocorticoids also confer resistance to stress but may damage body organs.
d. Cortical secretion is controlled by blood Na level and angiotensin (mineralocorticoids) and by ACTH (others).
e. Disorders of the cortex include Addison's disease (deficiency), Cushing's syndrome (excess), and the adrenogenital syndrome.

12. The ovaries and testes produce hormones that, with genetic and social factors, determine maleness and femaleness.
    a. Ovarian hormones include: estrogens, which develop female sex organs and characteristics, initiate puberty, and menses; progesterone, essential for ovulation, endometrial growth, and placenta formation.
    b. Ovarian activity is controlled by FSH and LH.
    c. The use of female hormones or materials having similar effects is the basis for oral contraception.
    d. "Fertility drugs" stimulate ovum development.
    e. The placenta produces estrogens, progesterone, LH-like materials, and other hormones.
    f. Testosterone, produced by the testes, ensures development of male sex organs and characteristics.
    g. Testosterone secretion is controlled by ICSH (LH).
    h. Both sexes may suffer deficiency of hormone(s) and loss of characteristics; excess hormone causes precocious puberty or accentuation of characteristics.
13. Climacterics are associated with decrease of gonadal hormone production, the development of nervous and constitutional symptoms, and atrophy of genital tissues.
14. Prostaglandins are derivates of fatty acids.
    a. They are widely distributed in animal cells.
    b. They appear to act to "turn off or on" certain chemical reactions in the body, notably cAMP production.
15. The pineal gland produces melatonin and may be involved in ovarian hormone production.

QUESTIONS

1. What features do endocrine glands have in common?
2. What hormones influence glucose metabolism? What is the effect of each on glucose metabolism?
3. What hormones are concerned with body growth, and what is the contribution of each to the process?
4. What hormones are involved with mineral metabolism? Discuss the effects of each.
5. How is sexual development controlled by hormones?
6. What hormones, and from where, ensure continuation of a pregnancy if the ovaries are removed?
7. What are the determinants of maleness and femaleness? Which are hormonally controlled?
8. Give an example for each of the basic control mechanisms for endocrine secretion.

READINGS

Bremner, William J., and David M. deKretser. "The Prospects for New, Reversible Male Contraceptives." *New Eng. J. Med. 295:*1111. Nov 11, 1976.

Chan, Lawrence, and Bert W. O'Malley. "Mechanism of Action of the Sex Steroid Hormones." *New Eng. J. Med. 294:*1322. June 10, 1976. *294:*1372. June 17, 1976. *294:*1430. June 24, 1976.

Chick, William L., et al. "Artificial Pancreas Using Living Beta Cells: Effects on Glucose Homeostasis in Rats." *Science 197:*780, 19 August 1977.

Dickmann, Z., and C. H. Spilman. "Prostaglandins in Rabbit Blastocysts." *Science 190:*997. 5 Dec 1975.

Edmondson, Hugh A., Brian Henderson, and Barbara Benton. "Liver-Cell Adenomas Associated with Use of Oral Contraceptives." *New Eng. J. Med. 294:*470. Feb 26, 1976.

Gerald, Park S. "Current Concepts: Sex Chromosome Disorders." *New Eng. J. Med. 294:*706. Mar 25, 1976.

Hays, Richard M. "Antidiuretic Hormone." *New Eng. J. Med. 295:*659. Sept 16, 1976.

Levine, Rachmiel. "Glucagon and the Regulation of Blood Sugar." *New Eng. J. Med. 294:*494. Feb 26, 1976.

Loeb, John N. "Corticosteroids and Growth." *New Eng. J. Med. 295:*547. Sept 2, 1976.

Marx, Jean L. "Estrogen Drugs: Do They Increase the Risk of Cancer?" *Science 191:*838. 27 Feb 1976.

Maugh, Thomas H. "Diabetes Commission: Problem Severe, Therapy Inadequate." *Science 191:*272. 23 Jan 1976.

Maugh, Thomas H. "Diabetes Therapy: Can New Techniques Halt Complications?" *Science 190:*1281. 26 Dec 1975.

Maugh, Thomas H. "Diabetic Retinopathy: New Ways to Prevent Blindness." *Science 192:*539. 7 May 1976.

Maugh, Thomas H. "Hormone Receptors: New Clues to the Cause of Diabetes." *Science 193:*220. 16 July 1976.

Miller, Orlando, and William R. Breg. "Current Concepts: Autosomal Chromosome Disorders and Variations." *New Eng. J. Med. 294:*596. Mar 11, 1976.

Mullen, Yoko S., William R. Clark. I. Gabriella Molnar, and Josiah Brown. "Complete Reversal of Experimental Diabetes Mellitus in Rats by a Single Fetal Pancreas." *Science 195:*68. 7 Jan 1977.

Needleman, Philip, Prasad S. Kulkarni, and Amiram Raz. "Coronary Tone Modulation: Formation and Actions of Prostaglandins, Endoperoxides, and Thromboxanes." *Science 195:*409. 28 Jan 1977.

Rosenberg, Lynn, Bruce Armstrong, Hershel Jick. "Myocardial Infarction and Estrogen Therapy in Post-Menopausal Women." *New Eng. J. Med. 294:*1256. June 3, 1976.

Shen, Shiao-Wei, and Rubin Bressler. "Clinical Pharmacology of Oral Antidiabetic Agents." *New Eng. J. Med. 296:*493. March 3, 1977.

Smith, Donald C., et al. "Association of Exogenous Estrogen and Endometrial Cancer." *New Eng. J. Med. 293:*1164. Dec 4, 1975.

Weiss, Gerson, E. M. O'Byrne, and B. G. Steinetz. "Relaxin: A Product of the Human Corpus Luteum of Pregnancy." *Science 194:*948. 26 Nov 1976.

Wurtman, Richard J., and Michael A. Moskowitz. "The Pineal Organ." *New Eng. J. Med. 296:*1329, June 9, 1977. *296:*1383, June 16, 1977.

Zonana, Jonathan, and David L. Rimoin. "Current Concepts: Inheritance of Diabetes Mellitus." *New Eng. J. Med. 295:*603. Sept 9, 1976.

27

INHERENT CONTROL; BIOLOGICAL RHYTHMS

OBJECTIVES

After studying this chapter, the reader should be able to:

☐ Explain the interrelationship between rhythms and maintenance of homeostasis.

☐ Define the terms cycle, period, frequency, amplitude, phase, and entrainment as related to rhythm.

☐ Explain what is meant by circannual, circamensual, circaseptan, and circadian rhythms, and give examples of each.

☐ Cite examples of hourly, second, and fraction-of-second rhythms.

☐ List some properties of rhythms, regardless of length of period.

☐ Discuss the hypotheses offered to explain the origin of cyclical alterations in function.

☐ Discuss the concept of "biological clocks" as determiners of cyclical function.

☐ List some possible external entraining stimuli that might be rhythm setters.

☐ Indicate, using examples, why some rhythms appear to be acquired after birth.

☐ Discuss the effects of aging on cyclical functions.

☐ Present, for the human, some examples of circadian rhythms.

☐ Indicate that both mental and physical performance shows cyclical patterns and relate this to internal and/or external stimuli.

RHYTHMS AND HOMEOSTASIS

Our study of body function has stressed the maintenance of that state of apparent constancy called homeostasis. However, a healthy individual's achievement of homeostasis masks a multitude of cyclical alterations in body function. In short, homeostasis is not a straight line running parallel to the abscissa of a graph, but a curving line showing periodic valleys and peaks. The upper and lower boundaries of those peaks and valleys are usually not allowed to stray beyond certain limits, and, as stated in Chapter 1, *controlled* deviation is really the essence of homeostasis, not undeviating adherence to a preset value. Thus, the concept of rhythm in life and nature compliments the concept of homeostasis.

THE NATURE OF RHYTHMS

DESCRIPTIVE TERMINOLOGY

The fluctuating nature of a function that repeats itself with time in the same order and at the same interval constitutes a CYCLE or RHYTHM (Fig. 27.1). Within a cycle, various of its parts may be named. The PERIOD is the time between similar parts of the cycle, as from peak-to-peak or trough-to-trough. FREQUENCY refers to the number of cycles per unit of time. AMPLITUDE refers to the height or intensity (the ordinate of a graph) of the swings of a function. PHASE refers to a particular part of a cycle, such as its upswing or downswing. The term ENTRAINMENT is used to refer to the assumption of a cycle as a result of an internal or external change. Particular terms may be applied when referring to the period. The prefix *circa-* (circa, about) is followed by a word reflecting length. For example *circannual* indicates a period of about a year (as in blood glucose and cholesterol levels that exhibit annual as well as daily fluctuations); *circamensual,* about a month (as in menstrual cycles); *circaseptan,* about a week; *circadian,* about a day (*see* Table 27.1); *diurnal,* a cycle peaking during the day; *nocturnal,* a cycle peaking during darkness. Shorter rhythms may be measured in hours (alternation of body temperature peaks from left side to right side), minutes (cell division), seconds (heartbeat), or fractions of seconds (EEG rhythms). These shorter periods do not generally receive special names. Additionally, many cycles appear to be interrelated in a consistent manner in the healthy individual; heartbeat and breathing rate usually exhibit about a 4:1 ratio. Many of the human rhythms appear to be circadian in nature.

Timing of such rhythms appears to reflect the presence of a "biological clock" that ensures the length of period and activity associated with that period.

PROPERTIES AND CHARACTERISTICS OF RHYTHMS

Among the properties and characteristics of a rhythm may be listed the following.

Physiological functions in any living organism exhibit periodicity as a function of time. The fluctuations persist even when an organism is removed from its natural or habitual environment, although the period may be altered somewhat.

Rhythms appear to be dependent on some internal mechanism, at least after they have been entrained.

Rhythm setting is probably the result of evolutionary processes that have been incorporated into the genetic mechanisms of the organism.

Rhythms have adaptive value to the organism in terms of survival.

Rhythms are generally resistant to a variety of chemical substances including stimulants and depressants. Their amplitude may be changed by such manipulations but period is not greatly altered, if at all.

FIGURE 27.1
The various aspects of a cycle.

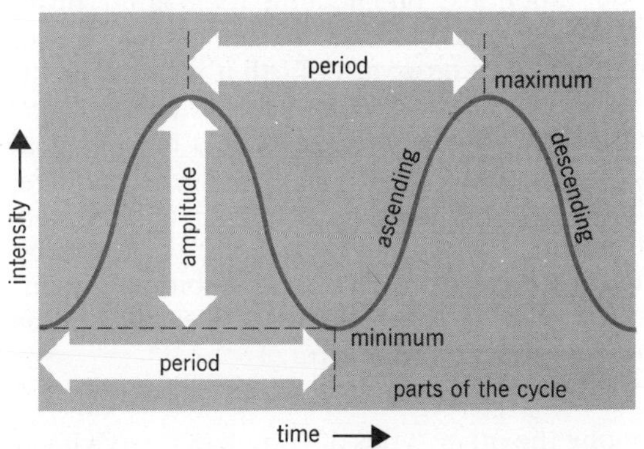

SETTING OF RHYTHMS

HYPOTHESES

As stated above, rhythms appear to be innate in the organism. Nevertheless, investigation designed to correlate rhythms with known alterations, either of exogenous or endogenous origin have been undertaken. Two general groups of hypotheses have evolved, named INTERNAL and EXTERNAL HYPOTHESES.

Adherents to the internal hypothesis point out that physiological function does not correlate exactly with external changes such as light-dark cycles, or temperature change, and that different individuals do not have the same periods in their identical functions. They conclude that the "clock" runs independently of external stimuli, perhaps depending on gene-determined alterations.

Supporters of the external hypothesis indicate that *exact* timing is not a necessary prerequisite for development of a rhythm; that the body possesses a variety of sensors that may have different sensitivities between individuals; that cycles do indeed exist in natural phenomena; and that these combine to provide an initial stimulus to rhythm setting.

Whatever explanation is correct has not yet been resolved and may reflect some combination of these two hypotheses.

Another argument that continues revolves

around whether there is a "master clock" that is set into motion by internal or external determinants, or whether there are "multiple clocks" that control one or a small number of interrelated functions. Again, two major hypotheses related to clocks have evolved.

The ESCAPEMENT-TYPE clock is described as being insensitive to external change, and is inherent within the cell. It measures off time independent of external forces.

The NONESCAPEMENT-TYPE clock is a responding clock, sensitive to environmental change, and measures time in response to some fundamental geophysical alteration.

Where the clock is located is also a problem. In the cerebrum? hypothalamus? eye? At any rate, the clock is said to be independent of the rhythm it controls, but is coupled to it by some as yet unknown mechanism.

If any exogenous stimulus is involved in setting a rhythm, what stimulus or stimuli are most effective, and by what means does the organism detect such stimuli? Many "rhythm-setters" have been investigated. Some of these are discussed below.

TEMPERATURE

During the course of a day, there is usually a point of highest environmental temperature, and a point of lowest environmental temperature. In those animals that do not maintain a nearly constant body temperature, chemical reactions vary in their rate according to temperature. Thus, periods of activity are often associated with elevated environmental temperature when body chemistry is accelerated, and periods of low activity may be associated with lower environmental temperature. An increase in "clock speed" as a result of higher temperature should be reflected in a shortening of the period of a cycle, and should occur in a predictable manner. A difficulty with temperature as a rhythm setter is that the clock itself should be affected by temperature change and would act like a thermometer, indicating temperature by the rate at which it runs. This, however, is not the case; the clock continues to run nearly at the same speed independent of ambient temperature. Several explanations have been offered to account for clock independence.

The clock depends for its accuracy on some type of passive process, these being more independent of temperature than active ones.

The basis of the clock is indeed chemical, but it has a compensating mechanism that ensures accuracy. The nature of such a compensating mechanism is unknown.

The clock's timing is set by some all-pervading force *outside* the organism; a force other than temperature change. Again, what that "cosmic force" may be is unknown.

For the human, temperature cannot be a significant clock setter because of the maintenance of a near-constant internal body temperature.

LIGHT-DARKNESS

Alternation of light and darkness as a result of the earth's rotation has an obvious periodicity. Time-lapse photography of plant blooming and leaf movement demonstrates a light-dark correlation. Birds migrate according to length of the light-dark periods. In humans, there is an obvious light-sensitive receptor, the eye. Nervous connections have been demonstrated between the eye and the pineal gland; the pineal cyclically produces several chemical substances including norepinephrine, serotonin, histamine, and melatonin. One or more of these substances may affect a part of the brain (hypothalamus?) or perhaps the pituitary, producing cyclical alteration of body functions. As described below in greater detail, human newborns do not show circadian rhythms in their body functions, but acquire them after birth. Before birth, the fetal environment is one of continual darkness; after birth it is one of light-dark cycles. Thus, for the human at least, light-dark cycles would appear to be a more logical cycle entraining device than temperature. The phenomenon of "jet lag," in which one's cycles of alertness and nonalertness are disrupted by rapid passage from one set of habituated light-dark cycles to nonhabituated ones, illustrates the operation of this factor.

OTHER STIMULI

Among the other types of stimuli that have been

investigated as entraining stimuli are cosmic radiation, geomagnetism, electromagnetism, electrical fields, sunspot cycles, lunar cycles (*lunatic*), and changes in barometric pressure. Folklore abounds with supposed examples of animals' ability to detect changes in one or more of these factors. Animals behave oddly just before earthquakes are due; crabs burrow in the sand two days before a storm; elk gather in the trees before a blizzard; "water dowsers" can find water using a forked stick. Can humans detect changes in such factors? If they can, they either possess receptors most of us do not have, or common receptors are far more sensitive in some individuals than in others. The water dowser is said to respond to small alterations in magnetism that may be associated with underground water; mental patients show changes that can be correlated with (among other things) lunar cycles. In summary, the role of such stimuli as listed at the beginning of this section in entrainment remains to be proven, at least for higher forms of life.

TIME OF ASSUMPTION OF A RHYTHM

Newborn humans seem to have only one predictable quality; unpredictability. Needs, wants, time of sleep and wakefulness are all "out of step" with the rhythm of family life. At about sixteen weeks of age a "settling down" appears to occur that results in a more comfortable situation for all.

Does this pattern of apparent lack of rhythm reflect the time required for entrainment by an external force or the maturation of an inherent clock? The answer is not known. Studies of infant physiological parameters such as hormone levels, REM sleep, electrolyte excretion, and urine volume indicate a noncircadian pattern for as long as three months after birth. It is obviously impossible to raise infants in constant darkness, or on 30 hour cycles to assess the effects of light-dark on entrainment of their rhythms, nor would one want to. Suffice it to say that adult rhythms in these same parameters are not evident until some weeks or months after birth, and that their entrainment does depend on the development of a clock that responds to some internal or external factor.

Of equal interest is the question: does aging have a demonstrable effect on the functioning of the clock? As we age, peaks in circadian rhythms appear to occur earlier. This may reflect alterations in sleep-waking patterns, for it *is* known that older individuals generally require less sleep than younger ones. More important in the aged is the dissociation or loss of correlation between rhythms. For example, in younger individuals, the peaks of body temperature, urine flow, heart rate and potassium excretion occur within a four hour time span. In older individuals, peak urine flow was up to 8 or more hours out of phase with potassium excretion. Degree of dissociation of various body functions may have diagnostic value as presagers of disease.

HUMAN RHYTHMS

METABOLIC PATTERNS

Table 27.1 presents several physiological parameters that demonstrate a circadian pattern in the adult human. To a degree, it is possible to "select" certain functions that may be determiners of others. For example, cyclical alteration in aldosterone secretion is a determinant of sodium and potassium excretion; cortisol elevation determines blood cell proliferation. In any event, a host of basic mechanisms show a circadian pattern.

SLEEP

It is the usual pattern for one to sleep when it is dark and to be awake when it is light. In those individuals whose work requires the reverse of this usual situation, the circadian nature of their processes is maintained, but peaks in body temperature, urinary excretion of hormones, and other parameters may be shifted by about 12 hours. The patient in a hospital offers a good example of an individual whose pattern may be disrupted to serve the hospital schedule. A classical and not so humorous event is waking a patient to receive a sleeping pill because routine demands it. Jet lag again illustrates the havoc wrought by the ability to pass through time zones at a rate greater than nature occasions it. Depression appears to be an increas-

TABLE 27.1
HUMAN CIRCADIAN RHYTHMS IN DAY-ADAPTED SUBJECTS

BODY CELL, TISSUE, ORGAN OR FLUID MONITORED	PARAMETER	TIME OF CYCLE PEAK[a]
Brain	Total EEG	1 pm
	Delta waves	12 m
	Theta waves	11 am
	Alpha waves	1 pm
	Beta waves	12 m
Epidermis	Mitosis	11:30 pm
Urine	Volume; rate of excretion	8:30 am
	Potassium excretion	1 pm
	Sodium excretion	9 am
	Hydroxycorticosteroid excretion	11 am
	Cortisol excretion	9:30 am
	17 ketosteroid excretion	11 am
	Epinephrine excretion	1 pm
	Norepinephrine excretion	1 pm
	Aldosterone excretion	1 pm
	Magnesium excretion	1 am
	Phosphate excretion	8 pm
	pH	1 pm
Blood	Neutrophils	12:30 pm
	Lymphocytes	11:30 pm
	Monocytes	3 am
	Eosinophils	11:30 pm
	Hematocrit	10 am
	Ca^{2+}	9:30 pm
	Na^{1+}	5 pm
	Pco_2	11 am
Plasma or serum	Proteins	2 pm
Whole body	Temperature (oral)	3 pm
	Physical vigor	3:30 pm
	Weight	6 pm
	Heart rate	4 pm
	Systolic blood pressure	7 pm
	Diastolic blood pressure	10 pm
	Respiratory rate	2:30 pm

[a] Times quoted reflect the average of the peak values of the particular function; the range of a function is not indicated, and the low point of a given function may be assumed to occur approximately 12 hours from the time indicated.

ingly common complaint among those whose normal sleep patterns are disrupted by voluntary or involuntary occurrences. Of particular importance to health maintenance is the time spent in REM sleep (*see* Chapter 9). This type of sleep is believed to serve useful functions in terms of "relieving" the brain of stress and "exercising" it. It is this phase of sleep that appears to undergo the most disruption due to travel or stress, and thus it is a phase in which one spends more time when recovering from sleep disruptions.

SENSITIVITY

We are all familiar with the "complaint" that we are too tired to do a job now, or that we will feel better later. Physiologically, human sensitivity to stress and resistance to a variety of factors show a circadian pattern. Resistance to drugs, stress, allergy, pain, and infection vary according to time of day.

Among examples of time related susceptibility may be listed the following.

Sodium pentobarbitol, a sedative, is most effective at the beginning of the dark phase of the light-dark cycle.

Amphetamines, stimulants, produce their greatest effect at the height of the activity cycle; in humans this corresponds to midafternoon.

Pain tolerance is greatest during the day and least at night.

Epileptic seizures are more common in the early morning.

Further examples could be cited. The conclusions to be drawn are:

Timing of the dose of a drug may influence its activity or effect on the body, and the same dose of a drug given at different times of the day may find the recipient more or less able to tolerate it.

Internal resistance changes as a function of time, and stimuli causing an effect at one time may be ineffective at others.

PERFORMANCE

Just as we are more susceptible to stressors during certain parts of the day, our mental and physical performance shows cyclical alterations.

Subjective evaluations of when a task is performed most efficiently indicate that performance is best in the morning of a working day. After a good night's sleep, we should feel most refreshed and eager to perform. More objective evaluations of performance of specific tasks in the laboratory indicate that performance of mental tasks is best in early afternoon, a time that correlates with the body temperature maximum in most individuals. There is, however, a "post-lunch dip" occurring about an hour after consumption of a regular meal about 12 noon. Some diversion of blood from brain to stomach may occur, or feelings of relaxation may contribute to such a "let down."

The effect of time of day on efficiency in jobs that require primarily physical exertion have not been as thoroughly investigated as mental-oriented tasks. A variety of subjective factors influence physical performance. Boredom, actual or apparent fatigue, and the practice of payment-by-results, all contribute to the *motivation* of a worker to carry out a task. The basic conclusion here appears to be that a program that minimizes worker boredom and emphasizes the inclusion of workers in decisions that affect their working conditions will result in increased output that does not show classical cyclical or circadian variation.

We must recognize that cyclical alterations in body function exist and that there is individual variation in resistance to a variety of internal and external factors. To ignore these facts is to invite disruption of our normally healthy status. The reader is referred to the readings at the end of this chapter for references that may enable a particular interest to be investigated to a greater degree.

SUMMARY

1. Within the general framework of homeostasis, a variety of body functions may be shown to exhibit cyclical alterations in their values.
2. The nature of the cyclical alteration of a function may be described using a series of terms.
   a. The repetitive nature of a function is termed a *cycle*.
   b. *Period* describes the length of the cycle.
   c. *Frequency* describes the number of cycles per unit of time.
   d. *Amplitude* describes the intensity of the change.
   e. *Phase* describes a particular part of a cycle.

f. *Entrainment* refers to the assumption of a cycle due to some demonstrable cause or stimulus.
g. A variety of terms are employed to indicate period length. Among these are circannual (yearly), circamensual (monthly), circaseptan (weekly), circadian (daily). Many human cycles are circadian in nature.
h. An important feature of human cycles is their interrelationship, such as a 4:1 ratio between heart and breathing rates.

3. Rhythms exhibit certain characteristics.
   a. Cyclical nature persists even when the organism is removed from its normal environment.
   b. Rhythms appear to depend on some sort of internal timing device, once they are entrained.
   c. They may depend on evolutionary processes to pass the cyclical nature to offspring.
   d. Rhythms usually have survival value.
   e. Chemical and other stimuli do not normally affect rhythm period, but may alter amplitude.
4. There are two hypotheses advanced to explain what causes the appearance of cyclical functions.
   a. Internal hypotheses suggest that genetic influences predominate in rhythm-setting.
   b. External hypotheses suggest that light-dark, temperature, or fluctuations in geophysical phenomena act as rhythm setters.
   c. Two types of "clocks" are described as controlling the period and amplitude of a cycle. One measures time independently of stimuli; the other responds to stimuli.
5. Temperature change, light-dark, and other stimuli are described as possible external rhythm-setting devices.
6. The human organism does not possess a full circadian set of rhythms at birth. It may entrain these by visual-pineal-nervous system pathways.
7. Several human circadian rhythms are described, including metabolic changes, sleep cycles, sensitivity, and performance, with the latter including both mental and physical performance.

QUESTIONS

1. Why is the concept of homeostasis not in dispute with the demonstration that many human physiological changes are cyclical in nature?
2. If a function was described as having a circadian periodicity, a low amplitude, an ascending phase, and nocturnal, what would be implied?
3. List and describe four properties of a rhythm.
4. What are some possible entraining stimuli that are external? Internal?
5. What characterizes an escapement-type biological clock?
6. If someone said "My arthritis (or rheumatism) kicks up just before it is going to rain," would this represent a logical statement in terms of external entraining factors?
7. What evidence exists to suggest that some human rhythms are acquired after birth? Does this argue for an internal or external entraining device?
8. List some human circadian rhythms.
9. Is it possible that you have "good days" or "bad days"? Explain in terms of cyclical patterns of physiological function.

READINGS

Colquhoun, W. P. (Ed). *Biological Rhythms and Human Performance*. Academic Press. New York, 1976.

Goodner, Charles J., Barbara C. Walike, Donna J. Koerker, John W. Ensinck, Arthur C. Brown, Elliott W. Chideckel, Jerry Palmer, and Lynne Kalnasy. "Insulin, Glucagon, and Glucose Exhibit Synchronous, Sustained Oscillations in Fasting Monkeys." *Science 195*:177. 14 Jan 1977.

Hedlund, Laurence, Michael M. Lischko, and Gordon D. Niswender. "Melatonin: Daily Cycle in Plasma and Cerebrospinal Fluid of Calves." *Science 195*:686. 18 Feb 1977.

Palmer, John D. *An Introduction to Biological Rhythms*. Academic Press. New York, 1976.

Palmer, John D. "Biological Clocks of the Tidal Zone." *Sci. Amer. 232*:70. (Feb) 1975.

Pengelley, Eric T. *Circannual Clocks*. Academic Press. New York, 1974.

Saunders, D. S. "The Biological Clock of Insects." *Sci. Amer. 234*:114. (Feb) 1976.

Siffre, Michel. "Six Months Alone in a Cave." *Nat'l. Geog.* (March) 1975.

Stetson, Milton H., and Marcia Watson-Whitmyre. "Nucleus Suprachiasmaticus: The Biological Clock in the Hamster?" *Science 191*:197. 16 Jan 1976.

Turek, Fred W., Joseph P. McMillan, and Michael Menaker. "Melatonin: Effects on the Circadian Locomotor Rhythm of Sparrows." *Science 194:*1441. 24 Dec 1976.

Underwood, Herbert. "Circadian Organization in Lizards." *Science 195:*587. 11 Feb 1977.

Wurtman, Richard J. "The Effects of Light on the Human Body." *Sci. Amer. 233:*68. (July) 1975.

FROM BEFORE CONCEPTION TO DEATH; THE BASICS OF THE ULTIMATE HUMAN CYCLE

OBJECTIVES

After studying this chapter, the reader should be able to:

☐ Describe, in general terms, the development of ova and spermatozoa in terms of results of the processes and length of cycle.

☐ Outline the roles of male and female in production of a new individual.

☐ Discuss the venereal diseases in terms of causative agent, mode of transmission, effects on the body, and susceptibility to treatment.

☐ Describe, in general terms, the developmental events that occur in the 1st, 2nd, 3rd, 4th, 5th to 8th, and 9th weeks to birth, of the new individual.

☐ List the factors involved in ensuring normal development of a new individual.

☐ Describe the events occurring at birth, and the dangers of the process.

☐ Show how the physiological status of the newborn may be assessed at birth.

☐ Describe the patterns of growth that occur from birth to maturity, what regulates that growth, and what characterizes the several "periods" in development.

☐ Explain some of the theories advanced to account for aging.

☐ List some changes in individual organs that occur as aging proceeds.

☐ Give a modern definition of death.

The ultimate human cycle is the time from conception to death of a given individual. Many factors operate to determine the development, length of life, and time of death of an individual. This chapter presents an overview of a human life, and offers the suggestion that our concern for health must begin before conception of a given individual and must terminate only with the death of that individual.

DEVELOPMENT OF OVA AND SPERMATOZOA

The stage for development of a new individual is set at the time of puberty of the potential parents, by ripening of ova and development of spermatozoa.

Ripening of ova (Fig. 28.1) occurs at 9 to 12 years of age from primordial follicles developed in the ovary before birth of the female. These ova remain in first meiotic prophase until puberty, when normally one mature ovum per month will be ovulated. Second meiotic division is completed after ovulation as fertilization occurs. Of four cells produced during ovum maturation, one will receive the bulk of cytoplasm as meiosis proceeds, and the remaining three cells become polar bodies. Thus, one *functional* cell results from ovum meiosis. As already indicated (Chapter 2), meiosis halves the chromosome number of the species, and introduces genetic variability in the gametes. The mature ovum contains, in the human, 22 autosomes plus an X sex chromosome.

Spermatozoa development (Fig. 28.2) begins at 13 to 15 years of age in the male. A cycle of

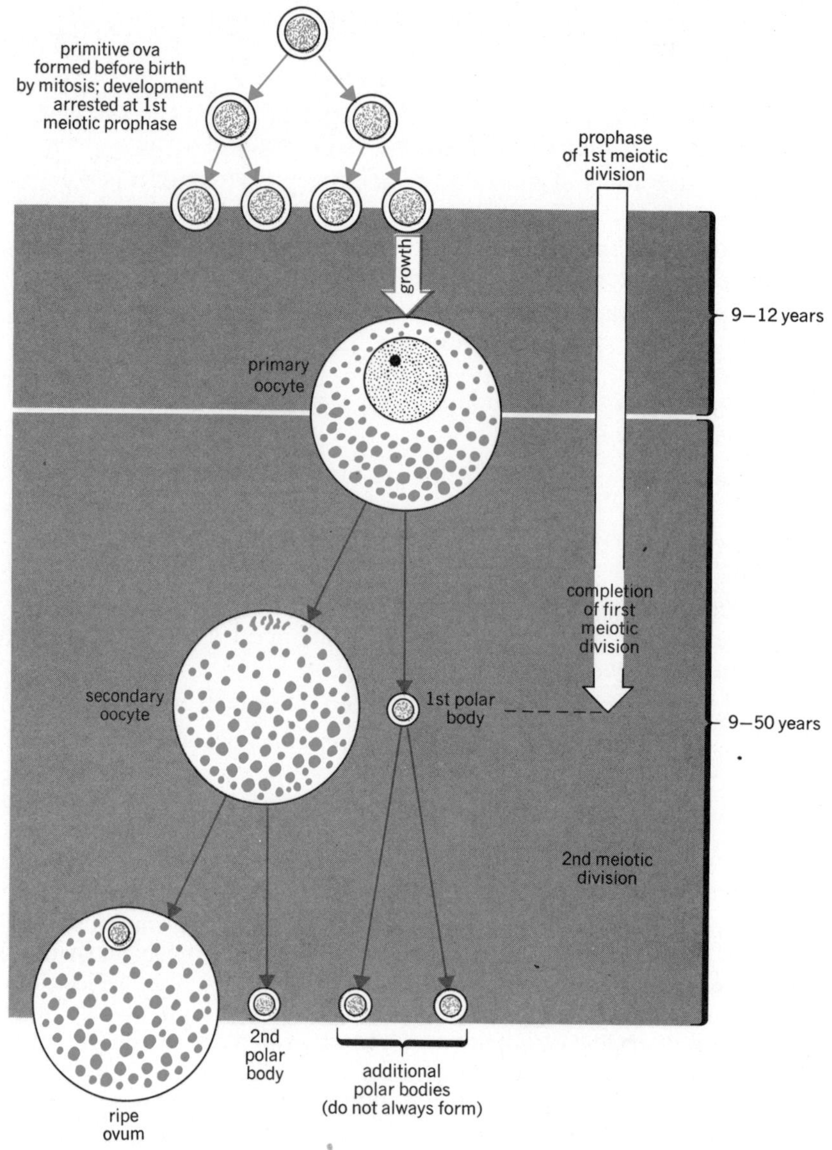

FIGURE 28.1
Oogenesis in the human female showing when the first and second meiotic divisions occur and the formation of one functional cell and the polar bodies.

sperm production has a period of about 74 days for a given seminiferous tubule, and not all tubules show the same stage of development at the same time. Four functional cells are produced, one-half of which will contain a chromosome complement of 22 + X, and one-half will contain a chromosome complement of 22 + Y.

Production of a new individual requires

fusion of a sperm and an ovum in the process known as fertilization. The time of fertilization is the time of conception.

SEXUAL INTERCOURSE

THE ROLE OF THE MALE

The organs of the male reproductive system and adnexa are depicted in Figure 28.3. The essential organs of the system are the testes; various tubular structures serve to conduct and/or store spermatozoa; accessory glands produce fluids to suspend, activate, and nourish the sperm. The penis serves as the organ of copulation and natural introduction (*insemination*) of sperm into the vagina of the female. To be an effective organ of copulation, the penis must become firm or erected.

Erection of the penis occurs as a result of psychic stimuli, or tactile stimuli applied to the genital area, or on the body's erogenous zones. Stimulation of the penis itself appears to be the primary stimulus leading to erection and eventual ejaculation. From touch corpuscles in the penis, nerve impulses pass over the pudendal nerve to the sacral spinal cord. Parasympathetic impulses pass to the penile arteries over the nervi erigentes. The smooth muscle of the arteries relaxes as a result of parasympathetic stimulation and a greater inflow of blood occurs into the penis. Its cavernous bodies fill with blood, compress the penile veins, and slow outflow from the organ. The penis becomes firm or erected. Penile stimulation causes secretion of mucus by the bulbourethral (Cowper's) glands that neutralize any urine remaining in the urethra. Ejaculation occurs when rhythmic sympathetic discharge from the lumbar spinal cord causes smooth muscle in the epididymis and vas deferens and skeletal muscle associated with the pelvic floor, to forcibly move sperm through the ducts and urethra to the outside. The sympathetic discharge also causes secretion of the prostate and seminal vesicles. The mixture of glandular secretions and sperm constitutes the ejaculate or semen.

The SEMEN has an average volume of 2 to 5 ml, a pH of 7.2 to 8.0, a sperm count of 60 to 150 million per ml, of which 80 percent or more should be motile and 80 to 90 percent of which should show normal morphology. About 60 percent of the semen volume is contributed by the seminal vesicles, 38 percent by the prostate gland, and the remainder is sperm. Vesicle secretion contains much fructose and ascorbic acid (vitamin C) and is yellowish in

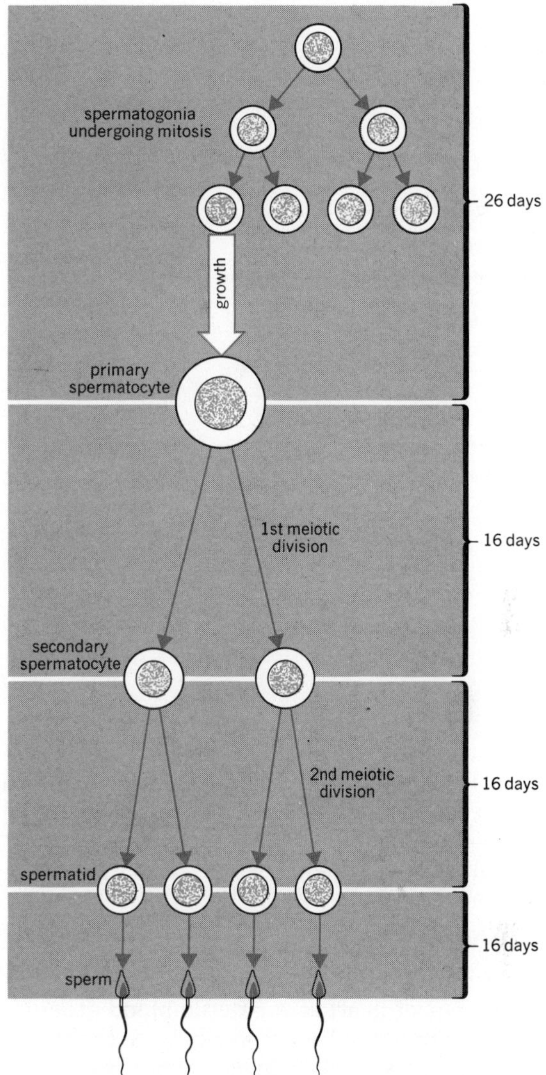

FIGURE 28.2
Spermatogenesis in the human male showing when the first and second meiotic divisions occur and the formation of four functional cells.

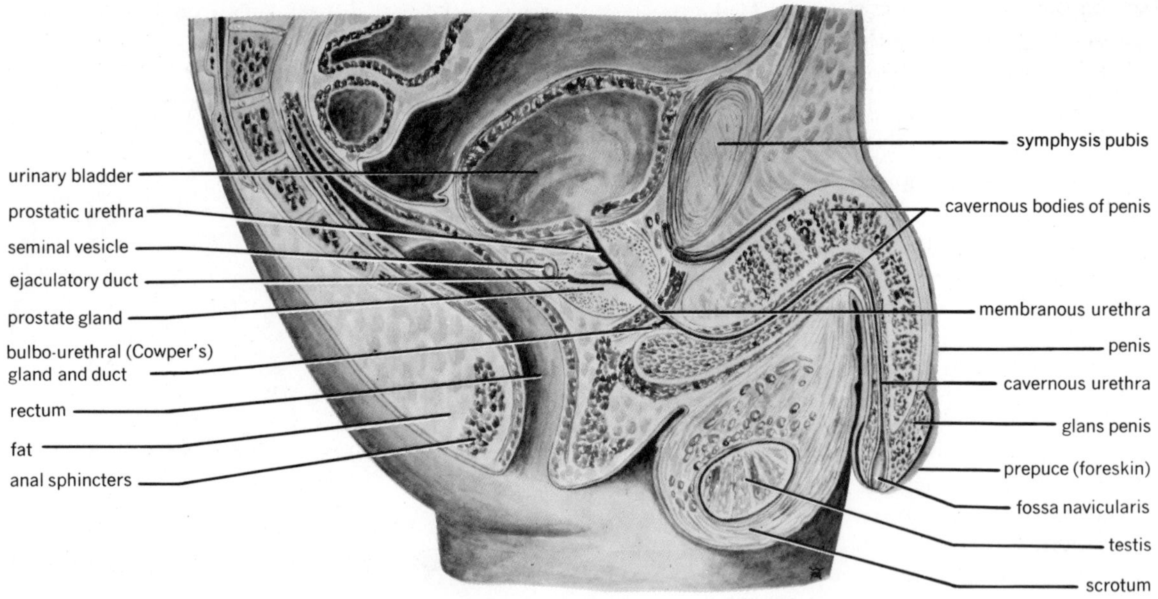

FIGURE 28.3
The male reproductive organs, and associated structures.

color. Prostatic secretion is odiferous and is rich in cholesterol, phospholipids, buffering salts, and prostaglandins. These substances provide nourishment for the sperm, a suspending medium, neutralization of vaginal acidity, and activation of the sperm (prostatic secretion).

Release of semen (ejaculation) is attended by widespread body movement, sensations, and elevations of heart rate, arterial blood pressure and respiration. These phenomena constitute the male orgasm, after which there is resumption of penile artery tone, and a return of the penis to the flaccid state. Thus, the role of the male in production of offspring is basically one of insemination of the female. Sperm deposited in the vagina must make their way to the outer portions of the uterine (*fallopian*) tubes before fertilization can occur.

THE ROLE OF THE FEMALE

The female reproductive organs and adenexa are depicted in Figure 28.4. Ovulation occurs about halfway through a menstrual cycle and is associated with an initial dip, and then a rise of body temperature (Fig. 28.5). If implantation occurs, body temperature remains elevated, with suspension of normal menses and this may be taken as a sign of probable pregnancy. Temperature then drops with resumption of menstruation. The vagina normally serves as the organ of reception of the penis and the area for sperm deposition. In artificial insemination, sperm are introduced by pipette directly into the uterus.

There appears to be little difference from that of the male in female response to sexual stimulation. The female erogenous zones include the breasts, perineum, and clitoris. As a result of psychic and/or tactile stimulation, the clitoris becomes erect, and the spongy tissues around the vaginal orifice also become engorged with blood. This latter phenomenon is believed to result in a tightening effect that speeds male orgasm. The impulses that originate from stimulation of the female genitalia pass over the pudendal nerve to the sacral cord, and parasympathetic impulses pass to the clitoral blood vessels, dilating them and causing erection, and to the Bartholin's glands that secrete fluids around the vaginal orifice. These secretions supply most of the lubrication for insertion of the penis into the vagina. Thrusting of the male alternately tightens and releases tension on the labia minora that end anteriorly around the clitoris. This action is believed to provide a massaging effect on the clitoris that aids in attainment of female orgasm. Also, during intercourse, the uterus shifts backward slightly, possibly because of an increased blood volume, and the vagina enlarges in the upper end to form a "pool" for the collection of semen. After female orgasm—attended by feelings of release of tension and body movement, elevations of heart rate, blood pressure, and respiration—the uterine orifice may be seen to open, and there are rhythmic contractions of the uterine muscle that may aid in sperm transport through the organ. The role of the female thus becomes one of reception of sperm, the site of FERTILIZATION, and the PROVISION OF NUTRIENTS to assure development of the offspring to a state where it can cope with the external environment on its own.

VENEREAL DISEASES

The venereal diseases (VD) are the number one communicable diseases in the United States today. The continuing increase in incidence of VD may be due to several basic causes.

There have been and will continue to be changes in attitudes toward sexual intercourse that separate it from its reproductive function. Increased sexual contact has therefore increased the possibility of spread of VD among the population.

The increasing number of younger people in

FIGURE 28.4
The female reproductive organs, and associated structures.

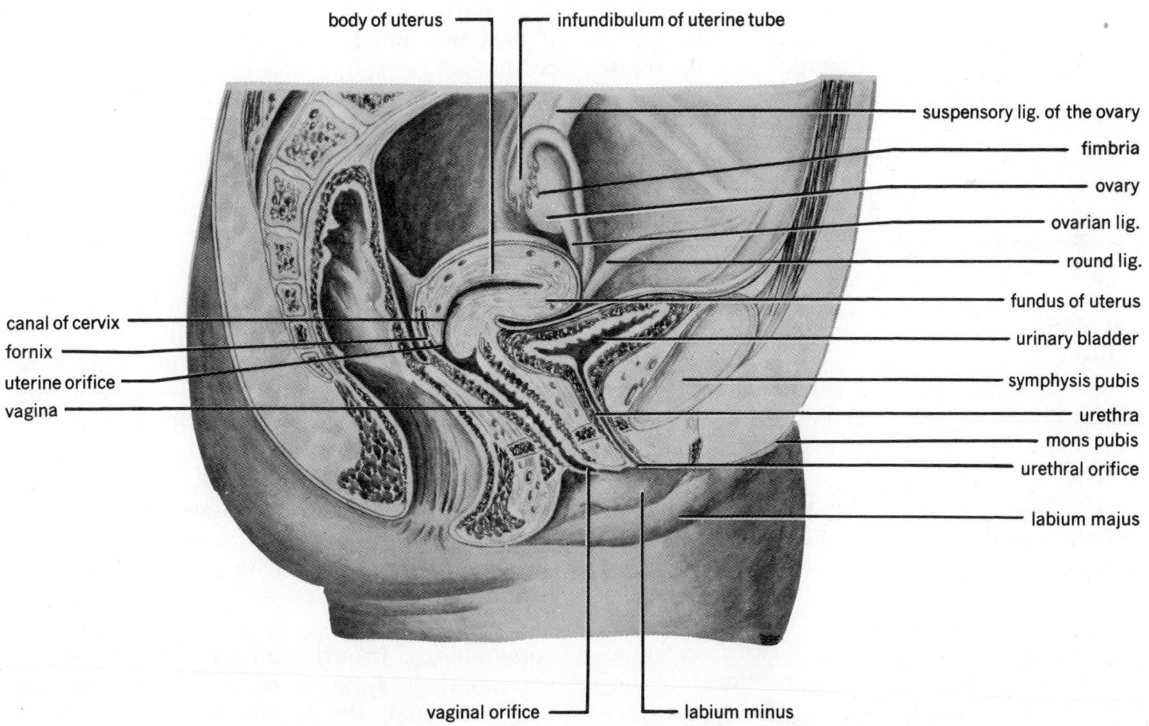

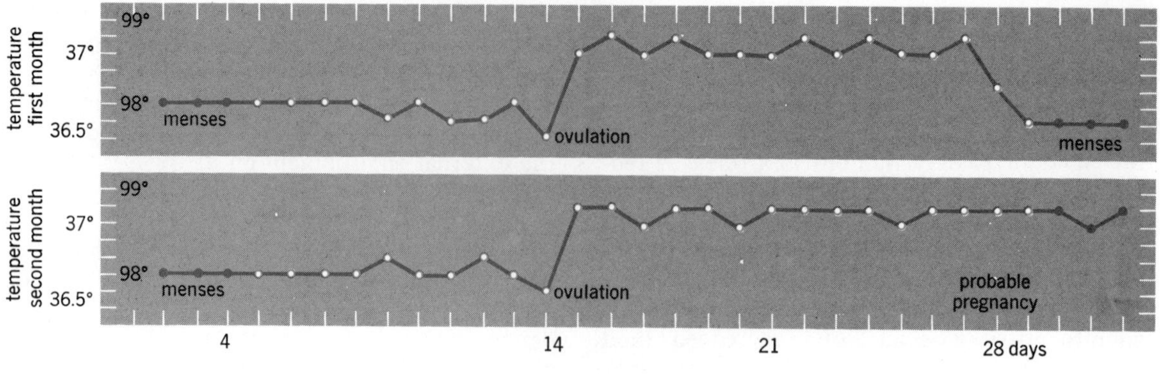

FIGURE 28.5
Change in basal body temperature of ovulation and with pregnancy.

the population has increased the incidence of sexual contact with greater possibility of spread of VD (the highest incidence is presently in the 15 to 30 age group).

A large part of the social stigma associated with VD has disappeared, and reporting of cases has increased. It may be hoped that increased reporting of cases has been an incentive for the establishment or increase in funding of programs designed for case finding and treatment of venereal diseases.

The discontinuance of indiscriminate use of antibiotics for "colds" and other infections that do not warrant their use has resulted in the removal of a "brake" on the spread of VD, in that many previously undiagnosed cases are no longer being treated inadvertently. Use of antibiotics has resulted in the development of antibiotic-resistant strains of the VD organisms.

Better epidemiological studies have resulted in the discovery of reservoirs of infection that may not have been brought to light previously.

The belief of many segments of the public that a "shot" of antibiotic will cure VD has perhaps led to the development of complacency, as well as a false sense of security. All patients must be rechecked at a certain interval following treatment to ensure that the tests for organisms are negative. Also, after an initial set of symptoms, several of the venereal diseases may enter an asymptomatic period that may persist for years, leading to the belief that the disease has disappeared. During this time, damage may continue to be wrought in various body systems, although no overt changes become apparent.

Reinfection is common since no lasting immunity is developed to the organisms. Contacts must be examined and adequately treated if they are infected to prevent further spread. Sexual partners, diagnosed as infected, must be treated at the same time or they will be playing "ping-pong" with the organism. Persons having a gonococcal infection must also be checked for syphilis because both infections may be present at the same time but the amount and type of treatment may differ.

Gonorrhea ("clap") and syphilis ("bad blood") are presently in epidemic proportions throughout the United States and some other countries. These diseases have no respect for race, creed, or color, and one cannot tell an infected person by just looking at him or her.

All personal information regarding cases, contacts, and suspects of VD is legally confidential, which should encourage infected persons to seek early treatment of the disease(s).

Free treatment is available through local health departments. Other clinic facilities, private physicians, or both are available in most communities. Information on a state's law regarding the treatment of minors may be obtained through the health department. Some states now treat minors without the parents' consent, since the child's future welfare is at stake, and as an effort to cut down the VD incidence in sexually active teenagers.

Some characteristics and effects of gonorrhea and syphilis are presented in Table 28.1.

A difficulty in the treatment of gonorrhea is the recent appearance of *penicillin-resistant* organisms in eleven different countries, including the United States. The appearance of such organisms obviously requires antibiotics other than penicillin to control them; a requirement of culturing for penicillin-resistant strains is a must before *and* after a course of penicillin

TABLE 28.1
SOME CHARACTERISTICS AND EFFECTS OF GONORRHEA AND SYPHILIS

	GONORRHEA	SYPHILIS
Causative organism	Neisseria gonorrhoeae	Treponema pallidum
Incubation period	2–14 days (usually 3 days)	7–90 days (usually 3 weeks)
Method of transmission	Sexual, oral, or physical contact with an infected person. Contaminated object up to 8 hours after organisms deposited. *Infants*—during birth through vagina.	Sexual, oral, or physical contact with an infected person. Blood transfusion. On contaminated objects, dies quickly by drying. *Infants*—may acquire during birth, or through the placenta if mother not treated before third trimester and adequately.
Contact examination *(all sex contacts exposed within the following time periods)*	2 weeks (male) 1 month (female)	*Primary.* 3 months (+ duration of symptoms). *Secondary.* 6 months (+ duration of symptoms). *Early latent.* 1 year. *All syphilis.* "Family" contacts as indicated.
Clinical characteristics	Discharge; burning, pain, swelling of genitals and glands. **Male.** Purulent urethral discharge, hematuria, chordee. Urethritis, prostatitis, seminal vesiculitis, epididymitis, occasional involvement of testes. If untreated, symptoms disappear in about 6 weeks, and organism persists in the prostate gland (Gc carrier). **Female.** *Child,* vaginitis. Leukorrhea, tubal abscess, urethritis, cervicitis, pelvic inflammation (Peritonitis). Tubal stricture, possible sterility due to closure of tubes. In both male and female, healing is by scar tissue formation. Strictures and closing of tubular structures may result. Arthritis, endocarditis, meningitis may occur.	**Primary.** Chancre present, solitary, nonpainful ulcer on genital or mucous membranes. **Secondary.** Rashes or mucous patches. Macules or papules on hands, feet, oral cavity, genitoanal area, trunk, extremities. **Tertiary:** *Latency.* No symptoms, positive serology. Profound changes are produced in the skin, mucous membranes, skeleton, GI tract, kidney, brain; heart and blood vessels show destructive changes (abscesses, scarring, tissue destruction). Tabes dorsalis in spinal cord destroys dorsal columns. Paresis and psychosis may result. *Relapse.* Recurrence of infectious lesions after disappearance of secondary lesions. *Late.* Cardiovascular, central nervous system, gummata. Obvious systemic damage appears.

TABLE 28.1 (continued)

	GONORRHEA	SYPHILIS
Diagnostic procedures	Culture of discharge. Smear. Fluorescent antibody test (FAT). Currettage to get issue containing cocci. Several cultures may be necessary as not all tests consistently show cocci. History, clinical and contacts. (Serologic test for syphilis).	Darkfield examination—microscopic for spirochete. Serologic (blood) test for antibody (reagin) to organism. [(Wasserman, VDRL—Venereal Disease Research Lab) False positive tests may be reported by smallpox antibodies, hepatitis, mononucleosis, and high fevers.] Spinal fluid test. X-rays of long bones of infants. History, clinical and contacts.
Treatment	Penicillin Broad spectrum antibiotics (e.g., sulfonamides, streptomycin). Organism must be sensitive to the drug of choice.	Penicillin Broad spectrum antibiotics (e.g., Erythromycin, Tetracycline). Organism must be sensitive to drug of choice. Reexamination at 6 months and 1 year to evaluate treatment results.

treatment to determine if the antibiotic is or was effective. If resistant organisms *are* discovered, other antibiotics must be administered. *Spectinomycin* is the antibiotic of choice if penicillin fails; however, a spectinomycin-resistant strain of *Neisseria gonorrhoeae* has been isolated in Denmark and in the United States. The U.S. patient was cured with tetracycline. The continued rise of gonorrhea in the world raises the spectre that strains resistant to all currently available antibiotics may appear.

The term venereal diseases today includes sexually or orally transmitted diseases other than gonorrhea and syphilis.

NONSPECIFIC URETHRITIS is caused by an unknow agent, or may be secondary to or associated with parasitic infections, fungus infections, or trauma. That there is an organism involved is demonstrated by transmissibility of the infection from one person to another.

GRANULOMA VENEREUM is caused by a gram-negative rod, *Donovania granulomatis*. Evidence of infection includes a skin lesion, usually on the glans penis, typically followed by ulceration. The lesions heal slowly and tend to spread. Antibiotic therapy will cure the disease.

LYMPHOGRANULOMA VENEREUM is caused by *Bedsonia sp.*, and is associated with ulcerative lesions on the genitals. The organism then invades nearby lymph nodes, causing them to enlarge to form a *bubo,* an enlarged matted mass of nodes. Again, appropriate antibiotics will cure the condition.

HERPES SIMPLEX VIRUS, type 2, produces vaginal, genital, and uterine cervix lesions. This virus may be transmitted to the fetus, either *in utero* or during birth. Herpes simplex encephalitis may occur that has a high mortality and common recurrences. "Treatment" is by acquisition of resistance or partial resistance through immune response; antibiotics are ineffective on viruses.

Recently, a compound designated Ara-A (for adenine arabinoside) has been shown to be effective against the herpes simplex organism causing herpes encephalitis. Perhaps the substance has promise in the therapy of venereal herpes infections.

FROM CONCEPTION TO BIRTH

THE FIRST WEEK

Ovulated egg cells enter the uterine (*fallopian*) tubes due to the action of cilia that line the tube. Sperm must fertilize the egg in the outer por-

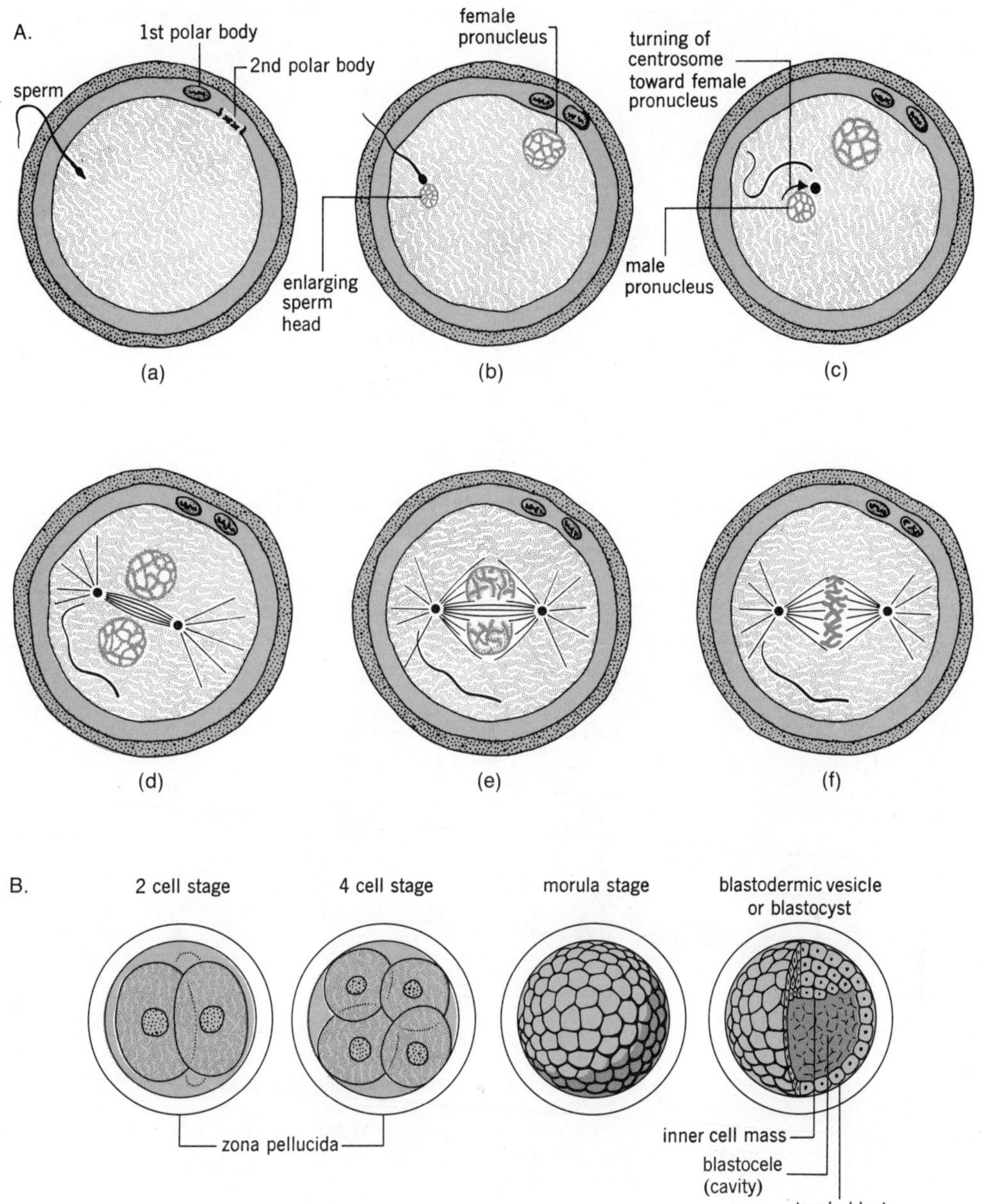

FIGURE 28.6

Development during the first week. A. The events of fertilization. (*a*) Sperm enters egg, (*b*) sperm head enlarges to form pronucleus, (*c*) rotation of centriole, (*d*) centriole divides and separates—spindle is formed. (*e*) pooling of chromosomes from pronuclei, (*f*) mitosis, metaphase. B. Cleavage and cell segregation.

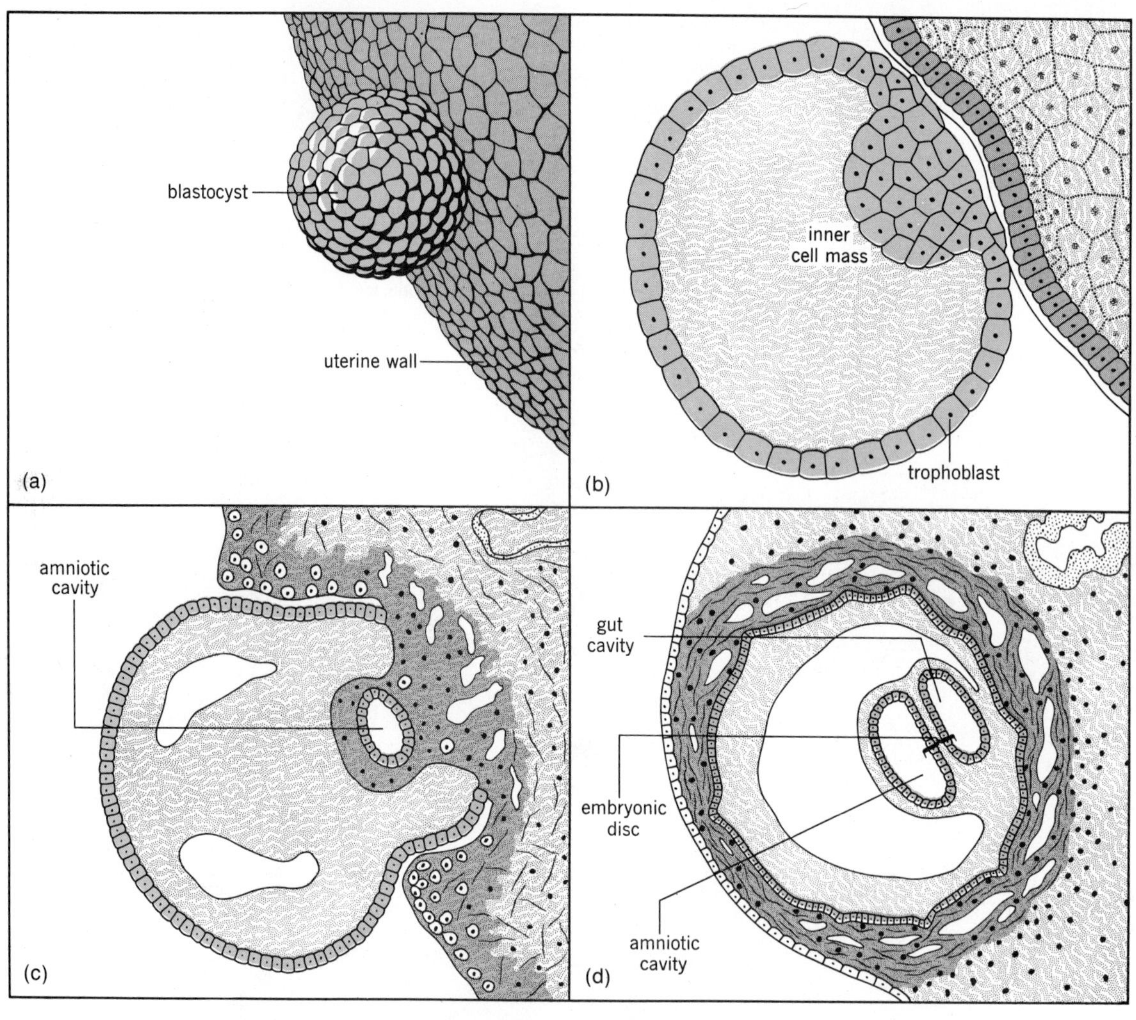

FIGURE 28.7
Development during the second week and implantation of the blastocyst. (*a*) Attachment of blastocyst to endometrium of uterus. (*b*) Appearance of sectioned blastocyst. (c) "Burrowing" of blastocyst and development of two-layered embryo. (*d*) Embryo enclosed by endometrium.

tion of the tube within 18 hours after release or no further development will occur. Fertilization (the time of conception) restores the normal chromosome complement and establishes the genetic sex of the human offspring (44 + XX = female; 44 + XY = male). As the fertilized egg or ZYGOTE passes down the uterine tube toward the uterus, a journey taking about 3 days, the zygote undergoes CLEAVAGE, a series of mitotic cell divisions. Cleavage results in a doubling of cell number for the first 8 to 10 divisions and thereafter becomes irregular. A solid mass of cells called a MORULA is formed. The morula adheres to the uterine wall and a reorganization of cells in the morula causes the formation of a BLASTOCYST, a hollow structure having a cavity, the *blastocele* and an *inner cell mass*. This reorganization takes 3 to 4 days. The events of the first week are presented in Figure 28.6.

THE SECOND WEEK

The blastocyst undergoes further reorganization with the appearance of two cavities in the inner cell mass. The upper one is the AMNIOTIC CAVITY; the lower one is the CAVITY OF THE YOLK SAC. A two-layered plate of cells separates the two cavities and is known as the EMBRYONIC PLATE or DISC. Development to this stage is said to result in the formation of a *two-layered embryo*. The upper layer of cells in this embryo is termed ECTODERM, the lower layer ENDODERM. These layers form two of the three "germ layers" from which all body structures will develop. The embryo then undergoes IMPLANTATION in the wall of the uterus. In this process, the blastocyst "digests" its way into the vascular and glandular endometrium of the uterus in order to establish a relationship with the mother's blood supply. This relationship assures the nutrients necessary for further growth and development. These processes are shown in Fig. 28.7.

THE THIRD WEEK

The major event occurring at this time is the formation of the third basic germ layer. By mitotic cell division, a layer of cells termed the MESODERM is formed between the ectoderm and endoderm of the two-layered embryo. Table 28.2 shows the ultimate derivatives of these three germ layers.

THE FOURTH WEEK

The fourth week is associated with the formation of the NEURAL TUBE, the beginning of the nervous system, and of SOMITES or blocks of mesoderm along the backbone of the embryo. These somites are the forerunners of the bones and muscles of the back. The embryo appears as shown in Fig. 28.8. It should be noted that development occurs in a head-to-tail (*cephalocaudal*) direction, as does maturation of function, once an organ or system is formed.

THE FIFTH THROUGH EIGHTH WEEKS

Completion of the embryo occurs during this

TABLE 28.2
TISSUES DERIVED FROM GERM LAYERS IN THE HUMAN

ECTODERM	MESODERM	ENDODERM
1. Outer layer (epidermis) of skin:	1. Muscle:	1. Epithelium of:
Skin glands	Skeletal,	Pharynx
Hair and nails	Cardiac,	Auditory (ear) tube
Lens of eye	Smooth	Tonsils
2. Lining tissue (epithelium) of:	2. Supporting (connective) tissue:	Thyroid
Nasal cavities	Cartilage	Parathyroid
Sinuses	Bone	Thymus
Mouth:	Blood	Larynx
Oral glands	3. Bone marrow	Trachea
Tooth enamel	4. Lymphoid tissue	Lungs
Sense organs	5. Epithelium of:	Digestive tube and its glands
Anal canal	Blood vessels	Bladder
3. Nervous tissues	Lymphatics	Vagina and vestibule
4. Pituitary (posterior lobe)	Celomic cavity	Urethra and glands
5. Adrenal medulla	Kidney and ureters	2. Pituitary (anterior and middle lobes)
	Gonads and ducts	
	Adrenal cortex	
	Joint cavities	

TABLE 28.3

A REFERENCE TABLE OF CORRELATED HUMAN DEVELOPMENT (from Arey)

AGE IN WEEKS	SIZE (C R) IN MM	BODY FORM	MOUTH	PHARYNX AND DERIVATIVES	DIGESTIVE TUBE AND GLANDS
2.5	1.5	Embryonic disc flat. Primitive streak prominent. Neural groove indicated.	—	—	Gut not distinct from yolk sac.
3.5	2.5	Neural groove deepens and closes (except ends). Somites 1 — 16 ± present. Cylindrical body constricting from yolk sac. Branchial arches 1 and 2 indicated.	Mandibular arch prominent. Stomodeum a definite pit. Oral membrane ruptures.	Pharynx broad and flat. Pharyngeal pouches forming. Thyroid indicated.	Fore- and hind-gut present. Yolk sac broadly attached at mid-gut. Liver bud present. Cloaca and cloacal membrane present.
4	5.0	Branchial arches completed. Flexed heart prominent. Yolk stalk slender. All somites present (40). Limb buds indicated. Eye and otocyst present. Body flexed; C shape.	Maxillary and mandibular processes prominent. Tongue primordia present. Rathke's pouch indicated.	Five pharyngeal pouches present. Pouches 1-4 have closing plates. Primary tympanic cavity indicated. Thyroid a stalked sac.	Esophagus short. Stomach spindle-shaped. Intestine a simple tube. Liver cords, ducts and gall bladder forming. Both pancreatic buds appear. Cloaca at height.
5	8.0	Nasal pits present. Tail prominent. Heart, liver and mesonephros protuberant. Umbilical cord organizes.	Jaws outlined. Rathke's pouch a stalked sac.	Phar. pouches gain dors. and vent. diverticula. Thyroid bilobed. Thyro-glossal duct atrophies.	Tail-gut atrophies. Yolk stalk detaches. Intestine elongates into a loop. Cecum indicated.
6	12.0	Upper jaw components prominent but separate. Lower jaw halves fused. Head becomes dominant in size. Cervical flexure marked. External ear appearing. Limbs recognizable as such.	Foramen caecum established. Labio-dental laminae appearing. Parotid and submaxillary buds indicated. Lingual primordia fusing	Thymic sacs, ultimobranchial sacs and solid parathyroids are conspicuous and ready to detach. Thyroid becomes solid and converts into plates.	Stomach rotating. Intestinal loop undergoes torsion. Hepatic lobes identifiable. Cloaca subdividing.
7	17.0	Branchial arches lost. Cervical sinus obliterates. Face and neck forming. Digits indicated. Back straightens. Heart and liver determine shape of body ventrally. Tail regressing.	Lingual primordia merge into single tongue. Separate labial and dental laminae distinguishable. Jaws formed and begin to ossify. Palate folds present and separated by tongue.	Thymi elongating and losing lumina. Parathyroids become trabeculate and associate with thyroid. Ultimobranchial bodies fuse with thyroid. Thyroid becoming crescentic.	Stomach attaining final shape and position. Duodenum temporarily occluded. Intestinal loops herniate into cord. Rectum separates from bladder-urethra. Anal membrane ruptures. Dorsal and ventral pancreatic primordia fuse.
8	23.0	Nose flat, eyes far apart. Digits well formed. Growth of gut makes body evenly rotund. Head elevating. Fetal state attained.	Tongue muscles well differentiated. Earliest taste buds indicated. Rathke's pouch detaches from mouth. Sublingual gland appearing.	Auditory tube and tympanic cavity distinguishable. Sites of tonsil and its fossae indicated. Thymic halves unite and become solid. Thyroid follicles forming.	Small intestine coiling within cord. Intestinal villi developing. Liver very large in relative size.
10	40.0	Head erect. Limbs nicely modeled. Nail folds indicated. Umbilical hernia reduced.	Fungiform and vallate papillae differentiating. Lips separate from jaws. Enamel organs and dental papillae forming. Palate folds fusing.	Thymic epithelium transforming into reticulum and thymic corpuscles. Ultimobranchial bodies disappear as such.	Intestines withdraw from cord and assume characteristic positions. Anal canal formed. Pancreatic alveoli present
12	56.0	Head still dominant. Nose gains bridge. Sex readily determined by external inspection.	Filiform and foliate papillae elevating. Tooth primordia form prominent cups. Cheeks represented. Palate fusion complete.	Tonsillar crypts begin to invaginate. Thymus forming medulla and becoming increasingly lymphoid. Thyroid attains typical structure.	Muscle layers of gut represented. Pancreatic islands appearing. Bile secreted.

RESPIRATORY SYSTEM	COELOM AND MESENTERIES	UROGENITAL SYSTEM	VASCULAR SYSTEM	SKELETAL SYSTEM
—	Extra-embryonic coelom present. Embryonic coelom about to appear.	Allantois present.	Blood islands appear on chorion and yolk sac. Cardiogenic plate reversing.	Head process (or notochordal plate) present
Respiratory primordium appearing as a groove on floor of pharynx.	Embryonic coelom a U-shaped canal, with a large pericardial cavity. Septum transversum indicated. Mesenteries forming. Mesocardium atrophying.	All pronephric tubules formed. Pronephric duct growing caudad as a blind tube. Cloaca and cloacal membrane present.	Primitive blood cells and vessels present. Embryonic vessels a paired symmetrical system. Heart tubes fuse, bend S-shape and beat begins.	Mesodermal segments appearing (1 – 16 ±). Older somites begin to show sclerotomes. Notochord a cellular rod.
Trachea and paired lung buds become prominent. Laryngeal opening a simple slit.	Coelom still a continuous system of cavities. Dorsal mesentery a complete median curtain. Omental bursa indicated.	Pronephros degenerated. Pronephric (mesonephric) duct reaches cloaca. Mesonephric tubules differentiating rapidly. Metanephric bud pushes into secretory primordium.	Hemopoiesis on yolk sac. Paired aortae fuse. Aortic arches and cardinal veins completed. Dilated heart shows sinus, atrium, ventricle, and bulbs.	All somites present (40). Sclerotomes massed as primitive vertebrae about notochord.
Bronchial buds presage future lung lobes. Arytenoid swellings and epiglottis indicated.	Pleuro-pericardial, and pleuro-peritoneal membranes forming. Ventral mesogastrium draws away from septum.	Mesonephros reaches its caudal limit. Ureteric and pelvic primordia distinct. Genital ridge bulges.	Primitive vessels extend into head and limbs. Vitelline and umbilical veins transforming. Myocardium condensing. Cardiac septa appearing. Spleen indicated.	Condensations of mesenchyme presage many future bones
Definitive pulmonary lobes indicated. Bronchi sub-branching. Laryngeal cavity temporarily obliterated.	Pleuro-pericardial communications close. Mesentery expands as intestine forms loop.	Cloaca subdividing. Pelvic anlage sprouts pole tubules. Sexless gonad and genital tubercle prominent. Mullerian duct appearing.	Hemopoiesis in liver. Aortic arches transforming. L. umbil. vein and d. venosus become important. Bulbus absorbed into right ventricle. Heart acquires its general definitive form.	First appearance of chondrification centers. Desmocranium.
Larynx and epiglottis well outlined; orifice T-shaped. Laryngeal and tracheal cartilages foreshadowed. Conchae appearing. Primary choanae rupturing.	Pericardium extended by splitting from body wall. Mesentery expanding rapidly as intestine coils. Ligaments of liver prominent.	Mesonephros at height of its differentiation. Metanephric collecting tubules begin branching. Earliest metanephric secretory tubules differentiating. Bladder-urethra separates from rectum. Urethral membrane rupturing	Cardinal veins transforming. Inf. vena cava outlined. Atrium, ventricle and bulbus partitioned. Cardiac valves present. Stem of pulm. vein absorbed into l. atrium. Spleen anlage prominent.	Chondrification more general. Chondrocranium.
Lung becoming gland-like by branching of bronchioles. Nostrils closed by epithelial plugs.	Pleuro-peritoneal communications close. Pericardium a voluminous sac. Diaphragm completed, including musculature. Diaphragm finishes its descent.	Testis and ovary distinguishable as such. Mullerian ducts, nearing urogenital sinus, are ready to unite as utero-vaginal primordium. Genital ligaments indicated.	Main blood vessels assume final plan. Primitive lymph sacs present. Sinus venosus absorbed into right atrium. Atrio-ventricular bundle represented.	First indications of ossification.
Nasal passages partitioned by fusion of septum and palate. Nose cartilaginous. Laryngeal cavity reopened; vocal folds appear.	Processus (saccus) vaginales forming. Intestine and its mesentery withdrawn from cord.	Kidney able to secrete. Bladder expands as sac. Genital duct of opposite sex degenerating. Bulbo-urethral and vestibular glands appearing. Vaginal sacs forming.	Thoracic duct and peripheral lymphatics developed. Early lymph glands appearing. Enucleated red cells predominate in blood.	Ossification centers more common. Chondrocranium at its height.
Conchae prominent. Nasal glands forming. Lungs acquire definitive shape.	Omentum as expansive apron partly fused with dorsal body wall. Mesenteries free but exhibit typical relations. Coelomic extension into umbilical cord obliterated.	Uterine horns absorbed. External genitalia attain distinctive features. Meson. and rete tubules complete male duct. Prostate and seminal vesicle appearing. Hollow viscera gaining muscular walls.	Blood formation beginning in bone marrow. Blood vessels acquire accessory coats.	Notochord degenerating rapidly. Ossification spreading. Some bones well outlined.

(continued)

TABLE 28.3 (continued)

AGE IN WEEKS	SIZE (C R) IN MM	BODY FORM	MOUTH	PHARYNX AND DERIVATIVES	DIGESTIVE TUBE AND GLANDS
16	112.0	Face looks human. Hair of head appearing. Muscles become spontaneously active. Body outgrowing head.	Hard and soft palates differentiating. Hypophysis acquiring definitive structure.	Lymphocytes accumulate in tonsils. Pharyngeal tonsil begins development.	Gastric and intestinal glands developing. Duodenum and colon affixing to body wall. Meconium collecting.
20–40 (5-10 mo)	160.0–350.0	Lanugo hair appears (5). Vernix caseosa collects (5). Body lean but better proportioned (6). Fetus lean, wrinkled and red; eyelids reopen (7). Testes invading scrotum (8). Fat collecting, wrinkles smoothing, body rounding (8-10).	Enamel and dentine depositing (5). Lingual tonsil forming (5). Permanent tooth primordia indicated (6-8). Milk teeth unerupted at birth	Tonsil structurally typical (5).	Lymph nodules and muscularis mucosae of gut present (5). Ascending colon becomes recognizable (6). Appendix lags behind caecum in growth (6). Deep esophageal glands indicated (7). Plicae circulares represented (8).

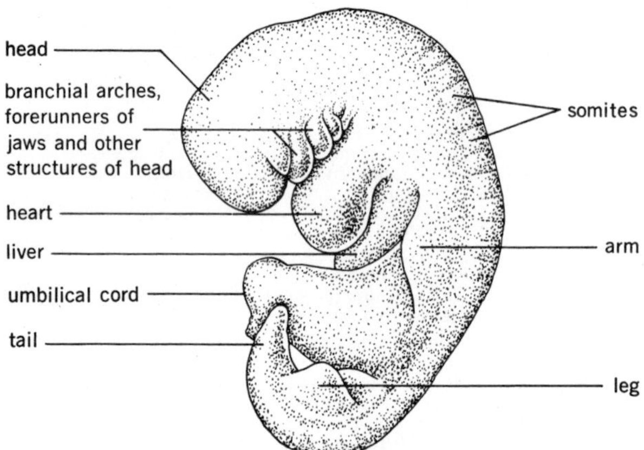

FIGURE 28.8
The human embryo at four weeks of development.

period, and it assumes a clearly human form. The brain is enclosed, the digestive system forms, a heart is formed, and circulation of blood in vessels is established. The limbs, eyes, ears, and other features are evident (Fig. 28.9).

THE NINTH WEEK TO BIRTH

This period constitutes the period of the FETUS, a creature with definite human form and all basic body systems. The fetus is, in some cases, capable of independent survival after about 26 weeks of development, a time which normally sees lung development far enough advanced to support life. Table 28.3 indicates that the fetal period is one mainly of growth of the body; by 8 to 12 weeks, all body systems are present and need only to grow and undergo final refinement.

The first 12 weeks of development are thus of critical importance in terms of establishing normal organs. Drugs used or diseases (e.g., Rubella) contracted during this period may lead to malformations in the development of body organs and systems. CONGENITAL (*born with*) ANOMALIES or malformations are the result of developmental errors. They may be evident at birth or cause disorders later in life. Such malformations are the third leading cause of death in the United States between birth through four years of age, and the fourth leading cause of death between the ages of five through fourteen years. Because several organs and systems develop at the same time, a process affecting one system may result in changes in a simultaneously developing system, so that malformations are usually seen in more than one system.

During fetal growth, and after birth, body proportions change (Fig. 28.10). Thus, a newborn has a relatively larger head and shorter legs than does the child and adult. Growth in size is the most obvious change (Fig. 28.11) occurring after about two months *in utero*.

THE ENVIRONMENT OF THE EMBRYO AND FETUS

From implantation to birth, the offspring is dependent on its mother for all sources of nutrients and for routes of elimination of its metabolic wastes. Four areas may be emphasized as primary determinants of the anatomical and physiological status of the fetus.

TABLE 28.3 (continued)

RESPIRATORY SYSTEM	COELOM AND MESENTERIES	UROGENITAL SYSTEM	VASCULAR SYSTEM	SKELETAL SYSTEM
Accessory nasal sinuses developing. Tracheal glands appear. Mesoderm still abundant between pulmonary alveoli. Elastic fibers appearing in lungs.	Greater omentum fusing with transverse mesocolon and colon. Mesoduodenum and ascending and descending mesocolon attaching to body wall.	Kidney attains typical shape and plan. Testis in position for later descent into scrotum. Uterus and vagina recognizable as such. Mesonephros involuted.	Blood formation active in spleen. Heart musculature much condensed.	Most bones distinctly indicated throughout body. Joint cavities appear.
Nose begins ossifying (5). Nostrils reopen (6). Cuboidal pulmonary epithelium disappearing from alveoli (6). Pulmonary branching only two-thirds completed (10). Frontal and sphenoidal sinuses still very incomplete (10).	Mesenterial attachments completed (5). Vaginal sacs passing into scrotum (7-9).	Female urogenital sinus becoming a shallow vestibule (5). Vagina regains lumen (5). Uterine glands appear (7). Scrotum solid until sacs and testes descend (7-9). Kidney tubules cease forming at birth.	Blood formation increasing in bone marrow and decreasing in liver (5-10). Spleen acquires typical structure (7). Some fetal blood passages discontinue (10).	Carpal, tarsal and sternal bones ossify late; some after birth. Most epiphyseal centers appear after birth; many during adolescence.

(continued)

The state of maternal nutrition.

The status of the placenta (structurally and functionally).

The genetic makeup of the offspring.

The presence of physical, chemical, or mechanical insults (other than during delivery) to mother and/or child during pregnancy.

MATERNAL NUTRITION

In general, with respect to nutritional and oxygen needs, the fetus will maintain itself at the

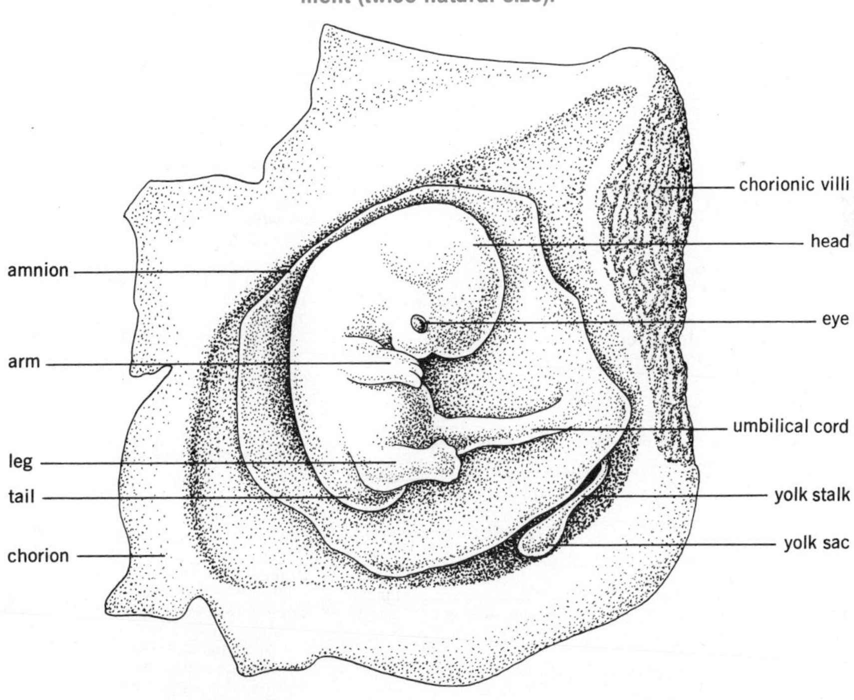

FIGURE 28.9
The human embryo at about eight weeks of development (twice natural size).

TABLE 28.3 (continued)

A REFERENCE TABLE OF CORRELATED HUMAN DEVELOPMENT

AGE IN WEEKS	SIZE (C R) IN MM	MUSCULAR SYSTEM	INTEGUMENTARY SYSTEM	NERVOUS SYSTEM	SENSE ORGANS
2.5	1.5	—	Ectoderm a single layer.	Neural groove indicated.	—
3.5	2.5	Mesodermal segments appearing (1 — 16 ±). Older somites show myotome plates.	—	Neural groove prominent; rapidly closing. Neural crest a continuous band.	Optic vesicle and auditory placode present. Acoustic ganglia appearing.
4	5.0	All somites present (40).	—	Neural tube closed. Three primary vesicles of brain represented. Nerves and ganglia forming. Ependymal, mantle and marginal layers present.	Optic cup and lens pit forming. Auditory pit becomes closed, detached otocyst. Olfactory placodes arise and differentiate nerve cells.
5	8.0	Premuscle masses in head, trunk and limbs.	Epidermis gaining a second layer (periderm).	Five brain vesicles. Cerebral hemispheres bulging. Nerves and ganglia better represented. (Suprarenal cortex accumulating).	Choroid fissure prominent. Lens vesicle free. Vitreous anlage appearing. Otocyst elongates and buds endolymph duct. Olfactory pits deepen.
6	12.0	Myotomes, fused into a continuous column, spread ventrad. Muscle segmentation largely lost.	Milk line present.	Three primary flexures of brain represented. Diencephalon large. Nerve plexuses present. Epiphysis recognizable. Sympathetic ganglia forming segmental masses. Meninges indicated.	Optic cup shows nervous and pigment layers. Lens vesicle thickens. Eyes set at 160°. Naso-lacrimal duct. Modeling of ext., med. and int. ear under way. Vomero-nasal organ.
7	17.0	Muscles differentiating rapidly throughout body and assuming final shapes and relations.	Mammary thickening lens-shaped	Cerebral hemispheres becoming large. Corpus striatum and thalamus prominent. Infundibulum and Rathke's pouch in contact. Choroid plexuses appearing. Suprarenal medulla begins invading cortex.	Choroid fissure closes, enclosing central artery. Nerve fibers invade optic stalk. Lens loses cavity by elongating lens fibers. Eyelids forming. Fibrous and vascular coats of eye indicated. Olfactory sacs open into mouth cavity
8	23.0	Definitive muscles of trunk, limbs and head well represented and fetus capable of some movement.	Mammary primordium a globular thickening	Cerebral cortex begins to acquire typical cells. Olfactory lobes visible. Dura and pia-arachnoid distinct. Chromaffin bodies appearing.	Eyes converging rapidly. Ext., mid. and int. ear assuming final form. Taste buds indicated. External nares plugged.
10	40.0	Perineal muscles developing tardily.	Epidermis adds intermediate cells. Periderm cells prominent. Nail field indicated. Earliest hair follicles begin developing on face.	Spinal cord attains definitive internal structure.	Iris and ciliary body organizing. Eyelids fused. Lacrimal glands budding. Spiral organ begins differentiating.
12	56.0	Smooth muscle layers indicated in hollow viscera.	Epidermis three-layered. Corium and subcutaneous now distinct.	Brain attains its general structural features. Cord shows cervical and lumbar enlargements. Cauda equina and filum terminale appearing. Neuroglial types begin to differentiate.	Characteristic organization of eye attained. Retina becoming layered. Nasal septum and palate fusions completed.

TABLE 28.3 (continued)

A REFERENCE TABLE OF CORRELATED HUMAN DEVELOPMENT

AGE IN WEEKS	SIZE (C R) IN MM	MUSCULAR SYSTEM	INTEGUMENTARY SYSTEM	NERVOUS SYSTEM	SENSE ORGANS
16	112.0	Cardiac muscle appearing in earlier weeks, now much condensed. Muscular movements in utero can be detected.	Epidermis begins adding other layers. Body hair starts developing. Sweat glands appear. First sebaceous glands differentiating.	Hemispheres conceal much of brain. Cerebral lobes delimited. Corpora quadrigemina appear. Cerebellum assumes some prominence.	Eye, ear and nose grossly approach typical appearance. General sense organs differentiating.
20–40 (5-10 mo)	160.0–350.0	Perineal muscles finish development (6).	Vernix caseosa seen (5). Epidermis cornifies (5). Nail plate begins (5). Hairs emerge (6). Mammary primordia budding (5); buds branch and hollow (8). Nail reaches finger tip (9). Lanugo hair prominent (7); sheds (10).	Commissures completed (5). Myelinization of cord begins (5). Cerebral cortex layered typically (6). Cerebral fissures and convolutions appearing rapidly (7). Myelinization of brain begins (10).	Nose and ear ossify (5). Vascular tunic of lens at height (7). Retinal layers completed and light perceptive (7). Taste sense present (8). Eyelids reopen (7-8). Mastoid cells unformed (10). Ear deaf at birth.

(From L. B. Arey, DEVELOPMENTAL ANATOMY, 7th ed., W. B. Saunders, 1965.)

expense of the mother. The fetus is thus a "parasite," maintaining itself at the expense of its host, its mother. The mother usually possesses reserves of necessary materials on which the offspring may draw. Chronic maternal malnutrition will, however, reduce both the quality and quantity of materials available for the offspring, and may, depending on when the shortage occurs, result in abnormal developmental processes.

As examples, we may cite the following.

Deficiency of iodine in the first two months may result in failure of the mother or fetus to synthesize sufficient thyroxin to assure normal nervous system development. Mental retardation may result.

Deficiency of iron, calcium, or phosphorus during the middle and last trimesters may result in inadequate blood and bone formation.

Inadequate protein intake at any time slows growth and reduces vigor of the fetus.

Nutritional deficiencies in maternal diet produce fetal deficiencies only if the maternal deficiency is severe. Differences between subjects on a poor or marginal diet and those on an adequate diet are minimal and difficult to find.

FETAL MEMBRANES

In vertebrates with internal embryonic development, the embryo acquires a series of protective, nutritive, and excretory structures comprising the fetal membranes (Fig. 28.12). These membranes include the yolk sac, amnion, chorion, allantois, umbilical cord, and placenta.

The yolk sac. No true yolk mass is present in human embryos, but a yolk sac appears as the endoderm-lined cavity below the embryonic disc. It is a transitory structure and is incorporated into the gut.

The amnion. The amnion overlies and surrounds the embryonic disc. As the embryo grows, the amnion comes to surround the embryo on all sides, and is filled with amniotic fluid. It forms the "bag of waters" that cushions the fetus. As the embryo develops, chemical substances and cells are shed into this fluid; the procedure designated *amniocentesis,* in which the amnion is punctured and fluid and these cells withdrawn, is used to determine the presence of abnormal processes in the fetus. The cells may be analyzed for chromosome makeup; metabolic disorders and erythroblastosis often result in the presence of abnormal chemicals in the fluid.

2 mo. fetus
3 mo. fetus
Newborn
2
5
13
22 years

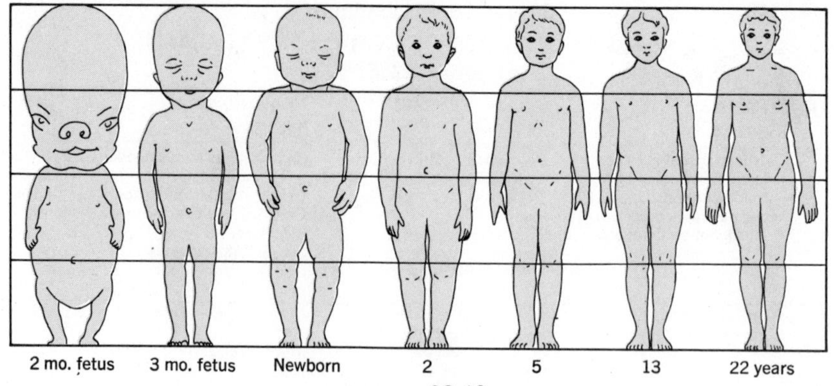

FIGURE 28.10
Changes in body proportions from before birth to adulthood.

FIGURE 28.11
Changes in body size during development in the uterus (all figures natural size).

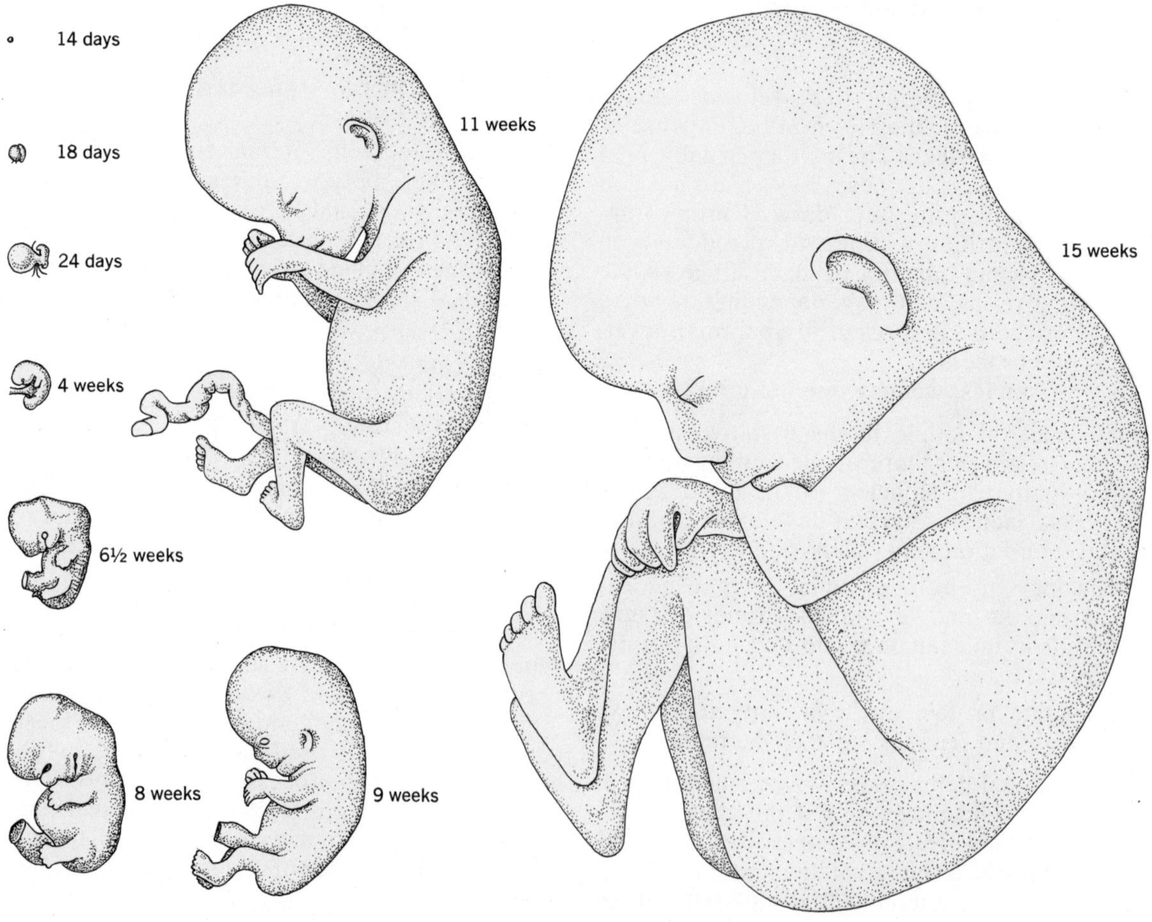

14 days
18 days
24 days
4 weeks
6½ weeks
8 weeks
9 weeks
11 weeks
15 weeks

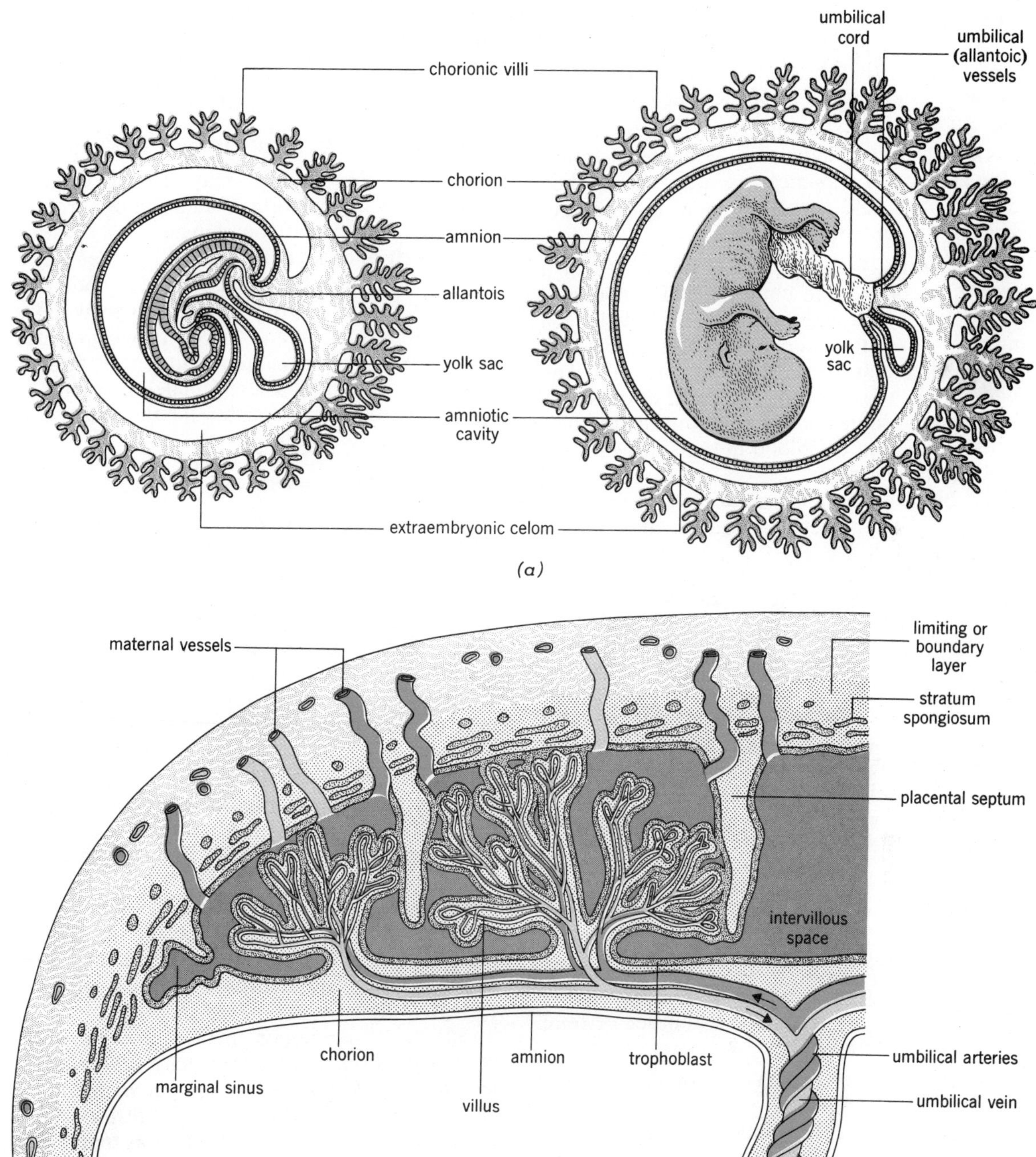

FIGURE 28.12
(*a*) The development of the fetal membranes. (*b*) The scheme of placental circulation.

The chorion. The chorion arises from the trophoblast of the blastocyst and the mesoderm associated with the trophoblast. It forms the major fetal components of the placenta.

The allantois. The allantois develops as an outpocketing of the hindgut of the embryo. In human embryos it becomes vascularized and then degenerates. The vessels remain as a component of the umbilical vessels.

The placenta. The placenta is composed of a fetal component, the chorion, and a maternal component, the endometrium. It develops a number of sections known as cotyledons, each of which contains countless placental villi. The villi act as the routes for exchange of materials between the fetus and the blood of the mother. The villi float in the blood in the sinuses of the endometrium.

At birth, the amnion ruptures, and the fetus emerges. The placenta detaches and constitutes the "afterbirth."

STATUS OF THE PLACENTA

The placenta serves as the means by which the fetus acquires its nutritional needs and eliminates wastes. Only if the placenta develops abnormally is it likely to lead to insufficient supplies of necessary nutrients. In a full-term fetus that is born obviously "malnourished" (as evidenced by weight and adiposity), the placenta is defective, typically showing one or more of the following characteristics:

It is small and thin for time of development. This reduces surface area for absorption and excretion.

It is usually fibrotic in texture. This reduces the ability of substances to diffuse through the organ and may reduce the number of fetal blood vessels.

It contains a smaller-than-normal umbilical cord. This may reflect single arteries rather than the usual two. It indicates a lower-than-normal blood flow.

It has fewer than the normal 12 to 15 cotyledons or sections. This usually means lessened surface area for exchange of materials.

As the major route for exchange of materials, large defects in the placenta must obviously influence the ability of the fetus to acquire those substances necessary for its development.

THE GENETIC STATUS OF THE OFFSPRING

Congenital abnormalities may result from gene mutation or the deletion or addition of chromosomes to the normal complement.

Growth and development will proceed at a rate determined by basic genetic potential. Some fetuses are "programmed" to grow more slowly than others, and low birth weight does not necessarily mean abnormality. Male fetuses grow more rapidly than female, resulting in higher average birth weights for the males.

INSULTS AND TRAUMA TO THE DEVELOPING INDIVIDUAL

A variety of trauma may be delivered to the fetus, either "externally" through the mother's surfaces or "internally" via the mother's bloodstream. PHYSICAL AGENTS such as radiation may injure or destroy cells or cause changes in genes that result in abnormal development. CHEMICAL AGENTS include virus or bacterial agents and their products as well as drugs that may interfere with a crucial developmental process. As examples of the action of chemical agents we may cite the congenital abnormalities (eye problems, deafness, heart defects) that result from the acquisition of measles (Rubella) by a female during the first trimester of pregnancy; the presumed effects of LSD in causing breakage of chromosomes and the higher incidence of skeletal abnormalities; the use of barbiturates and the development of hyperexcitability of brain neurons; the use of antiepileptics and the increased incidence of cleft lip and palate. An important fact to keep in mind is that *abnormalities are usually multiple depending on what processes are occurring at the time of the impingement of the insult.* Also, external symptoms may give clues as to internal problems. For example, deformities of the external ear may suggest kidney anomalies, since the two areas are developing at the same time. MECHANICAL INSULTS include such things as umbilical cord compression that may result in asphyxia of the fetus, and abnormal foot or leg positions that may produce a clubfoot or bowleg.

PARTURITION (BIRTH)

At some point, usually about 280 days from the last menstrual period (LMP) or some 267 days after conception, a new individual is born. The series of events preceding birth constitutes LABOR, and the appearance of the baby itself is termed DELIVERY or BIRTH.

CAUSES OF LABOR AND BIRTH

Ten to fourteen days before birth is imminent, the fetal head settles into the pelvis in the process called LIGHTENING. False labor pains (irregular contractions of the uterus) may occur 3 to 4 weeks before birth. REGULAR CONTRACTIONS of the uterus, discharge of the cervical mucus plug, and RUPTURE OF THE AMNION (bag of waters) with discharge of the amniotic fluid surrounding the fetus signal the approach of birth.

Among the factors suggested to be responsible for onset of labor are: stretching of the uterus with release of prostaglandins and subsequent stimulation of uterine contraction; increased sensitivity of the uterus to oxytocin; decrease of progesterone levels in the bloodstream. Although the reason is not definitely known, labor is usually initiated at a time when the fetus is sufficiently mature in body systems to sustain itself apart from the mother.

Duration of labor is usually longest in full-term first pregnancies (primiparous females)—about 14 hours—and decreases to 7 to 8 hours in second or more births (multiparous females).

Labor proceeds in three stages (Fig. 28.13). The first stage is EFFACEMENT and DILATION, in which the cervical canal is shortened and thinned (effacement) and the cervix opens to a maximum of about 10 cm (dilation). The second stage is BIRTH of the infant, due to uterine contractions 50 to 70 seconds in length and occurring every 2 to 3 minutes. Stage three involves separation and delivery of the placenta (*after-birth*). The uterus then normally contracts strongly, closing the uterine vessels and preventing hemorrhage.

DANGERS OF BIRTH

Prolonged labor, or DYSTOCIA, may result in separation of the placenta before birth, with asphyxia of the fetus. CORD COMPRESSION has the same effect. ABNORMAL PRESENTATION (breech) of the fetus may result in difficulty of passage of a body part through the birth canal. The normal "order" of birth is head, shoulders, trunk, and lower limbs. Attempts to extract the fetus in a breech position may result in trauma, as by the use of forceps, or by excessive traction on a body part.

ASPHYXIA may be signaled by a high heart rate (above 160 beats/min) or by slowing of rate (below 120 beats/min). Increased movement followed by depression of movement or the presence of colon contents (meconium) in the amniotic fluid all signal distress, with sympathetic nervous system discharge caused by the asphyxia.

ASSESSMENT OF STATUS OF THE NEWBORN

If birth proceeds normally, the newborn individual must be evaluated as to physiological status immediately after birth. The APGAR RATING SYSTEM looks at five basic physiological parameters and assigns points for each of three levels of function. The parameters and scoring are presented in Table 28.4. Good correlation between high Apgar scores and neonatal (newborn) survival have been demonstrated. Although the rating system is very general, it indicates general vigor and status of the organism as determined at one and five minutes after birth, and is an assessment for earliest possible medical intervention.

The neonate is usually born slightly acidotic as a result of accumulation of CO_2 in the bloodstream. With the assumption of breathing, the condition usually corrects itself. The ability of the infant to adjust body temperature precisely is not mature at birth, and fluctuations of temperature are more extreme. Kidney function is immature, with low filtration rates and lack of concentrating ability present. The neonate thus loses relatively more water per unit of solute than an older child, and water and electrolyte balances must be monitored more closely.

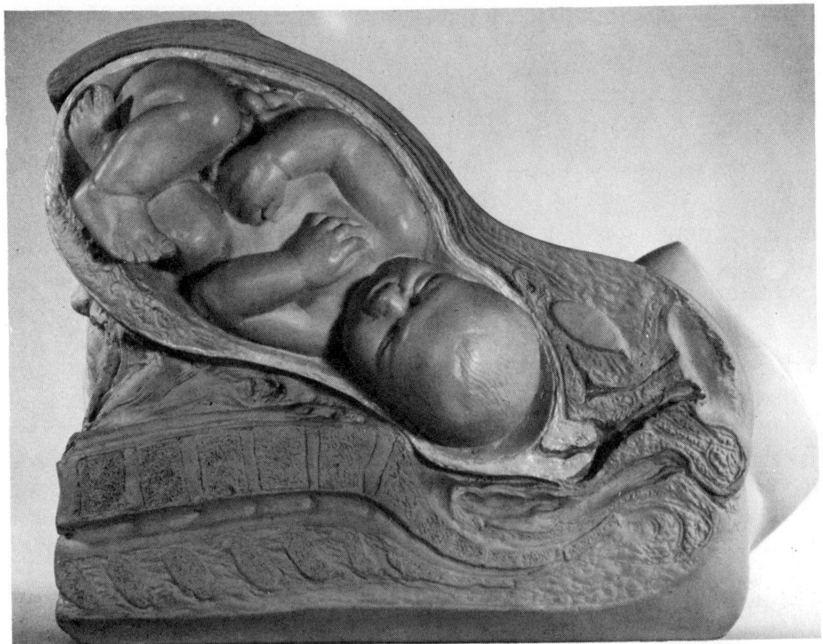

(a)

(b)

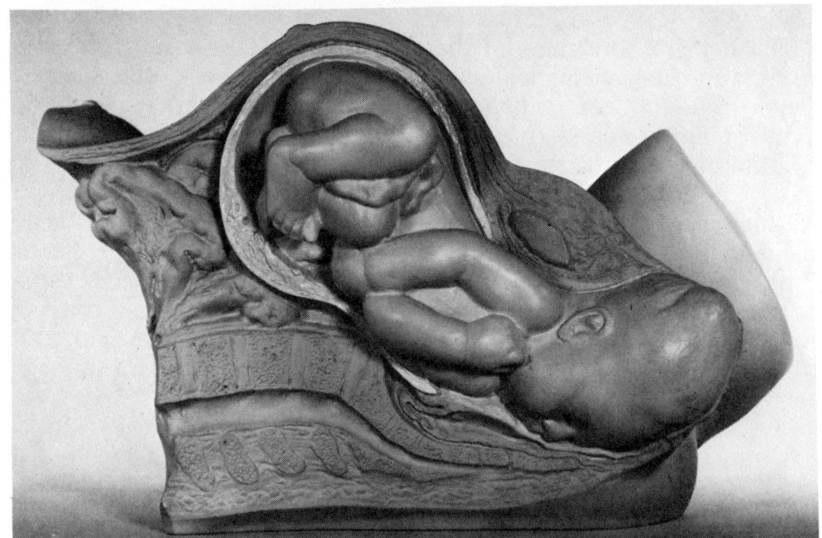

FIGURE 28.13

The three stages of labor. (*a*) First stage of labor. The rhythmic contractions of the uterus aid the progressive effacement and dilation of the cervix as well as the descent of the infant. (*b*) Second stage of labor. "Caput," or top of infant's head, begins to appear through the vulvar opening. (*c*) Third stage of labor. Separation of placenta. (Dickinson-Belskie models, courtesy Cleveland Health Education Museum.)

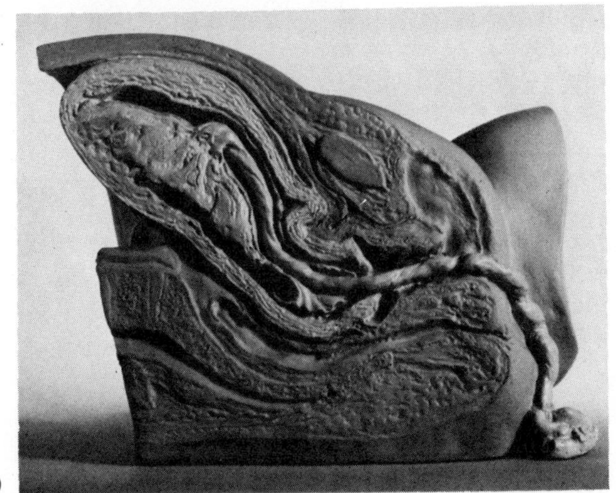

(c)

Fever indicates excessive metabolism as compared to heat loss or excessive loss of fluids. Immature immune response may be a source of inadequate reaction to bacterial or viral agents. Breast feeding is believed to provide the neonate with some maternal antibodies that may help protect the infant. The neonate therefore should be guarded against contact with infected persons.

The NEONATAL PERIOD ends after four weeks of life. The next period is INFANCY, extending from four weeks to two years of age. CHILDHOOD is the period between infancy and puberty. ADOLESCENCE extends from puberty to maturity. MATURITY extends from adolescence to SENESCENCE OR OLD AGE. Progressing

TABLE 28.4

APGAR SCORING OF NEWBORN INFANT

SIGN	SCORE		
	0	1	2
A Appearance (color)	Blue or pale	Body pink; extremities blue	All pink
P Pulse (heart rate)	None	Below 100	Over 100
G Grimace (reflex irritability in response to sole of foot stimulation)	No response	Some movement; grimace	Active with crying
A Activity (muscle tone)	Flaccid	Some flexion of extremities	Active motion
R Respiration (respiratory effort)	None	Slow and irregular; no crying	Good strong cry

TABLE 28.5

PHYSIOLOGICAL CHANGES DURING MATURATION

PARAMETER	STAGE OF DEVELOPMENT				
	NEONATE	INFANT	CHILD	ADOLESCENT	ADULT
Hemoglobin (grams %)	17–20	10.5–12.5	12–14	14–16	14–16
Heart rate, range (beats/min)	120–140	80–140	70–115	80–90	70–75
Blood pressure, average (systolic/diastolic, mm Hg)	75/40	80/50	85/60	112/70	120/80
Breathing rate (breaths/min)	30–50	20–30	16–20	14–16	14–16
Urine specific gravity	1.002–1.008	→	1.015–1.025	→	→
Daily urine volume (ml)	100–300	400–500	600–1000	800–1400	1000–1500
Motor development (landmarks)	Raise head	Sit; crawl; walk, run; talk	Hop; write; read	Fully developed gross and fine motor skills	→

through these states, the individual achieves physiological function equivalent to that of the adult, grows to adult stature, and becomes an independent instead of a dependent creature in the psychological and social realms. Table 28.5 shows some changes in physiological advancement.

GROWTH AND DEVELOPMENT

PATTERNS OF GROWTH

GROWTH is commonly defined as an increase in size and/or weight that occurs when synthetic or anabolic processes occur at a faster rate than catabolic ones. The definition stresses the concept that new protoplasmic substance is added to preexisting material to achieve growth. Mere increase in weight may reflect only water accumulation or addition of adipose tissue, and should not be used as a criterion of growth. Growth may occur by increase in size of cells, by increase in number of cells that grow between divisions, or by a combination of both processes. The latter option is the one employed to achieve the great increase in body size that occurs from conception to maturity.

It must further be emphasized that not all body organs or systems grow at the same rate, particularly after birth; some parts are growing while others are regressing. The progression, either up or down, is not a linear process, but is marked by plateaus, peaks, and valleys. The curve most representative of overall body growth is sigmoid in shape (Fig. 28.14).

REGULATION OF GROWTH; PRENATAL DEVELOPMENT

A wide variety of factors affect growth and development. HEREDITY appears to set a basic rate and direction of development that may be altered by environmental factors such as NUTRITION, ENDOCRINE STATUS, DRUG INTAKE, PARENTAL AGE, SMOKING, DISEASE, and

FIGURE 28.14
The human growth curve after birth. (From A. K. Laird, "Evolution of the Human Growth Curve," *Growth,* Vol. 31, p. 345, 1967.

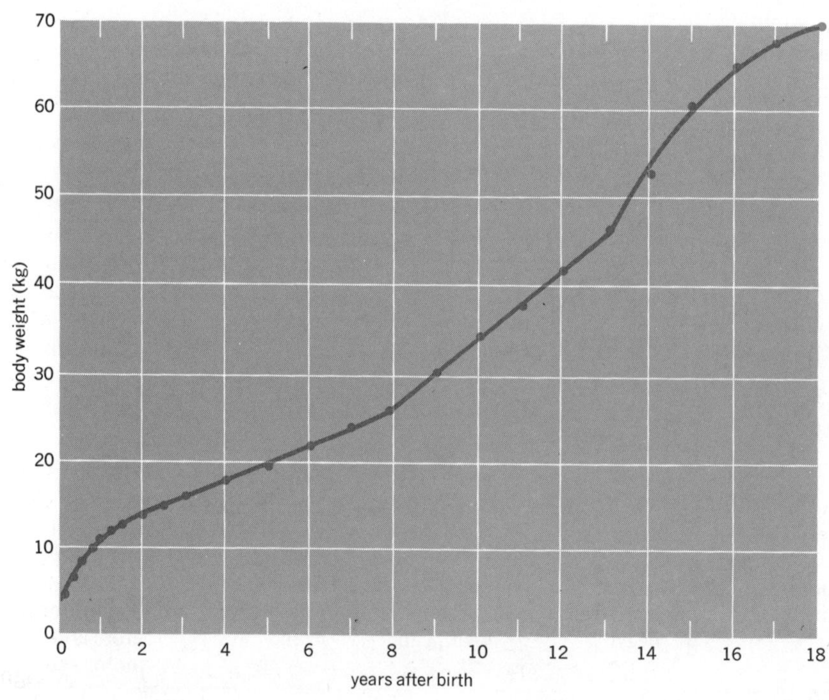

TABLE 28.6
SOME TERATOGENS AND THEIR EFFECTS

AGENT	TYPES OF EFFECTS ON EMBRYO/FETUS
Aminopterin (folic acid antagonist) used in: attempted abortion; treatment of leukemia.	Widespread anatomical malformations
Antibiotics (tetracyclines, actinomycin, streptomycin, others)	Deafness, growth retardation, hemolysis
Cortisone	Cleft palate
Dicumarol	Fetal bleeding
Estrogens (DES, diethylstilbesterol)	Masculinization of female fetus; rare vaginal cancer later in life
Herpes virus	Nervous system defects
Hypoxia	Patent ductus arteriosus
Progestational agents (e.g., nortestosterone)	Masculinization of female fetus (labial fusion, clitoral enlargement)
Quinine	Congenital deafness
Radiation	Skeletal deformities, microcephaly
Rubella virus (german measles)	Blindnes, deafness, congenital heart disease, inflammation of many body organs
Salicylates	Hemorrhage, anatomical malformations
Syphilis (untreated)	Blindness, deafness, congenital heart disease, others
Thiouracil derivatives (Iodine, Iodines)	Goiter
Tranquilizers (e.g., thalidomide)	Anatomic malformations, especially of the appendages

whether an individual is a member of a multiple birth. Environmental factors have a greater or lesser effect on the developing organism, depending on its maturity, the process involved, and the severity of the effect. The "critical-period concept" suggests that those times when changes are occurring rapidly represent points at which environmental factors may cause the greatest effect. The first trimester (three months) of pregnancy constitutes such a time, and drug effects, viral infections, and other insults wreak the greatest havoc on developmental processes at that time. The concept of critical periods may aid medical personnel in deciding if intervention to correct or modify an abnormal pattern of growth is warranted.

GENETIC FACTORS appear to be most important in the prenatal (before birth) period as the influence that directs the differentiation and organization of the offspring. Genetics also sets the general tone in terms of rate and direction of development. If there is a genetic defect that is not lethal, its presence may not be detected until birth, when the child will be said to present a congenital (born with) defect. TERATOGENS is the name given to substances that cause malformations. A partial list of teratogens and examples of the types of defects they produce are presented in Table 28.6. Exactly what effect is produced depends on what system was undergoing the most rapid change when the agent acted, and whether it affects the mother, placenta, or offspring to the greatest degree.

NUTRITION becomes a strong factor in development only if the mother is grossly deficient in intake of specific nutrients required for embryonic or fetal development. Recent theories advance the suggestion that the pregnant female should not limit her weight gain to some

TABLE 28.7

SOME NUTRIENTS ESSENTIAL FOR NORMAL DEVELOPMENT OF EMBRYO OR FETUS, AND EFFECTS OF THEIR DEFICIENCY

SUBSTANCE	DEFICIENCY EFFECTS
Vitamin D	Abnormal calcification of bones
Vitamin A	Visual disturbances
Vitamin E	Increased red cell fragility
Riboflavin	Defective H^+ transport
Fatty acids	Skin lesions, fatty liver
Iodine	Faulty nervous system development
Protein	General development retarded, small brain

specific number of pounds that might result in nutritional deficiencies for the fetus. Gain should not be so great to cause difficulty for the mother because of her size or the size of the child, and nutrition should be adequate at all times. Several important substances essential for embryonic and fetal development, and effects of their deficiency, are presented in Table 28.7.

ENDOCRINE STATUS may be reflected by changes that become evident at birth or later in life. ACTH given during the first trimester increases the incidence of cleft palate; adrenal sex hormone hypersecretion may cause adrenogenital syndrome; hyposecretion of thyroid hormone by the mother before the fetus produces its own (at about five months) retards brain growth.

DRUGS are usually thought of as materials that are abused. Any substance other than nutrients placed in the body for a therapeutic purpose may be considered a drug. Drugs act to produce a variety of effects on development, as shown in Table 28.8.

Any drug that must be taken in large doses during pregnancy should be checked for possible effects on the offspring.

Drug effects are generally more severe on the offspring *in utero* than on the mother because of the limited ability of the immature liver and kidneys to detoxify and excrete drugs.

AGE OF THE MOTHER exerts an influence on growth and development in that younger mothers tend to have smaller babies. A higher incidence of malformations is seen in very young and older mothers, and recent evidence suggests that mental development is slower in later children of larger families due to "dilution" of the learning environment by more persons.

Maternal SMOKING increases the incidence of premature births and results in the birth of smaller babies. The effect is believed to be due to the nicotine passed from mother to fetus.

As indicated earlier, INFECTIOUS AGENTS such as viruses may compromise placental function (e.g., mumps) to where life cannot be sustained, or affect the offspring directly (e.g., rubella). Congenital syphilis causes a variety of changes, many of which becomes evident only later in life.

TABLE 28.8

SOME DRUGS INFLUENCING GROWTH AND DEVELOPMENT

Analgesics (pain relievers) and antipyretics (fever reducers) (e.g., salicylates)	Hemorrhage, fetal death due to large doses
Antibiotics	Possible growth retardation
Penicillin	Deafness
Streptomycin	
Anticancer agents	Anatomical changes
Anticoagulants (e.g., dicumarol)	Hemorrhage
Antihistamines (e.g., meclizine)	Anatomical changes
Antihypertensives (e.g., reserpine, thiazides)	Impaired neonatal adjustment
Barbiturates	CNS depression or excitation, neurological changes

CHARACTERISTICS OF THE NEWBORN (NEONATE)

At birth the full-term fetus averages 3405 g ($7\frac{1}{2}$ lb) in weight and about 50 cm (20 in.) in length, and has a head circumference averaging $35\frac{1}{2}$ cm (range 33 to 38 cm). Five to ten percent of the body weight is lost in the first 24 to 48 hours after birth, representing a loss of water. Newborns usually appear chubby due to deposition of adipose tissue during the last months of pregnancy. Temperature regulation is imperfect and is handled by alterations in metabolic rate and not by sweating and shivering. The kidney lacks the ability to secrete a concentrated urine, and large volumes of dilute urine are produced; thus, dehydration poses a greater threat to the newborn than to an older child. The immune system is immature, and it makes sense to guard the neonate against contact with infected persons. Behavior at birth is almost entirely controlled by lower cerebral centers and the spinal cord. Many stimuli are received, but their localization is poor, and motor response is generally nondirected. Body functions are not circadian, and the newborn appears to be governed by sleep, hunger, and discomfort.

The neonatal period ends after four weeks of life, and infancy begins. Infancy extends until two years of age, followed by childhood extending from infancy to puberty.

GROWTH IN INFANCY AND CHILDHOOD

Three primary measures are employed to assess the "normality" of growth in this period of life: gain in weight, gain in height, and increase in head circumference. It must be emphasized that all children are individuals and are different in their rates of growth; thus, there is a wide range of "normality," and *average* or *mean* is the proper term to employ instead of *normal*.

Weight gain averages 200 g (7 oz) per week for the first three months, 140 g (5 oz) per week for months 4 to 6, 85 g (3 oz) per week for months 7 to 9, and 70 g ($2\frac{1}{2}$ oz) per week for months 10 to 12. At one year of age, the infant averages 20 lb, and at years 3, 5, and 7, weight averages 30, 40, and 50 lb. Food intake thus slows after the first half-year of life and should be of no concern to parents if advancement is steady. If the early rate of gain were maintained, an individual would weigh some 322 lb at age 14. Height increases by 25 to 30 cm between birth and one year of age, and by 10 to 15 cm the second year. Between 2 and 14 years of age, the formula:

$$\text{Ht(in.)} = (\text{age in years} \times 2.5) + 30$$

may be used to indicate average growth rate. A puberal growth spurt adds $3\frac{1}{2}$–4 in. to the height.

Head circumference indicates growth of the brain and may reveal rapid enlargement due to the presence of hydrocephalus. By the sixth month, the brain has reached 50 percent of the adult size, 60 percent by one year, and 75 percent by two years of age.

Figures 28.15 to 28.18 present curves depicting changes in the three parameters discussed.

In any one individual, increase in these parameters is subject to change according to genetic, nutritional, and disease factors as in the prenatal period.

Individual organ or system growth diverges as the person grows older. General, neural, lymphoid, and genital types of development are shown in Figure 28.19. Thus, enlargement of tonsils or lack of growth of genitals during certain time periods should not become causes for concern.

Mental and motor development also follows a more-or-less typical pattern, as shown in Figure 28.20. This figure presents a means of screening a child as to its standing relative to the average; it should not be used as a rigid guide to development. Again, all children are different; they may be behind in one area and ahead in another, development is continuous from conception to maturity, and order of development tends to remain the same in most children, but rate varies.

ADOLESCENCE

Adolescence is the period extending from puberty to maturity; it is primarily characterized by development of sexual organs, sec-

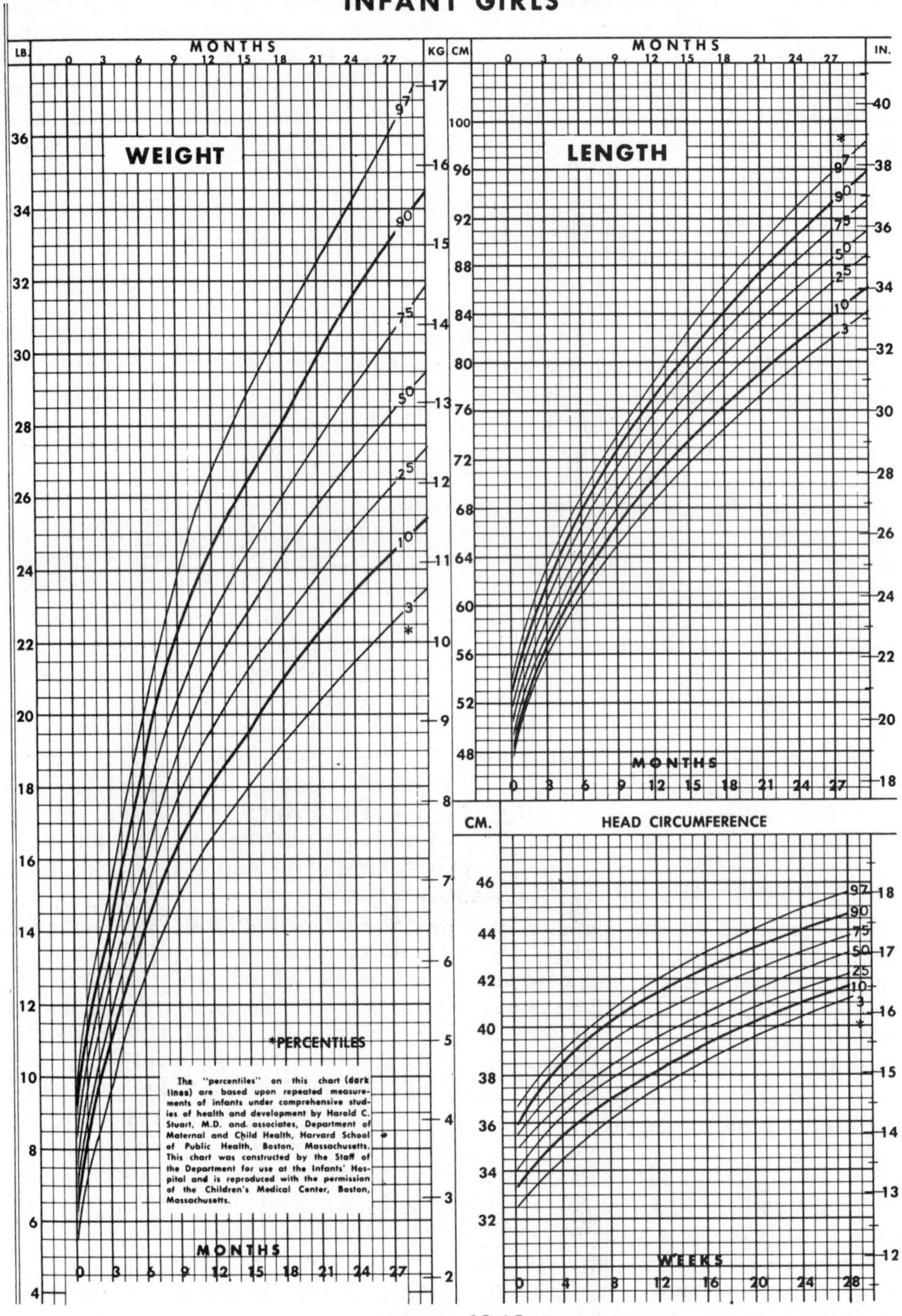

FIGURE 28.15
Growth curves for infant girls. (Courtesy Children's Medical Center, Boston, Mass.)

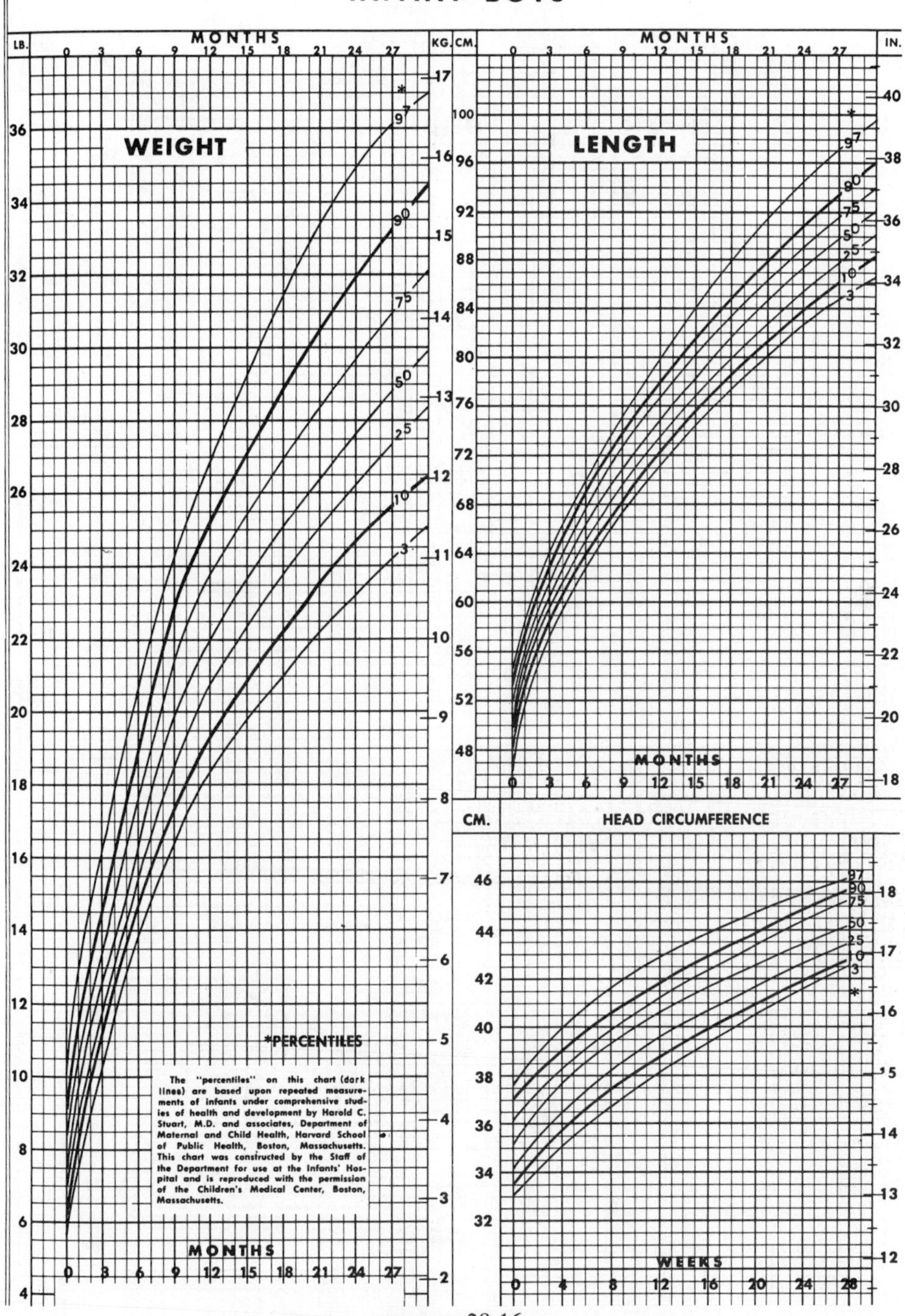

FIGURE 28.16
Growth curves for infant boys. (Courtesy Children's Medical Center, Boston, Mass.)

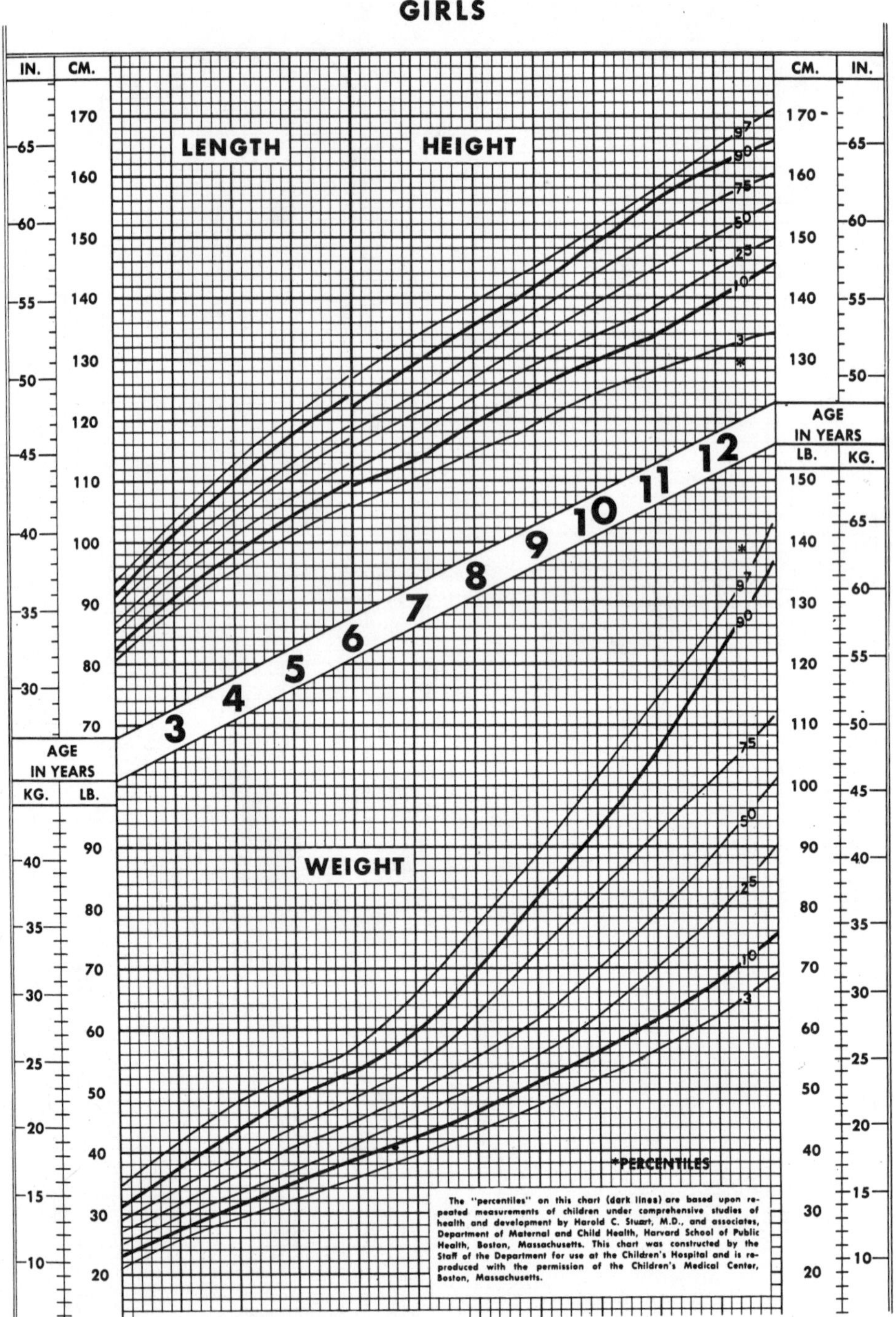

FIGURE 28.17
Growth curves for girls. (Courtesy Children's Medical Center, Boston, Mass.)

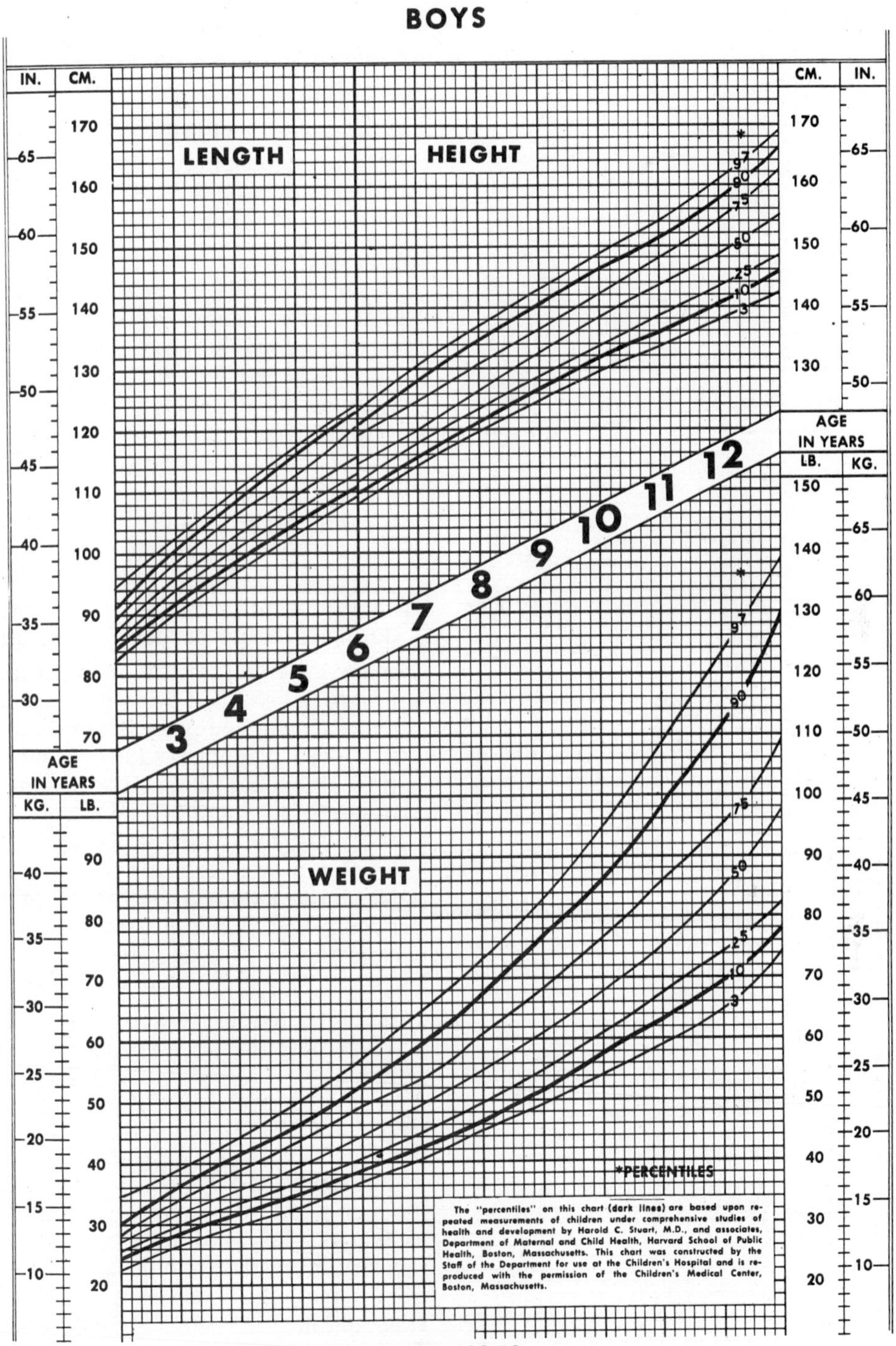

BOYS

FIGURE 28.18

Growth curves for boys. (Courtesy Children's Medical Center, Boston, Mass.)

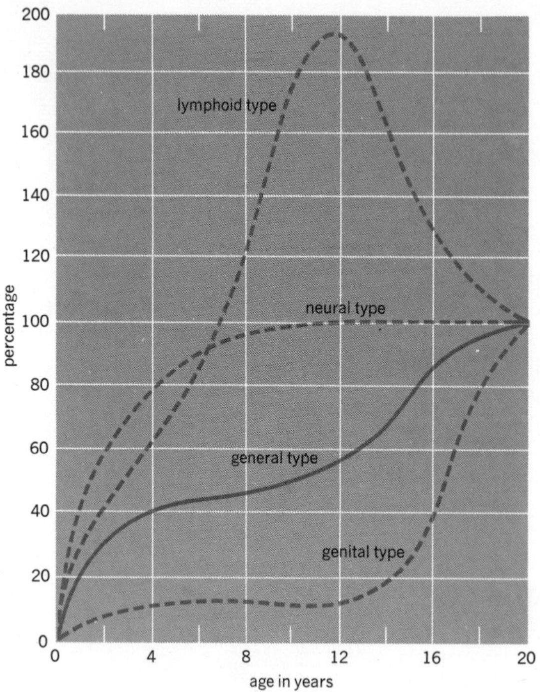

FIGURE 28.19
The major types of growth occurring in the body systems and organs. (From THE MEASUREMENT OF MAN by J. A. Harris. © Copyright 1930 by the University of Minnesota, University Press, Minneapolis.)

ondary sexual characteristics, and skeletal and muscular growth.

A growth spurt (Fig. 28.21) normally occurs between 10½ to 13 years of age in females and between 12½ to 15 years in males. Height growth usually stops at 17 years in females and 18 years in males. The peak of the growth spurt in females is usually followed, within two years, by menarche (menstruation). In females, the usual order of change is:

Rapid increase in height and weight.

Breast changes (pigmented areolae, enlargement of nipples, increase in mass of breast tissue).

Increase in pelvic girth.

Growth of pubic hair, becoming curly about a year after first appearance.

Function of the axillary apocrine sweat glands.

Growth of axillary hair.

Attainment of mature stature.

Menstruation.

The MENSTRUAL (*estrus*) CYCLE is hormonally controlled, depending on ovarian production of estrogen and progestin. The cycle is usually subdivided into four stages.

MENSTRUAL STAGE, the first three to five days of the cycle. This stage commences with the first external show of blood and is characterized by involution of the blood vessels of the thickened endometrium, tissue degeneration, and sloughing of the lining.

PROLIFERATIVE (*follicular*) STAGE, the next seven to fifteen days. Under the influence of estrogens from the growing follicle, epithelial repair is begun, and there is multiplication of connective tissue elements and proliferation of glands and blood vessels. A thickness of about 2 mm is attained. Ovulation occurs during the latter portion of this stage.

SECRETORY (*luteal*) STAGE, the next fourteen to fifteen days. After ovulation, corpus luteum development and progestin production cause increased glandular and vascular proliferation, and some glandular secretion. A thickness of 4 to 5 mm is attained preparatory to receiving the ovum if it is fertilized. Implantation of a fertilized ovum "signals" the luteum to maintain itself. If no implantation occurs, the luteum degenerates and precipitates the fourth stage.

The PREMENSTRUAL STAGE. Arteries begin to involute; tissue breakdown is initiated. The stage terminates with external show of blood.

In females who are commencing menstruation, the initial cycles are usually anovulatory.

The length of a cycle is variable according to the individual. It may be as short as 24 days or as long as 35 days. If there is one timespan that remains more or less constant, it is the time from ovulation to menstruation, 14 to 15 days. It should not be assumed that all cycles are 28 days in length, or that ovulation always occurs 14 days after menstruation.

DIRECTIONS DATE BIRTHDATE
NAME HOSP. NO.

1. Try to get child to smile by smiling, talking or waving to him. Do not touch him.
2. When child is playing with toy, pull it away from him. Pass if he resists.
3. Child does not have to be able to tie shoes or button in the back.
4. Move yarn slowly in an arc from one side to the other, about 6" above child's face. Pass if eyes follow 90° to midline. (Past midline; 180°)
5. Pass if child grasps rattle when it is touched to the backs or tips of fingers.
6. Pass if child continues to look where yarn disappeared or tries to see where it went. Yarn should be dropped quickly from sight from tester's hand without arm movement.
7. Pass if child picks up raisin with any part of thumb and a finger.
8. Pass if child picks up raisin with the ends of thumb and index finger using an over hand approach.

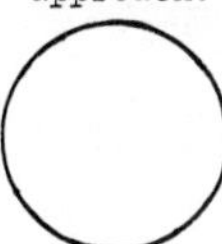

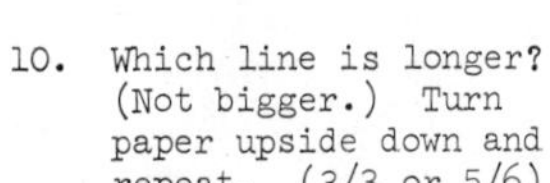

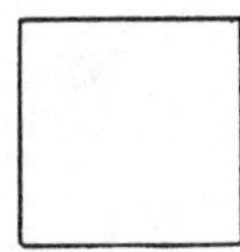

9. Pass any enclosed form. Fail continuous round motions.
10. Which line is longer? (Not bigger.) Turn paper upside down and repeat. (3/3 or 5/6)
11. Pass any crossing lines.
12. Have child copy first. If failed, demonstrate

When giving items 9, 11 and 12, do not name the forms. Do not demonstrate 9 and 11.

13. When scoring, each pair (2 arms, 2 legs, etc.) counts as one part.
14. Point to picture and have child name it. (No credit is given for sounds only.)

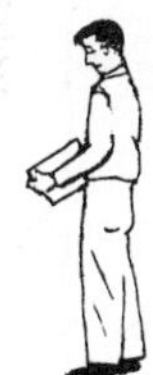

15. Tell child to: Give block to Mommie; put block on table; put block on floor. Pass 2 of 3. (Do not help child by pointing, moving head or eyes.)
16. Ask child: What do you do when you are cold? ..hungry? ..tired? Pass 2 of 3.
17. Tell child to: Put block <u>on</u> table; <u>under</u> table; <u>in front</u> of chair, <u>behind</u> chair. Pass 3 of 4. (Do not help child by pointing, moving head or eyes.)
18. Ask child: If fire is hot, ice is ?; Mother is a woman, Dad is a ?; a horse is big, a mouse is ?. Pass 2 of 3.
19. Ask child: What is a ball? ..lake? ..desk? ..house? ..banana? ..curtain? ..ceiling? ..hedge? ..pavement? Pass if defined in terms of use, shape, what it is made of or general category (such as banana is fruit, not just yellow). Pass 6 of 9.
20. Ask child: What is a spoon made of? ..a shoe made of? ..a door made of? (No other objects may be substituted.) Pass 3 of 3.
21. When placed on stomach, child lifts chest off table with support of forearms and/or hands.
22. When child is on back, grasp his hands and pull him to sitting. Pass if head does not hang back.
23. Child may use wall or rail only, not person. May not crawl.
24. Child must throw ball overhand 3 feet to within arm's reach of tester.
25. Child must perform standing broad jump over width of test sheet. (8-1/2 inches)
26. Tell child to walk forward, ⊂⊃⊂⊃⊂⊃⊂⊃→ heel within 1 inch of toe. Tester may demonstrate. Child must walk 4 consecutive steps, 2 out of 3 trials.
27. Bounce ball to child who should stand 3 feet away from tester. Child must catch ball with hands, not arms, 2 out of 3 trials.
28. Tell child to walk backward, ←⊂⊃⊂⊃⊂⊃⊂⊃ toe within 1 inch of heel. Tester may demonstrate. Child must walk 4 consecutive steps, 2 out of 3 trials.

<u>DATE AND BEHAVIORAL OBSERVATIONS</u> (how child feels at time of test, relation to tester, attention span, verbal behavior, self-confidence, etc,):

FIGURE 28.20
Denver Developmental Screening Test (see also p. 594).

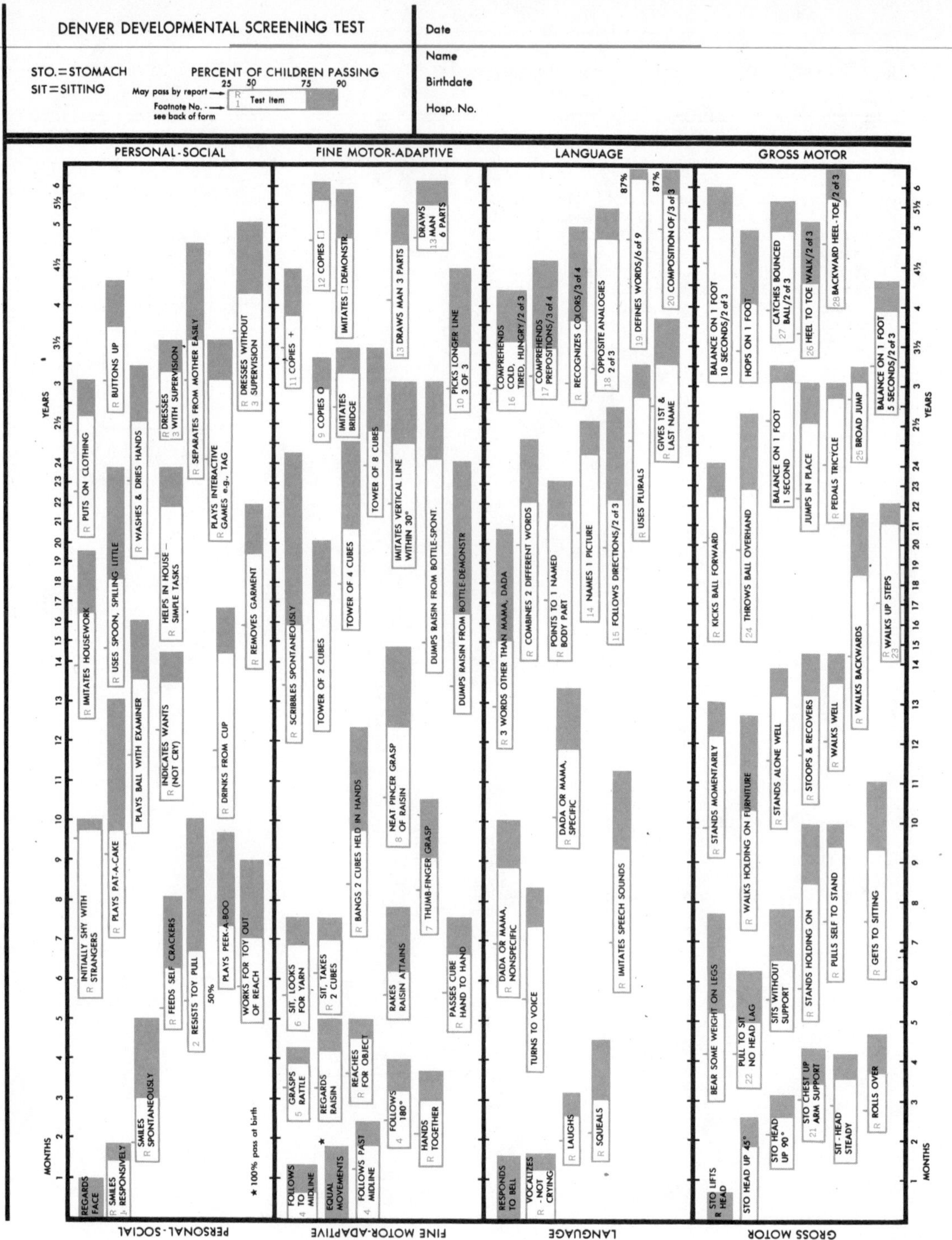

FIGURE 28.20
The Denver Developmental Screening Test for motor and mental development in infants and children. (Courtesy W. K. Frankenburg and J. B. Dodds, University of Colorado Medical Center.)

In males, as testosterone secretion increases, the usual order of changes is:

Increase in height and weight.

Enlargement of penis and testicles.

Development of pubic hair, becoming curly about one year after first appearance.

Growth of axillary hair.

Hair development on upper lip, groin, thighs, abdomen (on other parts of face about two years after pubic hair).

Voice changes due to larynx growth.

Nocturnal emissions.

Attainment of mature stature.

Several of these changes are summarized in Fig. 28.22.

Other physiological changes occurring during adolescence and extending into maturity include:

Progressive decrease of metabolic rate.

Leveling off of cardiovascular function, with attainment of adult levels of heart rate and blood pressure.

Blood cell formation reaches adult levels.

Breathing rates reach adult levels with increase of vital capacity and higher alveolar CO_2 levels.

Myelination of nervous structures continues into the 30s and possibly beyond (Fig. 28.23).

When most of these structural and functional changes have achieved adult values, the individual is considered to be "mature"; this is usually about 25 to 30 years of age.

FIGURE 28.21

Curves depicting the adolescent growth spurt in boys and girls. Ages represent averages and do not indicate range. (From J. M. Tanner, *Growth and Adolescence*, Second Edition, Blackwell Scientific Pubs., Oxford, 1962. Used with permission.)

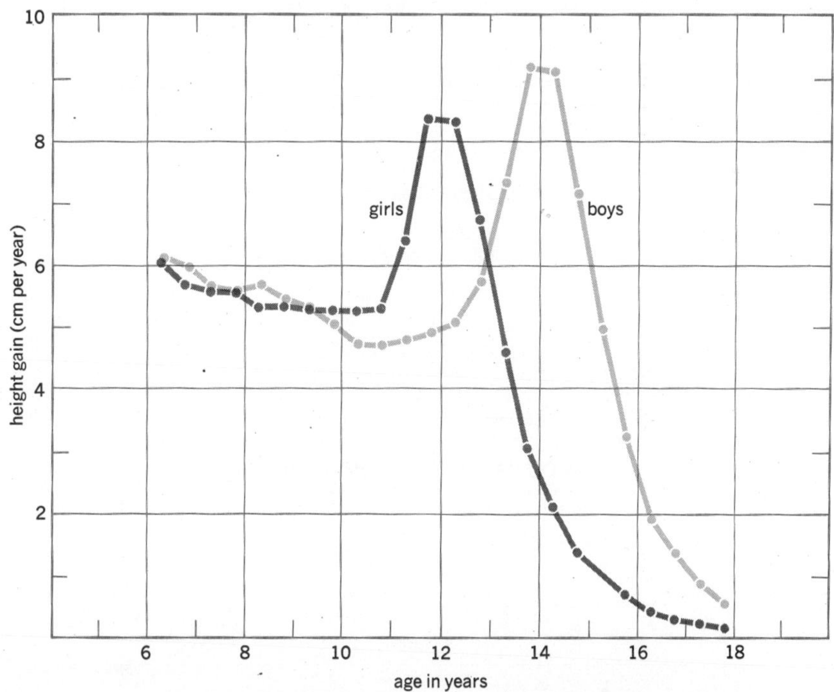

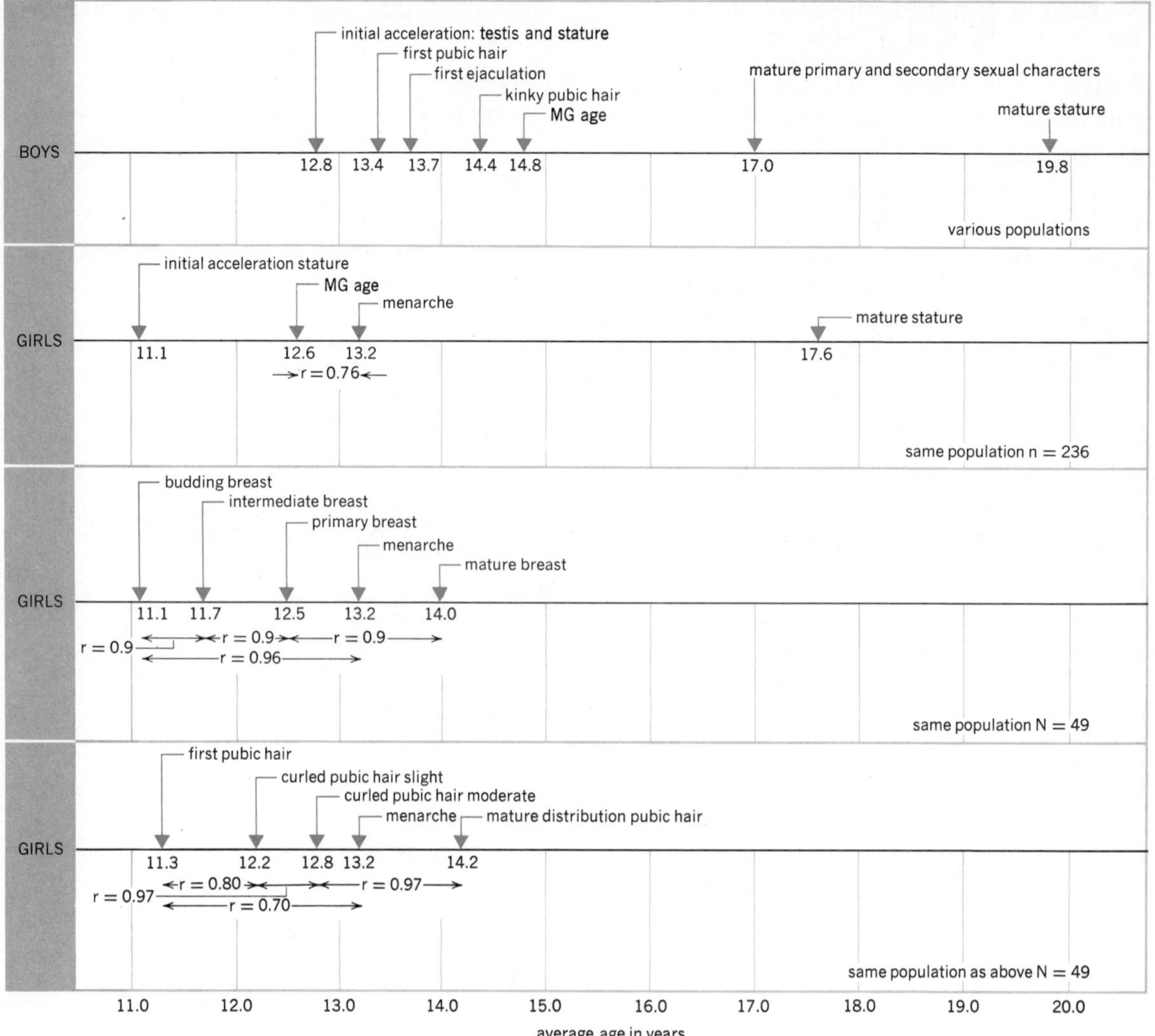

FIGURE 28.22
Average sequence of events occurring in development of secondary sex characteristics in boys and girls. MG, Mature Genital, implying production of ova or sperm. (From DEVELOPMENTAL PSYCHOLOGY by E. Hurlock. Copyright 1968 McGraw-Hill Book Company. Used with permission.)

MATURITY

Maturity is difficult to define, for it includes physical, mental, emotional, and chemical factors. Other than ability to reproduce, long considered a definition of (sexual) maturity, any definition should include mention of attainment of optimum integration between physiological processes and of a steady state in body functions.

AGING AND SENESCENCE

Aging is a term that should be used to denote the changes, some beginning in infancy, that occur as one grows older. *Senescence* is usually understood to mean the "state of old age encountered in the last years of life." There is no reason to associate either term with a particular age or mental state, as seems common. There is as yet no scientific validation that

FIGURE 28.23

The pattern of myelination in various parts of the nervous system. (From A. Minkowski (ed.), *Regional Development of the Brain in Early Life*, Blackwell Scientific Pubs., Oxford, 1967. Used with permission.)

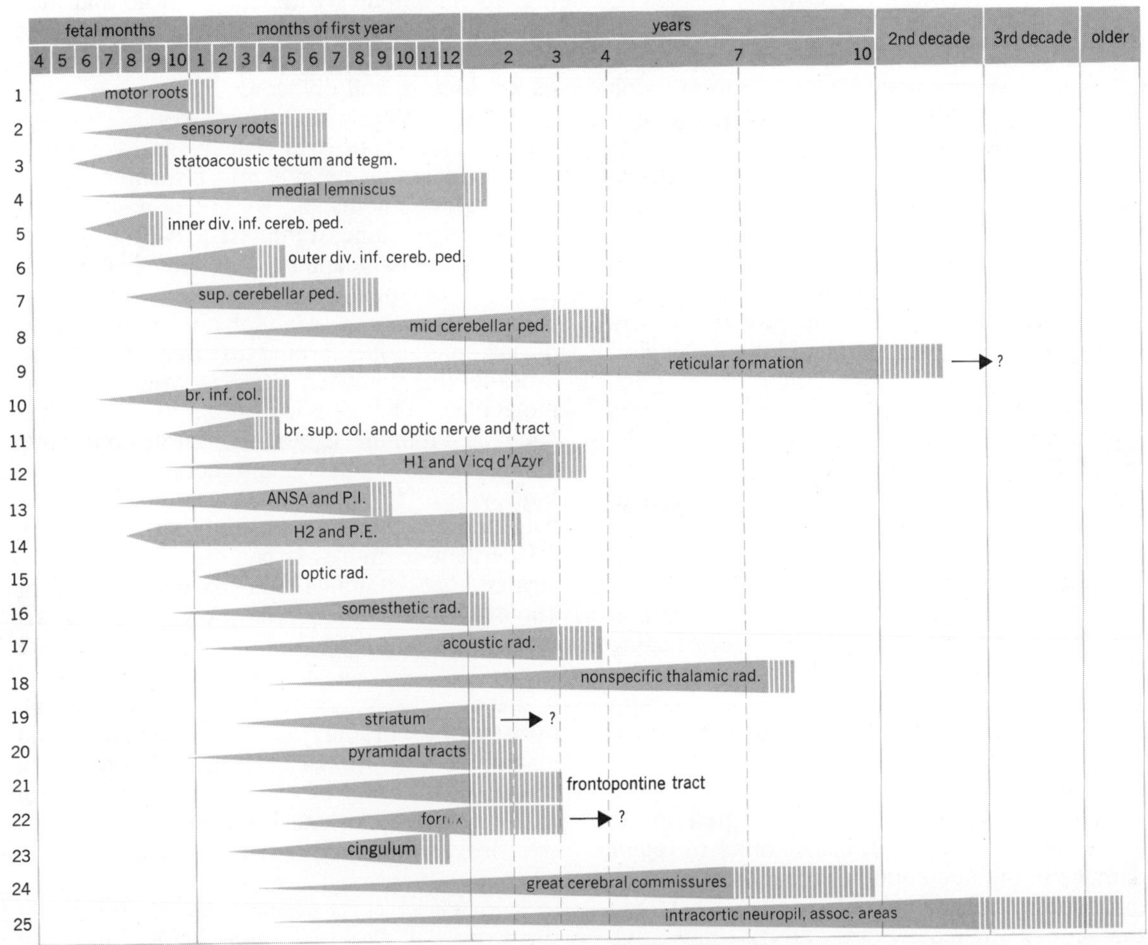

death is a *natural* event that must be accepted with stoic resignation. It appears that death may result from a compromising of functions that *may be* largely preventable. Adding years to life must also be accompanied by attempts to add life to years—continuing the use of the talents of productive individuals.

That body structure and function change with age is an obvious fact. Some changes, such as the deposition of lipids in arteries or changes in the eye, begin in infancy. They may ultimately comprise integration and function to where life can no longer be sustained. Systems age, as they also develop, at different rates; perhaps it is an inability to adapt that ultimately kills us.

THEORIES OF AGING

Two general categories of theories are advanced to explain why we age. EXTERNAL THEORIES suggest that environmental forces, such as radiation, pollutants, nutritional status, and disease are the causes of tissue aging. INTERNAL THEORIES suggest that changes in DNA, failure of the immune response, inability to detoxify harmful substances, or "using up" of a genetically determined "age substance" contribute to our eventual death. No one theory adequately explains the aging process or why we die. In the words of Dr. Hans J. Kugler, "There seem to be so many theories on aging, and each one makes some sense. One develops a preference for one or two of these theories, one looks for weakness in others, and one comes up with a new, ingenious approach to solve the aging problem."

A theory that combines various aspects of existing theories will probably prove to be the most correct of any. In any event, let us examine what some of the current theories of aging are.

CHANGES IN OR DAMAGE TO DNA AND RNA may alter the cell's ability to (ultimately) synthesize the proteins necessary to run its chemical machinery (enzymes) or to maintain its basic structure. DNA may become limited in its ability to replicate itself; in its ability to repair damage in the nucleotide sequences; it may be damaged by permanent cross-linkages between the coiled helices; viral DNA may give "wrong" instructions to the replication-transcription-translation mechanisms. A definite and direct correlation has been demonstrated between length of life and the efficiency of DNA repair mechanisms. Recently, it has been suggested that enzymes placed in the body may be able to break bonds that cross-link DNA helices and "turn on" the DNA repair mechanisms again. Life spans of 150 to 200 years may be "just around the corner."

In 1968, it was suggested that human cells have a "programmed" LIMIT OF ABOUT 50 CELL DIVISIONS. Theoretically then, when this limit is reached, repair of injuries will not occur, and the organism will die as a result of loss of cells. Administration of substances such as vitamin E (which combines with free radicals that may cross-link DNA) can increase the cell divisions to 100 times. Thus, it would seem that there is *no set limit* to cell division.

HIGH LEVELS OF MONAMINE OXIDASE (*MAO*) may occur, which destroys a variety of chemicals acting as synaptic transmitters and acceleratory chemicals. By giving a substance known as procaine (it inhibits MAO), improvement in mental capacity and physical ability has been demonstrated.

FREE RADICALS are substances produced during metabolic processes that can cause cross-linking between various types of linear molecules. The administration of substances such as vitamins C and E, sulfur-containing amino acids and selenium compounds combine with or lower the production of free radicals, and lengthen (in mice and rats) lifespan by 60 percent or more. It has also been demonstrated that vitamin C speeds the removal of cholesterol-containing lipoproteins from arterial walls and increases its conversion to bile acids.

Aldehydes, heavy metals, and radicals can CROSS-LINK not only DNA but collagen and elastin as well. Loss of strength of connective tissue, destruction of elastic fibers in the skin and elsewhere can cause the tissues to "sag," with loss of connective tissue strength and resiliency.

FAULTY NUTRITION, perhaps involving failure to supply essential amino acids, can result in failure to synthesize neural transmitters, and cause the production of abnormal enzymes.

RADIATION can cause mutations (breaks, deletions, etc.) in DNA and RNA synthesis that can lead to faulty protein synthesis.

FAILURE OF THE IMMUNE SYSTEM to adequately combat viruses that can "control" DNA and RNA synthesis can lead to deficiencies in normal body functioning. The thymus, in its role of aiding the B-lymphocyte to plasma cell transformation, may suffer decreased activity and reduce ability to trigger immune response.

Figure 28.24 summarizes the interrelationships of various factors in producing aging. Vertically oriented lines indicate secondary factors that result in primary (horizontal boxes) aging changes.

CHANGES DURING AGING

Physiological processes reach a peak during maturity (about 30 years of age) and thereafter undergo gradual decline. At some point, one or more functions decline below the point necessary to sustain life and death ensues. Heredity, environment, diet, and "speed of living" all influence the rate of decline and time of death.

Three general types of changes contribute to the aging process: SECULAR changes are the result of natural wear and tear; SENESCENT changes are the aging of tissues and organs, especially those with low mitotic rate; PATHOLOGIC complications are those resulting from disease processes developing in the aging organs.

Secular changes. As one grows older, tissues appear to demand more metabolic support than the body can supply. In short, secular changes appear to be the result of an IMBALANCE BETWEEN VASCULAR SUPPLY AND TISSUE DEMAND. Thus, the amount of active tissue declines in proportion to its vascularity, and body weight tends to be reduced. Little attempt appears to be made by the body to restore the balance. The changes are usually seen first in endocrine-dependent structures such as breast, prostate, internal reproductive organs, and thyroid.

Senescent changes. All body systems share in these changes. Some of the more noteworthy changes in systems are represented below.

Aging of the INTEGUMENT becomes apparent when the epidermis thins and becomes more translucent and dry, and the dermis becomes dehydrated and suffers loss of elastic fibers; the skin thus tends to "sag" on the body. There is usually loss or thinning of hair as a result of lowered sex-hormone levels and lowered synthesis of proteins. Sweat gland secretion diminishes, making the older individual more susceptible to the effects of high environmental temperature. Skin lesions (cancer, ruptured blood vessels, for example) are more common in the aged. Diabetic vasculopathy is also more common in the aged.

Aging of the EYE is reflected by a high incidence in the aged of cataracts, occlusion of retinal vessels with subsequent retinal changes, glaucoma, and changes in the transparency of the cornea. About one person in six in the over–65 age group exhibits some ocular pathology.

Aging of the EAR is evidenced by sclerotic alterations in the eardrum and ossicles that may result in loss of hearing. Vascular blockage may result in sudden hearing loss or loss of ability to maintain body equilibrium.

NEUROLOGICAL disorders are most commonly associated with changes in vascular supply, either gradually (Parkinson's disease) or suddenly ("stroke"). Peripheral loss of sensation acuity, neuritis, and neuralgia appear to be more common in the aged.

CIRCULATORY disorders may include hardening of arteries (arteriosclerosis) and deposition of lipids in the vessel walls (atherosclerosis). If deposition occurs in coronary vessels, the individual becomes a candidate for a "heart attack." Heart action lessens, with fall in cardiac output. Shortness of breath may develop as a result of low cardiac output and low pulmonary perfusion. Hypertension is common.

At the CELLULAR LEVEL, there is diminution of DNA and RNA synthesis, with failure to replace worn-out body proteins.

The ALIMENTARY TRACT undergoes changes in function to a greater degree than changes in structure. Over one-half (56 percent) of aged

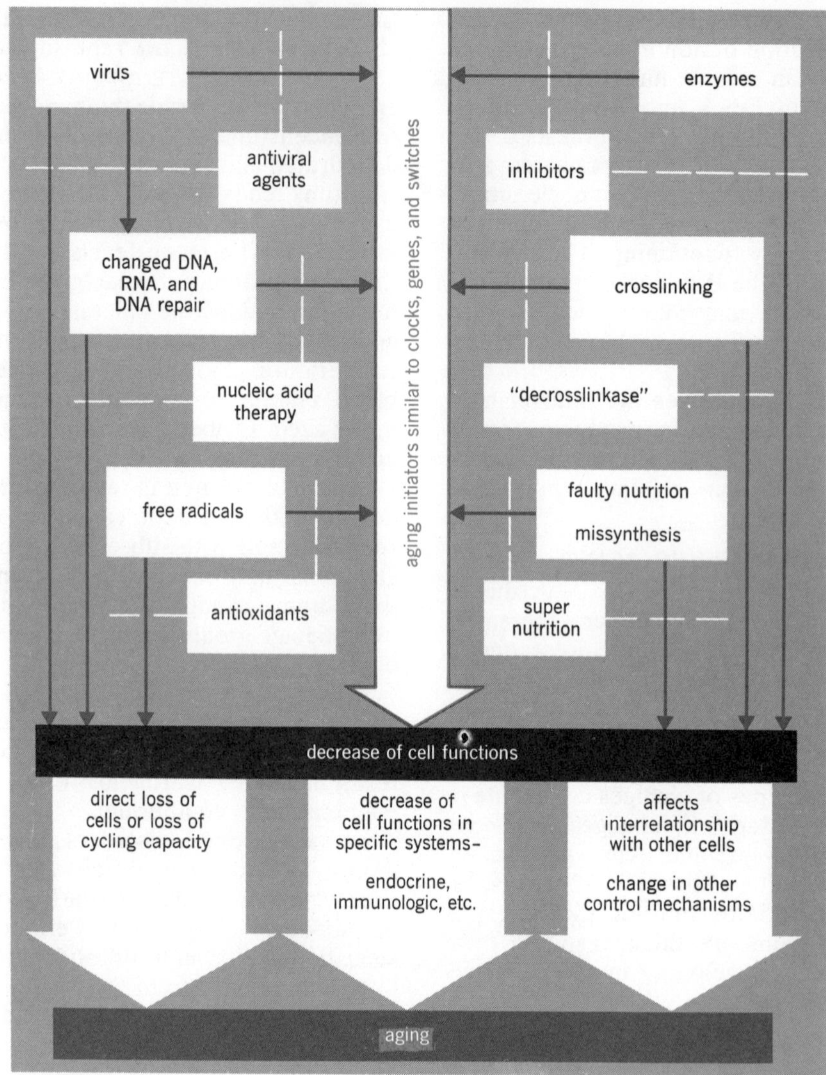

FIGURE 28.24
A scheme of causes of aging. (From "The Combination Theory of Aging," by H. J. Kugler, AMERICAN LABORATORY, November 1976, p. 24.)

persons show functional disorders such as heartburn, belching, nausea, diarrhea, constipation, and flatus. Many of these complaints have an emotional basis, based on fear of death and disease and loss of contact with offspring. An organic basis may be demonstrated in some cerebral arteriosclerosis patients. Malignancy of the lower tract is more common in elderly persons (11 percent).

RENAL DISORDERS include lowered filtration rates, reduced reabsorption, and renal hypertension. Pyelonephritis (infection) is the most common renal disease. Calculi (stones) in the kidney and ureter are also more common in the aged.

GONADAL FUNCTION decreases. The ovary atrophies, producing menopause in females; testicular function declines more slowly. Internal organs atrophy and often prolapse ("drop down"), as with the uterus entering the upper vagina, for example.

Loss of the inorganic component of the SKELETON produces osteoporosis and greater liability to fracture. The MUSCLES become smaller and weaker and are prone to cramps and effects of altered electrolyte balance (hypokalemia, low blood potassium).

ARTICULATIONS are affected by arthritis. Osteoarthritis is a noninflammatory condition in which joints undergo degenerative changes in cartilage and synovial membranes. Rheumatoid arthritis is typically inflammatory in nature. Gouty arthritis is a metabolic disorder in which uric acid crystals accumulate in the joints.

ENDOCRINES may increase activity (as in acromegaly) or decrease activity (diabetes, hypothyroidism). Gonadotropins appear to show the widest secretion ranges.

The CENTRAL NERVOUS SYSTEM shows the greatest change in the brain where neuron loss (estimated at 30 cells/min from 30 years onward) may cause increased central reaction times, loss of memory, and personality alterations.

Pathologic changes. The DISEASES OF OLD AGE are too numerous to consider in this text. Suffice it to say that the aged person is more susceptible to the diseases we all are heir to, and that the presence of a disease process tends to accelerate the aging changes described earlier.

DEATH

Although old age itself does not cause death, organs and systems age at different rates. If the function of a vital organ drops below the critical level for maintenance of life, life ceases. The operation of bodily systems is interrelated so that if one vital organ fails, all fail. "The chain is only as strong as its weakest link." Loss of recovery ability is obvious, so that alterations that would be rapidly returned to normal by a younger person become life-threatening to the aged.

Definitions of death are extremely varied, and a strict definition is almost impossible. Among the definitions considered are the following.

Total, irreversible cessation of cerebral function (as evidenced by a "flat" electroencephalogram) for at least 24 hours; no spontaneous heartbeat or respiration; permanent cessation of all vital functions. At one time, stoppage of kidney function, cardiac standstill, or stoppage of respiration may have been considered signs of death. The use of an artificial kidney, respirators, and heart-lung machines has caused a reevaluation of the definition of death, as has the problem of securing organs from a donor for transplant. At the present time, the spontaneous brain-activity criterion appears to be the most valid criterion of death and one to which specific legal definition has advanced in some states.

Although old age and eventual death are in every person's future, efforts must be made by all concerned not only to add years to life, but also life to years. Age alone should not determine when one ceases or is forced to cease as a productive member of society. It is becoming increasingly clear that care of the body must begin with conception, and that care should cease only with the death of a given individual. Prevention rather than treatment of disease and disorders should become the final goal for each of us.

SUMMARY

1. Ovum and spermatozoa production is necessary for creation of a new individual.
   a. Ova are present in the ovary before birth and begin to ripen at 9 to 12 years of age in most females. One functional haploid cell, having a chromosome complement of 22 + X results from maturation of a single primitive ovum.
   b. Spermatozoa are produced at 13 to 15 years of age in most males. Four functional haploid cells are produced from each primitive cell; one-half will contain a chromosome complement of 22 + X; one-half will contain 22 + Y.
   c. At fertilization, normal chromosome number is restored. A male results from the combination of 44 XY; a female from 44 XX.

2. The role of the male in sexual intercourse is basically one of insemination of the female.
   a. Erection of the penis occurs by vascular changes and ejaculation causes passage of sperm through the ducts and urethra to the exterior.
   b. Semen includes sperm and prostatic and seminal vesicle secretions. A normal ejaculate has a volume of 2 to 5 ml, a pH of 7.2 to 8.0, and contains 60 to 150 million sperm per ml.
   c. Semen provides for neutralization of acidity, suspension and nourishment of sperm, and sperm activation.
3. The role of the female in sexual intercourse is basically one of providing a reception area for sperm, an area for fertilization, and an area for development and nutrition of a new individual.
4. Venereal diseases are sexually or orally transmitted.
   a. Several possible reasons for the epidemic increase of venereal diseases are advanced.
   b. The characteristics and effects of gonorrhea and syphilis are presented (Table 28.1).
   c. Other microorganism-related, sexually transmitted diseases are described, including urethritis, granuloma and lymphogranuloma venereum, and herpes.
5. Development from conception to birth is described.
   a. The events of the first week include fertilization, cleavage, morula formation, and blastocyst formation.
   b. The second week sees implantation, formation of a two layered embryo with ectoderm and endoderm.
   c. The third week is associated with mesoderm formation and the commitment to specific tissue formation by the germ layers (Table 28.2).
   d. The fourth week establishes basic organ structure.
   e. The fifth through eighth weeks see human form established with systems organized.
   f. The ninth week to birth is a time of growth and refinement of form and function.
6. Embryonic and fetal development is influenced by maternal nutrition, placental status, genetic makeup of the individual, and insults delivered to the offspring.
   a. The fetal membranes protect, aid in formation of fetal blood vessels, and contribute to formation of the placenta.
   b. A normal placenta, in terms of size, texture, and number of sections, ensures adequate passage of nutrients to the new individual, and elimination of its wastes.
   c. Genetic influences determine direction and rate of development.
   d. Physical, chemical, and mechanical insults may influence the course of development.
7. Birth refers to appearance of the offspring in the world.
   a. Labor refers to effacement and dilation of the uterus, and the uterine contractions that expel the fetus.
   b. Dangers associated with birth include fetal asphyxia, breech presentation, or prolonged labor.
8. At birth, physiological status of the newborn may be assessed by the Apgar system that rates appearance, pulse rate, response to stimulation, activity, and respiration. High scores are associated with good infant survival.
9. The individual passes through several developmental stages after birth.
   a. Neonatal period: birth to four weeks.
   b. Infancy: 4 weeks to 2 years.
   c. Childhood: 2 years to puberty.
   d. Adolescence: puberty to maturity.
   e. Maturity: assumption of optimum function, mentally and physically.
   f. Senescence and old age: from maturity to death.
10. Growth is characterized by increase in body size.
   a. General growth displays a sigmoid-shaped curve.
   b. Growth is regulated by the same kinds of factors that govern prenatal development (nutrition, genetics, drug intake).
   c. The newborn averages 7.5 pounds and 20 inches long. Temperature regulation, kidney function, and immune response is immature.
   d. Growth during infancy and childhood increases height, weight and head circumference (Figs. 28.15 to 28.18).
   e. Adolescence is associated with assumption of gamete production and secondary sex characteristic appearance.
   f. The menstrual cycle is described according to its phases.
   g. Maturity sees integration and optimum physiological function.
   h. Old age is associated with deterioration of function.

11. Aging is explained by internal and external theories of cause.
    a. Internal theories involve changes in nucleic acid synthesis and repair, limits to cell division, chemical disorders, and cross-linking of nucleic acids and collagen.
    b. External theories involve physical agents, disease producing organisms and faulty nutrition.
12. Changes during aging involve
    a. Secular changes: vascularity falls below tissue demand.
    b. Senescent changes: changes in individual organs that compromise their function.
    c. Pathologic changes: diseases threaten the old to a greater degree than the young.
13. Death is perhaps best indicated by loss of brain function.
14. Prevention of change, slowing rate of change, and adding life to years should become the goal for a rewarding as well as longer life.

QUESTIONS

1. In what ways do the development of ova and sperm differ? Why is there only one functional ovum developed but four sperm per primitive cell? (*Clue:* How is cleavage sustained? What are some hazards sperm face in the female?)
2. What do the mature male and female contribute to creation and sustenance of a new individual?
3. Why do you suppose VD has reached epidemic proportions? What do you think could be done to reduce the incidence of the diseases, or to discover undiagnosed cases?
4. Compare the effects of untreated gonorrhea and syphilis on the body.
5. Give a "weekly summary" of developmental changes from conception to birth.
6. What are some of the hazards of birth?
7. What kind of an "Apgar score" would you give a newborn whose heart rate was below 100, who showed some movement, blue extremities, and irregular breathing? What does the rating indicate?
8. As you did for prenatal growth and development, list the "stages of life" after birth, and characterize each according to primary physiological changes.
9. Which theory of aging appeals to you? Defend your choice.
10. Pick three body systems and describe some aging changes that occur.
11. What would you do or not do to yourself or others, to ensure a long, productive, and rewarding life?

READINGS

Bower, T. G. R. "Repetitive Processes in Child Development." *Sci. Amer. 235:*38. (Nov) 1976.

Burnet, F. M. *Immunology, Aging, and Cancer.* Freeman. San Francisco, 1976.

Culliton, Barbara J. "Amniocentesis: HEW Backs Test for Prenatal Diagnosis of Disease." *Science 190:*537. 7 Nov 1975.

Culliton, Barbara J. "Penicillin-Resistant Gonorrhea: New Strain Spreading Worldwide." *Science 194:*1395. 24 Dec 1976.

Hayflick, Leonard. "The Cell Biology of Human Aging." *New Eng. J. Med. 295:*1302. Dec 2, 1976.

Kaufman, Richard E., et al. "Gonorrhea Therapy Monitoring: Treatment Results." *New Eng. J. Med. 294:*1. Jan 1, 1976.

Kolata, Gina Bari. "Gonorrhea: More of a Problem but Less of a Mystery." *Science 192:*244. 16 April 1976.

Marx, Jean L. "Cytomegalovirus: A Major Cause of Birth Defects." *Science 190:*1184. 19 Dec 1975.

Rabkin, Judith G., and Elmer L. Struening. "Life Events, Stress, and Illness." *Science 194:*1013. 3 Dec 1976.

Rafferty, Keen A. Jr. "Herpes Virus and Cancer." *Sci. Amer. 229:*26. (Oct) 1973.

Scientific American. (Entire issue on Medicine and Life of Man.) (Sept) 1973.

Segal, Sheldon J. "The Physiology of Human Reproduction." *Sci. Amer. 231:*52. (Sept) 1974.

Sobell, Henry M. "How Actinomycin Binds to DNA." *Sci. Amer. 231:*82. (Aug) 1974.

Whitley, Richard J., et al. "Adenosine Arabinoside Therapy of Biopsy-Proved Herpes Simplex Encephalitis." *New Eng. J. Med. 297:*289. Aug. 11, 1977.

GENERAL REFERENCES

Allen, Garland. *Life Sciences In the Twentieth Century*. Wiley. New York, 1975.

Arey, L. B. *Developmental Anatomy*. 7th ed. Saunders. Philadelphia, 1974.

Barr, Murray L. *The Human Nervous System*. Harper & Row. New York, 1974.

Beeson, Paul B., and Walsh McCermott (Editors). *Textbook of Medicine*. 14th ed. Saunders. Philadelphia, 1975.

Bhagavan, N. V. *Biochemistry, A Comprehensive Review*. Lippincott. Philadelphia, 1974.

Bloom, William, and Don W. Fawcett. *A Textbook of Histology*. 10th ed. Saunders. Philadelphia, 1976.

Cheek, Donald B. *Fetal and Postnatal Cellular Growth*. Wiley. New York, 1975.

Diem, K., and C. Lentner (Editors). *Scientific Tables*. 7th ed. J. R. Geigy. Basle, Switzerland, 1970.

Gray, Henry. *Anatomy of the Human Body*. Edited by C. M. Goss. 29th ed. Lea & Febiger. Philadelphia, 1973.

Guyton, Arthur C. *Textbook of Medical Physiology*. 5th ed. Saunders. Philadelphia, 1976.

Moore, Keith L. *Before We Are Born: Basic Embryology and Birth Defects*. Saunders. Philadelphia, 1974.

Moore, Keith L. *The Developing Human: Clinically Oriented Embryology*. Saunders. Philadelphia, 1973.

Mountcastle, Vernon B. (Ed.). *Medical Physiology*. Vols I and II. Mosby. St. Louis, Mo., 1974.

Netter, Frank H. *Ciba Collection of Medical Illustrations*. Vol. I. *Nervous System*. Vol. II. *Reproductive Systems*. Vol. III. (Parts I, II, III). *Digestive System*. Vol. IV. *Endocrine System and Selected Metabolic Diseases*. Vol. V. *Heart*. Ciba Pharmaceutical Products. Newark, N. J.

Ruch, Theodore C., and Harry D. Patton. *Physiology and Biophysics*. Vol. I. *The Nervous System*. 1977. Vol. II. *Circulation, Respiration, and Fluid Balance*. 1974. Vol. III *Digestion, Metabolism, Endocrine Function, and Reproduction*. 1973. Saunders. Philadelphia.

Scientific American—Readings. *Human Physiology and the Environment in Health and Disease*. Freeman. San Francisco, 1976.

Smith, Clement A., and Nicholas M. Nelson. *The Physiology of the Newborn Infant*. Thomas. Springfield, Ill., 1976.

Stryer, Lubert. *Biochemistry*. Freeman. San Francisco, 1975.

Timiras, Paola S. *Developmental Physiology and Aging*. Macmillan. New York, 1972.

Vaughn, Victor C., and R. James McKay. *Nelson Textbook of Pediatrics*. 10th ed. Saunders. Philadelphia, 1975.

APPENDIX

CALCULATION OF OSMOTIC PRESSURE

osmotic pressure in atmospheres $= P_{id} = 0.082055 \times T \times \frac{M}{V_m} \times m_2 \times \nu$

where

0.082055 = the gas constant R in liter atmospheres

T = absolute temperature in kelvin (K) = 273.15 + °C

$\frac{M}{V_m}$ = ratio of molar mass to molar volume for water (usually = 1)

m_2 = molality of the solute (number of moles of undissociated solute per 1000 g water)

ν = number of particles into which the solute dissociates

NERNST EQUATION

$$E = \frac{RT}{FZ} \log e \frac{[\text{ion}]_o}{[\text{ion}]_i}$$

where

E = potential in millivolts

R = the gas constant (8.317 joules/degree mole)

T = temperature in absolute degrees (273 + degrees centigrade)

F = the Faraday (96,520 coulomb/mole)

Z = the ion valence

$[\text{ion}]_o$ = ion concentration outside the cell

$[\text{ion}]_i$ = ion concentration inside the cell

Note: joule = volt coulomb, thus answer will be in volts (joule/coulomb).

terms and symptoms of drug abuse

This chart indicates the most common symptoms of drug abuse. However, all of the signs are not always evident, nor are they the only ones that may occur. The reaction produced by any drug will usually depend on the person, his mood, his environment, the dosage of the drug, and how the drug interacts with other drugs the abuser has taken, or contaminants within the drug. (Reprinted from "Drugs of Abuse," U.S. Government Printing Office, 1972. Courtesy of the Drug Enforcement Administration.)

	SLANG TERMS	DROWSINESS	EXCITATION & HYPERACTIVITY	IRRITABILITY & RESTLESSNESS	BELLIGERENCE	ANXIETY	EUPHORIA	DEPRESSION	HALLUCINATIONS	PANIC	IRRATIONAL BEHAVIOR	CONFUSION	TALKATIVENESS	RAMBLING SPEECH	SLURRED SPEECH	LAUGHTER
MORPHINE	M, dreamer, white stuff, hard stuff, morpho, unkie, Miss Emma, monkey, cube, morf, tab, emsel, hocus, morphie, melter	●		●		● ●	●	●	●	●		●			●	
HEROIN	Snow, stuff, H, junk, big Harry, caballo, Doojee, boy, horse, white stuff, Harry, hairy, joy powder, salt, dope, Duige, hard stuff, schmeek, shit, skag, thing,	●		●		●	●	●		●		●			●	
CODEINE	Schoolboy	●		●		● ●	●	●		●		●			●	
HYDROMORPHONE	Dilaudid, Lords	●		●		● ●	●	●		●		●			●	
MEPERIDINE	Demerol, Isonipecaine, Dolantol, Pethidine	●		●		● ●	●	●	●	●		●			●	
METHADONE	Dolophine, Dollies, dolls, amidone	●		●		● ●	●	●		●		●			●	
EXEMPT PREPARATIONS	P.G., P.O., blue velvet (Paregoric with anthistamine), red water, bitter, licorice	●		●		●	●	●		●		●				
COCAINE	The leaf, snow, C, cecil, coke, dynamite, flake, speedball (when mixed with Heroin), girl, happy dust, joy powder, white girl, gold dust, Corine, Bernies, Burese, gin, Bernice, Star dust, Carrie, Cholly, heaven dust, paradise		●	●		●	●		●				●			
MARIHUANA	Smoke, straw, Texas tea, jive, pod, mutah, splim, Acapulco Gold, Bhang, boo, bush, butter flower, Ganja, weed, grass, pot, muggles, tea, has, hemp, griffo, Indian hay, loco week, hay, herb, J, mu, giggles–smoke, love weed, Mary Warner, Mohasky, Mary Jane, joint sticks, reefers, sativa, roach,	●	●	●		●	●	●	●	●			●			●
AMPHETAMINES	Pep pills, bennies, wake–ups, eye–openers, lid poppers, co–pilots, truck drivers, peaches, roses, hearts, cart–wheels, whites, coast to coast, LA turnabouts, browns, footballs, greenies, bombido, oranges, dexies, jolly–beans, A's, jellie babies, sweets, beans, uppers		●	●		●	●		●	●			●			
METHAMPHETAMINE	Speed, meth, splash, crystal, bombita, Methedrine, Doe		●	●		●	●		●	●			●			
OTHER STIMULANTS	Pep pills, uppers		●	●			●		●				●			
BARBITURATES	Yellows, yellow jackets, nimby, nimbles, reds, pinks, red birds, red devils, seggy, seccy, pink ladies, blues, blue birds, blue devils, blue heavens, red and blues, double trouble, tooies, Christmas trees, phennies, barbs	●		● ●	●	●	●	●	●		●	●			●	●
OTHER DEPRESSANTS	Candy, goofballs, sleeping pills, peanuts	●		● ●	●	●	●	●	●		●	●			●	
LYSERGIC ACID DIETHYLAMIDE (LSD)	Acid, cubes, pearly gates, heavenly blue, royal blue, wedding bells, sugar, Big D, Blue Acid, the Chief, the Hawk, instant Zen, 25, Zen, sugar lump		●			●	●	●	●	●	●			●		
STP	Serenity, tranquility, peace, DOM, syndicate acid						●	●	●		●			●		
PHENCYCLIDINE (PCP)	PCP, peace pill, synthetic marihuana	●				●				●		●				●
PEYOTE	Mescal button, mescal beans, hikori, hikuli, huatari, seni, wokowi, cactus, the button, tops, a moon, half moon, P, the bad seed, Big Chief, Mesc.		●	●		●			●					●		
PSILOCYBIN	Sacred mushrooms, mushrooms		●	●		●	●	●	●		●			●		
DIETHYLTRYPTAMINE (DMT)	DMT, 45–minute psychosis, business–man's special		●			●	●	●	●		●			●		

● SYMPTOMS OF ABUSE

TREMOR	STAGGERING	IMPAIRMENT OF COORDINATION	DIZZINESS	HYPERACTIVE REFLEXES	DEPRESSED REFLEXES	INCREASED SWEATING	CONSTRICTED PUPILS	DILATED PUPILS	UNUSUALLY BRIGHT SHINY EYES	INFLAMED EYES	RUNNY EYES AND NOSE	LOSS OF APPETITE	INCREASED APPETITE	INSOMNIA	DISTORTION OF SPACE OR TIME	NAUSEA AND VOMITING	ABDOMINAL CRAMPS	DIARRHEA	CONSTIPATION	PHYSICAL DEPENDENCE	PSYCHOLOGICAL DEPENDENCE	TOLERANCE	CONVULSIONS	UNCONSCIOUSNESS	HEPATITIS	PSYCHOSIS	DEATH FROM WITHDRAWAL	DEATH FROM OVERDOSE	POSSIBLE CHROMOSOME DAMAGE	ORALLY	INJECTION	SNIFFED	SMOKED
●		●			●	●	●	●			●	●		●		●	●	●	●	●	●	●	●	●	●			●			●		
●		●			●	●	●	●			●	●		●		●	●	●	●	●	●	●	●	●	●			●			●	●	
●		●			●	●	●	●			●	●		●		●	●	●	●	●	●	●			●			●		●	●		
●		●			●	●	●	●			●	●		●		●	●	●	●	●	●	●		●	●			●		●	●		
●		●			●	●	●	●			●	●		●		●	●	●	●	●	●	●		●	●			●		●	●		
●		●			●	●	●	●			●	●		●		●	●	●	●	●	●	●		●	●			●		●	●		
●		●			●	●	●	●			●	●		●		●	●	●	●	●	●	●			●					●	●		
●				●				●				●		●							●		●		●			●			●	●	
		●								●			●		●						●									●			●
●			●	●		●		●	●			●		●							●	●			●	●		●		●	●		
●			●	●		●		●	●			●									●	●			●	●		●		●	●		
●				●				●													●	●			●			●		●	●		
●	●	●			●	●	●							●		●	●			●	●	●	●	●	●		●	●		●	●		
●	●	●			●									●		●	●			●	●	●	●		●			●		●	●		
●						●		●							●						●	●				●			●	●			
●								●							●						●				●					●	●		
		●	●																		●			●	●					●	●		
								●								●					●	●			●					●	●		
						●		●							●						●	●								●			
						●		●							●						●	●			●						●		●

● SYMPTOMS OF WITHDRAWAL ● DANGER OF ABUSE ● HOW TAKEN

GLOSSARY

A

a-, an-: prefix meaning without, away from, not.

abducent: to lead away from; the 6th cranial nerve serving the lateral rectus muscle of the eye.

aberration: deviation from normal; an imperfect eye.

ablate (ablation): to remove; removal.

absolute: considered by itself with no reference to anything else.

absorption: taking in, usually across a membrane.

accelerate: to increase speed of something.

acceleration: a change in speed.

acceptors: substances taking up hydrogen ion.

accessory: auxiliary or assisting; the 11th cranial nerve supplying the trapezius and sternocleidomastoid muscles and pharynx.

acclimatization: becoming accustomed to a different environment.

acetylcholine: an ester of choline serving as a synaptic transmitter; destroyed by cholinesterase.

acetyl coenzyme A: Ac CoA; a compound formed from all three basic foodstuffs, which is the compound combusted by the Krebs cycle.

accommodate: to adjust or adapt; decrease in nerve impulse discharge with continued stimulation.

achalasia: failure of sphincters to relax.

acid: any substance releasing H ion and neutralizing basic substances.

acidosis: lowering of pH of extracellular fluid caused by excess acid or loss of base.

acro- (acros): extremity or top.

actin: a protein of muscle composed of globular units in a double helix; is drawn together when muscle contracts.

activate: to make active.

activators: substances converting inactive materials to active ones, or causing activity.

active: functioning, working, doing something.

active transport: using carriers, energy, and enzymes of cell to cause a substance to cross a membrane.

actual: real or existing.

acuity: keenness (as of vision).

acute: sharp, severe; of rapid onset and short course.

adapt: to adjust, particularly to an environmental change, so as to maintain homeostasis.

adaptation: to adjust; change in pupil size with light intensity.

adeno-: prefix denoting a gland.

adenoma: a neoplasm of glandular epithelium.

adequate: equal to need; as, an adequate stimulus.

adipose: fatty; pertaining to fat.

adnexa: accessory structures or ones associated with a neighboring structure.

adolescence: the period of life from the beginning of puberty to maturity.

ADP: adenosine diphosphate; nitrogenous base, ribose sugar and two phosphoric acid radicals.

adrenergic: nerve fibers releasing norepinephrine or epinephrine at their endings; includes all sympathetic nerves except those to sweat glands.

adsorption: "sticking together" of a gas, liquid, or dissolved substance on a surface; no bonding.

aerobic: requiring the presence of oxygen.

afferent: carry something toward a center or toward a specific reference point.

agglomerate: to form a mass.

agglutinate: to cause a clumping or coming together.

agglutination: coming together (as of red cells) to form a mass.

agglutinin: an antibody of normal or immune serum causing clumping of antigens usually carried on red blood cells.

agglutinogen: a substance stimulating the production of a specific agglutinin (an antigen of the blood groups).

agnosia: inability to understand sensations although no loss of receptor function can be demonstrated.

alexia: word "blindness" due to a central lesion; a form of aphasia.

-algia: suffix denoting pain.

alimentary: pertaining to nutrition.

alkali: a substance that can neutralize acids; combines with acid to form soaps; basic in reaction (pH).

alkalosis: rise of pH of extracellular fluid caused by excess base or loss of acid.

allantois: one of the fetal membranes growing from the hindgut; its blood vessels become umbilical arteries and vein.

allele: one of two or more genes occurring at the same position on a specific pair of chromosomes and controlling the expression of a given characteristic.

allergy: a condition characterized by reaction of body

tissues to specific antigens without production of immunity.

alpha: the first letter of the Greek alphabet; implies the first carbon beyond a carboxyl group in a carbon chain such as in a fatty acid.

alveoli: a little hollow; air sacs of the lungs or sockets of the teeth.

ambient: encompassing on all sides; surrounding.

ameboid: like an ameba; ameboid movements imply cellular motion by use of pseudopods.

amenable: responsive; responsible.

amine: a nitrogen containing organic compound with one or more hydrogens of ammonia replaced by organic radicals.

amino acid: a compound containing both an amine group ($—NH_2$), and a carboxyl group ($—COOH$). Two groups: essential, not synthesized in body or not produced in amounts necessary for normal function; nonessential, synthesized in body.

amnion: a fetal membrane enclosing the fluid in which the embryo/fetus floats *in utero*.

AMP: adenosine monophosphate; nitrogenous base, ribose sugar, one phosphoric acid radical.

amphoteric: having properties of both an acid and a base.

amplitude: extent of movement; intensity.

ampulla: a sac-like enlargement of a tube or duct.

amygdaloid: almond-shaped; a nucleus of the cerebrum belonging to the limbic system.

anabolism: synthesis or constructive chemical reactions.

anaerobic: not requiring oxygen or air to exist or function.

analgesia: absence of the sense of pain.

analyze: to separate into parts to determine the function or meaning of the whole.

anaphase: stage in mitosis characterized by separation and movement of chromosomes toward poles of dividing cell.

anaphylaxis: a condition resulting from an antigen-antibody reaction. Characterized by circulatory collapse and respiratory difficulty.

androgen: substance producing or stimulating male characteristics (e.g., the male sex hormone, testosterone).

anemia: a condition where there is reduction of circulating red blood cells or of hemoglobin.

anesthesia: partial or complete loss of sensation.

angstrom: a metric unit; one ten-millionth of a millimeter.

anhydrous: containing no water.

anion: a negatively charged ion.

aniso-: prefix denoting abnormal shape.

annual: yearly.

annulus: a ring-shaped structure.

anomaly: an organ or structure that is abnormal as to form or structure.

anorexia: loss of appetite.

anti-: prefix meaning against.

antibody: protein (usually a globulin) developed against a specific foreign substance (antigen).

anticoagulant: a substance preventing blood coagulation or clotting.

anticodon: the three bases in the terminal loop of transfer RNA that codes for a particular amino acid and its position of attachment on messenger RNA.

antigen: a substance causing production of antibodies.

antigenic: capable of causing production of antibodies.

antitoxin: an antibody capable of neutralizing a specific chemical (toxin) produced by microorganisms.

apex: summit or peak; the narrow portion or tip of an organ shaped in pyramidal form.

aphasia: inability to express oneself properly through speech; loss of verbal understanding.

apnea: stoppage or cessation of breathing.

apneustic: sustained respiratory inspiratory effort.

apo-: prefix denoting separation or away from.

apoferritin: a cellular protein combining with ferric iron (Fe^{+++}) to form ferritin.

apraxia: inability to perform purposeful movements in coordinated fashion.

aqueous: watery; having the nature of water.

arachnoid: spiderlike; the middle of the three meninges around brain and spinal cord.

arcuate: shaped like a bow or arch.

argent (argentaffin): pertaining to silver (*Ag*); taking a silver-containing stain.

arrhythmia: lack of rhythm; irregularity.

arterial: pertaining to the arteries.

arterio-: prefix pertaining to arteries.

arthro-: prefix pertaining to the joints.

ascending: going upwards or higher, or in a superior direction as a sensory nerve tract going up the spinal cord to the brain.

ascites: accumulation of fluid in the abdominal cavity.

-ase: suffix used in forming the name of an enzyme.

asphyxia: lack of oxygen and excess of carbon dioxide in body due to some interference with breathing.

association: joining together, as a neuron between sensory and motor neurons.

asthenia: lack of muscular or mental strength.

astigmia (astigmatism): imperfect vision due to different curvatures in the horizontal or vertical meridians of the lens or cornea.

astro-: prefix indicating star-shaped.

ataxia: muscular incoordination.
atelectasis: collapse of lung alveoli; failure of lung to expand at birth.
athero-: prefix meaning porridge; used to refer to deposition of fatty substances in an organ wall.
athetosis: slow, repeated, wormlike movements of a body part.
atmosphere: the gases surrounding the earth; the pressure exerted by those gases at sea level, equal to about 15 lb/square in.
atopic: misplaced; out of normal position.
ATP: adenosine triphosphate; nitrogenous base, ribose sugar, and three phosphoric acid radicals.
atresia: pathological closure of a normal opening; degeneration of ovarian follicles.
atrium: a cavity; one of the two upper chambers of the heart.
atrophy: wasting of something, with decrease in size.
attenuated: diluted; decreased in strength (as an attenuated virus; one that is less virulent).
auditory: referring to the sense of hearing.
augment: to add to or increase.
auricle: the outer ear flap on the side of the head; a small conical portion of the upper chambers of the heart. Should NOT be used synonymously with atrium.
auto-: prefix meaning self.
autoimmune: an immune response to one's own chemicals.
autonomic: self controlling; refers to the involuntary or self-regulating portion of the peripheral nervous system.
autoregulation: control exerted by means of mechanisms within the organ controlled.
autosome(s): any of the chromosome(s) other than the sex chromosomes; 22 pairs in the human.
-axie: suffix denoting value or strength of something.
axoplasm: the cytoplasm of the axon, or efferent process of a neuron.

B

baro-: prefix referring to pressure or stretch.
baroreceptors: receptor organs sensitive to pressure or stretching.
basal: the base of an object; of primary importance; the lowest value.
base: lower or broader part of anything.
benign: mild; not malignant.
beta: second letter of the Greek alphabet; denotes a carbon second from the carboxyl carbon in a molecule containing a series of carbons (e.g., fatty acid); cells staining with basophilic dyes.
beta-oxidation: metabolic scheme that catabolizes fatty acids by removal of two carbon units.
bi-: prefix meaning two or double.
bilirubin: the orange or yellow pigment in bile; derived from breakdown of heme.
biliverdin: the green pigment of bile formed by oxidation of bilirubin.
-blast: suffix meaning immature or primtive.
blastocyst: stage in development of a mammalian embryo consisting of a hollow structure containing an inner mass of cells.
blood pressure: as commonly used, the pressures measured on a peripheral artery during contraction and relaxation of the heart.
bolus: a mass of food ready for swallowing.
bone marrow: tissue within certain bones of the skeleton responsible for production of certain blood cells.
booster: an additional dose of an immunizing agent.
brady-: prefix meaning to slow.
branchial: pertaining to gills.
breathe: to take in and release air from the lungs and respiratory system.
breathing: respiration; the act of taking in and releasing air from the lungs and respiratory system.
bronchial: pertaining to the lung tubes known as bronchi or bronchioles.
Brown-Sequard: Charles E. Brown-Sequard, French physician (1818–1894); associated with his syndrome occurring when spinal cord is cut halfway through.
buffer: a substance preserving pH upon addition of acids or bases.
buffer pair or buffer system: the combination of a weak acid and its conjugate base, which work together to maintain pH nearly constant.
bulk: size, mass, or volume.
bulk flow: movement together of water and associated solutes.

C

calorie: a unit of heat; amount of heat required to raise temperature of 1 gram of water 1°C.
Calorie: a large calorie or kilocalorie; 1000 times as large as a calorie; amount of heat required to raise temperature of 1 kg of water 1°C.
calorigenic: causing an increase in heat production.
canaliculus: a tiny canal or channel.
carbaminohemoglobin: the compound formed by the reaction of carbon dioxide and hemoglobin.
carbohydrate: a class of organic compounds composed of C, H, and O, with H and O in the ratio of 2:1; sugars and starches.
carbonic anhydrase: an enzyme catalyzing the re-

action of CO_2 and H_2O to form carbonic acid (H_2CO_3) and vice versa.

carcinogens: substances known to cause cancer or neoplasms.

carcinoma: a malignant cancer or neoplasm of epithelial tissues.

cardio-: prefix pertaining to the heart.

carotid: (*karos,* deep sleep) referring to the arteries of the neck and their immediate branches.

carrier: a large molecule transporting substances in active transport across cell membranes.

catabolism: destructive or energy releasing chemical reactions.

catalyst: a substance that alters the rates of chemical reactions without being altered in the process.

cataract: opacity of the cornea or lens.

catecholamine: biologically active molecules derived from the amino acid tyrosine; (e.g., epinephrine, norepinephrine, serotonin).

cation: a positively charged ion.

caudal: inferior in position; pertaining to the tail or a tail-like structure.

-cele: suffix denoting a swelling or cyst.

celiac: pertaining to the abdominal region.

center: usually refers to a group of nerve cells in the brain or spinal cord. It controls some specific activity.

central: refers to the center; a more important part, as central nervous system.

centrifugal: away from the center.

centrifuge: a machine designed to separate parts by spinning to create many times the force of gravity; to subject something to spinning to cause separation of parts.

centrioles: a cell organelle that is involved in cell division. Forms a spindle.

centripetal: toward the center.

centro-: pertaining to the center or middle portion of something; e.g., *centro*mere, the clear region of a chromosome where its arms join.

-cephalon: suffix pertaining to the head.

cerebrum (-cerebrate): the upper largest part of the brain.

cerumen: the waxlike secretion of glands in the external auditory canal.

chemo-: prefix referring to chemical substances.

chemoreceptors: receptors sensitive to chemical change.

chiasm: a crossing.

chloride shift: the movement of chloride ions into red blood cells as bicarbonate ions are formed and leave the cell during reaction of CO_2 and H_2O.

cholesterol: a sterol forming the basis of many lipid hormones in the body. Formula $C_{27}H_{45}OH$.

cholinergic: nerve endings liberating acetylcholine.

chorea: rapid, irregular, and wormlike movements of body parts.

chorion: a fetal membrane derived from the outer layer of the blastocyst and contributing to the formation of the placenta.

chromaffin: staining with chromium salts; cells of the adrenal medulla.

chromatid: one of the two bodies resulting from longitudinal separation of duplicated chromosomes.

chromatin: darkly staining substance within the nucleus of most cells. Consists mainly of DNA.

chromo-: prefix meaning colored or stained.

chromosome: a microscopic J- or V-shaped body developing from chromatin during cell division. Contains the hereditary determinants of body characteristics.

chrono-: prefix denoting time.

chronaxie: a value expressing sensitivity of nerve fibers to stimulation. It is the time that a current that is two times the threshold value must last to cause depolarization.

chronic: long drawn out; long duration.

chyle: the contents of intestinal lymph vessels. Consists mainly of absorbed products of fat digestion.

chylomicron: microscopically visible fat droplet found in the blood after absorption of fats in foods.

chyme: the mixture of partially digested foods and digestive juices found in the stomach and small intestine.

ciliary: pertaining to the eyelashes or any hairlike process.

circa-: about or approximately.

circadian: describes events that repeat in a length of time approximating a day (23 to 25 hours).

circulate (circulation): to cause to move in a circular course; movement in a circular course.

cistern: a cavity or enlargement in a tube or vessel.

clearance: elimination of a substance from the plasma by the kidney.

cleavage: cell division after fertilization; to split a molecule into smaller parts.

clone: a group of cells descended from a single cell; implies identical form and function to the original.

clot: the semisolid mass of fibrin threads plus trapped blood cells, which forms when the blood coagulates; to coagulate.

clump: to form a mass; synonym for agglutination.

Co A: a dinucleotide that activates many substances in metabolic cycles.

coagulation: the process by which liquid blood is changed into a gel to aid in preventing blood loss through injured vessels.

cochlea: referring to the cochlea or organ of hearing in the inner ear.
code: a set of symbols or rules used to specify meaning; to put into the form of a code.
codominant: in inheritance of characteristics, two genes exerting equal influence on expression of that character.
codon: the sequence of three nitrogenous bases of messenger RNA that specifies a given amino acid and its position in a protein.
cofactor: a substance essential to or necessary for the operation of another material such as an enzyme.
cognition: awareness; having understanding and memory.
collagen: a fibrous protein forming the main organic structures of connective tissues.
collateral: a side branch of a nerve or blood vessel; side by side or close to.
colliculus: a little mound or elevation; one of four elevations on the dorsal or back surface of the midbrain.
colloid: a system formed by large particles that remain suspended and do not settle. Size of particles lies between 1 and 100 nanometers.
coma: a deep stupor or state of unconsciousness from which external stimuli cannot arouse the patient.
combust: to burn or oxidize.
commissure: a band of nerve fibers crossing the midline of the central nervous system.
compartmentalize: to divide into noninterfering areas.
compensatory: serving to balance or offset; returning to a normal state.
competition: a striving for superiority; e.g., two substances trying for the same enzyme or transport system.
complement: a system of blood chemicals that interact to cause cell lysis.
compliance: in the lung, ease of expansion.
compound: a substance composed of two or more parts and having properties different from its parts.
concave: having a hollow or depressed surface.
concentration: amount of a solute per volume of solvent.
conception: the point in time when a sperm unites with an ovum to initiate formation of a new individual.
conductance: ability of a nerve fiber or membrane to conduct an electrical disturbance.
conductile: being capable of carrying or transmitting a nerve impulse.
conduction: transmission of a nerve impulse by exciting progressive segments of a nerve fiber.
conductivity: conducting ability.
congenital: present at birth; a condition evident at birth having causes operating before birth.
conjugate: paired to or with something else.
conjunctiva: a mucous membrane lining the eyelids and anterior portion of the eyeball.
conscious: awake, being aware.
consolidate: to become solid or firm.
constrict: to narrow or become smaller in diameter.
constriction: the act of narrowing or becoming smaller in diameter.
contra-: prefix meaning opposite or against.
contraction: shortening or tightening.
contralateral: originating in or affecting the opposite side of the body.
convection: transferrance of heat by currents in a liquid or gas.
converge (convergence): to come together; the inward rotation of both eyes to view nearby objects.
convex: having an outward curvature.
convulsion(s): involuntary and usually severe muscular contractions and relaxations.
coronary: pertaining to a crown or circle; the system of blood vessels supplying the heart.
corpus: body; the major portion of an organ.
correlation: a close relationship between two or more factors.
cortex (-corticate): an outer covering or layer of an organ or structure.
corticoid: one of many adrenal cortical steroid hormones.
cotyledon: a section or part; e.g., one of the parts of the placenta.
countercurrent: when flow of gases or fluids is in opposite directions in two closely spaced limbs of a bent tube.
cranial: pertaining to the cranium or part of the skull surrounding the brain.
crest: a ridge or prominence on a bone, especially on its edge(s).
crista: a crest or shelf.
criterion: a standard or means of judging something.
crystalloid: a substance that is capable of forming crystals, and that, in solution, diffuses rapidly through membranes.
cuneate: wedge-shaped.
cutaneous: referring to the skin.
cyano-: prefix meaning blue.
cyanosis: a condition of blue color given to skin and mucous membranes by excessive amounts of reduced hemoglobin in the blood stream.
cybernetic: referring to the comparative study of the nervous system and complex machines to discovering the possibility for and self-regulating nature of their operation.

cybernetics: the study of self-monitoring and regulating mechanisms for maintenance of near-constant values of a function.

cycle: a series of events; something that has a predictable period of repetition.

cyclic: occurring in cycles.

-cyte: suffix meaning cell.

cyto-: prefix meaning cell.

cytochrome: a yellow pigment that functions to transport electrons in cellular respiration.

cytokinesis: division of cytoplasm in mitosis.

cytoplasm: the cellular substance between the membrane and nucleus.

D

damp(ing): to reduce the amplitude of successive swings of a function; reduction of same.

de-: prefix meaning down or from.

deamination: enzymatic removal of an amine group ($—NH_2$) from an amino acid to form ammonia.

death: permanent cessation of all vital functions. Absence of life.

decarboxylation: enzymatic removal of CO_2 from the carboxyl group (—COOH) of an organic acid.

decelerate: to decrease velocity.

decibel: a unit indicating the relative difference in power in the intensity of two sounds.

decompression: removal of pressure from an organ or organism.

decrement: a stepwise decrease or step in the decrease of something.

decussate: to undergo crossing.

deficiency: lowered levels of a body constituent.

deficit: lack of or lacking in.

degenerate: to "go bad" or deteriorate.

deglutition: swallowing.

degradative: a process that breaks down or degrades something.

degrade: to tear down or break down.

dehydration: removal of water from something; the result of water removal.

delete: to remove or omit.

delta: the fourth letter of the Greek alphabet; denotes fourth in line of discovery or importance.

dendrite: a relatively short, highly branched process of a neuron conveying impulses toward the cell body.

denervate: to separate from nerve supply.

density: crowding; numbers of receptors per given area; ratio of mass to volume.

deoxy-: deficient in oxygen.

depolarization: loss of polarity or the polarized state.

depolarized: being in a state of depolarization.

depot: a storehouse.

depressant: an agent inhibiting or lowering the level of body function(s).

derived: coming from something else; a compound produced from another compound.

derma- (dermis): referring to the skin; the connective tissue layer of the skin under the epidermis or covering layer.

dermatome: a segmental skin area served by a particular pair of spinal sensory nerves; an instrument for cutting skin to be used in transplantation.

descending: going downwards, or lower, or in an inferior direction; motor nerve tract carrying impulses down the spinal cord from the brain.

detoxify: to remove the toxic or poisonous quality of something.

deuteranopia: color blindness to green.

di-: prefix meaning twice.

diabetes: to pass through, as a substance through the body.

dialysis: passage of a diffusible solute through a membrane that restricts passage of colloids.

diastasis: in the cardiac cycle, the time when there is little change in length of muscle fibers and filling of the chamber is very slow. It is followed by contraction of the muscle of the chamber.

diastole: the relaxation phase of cardiac activity, when fibers are elongating and filling of a chamber is rapid.

-dies: suffix referring to a day or approximately a day in length of a function.

diffusible: capable of diffusing.

diffusion: mixing of solute and solvent as a result of motion of molecules from regions of higher to lower concentration.

dilate: to become larger in size or diameter.

dilation: expansion of an organ or vessel.

dilute: to add solvent to a solute to render it less concentrated; the state of being less concentrated.

diopter: the refractive power of a lens or its ability to bend (refract) light rays.

diploid: cell having twice the number of chromosomes present in the egg or sperm of a given species.

-dipsia: suffix denoting thirst.

dissociate: to separate into components.

distal: farthest from the center of the body or point of attachment.

diuresis: secretion of abnormally large amounts of urine.

diuretic: a substance causing diuresis.

DNA: deoxyribonucleic acid; composed of nitrogenous bases, deoxyribose sugar and a phosphate group, it is believed to the site of determination of the body's hereditary characteristics.

dominance: in inheritance of characteristics, the

effects of one allele overshadow the effects of the other and the character determined by the dominant (stronger) gene prevails.

donor: one who gives.

dorsal (dorso-): referring to the upper or back portion of the body.

duplication: to double something; to form a copy.

dura: hard or tough.

dys-: prefix denoting bad, difficult, or painful.

dyspnea: labored or difficult breathing.

dystocia: difficult or prolonged labor at childbirth.

dystrophy: degeneration of an organ resulting from poor nutrition, abnormal development, infection, or unknown causes.

E

ECF: abbreviation for extracellular fluid, the fluids not contained within cells.

ectoderm: the outer layer of cells in the embryo, from which the nervous system, special sense organs, and certain endocrines develop.

-ectomy: suffix referring to surgical removal of an organ or gland.

ectopic: in an abnormal place or position.

edema: swelling due to increase of ECF volume; dropsy.

effacement: the thinning and shortening of the maternal uterine cervix at birth.

effector: the organ that responds to stimulation; a gland or muscle cell.

efferent: carrying away from a center or specific point of reference.

egest: to rid the body of something; used synonymously with eliminate.

elastance: the power of elastic recoil after being stretched.

electro-: prefix referring to electrical phenomena of one sort or another.

electrolyte: a substance that dissociates into electrically charged ions; a solution capable of conducting electricity because of the presence of ions.

element: a substance that cannot be separated into substances different from itself by ordinary chemical means.

eliminate: to rid the body of wastes by emptying of a hollow organ.

embolus (embolism): a floating undissolved mass or bubble of gas brought into a blood vessel by the fluid flow; -ism is the plugging of a vessel by an embolus.

embryo: a stage in development between fertilization and assumption of the form characteristic of the species. In the human, the term specifically refers to the period between 2 to 8 weeks.

emetic: a substance producing vomiting.

-emia: suffix referring to the blood.

emmetropic: normal in reference to vision.

emphysema: condition resulting from rupture or expansion of alveoli of the lungs.

emulsify: to form into an emulsion, that is, a nonsettling suspension of fat droplets in a watery solvent.

encephalo-: prefix referring to brain or cerebrum.

endo-: prefix denoting in, or inside of, or from within.

endocrine: internally secreting; a gland secreting into the blood stream.

endoderm: inner layer of cells of the embryo; gives rise to *linings* of gut and all outpocketings of the gut (e.g., lungs, pancreas, liver).

endogenous: arising from within the cell or organism.

endoplasmic reticulum: the series of tubular structures found in the cytoplasm of cells; transports substances through cells.

endothelium: the simple (1-layered) squamous (flat) cellular lining of the blood and lymph vessels and heart.

energy: capacity to do work; heat or chemical bond capacity for work. Energy is seen as motion, heat, and sound, for example.

enteroceptive: originating within body viscera.

-enteron (enteric): referring to the gut.

entrain(ment): to cause to follow, as in the assumption of a rhythm by an internal or external stimulus.

enzyme: organic catalyst causing alterations in rates of chemical reactions.

ependyma: cells forming the lining of the cavities within the central nervous system.

epi-: prefix meaning upon, at, or outside of.

epilepsy: recurrent transient attacks of disturbed brain function.

epinephrine (adrenalin): a catecholamine produced by the adrenal medulla.

epithelial: referring to epithelia, the covering or lining tissues of the body.

epitope: the part of the surface of an antigen that determines its antigenicity.

equilibrium: a state of balance or rest or in which opposite acting forces are equal.

equivalent: equal in force; the amount, by weight, of an element that will combine with 1 gm of Hydrogen or 8 gms of Oxygen.

erythro-: prefix meaning red; implies reference to blood or blood vessels.

erythropoietin: a substance produced by the kidney

that stimulates production of red blood cells.
essential: indispensable; in reference to amino acids, those the body cannot synthesize and must ingest.
estro-: prefix denoting female, as in estrous (menstrual) cycles or estrogen (female characteristic-producing hormone).
estrogen: a substance producing or stimulating female characteristics; the follicular hormone.
eu-: prefix meaning normal or well.
euphoria: exaggerated feeling of well-being.
eupnea: normal breathing.
eupneic: possessed of normal breathing.
Eustachio: Bartolomineo Eustachio, Italian anatomist (1524–1574). The tube connecting the middle ear to the throat bears his name (Eustachian tube).
evacuate: to empty.
evoke: to cause to happen, to call forth.
excitable: the property of being capable of responding to stimuli by alterations in ion separation.
excitatory: something that causes changes in state of excitability; something that stimulates a function.
excite: to stimulate or increase the activity of a physiological mechanism.
excrete: to separate and expel useless substances from the body.
exo-: prefix denoting out, or outside of, or from without.
exocrine: external secretion by a gland, through ducts, upon an epithelial surface.
exogenous: originating outside the cell or organism.
expiration: expulsion of air from the lungs.
expire: to breathe out; to die.
exponential: change in a function according to multiples of powers (e.g., square or cube), rather than numerically (e.g., one time, two times, etc.).
extension: a pulling apart, as in straightening a limb.
external: outside, toward the surface, or to the side.
exteroceptive: sensations or reflex acts originating from stimulation of receptors at or near a body surface.
extra-: prefix meaning outside of or in addition to.
extracellular: outside of cells.
extraneous: outside of and unrelated to something.
extrinsic: from without or coming from without; separate or outside of an organ or cell.

F

facial: pertaining to the face; the seventh cranial nerve.
facilitate: to make a process easier, as in facilitation of an impulse passage across a synapse.
facultative: implies ability to control a process.
fascicle: an arrangement like a bundle of rods.
fasciculate: to separate into smaller parts.
fecal: related to feces (stools).
feedback: detection of the nature of an output and using that to control the process producing the output.
ferritin: a compound formed by the union of ferrous iron (Fe^{++}) and a protein, apoferritin.
ferrous: the divalent form of the iron ion (Fe^{++}).
fertilization: the impregnation of an ovum by a sperm cell.
fetal hemoglobin: hemoglobin F, a hemoglobin present in the fetus; it loads O_2 more completely at lower partial pressures than does adult hemoglobin (A).
fetus: the term given to a developing human after it assumes clearly human form; the period from about six weeks gestation to birth.
fibril: a small filamentous structure, often a component of a body tissue.
fibrillar: having the shape of, or being composed of fibrils, long filamentous structures.
fibrillate: to quiver, as in cardiac muscle fibrillation.
fibrillation: the quivering of cardiac muscle; it destroys pumping action.
fibrin: a white or yellowish insoluble protein formed when blood clots.
fibrosis: abnormal formation of fibrous connective tissue in an organ.
fibrous: composed of fibers.
filter: to strain, or separate on the basis of size.
filtrate: the name given to the fluid that has passed through a filter.
filtration: the process of forming a filtrate by passing a fluid, under pressure, through a filter or selective membrane.
fissure: a deep groove or cleft.
flaccid: relaxed or flabby.
flexion: bending or a state of being bent.
follicle: a small hollow structure containing cells or secretion.
foramen: a hole through a bone or tissue usually allowing passage of blood, blood vessels, and/or nerves.
foreign: coming from, or having to do with another person or thing; not self.
frequency: number of repetitions of something per unit of time.
frontal: anterior; pertaining to the forehead or its bone.
-fusal: suffix referring to a spindle-shaped body.

G

ΔG: a symbol describing the amount of energy (usually expressed as calories) released from or required for a particular chemical reaction.

gamma: third letter of the Greek alphabet; used to denote the third category of something (e.g., globulins), or the third carbon beyond the carboxyl carbon in a carbon chain.

ganglion: a mass of nervous tissue containing cell bodies, lying outside the brain or spinal cord.

gastric: referring to the stomach.

gender: one's sex; male or female.

gene: the unit of heredity; believed to be a portion of a nucleic acid chain.

generation: an act of forming a new organism; to create an electric current; average time span between one generation of people and the next (about 25 years).

generator potential: small short-lived electrical potential that occurs when a receptor is first stimulated. It causes depolarization of the nerve fiber(s) supplying the receptor.

genesis: origin of anything, or, the act of generation.

genetic: referring to genesis or origin of something.

-genic: suffix denoting production.

genital: referring to the genitalia, that is, external sex organs.

genotype: the hereditary makeup of an individual as determined by his or her genes.

genu: a knee or kneelike bend in a structure.

glaucoma: an eye disease characterized by an increased intraocular pressure.

glia: the nonnervous tissues of the nervous system: supporting and nutritive cells of various types.

globin: the protein component of hemoglobin.

globular: having a rounded or spherical shape.

globulin: one of three major groups of the plasma proteins; contain most of the antibody activity of the body.

glomerulus (glomerulo-): a small round mass of cells or blood vessels.

glosso-: prefix referring to the tongue.

gluco-: prefix referring to glucose.

gluconeogenesis: the synthesis of glucose from noncarbohydrate sources.

-glyc-: referring to glucose.

glyco-: referring to glycogen.

glycogen: a polysaccharide composed of glucose molecules ("animal starch").

glycogenesis: synthesis of glycogen from glucose.

glycogenolysis: the breakdown of glycogen to release glucose.

glycolysis: the anaerobic breakdown of glucose to pyruvic acid.

goiter: enlargement of the thyroid gland.

Golgi apparatus: a cellular organelle consisting of localized sacs and vesicles; concerned with secretion.

gonad (gonado-): refers to the primary sex organs, that is, ovaries and testes.

gracile: slender or slight.

gradient: a slope or grade; a curve representing increase or decrease of a function.

-gram: combining form referring to the record drawn by a machine: cardiogram, telegram.

gram molecular weight (mol or mole): a quantity of a substance, expressed in grams, which is equal to its molecular weight.

grand mal: a form of epilepsy in which unconsciousness and muscular spasms occur.

-graph: the machine that draws a -gram: telegraph, electrocardiograph.

gross: large; anatomy seen with the naked eye.

growth: progressive increase in size of a cell, tissue, or whole organism.

gyrus: an upfolding or convolution of the cerebrum.

H

[H^+]: symbol for hydrogen ion concentration.

haploid: having half the normal number of chromosomes characteristic of the species; sperm and ova are haploid cells.

haptoglobin: a protein in the plasma that combines specifically with hemoglobin released from red blood cells.

helix (pl., helices): a coil or spiral.

hematocrit: the volume of red blood cells in a tube after centrifugation.

heme: the iron containing red pigment that, with globin, forms hemoglobin.

hemi-: prefix meaning half.

hemiplegia: paralysis of one-half of the body (in a right-left direction).

hemisect: to cut halfway through.

hemisphere: one-half of the cerebrum or cerebellum.

hemo-: prefix or combining form referring to the blood.

hemoglobin: the respiratory pigment of red blood cells that combines with O_2, CO_2, and acts as a buffer.

hemolysis: destruction of red blood cells with release of hemoglobin into the plasma.

hemorrhage: abnormal loss of blood from the blood vessels, either internally or externally.

hemostasis: stoppage of blood loss through a wound due to vascular constriction and coagulation.

hepar: the liver.

hepatic: referring to the liver.

hetero-: prefix referring to different or contrasting.

heterogeneous: composed of things of varied type or nature.
hex(a)-: prefix meaning six.
hexose: a six-carbon sugar (e.g., glucose, fructose).
histio- (also histo-): prefix referring to tissue.
homeodynamics: a term implying the constant adjustments necessary to maintain body conditions at near constant values.
homeokinesis: a synonym for homeodynamics.
homeostasis: the state of near constancy of body composition and function.
homeothermic: a "warm blooded" animal, or one capable of maintaining a near constant body temperature.
homo-: a prefix referring to sameness or likeness.
homograft: use of tissues of the same species for grafting purposes (e.g., skin).
hormone: the chemical produced by an endocrine gland.
humidify: increasing moisture content.
humor: any fluid or semifluid substance in the body.
humoral: pertaining to the body fluids or substances in them.
hyaline: clear, glassy, or translucent.
hyaluronidase: an enzyme that liquifies the intercellular cement holding cells together.
hydrate(d): to cause water to combine with a compound; the compound after water has been added.
hydro-: prefix referring to water.
hydrostatic pressure: pressure exerted by liquids.
hyper-: prefix meaning above, or excessive.
hyperemia: an extra amount of blood in an area.
hyperesthesia: excessive sensitivity to sensory stimulation.
hypermetropic: pertaining to farsightedness.
hyperphagia: excessive intake of food.
hyperplasia: increase in size due to increased numbers of cells derived by division.
hyperpnea: an increase in depth of breathing.
hypertension: high blood pressure.
hypertonic: a solution containing a greater solute concentration and thus osmotic pressure than the blood.
hypertrophy: increase in size by adding cellular substance and *not* by increasing numbers of cells.
hyperventilation: increased exchange of air by increasing both rate and depth of breathing.
hypnotic: an agent producing sleep or depression of the senses.
hypo-: prefix meaning less than or below.
hypokalemia: low blood potassium levels.
hypotension: low blood pressure.
hypothesis: an assumption unproven by experiment or observation.
hypotonic: a solution containing less solute, and thus having a lower osmotic pressure, than the blood.
hypovolemic: diminished volume.
hypoxemia: lowered blood oxygen levels.
hypoxia: low oxygen levels in inspired air or reduced tension in blood or tissues.
Hz (Hertz): a symbol denoting frequency, especially of sound waves or electromagnetic radiation (replaces *cycles per second, cps*).

I

iatrogenic: an abnormal condition caused by effects of treatment by a physician or surgeon.
ICF: abbreviation for intracellular fluid; fluid contained within cells.
idiopathic: refers to a condition without clear cause.
immune: protected from getting a given disease.
immunity: a state of being protected from certain noxious agents; resistance.
immuno-: a prefix denoting relationship to the immune process.
implantation: embedding of the blastocyst in the uterine lining; inserting something into a body organ.
impulse: a physicochemical or electrical change transmitted along nerve fibers or membranes.
inactivate: to make inactive or inert.
inclusion: a lifeless, usually temporary constituent of a cell's cytoplasm.
increment: a stepwise increase or step in the increase of something.
independent irritability: capable of reacting to external stimuli, or those delivered by other than the normal route.
induce (induction): to cause an effect; induction is used to express the effect that a gene has in causing an effect.
inducer: an agent causing induction.
infantile: pertaining to the infant, or a structure that fails to mature normally.
infectious: capable of being transmitted with or without contact. Usually used in connection with microorganism caused diseases.
infiltrate: to pass into or through a substance or space.
inflame: to cause to become warm, swollen, or red.
infundibulum: a funnel-shaped passage.
infusion: to introduce liquid in a vein.
ingest: to take foods into the body.
ingestion: the process of taking foods into the body.
inherent: contained within or belonging to anything naturally; not dependent on outside forces.
inhibit: to repress or slow down.

initial: related to the beginning or start.
inorganic: a chemical compound not containing both hydrogen and carbon; not associated with living things.
input: what is put into something.
insert: to place into.
insipid: lacking in taste or spirit; (*diabetes insipidus* is a loss of water due to failure of kidney reabsorption.)
insomnia: inability to sleep or premature termination of sleep.
inspiration: taking air into the lungs.
insufficiency: state of being inadequate for its purpose.
intensity: degree of strength, force, loudness, or activity.
inter-: prefix meaning between.
intercalated: inserted between, or in addition to.
intermediary: something occurring between two time periods. In metabolism, the compounds formed as foodstuffs are utilized.
internal: within the body or away from its surface.
internuncial: acting to connect, as an *internuncial neuron* between 2 other neurons.
interphase: the "resting state" of a cell when it is not dividing.
interstitial: lying between, as when interstitial fluid lies between vessels and cells or between cells.
intra-: prefix meaning within.
intracellular: within cells.
intravenous: within or into a vein.
intrinsic: originating within a structure.
in utero: within the uterus.
invert: to turn two items to a reversed position from that existing originally.
in vitro: in glass, as in a test tube, *not* within the body.
in vivo: within the body.
involuntary: occurring without an act of will.
involution: change in a backward or diminishing direction.
ion: one or more atoms carrying an electric charge.
ionized: a substance that has formed ions.
ipsilateral: on the same side, or affecting the same side of the body.
irritable: capable of reacting to a stimulus (synonym, excitable).
ischemia: temporary decrease of blood supply to a body part.
iso-: equal to.
isometric: refers to no change of length, as an isometric contraction.
isotonic: a solution having the same osmotic pressure as the blood; a muscular contraction in which shortening is allowed.
-itis: suffix meaning inflammation of.

J

jaundice: a condition in which the skin is yellowed due to excessive bile pigment (bilirubin) in the blood.
juxta-: close or near to; in close proximity.

K

kalium: referring to potassium (K).
kallikrein: an enzyme present in the bloodstream, urine or tissues in an inactive state.
karyoplasm: nuclear protoplasm.
karyotype: a grouping or arrangement of the 46 human chromosomes based on the size of individual chromosomes.
keratin: a protein found in hair, nails, and the epidermis; insoluble in water, and weak acids and bases.
keto-: prefix denoting the presence in an organic compound of the carbonyl or "keto" group

$$(\text{—}\overset{\text{O}}{\overset{\|}{\text{C}}}\text{—}).$$

keto acid: an organic acid having a carbonyl and a carboxyl group.
ketone: any substance (e.g., acetone) having the carbonyl group in its molecule.
ketosis: accumulation in the body of the ketone bodies (acetone, betahydroxy butyric acid, acetoacetic acid).
kilocalorie: a large calorie; the amount of heat required to raise the temperature of 1 kg of H_2O, 1°C.
kilogram: a metric measurement of weight; equals 1000 grams or 2.2 lb.
kinesthesia: the sensation concerned with appreciation of movement and body position.
kinetic: pertains to movement or work.
-kinin: suffix denoting causing motion or action.
kinin: a group of biologically active polypeptide enzymes; they are the active product of activation of kallikreins.
kininase: an enzyme destroying a kinin.
Krebs cycle: a metabolic scheme that degrades two carbon units (AcCoA) to CO_2 with production of ATP and H.
Kussmal breathing: deep and rapid breathing; seen in acidosis as an attempt to eliminate CO_2.

L

labile: not fixed; easily altered.
labyrinth: intricate and tortuous communicating channels.

lack: absence of something.
lacrimal: refers to the tears.
lactating: secreting milk.
-lemma: suffix meaning husk; denotes the surrounding layer or membrane of something.
lemniscus: a ribbonlike bundle of sensory nerve fibers in the brain stem.
lesion: a wound, injury, or infected area.
leuco-: prefix meaning white.
leukemia (leucemia): a disease of the bloodstream characterized by overproduction of white blood cells at the expense of other blood cells.
levo-: prefix meaning left.
limbic system: a group of nervous structures in the cerebrum and diencephalon serving emotional expression.
liminal: threshold; just perceptible.
linkage: in genetics, when two or more genes on a chromsome tend to remain together during sex-cell formation.
lipid: fats or fatlike substances that are not soluble in water.
liter: metric measure of fluid volume. Equals 1000 ml, 1.06 qt, or 61 cu in.
-lith (lithiasis): suffix meaning stone; lithiasis indicates presence of stones.
localize: to restrict to a small area.
locus: a place; in genetics the location or position of a gene in a chromosome.
logarithmic: progression of a function by powers (e.g., square or cube) and not by individual numbers.
L-tubule: longitudinally arranged tubules of the sarcoplasmic reticulum in muscle cells.
lumbar: referring to the loins or small of the back area of the body.
luteo-: prefix meaning yellow.
lympho-: prefix referring to lymph or lymphatic system.
lysis: to destroy or break down.
lysosome: cell organelle concerned with digestion of large molecules.

M

macro-: prefix meaning large or long.
malignancy: a severe form of something.
malignant: growing worse; resisting treatment.
malnutrition: literally, "poor nutrition"; most often used to refer to absence of essential foodstuffs in the diet.
mandible: the lower jaw.
marginal: referring to a border or outer layer.
marginate: the lining up or accumulation of white blood cells on the inner border of a capillary.
masticate: to chew.
mater: mother; a word used to refer to two of the three membranes (meninges) surrounding the central nervous system (dura mater, pia mater).
maxilla: a bone of the upper jaw.
maximal: highest; greatest possible.
mean: the average (sum of numbers divided by the number of numbers).
mechanistic: adhering to the view that body operations may be explained without recourse to metaphysical forces.
mediated: accomplished by indirect means.
medulla: inner or central portion of an organ; the medulla oblongata of the brain stem.
medullated: having a myelin sheath.
mega-: large; one million.
-megalo (megaly): suffix denoting enlargement.
Meibom: Heinrich Meibom, German anatomist (1638–1700); a sebaceous gland in the eyelid bears his name (Meibomian gland).
meiosis: a form of cell division that reduces chromosome number to haploid; occurs in formation of sex cells.
Meissner: George Meissner, German histologist (1829–1905). Touch corpuscles bear his name (Meissner's corpuscle).
melanin (melano-): a brown-to-black pigment. Melano- is a prefix used to refer to the pigment.
mellitus: honeylike or sweet; (*diabetes mellitus* is a condition involving flowing of glucose from the body through the urine).
membranous: resembling a membrane.
meninges: the membranes around the central nervous system. There are three: dura mater, arachnoid, pia mater.
menstrual: refers to menstruation or the sloughing of the uterine endometrium.
-mensual: suffix denoting a month or approximately a month.
-mere: suffix meaning a part or portion.
mero-: prefix denoting a part or portion.
mesentery (mesenteric): a membrane supporting or suspending an abdominal organ; the adjective is mesenteric.
mesoderm: the middle germ layer of the embryo. Gives rise to connective tissue, muscle, blood, and the cellular part of many organs.
meta-: a prefix denoting change, or following something in a series.
metabolic: refers to metabolism.
metabolite: any product of metabolism.
metaphase: a stage of mitosis in which chromosomes line up on the equator of the dividing cell.
metastasis: movement from one part of the body to another (e.g., cancer cells).

meten-: a prefix meaning after.

micelle: a small complete unit, usually of colloids.

micro-: small; one millionth.

micron: metric unit of length; equals one one-thousandth of a millimeter.

microtubules: microscopic tubules in the cytoplasm of cells.

milli-: metric prefix meaning one thousandth of.

milliequivalent: one one-thousandth of an equivalent weight. (An equivalent weight is the quantity, by weight, of a substance that will combine with one gram of H or 8 grams of O.)

millimicron: one one-thousandth of a micron.

milliosmol: one one-thousandth of an osmol (an osmol is the molecular weight of a substance in grams divided by the number of particles each molecule releases in solution).

millivolts: thousandths of volts.

mimetic: imitating or causing the same effects as something else.

miotic: an agent causing pupillary contraction.

mitochondria: a cell organelle producing most of the cell's ATP stores.

mitosis: a type of cell division that results in production of daughter cells like the parent.

mitotic: refers to mitosis.

modality: the nature or type of a stimulus; a property of a stimulus distinguishing it from all other stimuli.

molal: a solution in which one mol (molecular weight of a substance expressed in grams) is present in 1000 gm of solvent.

molar: a solution in which one mol of a substance is present in a liter of solution.

molar solution: a solution in which one gram molecular weight of a substance is present in each liter of the solution.

mole: (see gram molecular weight).

molecular weight: weight of a molecule obtained by adding the weights of its constituent atoms. It carries no units.

molecule: the smallest unit of a substance retaining the properties of that substance.

monitor: to watch, check, or keep track of; one or something that watches.

mono-: prefix meaning one or single.

morbidity: sickness or the state of being ill.

morphology: the science or study of form and structure.

mortality: death rate.

morula: mulberry; a solid mass of cells resulting from cleavage of a zygote and resembling a mulberry.

motor: refers to movement or those structures (e.g., nerves and muscles) that cause movement.

motor unit: one motor nerve fiber (axon) and the skeletal muscle fibers it supplies.

mucous: resembling, having the nature of, or secreting mucus.

mucus: a sticky, thick secretion of glands or membranes.

multicellular: composed of many cells.

mutation: a change or transformation of a gene; the evidence of a gene alteration.

mydriatic: an agent causing pupillary dilation.

myelin: a fatty substance forming a sheath or covering around nerve fibers.

myelo-: prefix referring to bone marrow.

myeloid: formed in bone marrow.

myo-: prefix referring to muscle.

myoglobin: a respiratory pigment of muscle that binds oxygen.

myopia: nearsightedness; synonym, hypometropia.

myosin: a muscle protein exhibiting enzyme activity and forming cross bridges with actin to shorten a muscle.

myotatic: refers to muscle or kinesthetic sense (kinesthesia).

myxedema: a hypothyroid condition in which a thick fluid forms beneath the skin.

N

narcosis: a state of unconsciousness produced by a narcotic (substance causing stupor or sleep).

natrium: referring to sodium (Na).

necrosis: death of cells or tissues.

negative: without positiveness; lacking results; marked by resistance.

neo-: prefix meaning new or recent.

neonate: a newborn individual.

neoplasm: a new and abnormal formation of tissue, as a cancer, tumor, or growth.

neph-: prefix referring to the kidney.

nephron: the functional unit of the kidney that forms urine and regulates blood composition.

net: that which remains after all factors influencing it have operated.

neurilemma: a thin living membrane around some nerve fibers; aids in myelin formation and fiber regeneration.

neuro-: prefix referring to nerves or nervous system.

neuron: the cell serving as the unit of structure and function of the nervous system.

neurosis: a disorder of the mind in which contact with the real world is maintained.

neutral: indifferent; having no positive or negative qualities; neither acidic nor basic in reaction.

neutralize: to counteract, make inert, or destroy the properties of something.
nigra: black.
nitro-: prefix referring to nitrogen or niter (sodium or potassium nitrate).
nitrogenous: referring to nitrogen; containing nitrogen.
nonessential: dispensable; in reference to amino acids, those that the body can synthesize and need not ingest.
noxious: harmful or injurious.
nucleic acids: large molecules formed of nucleotides. They form the basis of heredity and protein synthesis. DNA and RNA.
nucleolus: a spherical mass of nucleic acid within the nucleus. Site of t-RNA synthesis.
nucleoside: a compound formed by the combination of a purine or pyrimidine base with a pentose sugar.
nucleotide: a unit or compound formed of a nitrogenous base, a five-carbon sugar, and a phosphoric acid radical. Unit of structure of DNA and RNA.
nucleus: the controlling body of a cell; a functional grouping of neuron cell bodies in the central nervous system.
nutrient: a food substance necessary for normal body functioning.
nystagmus: involuntary cyclical movements of the eyeball.

O

obese: to be fat; overweight due to fat accumulation.
obligatory: carries the idea of not having a choice; bound to or fixed in function.
obstruct: to block or plug a vessel or tube.
occipital: referring to the back part of the head.
occlude: to close, obstruct or block something, such as, a blood vessel.
occlusion: the state of being closed.
ocular: referring to the eye or vision.
oculo-: prefix referring to the eye.
-ogen: suffix denoting an inactive form of an enzyme or chemical.
olfactory: referring to smell or the sense of smell.
oligo-: prefix meaning small or few.
oliguria: secretion of small amounts of urine.
-oma: suffix denoting a tumor.
oncotic: refers to the colloidal osmotic pressure created by the plasma proteins.
opaque: not transparent; does not allow light to pass; dark.
operator gene: a gene that turns on or off the activities of a structural gene.
operon: the combination of an operator gene and a structural gene.
operon concept: the current theory of how genes operate to control body function.
ophthalmic: referring to the eye.
-opsin: suffix referring to the protein component of the visual pigments.
optic: referring to the eye or sight; the second cranial nerve.
organ: two or more tissues organized to do a particular job.
organelle(s): submicroscopic formed structures within the cytoplasm that carry out particular functions.
organic: referring to organs; compounds containing carbon and hydrogen.
organic acids: an acid containing one or more carboxyl groups (—COOH).
-ose: a suffix referring to a carbohydrate, especially a simple or double sugar.
-osis: suffix meaning caused by, state of, disease or intensive.
osmo-: combining form referring to smell; referring to osmosis.
osmol: the molecular weight of a substance, in grams, divided by the number of particles it releases in solution.
osmolarity: number of osmols per liter of solution.
osmosis: the passage of solvent through a selective membrane.
osmotic: referring to osmosis.
osseous: bonelike, or bony.
osteo-: prefix referring to bone or bones.
oto-: combining form referring to the ear.
-otomy: suffix referring to opening or repair of an organ without its removal.
output: the result or product of the operation of an organ or a machine.
ovulatory: referring to release of an ovum from the ovary.
-oxia: suffix referring to oxygen or oxygen concentration.
oxidation: the combining of something with oxygen; loss of electrons.
oxidative phosphorylation: a metabolic scheme that transfers H ions and electrons to produce ATP and water.
oxy-: combining form referring to acidity or oxygen.
oxygenate: to supply with oxygen.
oxygen debt: a temporary shortage of oxygen necessary to combust products (lactic acid, pyruvic acid) of glucose catabolism.

oxyntic: acid producing.

P

Pacini: Filippo Pacini, Italian anatomist (1812–1883). The pressure corpuscle bears his name (Pacinian corpuscle).
pallid: lacking color, pale.
papilla: a small nipplelike elevation or protuberance.
para-: combining form meaning near, past, beyond, the opposite, abnormal, or irregular.
paradoxical: contrary to belief.
paralyze: to cause loss of muscle movement or sensation.
parameter: a value, constant or something measured.
paraplegia: paralysis of the lower part of the body, involving both legs.
parasympathetic: refers to the portion of the autonomic nervous system that controls normal body functions.
paresthesia: abnormal sensation without demonstrable cause.
parietal: refers to outer lining or covering, or wall of a cavity.
pars: part or portion.
parturition: the act of giving birth to offspring.
passive: not active.
pathogenic: causing disease.
pathological: diseased or abnormal.
Pco_2: symbol for partial pressure of carbon dioxide.
peduncle: a band of nervous tissue connecting parts of the brain.
-penia: suffix denoting a decrease in numbers.
pent(a)-: prefix referring to five.
peptide: a compound containing two or more amino acids.
perceive: to become aware of objects.
perforate: to make a hole through; puncture.
perfuse: to pass a fluid through something.
peri-: prefix meaning around, or about.
pericardial: around the heart.
period: an interval of time between two successive occurrences in a cyclic phenomenon; a cycle.
peripheral: located away from the center; the outside.
peristalsis: a progressive wavelike contraction of smooth muscle that propels materials through a tubular organ.
peritoneal: refers to the peritoneum or abdominal cavity or its lining.
peritoneum: the membranous lining of the abdominal and pelvic cavities.
permeability: being permeable.
permeable: allowing passage of solutes and solvents in solutions.
petit mal: a type of epileptic seizure in which there is momentary loss of consciousness.
pH: a symbol used to express the acidity or alkalinity of a solution. Strictly, the logarithm of the reciprocal of H ion concentration.
-phage: suffix meaning to eat or consume.
phago-: combining form meaning an eater, or ingesting, or engulfing.
phagocytosis: engulfing of particles by cells.
pharynx: the throat.
phase: a portion of a cycle or period.
phosphocreatine: an energy rich compound found in muscle, consisting of creatine and phosphate.
phospholipid: a fatty substance combined with some form of phosphorus.
phosphorylation: transfer of a phosphate group from ATP or ADP to another compound (e.g., a sugar).
photopic: a term referring to daylight and color vision.
pia: tender or delicate; the inner meninx carrying blood vessels to cord and brain (syn.: pia mater).
pigment: any coloring substance.
pino-: a prefix meaning to drink, absorb, or imbibe liquid.
pinocytosis: intake of solution by cells through "sinking in" of the cell membrane.
pitch: that quality of a sound making it high or low in a scale. Depends on frequency of vibration.
placenta: the structure attached to the inner uterine wall through which the fetus gets its nourishment and excretes its wastes.
-plasm: a suffix indicating something formed.
plasma: the liquid portion of the blood.
plasma cell: one capable of producing antibodies in response to antigenic challenge.
plasma membrane: the cell membrane.
pleural: referring to the membrane(s) lining the two lateral cavities of the thorax or covering the lung; or, the cavities themselves which house the lungs.
plexus: a network of nerves or blood vessels.
-pnea: suffix referring to breathing.
pneumo-: combining form referring to the lungs or air.
pneumonia: inflammation of the lungs.
pneumotaxic centers: groups of nerve cells in the brain stem that aid in inhibiting inspiration.
Po_2: symbol for partial pressure of oxygen.
-pod, poda: foot or feet.
poikilo-: a prefix denoting varied or variable.
poikilothermic: "cold-blooded" or not capable of maintaining a nearly constant body temperature.
polarized: state in which ions are in unequal concentrations on two sides of a membrane, with production of an electric potential across the membrane.

poly-: prefix meaning many or much or great.
polymerization: putting together similar units to form an aggregate of greater molecular weight.
polymerize: to put similar units together into a larger aggregate.
polypeptide: a compound containing more than 10 amino acids.
polypnea: rapid breathing.
polyuria: secretion and elimination of large quantities of urine.
pore: an opening of small size.
porosis: a condition characterized by cavity formation in an organ.
portal: an entryway; in the vascular system, a set of capillaries between veins.
positive: definite; having a positive electrical charge; having results.
post-: prefix meaning before or in front of.
postganglionic: beyond or past a ganglion or synapse.
potency: strength or power.
potential: an "electrical pressure." Implies a measurable electrical current flow or state between two areas of different electrical strength.
pre-: prefix meaning in front of or before.
precipitate: something usually insoluble that forms in a solution.
precocious: matured or developed earlier than normal.
preganglionic: in front of, or before a ganglion or synapse.
presbyopia: an eye condition arising in old age in which there is loss of ability to focus near objects due to loss of lens elasticity.
pro-: prefix meaning for or in favor of.
process: a projection from something; a method of action.
progesterone: hormone produced by the corpus luteum. It thickens the uterine lining and develops the mammary glands.
prognosis: prediction of the course and outcome of a disease or abnormal process.
projection: throwing forward; the efferent connection of a part of the brain.
propagation: carrying forward; act of reproducing or giving birth.
prophase: a stage in mitosis characterized by nuclear disorganization and formation of visible chromosomes.
proprioceptive: pertaining to awareness of posture, movement, equilibrium, and body position.
protanopia: red blindness.
protein: a large molecule composed of many amino acids.
proteinuria: the presence of protein in the urine.
proximal: closest to midline or point of attachment.
pseudo-: false.
psychosis: a disorder of the mind in which contact with reality is lost.
psychosomatic: an illness in which some part of the cause is related to emotional factors.
pulmo-: combining form meaning lung.
pulmonary: concerned with the lungs.
pulse: a throbbing caused in an artery by the shock wave resulting from ventricular contraction.
pupil: the opening in the center of the iris of the eye.
Purkinje: Johannes E. von Purkinje, Hungarian physiologist (1787–1869). His name is associated with cells of the cerebellum and heart.
pyrogen: an agent causing a rise in body temperature.

Q

quadriplegia: paralysis affecting all four limbs.
quality: the nature or characteristic(s) of something.
quantity: amount.
quantum: a unit, usually of light or light energy.

R

radiate: to spread outwards from a given point.
radical: a group of atoms acting as a single unit; a nonconservative approach to cure of illness.
ramus: a branch or division of a bone or vessel.
reabsorption: to absorb again (foods are absorbed from the intestine, excreted by the kidney and *reabsorbed* back into the bloodstream).
receptor: cells functioning in the reception of stimuli; a molecular group in or on a cell having a special affinity for other chemical groups.
recessive: to go back; in genetics, a characteristic that does not usually express itself because of suppression by a dominant gene (allele).
recipient: one who receives.
reciprocal: the opposite; interchangeable in nature.
recompression: to put pressure back on.
reduction: uptake of H by a compound; gain of electrons; restoring a broken bone to normal relationships.
referred: sent to another area.
reflex: an involuntary, stereotyped response to a stimulus.
refract: to bend or deflect a light ray.
refractory: resistant to stimulation.
reject: to show no affection; to ignore; the destruction of a transplant.
relative: considered in relation to something else.
relaxation: lessening of tension.
renal: referring to the kidney.
replication: to duplicate.

repolarization: the act of becoming repolarized.
repolarize: to return a muscle or nerve cell to its original resting state of separation of ions across the cell membrane and electrical potential.
repression: to put down or prevent expression of something.
repressor: something that causes repression, as a repressor gene.
resistance: to oppose, as flow of air or liquid; the sum total of the mechanisms conferring immunity to disease.
respiration: cellular metabolism; the act of exchanging gases between the body and its environment.
resting: not active.
resuscitation: to bring back to consciousness.
retain (retention): to hold; keeping something in the body that does not belong there, or in a greater quantity than is normal.
reticular: in the form of a network.
reticuloendothelial: referring to the tissues of the reticuloendothelial system, that is, the fixed phagocytic cells of the body and the plasma cells.
reticulum: a network.
retina(l): pertaining to the retina of the eye, the photosensitive tunic.
rheobase: the minimum strength of a stimulus required to produce a response.
rhin-, rhino-: combining form referring to the nose.
rhodopsin: a visual pigment found in rod cells; "visual purple."
rhythm: a regular recurrence of action or function.
ribo-: combining form referring to ribose, a five-carbon sugar.
RNA: ribonucleic acid, composed of nitrogenous bases, ribose sugar, and phosphate. The three types are messenger-RNA, transfer-RNA, and ribosomal-RNA.
ruber: red.

S

-saccharide: combining form referring to sugar.
sacral: pertaining to the sacrum, a bone at the inferior end of the vertebral column.
salt: NaCl (sodium chloride); a chemical compound that results from replacing a hydrogen in the carboxyl group of an acid with a metal or cation.
saltatory conduction: conduction of a nerve impulse down a myelinated nerve fiber by skipping from node to node.
sarco-: combining form meaning flesh.
sarcolemma: cell membrane of a skeletal muscle fiber.
sarcoma: a neoplasm or cancer arising from muscle, bone, or connective tissue.
sarcomere: the portion of a myofibril lying between two Z lines.
sarcoplasm: muscle protoplasm.
satiated: to be satisfied; "full."
saturated: holding all it can. A saturated fat has all the hydrogen it can hold on its chemical bonds.
saturation: the holding in a solution of all the solute possible without precipitation.
scheme: an orderly series of events or changes, as, a metabolic scheme.
sclerosis: hardening or toughening of a tissue or organ, usually by increase in fibrous tissue.
scotopic: referring to night vision.
secrete: to separate; to make a product different from that originally presented as a starting material; active movement of substances *into* a hollow organ.
secretion: the product of secretory activity; the act of producing a product actively.
secular: belonging to a particular age.
sedimentation: formation of a sediment; a settling out.
segmental: composed of segments, that is, individual parts or portions.
selective: exhibiting choice, as, a selective membrane passes some materials and not others.
semi-: a prefix meaning half.
semilunar: shaped like a half-moon; the valves of the aorta and pulmonary artery.
senescence: growing old; the period of old age.
sensibility: capacity to respond to stimuli.
sensitivity: the quality of being sensitive or able to receive and transmit sensory impressions.
sensitize: to make sensitive to an antigen by repeated exposure to that antigen.
sensory: refers to sensation or the afferent nerve fibers from the periphery to the central nervous system.
-septan: a suffix referring to seven, or about a week.
serum: the *fluid* remaining after blood has clotted.
shock: a state resulting from circulatory collapse (low blood pressure and weak heart action).
shunt: a "short-cut"; passage between arteries and veins which by-passes the capillaries; a scheme for metabolism of a foodstuff that eliminates certain steps found in a scheme that metabolizes the same substance.
sickle cell: an erythrocyte that is crescent-shaped because it contains an abnormal hemoglobin.
sigmoid: shaped like the letter *S*.
simultaneously: at the same time.
sinusoid: a small irregular blood vessel found in the liver and spleen.
smooth: not bumpy or knobby; lacking striations or stripes.

solubility: capable of being dissolved.
solute: the dissolved, suspended, or solid component of a solution.
solution: liquid containing dissolved substance(s).
solvent: a dissolving medium.
soma: referring to the body, as a cell body, or the body as a whole.
somatic: pertaining to the body generally; the nerves to and from skin and skeletal muscles.
-some: combining form meaning body.
somesthesia: the awareness of body sensations.
somite: blocks of embryonic tissue lying alongside the notochord (forerunner of part of the spinal column).
spasm: a sudden, involuntary, often painful contraction of a muscle.
spastic: contracted or in a state of continuous contraction.
spatial: referring to space.
specific: restricted or not generalized; a disease always caused by the same agent.
sphincter: a band of circularly arranged muscle that narrows an opening.
spirometer: a machine that measures air volumes of the lung.
spongio-: a prefix denoting sponge or spongy; used to refer to precursor cells of some types of glia.
spontaneous: occurring without apparent cause; activity seeming to occur "on its own."
stagnant: without motion, not flowing or moving.
stasis: stoppage of fluid flow.
stenosis: narrowing of an opening.
stereotyped: repeated, predictable response to stimulation.
steroid: a lipid substance having as its chemical skeleton the phenanthrene nucleus.
stimulate: to increase an activity of an organ or structure.
stimulus: an agent acting to cause a response by a living system.
strabismus: when the eyes do not both focus at the same point. Crossed eyes or "squint."
stressor: an agent tending to upset homeostasis and produce strain or stress on the organism.
striate: striped, or marked by bands.
stroke: the total set of symptoms resulting from a cerebral vascular disorder.
structural gene: a gene that directs the synthesis of a specific protein.
sub-: prefix meaning beneath or below.
subliminal: less than that required to get a response; not perceptible.
substitute: to replace a unit with another and different unit.
substrate: the substance an enzyme acts upon.
subthreshold: (see subliminal).
sulcus: a slight depression or groove, especially in the brain or spinal cord.
summate: to add together.
supplement: an addition to something, usually to restore it to normal.
surface tension (alveolus): the phenomenon whereby liquid droplets tend to assume the smallest area for their volume. The surface acts as though it had a "skin" on it due to cohesion among the surface water molecules.
surface-volume ratio: the relationship between the size of an external surface of a unit to the amount of material contained by that surface.
surfactant: a lipid-protein substance that reduces surface tension in the lung alveoli and thus reduces chances of alveolar collapse.
suspension: a state in which solute molecules are mixed in but not dissolved in a solvent. Solutes do not settle out of the mixture.
sympathetic: referring to the portion of the autonomic nervous system that controls response to stressful situations.
synapse: a point of junction between two neurons.
synchronous: occurring at the same time.
syncytium: a group of cells that have an anatomical or functional connection between themselves.
syndrome: a group of signs and symptoms characteristic of a particular disease.
syneresis: shrinking of a gel, as a clot.
synergistic: working together.
synonomy: names used to refer to the same process, characteristic or structure.
synovial: referring to the cavity or fluid in the space between bones of a freely movable joint.
synthesis: to form new or more complex substances from simple precursors.
system: an organized group of related structures.
systemic: referring to the whole or the greater part of the body.
systole: contraction of the muscle of a heart chamber.
systolic: referring to systole, as systolic pressure (highest pressure created by heart contraction).

T

tachy-: combining form meaning rapid or fast.
tachypnea: rapid rate of breathing.
target organ: the structure affected by a particular drug, hormone, or other agent.
taxis: response to an environmental change.
tegmentum: a roof or covering specifically of the midbrain.
telen-: a distant part; the anterior portion of a part

(e.g., *telencephalon,* the anterior portion of the developing brain that will form the cerebral hemispheres).

telo-: a prefix meaning end.

telophase: a stage of mitosis characterized by cytoplasmic division and reformation of nuclei.

temporal: referring to time.

tension: stretching or pressure; what a muscle develops when it contracts.

terminal: end or placed at the end.

terminate: to bring to an end or conclusion.

testosterone: the male sex hormone produced by the interstitial cells of the testes.

tetanus: a sustained contraction of a muscle; a disease caused by a bacterium characterized by sustained contraction of jaw muscles ("lockjaw").

therapy: the treatment of a condition or disease.

thermal: referring to heat.

theta: eighth letter of the Greek alphabet; a wave of the electroencephalogram indicating immaturity of malfunction of nervous tissue.

thorax: the chest.

-thermia: suffix referring to heat or temperature.

threshold: just perceptible; the lowest strength stimulus that results in a detectable response or reaction.

thrombus (thrombosis): a blood clot obstructing a vessel. Thrombosis is the formation of the clot or the result of blocking the vessel.

timbre: the resonance or quality of a sound, as distinguished from its pitch and intensity.

tissue: a group of cells similar in structure and function (e.g., epithelial, connective, muscular, and nervous tissues).

-tome: combining form meaning a cutting, or part.

tone: a state of slight constant tension or contraction exhibited by muscular tissue.

tonic: having tone; continual.

topography: the description of a particular area of the body.

toxic: poisonous.

toxin: a poisonous substance derived from plant or animal sources.

toxoid: a toxin that is treated to lower or decrease its toxicity but that will still cause antibody production.

tract: a bundle of nerve fibers in the central nervous system that carry particular kinds of motor or sensory impulses.

trans-: prefix meaning across, through, over, or beyond.

transamination: transfer of an amine (—NH_2) group from one compound to another without formation of ammonia.

transcellular: literally, through cells. Used to describe those body fluids separated from other fluids by an epithelial membrane (e.g., fluids in stomach, intestine).

transcription: the process of forming RNA by DNA.

transduce: to alter the nature of one form of energy to another form.

transferrin: the plasma protein (a beta-globulin) that binds iron and transports it.

transect: to cut completely across a structure.

transfuse: to inject a fluid into the veins of a person.

transient: temporary, not permanent.

translation: the process by which the code in messenger-RNA is utilized to synthesize a specific protein.

transplant: to take living tissue from one part of a given body, or from other bodies, and unite it with tissues already present, the object being to remedy defects or deformity.

trauma: an injury or wound.

tremor: an involuntary quivering or repetitive movement.

treppe: an increase in strength of muscular contraction when muscle is stimulated maximally and repeatedly.

tri-: a prefix signifying three.

tritanopia: color blindness to blue.

trochlea: a structure having the form of a pulley.

tropic (trophic): literally, nourishing. Used to refer to those hormones that control activity of other endocrine structures (e.g., ACTH, TSH, LH, FSH, ICSH).

tropomyosin: a filamentous muscle protein thought to lie twisted around and to cover active sites on actin, thus preventing contraction.

troponin: a globular protein of muscle having a strong affinity for calcium.

tympanum: the middle ear cavity.

T-tubule: transversely arranged microscopic tubules that convey ECF to the interior of muscle cells.

tumor: a cellular swelling or enlargement. Syn.: neoplasm.

tunic: a coat, covering, or layer.

turbulent: disturbed or not "smooth." Used to describe flow of blood in the circulatory system.

twitch: a single muscular contraction in response to a single stimulus.

U

ulcer: an open sore or lesion in the skin or a mucous membrane-lined organ (e.g., stomach, mouth, and intestine).

unconscious: insensible, not aware of surroundings.

uni-: prefix meaning one or single.

unicellular: one-celled.
unilateral: affecting or occurring on only one side of the body; one-sided.
unitary: related to a single unit.
unsaturated: a chemical compound (e.g., fatty acid) that has double C to C bonds (—C=C—) and is not holding all the hydrogen it is capable of holding.
uremia: a toxic condition in which blood urea levels are elevated above normal. Associated with kidney disease.
-uria: combining form referring to urine.
urobilinogen: the product of bacterial action on bilirubin in the intestine.
urogenital: referring to the urinary and reproductive systems or organs.

V

vaccinia: a disease resulting from cowpox virus. Immunization with cowpox virus confers immunity to smallpox in humans.
vacuole: a fluid filled space in a cell.
vagus: wandering; the tenth cranial nerve.
valence: the magnitude and sign of the electric charge on an ion.
variable: not constant; changing with time.
vascular: referring to, or having, blood vessels.
vaso-: combining form referring to a vessel.
vasomotor: referring to nervous control of muscular blood vessels.
velocity: speed or quickness of motion.
venereal: from Venus, goddess of love; refers to diseases commonly transferred by sexual intercourse.
venous: referring to the veins.
venous pressure: pressure of blood within veins.
ventilation: exchange of air between the respiratory system and the environment.
Alveolar ventilation: supplying of air to alveoli.
Pulmonary ventilation: supplying of air to the lung; includes alveoli and also the bronchi of the lungs.
ventral: referring to belly or front side of the body.
ventricle: one of the two lower chambers of the heart.
vermis: a worm, or wormlike structure; the central single lobe of the cerebellum.
vertebral: referring to the vertebral (spinal) column or a vertebra.
vesicle: a small, fluid filled sac.
vestibular: used to refer to the equilibrium structures of the inner ear and their nerves; literally, referring to a small space or cavity.
vestibule: a space or cavity at the beginning of a channel or channels.
viable: capable of living independently.
villus: a tuft of hair; a finger or tuftlike process of mucous membranes.
vis a tergo: a force that pushes; residual blood pressure in the venous system.
visceral: refers to the viscera, or organs of chest and abdomen; refers to an outer covering or membrane of an organ.
viscosity: thickness of a fluid; the state of being sticky.
viscous: being sticky, thick, or gummy.
visual: referring to vision or sight.
vitalistic: adhering to the view that life processes are not due to mechanical, physical, or chemical forces, but rather to metaphysical factors.
vitreous: glassy; the vitreous body of the eyeball.
volatile: easily vaporized or evaporated.
voluntary: referring to being under willful control.

W

waning: gradually growing less in extent, strength, or prominence.
waxing: gradually increasing in extent, strength, or prominence.
Willis: Thomas Willis, English anatomist (1621–1675). His name appears in connection with the arterial circle on the base of the brain.

Z

zonule: a band or girdle.
zygote: the cell produced by fusion of an ovum and a sperm; the fertilizer ovum.
zymogen: an inactive form of an enzyme.

INDEX

Italicized entries indicate that a *Figure* illustrates the point discussed in the text. A *t* following an entry indicates a Table that contains information pertinent to the topic discussed. The index is organized primarily by major topic, system, or organ.

A

C

D

E

H

J

K

L

M

T

U

V

W

Y

Z